Communications and Signal Processing

Stephen W. Director, Carnegie-Mellon University
Consulting Editor

DIGITAL
COMMUNICATIONS

Second Edition

John G. Proakis, Ph.D., P.E.

Department of Electrical and Computer Engineering
Northeastern University

McGraw-Hill Book Company

New York St. Louis San Francisco Auckland Bogotá Caracas
Colorado Springs Hamburg Lisbon London Madrid Mexico Milan
Montreal New Delhi Oklahoma City Panama Paris San Juan
São Paulo Singapore Sydney Tokyo Toronto

DIGITAL COMMUNICATIONS
INTERNATIONAL EDITION

Copyright © 1989
Exclusive rights by McGraw-Hill Book Co — Singapore for manufacture and export. This book cannot be re-exported from the country to which it is consigned by McGraw-Hill.

2 3 4 5 6 7 8 9 0 JJP PMP 9 4 3 2 1 0

This book was set in Times Roman by Science Typographers, Inc.
The editors were Alar E. Elken and John M. Morriss.
The drawings were done by Science Typographers, Inc.

Library of Congress Cataloging-in-Publication Data

Proakis, John G.
    Digital communications / John G. Proakis. — 2nd ed.
    p. cm. — (McGraw-Hill series in electrical engineering.
    Communications and signal processing)
    Includes bibliographies and index.
    ISBN 0-07-050937-9
    1. Digital communications. I. Title. II. Series
TK5103.7.P76   1989
621.38'0413–dc19   88-31455

When ordering this title use ISBN 0-07-100269-3

Printed in Singapore.

To
Felia, George, and Elena

CONTENTS

3   Characterization of Signals and Systems   148

4   Modulation and Demodulation for the Additive Gaussian Noise Channel   220

8   Spread Spectrum Signals for Digital Communications   800

References and Bibliography   878

Index   892

This second edition of Digital Communications is an expanded version of the first edition. New material has been added on source encoding methods including scalar and vector quantization, the spectral characteristics and performance characteristics of continuous-phase modulation (CPM), synchronization techniques; trellis-coded modulation, and the use of the cutoff rate as a criterion for modulation/demodulation and coding/decoding design.

As indicated in the first edition, the book is an outgrowth of my teaching, research, and consulting activities in the field of digital communications over the past 20 years. It is designed to serve as a text for a first-year graduate-level course for students in electrical engineering. It is also designed to serve as a text for self-study and as a reference book for the practicing engineer who is involved in the design of digital communications systems. As a background, I presume that the reader has a thorough understanding of basic calculus and elementary linear systems theory and some prior knowledge of probability and stochastic processes.

The field of digital communications has grown tremendously since the publication of C. E. Shannon's work over 40 years ago. It encompasses a broad range of topics that collectively may be called *information theory* and which include the theory of source coding and the theory of channel coding and decoding. An in-depth treatment and understanding of these subjects requires substantial mathematical sophistication. There are several excellent textbooks that treat these subjects thoroughly, e.g., R. G. Gallagher's *Information Theory and Reliable Communication*, E. R. Berlekamp's *Algebraic Coding Theory*, *Error-Correcting Codes*, by W. W. Peterson and E. J. Weldon, and *Principles of Digital Communication and Coding* by A. J. Viterbi and J. K. Omura.

In contrast to these more advanced books, the treatment of digital communications in this book is intended to serve as an introduction to the subject. The emphasis is on (1) the mathematical representation of digital signals, (2) the selection or design of digital signals for transmission of information over the communication channel, (3) the mathematical modeling and characterization of physical channels, and (4) the systems design of optimum and suboptimum receiver structures for demodulating and decoding of the channel-corrupted received signals. The selection of topics and the emphasis provided reflect my views on what should be contained in a first course on this subject. I have attempted to present the material at a level that is comprehensible to most graduate students who may have an interest in the subject.

Chapter 1 of the book contains a review of the basic elements of probability and stochastic processes. It includes a number of probability distribution functions and moments that are used throughout the book. It also includes the derivation of the Chernoff bound, which is useful in obtaining bounds on the performance of digital communications systems.

The second chapter introduces the reader to the elements of a digital communications system and to a number of basic results in information theory regarding the capacity of a channel. The encoding of sources is a topic that is also described. The emphasis is on the well-developed methods used for speech digitization, namely, pulse code modulation (PCM), differential PCM (DPCM), delta modulation (DM), and source modeling based on linear prediction. Emphasis is placed on scalar and vector quantization techniques, and comparisons are made with basic results from rate-distortion theory.

In Chapter 3, the reader is introduced to the representation of digitally modulated signals and to the characterization of narrowband signals and systems. Also treated in that chapter are the spectral characteristics of digitally modulated signals.

Chapter 4 treats the design of modulation and optimum demodulation methods for digital communications over an additive white gaussian noise channel. Emphasis is placed on the evaluation of the error rate performance for the various digital signaling techniques and on the channel bandwidth requirements of the corresponding signals. The important practical problems dealing with phase acquisition for coherent detection and time synchronization are also treated.

Channel encoding and decoding are introduced in Chapter 5. Of necessity, the treatment of this broad and rich subject is limited in scope. Coding is viewed as an efficient means for increasing the size of the signal alphabet and thus improving performance. Thus, coding and modulation are treated in an integrated manner. Both linear block codes and convolutional codes are considered. Soft-decision and hard-decision decoding are treated and their performance is evaluated for the AWGN channel. Although reference is made to a number of practical algorithms for soft-decision decoding and hard-decision decoding of block codes, for the sake of brevity, no attempt is made to describe the algorithms. Viterbi decoding of convolutional codes is described in detail, but sequential decoding, feedback decoding, and the stack algorithm are briefly treated. Random coding is also treated in this chapter and the cutoff rate parameter is introduced as a criterion for the design of coded modulation. Finally, trellis coded modulation is described and several examples are given on its use for bandwidth-constrained channels.

Chapter 6 treats the problem of digital communications over band-limited channels, most notably telephone channels. This involves the design of signals for band-limited channels, the characterization and the effect of channel distortion, which results in intersymbol interference, and a thorough discussion of adaptive equalization algorithms for mitigating intersymbol interference. The inclusion of the section on rapidly convergent algorithms based on recursive least-squares methods is particularly timely in view of the fact that such algorithms can now be implemented at a modest cost and with a reasonable amount of hardware. Also included in this chapter is a treatment of the demodulation and performance of continuous-phase modulated signals. The length of this chapter is indicative of my special interest and long-term involvement in the problem of intersymbol interference in high data rate transmission over band-limited channels.

Many channels, especially radio channels, exhibit time-variant transfer characteristics to transmitted signals. The characterization of such channels and the problem of signal design and receiver design is the topic of Chapter 7. The importance of diversity in the achievement of reliable transmission in the presence of signal fading is emphasized. Finally, the benefits of coding, which is viewed as a bandwidth-efficient means of obtaining diversity, are described and evaluated.

The last chapter of the book, Chapter 8, gives an introduction of direct sequence (DS) and frequency-hopped (FH) spread spectrum signals. Until a few years ago this topic was of interest primarily to engineers designing military communications systems. Today, however, spread spectrum signals are being considered in the design of commercial digital communications systems, primarily systems that accommodate many users of a channel in a multiple-access mode of transmission.

The material in Chapters 2 through 5 is basic and can be easily covered in a one-semester graduate course. The remaining three chapters are appropriate for a second course or a special topics course in digital communications. The homework problems at the end of each chapter are an essential element in learning the material covered in the text. Some of these problems are designed to supplement, and to extend, the results derived in the text. A solutions manual is available to instructors through the publisher.

Throughout my professional career I have had the opportunity to work with and learn from a number of people whom I would like to publicly acknowledge. I wish to express my appreciation to Dr. R. Price, P. R. Drouilhet, Jr., and Dr. P. E. Green, Jr., who introduced me to various aspects of digital communications through fading multipath channels and multichannel signal transmission during my employment at the MIT Lincoln Laboratory. I am also indebted to Professor D. W. Tufts, who supervised my Ph.D. dissertation at Harvard University and who introduced me to the problems of signal design and equalization for band-limited channels.

I have also benefited during the past 15 years from a close collaboration with a number of colleagues and friends at GTE Sylvania and Stein Associates. These two industrial connections have had a great influence on my selection of the topics covered in this book and on their treatment. I am especially indebted to Dr. S. Stein, Dr. B. Barrow, Dr. A. A. Giordano, Dr. A. H. Levesque, Dr. R. Greenspan, Dr. D. Freeman, Ph. H. Anderson, D. Gooding, and J. Lindholm. Special thanks are due to Dr. D. Freeman for reviewing the manuscript for the first edition and for several suggestions that have led to improvements in the presentation. Various parts of the manuscript for the first edition were also reviewed by R. Lin, T. Zhou, and H. P. Delgado, who made several useful suggestions. Dr. T. Schonhoff provided the graphs illustrating the spectral characteristics of CPFSK, H. Gibbons provided the data for the graphs in Chapter 7 that show the performance of PSK and DPSK with diversity, and F. Ling suggested the matrix inverse identity that simplified the derivation of the least-squares lattice equations in Chapter 6. The assistance of these colleagues is greatly appreciated.

McGraw-Hill and I would like to thank the reviewers of this second edition for their many useful comments and suggestions: David G. Daut, Rutgers University; Joseph D. DeLorenzo, Rochester Institute of Technology; Ronald Iltis, University of California at Santa Barbara; and H. W. Welch, Arizona State University.

I wish to thank my wife, Felia, for typing the entire 1000 page manuscript for the first edition. She accomplished this task while simultaneously holding a teaching position and fulfilling her role as a housewife and mother to our two small children. Finally, I wish to express my appreciation to Gloria Proakis for typing the manuscript for the second edition.

John G. Proakis

CHAPTER
1

PROBABILITY AND STOCHASTIC PROCESSES

The theory of probability and stochastic processes is an essential mathematical tool in the design of digital communication systems. This subject is important in the statistical modeling of sources that generate the information, in the digitization of the source output, in the characterization of the channel through which the digital information is transmitted, in the design of the receiver that processes the information-bearing signal from the channel, and in the evaluation of the performance of the communication system. Our coverage of this rich and interesting subject is brief and limited in scope. We present a number of definitions and basic concepts in the theory of probability and stochastic processes and we derive several results that are important in the design of efficient digital communication systems and in the evaluation of their performance.

We anticipate that most readers have had some prior exposure to the theory of probability and stochastic processes, so that our treatment serves primarily as a review. Some readers, however, who have had no previous exposure may find the presentation in this chapter extremely brief. These readers will benefit from additional reading of engineering-level treatments of the subject found in the texts by Davenport and Root (1958), Papoulis (1984), and Davenport (1970).

1.1 PROBABILITY

Let us consider an experiment, such as the rolling of a die, with a number of possible outcomes. The sample space S of the experiment consists of the set of all possible outcomes. In the case of the die,

$$S = \{1,2,3,4,5,6\} \qquad (1.1.1)$$

where the integers 1 through 6 represent the number of dots on the six faces of the die. These six possible outcomes are the sample points of the experiment. An event is a subset of S which may consist of any number of sample points. For example, the event A defined as

$$A = \{2,4\} \qquad (1.1.2)$$

consists of the outcomes 2 and 4. The complement of the event A, denoted as $\bar{A}$, consists of all the sample points in S that are not in A and, hence,

$$\bar{A} = \{1,3,5,6\} \qquad (1.1.3)$$

Two events are said to be mutually exclusive if they have no sample points in common—that is, if the occurrence of one event excludes the occurrence of the other. For example, if A is defined as in (1.1.2) and the event B is defined as

$$B = \{1,3,6\} \qquad (1.1.4)$$

then A and B are mutually exclusive events. Similarly, A and $\bar{A}$ are mutually exclusive events.

The union (sum) of two events is an event that consists of all the sample points in the two events. For example, if B is the event defined in (1.1.4) and C is the event defined as

$$C = \{1,2,3\} \qquad (1.1.5)$$

then, the union of B and C, denoted as $B \cup C$, is the event

$$D = B \cup C$$
$$= \{1,2,3,6\} \qquad (1.1.6)$$

Similarly, $A \cup \bar{A} = S$, where S is the entire sample space or the certain event. On the other hand, the intersection of two events is an event that consists of the points that are common to the two events. Thus, if $E = B \cap C$ represents the intersection of the events B and C, defined by (1.1.4) and (1.1.5), respectively, then

$$E = \{1,3\}$$

When the events are mutually exclusive, the intersection is the null event, denoted as ϕ. For example, $A \cap B = \phi$, and $A \cap \bar{A} = \phi$. The definitions of union and intersection are extended to more than two events in a straightforward manner.

Associated with each event A contained in S is its probability $P(A)$. In the assignment of probabilities to events, we adopt an axiomatic viewpoint. That is, we postulate that the probability of the event A satisfies the condition $P(A) \geq 0$. We also postulate that the probability of the sample space (certain event) is $P(S) = 1$. The third postulate deals with the probability of mutually exclusive events. Suppose that A_i, $i = 1,2\ldots$, are a (possibly infinite) number of events in the sample space S such that

$$A_i \cap A_j = \phi \qquad i \neq j = 1, 2, \ldots$$

Then the probability of the union of these mutually exclusive events satisfies the condition

$$P\left(\bigcup_i A_i\right) = \sum_i P(A_i) \tag{1.1.7}$$

For example, in a roll of a fair die, each possible outcome is assigned the probability $\frac{1}{6}$. The event A defined by (1.1.12) consists of two mutually exclusive subevents or outcomes and, hence, $P(A) = \frac{2}{6} = \frac{1}{3}$. Also, the probability of the event $A \cup B$, where A and B are the mutually exclusive events defined by (1.1.2) and (1.1.4), respectively, is $P(A) + P(B) = \frac{1}{3} + \frac{1}{2} = \frac{5}{6}$.

Joint events and joint probabilities. Instead of dealing with a single experiment, let us perform two experiments and consider their outcomes. For example, the two experiments may be two separate tosses of a single die or a single toss of two dice. In either case, the sample space S consists of the 36 two-tuples (i, j) where $i, j = 1, 2, \ldots, 6$. If the dice are fair, each point in the sample space is assigned the probability $\frac{1}{36}$. We may now consider joint events, such as $\{i$ is even, $j = 3\}$, and determine the associated probabilities of such events from knowledge of the probabilities of the sample points.

In general, if one experiment has the possible outcomes A_i, $i = 1, 2, \ldots, n$, and the second experiment has the possible outcomes B_j, $j = 1, 2, \ldots, m$, then the combined experiment has the possible joint outcomes (A_i, B_j), $i = 1, 2, \ldots, n$, $j = 1, 2, \ldots, m$. Associated with each joint outcome (A_i, B_j) is the joint probability $P(A_i, B_j)$ which satisfies the condition

$$0 \leq P(A_i, B_j) \leq 1$$

Assuming that the outcomes B_j, $j = 1, 2, \ldots, m$, are mutually exclusive, it follows that

$$\sum_{j=1}^{m} P(A_i, B_j) = P(A_i) \tag{1.1.8}$$

Similarly, if the outcomes A_i, $i = 1, 2, \ldots, n$, are mutually exclusive, we have

$$\sum_{i=1}^{n} P(A_i, B_j) = P(B_j) \tag{1.1.9}$$

Furthermore, if all the outcomes of the two experiments are mutually exclusive, then

$$\sum_{i=1}^{n} \sum_{j=1}^{m} P(A_i, B_j) = 1 \tag{1.1.10}$$

The generalization of the above treatment to more than two experiments is straightforward.

Conditional probabilities. Consider a combined experiment in which a joint event occurs with probability $P(A, B)$. Suppose that the event B has occurred and we wish to determine the probability of occurrence of the event A. This is called the *conditional probability* of the event A given the occurrence of the event B and is defined as

$$P(A|B) = \frac{P(A, B)}{P(B)} \tag{1.1.11}$$

provided $P(B) > 0$. In a similar manner, the probability of the event B conditioned on the occurrence of the event A is defined as

$$P(B|A) = \frac{P(A, B)}{P(A)} \tag{1.1.12}$$

provided $P(A) > 0$. The relations in (1.1.11) and (1.1.12) may also be expressed as

$$P(A, B) = P(A|B)P(B) = P(B|A)P(A) \tag{1.1.13}$$

The relations in (1.1.11), (1.1.12), and (1.1.13) also apply to a single experiment in which A and B are any two events defined on the sample space S and $P(A, B)$ is interpreted as the probability of the $A \cap B$. That is, $P(A, B)$ denotes the simultaneous occurrence of A and B. For example, consider the events B and C given by (1.1.4) and (1.1.5), respectively, for the single toss of a die. The joint event consists of the sample points $\{1, 3\}$. The conditional probability of the event C given that B occurred is

$$P(C|B) = \frac{\frac{2}{6}}{\frac{3}{6}} = \frac{2}{3}$$

In a single experiment, we observe that when two events A and B are mutually exclusive, $A \cap B = \phi$ and, hence, $P(A|B) = 0$. Also, if A is a subset of B, then $A \cap B = A$ and, hence,

$$P(A|B) = \frac{P(A)}{P(B)}$$

On the other hand, if B is a subset of A, we have $A \cap B = B$ and, hence,

$$P(A|B) = \frac{P(B)}{P(B)} = 1$$

An extremely useful relationship for conditional probabilities is Bayes' theorem, which states that if A_i, $i = 1, 2, \ldots, n$, are mutually exclusive events such that

$$\bigcup_{i=1}^{n} A_i = S$$

and B is an arbitrary event with nonzero probability, then

$$P(A_i|B) = \frac{P(A_i, B)}{P(B)}$$

$$= \frac{P(B|A_i)P(A_i)}{\sum_{j=1}^{n} P(B|A_j)P(A_j)} \qquad (1.1.14)$$

We use this formula in Chap. 4 to derive the structure of the optimum receiver for a digital communication system in which the events A_i, $i = 1, 2, \ldots, n$, represent the possible transmitted messages in a given time interval, $P(A_i)$ represent their *a priori* probabilities, B represents the received signal which consists of the transmitted message (one of the A_i) corrupted by noise, and $P(A_i|B)$ is the *a posteriori* probability of A_i conditioned on having observed the received signal B.

Statistical independence. The statistical independence of two or more events is another important concept in probability theory. It usually arises when we consider two or more experiments or repeated trials of a single experiment. To explain this concept, we consider two events A and B and their conditional probability $P(A|B)$, which is the probability of occurrence of A given that B has occurred. Suppose that the occurrence of A does not depend on the occurrence of B. That is,

$$P(A|B) = P(A) \qquad (1.1.15)$$

Substitution of (1.1.15) into (1.1.13) yields the result

$$P(A, B) = P(A)P(B) \qquad (1.1.16)$$

That is, the joint probability of the events A and B factors into the product of the elementary or marginal probabilities $P(A)$ and $P(B)$. When the events A and B satisfy the relation in (1.1.16), they are said to be *statistically independent*.

For example, consider two successive experiments in tossing a die. Let A represent the even-numbered sample points $\{2, 4, 6\}$ in the first toss and B represent the even-numbered possible outcomes $\{2, 4, 6\}$ in the second toss. In a fair die, we assign the probabilities $P(A) = \frac{1}{2}$ and $P(B) = \frac{1}{2}$. Now, the joint probability of the joint event "even-numbered outcome on the first toss and even-numbered outcome on the second toss" is just the probability of the nine pairs of outcomes (i, j), $i = 2, 4, 6$, $j = 2, 4, 6$, which is $\frac{1}{4}$. Also

$$P(A, B) = P(A)P(B) = \frac{1}{4}$$

Thus, the events A and B are statistically independent. Similarly, we may say that the outcomes of the two experiments are statistically independent.

The definition of statistical independence can be extended to three or more events. Three statistically independent events A_1, A_2, and A_3 must satisfy the following conditions:

$$P(A_1, A_2) = P(A_1)P(A_2)$$
$$P(A_1, A_3) = P(A_1)P(A_3)$$
$$P(A_2, A_3) = P(A_2)P(A_3) \tag{1.1.17}$$
$$P(A_1, A_2, A_3) = P(A_1)P(A_2)P(A_3)$$

In the general case, the events A_i, $i = 1, 2, \ldots, n$, are statistically independent provided that the probabilities of the joint events taken $2, 3, 4 \ldots$, and n at a time factor into the product of the probabilities of the individual events.

1.1.1   Random Variables, Probability Distributions, and Probability Densities

Given an experiment having a sample space S and elements $s \in S$, we define a function $X(s)$ whose domain is S and whose range is a set of numbers on the real line. The function $X(s)$ is called a *random variable*. For example, if we flip a coin the possible outcomes are head (H) and tail (T), so S contains two points labeled H and T. Suppose we define a function $X(s)$ such that

$$X(s) = \begin{cases} 1 & s = H \\ -1 & s = T \end{cases} \tag{1.1.18}$$

Thus we have mapped the two possible outcomes of the coin-flipping experiment into the two points (± 1) on the real line. Another experiment is the toss of a die with possible outcomes $S = \{1, 2, 3, 4, 5, 6\}$. A random variable defined on this sample space may be $X(s) = s$, in which case the outcomes of the experiment are mapped into the integers 1 through 6 or, perhaps, $X(s) = s^2$, in which case the possible outcomes are mapped into the integers $\{1, 4, 9, 16, 25, 36\}$. These are examples of discrete random variables.

Although we have used as examples experiments which have a finite set of possible outcomes, there are many physical systems (experiments) which generate outputs (outcomes), that are continuous. For example, the noise voltage generated by an electronic amplifier has a continuous amplitude. Consequently, the sample space S of voltage amplitudes $v \in S$ is continuous and so is the mapping $X(v) = v$. In such a case, the random variable† X is said to be a *continuous random variable*.

Given a random variable X, let us consider the event $\{X \leq x\}$ where x is any real number in the interval $(-\infty, \infty)$. We write the probability of this event as $P(X \leq x)$ and denote it simply as $F(x)$, i.e.,

$$F(x) = P(X \leq x) \qquad -\infty < x < \infty \tag{1.1.19}$$

The function $F(x)$ is called the *probability distribution function* of the random

---

† The random variable $X(s)$ will be written simply as X.

FIGURE 1.1.1
Examples of the cumulative distribution functions of two discrete random variables.

variable X. It is also called the *cumulative distribution function* (cdf). Since $F(x)$ is a probability, its range is limited to the interval $0 \leq F(x) \leq 1$. In fact, $F(-\infty) = 0$ and $F(\infty) = 1$. For example, the discrete random variable generated by flipping a fair coin and defined by (1.1.18) has the cdf shown in Fig. 1.1.1a. There are two discontinuities or jumps in $F(x)$, one at $x = -1$ and one at $x = 1$. Similarly, the random variable $X(s) = s$ generated by tossing a fair die has the cdf shown in Fig. 1.1.1b. In this case $F(x)$ has six jumps, one each at the points $x = 1, 2, 3, 4, 5, 6$.

The cdf of a continuous random variable typically appears as shown in Fig. 1.1.2. This is a smooth, nondecreasing function of x. In some practical problems, we may also encounter a random variable of a mixed type. The cdf of such a random variable is a smooth, nondecreasing function in certain parts of the real line and contains jumps at a number of discrete values of x. An example of such a cdf is illustrated in Fig. 1.1.3.

FIGURE 1.1.2
An example of the cumulative distribution function of a continuous random variable.

FIGURE 1.1.3
An example of the cumulative distribution function of a random variable of a mixed type.

The derivative of the cdf $F(x)$, denoted as $p(x)$, is called the *probability density function* (pdf) of the random variable X. Thus, we have

$$p(x) = \frac{dF(x)}{dx} \qquad -\infty < x < \infty \qquad (1.1.20)$$

or, equivalently,

$$F(x) = \int_{-\infty}^{x} p(u)\, du \qquad -\infty < x < \infty \qquad (1.1.21)$$

Since $F(x)$ is a nondecreasing function, it follows that $p(x) \geq 0$. When the random variable is discrete or of a mixed type, the pdf contains impulses at the points of discontinuity of $F(x)$. In such cases the discrete part of $p(x)$ may be expressed as

$$p(x) = \sum_{i=1}^{n} P(X = x_i)\delta(x - x_i) \tag{1.1.22}$$

where x_i, $i = 1, 2, \ldots, n$, are the possible discrete values of the random variable; $P(X = x_i)$, $i = 1, 2, \ldots, n$, are the probabilities; and $\delta(x)$ denotes an impulse at $x = 0$.

Often we are faced with the problem of determining the probability that a random variable X falls in an interval (x_1, x_2) where $x_2 > x_1$. To determine the probability of this event, let us begin with the event $\{X \le x_2\}$. This event can always be expressed as the union of two mutually exclusive events $\{X \le x_1\}$ and $\{x_1 < X \le x_2\}$. Hence the probability of the event $\{X \le x_2\}$ can be expressed as the sum of the probabilities of the mutually exclusive events. Thus we have

$$P(X \le x_2) = P(X \le x_1) + P(x_1 < X \le x_2)$$

$$F(x_2) = F(x_1) + P(x_1 < X \le x_2)$$

or, equivalently,

$$P(x_1 < X \le x_2) = F(x_2) - F(x_1)$$

$$= \int_{x_1}^{x_2} p(x)\,dx \tag{1.1.23}$$

In other words, the probability of the event $\{x_1 < X \le x_2\}$ is simply the area under the pdf in the range $x_1 < X \le x_2$.

Multiple random variables, joint probability distributions, and joint probability densities. In dealing with combined experiments or repeated trials of a single experiment, we encounter multiple random variables and their cdf's and pdf's. Multiple random variables are basically multidimensional functions defined on a sample space of a combined experiment. Let us begin with two random variables X_1 and X_2, each of which may be continuous, discrete, or mixed. The joint cumulative distribution function (joint cdf) for the two random variables is defined as

$$F(x_1, x_2) = P(X_1 \le x_1, X_2 \le x_2)$$

$$= \int_{-\infty}^{x_2} \int_{-\infty}^{x_2} p(u_1, u_2)\,du_1\,du_2 \tag{1.1.24}$$

where $p(x_1, x_2)$ is the joint probability density function (joint pdf). The latter may also be expressed in the form

$$p(x_1, x_2) = \frac{\partial^2}{\partial x_1 \, \partial x_2} F(x_1, x_2) \tag{1.1.25}$$

When the joint pdf $p(x_1, x_2)$ is integrated over one of the variables, we obtain the pdf of the other variable. That is,

$$\int_{-\infty}^{\infty} p(x_1, x_2)\, dx_1 = p(x_2)$$
$$\int_{-\infty}^{\infty} p(x_1, x_2)\, dx_2 = p(x_1)$$

(1.1.26)

The pdf's $p(x_1)$ and $p(x_2)$ obtained from integrating over one of the variables are called *marginal pdf*'s. Furthermore, if $p(x_1, x_2)$ is integrated over both variables, we obtain

$$\int_{-\infty}^{\infty}\int_{-\infty}^{\infty} p(x_1, x_2)\, dx_1\, dx_2 = F(\infty, \infty) = 1 \qquad (1.1.27)$$

We also note that $F(-\infty, -\infty) = F(-\infty, x_2) = F(x_1, -\infty) = 0$.

The generalization of the above expressions to multidimensional random variables is straightforward. Suppose that X_i, $i = 1, 2, \ldots, n$, are random variables with a joint cdf which is defined as

$$F(x_1, x_2, \ldots, x_n) = P(X_1 \le x_1, X_2 \le x_2, \ldots, X_n \le x_n)$$
$$= \int_{-\infty}^{x_1}\int_{-\infty}^{x_2} \cdots \int_{-\infty}^{x_n} p(u_1, u_2, \ldots, u_n)\, du_1\, du_2 \ldots du_n$$

(1.1.28)

where $p(x_1, x_2, \ldots, x_n)$ is the joint pdf. By taking the partial derivatives of $F(x_1, x_2, \ldots, x_n)$ given by (1.1.28), we obtain

$$p(x_1, x_2, \ldots, x_n) = \frac{\partial^n}{\partial x_1 \partial x_2 \cdots \partial x_n} F(x_1, x_2, \ldots, x_n) \qquad (1.1.29)$$

Any number of variables in $p(x_1, x_2, \ldots, x_n)$ can be eliminated by integrating over these variables. For example integration over x_2 and x_3 yields

$$\int_{-\infty}^{\infty}\int_{-\infty}^{\infty} p(x_1, x_2, x_3, \ldots, x_n)\, dx_2\, dx_3 = p(x_1, x_4, \ldots, x_n) \qquad (1.1.30)$$

It also follows that $F(x_1, \infty, \infty, x_4, \ldots, x_n) = F(x_1, x_4, x_5, \ldots, x_n)$ and

$$F(x_1, -\infty, -\infty, x_4, \ldots, x_n) = 0.$$

Conditional probability distribution functions. Let us consider two random variables X_1 and X_2 with joint pdf $p(x_1, x_2)$. Suppose that we wish to determine the probability that the random variable $X_1 \le x_1$ conditioned on

$$x_2 - \Delta x_2 < X_2 \le x_2,$$

where Δx_2 is some positive increment. That is, we wish to determine the probability of the event $(X_1 \le x_1 | x_2 - \Delta x_2 < X_2 \le x_2)$. Using the relations established earlier for the conditional probability of an event, the probability of the event $(X_1 \le x_1 | x_2 - \Delta x_2 < X_2 \le x_2)$ can be expressed as the probability of the joint event $(X_1 \le x_1, x_2 - \Delta x_2 < X_2 \le x_2)$ divided by the probability of the event $(x_2 - \Delta x_2 < X_2 \le x_2)$. Thus

$$P(X_1 \le x_1 | x_2 - \Delta x_2 < X_2 \le x_2) = \frac{\int_{-\infty}^{x_1} \int_{x_2 - \Delta x_2}^{x_2} p(u_1, u_2) \, du_1 \, du_2}{\int_{x_2 - \Delta x_2}^{x_2} p(u_2) \, du_2}$$

$$= \frac{F(x_1, x_2) - F(x_1, x_2 - \Delta x_2)}{F(x_2) - F(x_2 - \Delta x_2)} \quad (1.1.31)$$

Assuming that the pdf's $p(x_1, x_2)$ and $p(x_2)$ are continuous functions over the interval $(x_2 - \Delta x_2, x_2)$, we may divide both numerator and denominator in (1.1.31) by Δx_2 and take the limit as $\Delta x_2 \to 0$. Thus we obtain

$$P(X_1 \le x_1 | X_2 = x_2) \equiv F(x_1 | x_2) = \frac{\partial F(x_1, x_2) / \partial x_2}{\partial F(x_2) / \partial x_2}$$

$$= \frac{\partial \left[\int_{-\infty}^{x_1} \int_{-\infty}^{x_2} p(u_1, u_2) \, du_1 \, du_2 \right] / \partial x_2}{\partial \left[\int_{-\infty}^{x_2} p(u_2) \, du_2 \right] / \partial x_2}$$

$$= \frac{\int_{-\infty}^{x_1} p(u_1, x_2) \, du_1}{p(x_2)} \quad (1.1.32)$$

which is the conditional cdf of the random variable X_1 given the random variable X_2. We observe that $F(-\infty | x_2) = 0$ and $F(\infty | x_2) = 1$. By differentiating (1.1.32) with respect to x_1, we obtain the corresponding pdf $p(x_1 | x_2)$ in the form

$$p(x_1 | x_2) = \frac{p(x_1, x_2)}{p(x_2)} \quad (1.1.33)$$

Alternatively, we may express the joint pdf $p(x_1, x_2)$ in terms of the conditional pdf's, $p(x_1 | x_2)$ or $p(x_2 | x_1)$, as

$$p(x_1, x_2) = p(x_1 | x_2) p(x_2)$$

$$= p(x_2 | x_1) p(x_1) \quad (1.1.34)$$

The extension of the relations given above to multidimensional random variables is also easily accomplished. Beginning with the joint pdf of the random variables X_i, $i = 1, 2, \ldots, n$, we may write

$$p(x_1, x_2, \ldots, x_n) = p(x_1, x_2, \ldots, x_k | x_{k+1}, \ldots, x_n) p(x_{k+1}, \ldots, x_n) \quad (1.1.35)$$

where k is any integer in the range $1 < k < n$. The joint conditional cdf corresponding to the pdf $p(x_1, x_2, \ldots, x_k | x_{k+1}, \ldots, x_n)$ is

$$F(x_1, x_2, \ldots, x_k | x_{k+1}, \ldots, x_n)$$

$$= \frac{\int_{-\infty}^{x_1} \cdots \int_{-\infty}^{x_k} p(u_1, u_2, \ldots, u_k, x_{k+1}, \ldots, x_n) \, du_1 \, du_2 \cdots du_k}{p(x_{k+1}, \ldots, x_n)} \quad (1.1.36)$$

This conditional cdf satisfies the properties previously established for these functions, such as

$$F(\infty, x_2, \ldots, x_k | x_{k+1}, \ldots, x_n) = F(x_2, x_3, \ldots, x_k | x_{k+1}, \ldots, x_n)$$

$$F(-\infty, x_2, \ldots, x_k | x_{k+1}, \ldots, x_n) = 0$$

Statistically independent random variables. We have already defined statistical independence of two or more events of a sample space S. The concept of statistical independence can be extended to random variables defined on a sample space generated by a combined experiment or by repeated trials of a single experiment. If the experiments result in mutually exclusive outcomes, the probability of an outcome in one experiment is independent of an outcome in any other experiment. That is, the joint probability of the outcomes factors into a product of the probabilities corresponding to each outcome. Consequently, the random variables corresponding to the outcomes in these experiments are independent in the sense that their joint pdf factors into a product of marginal pdf's. Hence the multidimensional random variables are statistically independent if and only if

$$F(x_1, x_2, \ldots, x_n) = F(x_1)F(x_2) \cdots F(x_n) \qquad (1.1.37)$$

or, alternatively,

$$p(x_1, x_2, \ldots, x_n) = p(x_1)p(x_2) \cdots p(x_n) \qquad (1.1.38)$$

1.1.2 Functions of Random Variables

A problem that arises frequently in practical applications of probability is the following. Given a random variable X, which is characterized by its pdf $p(x)$, determine the pdf of the random variable $Y = g(X)$, where $g(X)$ is some given function of X. When the mapping g from X to Y is one to one, the determination of $p(y)$ is relatively straightforward. However, when the mapping is not one to one as is the case, for example, when $Y = X^2$, we must be very careful in our derivation of $p(y)$.

Example 1.1.1. Consider the random variable Y defined as

$$Y = aX + b \qquad (1.1.39)$$

where a and b are constants. We assume that $a > 0$. If $a < 0$, the approach is similar (see Prob. 1.3). We note that this mapping, illustrated in Fig. 1.1.4a is linear and monotonic. Let $F_X(x)$ and $F_Y(y)$ denote the cdf's for X and Y, respectively.† Then

$$F_Y(y) = P(Y \le y) = P(aX + b \le y) = P\left(X \le \frac{y - b}{a}\right)$$

$$= \int_{-\infty}^{(y-b)/a} p_X(x) \, dx = F_X\left(\frac{y - b}{a}\right) \qquad (1.1.40)$$

By differentiating (1.1.40) with respect to y, we obtain the relationship between the respective pdf's. It is

$$p_Y(y) = \frac{1}{a} p_X\left(\frac{y - b}{a}\right) \qquad (1.1.41)$$

---

† To avoid confusion in changing variables, subscripts are used in the respective pdf's and cdf's.

FIGURE 1.1.4
A linear transformation of a random variable X and an example of the corresponding pdf's of X and Y.

Thus (1.1.40) and (1.1.41) specify the cdf and pdf of the random variable Y in terms of the cdf and pdf of the random variable X for the linear transformation in (1.1.39). To illustrate this mapping for a specific pdf $p_X(x)$, consider the one shown in Fig. 1.1.4b. The pdf $p_Y(y)$ which results from the mapping in (1.1.39) is shown in Fig. 1.1.4c.

Example 1.1.2. Consider the random variable Y defined as

$$Y = aX^3 + b \qquad a > 0 \qquad (1.1.42)$$

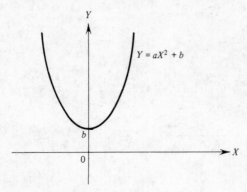

FIGURE 1.1.5
A quadratic transformation of the random variable X.

As in Example 1.1.1, the mapping between X and Y is one to one. Hence

$$F_Y(y) = P(Y \le y) = P(aX^3 + b \le y)$$

$$= P\left[X \le \left(\frac{y-b}{a}\right)^{1/3}\right] = F_X\left[\left(\frac{y-b}{a}\right)^{1/3}\right] \qquad (1.1.43)$$

Differentiation of (1.1.43) with respect to y yields the desired relationship between the two pdf's as

$$p_Y(y) = \frac{1}{3a[(y-b)/a]^{2/3}} p_X\left[\left(\frac{y-b}{a}\right)^{1/3}\right] \qquad (1.1.44)$$

Example 1.1.3. The random variable Y is defined as

$$Y = aX^2 + b \qquad a > 0 \qquad (1.1.45)$$

In contrast to Examples 1.1.1 and 1.1.2, the mapping between X and Y, illustrated in Fig. 1.1.5, is not one to one. To determine the cdf of Y, we observe that

$$F_Y(y) = P(Y \le y) = P(aX^2 + b \le y)$$

$$= P\left(|X| \le \sqrt{\frac{y-b}{a}}\right)$$

Hence

$$F_Y(y) = F_X\left(\sqrt{\frac{y-b}{a}}\right) - F_X\left(-\sqrt{\frac{y-b}{a}}\right) \qquad (1.1.46)$$

Differentiating (1.1.46) with respect to y, we obtain the pdf of Y in terms of the pdf of X in the form

$$p_Y(y) = \frac{p_X\left[\sqrt{(y-b)/a}\right]}{2a\sqrt{[(y-b)/a]}} + \frac{p_X\left[-\sqrt{(y-b)/a}\right]}{2a\sqrt{[(y-b)/a]}} \qquad (1.1.47)$$

In Example 1.1.3, we observe that the equation $g(x) = ax^2 + b = y$ has two real solutions,

$$x_1 = \sqrt{\frac{y - b}{a}}$$

$$x_2 = -\sqrt{\frac{y - b}{a}}$$

and that $p_Y(y)$ consists of two terms corresponding to these two solutions. That is,

$$p_Y(y) = \frac{p_X\left[x_1 = \sqrt{(y - b)/a}\right]}{\left|g'\left[x_1 = \sqrt{(y - b)/a}\right]\right|} + \frac{p_X\left[x_2 = -\sqrt{(y - b)/a}\right]}{\left|g'\left[x_2 = -\sqrt{(y - b)/a}\right]\right|} \tag{1.1.48}$$

where $g'(x)$ denotes the first derivative of $g(x)$.

In the general case, suppose that $x_1, x_2, \ldots, x_n$ are the real roots of the equation $g(x) = y$. Then the pdf of the random variable $Y = g(X)$ may be expressed as

$$p_Y(y) = \sum_{i=1}^{n} \frac{p_X(x_i)}{|g'(x_i)|} \tag{1.1.49}$$

where the roots x_i, $i = 1, 2, \ldots, n$, are functions of y.

Now let us consider functions of multidimensional random variables. Suppose that X_i, $i = 1, 2, \ldots, n$, are random variables with joint pdf $p_X(x_1, x_2, \ldots, x_n)$ and let Y_i, $i = 1, 2, \ldots, n$, be another set of n random variables related to the X_i by the functions

$$Y_i = g_i(X_1, X_2, \ldots, X_n) \qquad i = 1, 2, \ldots, n \tag{1.1.50}$$

We assume that the $g_i(X_1, X_2, \ldots, X_n)$, $i = 1, 2\ n$, are single-valued functions with continuous partial derivatives and invertible. By "invertible" we mean that the X_i, $i = 1, 2, \ldots, n$, can be expressed as functions of Y_i, $i = 1, 2, \ldots, n$, in the form

$$X_i = g_i^{-1}(Y_1, Y_2, \ldots, Y_n) \qquad i = 1, 2, \ldots, n \tag{1.1.51}$$

where the inverse functions are also assumed to be single-valued with continuous partial derivatives. The problem is to determine the joint pdf of Y_i, $i = 1, 2, \ldots, n$, denoted as $p_Y(y_1, y_2, \ldots, y_n)$, given the joint pdf $p_X(x_1, x_2, \ldots, x_n)$.

To determine the desired relation, let R_X be a region in the n-dimensional space of the random variables X_i, $i = 1, 2, \ldots, n$, and let R_Y be the (one-to-one) mapping of R_X defined by the functions $Y_i = g_i(X_1, X_2, \ldots, X_n)$. Clearly

$$\int\int_{R_Y} \cdots \int p_Y(y_1, y_2, \ldots, y_n) \, dy_1 \, dy_2 \, \cdots \, dy_n$$

$$= \int\int_{R_X} \cdots \int p_X(x_1, x_2, \ldots, x_n) \, dx_1 \, dx_2 \, \cdots \, dx_n \tag{1.1.52}$$

By making a change in variables in the multiple integral on the right-hand side of (1.1.52) with the substitution

$$x_i = g_i^{-1}(y_1, y_2, \ldots, y_n) \equiv g_i^{-1} \qquad i = 1, 2, \ldots, n$$

we obtain

$$\iint_{R_Y} \cdots \int p_Y(y_1, y_2, \ldots, y_n)\, dy_1\, dy_2\, \cdots\, dy_n$$

$$= \iint_{R_Y} \cdots \int p_X(x_1 = g_1^{-1}, x_2 = g_2^{-1}, \ldots, x_n = g_n^{-1})|J|\, dy_1\, dy_2\, \cdots\, dy_n \tag{1.1.53}$$

where J denotes the jacobian of the transformation, defined by the determinant

$$J = \begin{vmatrix} \dfrac{\partial g_1^{-1}}{\partial y_1} & \dfrac{\partial g_2^{-1}}{\partial y_1} & \cdots & \dfrac{\partial g_n^{-1}}{\partial y_1} \\ \cdots\cdots\cdots\cdots\cdots\cdots\cdots\cdots\cdots \\ \dfrac{\partial g_1^{-1}}{\partial y_n} & \dfrac{\partial g_2^{-1}}{\partial y_n} & \cdots & \dfrac{\partial g_n^{-1}}{\partial y_n} \end{vmatrix} \tag{1.1.54}$$

Consequently, the desired relation for the joint pdf of the Y_i, $i = 1, 2, \ldots, n$, is

$$p_Y(y_1, y_2, \ldots, y_n) = p_X(x_1 = g_1^{-1}, x_2 = g_2^{-1}, \ldots, x_n = g_n^{-1})|J| \tag{1.1.55}$$

Example 1.1.4. An important functional relation between two sets of n-dimensional random variables that frequently arises in practice is the linear transformation

$$Y_i = \sum_{j=1}^{n} a_{ij} X_j \qquad i = 1, 2, \ldots, n \tag{1.1.56}$$

where the $\{a_{ij}\}$ are constants. It is convenient to employ the matrix form for the transformation, which is

$$\mathbf{Y} = \mathbf{A}\mathbf{X} \tag{1.1.57}$$

where $\mathbf{X}$ and $\mathbf{Y}$ are n-dimensional vectors and $\mathbf{A}$ is an $(n \times n)$ matrix. We assume that $\mathbf{A}$ is nonsingular. Then $\mathbf{A}$ is invertible and, hence,

$$\mathbf{X} = \mathbf{A}^{-1}\mathbf{Y} \tag{1.1.58}$$

Equivalently, we have

$$X_i = \sum_{j=1}^{n} b_{ij} Y_j \qquad i = 1, 2, \ldots, n \tag{1.1.59}$$

where $\{b_{ij}\}$ are the elements of the inverse matrix $\mathbf{A}^{-1}$. The jacobian of this transformation is $J = 1/\det(\mathbf{A})$. Hence

$$p_Y(y_1, y_2, \ldots, y_n)$$

$$= p_X\left(x_1 = \sum_{j=1}^{n} b_{1j} y_j, x_2 = \sum_{j=1}^{n} b_{2j} y_j, \ldots, x_n = \sum_{j=1}^{n} b_{nj} y_j\right) \frac{1}{|\det(\mathbf{A})|} \tag{1.1.60}$$

1.1.3 Statistical Averages of Random Variables

Averages play an important role in the characterization of the outcomes of experiments and the random variables defined on the sample space of the experiments. Of particular interest are the first and second moments of a single random variable and the joint moments, such as the correlation and covariance, between any pair of random variables in a multidimensional set of random variables. Also of great importance are the characteristic function for a single random variable and the joint characteristic function for a multidimensional set of random variables. This section is devoted to the definition of these important statistical averages.

First we consider a single random variable X characterized by its pdf $p(x)$. The mean or expected value of X is defined as

$$E(X) \equiv m_x = \int_{-\infty}^{\infty} x p(x) \, dx \qquad (1.1.61)$$

where $E(\ )$ denotes expectation (statistical averaging). This is the first moment of the random variable X. In general, the nth moment is defined as

$$E(X^n) = \int_{-\infty}^{\infty} x^n p(x) \, dx \qquad (1.1.62)$$

Now, suppose that we define a random variable $Y = g(X)$, where $g(X)$ is some arbitrary function of the random variable X. The expected value of Y is

$$E(Y) = E[g(X)] = \int_{-\infty}^{\infty} g(x) p(x) \, dx \qquad (1.1.63)$$

In particular, if $Y = (X - m_x)^n$ where m_x is the mean value of X, then

$$E(Y) = E[(X - m_x)^n] = \int_{-\infty}^{\infty} (x - m_x)^n p(x) \, dx \qquad (1.1.64)$$

This expected value is called the nth *central moment* of the random variable X, because it is a moment taken relative to the mean. When $n = 2$, the central moment is called the *variance* of the random variable and denoted as σ_x^2. That is,

$$\sigma_x^2 = \int_{-\infty}^{\infty} (x - m_x)^2 p(x) \, dx \qquad (1.1.65)$$

This parameter provides a measure of the dispersion of the random variable X. By expanding the term $(x - m_x)^2$ in the integral of (1.1.65) and noting that the expected value of a constant is equal to the constant, we obtain an expression that relates the variance to the first and second moments, namely,

$$\sigma_x^2 = E(X^2) - [E(X)]^2$$
$$= E(X^2) - m_x^2 \qquad (1.1.66)$$

In the case of two random variables, X_1 and X_2, with joint pdf $p(x_1, x_2)$, we define the joint moment as

$$E(X_1^k X_2^n) = \int_{-\infty}^{\infty} \int_{-\infty}^{\infty} x_1^k x_2^n p(x_1, x_2) \, dx_1 \, dx_2 \qquad (1.1.67)$$

and the joint central moment as

$$E\left[(X_1 - m_1)^k (X_2 - m_2)^n\right]$$
$$= \int_{-\infty}^{\infty} \int_{-\infty}^{\infty} (x_1 - m_1)^k (x_2 - m_2)^n p(x_1, x_2) \, dx_1 \, dx_2 \quad (1.1.68)$$

where $m_i = E(X_i)$. Of particular importance to us are the joint moment and joint central moment corresponding to $k = n = 1$. These joint moments are called the *correlation* and the *covariance* of the random variables X_1 and X_2, respectively.

In considering multidimensional random variables, we can define joint moments of any order. However, the moments that are most useful in practical applications are the correlations and covariances between pairs of random variables. To elaborate, suppose that X_i, $i = 1, 2, \ldots, n$, are random variables with joint pdf $p(x_1, x_2, \ldots, x_n)$. Let $p(x_i, x_j)$ be the joint pdf of the random variables X_i and X_j. Then the correlation between X_i and X_j is given by the joint moment

$$E(X_i X_j) = \int_{-\infty}^{\infty} \int_{-\infty}^{\infty} x_i x_j p(x_i, x_j) \, dx_i \, dx_j \quad (1.1.69)$$

and the covariance of X_i and X_j is

$$\mu_{ij} \equiv E\left[(X_i - m_i)(X_j - m_j)\right]$$
$$= \int_{-\infty}^{\infty} \int_{-\infty}^{\infty} (x_i - m_i)(x_j - m_j) p(x_i, x_j) \, dx_i \, dx_j$$
$$= \int_{-\infty}^{\infty} \int_{-\infty}^{\infty} x_i x_j p(x_i, x_j) \, dx_i \, dx_j - m_i m_j$$
$$= E(X_i X_j) - m_i m_j \quad (1.1.70)$$

The $(n \times n)$ matrix with elements μ_{ij} is called the *covariance matrix* of the random variables X_i, $i = 1, 2, \ldots, n$. We shall encounter the covariance matrix in our discussion of jointly gaussian random variables in Sec. 1.1.4.

Two random variables are said to be uncorrelated if $E(X_i X_j) = E(X_i)E(X_j) = m_i m_j$. In that case, the covariance $\mu_{ij} = 0$. We note that when X_i and X_j are statistically independent, they are also uncorrelated. However, if X_i and X_j are uncorrelated, they are not necessarily statistically independent.

Two random variables are said to be orthogonal if $E(X_i X_j) = 0$. We note that this condition holds when X_i and X_j are uncorrelated and either one or both of the random variables have zero mean.

Characteristic functions. The characteristic function of a random variable X is defined as the statistical average

$$E(e^{jvX}) \equiv \psi(jv) = \int_{-\infty}^{\infty} e^{jvx} p(x) \, dx \quad (1.1.71)$$

where the variable v is real and $j = \sqrt{-1}$. We note that $\psi(jv)$ may be described as the Fourier transform† of the pdf $p(x)$. Hence the inverse Fourier transform is

$$p(x) = \frac{1}{2\pi} \int_{-\infty}^{\infty} \psi(jv) e^{-jvx} \, dv \qquad (1.1.72)$$

One useful property of the characteristic function is its relation to the moments of the random variable. We note that the first derivative of (1.1.71) with respect to v yields

$$\frac{d\psi(jv)}{dv} = j \int_{-\infty}^{\infty} x e^{jvx} p(x) \, dx$$

By evaluating the derivative at $v = 0$, we obtain the first moment (mean)

$$E(X) = m_x = -j \frac{d\psi(jv)}{dv} \bigg|_{v=0} \qquad (1.1.73)$$

The differentiation process can be repeated, so that the nth derivative of $\psi(jv)$ evaluated at $v = 0$ yields the nth moment

$$E(X^n) = (-j)^n \frac{d^n\psi(jv)}{dv^n} \bigg|_{v=0} \qquad (1.1.74)$$

Thus the moments of a random variable can be determined from the characteristic function. On the other hand, suppose that the characteristic function can be expanded in a Taylor series about the point $v = 0$. That is,

$$\psi(jv) = \sum_{n=0}^{\infty} \left[\frac{d^n\psi(jv)}{dv^n} \right]_{v=0} \frac{v^n}{n!} \qquad (1.1.75)$$

Using the relation in (1.1.74) to eliminate the derivative in (1.1.75), we obtain an expression for the characteristic function in terms of its moments in the form

$$\psi(jv) = \sum_{n=0}^{\infty} E(X^n) \frac{(jv)^n}{n!} \qquad (1.1.76)$$

The characteristic function provides a simple method for determining the pdf of a sum of statistically independent random variables. To illustrate this point, let X_i, $i = 1, 2, \ldots, n$, be a set of n statistically independent random variables and let

$$Y = \sum_{i=1}^{n} X_i \qquad (1.1.77)$$

The problem is to determine the pdf of Y. We shall determine the pdf of Y by first finding its characteristic function and then computing the inverse Fourier

---

† Usually the Fourier transform of a function $g(u)$ is defined as $G(v) = \int_{-\infty}^{\infty} g(u) e^{-juv} \, du$, which differs from (1.1.71) by the negative sign in the exponential. This is a trivial difference, however, so we call the integral in (1.1.71) a Fourier transform.

transform. Thus

$$\psi_Y(jv) = E(e^{jvY})$$

$$= E\left[\exp\left(jv \sum_{i=1}^{n} X_i\right)\right]$$

$$= E\left[\prod_{i=1}^{n} (e^{jvX_i})\right]$$

$$= \int_{-\infty}^{\infty} \cdots \int_{-\infty}^{\infty} \left(\prod_{i=1}^{n} e^{jvx_i}\right) p(x_1, x_2, \ldots, x_n) \, dx_1 \, dx_2 \cdots dx_n \quad (1.1.78)$$

Since the random variables are statistically independent, $p(x_1, x_2, \ldots, x_n) = p(x_1)p(x_2) \cdots p(x_n)$ and, hence, the nth-order integral in (1.1.78) reduces to a product of n single integrals, each corresponding to the characteristic function of one of the X_i. Hence,

$$\psi_Y(jv) = \prod_{i=1}^{n} \psi_{X_i}(jv) \quad (1.1.79)$$

If, in addition to their statistical independence, the X_i are identically distributed, then all the $\psi_{X_i}(jv)$ are identical. Consequently,

$$\psi_Y(jv) = [\psi_X(jv)]^n \quad (1.1.80)$$

Finally, the pdf of Y is determined from the inverse Fourier transform of $\psi_Y(jv)$, given by (1.1.72).

Since the characteristic function of the sum of n statistically independent random variables is equal to the product of the characteristic functions of the individual random variables X_i, $i = 1, 2, \ldots, n$, it follows that, in the transform domain, the pdf of Y is the n-fold convolution of the pdf's of the X_i. Usually the n-fold convolution is more difficult to perform when compared to the characteristic function method described above in determining the pdf of Y.

When working with n-dimensional random variables, it is appropriate to define an n-dimensional Fourier transform of the joint pdf. In particular, if X_i, $i = 1, 2, \ldots, n$, are random variables with pdf $p(x_1, x_2, \ldots, x_n)$, the n-dimensional characteristic function is defined as

$$\psi(jv_1, jv_2, \ldots, jv_n)$$

$$= E\left[\exp\left(j \sum_{i=1}^{n} v_i X_i\right)\right]$$

$$= \int_{-\infty}^{\infty} \cdots \int_{-\infty}^{\infty} \exp\left(j \sum_{i=1}^{n} v_i x_i\right) p(x_1, x_2, \ldots, x_n) \, dx_1 \, dx_2 \cdots dx_n \quad (1.1.81)$$

Of special interest is the two-dimensional characteristic function

$$\psi(jv_1, jv_2) = \int_{-\infty}^{\infty} \int_{-\infty}^{\infty} e^{j(v_1 x_1 + v_2 x_2)} p(x_1, x_2) \, dx_1 \, dx_2 \quad (1.1.82)$$

FIGURE 1.1.6
The probability distribution function of X.

We observe that the partial derivatives of $\psi(jv_1, jv_2)$ with respect to v_1 and v_2 can be used to generate the joint moments. For example, it is easy to show that

$$E(X_1 X_2) = -\frac{\partial^2 \psi(jv_1, jv_2)}{\partial v_1 \partial v_2}\bigg|_{v_1 = v_2 = 0} \tag{1.1.83}$$

Higher-order moments are generated in a straightforward manner.

1.1.4 Some Useful Probability Distributions

In subsequent chapters we shall encounter several different types of random variables. In this section we list these frequently encountered random variables, their pdf's, their cdf's, and their moments. We begin with the binomial distribution, which is the distribution of a discrete random variable, and then we present the distributions of several continuous random variables.

Binomial distribution. Let X be a discrete random variable that has two possible values, say $X = 1$ or $X = 0$, with probabilities p and $(1 - p)$, respectively. The pdf of X is shown in Fig. 1.1.6. Now, suppose that

$$Y = \sum_{i=1}^{n} X_i$$

where the X_i, $i = 1, 2, \ldots, n$, are statistically independent and identically distributed random variables with the pdf shown in Fig. 1.1.6. What is the probability distribution function of Y?

To answer this question, we observe that the range of Y is the set of integers from 0 to n. The probability that $Y = 0$ is simply the probability that all the $X_i = 0$. Since the X_i are statistically independent,

$$P(Y = 0) = (1 - p)^n$$

The probability that $Y = 1$ is simply the probability that one $X_i = 1$ and the rest of the $X_i = 0$. Since this event can occur in n different ways,

$$P(Y = 1) = np(1 - p)^{n-1}$$

To generalize, the probability that $Y = k$ is the probability that k of the X_i are equal to one and $(n - k)$ are equal to zero. Since there are

$$\binom{n}{k} \equiv \frac{n!}{k!(n - k)!} \tag{1.1.84}$$

different combinations which result in the event $\{Y = k\}$, it follows that

$$P(Y = k) = \binom{n}{k}p^k(1 - p)^{n-k} \tag{1.1.85}$$

where $\binom{n}{k}$ is the binomial coefficient. Consequently, the pdf of Y may be expressed as

$$p(y) = \sum_{k=0}^{n} P(Y = k)\delta(y - k)$$

$$= \sum_{k=0}^{n} \binom{n}{k}p^k(1 - p)^{n-k}\delta(y - k) \tag{1.1.86}$$

The cdf of Y is

$$F(y) = P(Y \leq y)$$

$$= \sum_{k=0}^{[y]} \binom{n}{k}p^k(1 - p)^{n-k} \tag{1.1.87}$$

where $[y]$ denotes the largest integer m such that $m \leq y$. The cdf in (1.1.87) characterizes a binomially distributed random variable.

The first two moments of Y are

$$E(Y) = np$$
$$E(Y^2) = np(1 - p) + n^2p^2 \tag{1.1.88}$$
$$\sigma^2 = np(1 - p)$$

and the characteristic function is

$$\psi(jv) = \left(1 - p + pe^{jv}\right)^n \tag{1.1.89}$$

Uniform distribution. The pdf and cdf of a uniformly distributed random variable X are shown in Fig. 1.1.7. The first two moments of X are

$$E(X) = \frac{a + b}{2}$$

$$E(X^2) = \frac{a^2 + b^2 + ab}{3} \tag{1.1.90}$$

$$\sigma^2 = \frac{(a - b)^2}{12}$$

and the characteristic function is

$$\psi(jv) = \frac{e^{jvb} - e^{jva}}{jv(b - a)} \tag{1.1.91}$$

FIGURE 1.1.7
The pdf and cdf of a uniformly distributed random variable.

Gaussian (normal) distribution. The pdf of a gaussian or normally distributed random variable is

$$p(x) = \frac{1}{\sqrt{2\pi}\,\sigma} e^{-(x-m)^2/2\sigma^2} \tag{1.1.92}$$

where m is the mean and σ^2 is the variance of the random variable. The cdf is

$$
\begin{aligned}
F(x) &= \int_{-\infty}^{x} p(u)\, du \\
&= \frac{1}{\sqrt{2\pi}\,\sigma} \int_{-\infty}^{x} e^{-(u-m)^2/2\sigma^2}\, du \\
&= \frac{1}{2}\,\frac{2}{\sqrt{\pi}} \int_{-\infty}^{(x-m)/\sqrt{2}\,\sigma} e^{-t^2}\, dt \\
&= \frac{1}{2} + \frac{1}{2}\operatorname{erf}\left(\frac{x-m}{\sqrt{2}\,\sigma}\right)
\end{aligned} \tag{1.1.93}
$$

where $\operatorname{erf}(x)$ denotes the error function which is defined as

$$\operatorname{erf}(x) = \frac{2}{\sqrt{\pi}} \int_{0}^{x} e^{-t^2}\, dt \tag{1.1.94}$$

The pdf and cdf are illustrated in Fig. 1.1.8.

(a)

(b)

FIGURE 1.1.8
The pdf and cdf of a gaussian-distributed random variable.

The cdf $F(x)$ may also be expressed in terms of the complementary error function. That is,

$$F(x) = 1 - \tfrac{1}{2}\,\mathrm{erfc}\left(\frac{x - m}{\sqrt{2}\,\sigma}\right)$$

where

$$\mathrm{erfc}(x) = \frac{2}{\sqrt{\pi}} \int_{x}^{\infty} e^{-t^2}\, dt$$

$$= 1 - \mathrm{erf}(x) \tag{1.1.95}$$

We note that $\mathrm{erf}(-x) = -\mathrm{erf}(x)$, $\mathrm{erfc}(-x) = 2 - \mathrm{erfc}(x)$, $\mathrm{erf}(0) = \mathrm{erfc}(\infty)$ $= 0$, and $\mathrm{erf}(\infty) = \mathrm{erfc}(0) = 1$.

The characteristic function of a gaussian random variable with mean m and variance σ^2 is

$$\psi(jv) = \int_{-\infty}^{\infty} e^{jvx}\left[\frac{1}{\sqrt{2\pi}\,\sigma} e^{-(x-m)^2/2\sigma^2}\right] dx$$

$$= e^{jvm - (1/2)v^2\sigma^2} \tag{1.1.96}$$

The central moments of a gaussian random variable are

$$E\left[(X - m)^k\right] \equiv \mu_k = \begin{cases} 1 \cdot 3 \cdots (k-1)\sigma^k & k \text{ even} \\ 0 & k \text{ odd} \end{cases} \tag{1.1.97}$$

and the ordinary moments may be expressed in terms of the central moments as

$$E(X^k) = \sum_{i=0}^{k} \binom{k}{i} m^i \mu_{k-i} \qquad (1.1.98)$$

The sum of n statistically independent gaussian random variables is also a gaussian random variable. To demonstrate this point, let

$$Y = \sum_{i=1}^{n} X_i \qquad (1.1.99)$$

where the X_i, $i = 1, 2, \ldots, n$, are statistically independent gaussian random variables with means m_i and variances σ_i^2. Using the result in (1.1.79), we find that the characteristic function of Y is

$$\begin{aligned}
\psi_Y(jv) &= \prod_{i=1}^{n} \psi_{X_i}(jv) \\
&= \prod_{i=1}^{n} e^{jvm_i - (1/2)v^2\sigma_i^2} \\
&= e^{jvm_y - (1/2)v^2\sigma_y^2} \qquad (1.1.100)
\end{aligned}$$

where

$$m_y = \sum_{i=1}^{n} m_i$$

$$\sigma_y^2 = \sum_{i=1}^{n} \sigma_i^2 \qquad (1.1.101)$$

Therefore, Y is gaussian-distributed with mean m_y and variance σ_y^2.

Chi-square distribution. A chi-square-distributed random variable is related to a gaussian-distributed random variable in the sense that the former can be viewed as a transformation of the latter. To be specific, let $Y = X^2$, where X is a gaussian random variable. Then Y has a chi-square distribution. We distinguish between two types of chi-square distributions. The first is called a *central chi-square distribution* and is obtained when X has zero mean. The second is called a *non-central chi-square distribution* and is obtained when X has a nonzero mean.

First we consider the central chi-square distribution. Let X be gaussian-distributed with zero mean and variance σ^2. Since $Y = X^2$, the result given in (1.1.47) applies directly with $a = 1$ and $b = 0$. Thus we obtain the pdf of Y in the form

$$p_Y(y) = \frac{1}{\sqrt{2\pi y}\,\sigma} e^{-y/2\sigma^2} \qquad y \ge 0 \qquad (1.1.102)$$

The cdf of Y is

$$F_Y(y) = \int_0^y p_Y(u)\, du$$

$$= \frac{1}{\sqrt{2\pi}\,\sigma} \int_0^y \frac{1}{\sqrt{u}} e^{-u/2\sigma^2}\, du \tag{1.1.103}$$

which cannot be expressed in a closed form. The characteristic function, however, can be determined in closed form. It is

$$\psi(jv) = \frac{1}{\left(1 - j2v\sigma^2\right)^{1/2}} \tag{1.1.104}$$

Now, suppose that the random variable Y is defined as

$$Y = \sum_{i=1}^n X_i^2 \tag{1.1.105}$$

where the X_i, $i = 1, 2, \ldots, n$, are statistically independent and identically distributed gaussian random variables with zero mean and variance σ^2. As a consequence of the statistical independence of the X_i, the characteristic function of Y is

$$\psi_Y(jv) = \frac{1}{\left(1 - j2v\sigma^2\right)^{n/2}} \tag{1.1.106}$$

The inverse transform of this characteristic function yields the pdf

$$p_Y(y) = \frac{1}{\sigma^n 2^{n/2} \Gamma(n/2)} y^{n/2-1} e^{-y/2\sigma^2} \qquad y \geq 0 \tag{1.1.107}$$

where $\Gamma(p)$ is the gamma function, defined as

$$\Gamma(p) = \int_0^\infty t^{p-1} e^{-t}\, dt \qquad p > 0$$

$$\Gamma(p) = (p-1)! \qquad p \text{ an integer, } p > 0 \tag{1.1.108}$$

$$\Gamma\!\left(\frac{1}{2}\right) = \sqrt{\pi} \qquad \Gamma\!\left(\frac{3}{2}\right) = \frac{\sqrt{\pi}}{2}$$

This pdf, which is a generalization of (1.1.102), is called a *chi-square* (or *gamma*) *pdf with n degrees of freedom*. It is illustrated in Fig. 1.1.9.

The first two moments of Y are

$$E(Y) = n\sigma^2$$

$$E(Y^2) = 2n\sigma^4 + n^2\sigma^4 \tag{1.1.109}$$

$$\sigma_y^2 = 2n\sigma^4$$

The cdf of Y is

$$F_Y(y) = \int_0^y \frac{1}{\sigma^n 2^{n/2} \Gamma(n/2)} u^{n/2-1} e^{-u/2\sigma^2}\, du \tag{1.1.110}$$

FIGURE 1.1.9
The pdf of a chi-square-distributed random variable for several degrees of freedom.

This integral can be easily manipulated into the form of the incomplete gamma function which is tabulated by Pearson (1965). When n is even, the integral in (1.1.110) can be expressed in closed form. Specifically, let $m = n/2$, where m is an integer. Then, by repeated integration by parts, we obtain

$$F_Y(y) = 1 - e^{-y/2\sigma^2} \sum_{k=0}^{m-1} \frac{1}{k!} \left(\frac{y}{2\sigma^2} \right)^k \qquad (1.1.111)$$

Let us now consider a noncentral chi-square distribution, which results from squaring a gaussian random variable having a nonzero mean. If X is gaussian with mean m and variance σ^2, the random variable $Y = X^2$ has the pdf

$$p_Y(y) = \frac{1}{\sqrt{2\pi y}\,\sigma} e^{-(y+m^2)/2\sigma^2} \cosh\left(\frac{\sqrt{y}\,m}{\sigma^2} \right) \qquad (1.1.112)$$

which is obtained by applying the result in (1.1.47) to the gaussian pdf given by (1.1.92). The characteristic function corresponding to this pdf is

$$\psi_Y(jv) = \frac{1}{\left(1 - j2v\sigma^2\right)^{1/2}} e^{jm^2v/(1-j2v\sigma^2)} \qquad (1.1.113)$$

To generalize these results, let Y be the sum of squares of gaussian random variables as defined by (1.1.105). The X_i, $i = 1, 2, \ldots, n$, are assumed to be statistically independent with means m_i, $i = 1, 2, \ldots, n$, and identical variances equal to σ^2. Then the characteristic function of Y, obtained from (1.1.113) by applying the relation in (1.1.77), is

$$\psi_Y(jv) = \frac{1}{(1 - j2v\sigma^2)^{n/2}} \exp\left(\frac{jv \sum_{i=1}^{n} m_i^2}{1 - j2v\sigma^2}\right) \qquad (1.1.114)$$

This characteristic function can be inverse-Fourier-transformed to yield the pdf

$$p_Y(y) = \frac{1}{2\sigma^2}\left(\frac{y}{s^2}\right)^{(n-2)/4} e^{-(s^2+y)/2\sigma^2} I_{n/2-1}\left(\sqrt{y}\,\frac{s}{\sigma^2}\right) \qquad y \geq 0 \quad (1.1.115)$$

where, by definition,

$$s^2 = \sum_{i=1}^{n} m_i^2 \qquad (1.1.116)$$

and $I_\alpha(x)$ is the αth-order modified Bessel function of the first kind, which may be represented by the infinite series

$$I_\alpha(x) = \sum_{k=0}^{\infty} \frac{(x/2)^{\alpha+2k}}{k!\,\Gamma(\alpha+k+1)} \qquad x \geq 0 \qquad (1.1.117)$$

The pdf given by (1.1.115) is called the *noncentral chi-square pdf with n degrees of freedom*. The parameter s^2 is called the *noncentrality parameter of the distribution*. The cdf of the noncentral chi square with n degrees of freedom is

$$F_Y(y) = \int_0^y \frac{1}{2\sigma^2}\left(\frac{u}{s^2}\right)^{(n-2)/4} e^{-(s^2+u)/2\sigma^2} I_{n/2-1}\left(\sqrt{u}\,\frac{s}{\sigma^2}\right) du \quad (1.1.118)$$

There is no closed-form expression for this integral. However, when $m = n/2$ is an integer, the cdf can be expressed in terms of the generalized Q function, which is defined as

$$Q_m(a,b) = \int_b^\infty x\left(\frac{x}{a}\right)^{m-1} e^{-(x^2+a^2)/2} I_{m-1}(ax)\, dx$$

$$= Q(a,b) + e^{(a^2+b^2)/2} \sum_{k=1}^{m-1} \left(\frac{b}{a}\right)^k I_k(ab) \qquad (1.1.119)$$

where

$$Q(a,b) \equiv Q_1(a,b) = e^{-(a^2+b^2)/2} \sum_{k=0}^{\infty} \left(\frac{a}{b}\right)^k I_k(ab) \qquad b > a > 0 \quad (1.1.120)$$

If we change the variable of integration in (1.1.118) from u to x where

$$x^2 = u/\sigma^2$$

and let $a^2 = s^2/\sigma^2$, then it is easily shown that

$$F_Y(y) = 1 - Q_m\left(\frac{s}{\sigma}, \frac{\sqrt{y}}{\sigma}\right) \qquad (1.1.121)$$

Finally, we state that the first two moments of a noncentral chi-square-distributed random variable are

$$E(Y) = n\sigma^2 + s^2$$

$$E(Y^2) = 2n\sigma^4 + 4\sigma^2 s^2 + \left(n\sigma^2 + s^2\right)^2 \qquad (1.1.122)$$

$$\sigma_y^2 = 2n\sigma^4 + 4\sigma^2 s^2$$

Rayleigh distribution. The Rayleigh distribution is closely related to the central chi-square distribution. To illustrate this point, let $Y = X_1^2 + X_2^2$ where X_1 and X_2 are zero mean statistically independent gaussian random variables each having a variance σ^2. From the discussion above, it follows that Y is chi-square-distributed with 2 degrees of freedom. Hence, the pdf of Y is

$$p_Y(y) = \frac{1}{2\sigma^2} e^{-y/2\sigma^2} \qquad y \geq 0 \qquad (1.1.123)$$

Now, suppose we define a new random variable R which is

$$R = \sqrt{X_1^2 + X_2^2} = \sqrt{Y} \qquad (1.1.124)$$

Making a simple change of variable in the pdf of (1.1.123), we obtain the pdf of R in the form

$$p_R(r) = \frac{r}{\sigma^2} e^{-r^2/2\sigma^2} \qquad r \geq 0 \qquad (1.1.125)$$

This is the pdf of a Rayleigh-distributed random variable. The corresponding cdf is

$$F_R(r) = \int_0^r \frac{u}{\sigma^2} e^{-u^2/2\sigma^2} \, du$$

$$= 1 - e^{-r^2/2\sigma^2} \qquad r \geq 0 \qquad (1.1.126)$$

The moments of R are

$$E(R^k) = (2\sigma^2)^{k/2} \Gamma\left(1 + \frac{k}{2}\right) \qquad (1.1.127)$$

and the variance is

$$\sigma_r^2 = \left(2 - \frac{\pi}{2}\right)\sigma^2 \qquad (1.1.128)$$

The characteristic function of the Rayleigh-distributed random variable is

$$\psi_R(jv) = \int_0^\infty \frac{r}{\sigma^2} e^{-r^2/2\sigma^2} e^{jvr} \, dr \qquad (1.1.129)$$

but unfortunately there is no closed-form expression. Instead we may use the Taylor series expansion given by (1.1.76) in which we substitute the moments given by (1.1.27).

As a generalization of the above expression, consider the random variable

$$R = \sqrt{\sum_{i=1}^{n} X_i^2} \tag{1.1.130}$$

where the X_i, $i = 1, 2, \ldots, n$, are statistically independent, identically distributed zero mean gaussian random variables. The random variable R has a generalized Rayleigh distribution. Clearly $Y = R^2$ is chi-square-distributed with n degrees of freedom. Its pdf is given by (1.1.107). A simple change in variable in (1.1.107) yields the pdf of R in the form

$$p_R(r) = \frac{r^{n-1}}{2^{(n-2)/2}\sigma^n \Gamma\left(\frac{n}{2}\right)} e^{-r^2/2\sigma^2} \qquad r \geq 0 \tag{1.1.131}$$

As a consequence of the functional relationship between the central chi-square and the Rayleigh distributions, the corresponding cdf's are similar. Thus, for any n, the cdf of R can be put in the form of the incomplete gamma function. In the special case when n is even, i.e., $n = 2m$, the cdf of R can be expressed in the closed form

$$F_R(r) = 1 - e^{-r^2/2\sigma^2} \sum_{k=0}^{m-1} \frac{1}{k!} \left(\frac{r^2}{2\sigma^2}\right)^k \qquad r \geq 0 \tag{1.1.132}$$

Finally, we state that the kth moment of R is

$$E(R^k) = (2\sigma^2)^{k/2} \frac{\Gamma[(n+k)/2]}{\Gamma(n/2)} \qquad k \geq 0 \tag{1.1.133}$$

which holds for any integer n.

Rice distribution. Just as the Rayleigh distribution is related to the central chi-square distribution, the Rice distribution is related to the noncentral chi-square distribution. To illustrate this relation, let $Y = X_1^2 + X_2^2$, where X_1 and X_2 are statistically independent gaussian random variables with means m_i, $i = 1, 2$, and common variance σ^2. From the previous discussion we know that Y has a noncentral chi-square distribution with noncentrality parameter $s^2 = m_1^2 + m_2^2$. The pdf of Y, obtained from (1.1.115) for $n = 2$, is

$$p_Y(y) = \frac{1}{2\sigma^2} e^{-(s^2+y)/2\sigma^2} I_0\left(\sqrt{y}\, \frac{s}{\sigma^2}\right) \qquad y \geq 0 \tag{1.1.134}$$

Now, we define a new random variable $R = \sqrt{Y}$. The pdf of R, obtained from (1.1.134) by a simple change in variable, is

$$p_R(r) = \frac{r}{\sigma^2} e^{-(r^2+s^2)/2\sigma^2} I_0\left(\frac{rs}{\sigma^2}\right) \qquad r \geq 0 \tag{1.1.135}$$

This is the pdf of a Ricean-distributed random variable. As will be shown in Chap. 4, this pdf characterizes the statistics of the envelope of a signal corrupted by additive narrowband gaussian noise. The cdf of R is easily obtained by specializing the results in (1.1.121) to the case $m = 1$. This yields

$$F_R(r) = 1 - Q\left(\frac{s}{\sigma}, \frac{r}{\sigma}\right) \tag{1.1.136}$$

where the Q function is defined by (1.1.120).

As a generalization of the expressions given above, let R be defined as in (1.1.130) where the X_i, $i = 1, 2, \ldots, n$ are statistically independent gaussian random variables with means m_i, $i = 1, 2, \ldots, n$, and identical variances equal to σ^2. The random variable $R^2 = Y$ has a noncentral chi-square distribution with n degrees of freedom and noncentrality parameter s^2 given by (1.1.116). Its pdf is given by (1.1.115). Hence the pdf of R is

$$p_R(r) = \frac{r^{n/2}}{\sigma^2 s^{(n-2)/2}} e^{-(r^2+s^2)/2\sigma^2} I_{n/2-1}\left(\frac{rs}{\sigma^2}\right) \tag{1.1.137}$$

and the corresponding cdf is

$$F_R(r) = P(R \le r) = P(\sqrt{Y} \le r) = P(Y \le r^2) = F_Y(r^2) \tag{1.1.138}$$

where $F_Y(r^2)$ is given by (1.1.118). In the special case where $m = n/2$ is an integer, we have

$$F_R(r) = 1 - Q_m\left(\frac{s}{\sigma}, \frac{r}{\sigma}\right) \tag{1.1.139}$$

which follows from (1.1.121). Finally, we state that the kth moment of R is

$$E(R^k) = (2\sigma^2)^{k/2} e^{-s^2/2\sigma^2} \frac{\Gamma[(n+k)/2]}{\Gamma(n/2)} {}_1F_1\left(\frac{n+k}{2}, \frac{n}{2}; \frac{s^2}{2\sigma^2}\right) \qquad k \ge 0 \tag{1.1.140}$$

where ${}_1F_1(\alpha, \beta; x)$ is the confluent hypergeometric function defined as

$$_1F_1(\alpha, \beta; x) = \sum_{k=0}^{\infty} \frac{\Gamma(\alpha+k)\Gamma(\beta)x^k}{\Gamma(\alpha)\Gamma(\beta+k)k!} \qquad \beta \ne 0, -1, -2, \ldots \tag{1.1.141}$$

Multivariate gaussian distribution. Of the many multivariate or multidimensional distributions that can be defined, the multivariate gaussian distribution is the most important and the one most likely to be encountered in practice. We shall briefly introduce this distribution and state its basic properties.

Let us assume that X_i, $i = 1, 2, \ldots, n$, are gaussian random variables with means m_i, $i = 1, 2, \ldots, n$ variances σ_i^2, $i = 1, 2, \ldots, n$ and covariances μ_{ij}, $i, j = 1, 2, \ldots, n$. Clearly, $\mu_{ii} = \sigma_i^2$, $i = 1, 2, \ldots, n$. Let $\mathbf{M}$ denote the $n \times n$ covariance matrix with elements $\{\mu_{ij}\}$, let $\mathbf{X}$ denote the $n \times 1$ column vector of random variables, and let $\mathbf{m}_x$ denote the $n \times 1$ column vector of mean values m_i, $i = 1, 2, \ldots, n$. The joint pdf of the gaussian random variables X_i, $i = 1, 2, \ldots, n$, is defined as

$$p(x_1, x_2, \ldots, x_n) = \frac{1}{(2\pi)^{n/2}[\det(\mathbf{M})]^{1/2}} \exp\left[-\tfrac{1}{2}(\mathbf{x} - \mathbf{m}_x)'\mathbf{M}^{-1}(\mathbf{x} - \mathbf{m}_x)\right] \tag{1.1.142}$$

where $\mathbf{M}^{-1}$ denotes the inverse of $\mathbf{M}$ and $\mathbf{x}'$ denotes the transpose of $\mathbf{x}$.

The characteristic function corresponding to this n-dimensional joint pdf is

$$\psi(j\mathbf{v}) = E(e^{j\mathbf{v}'\mathbf{x}})$$

where $\mathbf{v}$ is an n-dimensional vector with elements v_i, $i = 1, 2, \ldots, n$. Evaluation of this n-dimensional Fourier transform yields the result

$$\psi(j\mathbf{v}) = \exp\left(j\mathbf{m}_x'\mathbf{v} - \tfrac{1}{2}\mathbf{v}'\mathbf{M}\mathbf{v}\right) \tag{1.1.143}$$

An important special case of (1.1.142) is the bivariate or two-dimensional gaussian pdf. The mean $\mathbf{m}_x$ and the covariance matrix $\mathbf{M}$ for this case are

$$\mathbf{m}_x = \begin{bmatrix} m_1 \\ m_2 \end{bmatrix} \qquad \mathbf{M} = \begin{bmatrix} \sigma_1^2 & \mu_{12} \\ \mu_{12} & \sigma_2^2 \end{bmatrix} \tag{1.1.144}$$

where the joint central moment μ_{12} is defined as

$$\mu_{12} = E\left[(X_1 - m_1)(X_2 - m_2)\right]$$

It is convenient to define a normalized covariance

$$\rho_{ij} = \frac{\mu_{ij}}{\sigma_i\sigma_j} \qquad i \neq j \tag{1.1.145}$$

where ρ_{ij} satisfies the condition $0 \leq |\rho_{ij}| \leq 1$. When dealing with the two-dimensional case, it is customary to drop the subscripts on μ_{12} and ρ_{12}. Hence the covariance matrix is expressed as

$$\mathbf{M} = \begin{bmatrix} \sigma_1^2 & \rho\sigma_1\sigma_2 \\ \rho\sigma_1\sigma_2 & \sigma_2^2 \end{bmatrix} \tag{1.1.146}$$

Its inverse is

$$\mathbf{M}^{-1} = \frac{1}{\sigma_1^2\sigma_2^2(1 - \rho^2)} \begin{bmatrix} \sigma_2^2 & -\rho\sigma_1\sigma_2 \\ -\rho\sigma_1\sigma_2 & \sigma_1^2 \end{bmatrix} \tag{1.1.147}$$

and $\det(\mathbf{M}) = \sigma_1^2\sigma_2^2(1 - \rho^2)$. Substitution for $\mathbf{M}^{-1}$ into (1.1.142) yields the desired bivariate gaussian pdf in the form

$$p(x_1, x_2) = \frac{1}{2\pi\sigma_1\sigma_2\sqrt{1 - \rho^2}}$$

$$\times \exp\left(-\frac{\sigma_2^2(x_1 - m_1)^2 - 2\rho\sigma_1\sigma_2(x_1 - m_1)(x_2 - m_2) + \sigma_1^2(x_2 - m_2)^2}{2\sigma_1^2\sigma_2^2(1 - \rho^2)}\right) \tag{1.1.148}$$

We note that when $\rho = 0$, the joint pdf $p(x_1, x_2)$ in (1.1.148) factors into the product $p(x_1)p(x_2)$, where $p(x_i)$, $i = 1, 2$, are the marginal pdf's. Since ρ is a measure of the correlation between X_1 and X_2, we have shown that when the gaussian random variables X_1 and X_2 are uncorrelated, they are also statistically independent. This is an important property of gaussian random variables which does not hold in general for other distributions. It extends to n-dimensional gaussian random variables in a straightforward manner. That is, if $\rho_{ij} = 0$ for $i \neq j$, the random variables X_i, $i = 1, 2, \ldots, n$ are uncorrelated and, hence, statistically independent.

Now, let us consider a linear transformation of n gaussian random variables X_i, $i = 1, 2, \ldots, n$, with mean vector $\mathbf{m}_x$ and covariance matrix $\mathbf{M}$. Let

$$\mathbf{Y} = \mathbf{AX} \qquad (1.1.149)$$

where $\mathbf{A}$ is a nonsingular matrix. As shown previously, the jacobian of this transformation is $J = 1/\det(\mathbf{A})$. Since $\mathbf{X} = \mathbf{A}^{-1}\mathbf{Y}$, we may substitute for $\mathbf{X}$ in (1.1.142) and, thus, we obtain the joint pdf of $\mathbf{Y}$ in the form

$$p(\mathbf{y}) = \frac{1}{(2\pi)^{n/2}[\det(\mathbf{M})]^{1/2}\det(\mathbf{A})}\exp\left[-\tfrac{1}{2}(\mathbf{A}^{-1}\mathbf{y} - \mathbf{m}_x)'\mathbf{M}^{-1}(\mathbf{A}^{-1}\mathbf{y} - \mathbf{m}_x)\right]$$

$$= \frac{1}{(2\pi)^{n/2}[\det(\mathbf{Q})]^{1/2}}\exp\left[-\tfrac{1}{2}(\mathbf{y} - \mathbf{m}_y)'\mathbf{Q}^{-1}(\mathbf{y} - \mathbf{m}_y)\right] \qquad (1.1.150)$$

where the vector $\mathbf{m}_y$ and the matrix $\mathbf{Q}$ are defined as

$$\begin{aligned}\mathbf{m}_y &= \mathbf{Am}_x \\ \mathbf{Q} &= \mathbf{AMA}'\end{aligned} \qquad (1.1.151)$$

Thus we have shown that a linear transformation of a set of jointly gaussian random variables results in another set of jointly gaussian random variables.

Suppose that we wish to perform a linear transformation that results in n statistically independent gaussian random variables. How should the matrix $\mathbf{A}$ be selected? From our previous discussion, we know that the gaussian random variables are statistically independent if they are pairwise-uncorrelated, i.e., if the covariance matrix $\mathbf{Q}$ is diagonal. Therefore, we must have

$$\mathbf{AMA}' = \mathbf{D} \qquad (1.1.152)$$

where $\mathbf{D}$ is a diagonal matrix. The matrix $\mathbf{M}$ is a covariance matrix; hence it is positive definite. One solution is to select $\mathbf{A}$ to be an orthogonal matrix $(\mathbf{A}' = \mathbf{A}^{-1})$ consisting of columns which are the eigenvectors of the covariance matrix $\mathbf{M}$. Then $\mathbf{D}$ is a diagonal matrix with diagonal elements equal to the eigenvalues of $\mathbf{M}$.

Example 1.1.5. Consider the bivariate gaussian pdf with covariance matrix

$$\mathbf{M} = \begin{bmatrix} 1 & \tfrac{1}{2} \\ \tfrac{1}{2} & 1 \end{bmatrix}$$

Let us determine the transformation $\mathbf{A}$ that will result in uncorrelated random variables. First, we solve for the eigenvalues of $\mathbf{M}$. The characteristic equation is

$$\det(\mathbf{M} - \lambda \mathbf{I}) = 0$$

$$(1 - \lambda)^2 + \tfrac{1}{4} = 0$$

$$\lambda = \tfrac{3}{2}, \tfrac{1}{2}$$

Next we determine the two eigenvectors. If $\mathbf{a}$ denotes an eigenvector, we have

$$(\mathbf{M} - \lambda \mathbf{I})\mathbf{a} = 0$$

With $\lambda_1 = \tfrac{3}{2}$ and $\lambda_2 = \tfrac{1}{2}$, we obtain the eigenvectors

$$\mathbf{a}_1 = \begin{bmatrix} \dfrac{1}{\sqrt{2}} \\ \dfrac{1}{\sqrt{2}} \end{bmatrix} \qquad \mathbf{a}_2 = \begin{bmatrix} \dfrac{1}{\sqrt{2}} \\ \dfrac{-1}{\sqrt{2}} \end{bmatrix}$$

Therefore,

$$\mathbf{A} = \frac{1}{\sqrt{2}} \begin{bmatrix} 1 & 1 \\ 1 & -1 \end{bmatrix}$$

It is easily verified that $\mathbf{A}^{-1} = \mathbf{A}'$ and that

$$\mathbf{A}\mathbf{M}\mathbf{A}' = \mathbf{D}$$

where the diagonal elements of $\mathbf{D}$ are $\tfrac{3}{2}$ and $\tfrac{1}{2}$.

1.1.5 Upper Bounds on the Tail Probability

In evaluating the performance of a digital communication system, it is oftentimes necessary to determine the area under the tail of the pdf. We refer to this area as the *tail probability*. In this section we present two upper bounds on the tail probability. The first, obtained from the Chebyshev inequality, is rather loose. The second, called the *Chernoff bound*, is much tighter.

Chebyshev inequality. Suppose that X is an arbitrary random variable with finite mean m_x and finite variance σ_x^2. For any positive number δ,

$$P(|X - m_x| \geq \delta) \leq \frac{\sigma_x^2}{\delta^2} \qquad (1.1.153)$$

This relation is called the *Chebyshev inequality*. The proof of this bound is relatively simple. We have

$$\sigma_x^2 = \int_{-\infty}^{\infty} (x - m_x)^2 p(x)\, dx \geq \int_{|x - m_x| \geq \delta} (x - m_x)^2 p(x)\, dx$$

$$\geq \delta^2 \int_{|x - m_x| \geq \delta} p(x)\, dx = \delta^2 P(|X - m_x| \geq \delta)$$

Thus the validity of the inequality is established.

It is apparent that the Chebyshev inequality is simply an upper bound on the area under the tails of the pdf $p(y)$, where $Y = X - m_x$, i.e., the area of

$p(y)$ in the intervals $(-\infty, -\delta)$ and (δ, ∞). Hence the Chebyshev inequality may be expressed as

$$1 - \left[F_Y(\delta) - F_Y(-\delta) \right] \le \frac{\sigma_x^2}{\delta^2} \tag{1.1.154}$$

or, equivalently, as

$$1 - \left[F_X(m_x + \delta) - F_X(m_x - \delta) \right] \le \frac{\sigma_x^2}{\delta^2} \tag{1.1.155}$$

There is another way to view the Chebyshev bound. Working with the zero mean random variable $Y = X - m_x$, for convenience, suppose we define a function $g(Y)$ as

$$g(Y) = \begin{cases} 1 & \text{if } |Y| \ge \delta \\ 0 & \text{if } |Y| < \delta \end{cases} \tag{1.1.156}$$

Since $g(Y)$ is either 0 or 1 with probabilities $P(|Y| < \delta)$ and $P(|Y| \ge \delta)$, respectively, its mean value is

$$E[g(Y)] = P(|Y| \ge \delta) \tag{1.1.157}$$

Now suppose that we upper-bound $g(Y)$ by the quadratic $(Y/\delta)^2$, i.e.,

$$g(Y) \le \left(\frac{Y}{\delta} \right)^2 \tag{1.1.158}$$

The graph of $g(Y)$ and the upper bound are shown in Fig. 1.1.10. It follows that

$$E[g(Y)] \le E\left(\frac{Y^2}{\delta^2} \right) = \frac{E(Y^2)}{\delta^2} = \frac{\sigma_y^2}{\delta^2} = \frac{\sigma_x^2}{\delta^2}$$

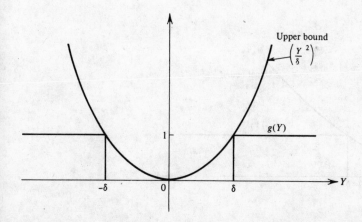

FIGURE 1.1.10
A quadratic upper bound on $g(Y)$ used in obtaining the tail probability (Chebyshev bound).

Since $E[g(Y)]$ is the tail probability, as seen from (1.1.157), we have obtained the Chebyshev bound.

For many practical applications, the Chebyshev bound is extremely loose. The reason for this may be attributed to the looseness of the quadratic $(Y/\delta)^2$ in overbounding $g(Y)$. There are certainly many other functions that can be used to overbound $g(Y)$. Below, we use an exponential bound to derive an upper bound on the tail probability that is extremely tight.

Chernoff bound. The Chebyshev bound given above involves the area under the two tails of the pdf. In some applications we are interested only in the area under one tail, either in the interval (δ, ∞) or in the interval $(-\infty, \delta)$. In such a case we can obtain an extremely tight upper bound by overbounding the function $g(Y)$ by an exponential having a parameter that can be optimized to yield as tight an upper bound as possible. Specifically, we consider the tail probability in the interval (δ, ∞). The function $g(Y)$ is overbounded as

$$g(Y) \le e^{\nu(Y-\delta)} \qquad (1.1.159)$$

where $g(Y)$ is now defined as

$$g(Y) = \begin{cases} 1 & Y \ge \delta \\ 0 & Y < \delta \end{cases} \qquad (1.1.160)$$

and $\nu \ge 0$ is the parameter to be optimized. The graph of $g(Y)$ and the exponential upper bound are shown in Fig. 1.1.11.

The expected value of $g(Y)$ is

$$E[g(Y)] = P(Y \ge \delta) \le E(e^{\nu(Y-\delta)}) \qquad (1.1.161)$$

This bound is valid for any $\nu \ge 0$. The tightest upper bound is obtained by selecting the value of ν which minimizes $E(e^{\nu(Y-\delta)})$. A necessary condition for a

FIGURE 1.1.11
An exponential upper bound on $g(Y)$ used in obtaining the tail probability (Chernoff bound).

minimum is

$$\frac{d}{d\nu}E(e^{\nu(Y-\delta)}) = 0 \tag{1.1.162}$$

But the order of differentiation and expectation can be interchanged, so that

$$\frac{d}{d\nu}E(e^{\nu(Y-\delta)}) = E\left(\frac{d}{d\nu}e^{\nu(Y-\delta)}\right)$$

$$= E\left[(Y-\delta)e^{\nu(Y-\delta)}\right]$$

$$= e^{-\nu\delta}\left[E(Ye^{\nu Y}) - \delta E(e^{\nu Y})\right] = 0$$

Therefore the value of ν that gives the tightest upper bound is the solution to the equation

$$E(Ye^{\nu Y}) - \delta E(e^{\nu Y}) = 0 \tag{1.1.163}$$

Let $\hat{\nu}$ be the solution of (1.1.163). Then, from (1.1.161), the upper bound on the one-sided tail probability is

$$P(Y \geq \delta) \leq e^{-\hat{\nu}\delta}E(e^{\hat{\nu}Y}) \tag{1.1.164}$$

This is the Chernoff bound for the upper tail probability for a discrete or a continuous random variable having a zero mean.†

An upper bound on the lower tail probability can be obtained in a similar manner, with the result that

$$P(Y \leq \delta) \leq e^{-\hat{\nu}\delta}E(e^{\hat{\nu}Y}) \tag{1.1.165}$$

where $\hat{\nu}$ is the solution to (1.1.163) and $\delta < 0$.

Example 1.1.6. Consider the (Laplace) pdf

$$p(y) = \tfrac{1}{2}e^{-|y|} \tag{1.1.166}$$

which is illustrated in Fig. 1.1.12. Let us evaluate the upper tail probability from the Chernoff bound and compare it to the true tail probability, which is

$$P(Y \geq \delta) = \int_{\delta}^{\infty} \tfrac{1}{2}e^{-y}\,dy = \tfrac{1}{2}e^{-\delta} \tag{1.1.167}$$

To solve (1.1.163) for $\hat{\nu}$, we must determine the moments $E(Ye^{\nu Y})$ and $E(e^{\nu Y})$. For the pdf in (1.1.166) we find that

$$E(Ye^{\nu Y}) = \frac{2\nu}{(\nu+1)^2(\nu-1)^2}$$

$$\tag{1.1.168}$$

$$E(e^{\nu Y}) = \frac{1}{(1+\nu)(1-\nu)}$$

Substituting these moments into (1.1.163), we obtain the quadratic equation

$$\nu^2\delta + 2\nu - \delta = 0$$

---

† Note that $E[e^{\nu Y}]$ for ν real is not the characteristic function of Y. It is called the *moment generating function* of Y.

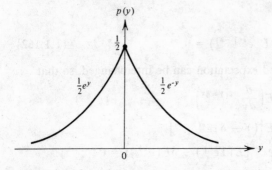

FIGURE 1.1.12
The pdf of a Laplace-distributed random variable.

which has the solutions

$$\hat{\nu} = \frac{-1 \pm \sqrt{1 + \delta^2}}{\delta} \qquad (1.1.169)$$

Since $\hat{\nu}$ must be positive, one of the two solutions is discarded. Thus

$$\hat{\nu} = \frac{-1 + \sqrt{1 + \delta^2}}{\delta} \qquad (1.1.170)$$

Finally, we evaluate the upper bound in (1.1.164) by eliminating $E(e^{\hat{\nu}Y})$ using the second relation in (1.1.168) and by substituting for $\hat{\nu}$ from (1.1.170). The result is

$$P(Y \geq \delta) \leq \frac{\delta^2}{2(-1 + \sqrt{1 + \delta^2})} e^{1 - \sqrt{1 + \delta^2}} \qquad (1.1.171)$$

For $\delta \gg 1$, (1.1.171) reduces to

$$P(Y \geq \delta) \leq \frac{\delta}{2} e^{-\delta} \qquad (1.1.172)$$

We note that the Chernoff bound decreases exponentially as δ increases. Consequently, it approximates closely the exact tail probability given by (1.1.167). In contrast, the Chebyshev upper bound for the upper tail probability obtained by taking one-half of the probability in the two tails (due to symmetry in the pdf) is

$$P(Y \geq \delta) \leq \frac{1}{\delta^2}$$

Hence this bound is extremely loose.

When the random variable has a nonzero mean, the Chernoff bound can be extended as we now demonstrate. If $Y = X - m_x$, we have

$$P(Y \geq \delta) = P(X - m_x \geq \delta) = P(X \geq m_x + \delta) = P(X \geq \delta_m)$$

where, by definition, $\delta_m = m_x + \delta$. Since $\delta > 0$, it follows that $\delta_m > m_x$. Let $g(X)$ be defined as

$$g(X) = \begin{cases} 1 & \text{if } X \geq \delta_m \\ 0 & \text{if } X < \delta_m \end{cases} \qquad (1.1.173)$$

and upper-bounded as

$$g(X) \leq e^{\nu(X-\delta_m)} \tag{1.1.174}$$

From this point, the derivation parallels the steps contained in (1.1.161) through (1.1.164). The final result is

$$P(X \geq \delta_m) \leq e^{-\hat{\nu}\delta_m} E(e^{\hat{\nu}X}) \tag{1.1.175}$$

where $\delta_m > m_x$ and $\hat{\nu}$ is the solution to the equation

$$E(Xe^{\nu X}) - \delta_m E(e^{\nu X}) = 0 \tag{1.1.176}$$

In a similar manner, we can obtain the Chernoff bound for the lower tail probability. For $\delta < 0$, we have

$$P(X - m_x \leq \delta) = P(X \leq m_x + \delta) = P(X \leq \delta_m) \leq E[e^{\nu(X-\delta_m)}] \tag{1.1.177}$$

From our previous development, it is apparent that (1.1.177) results in the bound

$$P(X \leq \delta_m) \leq e^{-\hat{\nu}\delta_m} E(e^{\hat{\nu}X}) \tag{1.1.178}$$

where $\delta_m < m_x$ and $\hat{\nu}$ is the solution to (1.1.176).

1.1.6 Sums of Random Variables and the Central Limit Theorem

We have previously considered the problem of determining the pdf of a sum of n statistically independent random variables. In this section, we again consider the sum of statistically independent random variables, but our approach is different and is independent of the particular pdf of the random variables in the sum. To be specific, suppose that X_i, $i = 1, 2, \ldots, n$, are statistically independent and identically distributed random variables, each having a finite mean m_x and a finite variance σ_x^2. Let Y be defined as the normalized sum, called the *sample mean*:

$$Y = \frac{1}{n} \sum_{i=1}^{n} X_i \tag{1.1.179}$$

First we shall determine upper bounds on the tail probabilities of Y and then we shall prove a very important theorem regarding the pdf of Y in the limit as $n \to \infty$.

The random variable Y defined in (1.1.179) is frequently encountered in estimating the mean of a random variable X from a number of observations X_i, $i = 1, 2, \ldots, n$. In other words, the X_i, $i = 1, 2, \ldots, n$, may be considered as independent samples drawn from a distribution $F_X(x)$, and Y is the estimate of the mean m_x.

The mean of Y is

$$E(Y) = m_y = \frac{1}{n} \sum_{i=1}^{n} E(X_i)$$

$$= m_x$$

The variance of Y is

$$\sigma_y^2 = E(Y^2) - m_y^2 = E(Y^2) - m_x^2$$

$$= \frac{1}{n^2} \sum_{i=1}^{n} \sum_{j=1}^{n} E(X_i X_j) - m_x^2$$

$$= \frac{1}{n^2} \sum_{i=1}^{n} E(X_i^2) + \frac{1}{n^2} \sum_{\substack{i=1 \\ i \neq j}}^{n} \sum_{j=1}^{n} E(X_i) E(X_j) - m_x^2$$

$$= \frac{1}{n}(\sigma_x^2 + m_x^2) + \frac{1}{n^2} n(n-1) m_x^2 - m_x^2$$

$$= \frac{\sigma_x^2}{n}$$

When Y is viewed as an estimate for the mean m_x, we note that its expected value is equal to m_x and its variance decreases inversely with the number of samples n. As n approaches infinity, the variance σ_y^2 approaches zero. An estimate of a parameter (in this case the mean m_x) which satisfies the conditions that its expected value converges to the true value of the parameter and the variance converges to zero as $n \to \infty$ is said to be a *consistent estimate*.

The tail probability of the random variable Y can be upper-bounded by use of the bounds presented in Sec. 1.1.5. The Chebyshev inequality applied to Y is

$$P(|Y - m_y| \geq \delta) \leq \frac{\sigma_y^2}{\delta^2}$$

$$P\left(\left|\frac{1}{n} \sum_{i=1}^{n} X_i - m_x\right| \geq \delta\right) \leq \frac{\sigma_x^2}{n\delta^2} \tag{1.1.180}$$

In the limit as $n \to \infty$, (1.1.180) becomes

$$\lim_{n \to \infty} P\left(\left|\frac{1}{n} \sum_{i=1}^{n} X_i - m_x\right| \geq \delta\right) = 0 \tag{1.1.181}$$

Therefore the probability that the estimate of the mean differs from the true mean m_x by more than δ ($\delta > 0$) approaches zero as n approaches infinity. This statement is a form of the law of large numbers. Since the upper bound converges to zero relatively slowly, i.e., inversely with n, the expression in (1.1.180) is called the *weak law of large numbers*.

The Chernoff bound applied to the random variable Y yields an exponential dependence on n and thus provides a tighter upper bound on the one-sided tail probability. Following the procedure developed in Sec. 1.1.5, we can determine that the tail probability for Y is

$$P(Y - m_y \geq \delta) = P\left(\frac{1}{n} \sum_{i=1}^{n} X_i - m_x \geq \delta \right)$$

$$= P\left(\sum_{i=1}^{n} X_i \geq n\delta_m \right) \leq E\left\{ \exp\left[\nu\left(\sum_{i=1}^{n} X_i - n\delta_m \right) \right] \right\} \quad (1.1.182)$$

where $\delta_m = m_x + \delta$ and $\delta > 0$. But the X_i, $i = 1, 2, \ldots, n$, are statistically independent and identically distributed. Hence

$$E\left\{ \exp\left[\nu\left(\sum_{i=1}^{n} X_i - n\delta_m \right) \right] \right\} = e^{-\nu n \delta_m} E\left[\exp\left(\nu \sum_{i=1}^{n} X_i \right) \right]$$

$$= e^{-\nu n \delta_m} \prod_{i=1}^{n} E(e^{\nu X_i})$$

$$= \left[e^{-\nu \delta_m} E(e^{\nu X}) \right]^n \quad (1.1.183)$$

where X denotes any one of the X_i. The parameter ν which yields the tightest upper bound is obtained by differentiating (1.1.183) and setting the derivative to zero. This yields the equation

$$E(Xe^{\nu X}) - \delta_m E(e^{\nu X}) = 0 \quad (1.1.184)$$

Let the solution of (1.1.184) be denoted as $\hat{\nu}$. Then, the bound on the upper tail probability is

$$P\left(\frac{1}{n} \sum_{i=1}^{n} X_i \geq \delta_m \right) \leq \left[e^{-\hat{\nu}\delta_m} E(e^{\hat{\nu} X}) \right]^n \quad \delta_m > m_x \quad (1.1.185)$$

In a similar manner we find that the lower tail probability is upper-bounded as

$$P(Y \leq \delta_m) \leq \left[e^{-\hat{\nu}\delta_m} E(e^{\hat{\nu} X}) \right]^n \quad \delta_m < m_x \quad (1.1.186)$$

where $\hat{\nu}$ is the solution to (1.1.184).

Example 1.1.7. Let X_i, $i = 1, 2, \ldots, n$, be a set of statistically independent random variables defined as

$$X_i = \begin{cases} 1 & \text{with probability } p < \frac{1}{2} \\ -1 & \text{with probability } 1 - p \end{cases}$$

We wish to determine a tight upper bound on the probability that the sum of the X_i is greater than zero. Since $p < \frac{1}{2}$, we note that the sum will have a negative value for the mean; hence we seek the upper tail probability. With $\delta_m = 0$ in (1.1.185) we have

$$P\left(\sum_{i=1}^{n} X_i \geq 0 \right) \leq \left[E(e^{\hat{\nu} X}) \right]^n \quad (1.1.187)$$

where $\hat{\nu}$ is the solution to the equation

$$E(Xe^{\nu X}) = 0 \tag{1.1.188}$$

Now

$$E(Xe^{\nu X}) = -(1 - p)e^{-\nu} + pe^{\nu} = 0$$

Hence

$$\hat{\nu} = \ln\left(\sqrt{\frac{1 - p}{p}}\right) \tag{1.1.189}$$

where $\ln(u)$ denotes the natural logarithm of u. Furthermore,

$$E(e^{\hat{\nu} X}) = pe^{\hat{\nu}} + (1 - p)e^{-\hat{\nu}}$$

Therefore the bound in (1.1.187) becomes

$$P\left(\sum_{i=1}^{n} X_i \geq 0\right) \leq [pe^{\hat{\nu}} + (1 - p)e^{-\hat{\nu}}]^n$$

$$\leq \left[p\sqrt{\frac{1 - p}{p}} + (1 - p)\sqrt{\frac{p}{1 - p}}\right]^n$$

$$\leq [4p(1 - p)]^{n/2} \tag{1.1.190}$$

We observe that the upper bound decays exponentially with n, as expected. In contrast, if the Chebyshev bound were evaluated, the tail probability would decrease inversely with n.

Central limit theorem. We conclude this section with an extremely useful theorem concerning the cdf of a sum of random variables in the limit as the number of terms in the sum approaches infinity. There are several versions of this theorem. We shall prove the theorem for the case in which the random variables X_i, $i = 1, 2, \ldots, n$, being summed are statistically independent and identically distributed, each having a finite mean m_x and a finite variance σ_x^2. For convenience, we define the normalized random variable

$$U_i = \frac{X_i - M_x}{\sigma_x} \qquad i = 1, 2, \ldots, n$$

Thus U_i has a zero mean and unit variance. Now, let

$$Y = \frac{1}{\sqrt{n}} \sum_{i=1}^{n} U_i \tag{1.1.191}$$

Since each term in the sum has a zero mean and unit variance, it follows that the normalized (by $1/\sqrt{n}$) random variable Y has zero mean and unit variance. We wish to determine the cdf of Y in the limit as $n \to \infty$.

The characteristic function of Y is

$$\psi_Y(jv) = E(e^{jvY}) = E\left[\exp\left(\frac{jv\sum_{i=1}^{n} U_i}{\sqrt{n}}\right)\right]$$

$$= \prod_{i=1}^{n} \psi_{U_i}\left(\frac{jv}{\sqrt{n}}\right)$$

$$= \left[\psi_U\left(\frac{jv}{\sqrt{n}}\right)\right]^n \tag{1.1.192}$$

where U denotes any of the U_i, which are identically distributed. Now, let us expand the characteristic function of U in a Taylor series. The expansion yields

$$\psi_U\left(j\frac{v}{\sqrt{n}}\right) = 1 + j\frac{v}{\sqrt{n}}E(U) - \frac{v^2}{n2!}E(U^2) + \frac{(jv)^3}{(\sqrt{n})^3 3!}E(U^3) - \cdots \tag{1.1.193}$$

Since $E(U) = 0$ and $E(U^2) = 1$, (1.1.193) simplifies to

$$\psi_U\left(\frac{jv}{\sqrt{n}}\right) = 1 - \frac{v^2}{2n} + \frac{1}{n}R(v, n) \tag{1.1.194}$$

where $R(v, n)/n$ denotes the remainder. We note that $R(v, n)$ approaches zero as $n \to \infty$. Substitution of (1.1.194) into (1.1.192) yields the characteristic function of Y in the form

$$\psi_Y(jv) = \left[1 - \frac{v^2}{2n} + \frac{R(v, n)}{n}\right]^n \tag{1.1.195}$$

Taking the natural logarithm of (1.1.195), we obtain

$$\ln\psi_Y(jv) = n\ln\left[1 - \frac{v^2}{2n} + \frac{R(v, n)}{n}\right] \tag{1.1.196}$$

For small values of x, $\ln(1 + x)$ can be expanded in the power series

$$\ln(1 + x) = x - \frac{x^2}{2} + \frac{x^3}{3} - \cdots$$

This expansion applied to (1.1.196) yields

$$\ln\psi_Y(jv) = n\left[-\frac{v^2}{2n} + \frac{R(v, n)}{n} - \frac{1}{2}\left(-\frac{v^2}{2n} + \frac{R(v, n)}{n}\right)^2 + \cdots\right] \tag{1.1.197}$$

Finally, when we take the limit as $n \to \infty$, (1.1.197) reduces to $\lim_{n \to \infty} \ln \psi_Y(jv) = -v^2/2$ or, equivalently,

$$\lim_{n \to \infty} \psi_Y(jv) = e^{-v^2/2} \qquad (1.1.198)$$

But, this is just the characteristic function of a gaussian random variable with zero mean and unit variance. Thus we have the important result that the sum of statistically independent and identically distributed random variables with finite mean and variance approaches a gaussian cdf as $n \to \infty$. This result is known as the *central limit theorem*.

Although we assumed that the random variables in the sum are identically distributed, the assumption can be relaxed provided that additional restrictions are imposed on the properties of the random variables. There is one variation of the theorem, for example, in which the assumption of identically distributed random variables is abandoned in favor of a condition on the third absolute moment of the random variables in the sum. For a discussion of this and other variations of the central limit theorem, the reader is referred to the book by Cramer (1946).

1.2 STOCHASTIC PROCESSES

Many of the random phenomena that occur in nature are functions of time. For example, the meteorological phenomena such as the random fluctuations in air temperature and air pressure are functions of time. The thermal noise voltages generated in the resistors of an electronic device such as a radio receiver are also a function of time. Similarly, the signal at the output of a source that generates information is characterized as a random signal that varies with time. An audio signal that is transmitted over a telephone channel is an example of such a signal. All these are examples of stochastic (random) processes. In our study of digital communications we encounter stochastic processes in the characterization and modeling of signals generated by information sources, in the characterization of communication channels used to transmit the information, in the characterization of noise generated in a receiver, and in the design of the optimum receiver for processing the received random signal.

At any given time instant the value of a stochastic process, whether it is the value of the noise voltage generated by a resistor or the amplitude of the signal generated by an audio source, is a random variable. Thus we may view a stochastic process as a random variable indexed by the parameter t. We shall denote such a process as $X(t)$. In general, the parameter t is continuous, whereas X may be either continuous or discrete depending on the characteristics of the source that generates the stochastic process.

The noise voltage generated by a single resistor or a single information source represents a single realization of the stochastic process. Hence it is called a *sample function* of the stochastic process. The set of all possible sample functions, e.g., the set of all noise voltage waveforms generated by resistors, constitute an ensemble of sample functions or, equivalently, the stochastic process $X(t)$. In general, the number of sample functions in the ensemble is assumed to be extremely large; often it is infinite.

Having defined a stochastic process $X(t)$ as an ensemble of sample functions, we may consider the values of the process at any set of time instants $t_1 > t_2 > t_3 > \cdots > t_n$ where n is any positive integer. In general, the random variables $X_{t_i} \equiv X(t_i)$, $i = 1, 2, \ldots, n$, are characterized statistically by their joint pdf $p(x_{t_1}, x_{t_2}, \ldots, x_{t_n})$. Furthermore, all the probabilistic relations defined in Sec. 1.1 for multidimensional random variables carry over to the random variables X_{t_i}, $i = 1, 2, \ldots, n$.

Stationary stochastic processes. As indicated above, the random variables X_{t_i}, $i = 1, 2, \ldots, n$, obtained from the stochastic process $X(t)$ for any set of time instants $t_1 > t_2 > t_3 > \cdots > t_n$ and any n are characterized statistically by the joint pdf $p(x_{t_1}, x_{t_2}, \ldots, x_{t_n})$. Let us consider another set of n random variables $X_{t_i + t} \equiv X(t_i + t)$, $i = 1, 2, \ldots, n$, where t is an arbitrary time shift. These random variables are characterized by the joint pdf $p(x_{t_1 + t}, x_{t_2 + t}, \ldots, x_{t_n + t})$. The joint pdf's of the random variables X_{t_i} and $X_{t_i + t}$, $i = 1, 2, \ldots, n$, may or may not be identical. When they are identical, i.e., when

$$p(x_{t_1}, x_{t_2}, \ldots, x_{t_n}) = p(x_{t_1 + t}, x_{t_2 + t}, \ldots, x_{t_n + t}) \qquad (1.2.1)$$

for all t and all n, the stochastic process is said to be *stationary in the strict sense*. That is, the statistics of a stationary stochastic process are invariant to any translation of the time axis. On the other hand, when the joint pdf's are different, the stochastic process is *nonstationary*.

1.2.1 Statistical Averages

Just as we have defined statistical averages for random variables, we may similarly define statistical averages for a stochastic process. Such averages are also called *ensemble averages*. Let $X(t)$ denote a random process and let $X_{t_i} \equiv X(t_i)$. The nth moment of the random variable X_{t_i} is defined as

$$E(X_{t_i}^n) = \int_{-\infty}^{\infty} x_{t_i}^n p(x_{t_i}) \, dx_{t_i} \qquad (1.2.2)$$

In general, the value of the nth moment will depend on the time instant t_i if the pdf of X_{t_i} depends on t_i. When the process is stationary, however, $p(x_{t_i + t}) = p(x_{t_i})$ for all t. Hence the pdf is independent of time and, as a consequence, the nth moment is independent of time.

Next we consider the two random variables $X_{t_i} \equiv X(t_i)$, $i = 1, 2$. The correlation between X_{t_1} and X_{t_2} is measured by the joint moment

$$E(X_{t_1} X_{t_2}) = \int_{-\infty}^{\infty} \int_{-\infty}^{\infty} x_{t_1} x_{t_2} p(x_{t_1}, x_{t_2}) \, dx_{t_1} \, dx_{t_2} \qquad (1.2.3)$$

Since this joint moment depends on the time instants t_1 and t_2, it is denoted as $\phi(t_1, t_2)$. The function $\phi(t_1, t_2)$ is called the *autocorrelation function* of the stochastic process. When the process $X(t)$ is stationary, the joint pdf of the pair (X_{t_1}, X_{t_2}) is identical to the joint pdf of the pair (X_{t_1+t}, X_{t_2+t}) for any arbitrary t. This implies that the autocorrelation function of $X(t)$ does not depend on the specific time instants t_1 and t_2 but, instead, it depends on the time difference $t_1 - t_2$. Thus, for a stationary stochastic process, the joint moment in (1.2.3) is

$$E(X_{t_1} X_{t_2}) = \phi(t_1, t_2) = \phi(t_1 - t_2) = \phi(\tau) \qquad (1.2.4)$$

where $\tau = t_1 - t_2$ or equivalently, $t_2 = t_1 - \tau$. If we let $t_2 = t_1 + \tau$, we have

$$\phi(-\tau) = E(X_{t_1} X_{t_1+t}) = E(X_{t_i'} X_{t_i'-\tau}) = \phi(\tau).$$

Therefore $\phi(\tau)$ is an even function. We also note that $\phi(0) = E(X_t^2)$ denotes the average power in the process $X(t)$.

There exist nonstationary processes with the property that the mean value of the process is independent of time (a constant) and where the autocorrelation function satisfies the condition that $\phi(t_1, t_2) = \phi(t_1 - t_2)$. Such a process is called *wide-sense stationary*. Consequently, wide-sense stationarity is a less stringent condition than strict-sense stationarity. When reference is made to a stationary stochastic process in any subsequent discussion in which correlation functions are involved, the less stringent condition (wide-sense stationarity) is implied.

Related to the autocorrelation function is the autocovariance function of a stochastic process, which is defined as

$$\mu(t_1, t_2) = E\{[X_{t_1} - m(t_1)][X_{t_2} - m(t_2)]\}$$
$$= \phi(t_1, t_2) - m(t_1)m(t_2) \qquad (1.2.5)$$

where $m(t_1)$ and $m(t_2)$ are the means of X_{t_1} and X_{t_2}, respectively. When the process is stationary, the autocovariance function simplifies to

$$\mu(t_1, t_2) = \mu(t_1 - t_2) = \mu(\tau) = \phi(\tau) - m^2 \qquad (1.2.6)$$

where $\tau = t_1 - t_2$.

Higher-order joint moments of two or more random variables derived from a stochastic process $X(t)$ are defined in an obvious manner. With the possible exception of the gaussian random process, for which higher-order moments can be expressed in terms of first and second moments, high-order moments are encountered very infrequently in practice.

Averages for a gaussian process. Suppose that $X(t)$ is a gaussian random process. Hence, at time instants $t = t_i$, $i = 1, 2, \ldots, n$, the random variables X_{t_i}, $i = 1, 2, \ldots, n$, are jointly gaussian with mean values $m(t_i)$, $i = 1, 2, \ldots, n$, and autocovariances

$$\mu(t_i, t_j) = E[(X_{t_i} - m(t_i))(X_{t_j} - m(t_j))] \qquad i, j = 1, 2, \ldots, n \quad (1.2.7)$$

If we denote the $(n \times n)$ covariance matrix with elements $\mu(t_i, t_j)$ as **M** and the vector of mean values by $\mathbf{m}_x$, then the joint pdf of the random variables X_{t_i}, $i = 1, 2, \ldots, n$ is given by (1.1.142).

If the gaussian process is stationary, then $m(t_i) = m$ for all t_i and $\mu(t_i, t_j) = \mu(t_i - t_j)$. We observe that the gaussian random process is completely specified by the mean and autocovariance functions. Since the joint gaussian pdf depends only on these two moments, it follows that if the gaussian process is wide-sense stationary, it is also strict-sense stationary. Of course, the converse is always true for any stochastic process.

Averages for joint stochastic processes. Let $X(t)$ and $Y(t)$ denote two stochastic processes and let $X_{t_i} \equiv X(t_i)$, $i = 1, 2, \ldots, n$, and $Y_{t'_j} \equiv Y(t'_j)$, $j = 1, 2, \ldots, m$, represent the random variables at times $t_1 > t_2 > t_3 > \cdots > t_n$ and $t'_1 > t'_2 > \cdots > t'_m$, respectively. The two processes are characterized statistically by their joint pdf

$$p\left(x_{t_1}, x_{t_2}, \ldots, x_{t_n}, y_{t'_1}, y_{t'_2}, \ldots, y_{t'_m}\right)$$

for any set of time instants $t_1, t_2, \ldots, t_n, t'_1, t'_2, \ldots, t'_m$ and for any positive integer values of n and m.

The cross-correlation function of $X(t)$ and $Y(t)$, denoted by $\phi_{xy}(t_1, t_2)$, is defined as the joint moment

$$\phi_{xy}(t_1, t_2) = E\left(X_{t_1} Y_{t_2}\right) = \int_{-\infty}^{\infty} \int_{-\infty}^{\infty} x_{t_1} y_{t_2} p\left(x_{t_1}, y_{t_2}\right) dx_{t_1} dy_{t_2} \qquad (1.2.8)$$

and the cross covariance is

$$\mu_{xy}(t_1, t_2) = \phi_{xy}(t_1, t_2) - m_x(t_1) m_y(t_2) \qquad (1.2.9)$$

When the processes are jointly and individually stationary we have $\phi_{xy}(t_1, t_2) = \phi_{xy}(t_1 - t_2)$ and $\mu_{xy}(t_1, t_2) = \mu_{xy}(t_1 - t_2)$. In this case we note that

$$\phi_{xy}(-\tau) = E\left(X_{t_1} Y_{t_1 + \tau}\right) = E\left(X_{t'_1 - \tau} Y_{t'_1}\right) = \phi_{yx}(\tau) \qquad (1.2.10)$$

The stochastic processes $X(t)$ and $Y(t)$ are said to be statistically independent if and only if

$$p\left(x_{t_1}, x_{t_2}, \ldots, x_{t_n}, y_{t'_1}, y_{t'_2}, \ldots, y_{t'_m}\right) = p\left(x_{t_1}, x_{t_2}, \ldots, x_{t_n}\right) p\left(y_{t'_1}, y_{t'_2}, \ldots, y_{t'_m}\right)$$

for all choices of t_i and t'_i and for all positive integers n and m. The processes are said to be uncorrelated if

$$\phi_{xy}(t_1, t_2) = E\left(X_{t_1}\right) E\left(Y_{t_2}\right)$$

Hence

$$\mu_{xy}(t_1, t_2) = 0$$

A complex-valued stochastic process $Z(t)$ is defined as

$$Z(t) = X(t) + jY(t) \qquad (1.2.11)$$

where $X(t)$ and $Y(t)$ are stochastic processes. The joint pdf of the random variables $Z_{t_i} \equiv Z(t_i)$, $i = 1, 2, \ldots$, is given by the joint pdf of the components

(X_{t_i}, Y_{t_i}), $i = 1, 2, \ldots, n$. Thus the pdf that characterizes Z_{t_i}, $i = 1, 2, \ldots, n$, is

$$p(x_{t_1}, x_{t_2}, \ldots, x_{t_n}, y_{t_1}, y_{t_2}, \ldots, y_{t_n})$$

The complex-valued stochastic process $Z(t)$ is encountered in the representation of narrowband bandpass noise in terms of its equivalent lowpass components. An important characteristic of such a process is its autocorrelation function. The function is defined as

$$\phi_{zz}(t_1, t_2) = \tfrac{1}{2} E\left(Z_{t_1} Z_{t_2}^*\right)$$

$$= \tfrac{1}{2} E\left[\left(X_{t_1} + jY_{t_1}\right)\left(X_{t_2} - jY_{t_2}\right)\right]$$

$$= \tfrac{1}{2}\left\{\phi_{xx}(t_1, t_2) + \phi_{yy}(t_1, t_2) + j\left[\phi_{yx}(t_1, t_2) - \phi_{xy}(t_1, t_2)\right]\right\}$$

$$(1.2.12)$$

where $\phi_{xx}(t_1, t_2)$ and $\phi_{yy}(t_1, t_2)$ are the autocorrelation functions of $X(t)$ and $Y(t)$, respectively, and $\phi_{yx}(t_1, t_2)$ and $\phi_{xy}(t_1, t_2)$ are the cross-correlation functions. The factor of $\tfrac{1}{2}$ in the definition of the autocorrelation function of a complex-valued stochastic process is an arbitrary but mathematically convenient normalization factor, as we will demonstrate in our treatment of such processes in Chap. 3.

When the processes $X(t)$ and $Y(t)$ are jointly and individually stationary, the autocorrelation function of $Z(t)$ becomes

$$\phi_{zz}(t_1, t_2) = \phi_{zz}(t_1 - t_2) = \phi_{zz}(\tau)$$

where $t_2 = t_1 - \tau$. Also, the complex conjugate of (1.2.12) is

$$\phi_{zz}^*(\tau) = \tfrac{1}{2} E\left(Z_{t_1}^* Z_{t_1 - \tau}\right) = \tfrac{1}{2} E\left(Z_{t_1' + \tau}^* Z_{t_1'}\right) = \phi_{zz}(-\tau) \qquad (1.2.13)$$

Hence $\phi_{zz}(\tau) = \phi_{zz}^*(-\tau)$.

Now, suppose that $Z(t) = X(t) + jY(t)$ and $W(t) = U(t) + jV(t)$ are two complex-valued stochastic processes. The cross-correlation function of $Z(t)$ and $W(t)$ is defined as

$$\phi_{zw}(t_1, t_2) = \tfrac{1}{2} E\left(Z_{t_1} W_{t_2}^*\right)$$

$$= \tfrac{1}{2} E\left[\left(X_{t_1} + jY_{t_1}\right)\left(U_{t_2} - jV_{t_2}\right)\right]$$

$$= \tfrac{1}{2}\left\{\phi_{xu}(t_1, t_2) + \phi_{yv}(t_1, t_2) + j\left[\phi_{yu}(t_1, t_2) - \phi_{xv}(t_1, t_2)\right]\right\}$$

$$(1.2.14)$$

When $X(t)$, $Y(t)$, $U(t)$, and $V(t)$ are pairwise-stationary, the cross-correlation functions in (1.2.14) become functions of the time difference $\tau = t_1 - t_2$. Furthermore,

$$\phi_{zw}^*(\tau) = \tfrac{1}{2} E\left(Z_{t_1}^* W_{t_1 - \tau}\right) = \tfrac{1}{2} E\left(Z_{t_1' + \tau}^* W_{t_1'}\right) = \phi_{wz}(-\tau) \qquad (1.2.15)$$

1.2.2 Power Density Spectrum

The frequency content of a signal is a very basic characteristic that distinguishes one signal from another. In general, a signal can be classified as having either a finite (nonzero) average power (infinite energy) or finite energy. The frequency content of a finite energy signal is obtained as the Fourier transform of the corresponding time function. If the signal is periodic, its energy is infinite and, consequently, its Fourier transform does not exist. The mechanism for dealing with periodic signals is to represent them in a Fourier series. With such a representation, the Fourier coefficients determine the distribution of power at the various discrete frequency components.

A stationary stochastic process is an infinite energy signal and, hence, its Fourier transform does not exist. The spectral characteristic of a stochastic signal is obtained by computing the Fourier transform of the autocorrelation function. That is, the distribution of power with frequency is given by the function

$$\Phi(f) = \int_{-\infty}^{\infty} \phi(\tau) e^{-j2\pi f\tau} \, d\tau \tag{1.2.16}$$

The inverse Fourier transform relationship is

$$\phi(\tau) = \int_{-\infty}^{\infty} \Phi(f) e^{j2\pi f\tau} \, df \tag{1.2.17}$$

We observe that

$$\phi(0) = \int_{-\infty}^{\infty} \Phi(f) \, df$$

$$= E(X_1^2) \geq 0 \tag{1.2.18}$$

Since $\phi(0)$ represents the average power of the stochastic signal, which is the area under $\Phi(f)$, then $\Phi(f)$ is the distribution of power as a function of frequency. Therefore, $\Phi(f)$ is called the *power density spectrum* of the stochastic process.

If the stochastic process is real, $\phi(\tau)$ is real and even and, hence, $\Phi(f)$ is real and even. On the other hand, if the process is complex, $\phi(\tau) = \phi^*(-\tau)$ and, hence,

$$\Phi^*(f) = \int_{-\infty}^{\infty} \phi^*(\tau) e^{j2\pi f\tau} \, d\tau = \int_{-\infty}^{\infty} \phi^*(-\tau) e^{-j2\pi f\tau} \, d\tau$$

$$= \int_{-\infty}^{\infty} \phi(\tau) e^{-j2\pi f\tau} \, d\tau = \Phi(f) \tag{1.2.19}$$

Therefore $\Phi(f)$ is real.

The definition of a power density spectrum can be extended to two jointly stationary stochastic processes $X(t)$ and $Y(t)$, which have a cross-correlation function $\phi_{xy}(\tau)$. The Fourier transform of $\phi_{xy}(\tau)$, i.e.,

$$\Phi_{xy}(f) = \int_{-\infty}^{\infty} \phi_{xy}(\tau) e^{-j2\pi f\tau} \, d\tau \tag{1.2.20}$$

is called the *cross-power density spectrum*. If we conjugate both sides of (1.2.20), we have

$$\Phi_{xy}^*(f) = \int_{-\infty}^{\infty} \phi_{xy}^*(\tau)e^{j2\pi f\tau}\,d\tau = \int_{-\infty}^{\infty} \phi_{xy}^*(-\tau)e^{-j2\pi f\tau}\,d\tau$$

$$= \int_{-\infty}^{\infty} \phi_{yx}(\tau)e^{-j2\pi f\tau}\,d\tau = \Phi_{yx}(f) \tag{1.2.21}$$

This relation holds in general. However, if $X(t)$ and $Y(t)$ are real stochastic processes,

$$\Phi_{xy}^*(f) = \int_{-\infty}^{\infty} \phi_{xy}(\tau)e^{j2\pi f\tau}\,d\tau = \Phi_{xy}(-f) \tag{1.2.22}$$

By combining the result in (1.2.21) with the result in (1.2.22), we find that the cross-power density spectrum of two real processes satisfies the condition

$$\Phi_{yx}(f) = \Phi_{xy}(-f) \tag{1.2.23}$$

1.2.3 Response of a Linear Time-Invariant System to a Random Input Signal

Consider a linear time-invariant system (filter) that is characterized by its impulse response $h(t)$ or, equivalently, by its frequency response $H(f)$, where $h(t)$ and $H(f)$ are a Fourier transform pair.† Let $x(t)$ be the input signal to the system and let $y(t)$ denote the output signal. The output of the system may be expressed in terms of the convolution integral as

$$y(t) = \int_{-\infty}^{\infty} h(\tau)x(t-\tau)\,d\tau \tag{1.2.24}$$

Now suppose that $x(t)$ is a sample function of a stationary stochastic process $X(t)$. Then, the output $y(t)$ is a sample function of a stochastic process $Y(t)$. We wish to determine the mean and autocorrelation functions of the output.

Since convolution is a linear operation performed on the input signal $x(t)$, the expected value of the integral is equal to the integral of the expected value. Thus the mean value of $Y(t)$ is

$$m_y = E[Y(t)] = \int_{-\infty}^{\infty} h(\tau)E[X(t-\tau)]\,d\tau$$

$$= m_x\int_{-\infty}^{\infty} h(\tau)\,d\tau = m_x H(0) \tag{1.2.25}$$

where $H(0)$ is the frequency response of the linear system at $f=0$. Hence the mean value of the output process is a constant.

---

† $h(t)$ is assumed to be real.

The autocorrelation function of the output is

$$\phi_{yy}(t_1, t_2) = E(Y_{t_1} Y_{t_2})$$

$$= \int_{-\infty}^{\infty} \int_{-\infty}^{\infty} h(\beta) h(\alpha) E[X(t_1 - \beta) X(t_2 - \alpha)] \, d\alpha \, d\beta$$

$$= \int_{-\infty}^{\infty} \int_{-\infty}^{\infty} h(\beta) h(\alpha) \phi_{xx}(t_1 - t_2 + \alpha - \beta) \, d\alpha \, d\beta$$

The last step indicates that the double integral is a function of the time difference $t_1 - t_2$. In other words, if the input process is stationary, the output is also stationary. Hence

$$\phi_{yy}(\tau) = \int_{-\infty}^{\infty} \int_{-\infty}^{\infty} h(\alpha) h(\beta) \phi_{xx}(\tau + \alpha - \beta) \, d\alpha \, d\beta \qquad (1.2.26)$$

By evaluating the Fourier transform of both sides of (1.2.26), we obtain the power density spectrum of the output process in the form

$$\Phi_{yy}(f) = \int_{-\infty}^{\infty} \phi_{yy}(\tau) e^{-j2\pi f\tau} \, d\tau$$

$$= \int_{-\infty}^{\infty} \int_{-\infty}^{\infty} \int_{-\infty}^{\infty} h(\alpha) h(\beta) \phi_{xx}(\tau + \alpha - \beta) e^{-j2\pi f\tau} \, d\tau \, d\alpha \, d\beta$$

$$= \Phi_{xx}(f) |H(f)|^2 \qquad (1.2.27)$$

Thus we have the important result that the power density spectrum of the output signal is the product of the power density spectrum of the input multiplied by the magnitude squared of the frequency response of the system.

When the autocorrelation function $\phi_{yy}(\tau)$ is desired, it is usually easier to determine the power density spectrum $\Phi_{yy}(f)$ and then to compute the inverse transform. Thus we have

$$\phi_{yy}(\tau) = \int_{-\infty}^{\infty} \Phi_{yy}(f) e^{j2\pi f\tau} \, df$$

$$= \int_{-\infty}^{\infty} \Phi_{xx}(f) |H(f)|^2 e^{j2\pi f\tau} \, df \qquad (1.2.28)$$

We observe that the average power in the output signal is

$$\phi_{yy}(0) = \int_{-\infty}^{\infty} \Phi_{xx}(f) |H(f)|^2 \, df \qquad (1.2.29)$$

Since $\phi_{yy}(0) = E(Y_t^2)$, it follows that

$$\int_{-\infty}^{\infty} \Phi_{xx}(f) |H(f)|^2 \, df \geq 0$$

Suppose we let $|H(f)|^2 = 1$ for any arbitrarily small interval $f_1 \leq f \leq f_2$, and $H(f) = 0$ outside this interval. Then

$$\int_{f_1}^{f_2} \Phi_{xx}(f) \, df \geq 0$$

But this is possible if and only if $\Phi_{xx}(f) \geq 0$ for all f.

FIGURE 1.2.1
An example of a lowpass filter.

Example 1.2.1. Suppose that the lowpass filter illustrated in Fig. 1.2.1 is excited by a stochastic process $x(t)$ having a power density spectrum

$$\Phi_{xx}(f) = \frac{N_0}{2} \quad \text{for all } f$$

A stochastic process having a flat power density spectrum is called *white noise*. Let us determine the power density spectrum of the output process. The transfer function of the lowpass filter is

$$H(f) = \frac{R}{R + j2\pi fL} = \frac{1}{1 + j2\pi fL/R}$$

and, hence,

$$|H(f)|^2 = \frac{1}{1 + (2\pi L/R)^2 f^2} \tag{1.2.30}$$

The power density spectrum of the output process is

$$\Phi_{yy}(f) = \frac{N_0}{2} \frac{1}{1 + (2\pi L/R)^2 f^2} \tag{1.2.31}$$

This power density spectrum is illustrated in Fig. 1.2.2. Its inverse Fourier transform yields the autocorrelation function

$$\phi_{yy}(\tau) = \int_{-\infty}^{\infty} \frac{N_0}{2} \frac{1}{1 + (2\pi L/R)^2 f^2} e^{j2\pi f\tau} \, df$$

$$= \frac{RN_0}{4L} e^{-(R/L)|\tau|} \tag{1.2.32}$$

The autocorrelation function $\phi_{yy}(\tau)$ is shown in Fig. 1.2.3. We observe that the second moment of the process $Y(t)$ is $\phi_{yy}(0) = RN_0/4L$.

As a final exercise, we determine the cross-correlation function between $y(t)$ and $x(t)$, where $x(t)$ denotes the input and $y(t)$ denotes the output of the linear

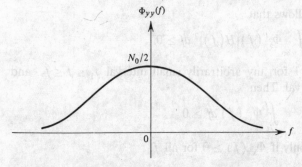

FIGURE 1.2.2
The power density spectrum of the lowpass filter output when the input is white noise.

process of efficiently converting the output of a source into a sequence of binary digits is called *source encoding or data compression*. We shall describe several source encoding techniques in Sec. 2.3.

The sequence of binary digits from the source encoder is to be transmitted through a channel to the intended receiver. For example, the real channel may be either a pair of wires, a coaxial cable, an optical fiber channel, a radio channel, a satellite channel, or some combination of these media. Such channels are basically waveform channels and, hence, they cannot be used to transmit directly the sequence of binary digits from the source. What is required is a device that converts the digital information sequence into waveforms that are compatible with the characteristics of the channel. Such a device is called a *digital modulator*, or, simply, a modulator. The modulator is a part of a larger device called a *channel encoder*, which serves a second function. The need for such a function is understood when we consider the characteristics of real channels.

In general, no real channel is ideal. First of all channels may have nonideal frequency response characteristics. In addition, there are noise disturbances and other interference that corrupt the signal transmitted through the channel. For example, there may be cross talk (interference) from signals being transmitted on adjacent channels. There is the thermal noise generated in the electronic equipment such as amplifiers and filters used in the transmitter and in the receiver. There may also be noise and interference of the impulsive type, caused by switching transients in wire line channels and thunderstorms on radio channels. Finally, there may be intentional jamming of the signal transmitted over the channel. Such disturbances corrupt the transmitted signal and cause errors in the received digital sequence.

In order to overcome such noise and interference and, thus, to increase the reliability of the data transmitted through the channel, it is often necessary to introduce in a controlled manner some redundancy in the binary sequence from the source. The introduction of redundancy in the information sequence for the

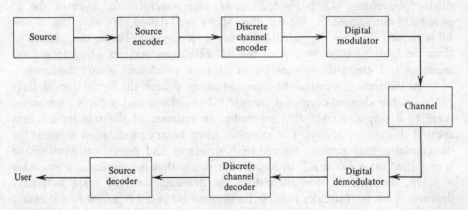

FIGURE 2.1.1
Block diagram of a digital communications system.

purpose of combating the detrimental effects of noise and interference in the channel is the second function of the channel encoder. The redundancy introduced at the transmitter aids the receiver in decoding the desired information-bearing sequence. For example, a (trivial) form of encoding of the binary information sequence is simply to repeat each binary digit m times, where m is some positive integer. More sophisticated (nontrivial) encoding involves taking k information bits at a time and mapping each k-bit sequence into a unique n-bit sequence, called a *code word*. The amount of redundancy introduced by encoding the data in this manner is measured by the ratio n/k. This is also the ratio by which the channel bandwidth must be increased to accommodate the added redundancy in the information sequence. The reciprocal of this ratio, namely, k/n, is called the *rate of the code* or, simply, the code rate.

An alternative to providing added redundancy in the information sequence as a means of overcoming the channel disturbances is to increase the power in the transmitted signal. However, transmitter power is often very expensive, so we usually design communications systems that are power efficient. In channels where bandwidth is readily available, we trade off transmitter power for channel bandwidth. In channels with limited bandwidth, clever use of redundancy allows us to transmit information efficiently without the need to expand the channel bandwidth. When such redundancy is introduced, it is both convenient and appropriate to view the channel encoder as consisting of two parts, a discrete channel encoder and a digital modulator. The former is discrete at both its input and its output. The latter has a discrete input but its output consists of waveforms.

To elaborate on the function performed by the modulator, suppose the information is to be transmitted 1 bit at a time at some uniform rate R bits/s. The modulator may simply map the binary digit 0 into a waveform $s_1(t)$ and the binary digit 1 into a waveform $s_2(t)$. In this manner, each bit from the channel encoder is transmitted separately. We call this *binary modulation*. Alternatively, the modulator may transmit k information bits at a time by using $M = 2^k$ distinct waveforms $s_i(t)$, $i = 1, 2, \ldots, M$, one waveform for each of the 2^k possible k-bit sequences. We call this M-ary modulation. We note that a new k-bit sequence enters the modulator every k/R seconds. Hence the amount of time available to transmit one of the M waveforms corresponding to a k-bit sequence is k times the time period in a system which uses binary modulation.

At the receiving end of the communications system, the digital demodulator processes the channel-corrupted transmitted waveform and reduces each waveform to a single number that represents an estimate of the transmitted data symbol (binary or M-ary). For example, when binary modulation is used, the demodulator may process the received waveform and decide on whether the transmitted bit is a 0 or a 1. In such a case we say that *the demodulator has made a binary decision*. As one alternative, the demodulator may make a ternary decision. That is, it decides that the transmitted bit is either a 0 or 1 or it makes no decision at all, depending on the apparent quality of the received signal. When no decision is made on a particular bit, we say that *the demodulator has inserted an erasure in the demodulated data.* Using the redundancy in the transmitted data, the decoder attempts to fill in the positions where erasures occurred. Viewing the decision process performed by the demodulator as a form of quantization, we observe that binary and ternary decisions are special cases of a demodulator that quantizes to Q levels, where $Q \geq 2$. In general, if the digital communications system employs M-ary modulation, where $m = 0, 1, \ldots, M - 1$ represent the M possible transmitted symbols, each corresponding to $k = \log_2 M$ bits, the demodulator may make a Q-ary decision, where $Q \geq M$. In the extreme case where no quantization is performed, $Q = \infty$.

When there is no redundancy in the transmitted information, the demodulator must decide which of the M waveforms was transmitted in any given time interval. Consequently $Q = M$ and, since there is no redundancy in the transmitted information, no discrete channel decoder is used following the demodulator. On the other hand, when there is redundancy introduced by a discrete channel encoder at the transmitter, the Q-ary output from the demodulator occurring every k/R seconds is fed to the decoder, which attempts to reconstruct the original information sequence from knowledge of the code used by the channel encoder and the redundancy contained in the received data.

A measure of how well the demodulator and encoder perform is the frequency with which errors occur in the decoded sequence. More precisely, the average probability of a bit error at the output of the decoder is a measure of the performance of the demodulator-decoder combination. In general, the probability of error is a function of the code characteristics, the types of waveforms used to transmit the information over the channel, the transmitter power, the characteristics of the channel, i.e., the amount of noise, the nature of the interference, etc., and the method of demodulation and decoding. These items and their effect on performance will be discussed in detail in subsequent chapters.

As a final step, when an analog output is desired, the source decoder accepts the output sequence from the channel decoder and, from knowledge of the source encoding method used, attempts to reconstruct the original signal from the source. Due to channel decoding errors and possible distortion introduced by the source encoder and, perhaps, the source decoder, the signal at the output of the source decoder is an approximation to the original source output. The difference or some function of the difference between the original signal and the reconstructed signal is a measure of the distortion introduced by the digital communications system illustrated in Fig. 2.1.1.

2.2 A LOGARITHMIC MEASURE FOR INFORMATION

Let X and Y be two discrete random variables with possible outcomes x_i, $i = 1, 2, \ldots, n$, and y_i, $i = 1, 2, \ldots, m$, respectively. Suppose we observe some outcome $Y = y_j$ and we wish to determine, quantitatively, the amount of information that the occurrence of the event $Y = y_j$ provides about the event $X = x_i$, $i = 1, 2, \ldots, n$. The problem is to select an appropriate measure for information.

We observe that when X and Y are statistically independent, the occurrence of $Y = y_j$ provides no information about the occurrence of the event $X = x_i$. On the other hand, when X and Y are fully dependent such that the occurrence of $Y = y_j$ determines the occurrence of $X = x_i$, the information content is simply that provided by the event $X = x_i$. A suitable measure that satisfies these conditions is the logarithm of the ratio of the conditional probability

$$P(X = x_i | Y = y_j) \equiv P(x_i | y_j)$$

divided by the probability

$$P(X = x_i) \equiv P(x_i)$$

That is, the information content provided by the occurrence of the event $Y = y_j$ about the event $X = x_i$ is defined as

$$I(x_i; y_j) = \log \frac{P(x_i | y_j)}{P(x_i)} \qquad (2.2.1)$$

$I(x_i; y_j)$ is called the *mutual information* between x_i and y_j.

The units of $I(x_i; y_j)$ are determined by the base of the logarithm, which is usually selected as either 2 or e. When the base of the logarithm is 2, the units of $I(x_i; y_j)$ are bits, and when the base is e, the units of $I(x_i; y_j)$ are called *nats* (natural units). (The standard abbreviation for $\log_e$ is ln.) Since

$$\ln a = \ln 2 \log_2 a = 0.69315 \log_2 a$$

the information measured in nats is equal to ln 2 times the information measured in bits.

When the random variables X and Y are statistically independent, $P(x_i | y_j) = P(x_i)$ and, hence, $I(x_i; y_j) = 0$. On the other hand, when the occurrence of the event $Y = y_j$ uniquely determines the occurrence of the event $X = x_i$, the conditional probability in the numerator of (2.2.1) is unity and, hence,

$$I(x_i; y_j) = \log \frac{1}{P(x_i)} = -\log P(x_i) \qquad (2.2.2)$$

But (2.2.2) is just the information of the event $X = x_i$. For this reason, it is called the *self-information* of the event $X = x_i$ and it is denoted as

$$I(x_i) = \log \frac{1}{P(x_i)} = -\log P(x_i) \qquad (2.2.3)$$

We note that a high-probability event conveys less information than a low-probability event. In fact, if there is only a single event x with probability $P(x) = 1$, then $I(x) = 0$. To demonstrate further that the logarithmic measure for information content is the appropriate one for digital communications, let us consider the following example.

Example 2.2.1. Suppose we have a discrete information source that emits a binary digit, either 0 or 1, with equal probability every τ_s seconds. The information content of each output from the source is

$$I(x_i) = -\log_2 P(x_i) \qquad x_i = 0, 1$$

$$= -\log_2 \tfrac{1}{2} = 1 \text{ bit}$$

Now suppose that successive outputs from the source are statistically independent. Let us consider a block of k binary digits from the source which occurs in a time interval $k\tau_s$. There are $M = 2^k$ possible k-bit blocks, each of which is equally probable with probability $1/M = 2^{-k}$. The self-information of a k-bit block is

$$I(x_i') = -\log_2 2^{-k} = k \text{ bits}$$

emitted in a time interval $k\tau_s$. Thus the logarithmic measure for information content possesses the desired additivity property when a number of source outputs is considered as a block.

Now let us return to the definition of mutual information given in (2.2.1) and multiply the numerator and denominator of the ratio of probabilities by $P(y_j)$. Since

$$\frac{P(x_i|y_j)}{P(x_i)} = \frac{P(x_i|y_j)P(y_j)}{P(x_i)P(y_j)} = \frac{P(x_i, y_j)}{P(x_i)P(y_j)} = \frac{P(y_j|x_i)}{P(y_j)}$$

we conclude that

$$I(x_i; y_j) = I(y_j; x_i) \tag{2.2.4}$$

Therefore the information provided by the occurrence of the event $Y = y_j$ about the event $X = x_i$ is identical to the information provided by the occurrence of the event $X = x_i$ about the event $Y = y_j$.

Example 2.2.2. Suppose that X and Y are binary-valued $\{0, 1\}$ random variables that represent the input and output of a binary-input, binary-output channel. The input symbols are equally likely and the output symbols depend on the input according to the conditional probabilities

$$P(Y = 0 | X = 0) = 1 - p_0$$

$$P(Y = 1 | X = 0) = p_0$$

$$P(Y = 1 | X = 1) = 1 - p_1$$

$$P(Y = 0 | X = 1) = p_1$$

Let us determine the mutual information about the occurrence of the events $X = 0$ and $X = 1$, given that $Y = 0$.

From the probabilities given above, we obtain

$$P(Y = 0) = P(Y = 0|X = 0)P(X = 0) + P(Y = 0|X = 1)P(X = 1)$$
$$= \tfrac{1}{2}(1 - p_0 + p_1)$$
$$P(Y = 1) = P(Y = 1|X = 0)P(X = 0) + P(Y = 1|X = 1)P(X = 1)$$
$$= \tfrac{1}{2}(1 - p_1 + p_0)$$

Then, the mutual information about the occurrence of the event $X = 0$, given that $Y = 0$ is observed, is

$$I(x_1; y_1) = I(0; 0) = \log_2 \frac{P(Y = 0|X = 0)}{P(Y = 0)} = \log_2 \frac{2(1 - p_0)}{1 - p_0 + p_1}$$

Similarly, given that $Y = 0$ is observed, the mutual information about the occurrence of the event $X = 1$ is

$$I(x_2; y_1) \equiv I(1; 0) = \log_2 \frac{2p_1}{1 - p_0 + p_1}$$

Let us consider some special cases: First, if $p_0 = p_1 = 0$, the channel is called *noiseless* and

$$I(0; 0) = \log_2 2 = 1 \text{ bit}$$

Hence, the output specifies the input with certainty. On the other hand, if $p_0 = p_1 = \tfrac{1}{2}$, the channel is *useless* because

$$I(0; 0) = \log_2 1 = 0$$

However, if $p_0 = p_1 = \tfrac{1}{4}$, then

$$I(0; 0) = \log_2 \tfrac{3}{2} = 0.587$$
$$I(0; 1) = \log_2 \tfrac{1}{2} = -1 \text{ bit}$$

We observe from this example that the mutual information between a pair of events can be either positive, negative, or zero.

In addition to the definition of mutual information and self-information, it is useful to define the *conditional self-information* as

$$I(x_i|y_j) = \log \frac{1}{P(x_i|y_j)} = -\log p(x_i|y_j) \qquad (2.2.5)$$

Then, by combining (2.2.1), (2.2.3), and (2.2.5) we obtain the relationship

$$I(x_i; y_j) = I(x_i) - I(x_i|y_j) \qquad (2.2.6)$$

We interpret $I(x_i|y_j)$ as the self-information about the event $X = x_i$ after having observed the event $Y = y_j$. Since both $I(x_i) \geq 0$ and $I(x_i|y_j) \geq 0$, it follows that $I(x_i; y_j) < 0$ when $I(x_i|y_j) > I(x_i)$, and $I(x_i; y_j) > 0$ when $I(x_i|y_j) < I(x_i)$.

2.2.1 Average Mutual Information and Entropy

Having defined the mutual information associated with the pair of events (x_i, y_j), which are possible outcomes of the two random variables X and Y, we can obtain the average value of the mutual information by simply weighting $I(x_i; y_j)$

by the probability of occurrence of the joint event and summing over all possible joint events. Thus, we obtain

$$I(X;Y) = \sum_{i=1}^{n} \sum_{j=1}^{m} P(x_i, y_j) I(x_i; y_j)$$

$$= \sum_{i=1}^{n} \sum_{j=1}^{m} P(x_i, y_j) \log \frac{P(x_i, y_j)}{P(x_i) P(y_j)} \quad (2.2.7)$$

as the average mutual information between X and Y. We observe that $I(X;Y) = 0$ when X and Y are statistically independent. An important characteristic of the average mutual information is that $I(X;Y) \geq 0$ (see Prob. 2.4).

Similarly, we define the average self-information, denoted by $H(X)$, as

$$H(X) = \sum_{i=1}^{n} P(x_i) I(x_i)$$

$$= - \sum_{i=1}^{n} P(x_i) \log P(x_i) \quad (2.2.8)$$

When X represents the alphabet of possible output letters from a source, $H(X)$ represents the average self-information per source letter, and it is called the *entropy*† of the source. In the special case in which the letters from the source are equally probable, $P(x_i) = 1/n$ for all i and, hence

$$H(X) = - \sum_{i=1}^{n} \frac{1}{n} \log \frac{1}{n}$$

$$= \log n \quad (2.2.9)$$

In general, $H(X) \leq \log n$ (see Prob. 2.5) for any given set of source letter probabilities. In other words, *the entropy of a discrete source is a maximum when the output letters are equally probable*.

Example 2.2.3. Consider a source that emits a sequence of statistically independent letters, where each output letter is either 0 with probability q or 1 with probability $1 - q$. The entropy of this source is

$$H(X) \equiv H(q) = -q \log q - (1 - q) \log (1 - q) \quad (2.2.10)$$

The binary entropy function $H(q)$ is illustrated in Fig. 2.2.1. We observe that the maximum value of the entropy function occurs at $q = \frac{1}{2}$ where $H(\frac{1}{2}) = 1$.

The average conditional self-information is called the *conditional entropy* and is defined

$$H(X|Y) = \sum_{i=1}^{n} \sum_{j=1}^{m} P(x_i, y_j) \log \frac{1}{P(x_i|y_j)} \quad (2.2.11)$$

---

† The term *entropy* is taken from statistical mechanics (thermodynamics) where a function similar to (2.2.8) is called (thermodynamic) entropy.

FIGURE 2.2.1
Binary entropy function.

We interpret $H(X|Y)$ as the information or uncertainty in X after Y is observed. By combining (2.2.7), (2.2.8), and (2.2.11) we obtain the relationship

$$I(X; Y) = H(X) - H(X|Y) \qquad (2.2.12)$$

Since $I(X;Y) \geq 0$. it follows that $H(X) \geq H(X|Y)$ with equality if and only if X and Y are statistically independent. If we interpret $H(X|Y)$ as the average amount of (conditional self-information) uncertainty in X after we observe Y, and $H(X)$ as the average amount of uncertainty (self-information) prior to the observation, then $I(X; Y)$ is the average amount of (mutual information) uncertainty provided about the set X by the observation of the set Y. Since $H(X) \geq H(X|Y)$, it is clear that conditioning on the observation Y does not increase the entropy.

Example 2.2.4. Let us evaluate the $H(X|Y)$ and $I(X; Y)$ for the binary-input, binary-output channel treated previously in Example 2.2.2 for the case where $p_0 = p_1 = p$, so that the channel is symmetric. Let the probabilities of the input symbols be $P(X = 0) = q$ and $P(X = 1) = 1 - q$. Then the entropy is

$$H(X) \equiv H(q) = -q \log q - (1 - q) \log (1 - q)$$

where $H(q)$ is the binary entropy function and the conditional entropy $H(X|Y)$ is

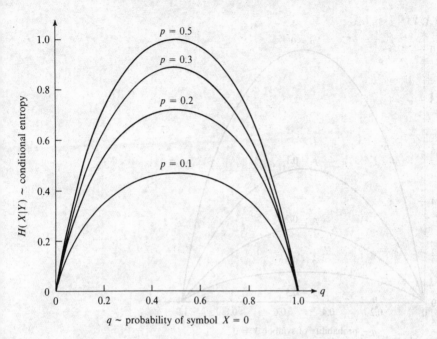

FIGURE 2.2.2
Conditional entropy for binary-input, binary-output symmetric channel.

defined by (2.2.11). A plot of $H(X|Y)$ as a function of q with p as a parameter is shown in Fig. 2.2.2. The average mutual information $I(X;Y)$ is plotted in Fig. 2.2.3.

As in the preceding example, when the conditional entropy $H(X|Y)$ is viewed in terms of a channel whose input is X and whose output is Y, then $H(X|Y)$ is called the *equivocation* and is interpreted as the amount of average uncertainty remaining in X after observation of Y.

The results given above can be generalized to more than two random variables. In particular, suppose we have a block of k random variables $(X_1, X_2, \ldots, X_k)$, with joint probability $P(x_1, x_2, \ldots, x_k) \equiv P(X_1 = x_1, X_2 = x_2, \ldots, X_k = x_k)$. Then, the entropy for the block is defined as

$$H(X_1 X_2 \cdots X_k) = -\sum_{j_1=1}^{n_1} \sum_{j_2=1}^{n_2} \cdots \sum_{j_k=1}^{n_k} P(x_1, x_2, \ldots, x_k) \log P(x_1, x_2, \ldots, x_k)$$

(2.2.13)

Since the joint probability $P(x_1, x_2, \ldots, x_k)$ can be factored as

$$P(x_1 x_2 \cdots x_k) = P(x_1) P(x_2|x_1) P(x_3|x_1 x_2) \cdots P(x_k|x_1, x_2, \ldots, x_{k-1})$$

(2.2.14)

FIGURE 2.2.3
Average mutual information for binary-input, binary-output symmetric channel.

it follows that

$$H(X_1X_2X_3 \cdots X_k) = H(X_1) + H(X_2|X_1) + H(X_3|X_1X_2)$$
$$+ \cdots + H(X_k|X_1 \cdots X_{k-1})$$
$$= \sum_{i=1}^{k} H(X_i|X_1X_2 \cdots H_{i-1}) \qquad (2.2.15)$$

By applying the result $H(X) \geq H(X|Y)$, where $X = X_m$ and $Y = X_1 X_2 \cdots X_{m-1}$ in (2.2.14) we obtain

$$H(X_1X_2 \cdots X_k) \leq \sum_{m=1}^{k} H(X_m) \qquad (2.2.16)$$

with equality if and only if the random variables $X_1, X_2, \ldots, X_k$ are statistically independent.

2.2.2   Information Measures for Continuous Random Variables

The definition of mutual information given above for discrete random variables may be extended in a straightforward manner to continuous random variables. In particular, if X and Y are random variables with joint pdf $p(x, y)$ and marginal pdf's $p(x)$ and $p(y)$, the average mutual information between X and Y is defined as

$$I(X; Y) = \int_{-\infty}^{\infty} \int_{-\infty}^{\infty} p(x)p(y|x)\log\frac{p(y|x)p(x)}{p(x)p(y)}\,dx\,dy \qquad (2.2.17)$$

Although the definition of the average mutual information carries over to continuous random variables, the concept of self-information does not. The problem is that a continuous random variable requires an infinite number of binary digits to represent it exactly. Hence, its self-information is infinite and, therefore, its entropy is also infinite. Nevertheless, we will define a quantity that we call the *differential entropy* of the continuous random variable X as

$$H(X) = -\int_{-\infty}^{\infty} p(x)\log p(x)\,dx \qquad (2.2.18)$$

We emphasize that this quantity does *not* have the physical meaning of self-information, although it may appear to be a natural extension of the definition of entropy for a discrete random variable (see Prob. 2.6).

By defining the average conditional entropy of X given Y as

$$H(X|Y) = -\int_{-\infty}^{\infty} \int_{-\infty}^{\infty} p(x|y)\log p(x|y)\,dx\,dy \qquad (2.2.19)$$

the average mutual information may be expressed as

$$I(X; Y) = H(X) - H(X|Y)$$

or, alternatively, as

$$I(X; Y) = H(Y) - H(Y|X)$$

In some cases of practical interest, the random variable X is discrete and Y is continuous. To be specific, suppose that X has possible outcomes x_i, $i = 1, 2, \ldots, n$, and Y is described by its marginal pdf $p(y)$. When X and Y are statistically dependent, we may express $p(y)$ as

$$p(y) = \sum_{i=1}^{n} p(y|x_i)P(x_i)$$

The mutual information provided about the event $X = x_i$ by the occurrence of the event $Y = y$ is

$$I(x_i; y) = \log\frac{p(y|x_i)P(x_i)}{p(y)P(x_i)}$$

$$= \log\frac{p(y|x_i)}{p(y)} \qquad (2.2.20)$$

Then the average mutual information between X and Y is

$$I(X; Y) = \sum_{i=1}^{n} \int_{-\infty}^{\infty} p(y|x_i)P(x_i)\log\frac{p(y|x_i)}{p(y)}\,dy \qquad (2.2.21)$$

Example 2.2.5. Suppose that X is a discrete random variable with two equally probable outcomes $x_1 = A$ and $x_2 = -A$. Let the conditional pdf's $p(y|x_i)$, $i = 1, 2$, be gaussian with mean x_i and variance σ^2. That is,

$$p(y|A) = \frac{1}{\sqrt{2\pi}\,\sigma} e^{-(y-A)^2/2\sigma^2}$$

$$p(y|-A) = \frac{1}{\sqrt{2\pi}\,\sigma} e^{-(y+A)^2/2\sigma^2}$$

$$(2.2.22)$$

The average mutual information obtained from (2.2.21) becomes

$$I(X; Y) = \frac{1}{2} \int_{-\infty}^{\infty} \left[p(y|A) \log \frac{p(y|A)}{p(y)} + p(y|-A) \log \frac{p(y|-A)}{p(y)} \right] dy$$

$$(2.2.23)$$

$$p(y) = \tfrac{1}{2}[\, p(y|A) + p(y|-A)\,] \qquad (2.2.24)$$

In Sec. 2.4, it will be shown that the average mutual information $I(X; Y)$ given by (2.2.23) represents the channel capacity of a binary-input additive white gaussian noise channel.

2.3 SOURCES, SOURCE MODELS, AND SOURCE ENCODING

In Sec. 2.2 we introduced a measure for the information content associated with a discrete random variable X. When X is the output of a discrete source, the entropy $H(X)$ of the source represents the average amount of information emitted by the source. In this section, we consider the process of encoding the output of a source, i.e., the process of representing the source output by a sequence of binary digits. A measure of the efficiency of a source encoding method can be obtained by comparing the average number of binary digits per output letter from the source to the entropy $H(X)$.

The encoding of a discrete source having a finite alphabet size may appear, at first glance, to be a relatively simple problem. However, this is true only when the source is memoryless, i.e., when successive symbols from the source are statistically independent and each symbol is encoded separately. The discrete memoryless source (DMS) is by far the simplest model that can be devised for a physical source. Few physical sources, however, closely fit this idealized mathematical model. For example, successive output letters from a teletype machine printing English text are expected to be statistically dependent. On the other hand, if the teletype output is a computer program coded in FORTRAN, the sequence of output letters is expected to exhibit a much smaller dependence. In any case, we shall demonstrate that it is always more efficient to encode blocks of symbol instead of encoding each symbol separately. By making the block size sufficiently large, the average number of binary digits per output letter from the source can be made arbitrarily close to the entropy of the source.

In contrast to a discrete-time source whose output is characterized by a discrete random variable X, a continuous-time source, which is also called an *analog source* or a *waveform source*, has an output $X(t)$ that is characterized as a stochastic process. In the encoding of the output of such a source, the first step usually involves sampling the source output periodically. In effect, the sampling process converts the continuous-time source into an equivalent discrete-time source. If the source output is a band-limited process, the sampling performed at the Nyquist rate or faster does not result in a loss of information. However, the amplitudes of the samples are continuous random variables and, hence, each sample cannot be represented by a finite sequence of binary digits. For this reason, each sample is quantized in amplitude to a finite number of levels. Thus an analog source is reduced to a discrete-time source having a finite number of levels (finite alphabet). The quantization process inherently results in the introduction of some distortion in the source output. The amount of this type of distortion is a function of the method by which the quantization is performed.

When each sample is quantized separately, the process is called *scalar quantization*. When a block of samples is quantized jointly as a single vector, the process is called *vector quantization* or *block quantization*. Both methods are considered in this section.

Sampling and quantization of the output of an analog source allow us to represent the analog output by a finite sequence of binary digits. Since the representation of the analog output requires an infinite number of bits, quantization results in *data compression*. Clearly, data compression is obtained at the cost of some signal distortion resulting from the quantization process. An objective in the encoding of an analog source output is to minimize the source bit rate for a given level of tolerable distortion. Alternatively, we may attempt to minimize the distortion for a specified bit rate. In either case, we will see that when the source encoding is performed on a block of outputs from the source, the efficiency of the encoding method generally improves as the block size is increased.

Considering the large amount of work that has been done on the problem of source encoding over the past three decades, our treatment of this subject is extremely brief. A thorough discussion of source encoding is beyond the scope of this book. More detailed treatments of this subject may be found in the texts by Blahut (1987), Gallager (1968), Viterbi and Omura (1979), and Berger (1971). Our discussion is focused on two aspects of source encoding: First we consider the efficient encoding of a DMS; then we describe several source encoding methods based on scalar quantization that have been used in practice to encode the output of analog sources. We consider as an example the encoding of speech signals and we compare the efficiency of the encoding methods in terms of the bit rate required to achieve acceptable telephone quality speech. Finally, we consider vector quantization for analog sources.

2.3.1 Coding for a Discrete Memoryless Source

Suppose that a DMS produces an output letter or symbol every τ_s seconds. Each symbol is selected from a finite alphabet of symbols x_i, $i = 1, 2, \ldots, L$, occurring with probabilities $P(x_i)$, $i = 1, 2, \ldots, L$. The entropy of the DMS in bits per source symbol is

$$H(X) = - \sum_{i=1}^{L} P(x_i) \log_2 P(x_i) \le \log_2 L \qquad (2.3.1)$$

where equality holds when the symbols are equally probable. The average number of bits per source symbol is $H(X)$ and the source rate in bits per second is defined as $H(X)/\tau_s$.

Fixed-length code words. First we consider a block encoding scheme that assigns a unique set of R binary digits to each symbol. Since there are L possible symbols, the number of binary digits per symbol required for unique encoding when L is a power of 2 is

$$R = \log_2 L \qquad (2.3.2)$$

and, when L is not a power of 2, it is

$$R = [\log_2 L] + 1 \qquad (2.3.3)$$

The notation $[x]$ denotes the largest integer contained in x. The code rate R in bits per symbol is now R and, since $H(X) \le \log_2 L$, it follows that $R \ge H(X)$.

The efficiency of the encoding for the DMS is defined as the ratio $H(X)/R$. We observe that when L is a power of 2 and the source letters are equally probable, $R = H(X)$. Hence a fixed-length code of R bits per symbol attains 100 percent efficiency. However, if L is not a power of 2 but the source symbols are still equally probable, R differs from $H(X)$ by at most 1 bit per symbol. When $\log_2 L \gg 1$, the efficiency of this encoding scheme is high. On the other hand, when L is small, the efficiency of the fixed-length code can be increased by encoding a sequence of J symbols at a time. To accomplish the desired encoding, we require L^J unique code words. By using sequences of N binary digits, we can accommodate 2^N possible code words. N must be selected such that

$$N \ge J \log_2 L$$

Hence the minimum integer value of N required is

$$N = [J \log_2 L] + 1 \qquad (2.3.4)$$

Now the average number of bits per source symbol is $N/J = R$ and, thus, the inefficiency has been reduced by approximately a factor of $1/J$ relative to the symbol-by-symbol encoding described above. By making J sufficiently large the efficiency of the encoding procedure, measured by the ratio $JH(X)/N$, can be made as close to unity as desired.

The encoding methods described above introduce no distortion since the encoding of source symbols or blocks of symbols into code words is unique. This type of encoding is called *noiseless*.

Now, suppose we attempt to reduce the code rate R by relaxing the condition that the encoding process be unique. For example, suppose that only a fraction of the L^J blocks of symbol is encoded uniquely. To be specific, let us select the $2^N - 1$ most probable J-symbol blocks and encode each of them uniquely, while the remaining $L^J - (2^N - 1)$ J-symbol blocks are represented by the single remaining code word. This procedure results in a decoding failure or (distortion) probability of error every time a low probability block is mapped into this single code word. Let P_e denote this probability of error. Based on this block encoding procedure, Shannon (1948a) proved the following source coding theorem.

> **Source coding theorem I.** Let X be the ensemble of letters from a DMS with finite entropy $H(X)$. Blocks of J symbols from the source are encoded into code words of length N from a binary alphabet. For any $\varepsilon > 0$, the probability P_e of a block decoding failure can be made arbitrarily small if
>
> $$R \equiv \frac{N}{J} \geq H(X) + \varepsilon \tag{2.3.5}$$
>
> and J is sufficiently large. Conversely, if
>
> $$R \leq H(X) - \varepsilon \tag{2.3.6}$$
>
> then, P_e becomes arbitrarily close to 1 as J is made sufficiently large.

From this theorem, we observe that the average number of bits per symbol required to encode the output of a DMS with arbitrarily small probability of decoding failure is lower bounded by the source entropy $H(X)$. On the other hand, if $R < H(X)$, the decoding failure rate approaches 100 percent as J is arbitrarily increased.

Variable-length code words. When the source symbols are not equally probable, a more efficient encoding method is to use variable-length code words. An example of such encoding is the Morse code, which dates back to the 1800s. In the Morse code the letters that occur more frequently are assigned short code words and the letters that occur infrequently are assigned long code words. Following this general philosophy, we may use the probabilities of occurrence of the different source letters in the selection of the code words. The problem is to devise a method for selecting and assigning the code words to source letters. This type of encoding is called *entropy coding*.

For example, suppose that a DMS with output letters a_1, a_2, a_3, a_4 and corresponding probabilities $P(a_1) = \frac{1}{2}$, $P(a_2) = \frac{1}{4}$, and $P(a_3) = P(a_4) = \frac{1}{8}$ is encoded as shown in Table 2.3.1. Code I is a variable-length code that has a basic

TABLE 2.3.1
Variable-length codes

Letter	$P(a_k)$	Code I	Code II	Code III
a_1	$\frac{1}{2}$	1	0	0
a_2	$\frac{1}{4}$	00	10	01
a_3	$\frac{1}{8}$	01	110	011
a_4	$\frac{1}{8}$	10	111	111

FIGURE 2.3.1
Code tree for code II in Table 2.3.1.

flaw. To see the flaw, suppose we are presented with the sequence $001001\ldots$. Clearly, the first symbol corresponding to 00 is a_2. However, the next four bits are ambiguous (not uniquely decodable). They may be decoded either as a_4a_3 or as $a_1a_2a_1$. Perhaps, the ambiguity can be resolved by waiting for additional bits, but such a decoding delay is highly undesirable. We shall only consider codes that are decodable *instantaneously*, that is, without any decoding delay.

Code II in Table 2.3.1 is *uniquely decodable* and *instantaneously decodable*. It is convenient to represent the code words in this code graphically as terminal nodes of a tree, as shown in Fig. 2.3.1. We observe that the digit 0 indicates the end of a code word for the first three code words. This characteristic plus the fact that no code word is longer than three binary digits makes this code instantaneously decodable. Note that no code word in this code is a prefix of any other code word. In general, the *prefix condition* requires that for a given code word $\mathbf{C}_k$ of length k having elements $(b_1, b_2, \ldots, b_k)$, there is no other code word of length $l < k$ with elements $(b_1, b_2, \ldots, b_l)$ for $1 \le l \le k - 1$. In other words, there is no code word of length $l < k$ that is identical to the first l binary digits of another code word of length $k > l$. This property makes the code words instantaneously decodable.

Code III given in Table 2.3.1 has the tree structure shown in Fig. 2.3.2. We note that in this case the code is uniquely decodable but it is *not* instantaneously decodable. Clearly, this code does *not* satisfy the prefix condition.

Our main objective is to devise a systematic procedure for constructing uniquely decodable variable-length codes that are efficient in the sense that the average number of bits per source letter, defined as the quantity-

$$\overline{R} = \sum_{k=1}^{L} n_k P(a_k) \tag{2.3.7}$$

is minimized. The conditions for the existence of a code that satisfies the prefix condition are given by the Kraft inequality.

Kraft inequality. A necessary and sufficient condition for the existence of a binary code with code words having lengths $n_1 \le n_2 \le \cdots \le n_L$ that satisfy the

FIGURE 2.3.2
Code tree for code III in Table 2.3.1.

prefix condition is

$$\sum_{k=1}^{L} 2^{-n_k} \leq 1 \qquad (2.3.8)$$

First, we prove that (2.3.8) is a sufficient condition for the existence of a code that satisfies the prefix condition. To construct such a code, we begin with a full binary tree of order $n = n_L$ that has 2^n terminal nodes and two nodes of order k stemming from each node of order $k - 1$, for each k, $1 \leq k \leq n$. Let us select any node of order n_1 as the first code word $\mathbf{C}_1$. This choice eliminates 2^{n-n_1} terminal nodes (or the fraction 2^{-n_1} of the 2^n terminal nodes). From the remaining available nodes of order n_2 we select one node for the second code word $\mathbf{C}_2$. This choice eliminates 2^{n-n_2} terminal nodes (or the fraction 2^{-n_2} of the 2^n terminal nodes). This process continues until the last code word is assigned at terminal node $n = n_L$. Since, at the node of order $j < L$, the fraction of the number of terminal nodes eliminated is

$$\sum_{k=1}^{j} 2^{-n_k} < \sum_{k=1}^{L} 2^{-n_k} \leq 1,$$

there is always a node of order $k > j$ available to be assigned to the next code word. Thus, we have constructed a code tree that is embedded in the full tree of 2^n nodes as illustrated in Fig. 2.3.3, for a tree having 16 terminal nodes and a source output consisting of five letters with $n_1 = 1$, $n_2 = 2$, $n_3 = 3$, and $n_4 = n_5 = 4$.

To prove that (2.3.8) is a necessary condition, we observe that in the code tree of order $n = n_L$, the number of terminal nodes eliminated from the total number of 2^n terminal nodes is

$$\sum_{k=1}^{L} 2^{n-n_k} \leq 2^n$$

Hence

$$\sum_{k=1}^{L} 2^{-n_k} \leq 1$$

and the proof of (2.3.8) is complete.

The Kraft inequality may be used to prove the following (noiseless) source coding theorem which applies to codes that satisfy the prefix condition.

Source coding theorem II. Let X be the ensemble of letters from a DMS with finite entropy $H(X)$, and output letters x_k, $1 \leq k \leq L$, with corresponding probabilities of occurrence p_k, $1 \leq k \leq L$. It is possible to construct a code that satisfies the prefix condition and has an average length $\bar{R}$ that satisfies the inequalities

$$H(X) \leq \bar{R} < H(X) + 1 \qquad (2.3.9)$$

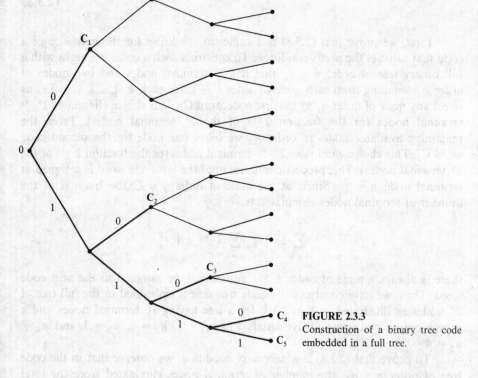

FIGURE 2.3.3
Construction of a binary tree code embedded in a full tree.

To establish the lower bound in (2.3.9) we note that for code words that have length n_k, $1 \leq k \leq L$, the difference $H(X) - \overline{R}$ may be expressed as

$$H(X) - \overline{R} = \sum_{k=1}^{L} p_k \log_2 \frac{1}{p_k} - \sum_{k=1}^{L} p_k n_k$$

$$= \sum_{k=1}^{L} p_k \log_2 \frac{2^{-n_k}}{p_k} \qquad (2.3.10)$$

Use of the inequality $\ln x \leq x - 1$ in (2.3.10) yields

$$H(X) - \overline{R} \leq (\log_2 e) \sum_{k=1}^{L} p_k \left[\frac{2^{-n_k}}{p_k} - 1 \right]$$

$$\leq (\log_2 e) \left[\sum_{k=1}^{L} 2^{-n_k} - 1 \right] \leq 0$$

where the last inequality follows from the Kraft inequality. Equality holds if and only if $p_k = 2^{-n_k}$ for $1 \leq k \leq L$.

The upper bound in (2.3.9) may be established under the constraint that n_k, $1 \le k \le L$, are integers, by selecting the $\{n_k\}$ such that $2^{-n_k} \le p_k < 2^{-n_k+1}$. But if the terms $p_k \ge 2^{-n_k}$ are summed over $1 \le k \le L$ we obtain the Kraft inequality for which we have demonstrated that there exists a code that satisfies the prefix condition. On the other hand, if we take the logarithm of $p_k < 2^{-n_k+1}$, we obtain

$$\log p_k < -n_k + 1$$

or equivalently,

$$n_k < 1 - \log p_k \tag{2.3.11}$$

If we multiply both sides of (2.3.11) by p_k and sum over $1 \le k \le L$, we obtain the desired upper bound given in (2.3.9). This completes the proof of (2.3.9).

We have now established that variable length codes that satisfy the prefix condition are efficient source codes for any DMS with source symbols that are not equally probable. Let us now describe an algorithm for constructing such codes.

Huffman (1952) devised a variable-length encoding algorithm, based on the source letter probabilities $P(x_i)$, $i = 1, 2, \ldots, L$. This algorithm is optimum in the sense that the average number of binary digits required to represent the source symbols is a minimum, subject to the constraint that the code words satisfy the prefix condition, as defined above, which allows the received sequence to be uniquely and instantaneously decodable. We illustrate this encoding algorithm by means of two examples.

Example 2.3.1. Consider a DMS with seven possible symbols $x_1, x_2, \ldots, x_7$ having the probabilities of occurrence illustrated in Fig. 2.3.4. We have ordered the source symbols in decreasing order of the probabilities, i.e., $P(x_1) > P(x_2) > \cdots > P(x_7)$. We begin the encoding process with the two least probable symbols x_6 and x_7. These two symbols are tied together as shown in Fig. 2.3.4, with the upper branch assigned a 0 and the lower branch assigned a 1. The probabilities of these two branches are added together at the node where the two branches meet to yield the probability 0.01. Now we have the source symbols x_1 through x_5 plus a new symbol, say x_6', obtained by combining x_6 and x_7. The next step is to join the two least probable symbols from the set $x_1, x_2, x_3, x_4, x_5, x_6'$. These are x_5 and x_6', which have a combined probability of 0.05. The branch from x_5 is assigned a 0 and the branch from x_6' is assigned a 1. This procedure continues until we exhaust the set of possible source letters. The result is a code tree with branches that contain the desired code words. The code words are obtained by beginning at the rightmost node in the tree and proceeding to the left. The resulting code words are listed in Fig. 2.3.4. The average number of binary digits per symbol for this code is $\bar{R} = 2.21$ bits/symbol. The entropy of the source is 2.11 bits/symbol.

We make the observation that the code is not necessarily unique. For example at the next to the last step in the encoding procedure we have a tie between x_1 and x_3', since these symbols are equally probable. At this point we chose to pair x_1 with x_2. An alternative is to pair x_2 with x_3'. If we choose this pairing, the resulting code is illustrated in Fig. 2.3.5. The average number of bits

Letter	Probability	Self-information	Code
x_1	0.35	1.5146	00
x_2	0.30	1.7370	01
x_3	0.20	2.3219	10
x_4	0.10	3.3219	110
x_5	0.04	4.6439	1110
x_6	0.005	7.6439	11110
x_7	0.005	7.6439	11111

$H(X) = 2.11$ $\bar{R} = 2.21$

FIGURE 2.3.4
An example of variable-length-source encoding for a DMS.

Letter	Code
x_1	0
x_2	10
x_3	110
x_4	1110
x_5	11110
x_6	111110
x_7	111111

$\bar{R} = 2.21$

FIGURE 2.3.5
An alternate code for the
DMS in Example 2.3.1.

FIGURE 2.3.6
Huffman code for Example 2.3.2.

Letter	Code
x_1	00
x_2	010
x_3	011
x_4	100
x_5	101
x_6	110
x_7	1110
x_8	1111

$$H(X) = 2.63 \qquad \overline{R} = 2.70$$

per source symbol for this code is also 2.21. Hence, the resulting codes are equally efficient. Secondly, the assignment of a 0 to the upper branch and a 1 to the lower (less probable) branch is arbitrary. We may simply reverse the assignment of a 0 and 1 and still obtain an efficient code satisfying the prefix condition.

Example 2.3.2. As a second example, let us determine the Huffman code for the output of a DMS illustrated in Fig. 2.3.6. The entropy of this source is $H(X) = 2.63$ bits per symbol. The Huffman code as illustrated in Fig. 2.3.6 has an average length of $\overline{R} = 2.70$ bits per symbol. Hence, its efficiency is 0.97.

The variable-length encoding (Huffman) algorithm described in the above examples generates a prefix code having an $\overline{R}$ that satisfies (2.3.9). However, instead of encoding on a symbol-by-symbol basis, a more efficient procedure is to encode blocks of J symbols at a time. In such a case, the bounds in (2.3.9) of source coding theorem II become

$$JH(X) \le \overline{R}_J < JH(X) + 1, \qquad (2.3.12)$$

since the entropy of a J-symbol block from a DMS is $JH(X)$, and $\overline{R}_J$ is the average number of bits per J-symbol blocks. If we divide (2.3.12) by J we obtain

$$H(X) \le \frac{\overline{R}_J}{J} < H(X) + \frac{1}{J} \qquad (2.3.13)$$

where $\overline{R}_J/J \equiv \overline{R}$ is the average number of bits per source symbol. Hence $\overline{R}$ can be made as close to $H(X)$ as desired by selecting J sufficiently large.

Example 2.3.3. The output of a DMS consists of *letters* x_1, x_2, x_3 with probabilities 0.45, 0.35, and 0.20, respectively. The entropy of this source is $H(X) = 1.518$ bits/ symbols. The Huffman code for this source, given in Table 2.3.2, requires $\overline{R}_1 = 1.55$ bits/symbol and results in an efficiency of 97.9 percent. If pairs of

TABLE 2.3.2
Huffman code for Example 2.8

Letter	Probability	Self-information	Code
x_1	0.45	1.156	1
x_2	0.35	1.520	00
x_3	0.20	2.330	01

$$H(X) = 1.518 \text{ bits/letter}$$
$$\bar{R}_1 = 1.55 \text{ bits/letter}$$
$$\text{Efficiency} = 97.9\%$$

symbols are encoded by means of the Huffman algorithm, the resulting code is given in Table 2.3.3. The entropy of the source output for pairs of letters is $2H(X) = 3.036$ bits/symbol pair. On the other hand, the Huffman code requires $\bar{R}_2 = 3.0675$ bits/symbol pair. Thus, the efficiency of the encoding increases to $2H(X)/\bar{R}_2 = 0.990$ or, equivalently, to 99.0 percent.

In summary, we have demonstrated that efficient encoding for a DMS may be done on a symbol-by-symbol basis using a variable-length code based on the Huffman algorithm. Furthermore, the efficiency of the encoding procedure is increased by encoding blocks of J symbols at a time. Thus, the output of a DMS with entropy $H(X)$ may be encoded by a variable-length code with an average number of bits per source letter that approaches $H(X)$ as closely as desired.

TABLE 2.3.3
Huffman code for encoding pairs of letters

Letter pair	Probability	Self-information	Code
$x_1 x_1$	0.2025	2.312	10
$x_1 x_2$	0.1575	2.676	001
$x_2 x_1$	0.1575	2.676	010
$x_2 x_2$	0.1225	3.039	011
$x_1 x_3$	0.09	3.486	111
$x_3 x_1$	0.09	3.486	0000
$x_2 x_3$	0.07	3.850	0001
$x_3 x_2$	0.07	3.850	1100
$x_3 x_3$	0.04	4.660	1101

$$2H(X) = 3.036 \text{ bits/letter pair}$$
$$\bar{R}_2 = 3.0675 \text{ bits/letter pair}$$
$$\frac{\bar{R}_2}{2} = 1.534 \text{ bits/letter}$$
$$\text{Efficiency} = 99.0\%$$

2.3.2 Discrete Stationary Sources

In the previous section we described the efficient encoding of the output of a DMS. In this section we consider discrete sources for which the sequence of output letters is statistically dependent. We limit our treatment to sources that are statistically stationary. If the source is stationary, the joint probabilities of the two sequences $x_1, x_2, \ldots, x_k$ and $x_{1+m}, x_{2+m}, \ldots, x_{k+m}$ are identical for all shifts m and for all $k \geq 1$. In other words, the joint probabilities of any sequence of outputs from the source are invariant to a shift in the time origin.

Let us evaluate the entropy of any sequence of letters from a stationary source. From the definition in (2.2.13) and the result given in (2.2.15), the entropy of a block of random variables $X_1 X_2 \cdots X_k$ is

$$H(X_1 X_2 \cdots X_k) = \sum_{i=1}^{k} H(X_i | X_1 X_2 \cdots X_{i-1}) \qquad (2.3.14)$$

where $H(X_i | X_1 X_2 \cdots X_{i-i})$ is the conditional entropy of the ith symbol from the source given the previous $i - 1$ symbols. The entropy per letter for the k-symbol block is defined as

$$H_k(X) = \frac{1}{k} H(X_1 X_2 \cdots X_k) \qquad (2.3.15)$$

We define the information content of a stationary source as the entropy per letter in (2.3.15) in the limit as $k \to \infty$. That is

$$H_\infty(X) = \lim_{k \to \infty} H_k(X) = \lim_{k \to \infty} \frac{1}{k} H(X_1 X_2 \cdots X_k) \qquad (2.3.16)$$

The existence of this limit is established below.

As an alternative, we may define the entropy per letter from the source in terms of the conditional entropy $H(X_k | X_1 X_2 \cdots X_{k-1})$ in the limit as k approaches infinity. Fortunately, this limit also exists and is identical to the limit in (2.3.16). That is,

$$H_\infty(X) = \lim_{k \to \infty} H(X_k | X_1 X_2 \cdots X_{k-1}) \qquad (2.3.17)$$

This result is also established below. Our development follows the approach in Gallager (1968).

First we show that

$$H(X_k | X_1 X_2 \cdots X_{k-1}) \leq H(X_{k-1} | X_1 X_2 \cdots X_{k-2}) \qquad (2.3.18)$$

for $k \geq 2$. From our previous result that conditioning on a random variable cannot increase entropy, we have

$$H(X_k | X_1 X_2 \cdots X_{k-1}) \leq H(X_k | X_2 X_3 \cdots X_{k-1}) \qquad (2.3.19)$$

From the stationarity of the source, we have

$$H(X_k | X_2 X_3 \cdots X_{k-1}) = H(X_{k-1} | X_1 X_2 \cdots X_{k-2}) \qquad (2.3.20)$$

Hence (2.3.18) follows immediately. This result demonstrates that $H(X_k | X_1 X_2 \cdots X_{k-1})$ is a nonincreasing sequence in k.

Second, we have the result

$$H_k(X) \geq H(X_k|X_1X_2 \cdots X_{k-1}) \qquad (2.3.21)$$

which follows immediately from (2.3.14) and (2.3.15) and the fact that the last term in the sum of (2.3.14) is a lower bound on each of the other $(k-1)$ terms.

Third, from the definition of $H_k(X)$, we may write

$$H_k(X) = \frac{1}{k}\left[H(X_1X_2 \cdots X_{k-1}) + H(X_k|X_1 \cdots X_{k-1})\right]$$

$$= \frac{1}{k}\left[(k-1)H_{k-1}(X) + H(X_k|X_1 \cdots X_{k-1})\right]$$

$$\leq \frac{k-1}{k}H_{k-1}(X) + \frac{1}{k}H_k(X)$$

which reduces to

$$H_k(X) \leq H_{k-1}(X) \qquad (2.3.22)$$

Hence, $H_k(X)$ is a nonincreasing sequence in k.

Since $H_k(X)$ and the conditional entropy $H(X_k|X_1 \cdots X_{k-1})$ are both nonnegative and nonincreasing with k, both limits must exist. Their limiting forms can be established by using (2.3.14) and (2.3.15) to express $H_{k+j}(X)$ as

$$H_{k+j}(X) = \frac{1}{k+j}H(X_1X_2 \cdots X_{k-1})$$

$$+ \frac{1}{k+j}\Big[H(X_k|X_1 \cdots X_{k-1}) + H(X_{k+1}|X_1 \cdots X_k)$$

$$+ \cdots + H(X_{k+j}|X_1 \cdots X_{k+j-1})\Big]$$

Since the conditional entropy is nonincreasing, the first term in the brackets serves as an upper bound on the other terms. Hence,

$$H_{k+j}(X) \leq \frac{1}{k+j}H(X_1X_2 \cdots X_{k-1}) + \frac{j+1}{k+j}H(X_k|X_1X_2 \cdots X_{k-1})$$

$$\qquad (2.3.23)$$

For a fixed k, the limit of (2.3.23) as $j \to \infty$ yields

$$H_\infty(X) \leq H(X_k|X_1X_2 \cdots X_{k-1}) \qquad (2.3.24)$$

But (2.3.24) is valid for all k; hence it is valid for $k \to \infty$. Therefore,

$$H_\infty(X) \leq \lim_{k \to \infty} H(X_k|X_1X_2 \cdots X_{k-1}) \qquad (2.3.25)$$

On the other hand, from (2.3.21) we obtain in the limit as $k \to \infty$,

$$H_\infty(X) \geq \lim_{k \to \infty} H(X_k|X_1X_2 \cdots X_{k-1}) \qquad (2.3.26)$$

which establishes (2.3.17).

Now suppose we have a discrete stationary source that emits J letters with $H_J(X)$ as the entropy per letter. We can encode the sequence of J letters with a variable-length Huffman code that satisfies the prefix condition by following the procedure described in the previous section. The resulting code has an average number of bits for the J-letter block which satisfies the condition

$$H(X_1 \cdots X_J) \leq \overline{R}_J < H(X_1 \cdots X_J) + 1 \qquad (2.3.27)$$

By dividing each term of (2.3.27) by J we obtain the bounds on the average number $\overline{R} = \overline{R}_J/J$ of bits per source letter as

$$H_J(X) \leq \overline{R} < H_J(X) + \frac{1}{J} \qquad (2.3.28)$$

By increasing the block size J, we can approach $H_J(X)$ arbitrarily close and in the limit as $J \to \infty$, $\overline{R}$ satisfies

$$H_\infty(X) \leq \overline{R} < H_\infty(X) + \varepsilon \qquad (2.3.29)$$

where ε approaches zero as $1/J$. Thus, efficient encoding of stationary sources is accomplished by encoding large blocks of symbols into code words. We should emphasize, however, that the design of the Huffman code requires knowledge of the joint pdf for the J-symbol blocks.

2.3.3 Coding for Analog Sources

An analog source emits a message waveform $x(t)$ that is a sample function of a stochastic process $X(t)$. Suppose that $X(t)$ is a stationary stochastic process with autocorrelation function $\phi_{xx}(\tau)$ and power spectral density function $\Phi_{xx}(f)$. Furthermore, suppose that $X(t)$ is a band-limited stochastic process, i.e., $\Phi_{xx}(f) = 0$ for $|f| > W$. Then, by use of the sampling theorem, $X(t)$ may be represented as

$$X(t) = \sum_{n=-\infty}^{\infty} X\left(\frac{n}{2W}\right) \frac{\sin 2\pi W(t - n/2W)}{2\pi W(t - n/2W)} \qquad (2.3.30)$$

where $\{ X(n/2W) \}$ denote the samples of the process $X(t)$ taken at the sampling rate $f_s = 2W$ samples per second.

By applying the sampling theorem, the output of an analog source is converted to an equivalent discrete-time sequence of samples. The samples are then quantized in amplitude and encoded. One type of simple encoding is to represent each discrete amplitude level by a sequence of binary digits. Hence, if we have L levels, we need $R = \log_2 L$ bits per sample if L is a power of 2, or $R = [\log_2 L] + 1$ if L is not a power of 2. On the other hand, if the levels are not equally probable, and the probabilities of the output levels are known, we may use Huffman coding (also called *entropy coding*) to improve the efficiency of the encoding process. Even when the probabilities are not known it is possible to obtain an estimate of the probabilities from the observed data.

Quantization of the amplitudes of the sampled signal results in data compression but it also introduces some distortion of the waveform or a loss of signal fidelity. The minimization of this distortion is considered in the following two sections. In this section we describe several waveform quantization and encoding techniques that have been applied to encoding of speech and images. In particular, we consider pulse code modulation (PCM), differential PCM (DPCM), and delta modulation (DM), and assess the distortion resulting from quantization. We also introduce a model-based method, called *linear-predictive coding* (LPC), for source encoding. Typical bit rates that result from the use of these methods in the encoding of speech signals are given.

Pulse code modulation (PCM). Let $x(t)$ denote a sample function emitted by a source and let x_n denote the samples taken at a sampling rate $f_s \geq 2W$ where W is the highest frequency in the spectrum of $x(t)$. In PCM each sample of the signal is quanitized to one of 2^R amplitude levels, where R is the number of binary digits used to represent each sample. Thus the rate from the source is Rf_s bits/s.

The quantization process may be modeled mathematically as

$$\tilde{x}_n = x_n + q_n \qquad (2.3.31)$$

where $\tilde{x}_n$ represents the quantized value of x_n and q_n represents the quantization error which we treat as an additive noise. Assuming that a uniform quantizer is used, having the input-output characteristic illustrated in Fig. 2.3.7, the quantization noise is well characterized statistically by the uniform pdf

$$p(q) = \frac{1}{\Delta} \qquad -\frac{\Delta}{2} \leq q \leq \frac{\Delta}{2} \qquad (2.3.32)$$

where the step size of the quantizer is $\Delta = 2^{-R}$. The mean square value of the quantization error is

$$E(q^2) = \frac{\Delta^2}{12} = \frac{2^{-2R}}{12} \qquad (2.3.33)$$

Measured in decibels, the mean square value of the noise is

$$10 \log \frac{\Delta^2}{12} = 10 \log \frac{2^{-2R}}{12} = -6R - 10.8 \text{ dB} \qquad (2.3.34)$$

We observe that the quantization noise decreases by 6 dB/bit used in the quantizer. For example, a 7-bit quantizer results in a quantization noise power of -52.8 dB.

Many source signals such as speech waveforms have the characteristic that small signal amplitudes occur more frequently than large signal amplitudes. However, a uniform quantizer provides the same spacing between successive levels throughout the entire dynamic range of the signal. A better approach is to have more closely spaced levels at the low signal amplitudes and more widely spaced levels at the large signal amplitudes. For a quantizer with R bits, this approach yields a (nonuniform) quantization error having a mean square value

FIGURE 2.3.7
Input-output characteristic for a uniform quantizer.

that is smaller than (2.3.33). A nonuniform quantizer characteristic is usually obtained by passing the signal through a nonlinear device that compresses the signal amplitude, followed by a uniform quantizer. For example, a logarithmic compressor has an input-output magnitude characteristic of the form

$$|y| = \frac{\log{(1 + \mu|x|)}}{\log{(1 + \mu)}} \tag{2.3.35}$$

where $|x|$ is the magnitude of the input, $|y|$ is the magnitude of the output, and μ is a parameter that is selected to give the desired compression characteristic. Figure 2.3.8 illustrates this compression relationship for several values of μ. The value $\mu = 0$ corresponds to no compression.

In the encoding of speech waveforms, for example, the value of $\mu = 255$ is often used. This value results in about a 24-dB reduction in the quantization noise power relative to uniform quantization, as shown by Jayant (1974). Consequently, a 7-bit quantizer used in conjunction with a $\mu = 255$ logarithmic compressor produces a quantization noise power of approximately -77 dB compared with the -53 dB for uniform quantization.

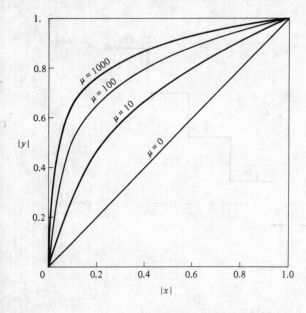

FIGURE 2.3.8
Input-output magnitude characteristic for a logarithmic compressor.

In the reconstruction of the signal from the quantized values, the inverse logarithmic relation is used to expand the signal amplitude. The combined compressor-expandor pair is termed a *compandor*.

Differential pulse code modulation (DPCM). In PCM each sample of the waveform is encoded independently of all the other samples. However, most source signals sampled at the Nyquist rate or faster exhibit significant correlation between successive samples. In other words, the average change in amplitude between successive samples is relatively small. Consequently an encoding scheme that exploits the redundancy in the samples will result in a lower bit rate for the source output.

A relatively simple solution is to encode the differences between successive samples rather than the samples themselves. Since differences between samples are expected to be smaller than the actual sampled amplitudes, fewer bits are required to represent the differences. A refinement of this general approach is to predict the current sample based on the previous p samples. To be specific, let x_n denote the current sample from the source and let $\hat{x}_n$ denote the predicted value of x_n, defined as

$$\hat{x}_n = \sum_{i=1}^{p} a_i x_{n-i} \qquad (2.3.36)$$

Thus $\hat{x}_n$ is a weighted linear combination of the past p samples and the $\{a_i\}$ are the predictor coefficients. The $\{a_i\}$ are selected to minimize some function of the error between x_n and $\hat{x}_n$.

A mathematically and practically convenient error function is the mean square error (MSE). With the MSE as the performance index for the predictor, we select the $\{a_i\}$ to minimize

$$\mathcal{E}_p = E\left(e_n^2\right) = E\left[\left(x_n - \sum_{i=1}^{p} a_i x_{n-i}\right)^2\right]$$

$$= E\left(x_n^2\right) - 2\sum_{i=1}^{p} a_i E(x_n x_{n-i}) + \sum_{i=1}^{p}\sum_{j=1}^{p} a_i a_j E(x_{n-i} x_{n-j}) \quad (2.3.37)$$

Assuming that the source output is (wide-sense) stationary, we may express (2.3.37) as

$$\mathcal{E}_p = \phi(0) - 2\sum_{i=1}^{p} a_i \phi(i) + \sum_{i=1}^{p}\sum_{j=1}^{p} a_i a_j \phi(i - j) \quad (2.3.38)$$

where $\phi(m)$ is the autocorrelation function of the sampled signal sequence x_n. Minimization of $\mathcal{E}_p$ with respect to the predictor coefficients $\{a_i\}$ results in the set of linear equations

$$\sum_{i=1}^{p} a_i \phi(i - j) = \phi(j) \qquad j = 1, 2, \ldots, p \quad (2.3.39)$$

Thus the values of the predictor coefficients are established. When the autocorrelation function $\phi(n)$ is not known *a priori*, it may be estimated from the samples $\{x_n\}$ using the relation†

$$\hat{\phi}(n) = \frac{1}{N}\sum_{i=1}^{N-n} x_i x_{i+n} \qquad n = 0, 1, 2, \ldots, p \quad (2.3.40)$$

and the estimate $\hat{\phi}(n)$ is used in (2.3.39) to solve for the coefficients $\{a_i\}$. Note that the normalization factor of $1/N$ in (2.3.40) drops out when $\hat{\phi}(n)$ is substituted in (2.3.39).

The linear equations in (2.3.39) for the predictor coefficients are called the *normal equations* or the *Yule-Walker equations*. There is an algorithm developed by Levinson (1947) and Durbin (1959) for solving these equations efficiently. It is described in Appendix 2A. We shall deal with the solution in greater detail in the subsequent discussion on linear predictive coding.

Having described the method for determining the predictor coefficients, let us now consider the block diagram of a practical DPCM system, shown in Fig. 2.3.9a. In this configuration, the predictor is implemented with the feedback loop around the quantizer. The input to the predictor is denoted as $\tilde{x}_n$, which represents the signal sample x_n modified by the quantization process, and the

---

† The estimation of the autocorrelation function from a finite number of observations $\{x_i\}$ is a separate issue which is beyond the scope of this discussion. The estimate in (2.3.40) is one that is frequently used in practice.

$$e_n = x_n - \hat{\tilde{x}}_n = x_n - \sum_{k=1}^{p} a_k \tilde{x}_{n-k}$$

$$\tilde{e}_n - e_n = \tilde{e}_n - \left(x_n - \hat{\tilde{x}}_n\right) = \tilde{e}_n + \hat{\tilde{x}}_n - x_n$$

$$= \tilde{x}_n - x_n$$

$$= q_n = \text{quantization error}$$

(a) Encoder

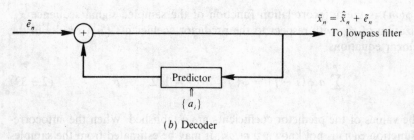

(b) Decoder

FIGURE 2.3.9
(a) block diagram of a DPCM encoder; (b) DPCM decoder at the receiver.

output of the predictor is

$$\hat{\tilde{x}}_n = \sum_{i=1}^{p} a_i \tilde{x}_{n-i} \qquad (2.3.41)$$

The difference

$$e_n = x_n - \hat{\tilde{x}}_n \qquad (2.3.42)$$

is the input to the quantizer and $\tilde{e}_n$ denotes the output. Each value of the quantized prediction error $\tilde{e}_n$ is encoded into a sequence of binary digits and transmitted over the channel to the destination. The quantized error $\tilde{e}_n$ is also added to the predicted value $\hat{\tilde{x}}_n$ to yield $\tilde{x}_n$.

At the destination the same predictor that was used at the transmitting end is synthesized and its output $\hat{\tilde{x}}_n$ is added to $\tilde{e}_n$ to yield $\tilde{x}_n$. The signal $\tilde{x}_n$ is the desired excitation for the predictor and also the desired output sequence from which the reconstructed signal $\tilde{x}(t)$ is obtained by filtering, as shown in Fig. 2.3.9b.

The use of feedback around the quantizer, as described above, ensures that the error in $\tilde{x}_n$ is simply the quantization error $q_n = \tilde{e}_n - e_n$ and that there is no accumulation of previous quantization errors in the implementation of the

decoder. That is,

$$q_n = \tilde{e}_n - e_n$$
$$= \tilde{e}_n - \left(x_n - \hat{\tilde{x}}_n \right)$$
$$= \tilde{x}_n - x_n \tag{2.3.43}$$

Hence $\tilde{x}_n = x_n + q_n$. This means that the quantized sample $\tilde{x}_n$ differs from the input x_n by the quantization error q_n independent of the predictor used. Therefore the quantization errors do not accumulate.

In the DPCM system illustrated in Fig. 2.3.9, the estimate or predicted value $\hat{\tilde{x}}_n$ of the signal sample x_n is obtained by taking a linear combination of past values $\tilde{x}_{n-k}$, $k = 1, 2, \dots, p$, as indicated by (2.3.41). An improvement in the quality of the estimate is obtained by including linearly filtered past values of the quantized error. Specifically, the $\hat{\tilde{x}}_n$ estimate may be expressed as

$$\tilde{x}_n = \sum_{i=1}^{p} a_i \tilde{x}_{n-i} + \sum_{i=1}^{m} b_i \tilde{e}_{n-i} \tag{2.3.44}$$

where $\{b_i\}$ are the coefficients of the filter for the quantized error sequence $\tilde{e}_n$. The block diagram of the encoder at the transmitter and the decoder at the receiver are shown in Fig. 2.3.10. The two sets of coefficients $\{a_i\}$ and $\{b_i\}$ are

(a) Encoder

(b) Decoder

FIGURE 2.3.10
DPCM modified by the addition of linearly filtered error sequence.

selected to minimize some function of the error $e_n = x_n - \hat{\tilde{x}}_n$, such as the mean square error.

Adaptive PCM and DPCM. Many real sources are quasistationary in nature. One aspect of the quasistationary characteristic is that the variance and the autocorrelation function of the source output vary slowly with time. PCM and DPCM encoders, however, are designed on the basis that the source output is stationary. The efficiency and performance of these encoders can be improved by having them adapt to the slowly time-variant statistics of the source.

In both PCM and DPCM, the quantization error q_n resulting from a uniform quantizer operating on a quasistationary input signal will have a time-variant variance (quantization noise power). One improvement which reduces the dynamic range of the quantization noise is the use of an adaptive quantizer. Although the quantizer can be made adaptive in different ways, a relatively simple method is to use a uniform quantizer which varies its step size in

FIGURE 2.3.11
Example of a quantizer with an adaptive step size (Jayant 1974)).

TABLE 2.3.4
Multiplication factors for adaptive step size adjustment (Jayant (1974))

	PCM			DPCM		
	2	3	4	2	3	4
$M(1)$	0.60	0.85	0.80	0.80	0.90	0.90
$M(2)$	2.20	1.00	0.80	1.60	0.90	0.90
$M(3)$		1.00	0.80		1.25	0.90
$M(4)$		1.50	0.80		1.70	0.90
$M(5)$			1.20			1.20
$M(6)$			1.60			1.60
$M(7)$			2.00			2.00
$M(8)$			2.40			2.40

accordance with the variance of the past signal samples. For example, a short-term running estimate of the variance of x_n can be computed from the input sequence $\{x_n\}$ and step size can be adjusted on the basis of such an estimate. In its simplest form, the algorithm for the step-size adjustment employs only the previous signal sample. Such an algorithm has been successfully used by Jayant (1974) in the encoding of speech signals. Figure 2.3.11 illustrates such a (3-bit) quantizer in which the step size is adjusted recursively according to the relation

$$\Delta_{n+1} = \Delta_n \cdot M(n) \qquad (2.3.45)$$

where $M(n)$ is a multiplication factor whose value depends on the quantizer level for the sample x_n, and Δ_n is the step size of the quantizer for processing x_n. Values of the multiplication factors optimized for speech encoding have been given by Jayant (1974). These values are displayed in Table 2.3.4 for 2-, 3-, and 4-bit adaptive quantization.

   In DPCM, the predictor can also be made adaptive when the source output is quasistationary. The coefficients of the predictor can be changed periodically to reflect the changing signal statistics of the source. The linear equations given by (2.3.39) still apply, with the short-term estimate of the autocorrelation function of x_n substituted in place of the ensemble correlation function. The predictor coefficients thus determined may be transmitted along with the quantized error $\tilde{e}(n)$ to the receiver, which implements the same predictor. Unfortunately, the transmission of the predictor coefficients results in a higher bit rate over the channel, offsetting, in part, the lower data rate achieved by having a quantizer with fewer bits (fewer levels) to handle the reduced dynamic range in the error e_n resulting from adaptive prediction.

   As an alternative, the predictor at the receiver may compute its own production coefficients from $\tilde{e}_n$ and $\tilde{x}_n$, where

$$\tilde{x}_n = \tilde{e}_n + \sum_{i=1}^{p} a_i \tilde{x}_{n-i} \qquad (2.3.46)$$

If we neglect the quantization noise, $\tilde{x}_n$ is equivalent to x_n. Hence $\tilde{x}_n$ may be used to estimate the autocorrelation function $\phi(n)$ at the receiver and the resulting estimates can be used in (2.3.39) in place of $\phi(n)$ to solve for the predictor coefficients. For sufficiently fine quantization, the difference between x_n and $\tilde{x}_n$ is very small. Hence the estimate of $\phi(n)$ obtained from $\tilde{x}_n$ is usually adequate for determining the predictor coefficients. Implemented in this manner, the adaptive predictor results in a lower source data rate.

Instead of using the block processing approach for determining the predictor coefficients $\{a_i\}$ as described above, we may adapt the predictor coefficients on a sample-by-sample basis by using a gradient-type algorithm, similar in form to the adaptive gradient equalization algorithm that is described in Chapter 6. Similar gradient-type algorithms have also been devised for adapting the filter coefficients $\{a_i\}$ and $\{b_i\}$ of the DPCM system shown in Fig. 2.3.10. For details on such algorithms the reader may refer to the book by Jayant and Noll (1984).

Delta modulation (DM) and adaptive delta modulation (ADM). Delta modulation may be viewed as a simplified form of DPCM in which a two-level (1-bit) quantizer is used in conjunction with a fixed first-order predictor. The block diagram of a DM encoder-decoder is shown in Fig. 2.3.12a. We note that

$$\hat{\tilde{x}}_n = \tilde{x}_{n-1} = \hat{\tilde{x}}_{n-1} + \tilde{e}_{n-1} \qquad (2.3.47)$$

Since

$$q_n = \tilde{e}_n - e_n$$

$$= \tilde{e}_n - \left(x_n - \hat{\tilde{x}}_n\right)$$

It follows that

$$\hat{\tilde{x}}_n = x_{n-1} + q_{n-1}$$

Thus the estimated (predicted) value of x_n is really the previous sample x_{n-1} modified by the quantization noise q_{n-1}. We also note that the difference equation in (2.3.47) represents an integrator with an input $\tilde{e}_n$. Hence an equivalent realization of the one-step predictor is an accumulator with an input equal to the quantized error signal $\tilde{e}_n$. In general, the quantized error signal is scaled by some value, say Δ_1, which is called the *step size*. This equivalent realization is illustrated in Fig. 2.3.12b. In effect, the encoder shown in Fig. 2.3.12 approximates a waveform $x(t)$ by a linear staircase function. In order for the approximation to be relatively good, the waveform $x(t)$ must change slowly relative to the sampling rate. This requirement implies that the sampling rate must be several (a factor of at least 5) times the Nyquist rate.

At any given sampling rate, the performance of the DM encoder is limited by two types of distortion, as illustrated in Fig. 2.3.13. One is called *slope-overload distortion*. It is due to the use of a step size Δ_1 that is too small to follow portions of the waveform that have a steep slope. The second type of distortion,

(a)

(b)

FIGURE 2.3.12
(a) Block diagram of a delta modulation system; (b) an equivalent realization of a delta modulation system.

called *granular noise*, results from using a step size that is too large in parts of the waveform having a small slope. The need to minimize both of these two types of distortion results in conflicting requirements in the selection of the step size Δ_1. One solution is to select Δ_1 to minimize the sum of the mean square values of these two distortions.

Even when Δ_1 is optimized to minimize the total mean square value of the slope-overload distortion and the granular noise, the performance of the DM

FIGURE 2.3.13
An example of slope overload distortion and granular noise in a delta modulation encoder.

encoder may still be less than satisfactory. An alternative solution is to employ a variable step size that adapts itself to the short-term characteristics of the source signal. That is, the step size is increased when the waveform has a steep slope and decreased when the waveform has a relatively small slope. This adaptive characteristic is illustrated in Fig. 2.3.14.

A variety of methods can be used to set adaptively the step size in every iteration. The quantized error sequence $\tilde{e}_n$ provides a good indication of the slope characteristics of the waveform being encoded. When the quantized error $\tilde{e}_n$ is changing signs between successive iterations, this is an indication that the slope of the waveform in that locality is relatively small. On the other hand, when the waveform has a steep slope, successive values of the error $\tilde{e}_n$ are expected to have

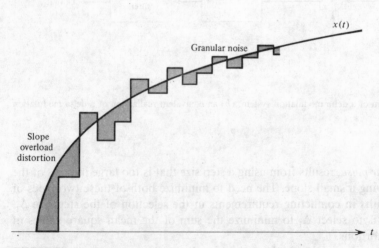

FIGURE 2.3.14
An example of variable-step-size delta modulation encoding.

identical signs. From these observations it is possible to devise algorithms which decrease or increase the step size depending on successive values of $\tilde{e}_n$. A relatively simple rule devised by Jayant (1970) is to vary adaptively the step size according to the relation

$$\Delta_n = \Delta_{n-1} K^{\tilde{e}_n \cdot \tilde{e}_{n-1}} \qquad n = 1, 2, \ldots$$

where $K \geq 1$ is a constant that is selected to minimize the total distortion. A block diagram of a DM encoder-decoder that incorporates this adaptive algorithm is illustrated in Fig. 2.3.15.

Several other variations of adaptive DM encoding have been investigated and described in the technical literature. A particularly effective and popular technique first proposed by Greefkes (1970) is called *continuously variable slope delta modulation* (CVSD). In CVSD the adaptive step-size parameter may be expressed as

$$\Delta_n = \alpha \Delta_{n-1} + k_1$$

if $\tilde{e}_n$, $\tilde{e}_{n-1}$, and $\tilde{e}_{n-2}$ have the same sign; otherwise,

$$\Delta_n = \alpha \Delta_{n-1} + k_2$$

The parameters α, k_1, and k_2 are selected such that $0 < \alpha < 1$ and $k_1 \gg k_2 > 0$. For more discussion on this and other variations of adaptive DM, the interested reader is referred to the papers by Jayant (1974) and Flanagan et al. (1979) and to the extensive references contained in these papers.

PCM, DPCM, adaptive PCM, and adaptive DPCM and DM are all source encoding techniques that attempt to faithfully represent the output waveform from the source. Consequently the methods are classified as waveform encoding techniques. The linear predictive coding method described next is an example of another class of source coding techniques that estimate and communicate a linear model of the source.

Linear predictive coding (LPC). In contrast to the waveform encoding methods described above, linear predictive coding (LPC) represents a completely different approach to the problem of source encoding. In LPC the source is modeled as a linear system (filter) which, when excited by an appropriate input signal results in the observed source output. Instead of transmitting the samples of the source waveform to the receiver, the parameters of the linear system are transmitted along with the appropriate excitation signal.

To be specific, suppose the source output is sampled at a rate equal to or exceeding the Nyquist rate and let the samples taken over some time interval be denoted as x_n, $n = 0, 1, \ldots, N - 1$. In LPC the sampled sequence is assumed to have been generated by an all-pole (discrete-time) filter having the transfer function

$$H(z) = \frac{G}{1 - \displaystyle\sum_{k=1}^{p} a_k z^{-k}} \qquad (2.3.48)$$

FIGURE 2.3.15
An example of a delta modulation system with adaptive step size.

Appropriate excitation functions are an impulse, a sequence of impulses, or a sequence of white noise with unit variance. In any case, suppose that the input sequence is denoted as v_n, $n = 0, 1, 2, \ldots$. Then the output sequence of the all-pole model satisfies the difference equation

$$x_n = \sum_{k=1}^{p} a_k x_{n-k} + G v_n \qquad n = 0, 1, 2, \ldots \qquad (2.3.49)$$

In general, the observed source output x_n, $n = 0, 1, 2, \ldots, N - 1$, does not satisfy the difference equation in (2.3.49), but only its model does. If the input is a white-noise sequence or an impulse, we may form an estimate (or prediction) of x_n by the weighted linear combination

$$\hat{x}_n = \sum_{k=1}^{p} a_k x_{n-k} \qquad n > 0 \qquad (2.3.50)$$

The difference between x_n and $\hat{x}_n$, namely,

$$e_n = x_n - \hat{x}_n$$

$$= x_n - \sum_{k=1}^{p} a_k x_{n-k} \qquad (2.3.51)$$

represents the error between the observed value x_n and the estimated (predicted) value $\hat{x}_n$. The filter coefficients $\{a_k\}$ can be selected to minimize the mean square value of this error.

Suppose for the moment that the input $\{v_n\}$ is a white-noise sequence. Then the filter output x_n is a random sequence and so is the difference $e_n = x_n - \hat{x}_n$. The ensemble average of the squared error is

$$\mathscr{E}_p = E\left(e_n^2\right)$$

$$= E\left[\left(x_n - \sum_{k=1}^{p} a_k x_{n-k}\right)^2\right]$$

$$= \phi(0) - 2\sum_{k=1}^{p} a_k \phi(k) + \sum_{k=1}^{p} \sum_{m=1}^{p} a_k a_m \phi(k-m) \qquad (2.3.52)$$

where $\phi(m)$ is the autocorrelation function of the sequence x_n, $n = 0, 1, \ldots,$ $N - 1$. But $\mathscr{E}_p$ is identical to the MSE given by (2.3.38) for a predictor used in DPCM. Consequently minimization of $\mathscr{E}_p$ in (2.3.52) yields the set of normal equations given previously by (2.3.39). To completely specify the filter $H(z)$, we must also determine the filter gain G. From (2.3.49) we have

$$E\left[(Gv_n)^2\right] = G^2 E\left(v_n^2\right) = G^2 = E\left[\left(x_n - \sum_{k=1}^{p} a_k x_{n-k}\right)^2\right] = \mathscr{E}_p \quad (2.3.53)$$

where $\mathscr{E}_p$ is the residual MSE obtained from (2.3.52) by substituting the optimum prediction coefficients, which result from the solution of (2.3.39). With this substitution, the expression for $\mathscr{E}_p$ and, hence, G^2 simplifies to

$$\mathscr{E}_p = G^2 = \phi(0) - \sum_{k=1}^{p} a_k \phi(k) \qquad (2.3.54)$$

In practice, we do not usually know *a priori* the true autocorrelation function of the source output. Hence, in place of $\phi(n)$, we substitute an estimate $\hat{\phi}(n)$ as given by (2.3.40), which is obtained from the set of samples x_n, $n = 0, 1, \ldots, N - 1$, emitted by the source.

It is convenient to express the equation in (2.3.39), with $\hat{\phi}(n)$ substituted for $\phi(n)$, in the matrix form

$$\mathbf{\Phi a} = \mathbf{\phi} \tag{2.3.55}$$

where $\mathbf{\Phi}$ is a $p \times p$ matrix with elements $\hat{\phi}_{ij} \equiv \hat{\phi}(i-j)$, $\mathbf{a}$ is a $p \times 1$ column vector of the predictor coefficients, and $\mathbf{\phi}$ is a $p \times 1$ column vector with elements $\hat{\phi}(i)$, $i = 1, 2, \ldots, p$. The matrix $\mathbf{\Phi}$ has the characteristic that the elements along each diagonal are equal. For example, a fourth-order ($p = 4$) predictor results in the 4×4 matrix

$$\mathbf{\Phi} = \begin{bmatrix} \hat{\phi}(0) & \hat{\phi}(1) & \hat{\phi}(2) & \hat{\phi}(3) \\ \hat{\phi}(1) & \hat{\phi}(0) & \hat{\phi}(1) & \hat{\phi}(2) \\ \hat{\phi}(2) & \hat{\phi}(1) & \hat{\phi}(0) & \hat{\phi}(1) \\ \hat{\phi}(3) & \hat{\phi}(2) & \hat{\phi}(1) & \hat{\phi}(0) \end{bmatrix} \tag{2.3.56}$$

A matrix that satisfies this property is called a *Toeplitz matrix*. The Toeplitz property allows us to invert $\mathbf{\Phi}$ very efficiently by means of a recursive algorithm originally proposed by Levinson (1947) and modified by Durbin (1959).

As indicated previously, the Levinson–Durbin algorithm is derived in detail in Appendix 2A. This algorithm determines the predictor coefficients $\{a_k\}$ recursively, beginning with a first-order predictor and iterating the order of the predictor up to order p. The recursive equations for the $\{a_k\}$ may be expressed as

$$a_{ii} = \frac{\hat{\phi}(i) - \sum_{k=1}^{i-1} a_{i-1k}\hat{\phi}(i-k)}{\hat{\mathscr{E}}_{i-1}}$$

$$a_{ik} = a_{i-1k} - a_{ii}a_{i-1i-k} \qquad \begin{array}{l} 1 \le k \le i-1 \\ i = 2, 3, \ldots, p \end{array} \tag{2.3.57}$$

$$\hat{\mathscr{E}}_i = (1 - a_{ii})\hat{\mathscr{E}}_{i-1}$$

$$a_{11} = \frac{\hat{\phi}(1)}{\hat{\phi}(0)} \qquad \hat{\mathscr{E}}_0 = \hat{\phi}(0)$$

where a_{ik}, $k = 1, 2, \ldots, i$, represents the coefficients of the ith-order predictor. The desired coefficients for the predictor of order p are

$$a_k \equiv a_{pk} \qquad k = 1, 2, \ldots, p \tag{2.3.58}$$

and the residual MSE is

$$\hat{\mathscr{E}}_p = G^2 = \hat{\phi}(0) - \sum_{k=1}^{p} a_k\hat{\phi}(k)$$

$$= \hat{\phi}(0) \prod_{i=1}^{p} (1 - a_{ii}^2) \tag{2.3.59}$$

We observe that the recursive relations in (2.3.57) not only give us the coefficients of the predictor for order p, but they also give us the predictor coefficients of all orders less than p.

The residual MSE $\hat{\mathscr{E}}_i$, $i = 1, 2, \ldots, p$, forms a monotone decreasing sequence, i.e., $\hat{\mathscr{E}}_p \leq \hat{\mathscr{E}}_{p-1} \leq \cdots \leq \hat{\mathscr{E}}_1 \leq \hat{\mathscr{E}}_0$, and the prediction coefficients a_{ii} satisfy the condition

$$|a_{ii}| < 1 \qquad i = 1, 2, \ldots, p \tag{2.3.60}$$

This condition is necessary and sufficient for all the poles of $H(z)$ to be inside the unit circle. Thus (2.3.60) ensures that the model is stable.

LPC has been successfully used in the modeling of a speech source. In this case, the coefficients a_{ii}, $i = 1, 2, \ldots, p$, are called *reflection coefficients* as a consequence of their correspondence to the reflection coefficients in the acoustic tube model of the vocal tract [for reference, see Rabiner and Schafer (1978)].

An alternative approach to the method used above, which yields a similar set of linear equations for the prediction coefficients, is based on the method of least squares. In the method of least squares, the prediction coefficients a_k, $k = 1, 2, \ldots, p$, are determined from the source output sequence x_i, $i = 0, 1, \ldots, N - 1$, by minimizing the time-averaged squared (least-squares) error

$$\tilde{\mathscr{E}}_p = \frac{1}{N - p} \sum_{n=p}^{N-1} e_n^2$$

$$= \frac{1}{N - p} \sum_{n=p}^{N-1} \left[x_n - \sum_{k=1}^{p} a_k x_{n-k} \right]^2$$

$$= \phi_1(0, 0) - 2 \sum_{k=1}^{p} a_k \phi_1(k, 0) + \sum_{k=1}^{p} \sum_{m=1}^{p} a_k a_m \phi_1(k, m) \tag{2.3.61}$$

where, by definition,

$$\phi_1(k, m) = \frac{1}{N - p} \sum_{n=p}^{N-1} x_{n-k} x_{n-m} \tag{2.3.62}$$

The minimization of $\tilde{\mathscr{E}}_p$ with respect to the predictor coefficients results in the set of linear equations

$$\sum_{k=1}^{p} a_k \phi_1(m, k) = \phi_1(m, 0) \qquad m = 1, 2, \ldots, p \tag{2.3.63}$$

The gain parameter G is determined from the equation for the minimum value of $\tilde{\mathscr{E}}_p$, namely,

$$G^2 = \tilde{\mathscr{E}}_p = \phi_1(0, 0) - \sum_{k=1}^{p} a_k \phi_1(k, 0) \tag{2.3.64}$$

FIGURE 2.3.16
Block diagram of waveform synthesizer (source decoder) for an LPC system.

In matrix form, the equations in (2.3.63) become

$$\boldsymbol{\Phi}_1 \mathbf{a} = \boldsymbol{\phi}_1 \qquad (2.3.65)$$

where $\boldsymbol{\Phi}_1$ is a $p \times p$ symmetric matrix with elements $\phi_1(k, m)$ and $\boldsymbol{\phi}_1$ is a $p \times 1$ vector with elements $\phi_1(m, 0)$, $m = 1, 2, \ldots, p$. A comparison of the equations in (2.3.63) with the linear equations given in (2.3.55) reveals that, unlike $\boldsymbol{\Phi}$, the matrix $\boldsymbol{\Phi}_1$ is not Toeplitz. As a consequence, the Levinson–Durbin algorithm cannot be applied to solve (2.3.65) for the predictor coefficients. This is not a major problem, however, because $\boldsymbol{\Phi}_1$ is sufficiently similar to a Toeplitz matrix that other algorithms have been developed which are almost as efficient as the Levinson–Durbin algorithm [for reference, see Marple (1980)]. Perhaps a more serious problem is that the prediction coefficients obtained from the solution of (2.3.65) may not correspond to a stable all-pole filter. In some applications this may be a serious shortcoming of the least-squares method.

Once the predictor coefficients and the gain G have been estimated from the source output $\{x_n\}$ by one of the two methods described above, each parameter is coded into a sequence of binary digits and transmitted to the receiver. Source decoding or waveform synthesis may be accomplished at the receiver as illustrated in Fig. 2.3.16. The signal generator is used to produce the excitation function $\{v_n\}$, which is scaled by G to produce the desired input to the all-pole filter model $H(z)$ synthesized from the received prediction coefficients. The analog signal may be reconstructed by passing the output sequence from $H(z)$ through an analog filter that basically performs the function of interpolating the signal between sample points. In this realization of the waveform synthesizer, the excitation function and the gain parameter must be transmitted along with the prediction coefficients to the receiver.

When the source output is stationary, the filter parameters need to be determined only once. However, the statistics of most sources encountered in practice are at best quasistationary. Under these circumstances, it is necessary periodically to obtain new estimates of the filter coefficients, the gain G, and the type of excitation function and transmit these estimates to the receiver.

FIGURE 2.3.17
Block diagram model of the generation of a speech signal.

Example 2.3.4. The block diagram shown in Fig. 2.3.17 illustrates a model for a speech source. There are two mutually exclusive excitation functions to model voiced and unvoiced speech sounds. On a short-time basis, voiced speech is periodic with a fundamental frequency f_0 or a pitch period $1/f_0$ which depends on the speaker. Thus voiced speech is generated by exciting an all-pole filter model of the vocal tract by a periodic impulse train with a period equal to the desired pitch period. Unvoiced speech sounds are generated by exciting the all-pole filter model by the output of a random-noise generator. The speech encoder at the transmitter must determine the proper excitation function, the pitch period for voiced speech, the gain parameter G, and the prediction coefficients. These parameters are encoded into binary digits and transmitted to the receiver. Typically, the voiced and unvoiced information requires 1 bit, the pitch period is adequately represented by 6 bits, and the gain parameter may be represented by 5 bits after its dynamic range is compressed logarithmically. The prediction coefficients require between 8 to 10 bits/coefficient for adequate representation [for reference, see Rabiner and Schafer (1978)]. The reason for such a high accuracy is that relatively small changes in the prediction coefficients result in a large change in the pole positions of the filter model $H(z)$. The accuracy requirements may be lessened by transmitting the reflection coefficients a_{ii} which have a smaller dynamic range. These are adequately represented by 6 bits. Thus, for a predictor of $p = 10$ order [five poles in $H(z)$], the total number of bits is 72. Due to the quasistationary nature of the speech signal, the linear system model must be changed periodically, typically once every 15 to 30 ms. Consequently, the bit rate from the source encoder is in the range of 4800 to 2400 bits/s.

When the reflection coefficients are transmitted to the decoder, it is not necessary to recompute the prediction coefficients in order to realize the speech synthesizer. Instead, the synthesis is performed by realizing a lattice filter, shown in Fig. 2.3.18, which utilizes the reflection coefficients directly and which is equivalent to the linear prediction filter.

Another variation of the source encoder-decoder based on LPC is illustrated in Fig. 2.3.19. In this scheme, the source signal is synthesized at the transmitter from the estimated filter parameters and excitation and it is com-

FIGURE 2.3.18
All-pole lattice filter for synthesizing the speech signal.

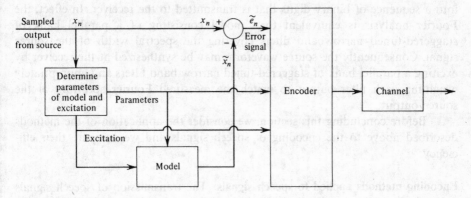

FIGURE 2.3.19
An alternative type of source encoder based on LPC.

pared with the source output to generate an error signal. The error signal is quantized and transmitted along with the filter parameters and the excitation function to the receiver. The source decoder at the receiver synthesizes the source output by adding the error to the output of the all-pole filter. The addition of the error signal reduces the distortion in the reconstructed signal at the receiver.

The linear all-pole filter model for which the filter coefficients are estimated via linear prediction is by far the simplest linear model for a source. A more general source model is a linear filter that contains both poles and zeros. In a pole-zero model, the source output x_n satisfies the difference equation

$$x_n = \sum_{k=1}^{p} a_k x_{n-k} + \sum_{k=0}^{q} b_k v_{n-k}$$

where v_n is the input excitation sequence. The problem now is to estimate the filter parameters $\{a_k\}$ and $\{b_k\}$ from the data x_i, $i = 0, 1, \ldots, N - 1$, emitted by the source. However, the MSE criterion applied to the minimization of the error $e_n = x_n - \hat{x}_n$, where $\hat{x}_n$ is an estimate of x_n, results in a set of nonlinear equations for the parameters $\{a_k\}$ and $\{b_k\}$. Consequently the evaluation of the $\{a_k\}$ and $\{b_k\}$ becomes tedious and difficult mathematically. To avoid having to solve the non-linear equations, a number of suboptimum methods have been devised for pole-zero modeling. A discussion of these techniques would lead us too far afield, however.

In addition to the all-pole and pole-zero linear filter models, there are other source encoding techniques that are based on modeling the spectral characteristics of the source output. A Fourier transform analysis is usually performed on short-time segments of the source waveform to determine the spectral components of each segment. In addition to the spectrum, the appropriate excitation function is determined for each segment. The spectral components and the excitation function corresponding to each segment are then converted (encoded) into a sequence of binary digits that is transmitted to the receiver. In effect, the Fourier analysis is equivalent to a model consisting of a parallel bank of staggered-tuned narrowband filters covering the spectral width of the source signal. Consequently the source waveform may be synthesized at the receiver by exciting a parallel bank of staggered-tuned narrow-band filters and appropriately weighting each filter output to match the measured Fourier spectrum of the source output.

Before concluding this section, we consider the application of the methods described above to the encoding of speech signals and we compare their efficiency.

Encoding methods applied to speech signals. The transmission of speech signals over telephone lines, radio channels, and satellite channels constitutes by far the largest part of our daily communications. It is understandable, therefore, that over the past three decades more research has been performed on speech encoding than on any other type of information-bearing signal. In fact, all the encoding techniques described in this section have been applied to the encoding of speech signals. It is appropriate, therefore, to compare the efficiency of these methods in terms of the bit rate required to transmit the speech signal.

The speech signal is assumed to be band-limited to the frequency range 200 to 3200 Hz and sampled at a nominal rate of 8000 samples/s for all encoders except DM where the sampling rate is f_s identical to the bit rate. For an LPC encoder, the parameters given in Example 2.3.4 are assumed.

Table 2.3.5 summarizes the main characteristics of the encoding methods described in this section and the required bit rate. In terms of the quality of the speech signal synthesized at the receiver from the (error-free) binary sequence, all the waveform encoding methods (PCM, DPCM, ADPCM, DM, ADM) provide telephone (toll) quality speech. In other words, a listener would have difficulty discerning the difference between the digitized speech and the analog speech waveform. ADPCM and ADM are particularly efficient waveform encoding techniques. With CVSD it is possible to operate down to 9600 bits/s with some noticeable waveform distortion. In fact, at rates below 16,000 bits/s the distor-

TABLE 2.3.5
Encoding techniques applied to speech signals

Encoding method	Quantizer	Coder	Transmission rate, bits / s
PCM	Linear	12 bits	96,000
Log PCM	Logarithamic	7–8 bits	56,000–64,000
DPCM	Logarithmic	4–6 bits	32,000–48,000
ADPCM	Adaptive	3–4 bits	24,000–32,000
DM	Binary	1 bit	64,000
ADM	Adaptive binary	1 bit	16,000–32,000
LPC			2,400–4,800

tion produced by waveform encoders increases significantly. Consequently, these techniques are not used below 9600 bits/s.

For rates below 9600 bits/s, encoding techniques, such as LPC, which are based on linear models of the source are usually employed. The synthesized speech obtained from this class of encoding techniques is intelligible. However, the speech signal has a synthetic quality and there is noticeable distortion.

2.3.4 Performance Characteristics of Optimum Scalar Quantizers

In the previous section we described some practical methods for encoding the output of analog sources. In this and the next section we consider the problem of quantization from a theoretical viewpoint and present some performance results from information theory on basic limits that relate the minimum achievable source code rate to the distortion introduced in the encoding process.

Many of the results given in this section apply directly to the discrete time, continuous-amplitude, memoryless gaussian source. Such a source serves as a good model for the residual error in speech coding and, more generally, in source coding based on linear prediction such as in adaptive DPCM and LPC. In these methods, the linear predictor not only removes linear dependencies in the data but it also whitens the output sequence and reduces the dynamic range of the signal samples. The residual error at the output of the predictor which is to be quantized is then well modeled as gaussian with zero mean.

In the PCM and DPCM waveform encoding methods described in the previous section, the quantizer can be optimized if we know the probability density function of the signal amplitude at the input to the quantizer. For example, suppose that the sequence $\{x_n\}$ at the input to the quantizer has a pdf $p(x)$ and let $L = 2^R$ be the desired number of levels. We wish to design the optimum scalar quantizer that minimizes some function of the quantization error $q = \tilde{x} - x$, where $\tilde{x}$ is the quantized value of x. To elaborate, suppose that $f(\tilde{x} - x)$ denotes the desired function of the error. Then, the average distortion resulting from quantization of the signal amplitude is

$$D = \int_{-\infty}^{\infty} f(\tilde{x} - x) p(x) \, dx \qquad (2.3.66)$$

In general, an optimum quantizer is one that minimizes D by optimally selecting the output levels and the corresponding input range of each output level. This optimization problem has been considered by Lloyd (1982) and Max (1960), and the resulting optimum quantizer is usually called the *Lloyd–Max quantizer*.

For a uniform quantizer, the output levels are specified as $\tilde{x}_k = (2k - 1)\Delta/2$, corresponding to an input signal amplitude in the range $(k - 1)\Delta \leq x < k\Delta$, where Δ is the step size. When the uniform quantizer is symmetric with an even number of levels, the average distortion in (2.3.66) may be expressed as

$$D = 2 \sum_{k=1}^{L/2-1} \int_{(k-1)\Delta}^{k\Delta} f\left[\frac{(2k-1)\Delta}{2} - x\right] p(x)\, dx$$

$$+ 2 \int_{(L/2-1)\Delta}^{\infty} f\left[\frac{(2k-1)\Delta}{2} - x\right] p(x)\, dx \quad (2.3.67)$$

In this case, the minimization of D is carried out with respect to the step-size parameter Δ. By differentiating D with respect to Δ we obtain

$$\sum_{k=1}^{L/2-1} (2k-1) \int_{(k-1)\Delta}^{k\Delta} f\left[\frac{(2k-1)\Delta}{2} - x\right] p(x)\, dx$$

$$+ (L-1) \int_{-(L/2-1)\Delta}^{\infty} f'\left[\frac{(L-1)\Delta}{2} - x\right] p(x)\, dx = 0 \quad (2.3.68)$$

where $f'(x)$ denotes the derivative of $f(x)$.

By selecting the error criterion function $f(x)$, the solution of (2.3.68) for the optimum step size can be obtained numerically on a digital computer for any given pdf $p(x)$. For the mean-square-error criterion, for which $f(x) = x^2$, Max (1960) evaluated the optimum step size Δ_{opt} and the minimum mean square error when the pdf $p(x)$ is zero mean gaussian with unit variance. Some of these results are given in Table 2.3.6. We observe that the minimum mean square distortion D_{min} decreases by a little more than 5 dB for each doubling of the number of levels L. Hence, each additional bit that is employed in a uniform quantizer with optimum step size Δ_{opt} for a gaussian distributed signal amplitude reduces the distortion by more than 5 dB.

By relaxing the constraint that the quantizer be uniform, the distortion can be reduced further. In this case, we let the output level be $\tilde{x} = y_k$ when the input signal amplitude is in the range $x_{k-1} \le x < x_k$. For an L-level quantizer, the end points are $x_0 = -\infty$ and $x_L = \infty$. The resulting distortion is

$$D = \sum_{k=1}^{L} \int_{x_{k-1}}^{x_k} f(y_k - x) p(x)\, dx \quad (2.3.69)$$

which is now minimized by optimally selecting the $\{y_k\}$ and $\{x_k\}$.

TABLE 2.3.6
Optimum step sizes for uniform quantization of a gaussian random variable

Number of output levels	Optimum step size Δ_{opt}	Minimum MSE D_{min}	$10 \log D_{min}$ (dB)
2	1.596	0.3634	−4.4
4	0.9957	0.1188	−9.25
8	0.5860	0.03744	−14.27
16	0.3352	0.01154	−19.38
32	0.1881	0.00349	−24.57

TABLE 2.3.7
Optimum four-level quantizer for a gaussian random variable

Level k	x_k	y_k
1	-0.9816	-1.510
2	0.0	-0.4528
3	0.9816	0.4528
4	∞	1.510

$$D_{\min} = 0.1175$$
$$10 \log D_{\min} = -9.3 \text{ dB}$$

The necessary conditions for a minimum distortion are obtained by differentiating D with respect to the $\{x_k\}$ and $\{y_k\}$. The result of this minimization is the pair of equations

$$f(y_k - x_k) = f(y_{k+1} - x_k) \qquad k = 1, 2, \ldots, L-1 \qquad (2.3.70)$$

$$\int_{x_{k-1}}^{x_k} f'(y_k - x) p(x) \, dx = 0 \qquad k = 1, 2, \ldots, L \qquad (2.3.71)$$

As a special case, we again consider minimizing the mean square value of the distortion. In this case, $f(x) = x^2$ and, hence, (2.3.70) becomes

$$x_k = \frac{y_k + y_{k-1}}{2} \qquad k = 1, 2, \ldots, L-1 \qquad (2.3.72)$$

which is the midpoint between y_k and y_{k-1}. The corresponding equations determining $\{y_k\}$ are

$$\int_{x_{k-1}}^{x_k} (y_k - x) p(x) \, dx = 0 \qquad k = 1, 2, \ldots, L \qquad (2.3.73)$$

Thus, y_k is the centroid of the area of $p(x)$ between x_{k-1} and x_k. These equations may be solved numerically for any given $p(x)$.

Tables 2.3.7 and 2.3.8 give the results of this optimization obtained by Max (1960) for the optimum four-level and eight-level quantizers of a gaussian distributed signal amplitude having zero mean and unit variance. In Table 2.3.9 we compare the minimum mean square distortion of a uniform quantizer to that of a nonuniform quantizer for the gaussian distributed signal amplitude. From the results of this table, we observe that the difference in the performance of the two types of quantizers is relatively small for small values of R (less than 0.5 dB for $R \le 3$) but it increases as R increases. For example, at $R = 5$, the nonuniform quantizer is approximately 1.5 dB better than the uniform quantizer.

It is instructive to plot the minimum distortion of a function of the bit rate $R = \log_2 L$ bits per source sample (letter) for both the uniform and nonuniform quantizers. These curves are illustrated in Fig. 2.3.20. The functional dependence of the distortion D on the bit rate R may be expressed as $D(R)$. This function is

TABLE 2.3.8
Optimum eight-level quantizer for a gaussian random variable [Max (1960)]

Level k	x_k	y_k
1	-1.748	-2.152
2	-1.050	-1.344
3	-0.5006	-0.7560
4	0	-0.2451
5	0.5006	0.2451
6	1.050	-0.7560
7	1.748	1.344
8	∞	2.152

$$D_{\min} = 0.03454$$
$$10 \log D_{\min} = -14.62 \text{ dB}$$

called the *distortion rate function* for the corresponding quantizer. We observe that the distortion rate function for the optimum nonuniform quantizer falls below that of the optimum uniform quantizer.

Since any quantizer reduces a continuous amplitude source into a discrete amplitude source, we may treat the discrete amplitudes as letters, say $Y = \{y_k, 1 \le k \le L\}$, with associated probabilities $\{p_k\}$. If the signal amplitudes are statistically independent, the discrete source is memoryless and, hence, its entropy is

$$H(Y) = -\sum_{k=1}^{L} p_k \log_2 p_k$$

For example, the optimum four-level nonuniform quantizer for the gaussian distributed signal amplitude results in the probabilities $p_1 = p_4 = 0.1635$ for the

TABLE 2.3.9
Comparison of optimum uniform and nonuniform quantizers for a gaussian random variable [Max (1960) and Paez and Glisson (1972)]

R (bits / samples)	$10 \log_{10} D_{\min}$	
	Uniform (dB)	Nonuniform (dB)
1	-4.4	-4.4
2	-9.25	-9.30
3	-14.27	-14.62
4	-19.38	-20.22
5	-24.57	-26.02
6	-29.83	-31.89
7	-35.13	-37.81

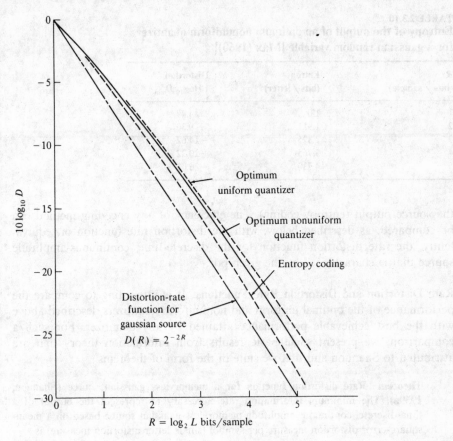

FIGURE 2.3.20
Distortion versus rate curves for discrete-time memoryless gaussian source.

two outer levels and $p_2 = p_3 = 0.3365$ for the two inner levels. The entropy for this discrete source is $H(Y) = 1.911$ bits per letter. Hence, with entropy coding (Huffman coding) of blocks of output letters, we can achieve the minimum distortion of -9.30 dB with 1.911 bits per letter instead of two bits per letter. Max (1960) has given the entropy for the discrete source letters resulting from quantization. Table 2.3.10 lists the values of the entropy for the nonuniform quantizer. These values are also plotted in Fig. 2.3.20 and labeled *entropy coding*.

From this discussion we conclude that the quantizer can be optimized when the pdf of the continuous source output is known. The optimum quantizer of $L = 2^R$ levels results in a minimum distortion of $D(R)$ where $R = \log_2 L$ bits per sample. Thus, this distortion can be achieved by simply representing each quantized sample by R bits. However, more efficient encoding is possible. The discrete source output that results from quantization is characterized by a set of probabilities $\{ p_k \}$ that can be used to design efficient variable-length codes for

TABLE 2.3.10
Entropy of the output of an optimum nonuniform quantizer for a gaussian random variable [Max (1960)]

$\bar{R}$ (bits / sample)	Entropy (bits / letter)	Distortion $10 \log_{10} D_{min}$
1	1.0	−4.4
2	1.911	−9.30
3	2.825	−14.62
4	3.765	−20.22
5	4.730	−26.02

the source output (entropy coding). The efficiency of any encoding method can be compared, as described below, with the distortion rate function or, equivalently, the rate distortion function for the discrete-time, continuous-amplitude source that is characterized by the given pdf.

Rate Distortion and Distortion Rate Functions. It is interesting to compare the performance of the optimal uniform and nonuniform quantizers described above with the best achievable performance attained by any quantizer. For such a comparison, we present some basic results from information theory that are attributed to Shannon and that we state in the form of theorems.

Theorem   Rate distortion function for a memoryless gaussian source (Shannon, 1959a). The minimum information rate necessary to represent the output of a time-discrete, continuous-amplitude memoryless gaussian source based on a mean-square-error distortion measure per symbol (single letter distortion measure) is

$$R_g(D) = \begin{cases} \frac{1}{2} \log_2 \dfrac{\sigma_x^2}{D} & 0 \le D \le \sigma_x^2 \\ 0 & D > \sigma_x^2 \end{cases} \tag{2.3.74}$$

where σ_x^2 is the variance of the gaussian source output. The function $R(D)$ is called the *rate distortion function* for the source.

We should note that (2.3.74) implies that no information needs to be transmitted when the distortion $D \ge \sigma_x^2$. Specifically, $D = \sigma_x^2$ can be obtained by using zeros in the reconstruction of the signal. For $D > \sigma_x^2$, we can use statistically independent, zero mean gaussian noise samples with a variance of $D - \sigma_x^2$ for the reconstruction. $R_g(D)$ is plotted in Fig. 2.3.21.

The rate distortion function $R(D)$ of a source is associated with the following basic source coding theorem in information theory.

Theorem   Source coding with a distortion measure (Shannon, 1959a). There exists an encoding scheme that maps the source output into code words such that for any given distortion D, the minimum rate $R(D)$ bits per symbol (sample) is sufficient to reconstruct the source output with an average distortion that is arbitrarily close to D.

FIGURE 2.3.21
Rate distortion function for a continuous-amplitude, memoryless gaussian source.

It is clear, therefore, that the rate distortion function $R(D)$ for any source represents a lower bound on the source rate that is possible for a given level of distortion.

Let us return to the result in (2.3.74) for the rate distortion function of a memoryless gaussian source. If we reverse the functional dependence between D and R, we may express D in terms of R as

$$D_g(R) = 2^{-2R} \sigma_x^2 \qquad (2.3.75)$$

This function is called the *distortion rate function* for the discrete-time, memoryless gaussian source.

When we express the distortion in (2.3.75) in dB we obtain

$$10 \log_{10} D_g(R) = -6R + 10 \log_{10} \sigma_x^2 \qquad (2.3.76)$$

Note that the mean square distortion decreases at a rate of 6 dB per bit. The graph of (2.3.76) is plotted in Fig. 2.3.20 and may be used as a basis for comparing the performance of the optimal quantizers against the minimum achievable distortion for the gaussian source given by (2.3.76).

If we compare the performance of the optimum nonuniform quantizer to the distortion rate function we find, for example, that at a distortion of -26 dB, entropy coding is 0.41 bits per sample more than the minimum rate given by (2.3.76) and simple block coding of each letter requires 0.68 bits per sample more than the minimum rate. We also observe that the distortion rate functions for the optimal uniform and nonuniform quantizers for the gaussian source approach the slope of -6 dB per bit asymptotically for large R.

Explicit results on the rate distortion functions for memoryless non-gaussian sources are not available. However, there are useful upper and lower bounds on the rate distortion function for any discrete-time continuous-amplitude memoryless source. An upper bound is given by the following theorem.

Theorem Upper bound on $R(D)$. The rate distortion function of a memoryless, continuous amplitude source with zero mean and finite variance σ_x^2 with respect to the mean-square-error distortion measure is upper bounded as

$$R(D) \leq \tfrac{1}{2} \log_2 \frac{\sigma_x^2}{D} \qquad 0 \leq D \leq \sigma_x^2$$

A proof of this theorem is given by Berger (1971). It implies that the gaussian source requires the maximum rate among all other sources for a specified level of mean square distortion. Thus, the rate distortion $R(D)$ of any continuous-amplitude, memoryless source with zero mean and finite variance σ_x^2 satisfies the condition $R(D) \leq R_g(D)$. Similarly, the distortion rate function of the same source satisfies the condition

$$D(R) \leq D_g(R) = 2^{-2R} \sigma_x^2 \tag{2.3.77}$$

A lower bound on the rate distortion function also exists. This is called the *Shannon lower bound* for a mean-square-error distortion measure, and is given as

$$R^*(D) = H(X) - \tfrac{1}{2} \log_2 2\pi e D \tag{2.3.78}$$

where $H(X)$ is the differential entropy of the continuous-amplitude memoryless source. The distortion rate function corresponding to (2.3.78) is

$$D^*(R) = \frac{1}{2\pi e} 2^{-2[R - H(X)]} \tag{2.3.79}$$

Therefore, the rate distortion function for any continuous-amplitude memoryless source is bounded from above and below as

$$R^*(D) \leq R(D) \leq R_g(D) \tag{2.3.80}$$

and the corresponding distortion rate function is bounded as

$$D^*(R) \leq D(R) \leq D_g(R) \tag{2.3.81}$$

The differential entropy of the memoryless gaussian source is

$$H_g(X) = \tfrac{1}{2} \log_2 2\pi e \sigma_x^2 \tag{2.3.82}$$

so that the lower bound $R^*(D)$ in (2.3.78) reduces to $R_g(D)$. Now, if we express $D^*(R)$ in terms of decibels and normalize it by setting $\sigma_x^2 = 1$ [or dividing $D^*(R)$ by σ_x^2], we obtain from (2.3.79)

$$10 \log_{10} D^*(R) = -6R - 6\left[H_g(X) - H(X) \right] \tag{2.3.83}$$

TABLE 2.3.11
Differential entropies and rate distortion comparisons of four common pdf's for signal models

pdf	$p(x)$	$H(X)$	$R_g(D) - R^*(D)$ (bits / sample)	$D_g(R) - D^*(R)$ (dB)
Gaussian	$\dfrac{1}{\sqrt{2\pi}\,\sigma_x}e^{-x^2/2\sigma_x^2}$	$\frac{1}{2}\log_2(2\pi e\sigma_x^2)$	0	0
Uniform	$\dfrac{1}{2\sqrt{3}\,\sigma_x}, \; |x| \le \sqrt{3}\,\sigma_x$	$\frac{1}{2}\log_2(12\sigma_x^2)$	0.255	1.53
Laplacian	$\dfrac{1}{\sqrt{2}\,\sigma_x}e^{-\sqrt{2}|x|/\sigma_x}$	$\frac{1}{2}\log_2(2e^2\sigma_x^2)$	0.104	0.62
Gamma	$\dfrac{\sqrt[4]{3}}{\sqrt{8\pi\sigma_x|x|}}e^{-\sqrt{3}|x|/2\sigma_x}$	$\frac{1}{2}\log_2(4\pi e^{0.423}\sigma_x^2/3)$	0.709	4.25

or, equivalently,

$$10\log_{10}\frac{D_g(R)}{D^*(R)} = 6\big[H_g(X) - H(X)\big] \text{ dB}$$

$$= 6\big[R_g(D) - R^*(D)\big] \text{ dB} \qquad (2.3.84)$$

The relations in (2.3.83) and (2.3.84) allow us to compare the lower bound in the distortion with the upper bound which is the distortion for the gaussian source, We note that $D^*(R)$ also decreases at -6 dB per bit. We should also mention that the differential entropy $H(X)$ is upper bounded by $H_g(X)$ as shown by Shannon (1948b).

Table 2.3.11 lists four pdf's that are models commonly used for source signal distributions. Shown in the table are the differential entropies, the differences in rates in bits per sample, and the difference in distortion between the upper and lower bounds. Note that the gamma pdf shows the greatest deviation from the gaussian. The Laplacian pdf is the most similar to the gaussian and the uniform pdf ranks second of the pdf's shown in the table. These results provide some benchmarks on the difference between the upper and lower bounds on distortion and rate.

Before concluding this section, let us consider a band-limited gaussian source with spectral density

$$\Phi(f) = \begin{cases} \sigma_x^2/2W & |f| \le W \\ 0 & |f| > W \end{cases} \qquad (2.3.85)$$

When the output of this source is sampled at the Nyquist rate, the samples are uncorrelated and, since the source is gaussian, the samples are also statistically independent. Hence, the equivalent discrete-time gaussian source is memoryless.

The rate distortion function for each sample is given by (2.3.74). Therefore, the rate distortion function for the band-limited white gaussian source in bits per second is

$$R_g(D) = W \log_2 \frac{\sigma_x^2}{D} \qquad 0 \le D \le \sigma_x^2 \tag{2.3.86}$$

The corresponding distortion rate function is

$$D_g(R) = 2^{-R/W}\sigma_x^2 \tag{2.3.87}$$

which, when expressed in decibels and normalized by σ_x^2 becomes

$$10 \log D_g(R)/\sigma_x^2 = -3R/W \tag{2.3.88}$$

The more general case in which the gaussian process is neither white nor band-limited has been treated by Gallager (1968) and Goblick and Holsinger (1967).

2.3.5 Vector Quantization

In the previous section we considered the quantization of the output signal from a continuous-amplitude source when the quantization is performed on a sample-by-sample basis, i.e., by scalar quantization. In this section we consider the joint quantization of a block of signal samples or a block of signal parameters. This type of quantization is called *block* or *vector quantization*.

A fundamental result of rate distortion theory is that better performance can be achieved by quantizing vectors instead of scalars, even if the continuous-amplitude source is memoryless. If, in addition, the signal samples or signal parameters are statistically dependent, we can exploit the dependency by jointly quantizing blocks of samples or parameters and thus, achieve an even greater efficiency (lower bit rate) compared with that which is achieved by scalar quantization.

The vector quantization problem may be formulated as follows. We have an n-dimensional vector $\mathbf{X} = (x_1 x_2 \cdots x_n)$ with real-valued, continuous-amplitude components $\{x_k, 1 \le k \le n\}$ that are described by a joint pdf $p(x_1 x_2 \cdots x_n)$. The vector $\mathbf{X}$ is quantized into another n-dimensional vector $\mathbf{Y}$ with components $\{y_k, 1 \le k \le n\}$. We express the quantization as $Q(\cdot)$ so that

$$\mathbf{Y} = Q(\mathbf{X})$$

where $\mathbf{Y}$ is the output of the vector quantizer when the input vector is $\mathbf{X}$.

Basically, vector quantization of blocks of data may be viewed as a pattern recognition problem involving the classification of the blocks of data into a discrete number of categories or *cells* in a way that optimizes some fidelity criterion, such as mean square distortion. For example, let us consider the quantization of two-dimensional vectors $\mathbf{X} = (x_1 x_2)$. The two-dimensional space is partitioned into cells as illustrated in Fig. 2.3.22, where we have arbitrarily selected hexagonal-shaped cells $\{C_k\}$. All input vectors that fall in cell C_k are quantized into the vector $\mathbf{Y}_k$, which is shown in Fig. 2.3.22 as the center of the hexagon. In this example, there are $L = 37$ vectors, one for each of the 37 cells

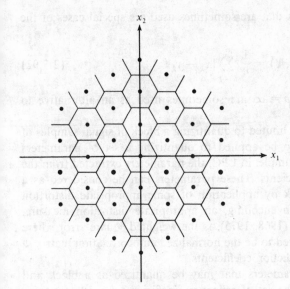

FIGURE 2.3.22
An example of quantization in two-dimensional space.

into which the two-dimensional space has been partitioned. We denote the set of possible output vectors as $\{Y_k, 1 \le k \le L\}$.

In general, quantization of the n-dimensional vector X into an n-dimensional vector Y introduces a quantization error or a distortion $d(X, Y)$. The average distortion over the set of input vectors X is

$$D = \sum_{k=1}^{L} p(X \in C_k) E\left[d(X, Y_k) | X \in C_k \right]$$

$$= \sum_{k=1}^{L} P(X \in C_k) \int_{X \in C_k} d(X, Y_k) p(X) \, dX \qquad (2.3.89)$$

where $P(X \in C_k)$ is the probability that the vector X falls in the cell C_k and $p(X)$ is the joint pdf of the n random variables. As in the case of scalar quantization, we can minimize D by selecting the cells $\{C_k, 1 \le k \le L\}$ for a given pdf $p(X)$.

The most common distortion measure that is adopted is the mean square error (l_2 norm) defined as

$$d_2(X, Y) = \frac{1}{n}(X - Y)(X - Y) = \frac{1}{n} \sum_{k=1}^{n} (x_k - y_k)^2 \qquad (2.3.90)$$

or, more generally, the weighted mean square error

$$d_{2W}(X, Y) = (X - Y')W(X - Y) \qquad (2.3.91)$$

where W is a positive-definite weighting matrix. Usually, W is selected to be the inverse of the covariance matrix of the input data vector X.

Other distortion measures that are sometimes used are special cases of the l_p norm defined as

$$d_p(\mathbf{X}, \mathbf{Y}) = \frac{1}{n} \sum_{k=1}^{n} |x_k - y_k|^p \qquad (2.3.92)$$

The special cases $p = 1$ and $p = \infty$ are sometimes used as an alternative to $p = 2$.

Vector quantization is not limited to quantizing a block of signal samples of a source waveform. It can also be applied to quantizing a set of parameters extracted from the data. For example, in LPC, the parameters extracted from the signal are the prediction coefficients. These parameters can be considered as a block and quantized as a block by application of some appropriate distortion measure. In the case of speech encoding, an appropriate distortion measure, proposed by Itakura and Saito (1968, 1975), is the weighted square error where the weighting matrix $\mathbf{W}$ is selected to be the normalized autocorrelation matrix $\mathbf{\Phi}$ used to solve for the linear prediction coefficients.

An alternative set of parameters that may be quantized as a block and transmitted to the receiver is the set of reflection coefficients $\{a_{ii}, 1 \le i \le p\}$. Yet another set of parameters that is sometimes used for vector quantization in linear predictive coding of speech is the log-area ratios $\{r_k\}$, which are defined in terms of the reflection coefficients as

$$r_k = \log \frac{1 + a_{kk}}{1 - a_{kk}} \qquad 1 \le k \le p \qquad (2.3.93)$$

Now, let us return to the mathematical formulation of vector quantization and let us consider the partitioning of the n-dimensional space into L cells $\{C_k, 1 \le k \le L\}$ so that the average distortion is minimized over all L-level quantizers. There are two conditions for optimality. The first is that the optimal quantizer employs a nearest neighbor selection rule, which may be expressed mathematically as

$$Q(\mathbf{X}) = \mathbf{Y}_k$$

if and only if

$$D(\mathbf{X}, \mathbf{Y}_k) \le D(\mathbf{X}, \mathbf{Y}_j) \qquad k \ne j, 1 \le j \le L \qquad (2.3.94)$$

The second condition necessary for optimality is that each output vector $\mathbf{Y}_k$ be chosen to minimize the average distortion in cell C_k. In other words, $\mathbf{Y}_k$ is the vector in C_k that minimizes

$$D_k = E\left[d(\mathbf{X}, \mathbf{Y}) | \mathbf{X} \in C_k\right] = \int_{\mathbf{X} \in C_k} d(\mathbf{X}, \mathbf{Y}) p(\mathbf{X}) \, d\mathbf{X} \qquad (2.3.95)$$

The vector $\mathbf{Y}_k$ that minimizes D_k is called the *centroid* of the cell. Thus, these conditions for optimality can be applied to partition the n-dimensional space into cells $\{C_k, 1 \le k \le L\}$ when the joint pdf $p(\mathbf{X})$ is known. It is clear that these two conditions represent the generalization of the optimum scalar quantization problem to the n-dimensional vector quantization problem. In general, we expect the code vectors to be closer together in regions where the joint pdf is large and farther apart in regions where $p(\mathbf{X})$ is small.

As an upper bound on the performance of a vector quantizer, we may use the performance of the optimal scalar quantizer which can be applied to each component of the vector as described in the previous section. On the other hand, the best performance that can be achieved by optimum vector quantization is given by the rate distortion function or, equivalently, the distortion rate function.

The distortion rate function which was introduced in the previous section may be defined in the context of vector quantization as follows. Suppose we form a vector $\mathbf{X}$ of dimension n from n consecutive samples $\{x(m)\}$. The vector $\mathbf{X}$ is then quantized to form $\mathbf{Y} = Q(\mathbf{X})$, where $\mathbf{Y}$ is a vector from the set $\{\mathbf{Y}_k, 1 \le k \le L\}$. As described above, the average distortion D resulting from representing $\mathbf{X}$ by $\mathbf{Y}$ is $E[d(\mathbf{X}, \mathbf{Y})]$ where $d(\mathbf{X}, \mathbf{Y})$ is the distortion per dimension, e.g.,

$$d(\mathbf{X}, \mathbf{Y}) = \frac{1}{n} \sum_{k=1}^{n} (x_k - y_k)^2$$

The vectors $\{\mathbf{Y}_k, 1 \le k \le L\}$ can be transmitted at an average bit rate of

$$R = \frac{H(\mathbf{Y})}{n} \text{ bits/sample} \tag{2.3.96}$$

where $H(\mathbf{Y})$ is the entropy of the quantized source output defined as

$$H(\mathbf{Y}) = - \sum_{i=1}^{L} p(\mathbf{Y}_i) \log_2 P(\mathbf{Y}_i) \tag{2.3.97}$$

For a given average rate R, the minimum achievable distortion $D_n(R)$ is

$$D_n(R) = \min_{Q(\mathbf{X})} E[d(X, Y)] \tag{2.3.98}$$

where $R \ge H(\mathbf{Y})/n$ and the minimum in (2.3.98) is taken over all possible mappings $Q(\mathbf{X})$. In the limit as the number of dimensions n is allowed to approach infinity, we obtain

$$D(R) = \lim_{n \to \infty} D_n(R) \tag{2.3.99}$$

where $D(R)$ is the distortion rate function that was introduced in the previous section. It is apparent from this development that the distortion rate function can be approached arbitrarily closely by increasing the size n of the vectors.

The development above is predicated on the assumption that the joint pdf $p(\mathbf{X})$ of the data vector is known. However, in practice, the joint pdf $p(\mathbf{X})$ of the data may not be known. In such a case, it is possible to select the quantized output vectors adaptively from a set of training vectors $\mathbf{X}(m)$. Specifically, suppose that we are given a set of M training vectors where M is much greater than L ($M \gg L$). An iterative clustering algorithm, called the *K means algorithm*, where in our case $K = L$, can be applied to the training vectors. This algorithm iteratively subdivides the M training vectors into L clusters such that the two necessary conditions for optimality are satisfied. The K means algorithm may be described as follows [Makhoul et al. (1985)].

K means algorithm.

Step 1. Initialize by setting the iteration number $i = 0$. Choose a set of output vectors $\mathbf{Y}_k(0)$, $1 \le k \le L$.

Step 2. Classify the training vectors $\{\mathbf{X}(m), 1 \le m \le M\}$ into the clusters $\{C_k\}$ by applying the nearest neighbor rule

$$\mathbf{X} \in C_k(i) \quad \text{iff} \quad D\big(\mathbf{X}, \mathbf{Y}_k(i)\big) \le D\big(\mathbf{X}, \mathbf{Y}_j(i)\big) \quad \text{for all } k \ne j$$

Step 3. Recompute the output vectors of every cluster by computing the centroid

$$\mathbf{Y}_k(i) = \frac{1}{M_k} \sum_{\mathbf{X} \in C_k} \mathbf{X}(m) \qquad 1 \le k \le L$$

of the training vectors that fall in each cluster. Also compute the resulting distortion $D(i)$ at the ith iteration.

Step 4. Terminate the test if the change $D(i-1) - D(i)$ in the average distortion is relatively small. Otherwise, go to Step 2.

The K means algorithm converges to a local minimum [for reference, see Anderberg (1973) and Linde et al. (1980)]. By beginning the algorithm with different sets of initial output vectors $\{Y_k(0)\}$ and each time performing the optimization described in the K means algorithm it is possible to find a global optimum. However, the computational burden of this search procedure may limit the search to a few initializations.

Once we have selected the output vectors $\{\mathbf{Y}_k, 1 \le k \le L\}$ each signal vector $\mathbf{X}(m)$ is quantized to the output vector that is nearest to it according to the distortion measure that is adopted. If the computation involves evaluating the distance between $X(m)$ and each of the L possible output vectors $\{Y_k\}$, the procedure constitutes a *full search*. If we assume that each computation requires n multiplications and additions, the computational requirement for a full search is

$$\mathscr{C} = nL \tag{2.3.100}$$

multiplications and additions per input vector.

If we select L to be a power of 2, then $\log_2 L$ is the number of bits required to represent each vector. Now, if R denotes the bit rate per sample [per component or dimension of $\mathbf{X}(m)$], we have $nR = \log_2 L$ and, hence, the computational cost is

$$\mathscr{C} = n2^{nR} \tag{2.3.101}$$

Note that the number of computations grows exponentially with the dimensionality parameter n and the bit rate R per dimension. Because of this exponential increase of the computational cost, vector quantization has been applied to low bit source encoding, such as coding the reflection coefficients or log area ratios in LPC.

The computational cost associated with full search can be reduced by slightly suboptimum algorithms [for reference, see Cheng et al. (1984) and Gersho (1982)].

In order to demonstrate the benefits of vector quantization compared with scalar quantization, we present the following example taken from the paper of Makhoul et al. (1985).

Example 2.3.5. Let x_1 and x_2 be two random variables with a uniform joint pdf

$$p(x_1, x_2) \equiv p(\mathbf{X}) = \begin{cases} \dfrac{1}{ab} & \mathbf{X} \in C \\ 0 & \text{otherwise} \end{cases} \tag{2.3.102}$$

where C is the rectangular region illustrated in Fig. 2.3.23. Note that the rectangle is rotated by $45°$ relative to the horizontal axis. Also shown in Fig. 2.3.23 are the marginal densities $p(x_1)$ and $p(x_2)$.

If we quantize x_1 and x_2 separately by using uniform intervals of length Δ, the number of levels needed is

$$L_1 = L_2 = \frac{a + b}{\sqrt{2}\,\Delta} \tag{2.3.103}$$

Hence, the number of bits needed for coding the vector $\mathbf{X} = (x_1, x_2)$ is

$$R_x = R_1 + R_2 = \log_2 L_1 + \log_2 L_2$$

$$R_x = \log_2 \frac{(a + b)^2}{2\Delta^2} \tag{2.3.104}$$

Thus, scalar quantization of each component is equivalent to vector quantization with the total number of levels

$$L_x = L_1 L_2 = \frac{(a + b)^2}{2\Delta^2} \tag{2.3.105}$$

We observe that this approach is equivalent to covering the large square that encloses the rectangle by square cells where each cell represents one of the L_x quantized regions. Since $p(\mathbf{X}) = 0$ except for $\mathbf{X} \in C$, this encoding is wasteful and results in an increase of the bit rate.

If we were to cover only the region for which $p(\mathbf{X}) \neq 0$ with squares having area Δ^2, the total number of levels that will result is the area of the rectangle divided by Δ^2, i.e.,

$$L_x' = \frac{ab}{\Delta^2} \tag{2.3.106}$$

therefore, the difference in bit rate between the scalar and vector quantization methods is

$$R_x - R_x' = \log_2 \frac{(a + b)^2}{2ab} \tag{2.3.107}$$

FIGURE 2.3.23
A uniform pdf in two dimensions [Makhoul et al. (1985)].

For instance, if $a = 4b$, the difference in bit rate is

$$R_x - R'_x = 1.64 \text{ bits/vector}$$

Thus, vector quantization is 0.82 bits per sample better for the same distortion.

It is interesting to note that a linear transformation (rotation by 45°) will decorrelate x_1 and x_2 and render the two random variables statistically independent. Then scalar quantization and vector quantization achieve the same efficiency. Although a linear transformation can decorrelate a vector of random variables, it does not result in statistically independent random variables, in general. Consequently, vector quantization will always equal or exceed the performance of scalar quantization (see Prob. 2.27).

Vector quantization has been applied to several types of speech encoding methods including both waveform and model-based methods. In model-based methods such as LPC, vector quantization has made possible the coding of speech at rates below 1000 bits per second [see papers by Buzo et al. (1980),

Roucos et al. (1982), and Paul (1983)]. When applied to waveform encoding methods, it is possible to obtain good quality encoded speech at 16,000 bits per second or, equivalently, at $R = 2$ bits per sample. With additional computational complexity, it may be possible in the future to implement waveform encoders producing good quality speech at a rate of $R = 1$ bit per sample.

2.4 CHANNEL MODELS AND CHANNEL CAPACITY

In the model for a digital communications system described in Sec. 2.1, we found it convenient to separate the channel encoder into two parts, the discrete channel encoder and the digital modulator. The channel itself is a waveform channel and, hence, the modulator serves as the interface that accepts a discrete-time information sequence at its input and puts out a set of corresponding waveforms. Similarly, the demodulator at the receiving end serves as the interface between the waveform channel and the discrete-time channel decoder. Hence the demodulator accepts waveforms at its input, processes the waveforms, and delivers to the channel decoder a sequence of discrete-time symbols.

Suppose that the modulator and demodulator are considered to be parts of the channel. Thus we have a composite channel having a discrete-time input and a discrete-time output. Such a composite channel is characterized by the set of possible inputs, the set of possible outputs, and a set of conditional probabilities relating the possible outputs to the possible inputs.

Example 2.4.1. Suppose the input to the composite channel is a binary sequence and the output is also a binary sequence. Consequently the inputs are selected from the alphabet $X = \{0,1\}$ and the outputs are selected from the alphabet $Y = \{0,1\}$. Also, suppose that the channel noise and other disturbances cause statistically independent errors in the transmitted binary sequence with average probability p. That is,

$$\begin{aligned} P(Y = 0|X = 1) = P(Y = 1|X = 0) = p \\ P(Y = 1|X = 1) = P(Y = 0|X = 0) = 1 - p \end{aligned} \quad (2.4.1)$$

Thus we have reduced the cascade of the binary modulator, the waveform channel, and the binary demodulator into an equivalent discrete-time channel which is represented by the diagram shown in Fig. 2.4.1. This binary-input, binary-output, symmetric channel is simply called a *binary symmetric channel* (BSC). Since each output bit from the channel depends only on the corresponding input bit, we say that the channel is memoryless.

The BSC is a special case of a more general discrete memoryless channel (DMC) model which is characterized by an input alphabet $X = \{x_0, x_1, \ldots, x_{q-1}\}$, an output alphabet $Y = \{y_0, y_1, \ldots, y_{Q-1}\}$, and a set of qQ conditional probabilities

$$P(Y = y_i|X = x_j) \equiv P(y_i|x_j) \quad (2.4.2)$$

where $i = 0, 1, \ldots, Q - 1$ and $j = 0, 1, \ldots, q - 1$. If the input to a DMC is a

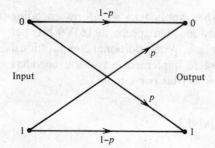

FIGURE 2.4.1
Binary symmetric channel.

sequence of n symbols $u_1, u_2, \ldots, u_n$ selected from the alphabet X and the corresponding output is the sequence $v_1, v_2, \ldots, v_n$ of symbols from the alphabet Y, the joint conditional probability is

$$P(Y_1 = v_1, Y_2 = v_2, \ldots, Y_n = v_n | X = u_1, X = u_2, \ldots, X = u_n)$$

$$= \prod_{k=1}^{n} P(Y = v_k | X = u_k) \qquad (2.4.3)$$

This expression is simply a mathematical statement of the memoryless condition.

In general, the conditional probabilities $\{P(y_i|x_j)\}$ that characterize a DMC can be arranged in the matrix form $\mathbf{P} = [p_{ij}]$, where by definition $p_{ij} \equiv P(y_i|x_j)$. $\mathbf{P}$ is called the *probability transition matrix* for the channel.

In some cases it is appropriate to model the composite discrete-time channel as having a finite input alphabet $X = \{x_0, x_1, \ldots, x_{q-1}\}$ and an output Y that can assume any value on the real line, i.e., $Y = \{-\infty, \infty\}$. This leads us to define a discrete-time memoryless channel that is characterized by the discrete input X, the continuous output Y, and the set of conditional probability density functions

$$p(y|X = x_k) \qquad k = 0, 1, \ldots, q - 1$$

The most important channel of this type is the additive white gaussian noise channel (AWGN) for which

$$Y = X + G \qquad (2.4.4)$$

where G is a zero mean gaussian random variable with variance σ^2 and $X = x_k$, $k = 0, 1, \ldots, q - 1$. For a given X, it follows that Y is gaussian with mean x_k and variance σ^2. That is,

$$p(y|X = x_k) = \frac{1}{\sqrt{2\pi}\,\sigma} e^{-(y-x_k)^2/2\sigma^2} \qquad (2.4.5)$$

For any given input sequence, X_i, $i = 1, 2, \ldots, n$, there is a corresponding output sequence

$$Y_i = X_i + G_i \qquad i = 1, 2, \ldots, n \qquad (2.4.6)$$

The condition that the channel is memoryless may be expressed as

$$p(y_1, y_2, \ldots, y_n | X_1 = u_1, X_2 = u_2, \ldots, X_n = u_n) = \prod_{i=1}^{n} p(y_i | X_i = u_i) \quad (2.4.7)$$

A third channel model that we wish to consider is a waveform channel which has a given bandwidth W and is corrupted by an additive gaussian noise process. Suppose that $x(t)$ is a band-limited input to such a channel and $y(t)$ is the corresponding output. For the additive gaussian noise waveform channel, the output is

$$y(t) = x(t) + n(t) \quad (2.4.8)$$

where $n(t)$ represents a sample function of the additive noise process. A suitable method for defining a set of probabilities that characterizes the channel is to expand $x(t)$, $y(t)$, and $n(t)$ into a complete set of orthonormal functions. That is, we express $x(t)$, $y(t)$, and $n(t)$ in the form

$$y(t) = \sum_i y_i f_i(t)$$

$$x(t) = \sum_i x_i f_i(t) \quad (2.4.9)$$

$$n(t) = \sum_i n_i f_i(t)$$

where $\{y_i\}$, $\{x_i\}$, and $\{n_i\}$ are the sets of coefficients in the corresponding expansions, e.g.,

$$\begin{aligned} y_i &= \int_0^T y(t) f_i^*(t)\, dt \\ &= \int_0^T [x(t) + n(t)] f_i^*(t)\, dt \\ &= x_i + n_i \end{aligned} \quad (2.4.10)$$

The functions $\{f_i(t)\}$ form a complete orthonormal set over the interval $(0, T)$, i.e.,

$$\int_0^T f_i(t) f_j^*(t)\, dt = \delta_{ij} = \begin{cases} 1 & i = j \\ 0 & i \neq j \end{cases} \quad (2.4.11)$$

where δ_{ij} is the Kronecker delta function. Such an expansion, called the Karhunen–Loève expansion, is used in Chap. 4 in the derivation of the optimum demodulator for the waveform channel when the additive gaussian noise is colored. When the gaussian noise is white, any complete set of orthonormal functions may be used in the expansions given by (2.4.9).

We may now use the coefficients in the expansions for characterizing the channel. Since

$$y_i = x_i + n_i$$

where n_i is gaussian, it follows that

$$p(y_i|x_i) = \frac{1}{\sqrt{2\pi}\,\sigma_i} e^{-(y_i-x_i)^2/2\sigma_i^2} \qquad i = 1, 2, \ldots \qquad (2.4.12)$$

As shown in Chap. 4, it is possible to select the functions $\{f_i(t)\}$ in the expansion such that the $\{n_i\}$ are uncorrelated. Since they are gaussian, they are also statistically independent. Hence

$$p(y_1, y_2, \ldots, y_N | x_1, x_2, \ldots, x_N) = \prod_{i=1}^{N} p(y_i|x_i) \qquad (2.4.13)$$

for any N. In this manner, the waveform channel is reduced to an equivalent discrete-time channel characterized by the conditional pdf given in (2.4.12).

When the additive noise is white and gaussian with spectral density $N_0/2$, the variances $\sigma_i^2 = N_0/2$ for all i in (2.4.12). In this case, samples of $x(t)$ and $y(t)$ may be taken at the Nyquist rate of $2W$ samples/s, so that $x_i = x(i/2W)$ and $y_i = y(i/2W)$. Since the noise is white, the noise samples are statistically independent. Thus (2.4.12) and (2.4.13) describe the statistics of the sampled signal. We note that in a time interval of length T, there are $N = 2WT$ samples. This parameter is used below in obtaining the capacity of the band-limited AWGN waveform channel.

The choice of which channel model to use at any one time depends on our objectives. If we are interested in the design and analysis of the performance of the discrete channel encoder and decoder, it is appropriate to consider channel models in which the modulator and demodulator are a part of the composite channel. On the other hand, if our intent is to design and analyze the performance of the digital modulator and digital demodulator, we use a channel model for the waveform channel.

Now let us consider a DMC having an input alphabet $X = \{x_0, x_1, \ldots, x_{q-1}\}$, an output alphabet $Y = \{y_0, y_1, \ldots, y_{Q-1}\}$, and the set of transition probabilities $P(y_i|x_j)$ as defined in (2.4.2). Suppose that the symbol x_j is transmitted and the symbol y_i is received. The mutual information provided about the event $X = x_j$ by the occurrence of the event $Y = y_i$ is $\log P(y_i|x_j)/P(y_i)$, where

$$P(y_i) \equiv P(Y = y_i) = \sum_{k=0}^{q-1} P(x_k)P(y_i|x_k) \qquad (2.4.14)$$

Hence the average mutual information provided by the output Y about the input X is

$$I(X; Y) = \sum_{j=0}^{q-1} \sum_{i=0}^{Q-1} P(x_j)P(y_i|x_j) \log \frac{P(y_i|x_j)}{P(y_i)} \qquad (2.4.15)$$

The channel characteristics determine the transition probabilities $P(y_i|x_j)$, but the probabilities of the input symbols are under the control of the discrete channel encoder. The value of $I(X; Y)$ maximized over the set of input symbol probabilities $P(x_j)$ is a quantity that depends only on the characteristics of the

DMC through the conditional probabilities $P(y_i|x_j)$. This quantity is called the *capacity* of the channel and is denoted by C. That is, the capacity of a DMC is defined as

$$C = \max_{P(x_j)} I(X; Y)$$

$$= \max_{P(x_j)} \sum_{j=0}^{q-1} \sum_{i=0}^{Q-1} P(x_j) P(y_i|x_j) \log \frac{P(y_i|x_j)}{P(y_i)} \qquad (2.4.16)$$

The maximization of $I(X; Y)$ is performed under the constraints that

$$P(x_j) \geq 0$$

$$\sum_{j=0}^{q-1} P(x_j) = 1$$

The units of C are bits per input symbol into the channel (bits per use of the channel) when the logarithm is base 2 and nats per input symbol when the natural logarithm (base e) is used. If a symbol enters the channel every τ_s seconds, the channel capacity in bits per second or nats per second is C/τ_s.

Example 2.4.2. For the BSC with transition probabilities

$$P(0|1) = P(1|0) = p$$

the average mutual information is maximized when the input probabilities $P(0) = P(1) = \frac{1}{2}$. Thus the capacity of the BSC is

$$C = p \log 2p + (1 - p) \log 2(1 - p) = 1 - H(p) \qquad (2.4.17)$$

where $H(p)$ is the binary entropy function. A plot of C versus p is illustrated in Fig. 2.4.2. Note that for $p = 0$, the capacity is 1 bit per channel use. On the other hand, for $p = \frac{1}{2}$, the mutual information between input and output is zero. Hence the channel capacity is zero. For $\frac{1}{2} < p \leq 1$, we may reverse the position of 0 and 1 at the output of the BSC, so that C becomes symmetric with respect to the point $p = \frac{1}{2}$. In our treatment of binary modulation and demodulation given in Chap. 4, we show that p is a monotonic function of the signal-to-noise ratio (SNR) as illustrated in Fig. 2.4.3a. Consequently when C is plotted as a function of the SNR,

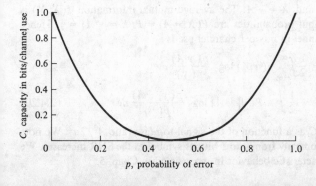

FIGURE 2.4.2
The capacity of a BSC as a function of the error probability p.

FIGURE 2.4.3
General behavior of error probability and channel capacity as a function of SNR.

it increases monotonically as the SNR increases. This characteristic behavior of C versus SNR is illustrated in Fig. 2.4.3b.

Next let us consider the discrete-time AWGN memoryless channel described by the transition probability density functions defined by (2.4.5). The average mutual information between the discrete input $X = \{x_0, x_1, \ldots, x_{q-1}\}$ and the output $Y = \{-\infty, \infty\}$ is given by (2.2.21). The capacity of this channel in bits per channel use is

$$C = \max_{P(x_i)} \sum_{i=0}^{q-1} \int_{-\infty}^{\infty} p(y|x_i) P(x_i) \log_2 \frac{p(y|x_i)}{p(y)} \, dy \qquad (2.4.18)$$

where

$$p(y) = \sum_{k=0}^{q-1} p(y|x_k) P(x_k) \qquad (2.4.19)$$

Example 2.4.3. Let us consider a binary-input AWGN memoryless channel with possible inputs $X = A$ and $X = -A$. The average mutual information $I(X;Y)$ is maximized when the input probabilities are $P(X = A) = P(X = -A) = \frac{1}{2}$. Hence the capacity of this channel in bits per channel use is

$$C = \frac{1}{2} \int_{-\infty}^{\infty} p(y|A) \log_2 \frac{p(y|A)}{p(y)} \, dy$$

$$+ \frac{1}{2} \int_{-\infty}^{\infty} P(y|-A) \log_2 \frac{p(y|-A)}{p(y)} \, dy \qquad (2.4.20)$$

Figure 2.4.4 illustrates C as a function of the signal-to-noise ratio $A^2/2\sigma^2$. We note that C increases monotonically from 0 to 1 bit per symbol as the SNR increases. We shall consider this characteristic behavior in more detail in Chap. 5.

FIGURE 2.4.4
Channel capacity as a function of SNR for binary-input AWGN memoryless channel.

It is interesting to note that in the two channel models described above, the choice of equally probable input symbols maximizes the average mutual information. Thus the capacity of the channel is obtained when the input symbols are equally probable. This is not always the solution for the capacity formulas given in (2.4.16) and (2.4.18), however. Nothing can be said in general about the input probability assignment that maximizes the average mutual information. However, in the two channel models considered above, the channel transition probabilities exhibit a form of symmetry that results in the maximum of $I(X; Y)$ being obtained when the input symbols are equally probable. The symmetry condition can be expressed in terms of the elements of the probability transition matrix $\mathbf{P}$ of the channel. When each row of this matrix is a permutation of any other row and each column is a permutation of any other column, the probability transition matrix is symmetric and input symbols with equal probability maximize $I(X; Y)$.

In general, necessary and sufficient conditions for the set of input probabilities $\{P(x_j)\}$ to maximize $I(X; Y)$ and, thus, to achieve capacity on a DMC are that (Prob. 2.28)

$$
\begin{aligned}
I(x_j; Y) &= C \quad \text{for all } j \text{ with } P(x_j) > 0 \\
I(x_j; Y) &\leq C \quad \text{for all } j \text{ with } P(x_j) = 0
\end{aligned}
\tag{2.4.21}
$$

where C is the capacity of the channel and

$$
I(x_j; Y) = \sum_{i=0}^{Q-1} P(y_i|x_j) \log \frac{P(y_i|x_j)}{P(y_i)}
\tag{2.4.22}
$$

Usually it is relatively easy to check if the equally probable set of input symbols satisfy the conditions in (2.4.21). If they do not, then one must determine the set of unequal probabilities $\{P(x_j)\}$ that satisfy (2.4.21).

Now let us consider a band-limited waveform channel with additive white gaussian noise. Formally the capacity of the channel per unit time has been defined by Shannon (1948b) as

$$C = \lim_{T \to \infty} \max_{p(x)} \frac{1}{T} I(X; Y) \tag{2.4.23}$$

where the average mutual information $I(X; Y)$ is given in (2.2.17). Alternatively, we may use the samples or the coefficients $\{y_i\}$, $\{x_i\}$, and $\{n_i\}$ in the series expansions of $y(t)$, $x(t)$, and $n(t)$, respectively, to determine the average mutual information between $\mathbf{x}_N = (x_1, x_2, \ldots, x_N)$ and $\mathbf{y}_N = (y_1, y_2, \ldots, y_N)$ where $N = 2WT$, $y_i = x_i + n_i$, and $p(y_i|x_i)$ is given by (2.4.12). The average mutual information between $\mathbf{x}_N$ and $\mathbf{y}_N$ for the AWGN channel is

$$
I(\mathbf{X}_N; \mathbf{Y}_N) = \int_{\mathbf{x}_N} \cdots \int\int_{\mathbf{y}_N} \cdots \int p(\mathbf{y}_N|\mathbf{x}_N) p(\mathbf{x}_N) \log \frac{p(\mathbf{y}_N|\mathbf{x}_N)}{p(\mathbf{y}_N)} \, d\mathbf{x}_N \, d\mathbf{y}_N
$$

$$
= \sum_{i=1}^{N} \int_{-\infty}^{\infty} \int_{-\infty}^{\infty} p(y_i|x_i) p(x_i) \log \frac{p(y_i|x_i)}{p(y_i)} \, dy_i \, dx_i \tag{2.4.24}
$$

where

$$p(y_i|x_i) = \frac{1}{\sqrt{\pi N_0}} e^{-(y_i - x_i)^2 / N_0} \tag{2.4.25}$$

The maximum of $I(X; Y)$ over the input pdf's $p(x_i)$ is obtained when the $\{x_i\}$ are statistically independent zero mean gaussian random variables, i.e.,

$$p(x_i) = \frac{1}{\sqrt{2\pi}\,\sigma_x} e^{-x_i^2 / 2\sigma_x^2} \tag{2.4.26}$$

where σ_x^2 is the variance of each x_i. Then, it follows from (2.4.24) that

$$
\max_{p(x)} I(\mathbf{X}_N; \mathbf{Y}_N) = \sum_{i=1}^{N} \frac{1}{2} \log\left(1 + \frac{2\sigma_x^2}{N_0}\right)
$$

$$
= \frac{N}{2} \log\left(1 + \frac{2\sigma_x^2}{N_0}\right)
$$

$$
= WT \log\left(1 + \frac{2\sigma_x^2}{N_0}\right) \tag{2.4.27}
$$

Suppose we put a constraint on the average power in $x(t)$. That is,

$$
P_{\text{av}} = \frac{1}{T} \int_0^T E[x^2(t)] \, dt
$$

$$
= \frac{1}{T} \sum_{i=1}^{N} E(x_i^2)
$$

$$
= \frac{N\sigma_x^2}{T} \tag{2.4.28}
$$

FIGURE 2.4.5
Normalized channel capacity as a function of SNR for band-limited AWGN channel.

Hence

$$\sigma_x^2 = \frac{TP_{av}}{N}$$

$$= \frac{P_{av}}{2W} \tag{2.4.29}$$

Substitution of this result into (2.4.27) for σ_x^2 yields

$$\max_{p(x)} I(\mathbf{X}_N; \mathbf{Y}_N) = WT \log\left(1 + \frac{P_{av}}{WN_0}\right) \tag{2.4.30}$$

Finally, the channel capacity per unit time is obtained by dividing the result in (2.4.30) by T. Thus

$$C = W \log\left(1 + \frac{P_{av}}{WN_0}\right) \tag{2.4.31}$$

This is the basic formula for the capacity of the band-limited AWGN waveform channel with a band-limited and average power-limited input. It was originally derived by Shannon (1948b).

A plot of the capacity in bits per second normalized by the bandwidth W is plotted in Fig. 2.4.5 as a function of the ratio of signal power (P_{av}) to noise power (WN_0). We note that the capacity increases monotonically with an increase

FIGURE 2.4.6
Channel capacity as a function of bandwidth with a fixed transmitted average power.

in SNR. Thus, for a fixed bandwidth, the capacity of the waveform channel increases with an increase in the transmitted signal power. On the other hand, if P_{av} is fixed, the capacity can be increased by increasing the bandwidth W. Figure 2.4.6 illustrates a graph of C versus W. We note that as W approaches infinity the capacity of the channel approaches the asymptotic value

$$C_\infty = \frac{P_{av}}{N_0} \log_2 e = \frac{P_{av}}{N_0 \ln 2} \text{ bits/s} \qquad (2.4.32)$$

Thus, we have derived the channel capacities of three important channel models that are considered in this book. The first is the discrete-input, discrete-output channel, of which the BSC is a special case. The second is a discrete-input, continuous output memoryless additive white gaussian noise channel. From these two channel models we can obtain benchmarks for the coded performance with hard- and soft-decision decoding in digital communications systems.

The third channel model focuses on the capacity in bits per second of a waveform channel. In this case, we assumed that we have a bandwidth limitation on the channel, an additive gaussian noise that corrupts the signal, and an average power constraint at the transmitter. Under these conditions we derived the result given in (2.4.31). We shall use this result as a benchmark on the performance of various modulation/demodulation techniques that are described in Chap. 4.

The major significance of the channel capacity formulas given above stem from Shannon's basic coding theorem, which states that with a sufficiently sophisticated channel encoder and decoder, we can transmit digital information over the channel at a rate up to the channel capacity with arbitrarily small probability of error. On the other hand, if the transmission rate exceeds the channel capacity, the probability of error cannot be reduced to zero.

2.5 BIBLIOGRAPHICAL NOTES AND REFERENCES

As indicated in the opening paragraph of this chapter, Shannon, in his two classic papers, laid the mathematical foundations for information theory, which is a discipline that is concerned with the transmission and processing of information. Information theory has developed rapidly over the last four decades along with the practical applications to digital communications. The modern digital communications systems and communications networks that exist today and those currently under development reflect these recent advances in information theory, as well as the advances in integrated circuit technology which have made it possible to implement complex modulation/demodulation and coding/decoding techniques that efficiently utilize the various communication channels.

We have presented in brief some of the basic concepts in information theory but did not attempt to delve deeply into the subject. A number of books treat information theory in more depth, including the texts by Fenstein (1958), Fano (1961), Abramson (1963), Ash (1965), Gallager (1968), Jelinek (1968), Viterbi and Omura (1979), Hamming (1986), and Blahut (1987). A number of key papers published during the early development of information theory have been reprinted by the IEEE Press in a book edited by Slepian (1974).

Treatments of rate distortion theory are found in the books by Gallager (1968), Berger (1971), Viterbi and Omura (1979), and Blahut (1987).

Much work has been done over the past several decades on speech encoding methods. Our treatment provides an overview of this important topic. A more comprehensive treatment is given in the books by Rabiner and Schafer (1978) and Jayant and Noll (1984). In addition to these texts, there have been special issues of the *IEEE Transactions on Communications* (April 1979, April 1982) and, more recently, the *IEEE Journal on Selected Areas in Communications* (February 1988) devoted to speech encoding. We should also mention the publication by IEEE Press of a book containing reprints of published papers on waveform quantization and coding, edited by Jayant (1976).

Over the past decade we have also seen a number of important developments in vector quantization. Our treatment of this topic was based on the comprehensive tutorial paper by Makhoul et al. (1985). This paper contains an extensive list of references to published work.

APPENDIX 2A
THE LEVINSON–DURBIN ALGORITHM

The Levinson–Durbin algorithm is an order-recursive method for determining the solution to the set of linear equations

$$\Phi_p \mathbf{a}_p = \phi_p \tag{2A.1}$$

where Φ_p is a $p \times p$ Toeplitz matrix, $\mathbf{a}_p$ is the vector of predictor coefficients expressed as

$$\mathbf{a}'_p = [a_{p1} a_{p2} \cdots a_{pp}]$$

and $\boldsymbol{\phi}_p$ is a p-dimensional vector with elements

$$\boldsymbol{\phi}'_p = [\phi(1)\phi(2) \cdots \phi(p)]$$

For a first-order ($p = 1$) predictor, we have the solution

$$\phi(0)a_{11} = \phi(1)$$

$$a_{11} = \frac{\phi(1)}{\phi(0)} \tag{2A.2}$$

The residual mean square error (MSE) for the first-order predictor is

$$\begin{aligned}
\mathcal{E}_1 &= \phi(0) - a_{11}\phi(1) \\
&= \phi(0) - a_{11}^2\phi(0) \\
&= \phi(0)(1 - a_{11}^2)
\end{aligned} \tag{2A.3}$$

In general, we may express the solution for the coefficients of an mth-order predictor in terms of the coefficients of the $(m-1)$st-order predictor. Thus we express $\mathbf{a}_m$ as the sum of two vectors, namely,

$$\mathbf{a}_m = \begin{bmatrix} a_{m1} \\ a_{m2} \\ \vdots \\ a_{mm} \end{bmatrix} = \begin{bmatrix} \mathbf{a}_{m-1} \\ \hline 0 \end{bmatrix} + \begin{bmatrix} \mathbf{d}_{m-1} \\ \hline k_m \end{bmatrix} \tag{2A.4}$$

where the vector $\mathbf{d}_{m-1}$ and the scalar k_m are to be determined. Also, $\boldsymbol{\Phi}_m$ may be expressed as

$$\boldsymbol{\Phi}_m = \begin{bmatrix} \boldsymbol{\Phi}_{m-1} & \boldsymbol{\phi}^r_{m-1} \\ \hline \boldsymbol{\phi}^{r\prime}_{m-1} & \phi(0) \end{bmatrix} \tag{2A.5}$$

where $\boldsymbol{\phi}^r_{m-1}$ is just the vector $\boldsymbol{\phi}_{m-1}$ in reverse order.

Now

$$\begin{bmatrix} \boldsymbol{\Phi}_{m-1} & \boldsymbol{\phi}^r_{m-1} \\ \hline \boldsymbol{\phi}^{r\prime}_{m-1} & \phi(0) \end{bmatrix} \left\{ \begin{bmatrix} \mathbf{a}_{m-1} \\ \hline 0 \end{bmatrix} + \begin{bmatrix} \mathbf{d}_{m-1} \\ \hline k_m \end{bmatrix} \right\} = \begin{bmatrix} \boldsymbol{\phi}_{m-1} \\ \hline \phi(m) \end{bmatrix} \tag{2A.6}$$

From (2A.6) we obtain two equations. The first is the matrix equation

$$\boldsymbol{\Phi}_{m-1}\mathbf{a}_{m-1} + \boldsymbol{\Phi}_{m-1}\mathbf{d}_{m-1} + k_m\boldsymbol{\phi}^r_{m-1} = \boldsymbol{\phi}_{m-1} \tag{2A.7}$$

But $\boldsymbol{\Phi}_{m-1}\mathbf{a}_{m-1} = \boldsymbol{\phi}_{m-1}$. Hence (2A.7) simplifies to

$$\boldsymbol{\Phi}_{m-1}\mathbf{d}_{m-1} + k_m\boldsymbol{\phi}^r_{m-1} = \mathbf{0} \tag{2A.8}$$

This equation has the solution

$$\mathbf{d}_{m-1} = -k_m \Phi_{m-1}^{-1} \phi_{m-1}^r \qquad (2A.9)$$

But ϕ_{m-1}^r is just ϕ_{m-1} in reverse order. Hence the solution in (2A.9) is simply $\mathbf{a}_{m-1}$ in reverse order multiplied by $-k_m$. That is,

$$\mathbf{d}_{m-1} = -k_m \begin{bmatrix} a_{m-1\,m-1} \\ a_{m-1\,m-2} \\ \vdots \\ a_{m-1\,1} \end{bmatrix} \qquad (2A.10)$$

The second equation obtained from (2A.6) is the scalar equation

$$\phi_{m-1}^{r'}\mathbf{a}_{m-1} + \phi_{m-1}^{r'}\mathbf{d}_{m-1} + \phi(0)k_m = \phi(m) \qquad (2A.11)$$

We elimiante $\mathbf{d}_{m-1}$ from (2A.11) by use of (2A.10). The resulting equation gives us k_m. That is,

$$
\begin{aligned}
k_m &= \frac{\phi(m) - \phi_{m-1}^{r'}\mathbf{a}_{m-1}}{\phi(0) - \phi_{m-1}^{r'}\Phi_{m-1}^{-1}\phi_{m-1}^r} \\[2mm]
&= \frac{\phi(m) - \phi_{m-1}^{r'}\mathbf{a}_{m-1}}{\phi(0) - a_{m-1}'\phi_{m-1}} \\[2mm]
&= \frac{\phi(m) - \phi_{m-1}^{r'}\mathbf{a}_{m-1}}{\mathscr{E}_{m-1}}
\end{aligned} \qquad (2A.12)
$$

where $\mathscr{E}_{m-1}$ is the residual MSE given as

$$\mathscr{E}_{m-1} = \phi(0) - \mathbf{a}_{m-1}'\phi_{m-1} \qquad (2A.13)$$

By substituting (2A.10) for $\mathbf{d}_{m-1}$ in (2A.4) we obtain the order-recursive relation

$$a_{mk} = a_{m-1\,k} - k_m a_{m-1\,m-k} \qquad \begin{aligned} k &= 1, 2, \ldots, m-1 \\ m &= 1, 2, \ldots, p \end{aligned} \qquad (2A.14)$$

and

$$a_{mm} = k_m$$

The minimum MSE may also be computed recursively. We have

$$\mathscr{E}_m = \phi(0) - \sum_{k=1}^{m} a_{mk}\phi(k) \qquad (2A.15)$$

Using (2A.14) in (2A.15), we obtain

$$\mathscr{E}_m = \phi(0) - \sum_{k=1}^{m-1} a_{m-1\,k}\phi(k) - a_{mm}\left[\phi(m) - \sum_{k=1}^{m-1} a_{m-1\,m-k}\phi(k)\right] \qquad (2A.16)$$

But the term in the brackets in (2A.16) is just the numerator of k_m in (2A.12). Hence

$$
\begin{aligned}
\mathscr{E}_m &= \mathscr{E}_{m-1} - a_{mm}^2 \mathscr{E}_{m-1} \\
&= \mathscr{E}_{m-1}\left(1 - a_{mm}^2\right)
\end{aligned} \qquad (2A.17)
$$

PROBLEMS

2.1. Consider the joint experiment described in Prob. 1.1 (Chap. 1) with the given joint probabilities $P(A_i, B_j)$. Suppose we observe the outcomes A_i, $i = 1, 2, 3, 4$ of experiment A.
   (*a*) Determine the mutual information $I(B_j; A_i)$ for $j = 1, 2, 3$ and $i = 1, 2, 3, 4$, in bits.
   (*b*) Determine the average mutual information $I(B; A)$.

2.2. Suppose the outcomes B_j, $j = 1, 2, 3$ in Prob. 2.1 represent the three possible output letters from the DMS. Determine the entropy of the source.

2.3. Prove that $\ln u \le u - 1$ and also demonstrate the validity of this inequality by plotting $\ln u$ and $u - 1$ on the same graph.

2.4. X and Y are two discrete random variables with probabilities

$$P(X = x, Y = y) \equiv P(x, y)$$

Show that $I(X; Y) \ge 0$, where the equality holds if and only if X and Y are statistically independent.
   [*Hint:* Use the inequality $\ln u < u - 1$, for $0 < u < 1$, to show that $-I(X; Y) \le 0$.]

2.5. The output of a DMS consists of the possible letters $x_1, x_2, \ldots, x_n$, which occur with probabilities $p_1, p_2, \ldots, p_n$, respectively. Prove that the entropy $H(X)$ of the source is at most $\log n$.

2.6. Determine the differential entropy $H(X)$ of the uniformly distributed random variable X with pdf

$$p(x) = \begin{cases} \dfrac{1}{a} & 0 \le x \le a \\ 0 & \text{otherwise} \end{cases}$$

for the following three cases:
   (*a*) $a = 1$.
   (*b*) $a = 4$.
   (*c*) $a = \frac{1}{4}$.
Observe from these results that $H(X)$ is not an absolute measure, but only a relative measure of randomness.

2.7. A DMS has an alphabet of eight letters, x_i, $i = 1, 2, \ldots, 8$, with probabilities 0.25, 0.20, 0.15, 0.12, 0.10, 0.08, 0.05, and 0.05.
   (*a*) Use the Huffman encoding procedure to determine a binary code for the source output.
   (*b*) Determine the average number $\overline{R}$ of binary digits per source letter.
   (*c*) Determine the entropy of the source and compare it with $\overline{R}$.

2.8. A DMS has an alphabet of five letters, x_i, $i = 1 = 1, 2, 3, 4, 5$, each occurring with probability $\frac{1}{5}$. Evaluate the efficiency of a fixed-length binary code in which
   (*a*) Each letter is encoded separately into a binary sequence.
   (*b*) Two letters at a time are encoded into a binary sequence.
   (*c*) Three letters at a time are encoded into a binary sequence.

2.9. In (2.2.6) it was shown that

$$I(x_i; y_j) = I(x_i) - I(x_i|y_j)$$

Prove that

(a) $I(x_i; y_j) = I(y_j) - I(y_j|x_i)$.

(b) $I(x_i; y_j) = I(x_i) - I(y_j|x_i)$.

2.10. Show that $I(X; Y) = H(X) + H(Y) - H(XY)$.

2.11. Show that for statistically independent events

$$H(X_1 X_2 \cdots X_n) = \sum_{i=1}^{n} H(X_i)$$

2.12. For a noiseless channel, show that $H(X|Y) = 0$.

2.13. Show that

$$I(X_3; X_2|X_1) = H(X_3|X_1) - H(X_3|X_1 X_2)$$

and that

$$H(X_3|X_1) \geq H(X_3|X_1 X_2)$$

2.14. Let X be a random variable with pdf $p_X(x)$ and let $Y = aX + b$ be a linear transformation of X, where a and b are two constants. Determine the differential entropy $H(Y)$ in terms of $H(X)$.

2.15. The outputs x_1, x_2, x_3 of a DMS with corresponding probabilities $p_1 = 0.45$, $p_2 = 0.35$, and $p_3 = 0.20$ are transformed by the linear transformation $Y = aX + b$, where a and b are constants. Determine the entropy $H(Y)$ and comment on what effect the transformation has had on the entropy of X.

2.16. The optimum four-level nonuniform quantizer for a gaussian distributed signal amplitude results in the four levels a_1, a_2, a_3, a_4 with corresponding probabilities of occurrence $p_1 = p_2 = 0.3365$ and $p_3 = p_4 = 0.1635$.

(a) Design a Huffman code that encodes a single level at a time and determine the average bit rate.

(b) Design a Huffman code that encodes two output levels at a time and determine the average bit rate.

(c) What is the minimum rate obtained by encoding J outputs levels at a time as $J \to \infty$?

2.17. A first-order Markov source is characterized by the state probabilities $P(x_i)$, $i = 1, 2, \ldots, L$, and the transition probabilities $P(x_k|x_i)$, $k = 1, 2, \ldots, L$, and $k \neq i$. The entropy of the Markov source is

$$H(X) = \sum_{k=1}^{L} p(x_k) H(X|x_k)$$

where $H(X|x_k)$ is the entropy conditioned on the source being in state x_k.

Determine the entropy of the binary, first-order Markov source shown in Fig. P2.17 which has the transition probabilities $P(x_2|x_1) = 0.2$ and $P(x_1|x_2) = 0.3$. [Note that the conditional entropies $H(X|x_1)$ and $H(X|x_2)$ are given by the binary entropy functions $H[P(x_2|x_1)]$ and $H[P(x_1|x_2)]$, respectively.] How does the entropy of the Markov source compare with the entropy of a binary DMS with the same output letter probabilities $P(x_1)$ and $P(x_2)$?

$$P(x_2/x_1)$$

$$1 - P(x_2/x_1) \qquad\qquad 1 - P(x_1/x_2)$$

$$x_1 \qquad\qquad x_2$$

$$P(x_1/x_2)$$

FIGURE P2.17

2.18. An additive white gaussian noise channel has the output $Y = X + G$ where X is the channel input and G is the noise with probability density function

$$p(n) = \frac{1}{\sqrt{2\pi}\,\sigma_n} e^{-n^2/2\sigma_n^2}$$

If X is a white gaussian input with $E(X) = 0$ and $E(X^2) = \sigma_x^2$:
(a) Determine the conditional differential entropy $H(X|G)$.
(b) Determine the average mutual information $I(X; Y)$.

2.19. A DMS has an alphabet of eight letters, x_i, $i = 1, 2, \ldots, 8$, with probabilities given in Prob. 2.7. Use the Huffman encoding procedure to determine a ternary code (using symbols $0, 1, 2$) for encoding the source output.
[*Hint*: Add a symbol x_9 with probability $p_9 = 0$ and group three symbols at a time.]

2.20. Determine if there exists a binary code with code word lengths $(n_1, n_2, n_3, n_4) = (1, 2, 2, 3)$ that satisfy the prefix condition.

2.21. Consider a binary block code with 2^n code words of the same length n. Show that the Kraft inequality is satisfied for such a code.

2.22. Show that the entropy of an n-dimensional gaussian vector $\mathbf{X} = (x_1, x_2, \ldots, x_n)$ with zero mean and covariance matrix $\mathbf{M}$ is

$$H(\mathbf{X}) = \tfrac{1}{2}\log_2 (2\pi e)^n |M^{-1}|$$

2.23. Consider a DMS with output bits $\{0, 1\}$ that are equiprobable. Define the distortion measure as $D = P_e$, where P_e is the probability of error in transmitting the binary symbols to the user over a BSC. Then the rate distortion function is [Berger (1971)]

$$R(D) = 1 + D\log_2 D + (1 - D)\log_2 (1 - D) \qquad 0 \le D = P_e \le \tfrac{1}{2}$$

Plot $R(D)$ for $0 \le D \le \tfrac{1}{2}$. Note that this is also the expression for the capacity of a BSC.

2.24. Evaluate the rate distortion function for an M-ary symmetric channel where $D = P_M$ and

$$R(D) = \log_2 M + D\log_2 D + (1 - D)\log_2 \frac{1 - D}{M - 1}$$

for $M = 2, 4, 8, 16$. P_M is the probability of error.

2.25. Consider the use of the weighted mean-square-error (MSE) distortion measure defined as

$$d_w(\mathbf{X}, \mathbf{Y}) = (\mathbf{X} - \mathbf{Y})\mathbf{W}(\mathbf{X} - \mathbf{Y})$$

where $\mathbf{W}$ is a symmetric, positive-definite weighting matrix. By factoring $\mathbf{W}$ as $\mathbf{W} = \mathbf{P}'\mathbf{P}$ show that $d_w(\mathbf{X}, \mathbf{Y})$ is equivalent to an unweighted MSE distortion measure $d_2(\tilde{\mathbf{X}}, \tilde{\mathbf{Y}})$ involving transformed vectors $\tilde{\mathbf{X}}$ and $\tilde{\mathbf{Y}}$.

FIGURE P2.27

2.26. Consider a stationary stochastic signal sequence $\{X(n)\}$ with zero mean and autocorrelation sequence

$$\phi(n) = \begin{cases} 1 & n = 0 \\ \frac{1}{2} & n = \pm 1 \\ 0 & \text{otherwise} \end{cases}$$

(a) Determine the prediction coefficient of the first-order minimum MSE predictor for $\{X(n)\}$ given as

$$\hat{x}(n) = a_1 x(n-1)$$

and the corresponding minimum mean square error $\mathcal{E}_1$.

(b) Repeat (a) for the second-order predictor

$$\hat{x}(n) = a_1 x(n-1) + a_2 x(n-2)$$

2.27. Consider the encoding of the random variables x_1 and x_2 that are characterized by the joint pdf $p(x_1, x_2)$ given as

$$p(x_1, x_2) = \begin{cases} \dfrac{10}{7ab} & x_1, x_2 \in C \\ 0 & \text{otherwise} \end{cases}$$

as shown in Fig. P2.27. Evaluate the bit rates required for uniform quantization of x_1 and x_2 separately (scalar quantization) and combined (vector) quantization of (x_1, x_2). Determine the difference in bit rate when $a = 4b$.

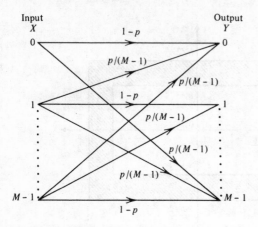

FIGURE P2.29

2.28. Show that the following two relations are necessary and sufficient conditions for the set of input probabilities $\{P(x_j)\}$ to maximize $I(X; Y)$ and, thus, to achieve capacity for a DMC:

$$I(x_j; Y) = C \quad \text{for all } j \text{ with } P(x_j) > 0$$

$$I(x_j; Y) \leq C \quad \text{for all } j \text{ with } P(x_j) = 0$$

where C is the capacity of the channel and

$$I(x_j; Y) = \sum_{i=0}^{Q-1} P(y_i|x_j) \log \frac{P(y_i|x_j)}{P(y_i)}$$

2.29. Figure P2.29 illustrates an M-ary symmetric DMC with transition probabilities $P(y|x) = 1 - p$ when $x = y = k$ for $k = 0, 1, \ldots, M - 1$, and $P(y|x) = p/(M - 1)$ when $x \neq y$.
   (*a*) Show that this channel satisfies the condition given in Prob. 2.28 when $P(x_k) = 1/M$.
   (*b*) Determine and plot the channel capacity as a function of p.
2.30. Determine the capacities of the channels shown in Fig. P2.30.
2.31. Consider the two channels with the transition probabilities as shown in fig. P2.31. Determine if equally probable input symbols maximize the information rate through the channel.
2.32. A telephone channel has a bandwidth $W = 3000$ Hz and a signal-to-noise power ratio of 400 (26 dB). Suppose we characterize the channel as a band-limited AWGN waveform channel with $P_{av}/WN_0 = 400$.
   (*a*) Determine the capacity of the channel in bits per second.
   (*b*) Is the capacity of the channel sufficient to support the transmission of a speech signal that has been sampled and encoded by means of logarithmic PCM?
   (*c*) Usually channel impairments other than additive noise limit the transmission rate over the telephone channel to less than the channel capacity of the equivalent band-limited AWGN channel considered in (*a*). Suppose that a transmission rate of $0.7C$ is achievable in practice without channel encoding.

FIGURE P2.30

Which of the speech source encoding methods described in Sec. 2.3 provide sufficient compression to fit the bandwidth restrictions of the telephone channel?

2.33. Consider the binary-input, quaternary-output DMC shown in Fig. P2.33.

(*a*) Determine the capacity of the channel.

(*b*) Show that this channel is equivalent to a BSC.

2.34. Determine the channel capacity for the channel shown in Fig. P2.34.

FIGURE P2.31

FIGURE P2.33

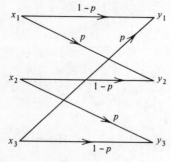

FIGURE P2.34

2.35. Consider a BSC with crossover probability of error p. Suppose that R is the number of bits in a source code word that represents one of 2^R possible levels at the output of a quantizer. Determine:

(a) The probability that a code word transmitted over the BSC is received correctly.

(b) The probability of having at least one bit error in a code word transmitted over the BSC.

(c) The probability of having n_e or less bit errors in a code word.

(d) Evaluate the probability in (a), (b), and (c) for $R = 5$, $p = 0.01$, and $n_e = 5$.

2.36. Show that for a DMC the average mutual information between a sequence $X_1 X_2 \cdots X_n$ of channel inputs and the corresponding channel outputs satisfy the

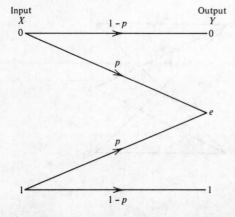

FIGURE P2.37

condition

$$I(X_1 X_2 \cdots X_n; Y_1, Y_2, \ldots, Y_n) \le \sum_{i=1}^{n} I(X_i; Y_i)$$

with equality if and only if the set of input symbols is statistically independent.

2.37. Figure P2.37 illustrates a binary erasure channel with transition probabilities $P(0|0) = P(1|1) = 1 - p$ and $P(e|0) = P(e|1) = p$. The probabilities for the input symbols are $P(X = 0) = \alpha$ and $P(X = 1) = 1 - \alpha$.

   (*a*) Determine the average mutual information $I(X; Y)$ in bits.

   (*b*) Determine the value of α which maximizes $I(X; Y)$, i.e., the channel capacity C in bits per channel use, and plot C as a function of p for the optimum value of α.

   (*c*) For the value of α found in (*b*), determine the mutual information $I(x; y) = I(0; 0)$, $I(1; 1)$, $I(0; e)$, and $I(1; e)$.

CHAPTER
3

CHARACTERIZATION OF SIGNALS AND SYSTEMS

Signals can be categorized in a number of different ways such as random versus deterministic, discrete time versus continuous time, discrete amplitude versus continuous amplitude, lowpass versus bandpass, finite energy versus infinite energy, finite average power versus infinite average power, etc. In this chapter we treat the characterization of signals and systems that are usually encountered in the transmission of digital information over a communication channel. In particular, we introduce the representation of various forms of digitally modulated signals and describe their spectral characteristics.

We begin with the characterization of bandpass signals and systems, including the mathematical representation of bandpass stationary stochastic processes. Then, we present a vector space representation of signals. We conclude with the representation of digitally modulated signals and their spectral characteristics.

3.1 REPRESENTATION OF BANDPASS SIGNALS AND SYSTEMS

Many digital information-bearing signals are transmitted by some type of carrier modulation. The channel over which the signal is transmitted is limited in bandwidth to an interval of frequencies centered about the carrier, as in double-sideband modulation, or adjacent to the carrier, as in single-sideband modulation. Signals and channels (systems) which satisfy the condition that their bandwidth is much smaller than the carrier frequency are termed *narrowband bandpass signals and channels* (*systems*). The modulation performed at the transmitting end of the communication system to generate the bandpass signal and the demodulation performed at the receiving end to recover the digital information involve frequency translations. With no loss in generality and for mathematical convenience, it is desirable to reduce all bandpass signals and channels to equivalent lowpass signals and channels. As a consequence, the results of the performance of the various modulation and demodulation techniques presented in the subsequent chapters are independent of carrier frequencies and channel frequency bands. The representation of bandpass signals and systems in terms of equivalent lowpass waveforms and the characterization of bandpass stationary stochastic processes are the main topics of this section.

3.1.1 Representation of Bandpass Signals

A real-valued signal $s(t)$ with a frequency content concentrated in a narrow band of frequencies in the vicinity of a frequency f_c can be expressed in the form

$$s(t) = a(t)\cos[2\pi f_c t + \theta(t)] \tag{3.1.1}$$

where $a(t)$ denotes the amplitude (envelope) of $s(t)$, and $\theta(t)$ denotes the phase of $s(t)$. The frequency f_c is usually called the *carrier* of $s(t)$ and may be any convenient frequency within or near the frequency band occupied by the signal. When the band of frequencies occupied by $s(t)$ is small relative to f_c, the signal is called a *narrowband bandpass signal* or, simply, a *bandpass signal*.

By expanding the cosine function in (3.1.1) a second representation for $s(t)$ is obtained, namely,

$$\begin{aligned} s(t) &= a(t)\cos\theta(t)\cos 2\pi f_c t - a(t)\sin\theta(t)\sin 2\pi f_c t \\ &= x(t)\cos 2\pi f_c t - y(t)\sin 2\pi f_c t \end{aligned} \tag{3.1.2}$$

where the signals $x(t)$ and $y(t)$, termed the *quadrature components* of $s(t)$, are defined as

$$\begin{aligned} x(t) &= a(t)\cos\theta(t) \\ y(t) &= a(t)\sin\theta(t) \end{aligned} \tag{3.1.3}$$

The frequency content of the quadrature components $x(t)$ and $y(t)$ is concentrated at low frequencies (around $f = 0$, as shown below) and, hence, these components are appropriately called *lowpass signals*. Finally, a third representation for $s(t)$ is obtained from (3.1.1) by defining the complex envelope $u(t)$ as

$$\begin{aligned} u(t) &= a(t)e^{j\theta(t)} \\ &= x(t) + jy(t) \end{aligned} \tag{3.1.4}$$

so that

$$s(t) = \text{Re}\left[u(t)e^{j2\pi f_c t}\right] \tag{3.1.5}$$

where Re[ ] denotes the real part of the complex-valued quantity in the brackets. Thus a real bandpass signal is completely described by any one of the three equivalent forms given in (3.1.1), (3.1.2), or (3.1.5).

The Fourier transform of $s(t)$ is

$$S(f) = \int_{-\infty}^{\infty} s(t) e^{-j2\pi ft} \, dt$$

$$= \int_{-\infty}^{\infty} \left\{ \mathrm{Re}\left[u(t) e^{j2\pi f_c t} \right] \right\} e^{-j2\pi ft} \, dt \qquad (3.1.6)$$

Use of the identity

$$\mathrm{Re}\,(\xi) = \tfrac{1}{2}(\xi + \xi^*) \qquad (3.1.7)$$

in (3.1.6) yields the result

$$S(f) = \frac{1}{2} \int_{-\infty}^{\infty} \left[u(t) e^{j2\pi f_c t} + u^*(t) e^{-j2\pi f_c t} \right] e^{-j2\pi ft} \, dt$$

$$= \tfrac{1}{2}\left[U(f - f_c) + U^*(-f - f_c) \right] \qquad (3.1.8)$$

where $U(f)$ is the Fourier transform of $u(t)$. Since the frequency content of the bandpass signal $s(t)$ is concentrated in the vicinity of the carrier f_c, the result in (3.1.8) indicates that the frequency content of $u(t)$ is concentrated in the vicinity of $f = 0$. Consequently, the complex-valued waveform $u(t)$ is basically a low-pass signal waveform and, hence, is called the *equivalent lowpass signal*.

The energy in the signal $s(t)$ is defined as

$$\mathscr{E} = \int_{-\infty}^{\infty} s^2(t) \, dt$$

$$= \int_{-\infty}^{\infty} \left\{ \mathrm{Re}\left[u(t) e^{j2\pi f_c t} \right] \right\}^2 \, dt \qquad (3.1.9)$$

When the identity in (3.1.7) is used in (3.1.9), we obtain the following result:

$$\mathscr{E} = \frac{1}{2} \int_{-\infty}^{\infty} |u(t)|^2 \, dt$$

$$+ \frac{1}{2} \int_{-\infty}^{\infty} |u(t)|^2 \cos\left[4\pi f_c t + 2\theta(t) \right] \, dt \qquad (3.1.10)$$

Consider the second integral in (3.1.10). Since the signal $s(t)$ is narrowband, the real envelope $a(t) \equiv |u(t)|$ or, equivalently, $a^2(t)$ varies slowly relative to the rapid variations exhibited by the cosine function. A graphical illustration of the integrand in the second integral of (3.1.10) is shown in Fig. 3.1.1. The value of the integral is just the net area under the cosine function modulated by $a^2(t)$. Since the modulating waveform $a^2(t)$ varies slowly relative to the cosine function, the net area contributed by the second integral is very small relative to the value of the first integral in (3.1.10) and, hence, it can be neglected. Thus, for all practical purposes, the energy in the bandpass signal $s(t)$, expressed in terms of the equivalent lowpass signal $u(t)$, is

$$\mathscr{E} = \frac{1}{2} \int_{-\infty}^{\infty} |u(t)|^2 \, dt \qquad (3.1.11)$$

where $|u(t)|$ is just the envelope $a(t)$ of $s(t)$.

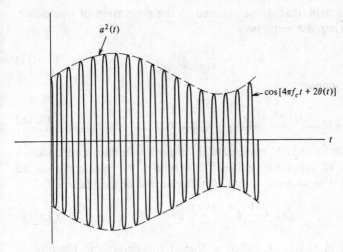

FIGURE 3.1.1
The signal $a^2(t)\cos[4\pi f_c t + 2\theta(t)]$.

3.1.2 Representation of Linear Bandpass Systems

A linear filter or system may be described either by its impulse response $h(t)$ or by its frequency response $H(f)$, which is the Fourier transform of $h(t)$. Since $h(t)$ is real,

$$H^*(-f) = H(f) \tag{3.1.12}$$

Let us define $C(f - f_c)$ as

$$C(f - f_c) = \begin{cases} H(f) & f > 0 \\ 0 & f < 0 \end{cases} \tag{3.1.13}$$

Then

$$C^*(-f - f_c) = \begin{cases} 0 & f > 0 \\ H^*(-f) & f < 0 \end{cases} \tag{3.1.14}$$

Using (3.1.12), we have

$$H(f) = C(f - f_c) + C^*(-f - f_c) \tag{3.1.15}$$

The inverse transform of $H(f)$ in (3.2.4) yields $h(t)$ in the form

$$h(t) = c(t)e^{j2\pi f_c t} + c^*(t)e^{-j2\pi f_c t}$$

$$= 2\operatorname{Re}\left[c(t)e^{j2\pi f_c t}\right] \tag{3.1.16}$$

where $c(t)$ is the inverse Fourier transform of $C(f)$. In general, the impulse response $c(t)$ of the equivalent lowpass system is complex-valued.

A filter characteristic that is encountered in the generation of single-side-band signals has the impulse response

$$h(t) = \frac{1}{\pi t} \tag{3.1.17}$$

and the frequency response

$$H(f) = \begin{cases} -j & f > 0 \\ j & f < 0 \end{cases} \tag{3.1.18}$$

The filter characteristic in (3.1.18) represents an all-pass filter which introduces a $-90°$ phase shift to all input frequencies $f > 0$ (and a $90°$ phase shift to all negative frequencies). For an input signal $s(t)$, the output of the filter is

$$r(t) = \frac{1}{\pi} \int_{-\infty}^{\infty} \frac{s(\tau)}{t - \tau} d\tau \tag{3.1.19}$$

The integral relation in (3.1.19) is called a *Hilbert transform*, and the filter is often called a *Hilbert transformer*. When the input is $s(t)$, its output is usually denoted as $\hat{s}(t)$.

3.1.3   Response of a Bandpass System to a Bandpass Signal

In Secs. 3.1.1 and 3.1.2 we have shown that narrowband bandpass signals and systems can be represented by equivalent lowpass signals and systems. In this section we demonstrate that the output of a bandpass system to a bandpass input signal is simply obtained from the equivalent lowpass input signal and the equivalent lowpass impulse response of the system.

Suppose that $s(t)$ is a narrowband bandpass signal and $u(t)$ is the equivalent lowpass signal. This signal excites a narrowband bandpass system characterized by its bandpass impulse response $h(t)$ or by its equivalent lowpass impulse response $c(t)$. The output of the bandpass system is also a bandpass signal and, therefore, it can be expressed in the form

$$r(t) = \text{Re}\left[v(t)e^{j2\pi f_c t}\right] \tag{3.1.20}$$

where $r(t)$ is related to the input signal $s(t)$ and the impulse response $h(t)$ by the convolution integral

$$r(t) = \int_{-\infty}^{\infty} s(\tau)h(t - \tau) d\tau \tag{3.1.21}$$

Equivalently, the output of the system, expressed in the frequency domain, is

$$R(f) = S(f)H(f) \tag{3.1.22}$$

Substituting from (3.1.8) for $S(f)$ and from (3.1.15) for $H(f)$, we obtain the result

$$R(f) = \tfrac{1}{2}\left[U(f - f_c) + U^*(-f - f_c)\right]\left[C(f - f_c) + C^*(-f - f_c)\right] \tag{3.1.23}$$

When $s(t)$ is narrowband signal and $h(t)$ is the impulse response of a narrowband system, $U(f - f_c) \approx 0$ and $C(f - f_c) = 0$ for $f < 0$. It follows from this narrowband condition that $U(f - f_c)C^*(-f - f_c) = 0$ and $U^*(-f - f_c) \times C(f - f_c) = 0$. Therefore, (3.1.23) simplifies to

$$R(f) = \tfrac{1}{2}[U(f - f_c)C(f - f_c) + U^*(-f - f_c)C^*(-f - f_c)]$$
$$= \tfrac{1}{2}[V(f - f_c) + V^*(-f - f_c)] \tag{3.1.24}$$

where

$$V(f) = U(f)C(f) \tag{3.1.25}$$

is the output spectrum of the equivalent lowpass system excited by the equivalent lowpass signal. It is clear that the time domain relation for the output $v(t)$ is given by the convolution of $u(t)$ with $c(t)$. That is,

$$v(t) = \int_{-\infty}^{\infty} u(\tau)c(t - \tau)\, d\tau \tag{3.1.26}$$

The combination of (3.1.26) with (3.1.20) gives the relationship between the bandpass output signal $r(t)$ and the equivalent lowpass time functions $u(t)$ and $c(t)$. This simple relationship allows us to ignore any linear frequency translations encountered in the modulation of a signal for purposes of matching its spectral content to the frequency allocation of a particular channel. Thus, for mathematical convenience, we shall deal only with the transmission of equivalent lowpass signals through equivalent lowpass channels.

3.1.4 Representation of Bandpass Stationary Stochastic Processes

The representation of bandpass signals presented in Sec. 3.1.1 applied to deterministic signals. In this section we extend the representation to sample functions of a bandpass stationary stochastic process. In particular, we derive the important relations between the correlation functions and power spectra of the bandpass signal and the correlation functions and power spectra of the equivalent lowpass signal.

Suppose that $n(t)$ is a sample function of a wide-sense stationary stochastic process with zero mean and power spectral density $\Phi_{nn}(f)$. The power spectral density is assumed to be zero outside of an interval of frequencies centered about $\pm f_c$, where f_c is termed the *carrier frequency*. The stochastic process $n(t)$ is said to be a *narrowband bandpass process* if the width of the spectral density is much smaller than f_c. Under this condition, a sample function of the process $n(t)$ can be represented by any of the three equivalent forms given in Sec. 3.1.1, namely,

$$n(t) = a(t)\cos[2\pi f_c t + \theta(t)] \tag{3.1.27}$$

$$= x(t)\cos 2\pi f_c t - y(t)\sin 2\pi f_c t \tag{3.1.28}$$

$$= \operatorname{Re}[z(t)e^{j2\pi f_c t}] \tag{3.1.29}$$

where $a(t)$ is the envelope and $\theta(t)$ is the phase of the real-valued signal, $x(t)$ and $y(t)$ are the quadrature components of $n(t)$, and $z(t)$ is called the *complex envelope of $n(t)$*.

Let us consider the form given by (3.1.28) in more detail. First we observe that if $n(t)$ is zero mean, then $x(t)$ and $y(t)$ must also have zero mean values. In addition, the stationarity of $n(t)$ implies that the autocorrelation and cross-correlation functions of $x(t)$ and $y(t)$ satisfy the following properties:

$$\phi_{xx}(\tau) = \phi_{yy}(\tau) \qquad (3.1.30)$$

$$\phi_{xy}(\tau) = -\phi_{yx}(\tau) \qquad (3.1.31)$$

That these two properties follow from the stationarity of $n(t)$ is now demonstrated. The autocorrelation function $\phi_{nn}(\tau)$ of $n(t)$ is

$$\begin{aligned}
E[n(t)n(t+\tau)] &= E\{[x(t)\cos 2\pi f_c t - y(t)\sin 2\pi f_c t] \\
&\quad \times [x(t+\tau)\cos 2\pi f_c(t+\tau) \\
&\quad\quad - y(t+\tau)\sin 2\pi f_c(t+\tau)]\} \\
&= \phi_{xx}(\tau)\cos 2\pi f_c t \cos 2\pi f_c(t+\tau) \\
&\quad + \phi_{yy}(\tau)\sin 2\pi f_c t \sin 2\pi f_c(t+\tau) \\
&\quad - \phi_{xy}(\tau)\sin 2\pi f_c t \cos 2\pi f_c(t+\tau) \\
&\quad - \phi_{yx}(\tau)\cos 2\pi f_c t \sin 2\pi f_c(t+\tau) \qquad (3.1.32)
\end{aligned}$$

Use of the following trigonometric identities

$$\begin{aligned}
\cos A \cos B &= \tfrac{1}{2}[\cos(A-B)+\cos(A+B)] \\
\sin A \sin B &= \tfrac{1}{2}[\cos(A-B)-\cos(A+B)] \qquad (3.1.33)\\
\sin A \cos B &= \tfrac{1}{2}[\sin(A-B)+\sin(A+B)]
\end{aligned}$$

in (3.1.32) yields the result

$$\begin{aligned}
E[n(t)n(t+\tau)] &= \tfrac{1}{2}[\phi_{xx}(\tau)+\phi_{yy}(\tau)]\cos 2\pi f_c\tau \\
&\quad + \tfrac{1}{2}[\phi_{xx}(\tau)-\phi_{yy}(\tau)]\cos 2\pi f_c(2t+\tau) \\
&\quad - \tfrac{1}{2}[\phi_{yx}(\tau)-\phi_{xy}(\tau)]\sin 2\pi f_c\tau \\
&\quad - \tfrac{1}{2}[\phi_{yx}(\tau)+\phi_{xy}(\tau)]\sin 2\pi f_c(2t+\tau) \qquad (3.1.34)
\end{aligned}$$

Since $n(t)$ is stationary, the right-hand side of (3.1.34) must be independent of t. But this condition can only be satisfied if (3.1.30) and (3.1.31) hold. As a consequence, (3.1.34) reduces to

$$\phi_{nn}(\tau) = \phi_{xx}(\tau)\cos 2\pi f_c\tau - \phi_{yx}(\tau)\sin 2\pi f_c\tau \qquad (3.1.35)$$

We note that the relation between the autocorrelation function $\phi_{nn}(\tau)$ of the bandpass process and the autocorrelation and cross-correlation functions $\phi_{xx}(\tau)$

and $\phi_{yx}(\tau)$ of the quadrature components is identical in form to (3.1.28), which expresses the bandpass process in terms of the quadrature components.

The autocorrelation function of the equivalent lowpass process

$$z(t) = x(t) + jy(t) \tag{3.1.36}$$

is defined as

$$\phi_{zz}(\tau) = \tfrac{1}{2}E\left[z^*(t)z(t+\tau)\right] \tag{3.1.37}$$

Substituting (3.1.36) into (3.1.37) and performing the expectation operation, we obtain

$$\phi_{zz}(\tau) = \tfrac{1}{2}\left[\phi_{xx}(\tau) + \phi_{yy}(\tau) - j\phi_{xy}(\tau) + j\phi_{yx}(\tau)\right] \tag{3.1.38}$$

Now if the symmetry properties given in (3.1.30) and (3.1.31) are used in (3.1.38), we obtain

$$\phi_{zz}(\tau) = \phi_{xx}(\tau) + j\phi_{yx}(\tau) \tag{3.1.39}$$

which relates the autocorrelation function of the complex envelope to the autocorrelation and cross-correlation functions of the quadrature components. Finally, we incorporate the result given by (3.1.39) into (3.1.35) and we have

$$\phi_{nn}(\tau) = \mathrm{Re}\left[\phi_{zz}(\tau)e^{j2\pi f_c \tau}\right] \tag{3.1.40}$$

Thus the autocorrelation function $\phi_{nn}(\tau)$ of the bandpass stochastic process is uniquely determined from the autocorrelation function $\phi_{zz}(\tau)$ of the equivalent lowpass process $z(t)$ and the carrier frequency f_c.

The power density spectrum $\Phi_{nn}(f)$ of the stochastic process $n(t)$ is the Fourier transform of $\phi_{nn}(\tau)$. Hence

$$\begin{aligned}\Phi_{nn}(f) &= \int_{-\infty}^{\infty}\left\{\mathrm{Re}\left[\phi_{zz}(\tau)e^{j2\pi f_c \tau}\right]\right\}e^{-j2\pi f \tau}\,d\tau \\ &= \tfrac{1}{2}\left[\Phi_{zz}(f - f_c) + \Phi_{zz}(-f - f_c)\right]\end{aligned} \tag{3.1.41}$$

where $\Phi_{zz}(f)$ is the power density spectrum of the equivalent lowpass process $z(t)$. Since the autocorrelation function of $z(t)$ satisfies the property $\phi_{zz}(\tau) = \phi_{zz}^*(-\tau)$, it follows that $\Phi_{zz}(f)$ is a real-valued function of frequency.

Properties of the quadrature components. It was just demonstrated above that the cross-correlation function of the quadrature components $x(t)$ and $y(t)$ of the bandpass stationary stochastic process $n(t)$ satisfies the symmetry condition in (3.1.31). Furthermore, any cross correlation function satisfies the condition

$$\phi_{yx}(\tau) = \phi_{xy}(-\tau) \tag{3.1.42}$$

From these two conditions we conclude that

$$\phi_{xy}(\tau) = -\phi_{xy}(-\tau) \tag{3.1.43}$$

That is, $\phi_{xy}(\tau)$ is an odd function of τ. Consequently $\phi_{xy}(0) = 0$ and, hence, $x(t)$ and $y(t)$ are uncorrelated (for $\tau = 0$, only). Of course this does not mean that the processes $x(t)$ and $y(t+\tau)$ are uncorrelated for all τ, since that would imply that $\phi_{xy}(\tau) = 0$ for all τ. If, indeed, $\phi_{xy}(\tau) = 0$ for all τ, then $\phi_{zz}(\tau)$ is real and the power spectral density $\Phi_{zz}(f)$ satisfies the condition

$$\Phi_{zz}(f) = \Phi_{zz}(-f) \qquad (3.1.44)$$

and vice versa. That is, $\Phi_{zz}(f)$ is symmetric about $f = 0$.

In the special case in which the stationary stochastic process $n(t)$ is gaussian, the quadrature components $x(t)$ and $y(t + \tau)$ are jointly gaussian. Moreover, for $\tau = 0$ they are statistically independent and, hence, their joint probability density function is

$$p(x, y) = \frac{1}{2\pi\sigma^2} e^{-(x^2+y^2)/2\sigma^2} \qquad (3.1.45)$$

where the variance σ^2 is defined as $\sigma^2 = \phi_{xx}(0) = \phi_{yy}(0) = \phi_{nn}(0)$.

Representation of white noise. White noise is a stochastic process that is defined to have a flat (constant) power spectral density over the entire frequency range. This type of noise cannot be expressed in terms of quadrature components, as a result of its wideband character.

In problems concerned with the demodulation of narrowband signals in noise, it is mathematically convenient to model the additive noise process as white and to represent the noise in terms of quadrature components. This can be accomplished by postulating that the signals and noise at the receiving terminal have passed through an ideal bandpass filter, having a passband that includes the spectrum of the signals but is much wider. Such a filter will introduce negligible, if any, distortion on the signal but it does eliminate the noise frequency components outside of the passband.

The noise resulting from passing the white noise process through a spectrally flat (ideal) bandpass filter is termed *bandpass white noise* and has the power spectral density depicted in Fig. 3.1.2. Bandpass white noise can be represented by any of the forms given in (3.1.27), (3.1.28), and (3.1.29). The equivalent

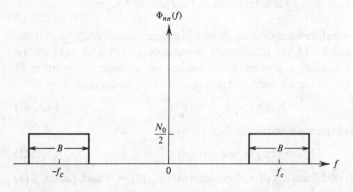

FIGURE 3.1.2
Bandpass noise with a flat spectrum.

lowpass noise $z(t)$ has a power spectral density

$$\Phi_{zz}(f) = \begin{cases} N_0 & |f| \leq B/2 \\ 0 & |f| > B/2 \end{cases} \tag{3.1.46}$$

and its autocorrelation function is

$$\phi_{zz}(\tau) = N_0 \frac{\sin \pi B \tau}{\pi \tau} \tag{3.1.47}$$

The limiting form of $\phi_{zz}(\tau)$ as B approaches infinity is

$$\phi_{zz}(\tau) = N_0 \delta(\tau) \tag{3.1.48}$$

The power spectral density for white noise and bandpass white noise is symmetric about $f = 0$, so $\phi_{yx}(\tau) = 0$ for all τ. Therefore

$$\phi_{zz}(\tau) = \phi_{xx}(\tau) = \phi_{yy}(\tau) \tag{3.1.49}$$

That is, the quadrature components $x(t)$ and $y(t)$ are uncorrelated for all time shifts τ and the autocorrelation functions of $z(t)$, $x(t)$, and $y(t)$ are all equal.

3.2 REPRESENTATION OF FINITE ENERGY SIGNALS BY ORTHONORMAL EXPANSIONS

Let $s(t)$ denote a deterministic, possibly complex-valued signal with finite energy

$$\mathcal{E}_s = \int_{-\infty}^{\infty} |s(t)|^2 dt \tag{3.2.1}$$

Furthermore, suppose that there exists a set of functions $\{ f_n(t), n = 1, 2, \ldots, N \}$ that are orthogonal in the sense that

$$\int_{-\infty}^{\infty} f_n(t) f_m^*(t) \, dt = \begin{cases} 0 & m \neq n \\ 1 & m = n \end{cases} \tag{3.2.2}$$

When the functions $\{ f_n(t) \}$ are normalized to have unity energy as in (3.2.2), they are called *orthonormal*.

We may approximate the signal $s(t)$ by a weighted linear combination of these functions, i.e.,

$$\hat{s}(t) = \sum_{k=1}^{K} s_k f_k(t) \tag{3.2.3}$$

where $\{ s_k, 1 \leq k \leq n \}$ are the coefficients in the approximation of $s(t)$. The approximation error incurred is

$$e(t) = s(t) - \hat{s}(t) \tag{3.2.4}$$

Let us select the coefficients $\{ s_k \}$ so as to minimize the energy $\mathcal{E}_e$ of the approximation error. Thus,

$$\mathcal{E}_e = \int_{-\infty}^{\infty} |s(t) - \hat{s}(t)|^2 dt$$

$$= \int_{-\infty}^{\infty} \left| s(t) - \sum_{k=1}^{K} s_k f_k(t) \right|^2 dt \tag{3.2.5}$$

The optimum coefficients in the series expansion of $s(t)$ may be found by differentiating (3.2.5) with respect to each of the coefficients $\{s_k\}$ and setting the first derivatives to zero. Alternatively, we may use the well known result in estimation theory based on the mean-square-error criterion which, simply stated, is that the minimum of $\mathscr{E}_e$ with respect to the $\{s_k\}$ is obtained when the error is orthogonal to each of the functions in the series expansion. Thus,

$$\int_{-\infty}^{\infty}\left[s(t) - \sum_{k=1}^{K} s_k f_k(t)\right] f_n^*(t)\, dt = 0 \qquad n = 1, 2, \ldots, K \qquad (3.2.6)$$

Since the functions $\{f_n(t)\}$ are orthonormal. (3.2.6) reduces to

$$s_n = \int_{-\infty}^{\infty} s(t) f_n^*(t)\, dt \qquad n = 1, 2, \ldots, K \qquad (3.2.7)$$

Thus, the coefficients are obtained by projecting the signal $s(t)$ onto each of the functions $\{f_n(t)\}$. Consequently, $\hat{s}(t)$ is the projection of $s(t)$ onto the K-dimensional signal space spanned by the functions $\{f_n(t)\}$. The minimum mean square approximation error is

$$\begin{aligned}
\mathscr{E}_{\min} &= \int_{-\infty}^{\infty} e(t) s^*(t)\, dt \\
&= \int_{-\infty}^{\infty} |s(t)|^2\, dt - \int_{-\infty}^{\infty} \sum_{k=1}^{K} s_k f_k(t) s^*(t)\, dt \\
&= \mathscr{E}_s - \sum_{k=1}^{K} |s_k|^2 \qquad (3.2.8)
\end{aligned}$$

which is nonnegative, by definition.

When the minimum mean square approximation error $\mathscr{E}_{\min} = 0$, then

$$\mathscr{E}_s = \sum_{k=1}^{K} |s_k|^2 = \int_{-\infty}^{\infty} |s(t)|^2\, dt \qquad (3.2.9)$$

Under the condition that $\mathscr{E}_{\min} = 0$, we may express $s(t)$ as

$$s(t) = \sum_{k=1}^{K} s_k f_k(t) \qquad (3.2.10)$$

where it is understood that equality of $s(t)$ to its series expansion holds in the sense that the approximation error has zero energy.

When every finite energy signal can be represented by a series expansion of the form in (3.2.10) for which $\mathscr{E}_{\min} = 0$, the set of orthonormal functions $\{f_n(t)\}$ is said to be *complete*.

Example 3.2.1 Exponential Fourier series. A finite energy signal $s(t)$ that is zero everywhere except in the range $0 \le t \le T$ and has a finite number of discontinuities in this interval, can be represented in an exponential Fourier series as

$$s(t) = \sum_{k=-\infty}^{\infty} s_k e^{j2\pi kt/T} \qquad (3.2.11)$$

where the coefficients that minimize the mean square error are given by

$$s_k = \frac{1}{T} \int_0^T s(t) e^{-j2\pi kt/T} dt \qquad (3.2.12)$$

The set of exponential functions $\{f_n(t) = e^{j2\pi nt/T}\}$ is complete and, hence, the series expansion results in zero mean square error. These properties are easily established from the development given above.

Now suppose that we have a set of finite energy signal waveforms $\{s_i(t),$ $= 1, 2, \ldots, M\}$ and we wish to construct a set of orthonormal waveforms. The am–Schmidt orthogonalization procedure allows us to construct such a set.

am–Schmidt procedure. We begin with the first waveform $s_1(t)$, which is sumed to have energy $\mathscr{E}_1$. The first waveform is simply constructed as

$$f_1(t) = \frac{s_1(t)}{\sqrt{\mathscr{E}_1}} \qquad (3.2.13)$$

us, $f_1(t)$ is simply $s_1(t)$ normalized to unit energy.

The second waveform is constructed from $s_2(t)$ by first computing the ojection of $f_1(t)$ onto $s_2(t)$, which is

$$c_{12} = \int_{-\infty}^{\infty} s_2(t) f_1^*(t) dt \qquad (3.2.14)$$

en, $c_{12} f_1(t)$ is subtracted from $s_2(t)$ to yield

$$d_2(t) = s_2(t) - c_{12} f_1(t) \qquad (3.2.15)$$

is waveform is orthogonal to $f_1(t)$ but it does not have unit energy. If $\mathscr{E}_2$ notes the energy of $d_2(t)$, the normalized waveform that is orthogonal to $f_1(t)$

$$f_2(t) = \frac{d_2(t)}{\sqrt{\mathscr{E}_2}} \qquad (3.2.16)$$

In general, the orthogonalization of the kth function leads to

$$f_k(t) = \frac{d_k(t)}{\sqrt{\mathscr{E}_k}} \qquad (3.2.17)$$

here

$$d_k(t) = s_k(t) - \sum_{i=1}^{k-1} c_{ik} f_i(t) \qquad (3.2.18)$$

nd

$$c_{ik} = \int_{-\infty}^{\infty} s_k(t) f_i^*(t) dt \qquad i = 1, 2, \ldots, k-1 \qquad (3.2.19)$$

hus, the orthogonalization process is continued until all the M signal waveforms $s_i(t)\}$ have been exhausted and $N \leq M$ orthonormal waveforms have been

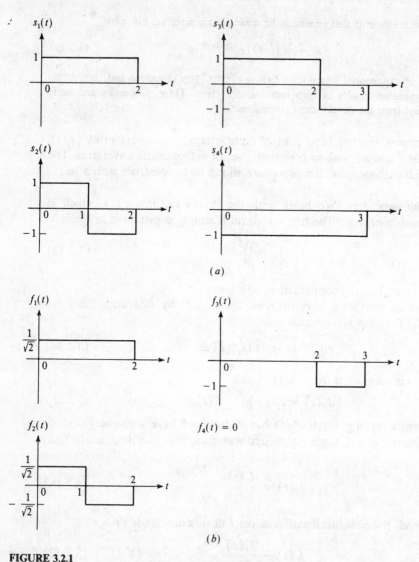

FIGURE 3.2.1
Gram–Schmidt orthogonalization of the signals $\{s_i(t), i = 1, 2, 3, 4\}$ and the corresponding orthogonal signals.

constructed. The dimensionality N of the signal space will be equal to M if all the signal waveforms are linearly independent, i.e., none of the signals waveforms is a linear combination of the other signal waveforms.

Example 3.2.2. Let us apply the Gram–Schmidt procedure to the set of four waveforms illustrated in Fig. 3.2.1a. The waveform $s_1(t)$ has energy $\mathscr{E}_1 = 2$, so that $f_1(t) = s_1(t)/\sqrt{2}$. Next, we observe that $c_{12} = 0$; hence, $s_2(t)$ and $f_1(t)$ are orthogonal. Therefore, $f_2(t) = s_2(t)/\sqrt{\mathscr{E}_2} = s_2(t)/\sqrt{2}$. To obtain $f_3(t)$ we compute c_{13} and c_{23}, which are $c_{13} = \sqrt{2}$ and $c_{23} = 0$. Thus,

$$d_3(t) = s_3(t) - \sqrt{2} f_1(t) = \begin{cases} -1 & 2 \leq t \leq 3 \\ 0 & \text{otherwise} \end{cases}$$

Since $d_3(t)$ has unit energy, it follows that $f_3(t) = d_3(t)$. In determining $f_4(t)$, we find that $c_{14} = -\sqrt{2}$, $c_{24} = 0$, and $c_{34} = 1$. Hence,

$$d_4(t) = s_4(t) - \sqrt{2} f_1(t) + f_3(t) = 0$$

Consequently, $s_4(t)$ is a linear combination of $f_1(t)$ and $f_3(t)$ and, hence, $f_4(t) = 0$. The three orthonormal functions are illustrated in Fig. 3.2.1*b*.

Once we have constructed the set of orthonormal waveforms $\{f_n(t)\}$, we can express the M signals $\{s_n(t)\}$ as linear combinations of the $\{f_n(t)\}$. Thus, we may write

$$s_k(t) = \sum_{n=1}^{N} s_{kn} f_n(t) \qquad k = 1, 2, \ldots, M \tag{3.2.20}$$

and

$$\mathscr{E}_k = \int_{-\infty}^{\infty} |s_k(t)|^2 \, dt = \sum_{n=1}^{N} |s_{kn}|^2 \tag{3.2.21}$$

Based on the expression in (3.2.20), each signal may be represented by the vector

$$\mathbf{s}_k = (s_{k1} s_{k2} \cdots s_{kN}) \tag{3.2.22}$$

or, equivalently, as a point in the N-dimensional signal space with coordinates $\{s_{ki}, i = 1, 2, \ldots, N\}$. The energy in the kth signal is simply the square of the length of the vector or, equivalently, the square of the Euclidean distance from the origin to the point in the N-dimensional space. Thus, any signal can be represented geometrically as a point in the signal space spanned by the orthonormal functions $\{f_n(t)\}$.

Example 3.3.3. Let us obtain the vector representation of the four signals shown in Fig. 3.2.1*a* by using the orthonormal set of functions in Fig. 3.2.1*b*. Since the dimensionality of the signal space is $N = 3$, each signal is described by three components. The signal $s_1(t)$ is characterized by the vector $\mathbf{s}_1 = (\sqrt{2}, 0, 0)$. Similarly, the signals $s_2(t)$, $s_3(t)$, and $s_4(t)$ are characterized by the vectors $\mathbf{s}_2 = (0, \sqrt{2}, 0)$, $\mathbf{s}_3 = (\sqrt{2}, 0, 1)$, and $\mathbf{s}_4 = (-\sqrt{2}, 0, 1)$, respectively. These vectors are shown in Fig. 3.2.2. Their lengths are $|\mathbf{s}_1| = \sqrt{2}$, $|\mathbf{s}_2| = \sqrt{2}$, $|\mathbf{s}_3| = \sqrt{3}$, and $|\mathbf{s}_4| = \sqrt{3}$ and the corresponding signal energies are $\mathscr{E}_k = |\mathbf{s}_k|^2$, $k = 1, 2, 3, 4$.

We have demonstrated that a set of M finite energy waveforms $\{s_n(t)\}$ can be represented by a weighted linear combination of orthonormal functions $\{f_n(t)\}$ of dimensionality $N \leq M$. The functions $\{f_n(t)\}$ are obtained by

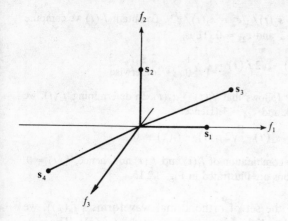

FIGURE 3.2.2
The four signal vectors represented as points in three-dimensional function space.

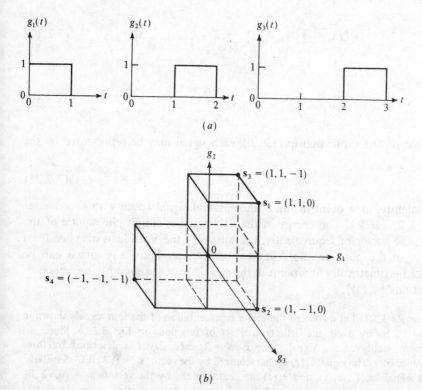

(a)

(b)

FIGURE 3.2.3
An alternative set of orthonormal functions for the four signals in Fig. 3.2.1*a*, and the corresponding signal points.

applying the Gram–Schmidt orthogonalization procedure on $\{s_n(t)\}$. It should be emphasized, however, that the functions $\{f_n(t)\}$ obtained from the Gram–Schmidt procedure are not unique. If we alter the order in which the orthogonalization of the signals $\{s_n(t)\}$ is performed, the orthonormal waveforms will be different and the corresponding vector representation of the signals $\{s_n(t)\}$ will depend on the choice of the orthonormal functions $\{f_n(t)\}$. Nevertheless, the vectors $\{s_n\}$ will retain their geometrical configuration and their lengths will be invariant to the choice of orthonormal functions $\{f_n(t)\}$.

Example 3.2.4. An alternative set of orthonormal functions for the four signals in Fig. 3.2.1 are illustrated in Fig. 3.2.3a. By using these functions to expand $\{s_n(t)\}$, we obtain the corresponding vectors $s_1 = (1, 1, 0)$, $s_2 = (1, -1, 0)$, $s_3 = (1, 1, -1)$, and $s_4 = (-1, -1, -1)$, which are shown in Fig. 3.2.3b. Note that the vector lengths are identical to those obtained from the orthonormal functions $\{f_n(t)\}$.

In the following section we describe digitally modulated signals and make use of the signal space representation for such signals. The most commonly used digitally modulated signals are basically two-dimensional signals and, hence, the waveforms may be represented as points in two-dimensional space.

3.3 REPRESENTATION OF DIGITALLY MODULATED SIGNALS

In the transmission of digital information over a communications channel, the modulator is the interface device that maps the digital information into analog waveforms that match the characteristics of the channel. The mapping is generally performed by taking blocks of $k = \log_2 M$ binary digits at a time from the information sequence $\{a_n\}$ and selecting one of $M = 2^k$ deterministic, finite energy waveforms $\{s_m(t), m = 1, 2, \ldots, M\}$ for transmission over the channel.

When the mapping from the digital sequence $\{a_n\}$ to waveforms is performed under the constraint that a waveform transmitted in any time interval depends on one or more previously transmitted waveforms, the modulator is said to have *memory*. On the other hand, when the mapping from the sequence $\{a_n\}$ to the waveforms $\{s_m(t)\}$ is performed without any constraint on previously transmitted waveforms, the modulator is called *memoryless*.

In addition to classifying the modulator as either memoryless or having memory, we may classify it as either *linear* or *nonlinear*. Linearity of a modulation method requires that the principle of superposition applies in the mapping of the digital sequence into successive waveforms. In nonlinear modulation, the superposition principle does not apply to signals transmitted in successive time intervals. We shall begin by describing linear modulation methods.

3.3.1 Linear, Memoryless Modulation Methods

Recall that in the transmission of analog signals the analog information-bearing signal is impressed onto a carrier frequency by modulating either the carrier amplitude (amplitude modulation) or the carrier phase (phase modulation) or the carrier frequency (frequency modulation). Similarly, the modulator in a digital communications system may map a sequence of binary digits into a corresponding set of *discrete amplitudes* or a set of *discrete phases* or a set of *discrete frequencies*. In general, the output of a carrier modulator is a bandpass signal that can be represented as

$$s(t) = \text{Re}\left[v(t)e^{j2\pi f_c t}\right]$$

where f_c denotes the carrier frequency and $v(t)$ is the equivalent lowpass information-bearing signal.

The signal waveform $v(t)$ depends on the method by which the binary information sequence is mapped into either the set of discrete amplitudes or the set of discrete phases or the set of discrete frequencies. To be specific, suppose that the sequence of binary digits (bits) at the input to the modulator occurs at a rate R bits per second. Then, if the modulator maps the information sequence into a corresponding set of discrete amplitudes, the modulation method is called *digital pulse amplitude modulation* (PAM) or *amplitude shift keying* (ASK), and the signal waveforms may be represented as

$$s_m(t) = \text{Re}\left[A_m u(t)e^{j2\pi f_c t}\right] \qquad m = 1, 2, \ldots, M \qquad (3.3.1)$$

where $\{A_m, m = 1, 2, \ldots, M\}$ denote the M possible signal amplitudes corresponding to the $M = 2^k$ possible k-bit blocks or *symbols*. The waveform $u(t)$ is a real-valued signal pulse whose shape constitutes an important signal design problem when there is a bandwidth limitation on the channel. (This problem is treated in detail in Chapter 6.) The symbol rate for the signal in (3.3.1) is R/k. This is the rate at which changes occur in the amplitude of the carrier to reflect the transmission of new information.

The *signal space diagram* that represents PAM (ASK) is illustrated in Fig. 3.3.1 for $M = 2$, 4, and 8. We make two observations concerning the PAM signal space constellation. One is that the signal has a one-dimensional (vector) representation, namely, the signal amplitude. Second, the signal points are selected to be symmetrical with respect to the origin on the real line. The reason for this choice is explained in Sec. 3.4, where we derive the spectral content of the PAM signal.

The digital PAM signal is also appropriate for transmission over a channel that does not require carrier modulation. In this case, the signal waveforms may be simply represented as

$$A_m(t) = A_m u(t) \qquad m = 1, 2, \ldots, M \qquad (3.3.2)$$

This is now called a *baseband* signal. For example, a four-amplitude level baseband PAM signal is illustrated in Fig. 3.3.2a. The carrier-modulated version of the signal is shown in Fig. 3.3.2b.

The PAM carrier-modulated signal represented by (3.3.1) is a double-sideband (DSB) signal and requires twice the channel bandwidth of the equivalent lowpass signal for transmission. Alternatively, we may use single-sideband (SSB)

(a) $M = 2$

(b) $M = 4$

(c) $M = 8$

FIGURE 3.3.1
Signal space diagram for digital PAM (ASK) signals.

PAM which has the representation (lower or upper sideband)

$$s_m(t) = \mathrm{Re}\left\{ A_m[u(t) \pm j\hat{u}(t)] e^{j2\pi f_c t} \right\} \qquad m = 1, 2, \ldots, M \quad (3.3.3)$$

where $\hat{u}(t)$ denotes the Hilbert transform of $u(t)$. Thus, the spectrum of the single-sideband signal is reduced by a factor of 2 in spectral width.

The bandwidth efficiency of PAM/SSB can also be obtained by simultaneously impressing two separate k-bit symbols from the sequence $\{a_n\}$ on two quadrature carriers $\cos 2\pi f_c t$ and $\sin 2\pi f_c t$. The resulting modulation technique is called quadrature PAM, or QAM, and the corresponding signal waveforms are

$$
\begin{aligned}
s_m(t) &= \mathrm{Re}\left[(A_{mc} + jA_{ms}) u(t) e^{j2\pi f_c t} \right] \\
&= A_{mc} u(t) \cos 2\pi f_c t - A_{ms} u(t) \sin 2\pi f_c t \qquad m = 1, 2, \ldots, M \quad (3.3.4)
\end{aligned}
$$

Signal
amplitude

Data: 11   10   00   01   11   00

(*a*) Baseband PAM signal

(*b*) Bandpass PAM signal

FIGURE 3.3.2
Baseband and bandpass PAM signals.

where A_{mc} is the signal amplitude for the information conveyed by the cosine of the carrier and A_{ms} is the signal amplitude for the information conveyed by the sine of the carrier.

Alternatively, the signal waveforms in (3.3.4) can be expressed as

$$s_m(t) = \text{Re}\left[V_m e^{j\theta_m} u(t) e^{j2\pi f_c t}\right]$$
$$= V_m u(t) \cos(2\pi f_c t + \theta_m) \qquad (3.3.5)$$

where $V_m = \sqrt{A_{mc}^2 + A_{ms}^2}$ and $\theta_m = \tan^{-1} A_{ms}/A_{mc}$. From this expression, it is

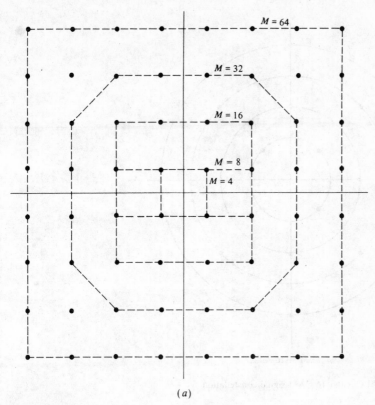

(a)

FIGURE 3.3.3
Several signal-space diagrams for QAM.

apparent that the QAM signal waveforms are a form of combined amplitude and phase modulation which is further described below.

Several different signal space diagrams for QAM are illustrated in Fig. 3.3.3 for $M = 4, 8, 16, 32,$ and 64. We observe that the QAM signal waveforms in (3.3.4) result in rectangular signal space constellations when the amplitude values $\{A_{mc}\}$ and $\{A_{ms}\}$ on the two quadrature carriers are equally spaced. If the waveforms $u(t)\cos 2\pi f_c t$ and $u(t)\sin 2\pi f_c t$ are selected to have unit energy, then the signal space vector representation of the QAM signal is

$$\mathbf{s}_m = (A_{mc}, A_{ms}) \qquad m = 1, 2, \ldots, M \tag{3.3.6}$$

where M is the total number of signal points in the two-dimensional space.

Digital phase modulation of the carrier results when the binary digits from the information sequence $\{a_n\}$ are mapped into a set of discrete phases of the carrier. Thus, an M-phase-shift-keyed (PSK or MPSK) signal is generated by mapping a block of $k = \log_2 M$ binary digits of the sequence $\{a_n\}$ into one of the M corresponding phases $\theta_m = 2\pi(m-1)/M,\ m = 1, 2, \ldots, M$. The result-

$M = 16$

(*b*) Circular 16-QAM signal constellation

FIGURE 3.3.3 (*Continued*).

ing waveforms are

$$s_m(t) = \text{Re}\left[u(t)e^{j\theta_m}e^{j2\pi f_c t}\right]$$

$$= u(t)\cos\left[2\pi f_c t + \frac{2\pi}{M}(m-1)\right] \qquad m = 1, 2, \ldots, M \quad (3.3.7)$$

where $u(t)$ is a pulse that serves the purpose of shaping the spectrum of the transmitted signal. Note that when $u(t)$ is a constant, say $u(t) = A$, over the time interval k/R seconds, the PSK signal is a constant amplitude signal. This is often a constraint in the implementation of PSK modulation for radio communication systems at carrier frequencies above the AM broadcast frequency band.

Signal space diagrams for PSK modulation are illustrated in Fig. 3.3.4 for $M = 2$, 4, and 8. As in the case of PAM, the signal points in the constellation have a two-dimensional vector representation.

Another form of digital modulation is combined PAM and PSK. For example, we may select a combination of M_1-level PAM and M_2-phase PSK. In

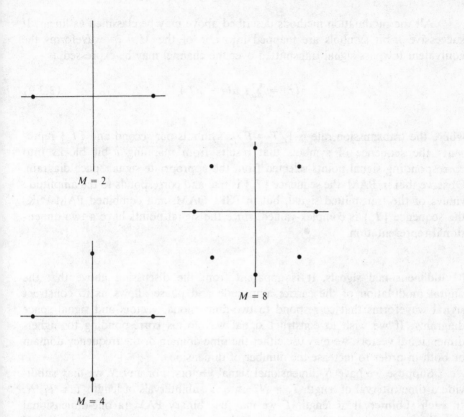

FIGURE 3.3.4
Signal-space diagrams for PSK signals.

such a case we have a set of $M = M_1 M_2$ possible distinct amplitude phase values that may be represented in functional form as

$$A_{mi}(t) = \text{Re}\left[A_m e^{j\theta_i} u(t) e^{j2\pi f_c t} \right]$$

$$= A_m u(t) \cos(2\pi f_c t + \theta_i) \qquad m = 1, 2, \ldots, M_1, i = 1, 2, \ldots, M_2 \quad (3.3.8)$$

If $M_1 = 2^n$ and $M_2 = 2^m$, the combined PAM-PSK modulation method results in the simultaneous transmission of $m + n = \log M_1 M_2$ binary digits occurring at a symbol rate $R/(m + n)$.

Examples of signal space diagrams for combined PAM-PSK modulation are shown in Fig. 3.3.5. It is apparent that the rectangular QAM diagrams shown in Fig. 3.3.3 are special cases of combined PAM-PSK modulation.

All the modulation methods described above may be classified as linear. If successive k-bit symbols are mapped into one of the $M = 2^k$ waveforms the equivalent lowpass signal transmitted over the channel may be expressed as

$$v(t) = \sum_n I_n u(t - nT) \tag{3.3.9}$$

where the transmission rate is $1/T = R/k$ symbols per second and $\{I_n\}$ represents the sequence of symbols that results from mapping k-bit blocks into corresponding signal points selected from the appropriate signal space diagram. Observe that in PAM, the sequence $\{I_n\}$ is real and corresponds to the amplitude values of the transmitted signal, but in PSK, QAM, and combined PAM-PSK, the sequence $\{I_n\}$ is complex-valued, since the signal points have a two-dimensional representation.

Multidimensional signals. It is apparent from the discussion above that the digital modulation of the carrier amplitude and phase allows us to construct signal waveforms that correspond to two-dimensional vectors and signal space diagrams. If we wish to construct signal waveforms corresponding to higher-dimensional vectors, we may use either the time domain or the frequency domain or both in order to increase the number of dimensions.

Suppose we have N-dimensional signal vectors. For any N we may subdivide a time interval of length $T_1 = NT$ into N subintervals of length $T = T_1/N$. In each subinterval of length T we may use binary PAM (a one-dimensional signal) to transmit an element of the N-dimensional signal vector. Thus, the N time slots are used to transmit the N-dimensional signal vector. If N is even, a time slot of length T may be used to simultaneously transmit two components of the N-dimensional vector by modulating the amplitude of quadrature carriers independently by the corresponding components. In this manner, the N-dimensional signal vector is transmitted in $NT/2$ seconds ($N/2$ time slots).

Alternatively, a frequency interval of length $N\Delta f$ may be subdivided into N frequency slots each of width Δf. An N-dimensional signal vector can be transmitted over the channel by simultaneously modulating the amplitude of N carriers, one in each of the N frequency slots. Care must be taken to provide sufficient frequency separation Δf between successive carriers so that there is no cross talk interference among the signals on the N carriers. If quadrature carriers are used in each frequency slot, the N-dimensional vector (N even) may be transmitted in $N/2$ frequency slots, thus reducing the channel bandwidth utilization by a factor of 2.

More generally, we may use both the time and frequency domains jointly to transmit an N-dimensional signal vector. For example, Fig. 3.3.6 illustrates a subdivision of the time and frequency axes into 12 slots. Thus, an $N = 12$-dimensional signal vector may be transmitted by PAM or an $N = 24$-dimensional

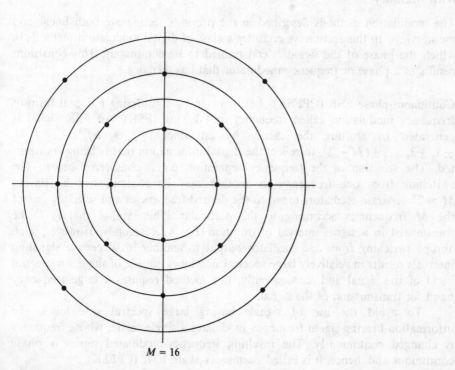

FIGURE 3.3.5
Examples of combined PAM-PSK signal-space diagrams.

FIGURE 3.3.6
Subdivision of time and frequency axes into distinct slots.

signal vector may be transmitted by use of two quadrature carriers (QAM) in each slot.

3.3.2 Nonlinear Modulation Methods with Memory

The modulation methods described in the previous section are both linear and memoryless. In this section we consider a class of digital modulation methods in which the phase of the signal is constrained to be continuous. This constraint results in a phase or frequency modulator that has memory.

Continuous-phase FSK (CPFSK). Let us begin by considering a digital form of frequency modulation called frequency-shift keying (FSK). An FSK signal is generated by shifting the carrier by an amount $f_n = (\Delta f/2)I_n$, $I_n = \pm 1, \pm 3, \ldots, \pm (M - 1)$, to reflect the digital information that is being transmitted. The selection of the frequency separation Δf is considered below. The switching from one frequency to another may be accomplished by having $M = 2^k$ separate oscillators tuned to the desired frequencies and selecting one of the M frequencies according to the particular k-bit symbol that is to be transmitted in a signal interval of duration $T = k/R$ seconds. However, such abrupt switching from one oscillator output to another in successive signaling intervals results in relatively large spectral side lobes outside of the main spectral band of the signal and, consequently, this method requires a large frequency band for transmission of the signal.

To avoid the use of signals having large spectral side lobes, the information-bearing signal frequency modulates a single carrier whose frequency is changed continuously. The resulting frequency-modulated signal is phase continuous and, hence, it is called *continuous-phase* FSK (CPFSK).

In order to represent a CPFSK signal, we begin with a PAM signal

$$d(t) = \sum_n I_n u(t - nT) \tag{3.3.10}$$

where the amplitudes $\{I_n\}$ are obtained by mapping k-bit blocks of binary digits from the information sequence $\{a_n\}$ into the amplitude levels $\pm 1, \pm 3, \ldots, \pm (M - 1)$ and $u(t)$ is a rectangular pulse of amplitude $1/2T$ and duration T seconds. The signal $d(t)$ is used to frequency-modulate the carrier. Consequently, the equivalent lowpass waveform $v(t)$ is expressed as

$$v(t) = A \exp\left\{ j\left[4\pi T f_d \int_{-\infty}^{t} d(\tau)\, d\tau + \phi_0 \right] \right\} \tag{3.3.11}$$

where f_d is the *peak frequency deviation* and ϕ_0 is an initial phase of the carrier.

The carrier-modulated signal corresponding to (3.3.11) may be expressed as

$$s(t) = A \cos\left[2\pi f_c t + \phi(t; \mathbf{I}) + \phi_0 \right] \tag{3.3.12}$$

where $\phi(t; \mathbf{I})$ represents the time-varying phase of the carrier, which is defined as

$$\phi(t; \mathbf{I}) = 4\pi T f_d \int_{-\infty}^{t} d(\tau)\, d\tau$$

$$= 4\pi T f_d \int_{-\infty}^{t} \left[\sum_n I_n u(\tau - nT) \right] d\tau \tag{3.3.13}$$

Note that, although $d(t)$ contains discontinuities, the integral of $d(t)$ is continuous. Hence, we have a continuous-phase signal. The phase of the carrier in the interval $nT \le t \le (n + 1)T$ is determined by integrating (3.3.13). Thus,

$$\phi(t; \mathbf{I}) = 2\pi f_d T \sum_{k=-\infty}^{n-1} I_k + 2\pi f_d (t - nT) I_n$$

$$= \theta_n + 2\pi h I_n q(t - nT) \tag{3.3.14}$$

where h, θ_n, and $q(t)$ are defined as

$$h = 2 f_d T \tag{3.3.15}$$

$$\theta_n = \pi h \sum_{k=-\infty}^{n-1} I_k \tag{3.3.16}$$

$$q(t) = \begin{cases} 0 & t < 0 \\ t/2T & 0 \le t \le T \\ 1/2 & t > T \end{cases} \tag{3.3.17}$$

We observe that θ_n represents the accumulation (memory) of all symbols up to time $(n - 1)T$. The parameter h is called the *modulation index*.

Continuous-phase modulation (CPM). When expressed in the form of (3.3.14), CPFSK becomes a special case of a general class of continuous-phase modulated (CPM) signals in which the carrier phase is

$$\phi(t; \mathbf{I}) = 2\pi \sum_{k=-\infty}^{n} I_k h_k q(t - kT) \qquad nT \le t \le (n + 1)T \tag{3.3.18}$$

where $\{I_k\}$ is the sequence of M-ary information symbols selected from the

FIGURE 3.3.7
Pulse shapes for full response CPM (a, b) and partial response CPM (c, d).

alphabet $\pm 1, \pm 3, \ldots, \pm(M-1)$, $\{h_k\}$ is a sequence of modulation indices, $q(t)$ is some normalized waveform shape.

When $h_k = h$ for all k, the modulation index is fixed for all symbols. W' the modulation index varies from one symbol to another, the CPM signa' called *multi-h*. In such a case, the $\{h_k\}$ are made to vary in a cyclic man through a set of indices.

The waveform $q(t)$ may be represented in general as the integral of so pulse $u(t)$, i.e.,

$$q(t) = \int_0^t u(\tau)\, d\tau \tag{3.3.}$$

If $u(t) = 0$ for $t > T$, the CPM signal is called *full response CPM*. If $u(t)$ for $t > T$, the modulated signal is called *partial response CPM*. Figure 3 illustrates several pulse shapes for $u(t)$, and the corresponding $q(t)$. I apparent that an infinite variety of CPM signals can be generated by choo' different pulse shapes $u(t)$ and by varying the modulation index h and alphabet size M.

(c)

$$u(t) = \frac{1}{4T}\left[1 - \cos\frac{\pi t}{T}\right]$$

(d)

FIGURE 3.3.7 (*Continued*).

It is instructive to sketch the set of phase trajectories $\phi(t; \mathbf{I})$ generated by all possible values of the information sequence $\{I_n\}$. For example, in the case of CPFSK with binary symbols $I_n = \pm 1$, the set of phase trajectories beginning at time $t = 0$ are shown in Fig. 3.3.8. For comparison, the phase trajectories for quaternary CPFSK are illustrated in Fig. 3.3.9. These phase diagrams are called *phase trees*. We observe that the phase trees for CPFSK are piecewise linear as a consequence of the fact that the pulse $u(t)$ is rectangular. Smoother phase trajectories and phase trees are obtained by using pulses that do not contain discontinuities, such as the class of raised cosine pulses. For example, a phase trajectory generated by the sequence $(1, -1, -1, -1, 1, 1, -1, 1)$ for a partial response, raised cosine pulse of length $3T$ is illustrated in Fig. 3.3.10. For comparison, the corresponding phase trajectory generated by CPFSK is also shown.

The phase trees shown in these figures grow with time. However, the phase of the carrier is unique only in the range from $\phi = 0$ to $\phi = 2\pi$ or, equivalently, from $\phi = -\pi$ to $\phi = \pi$. When the phase trajectories are plotted modulo 2π, say in the range $(-\pi, \pi)$, the phase tree collapses into a structure called a *phase trellis*. To properly view the phase trellis diagram, we may plot the two quadrature components $x_c(t; \mathbf{I}) = \cos\phi(t; \mathbf{I})$ and $x_s(t; \mathbf{I}) = \sin\phi(t; \mathbf{I})$ as a function of

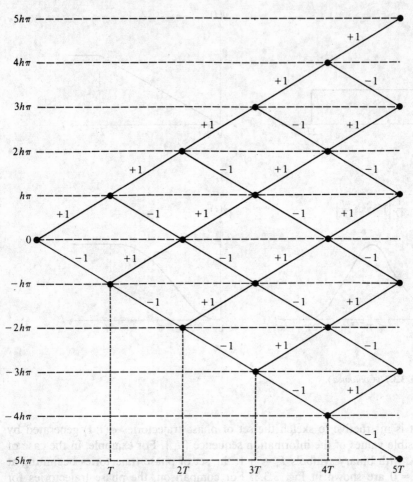

FIGURE 3.3.8
Phase trajectory for binary CPFSK.

time. Thus, we generate a three-dimensional plot in which the quadrature components x_c and x_s appear on the surface of a cylinder of unit radius. For example, Fig. 3.3.11 illustrates the phase trellis or phase cylinder obtained with binary modulation, a modulation index $h = \frac{1}{2}$, and a raised cosine pulse of length $3T$.

Simpler representations for the phase trajectories can be obtained by displaying only the terminal values of the signal phase at the time instants $t = nT$. In this case, we restrict the modulation index of the CPM signal to be rational. In particular, let us assume that $h = m/p$ where m and p are relatively prime integers. Then, a full response CPM signal at the time instants $t = nT$ will

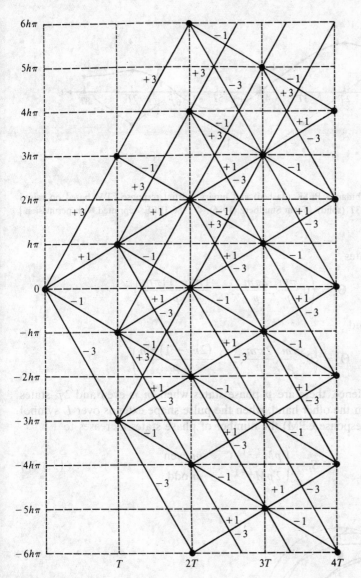

FIGURE 3.3.9
Phase trajectory for quaternary CPFSK.

FIGURE 3.3.10
Phase trajectories for binary CPFSK (dashed) and binary, partial response CPM based on raised cosine pulse of length $3T$ (solid). [From Sundberg (1986), © 1986 IEEE; reprinted with permission.]

have the phase states

$$\Theta_s = \left\{ 0, \frac{\pi m}{p}, \frac{2\pi m}{p}, \ldots, \frac{(m-1)\pi m}{p} \right\}$$

when m is even and

$$\Theta_s = \left\{ 0, \frac{\pi m}{p}, \frac{2\pi m}{p}, \ldots, \frac{(2m-1)\pi m}{p} \right\}$$

when m is odd. Hence, there are p phase states when m is even and $2p$ states when m is odd. On the other hand, when the pulse shape extends over L symbol intervals (partial response CPM) the number of phase states increases to

$$S_t = \begin{cases} pM^{L-1} & m \text{ even} \\ 2pM^{L-1} & m \text{ odd} \end{cases} \tag{3.3.20}$$

FIGURE 3.3.11
Phase cylinder for binary CPM with $h = \frac{1}{2}$ and a raised cosine pulse of length $3T$. [From Sundberg (1986), © 1986 IEEE; reprinted with permission.]

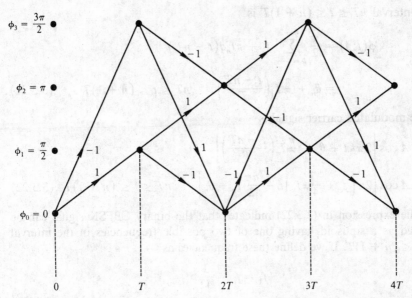

FIGURE 3.3.12
State trellis for binary CPFSK with $h = \frac{1}{2}$.

where M is the alphabet size. For example, the binary CPFSK signal (full response, rectangular pulse) with $h = \frac{1}{2}$, has $S_t = 4$ (terminal) phase states. The *state trellis* for this signal is illustrated in Fig. 3.3.12. We emphasize that the phase transitions from one state to another are not true phase trajectories. They represent phase transitions for the (terminal) states at the time instants $t = nT$.

An alternative representation to the state trellis is the state diagram, which also illustrates the state transitions at the time instants $t = nT$. This is an even more compact representation of the CPM signal characteristics. Only the possible (terminal) phase states and their transitions are displayed in the state diagram. Time does not appear explicitly as a variable. For example, the state diagram for the CPFSK signal with $h = \frac{1}{2}$ is shown in Fig. 3.3.13.

Minimum-shift keying (MSK). MSK is a special form of binary CPFSK (and, therefore, CPM) in which the modulation index $h = \frac{1}{2}$. The phase of the carrier

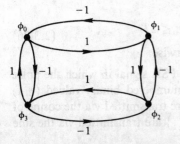

FIGURE 3.3.13
State diagrams for binary CPFSK with $h = \frac{1}{2}$.

in the interval $nT \le t \le (n + 1)T$ is

$$\phi(t; \mathbf{I}) = \frac{\pi}{2} \sum_{k=-\infty}^{n-1} I_k + \pi I_n q(t - nT)$$

$$= \theta_n + \frac{\pi}{2} I_n \left(\frac{t - nT}{T} \right) \qquad nT \le t \le (n + 1)T \qquad (3.3.21)$$

and the modulated carrier signal is

$$s(t) = A \cos \left[2\pi f_c t + \theta_n + \frac{\pi}{2} I_n \left(\frac{t - nT}{T} \right) \right]$$

$$= A \cos \left[2\pi \left(f_c + \frac{1}{4T} I_n \right) t - \frac{n\pi}{2} I_n + \theta_n \right] \qquad nT \le t \le (n + 1)T \quad (3.3.22)$$

The expression in (3.3.22) indicates that the binary CPFSK signal can be expressed as a sinusoid having one of two possible frequencies in the interval $nT \le t \le (n + 1)T$. If we define these frequencies as

$$f_1 = f_c - \frac{1}{4T}$$

$$f_2 = f_c + \frac{1}{4T} \qquad (3.3.23)$$

then the binary CPFSK signal given by (3.3.22) may be written in the form

$$s_i(t) = A \cos \left[2\pi f_i t + \theta_n + \frac{n\pi}{2} (-1)^{i-1} \right] \qquad i = 1, 2 \qquad (3.3.24)$$

The frequency separation $\Delta f = f_2 - f_1 = 1/2T$. It will be shown in the next chapter that $\Delta f = 1/2T$ is the minimum frequency separation that is necessary to ensure the orthogonality of the signals $s_1(t)$ and $s_2(t)$ over a signaling interval of length T. This explains why binary CPFSK with $h = \frac{1}{2}$ is called minimum-shift keying (MSK). The phase in the nth signaling interval is the phase state of the signal that results in phase continuity between adjacent intervals.

MSK may also be represented as a form of four-phase PSK. Specifically, we may express the equivalent lowpass digitally modulated signal in the form (see Prob. 3.6)

$$v(t) = \sum_{n=-\infty}^{\infty} \left[I_{2n} u(t - 2nT) - j I_{2n+1} u(t - 2nT - T) \right] \qquad (3.3.25)$$

where $u(t)$ is a sinusoidal pulse defined as

$$u(t) = \begin{cases} \sin \dfrac{\pi t}{2T} & 0 \le t \le 2T \\ 0 & \text{otherwise} \end{cases} \qquad (3.3.26)$$

Thus, this type of signal is viewed as a four-phase PSK signal in which the pulse shape is one-half cycle of a sinusoid. The even-numbered binary-valued (± 1) symbols $\{ I_{2n} \}$ of the information sequence $\{ I_n \}$ are transmitted via the cosine of the carrier while the odd-numbered symbols $\{ I_{2n+1} \}$ are transmitted via the sine

(*a*) In-phase signal component

(*b*) Quadrature signal component

(*c*) MSK signal [sum of (*a*) and (*b*)]

FIGURE 3.3.14
Representation of MSK signal as a form of two staggered binary PSK signals, each with a sinusoidal envelope.

of the carrier. The transmission rate on the two orthogonal carrier components is $1/2T$ bits per second so that the combined transmission rate is $1/T$ bits per second. Note that the bit transitions on the sine and cosine carrier components are staggered or offset in time by T seconds. For this reason, the signal

$$s(t) = A\left\{\left[\sum_{n=-\infty}^{\infty} I_{2n}u(t - 2nT)\right]\cos 2\pi f_c t\right.$$
$$\left. + \left[\sum_{n=-\infty}^{\infty} I_{2n+1}u(t - 2nT - T)\right]\sin 2\pi f_c t\right\} \quad (3.3.27)$$

is called *offset quadrature PSK (OQPSK)* or *staggered quadrature PSK (SQPSK)*.

Figure 3.3.14 illustrates the representation of the MSK signals as two staggered quadrature-modulated binary PSK signals. The corresponding sum of the two quadrature signals is a constant amplitude, frequency-modulated signal.

It is also interesting to compare the waveforms for MSK with offset QPSK in which the pulse $u(t)$ is rectangular for $0 \le t \le 2T$, and with conventional quadrature (four-phase) PSK (QPSK) in which the pulse $u(t)$ is rectangular for $0 \le t \le T$. Clearly, all three of the modulation methods result in identical data rates. The MSK signal has continuous phase. The offset QPSK signal with a rectangular pulse is basically two binary PSK signals for which the phase

FIGURE 3.3.15
Signal waveforms for (a) MSK, (b) offset QPSK (rectangular pulse), and (c) conventional QPSK (rectangular pulse). [From Gronemeyer and McBride (1976); © 1976 IEEE; reprinted with permission.]

transitions are staggered in time by T seconds. Thus, the signal contains phase jumps of $\pm 90°$ that may occur as often as every T seconds. On the other hand, the conventional four-phase PSK signal with constant amplitude will contain phase jumps of $\pm 180°$ or $\pm 90°$. An illustration of these three signal types is given in Fig. 3.3.15.

Signal space diagrams for CPM. In general, continuous-phase signals cannot be represented by discrete points in signal space as in the case of PAM, PSK, and QAM, because the phase of the carrier is time-variant. Instead, a continuous-phase signal is described by the various paths or trajectories from one phase state to another. For a constant amplitude CPM signal, the various trajectories form a circle.

For example, Fig. 3.3.16 illustrates the signal space (phase trajectory) diagram for CPFSK signals with $h = \frac{1}{4}$, $h = \frac{1}{3}$, $h = \frac{1}{2}$, and $h = \frac{2}{3}$. The beginning and ending points of these phase trajectories are marked in the figure by

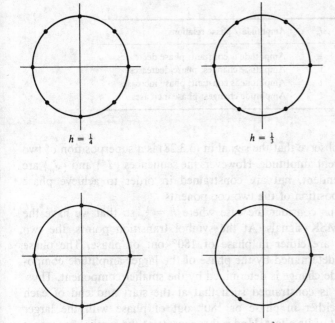

FIGURE 3.3.16
Signal-space diagram for CPFSK.

a dot. Note that the length of the phase trajectory increases with an increase in h. An increase in h also results in an increase of the signal bandwidth, as demonstrated in the following section.

Multiamplitude CPM. Multiamplitude CPM is a generalization of ordinary CPM in which the signal amplitude is allowed to vary over a set of amplitude values while the phase of the signal is constrained to be continuous. For example, let us consider a two-amplitude CPFSK signal, which may be represented as

$$s(t) = 2A \cos\left[2\pi f_c t + \phi_2(t; \mathbf{I})\right] + A \cos\left[2\pi f_c t + \phi_1(t; \mathbf{J})\right] \quad (3.3.28)$$

where

$$\phi_2(t; \mathbf{I}) = \pi h \sum_{k=-\infty}^{n-1} I_k + \pi h I_n(t - nT)/T \qquad nT \le t \le (n+1)T \quad (3.3.29)$$

$$\phi_1(t; \mathbf{J}) = \pi h \sum_{k=-\infty}^{n-1} J_k + \pi h J_n(t - nT)/T \qquad nT \le t \le (n+1)T \quad (3.3.30)$$

The information is conveyed by the symbol sequences $\{I_n\}$ and $\{J_n\}$, which are related to two independent binary information sequences $\{a_n\}$ and $\{b_n\}$ that

184 DIGITAL COMMUNICATIONS

TABLE 3.3.1

a_n	b_n	I_n	J_n	Amplitude / phase relations
0	0	-1	-1	Amplitude is constant; phase decreases
0	1	-1	1	Amplitude changes; phase decreases
1	0	1	1	Amplitude is constant; phase increases
1	1	1	-1	Amplitude changes; phase increases

take values $\{0, 1\}$. We observe that the signal in (3.3.28) is a superposition of two CPFSK signals of different amplitude. However, the sequences $\{I_n\}$ and $\{J_n\}$ are not statistically independent, but are constrained in order to achieve phase continuity in the superposition of the two components.

To elaborate, let us consider the case where $h = \frac{1}{2}$, so that we have the superposition of two MSK signals. At the symbol transition points, the two amplitude components are either in phase or 180° out of phase. The phase change in the signal is determined by the phase of the larger amplitude component, while the amplitude change is determined by the smaller component. Thus, the smaller component is constrained such that at the start and end of each symbol interval, it is either in phase or 180° out of phase with the larger component, independent of its phase. Under this constraint the symbol sequences $\{I_n\}$ and $\{J_n\}$ may be expressed as

$$I_n = 2a_n - 1$$
$$J_n = I_n(1 - 2b_n) = I_n\left(1 - \frac{b_n}{h}\right) \tag{3.3.31}$$

These relationships are summarized in Table 3.3.1.

As a generalization, a multiamplitude CPFSK signal with N components may be expressed as

$$s(t) = 2^{N-1}\cos\left[2\pi f_c t + \phi_N(t; \mathbf{I})\right] + \sum_{m=1}^{N-1} 2^{m-1}\cos\left[2\pi f_c t + \phi_m(t; \mathbf{J}_m)\right] \tag{3.3.32}$$

where

$$\phi_N(t; \mathbf{I}) = \pi h I_n \frac{t - nT}{T} + \pi h \sum_{k=-\infty}^{n-1} I_k \qquad nT \le t \le (n+1)T \tag{3.3.33}$$

and

$$\phi_m(t; \mathbf{J}_m) = I_n\pi\left(h + \frac{J_{mn}+1}{2}\right)\frac{t-nT}{T} + \sum_{k=-\infty}^{n-1}\pi I_k\left(h + \frac{J_{mk}+1}{2}\right)$$
$$nT \le t \le (n+1)T \tag{3.3.34}$$

The sequences $\{I_n\}$ and $\{J_{mn}\}$ are statistically independent, binary-valued sequences that take values from the set $\{1, -1\}$.

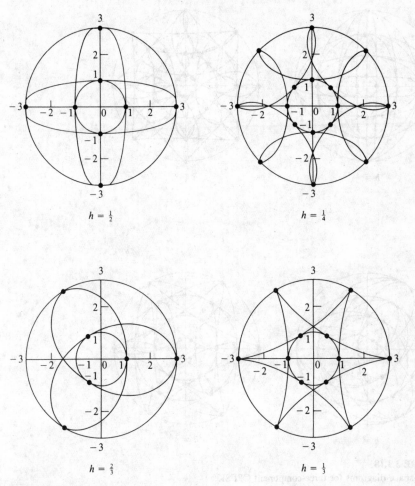

FIGURE 3.3.17
Signal-space diagrams for two-component CPFSK.

From (3.3.33) and (3.3.34) we observe that each component in the sum will be either in phase or 180° out of phase with the largest component at the end of the nth symbol interval, i.e., at $t = (n + 1)T$. Thus, the signal states are specified by an amplitude level from the set of amplitudes $\{1, 3, 5, \ldots, 2^N - 1\}$ and a phase level from the set $\{0, \pi\theta, 2\pi\theta, \ldots, 2\pi - \pi h\}$. The phase constraint is required to maintain the phase continuity of the CPM signal.

Figure 3.3.17 illustrates the signal space diagrams for two-amplitude ($N = 2$) CPFSK with $h = \frac{1}{4}$, $\frac{1}{3}$, $\frac{1}{2}$, and $\frac{2}{3}$. The signal space diagrams for three-component ($N = 3$) CPFSK are shown in Fig. 3.3.18. In this case there are *four* amplitude levels. The number of states depends on the modulation index h as well as N. Note that the beginning and ending points of the phase trajectories are marked by a dot.

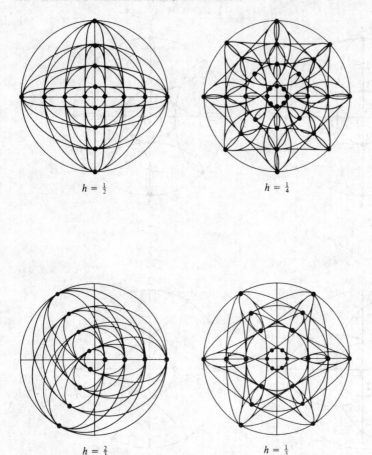

$h = \frac{1}{2}$

$h = \frac{1}{4}$

$h = \frac{2}{3}$

$h = \frac{1}{3}$

FIGURE 3.3.18
Signal-space diagrams for three-component CPFSK.

Additional multiamplitude CPM signal formats may be obtained by using pulse shapes other than rectangular as well as signal pulses that span more than one symbol interval (partial response).

3.3.3 Baseband Signals

Although a large amount of digital information is transmitted by means of carrier modulation, there are also communication channels that support non-carrier-type or *baseband* (lowpass) signal transmission. Some telephone channels and digital magnetic recording channels fall in this category.

Four different types of baseband signals often used in practice for digital transmission over such channels are illustrated and described in Fig. 3.3.19. Let us briefly consider the characteristics of these signals.

NRZ

NRZI

Biphase
(Manchester
code)

Delay
modulation
(Miller
code)

Data   1   0   1   1   0   0   0   1

FIGURE 3.3.19
Baseband signals.

The NRZ signal is the simplest. The binary information digit 1 is represented by a rectangular pulse of polarity A and the binary digit zero is represented by a rectangular pulse of polarity $-A$. Hence, the NRZ modulation is memoryless and is equivalent to a binary PAM or a binary PSK signal in a carrier-modulated system.

The NRZI signal is different from the NRZ signal in that transitions from one amplitude level to another occur only when a 1 is transmitted. The amplitude level remains unchanged when a zero is transmitted. This type of signal encoding is called *differential encoding*. The encoding operation is described mathematically by the relation

$$b_k = a_k \oplus b_{k-1} \qquad (3.3.35)$$

where $\{a_k\}$ is the binary information sequence into the encoder, $\{b_k\}$ is the output sequence of the encoder, and $\oplus$ denotes modulo-2 addition. When $b_k = 1$, the transmitted waveform is a rectangular pulse of amplitude A and when $b_k = 0$, the transmitted waveform is a rectangular pulse of amplitude $-A$. Hence,

FIGURE 3.3.20
State diagram for the NRZI signal.

the output of the encoder is mapped into one of two waveforms in exactly the same manner as for the NRZ signal.

The differential encoding operation introduces memory in the signal. The combination of the encoder and the modulator operations may be represented by a state diagram (a Markov chain) as shown in Fig. 3.3.20. The state diagram may be described by two transition matrices corresponding to the two possible input bits $\{0, 1\}$. We note that when $a_k = 0$, the encoder stays in the same state. Hence, the state transition matrix for a zero is simply

$$T_1 = \begin{bmatrix} 1 & 0 \\ 0 & 1 \end{bmatrix} \qquad (3.3.36)$$

where $t_{ij} = 1$ if a_k results in a transition from state i to state j, $i = 1, 2$, and $j = 1, 2$; otherwise $t_{ij} = 0$. Similarly, the state transition matrix for $a_k = 1$ is

$$T_2 = \begin{bmatrix} 0 & 1 \\ 1 & 0 \end{bmatrix} \qquad (3.3.37)$$

Thus, these two state transition matrices characterize the NRZI signal.

The biphase signal shown in Fig. 3.3.19 may be described as a baseband signal that is generated from the pulse $s(t)$ shown in Fig. 3.3.21. When the information symbol is a 1, the pulse $s(t)$ is transmitted. When the information symbol is a 0 the transmitted waveform is $-s(t)$. Hence, there is no memory in the biphase-modulated signal.

The signal generated by delay modulation or "Miller encoding" does have memory. This type of digital modulation has been used extensively for digital

FIGURE 3.3.21
Pulse shape for biphase signal.

(a)

$s_3(t) = -s_2(t) \quad 0 \le t \le T$

$s_4(t) = -s_1(t) \quad 0 \le t \le T$

(b)

FIGURE 3.3.22
State diagram (a) and basic waveforms (b) for delay modulated (Miller encoded) signal.

magnetic recording and in carrier modulation systems employing binary PSK. The signal may be described by a state diagram that has four states as shown in Fig. 3.3.22a. There are two elementary waveforms $s_1(t)$ and $s_2(t)$ and their negatives $-s_1(t)$ and $-s_2(t)$, which are used for transmitting the binary information. These waveforms are illustrated in Fig. 3.3.22b. The mapping from bits to corresponding waveforms is illustrated in the state diagram. The state transition matrices that characterize the memory of this encoding and modulation method are easily obtained from the state diagram in Fig. 3.3.22. When $a_k = 0$, we have

$$ T_1 = \begin{bmatrix} 0 & 0 & 0 & 1 \\ 0 & 0 & 0 & 1 \\ 1 & 0 & 0 & 0 \\ 1 & 0 & 0 & 0 \end{bmatrix} \qquad (3.3.38) $$

and when $a_k = 1$ the transition matrix is

$$T_2 = \begin{bmatrix} 0 & 1 & 0 & 0 \\ 0 & 0 & 1 & 0 \\ 0 & 1 & 0 & 0 \\ 0 & 0 & 1 & 0 \end{bmatrix} \qquad (3.3.39)$$

Thus, these two 4×4 state transition matrices characterize the state diagram for the Miller-encoded signal.

Modulation techniques with memory such as NRZI and Miller coding are generally characterized by a K-state Markov chain with *stationary state probabilities* $\{ p_i, i, j = 1, 2, \ldots, K \}$. Associated with each transition is a signal waveform $s_j(t)$, $j = 1, 2, \ldots, K$. Thus, the transition probability p_{ij} denotes the probability that signal waveform $s_j(t)$ is transmitted in a given signaling interval after the transmission of the signal waveform $s_i(t)$ in the previous signaling interval. The transition probabilities may be arranged in matrix form as

$$\mathbf{P} = \begin{bmatrix} p_{11} & p_{12} & \cdots & p_{1K} \\ p_{21} & p_{22} & \cdots & p_{2K} \\ \vdots & & & \\ p_{K1} & p_{K2} & \cdots & p_{KK} \end{bmatrix} \qquad (3.3.40)$$

where $\mathbf{P}$ is called the *transition probability matrix*.

The transition probability matrix is easily obtained from the transition matrices $\{ T_i \}$ and the corresponding probabilities of occurrence of the input bits (or, equivalently, the stationary state transition probabilities $\{ p_i \}$). The general relationship may be expressed as

$$\mathbf{P} = \sum_{i=1}^{2} q_i T_i \qquad (3.3.41)$$

where $q_1 = P(a_k = 0)$ and $q_2 = P(a_k = 1)$.

For the NRZI signal with equal state probabilities $p_1 = p_2 = \frac{1}{2}$ and transition matrices given by (3.3.36) and (3.3.37), the transition probability matrix is

$$\mathbf{P} = \begin{bmatrix} \frac{1}{2} & \frac{1}{2} \\ \frac{1}{2} & \frac{1}{2} \end{bmatrix} \qquad (3.3.42)$$

Similarly, the transition probability matrix for the Miller-coded signal with equally likely symbols ($q_1 = q_2 = \frac{1}{2}$ or, equivalently, $p_1 = p_2 = p_3 = p_4 = \frac{1}{4}$) is

$$\mathbf{P} = \begin{bmatrix} 0 & \frac{1}{2} & 0 & \frac{1}{2} \\ 0 & 0 & \frac{1}{2} & \frac{1}{2} \\ \frac{1}{2} & \frac{1}{2} & 0 & 0 \\ \frac{1}{2} & 0 & \frac{1}{2} & 0 \end{bmatrix} \qquad (3.3.43)$$

The transition probability matrix is useful in the determination of the spectral characteristics of digital modulation techniques with memory, as we shall observe in Sec. 3.4.

3.4 SPECTRAL CHARACTERISTICS OF DIGITALLY MODULATED SIGNALS

In most digital communications systems, the available channel bandwidth is limited. Consequently, the system designer must consider the constraints imposed by the channel bandwidth limitation in the selection of the modulation technique used to transmit the information. For this reason, it is important for us to determine the spectral content of the digitally modulated signals described in Sec. 3.3.

Since the information sequence is random, a digitally modulated signal is a stochastic process. We are interested in determining the power density spectrum of such a process. From the power density spectrum we can determine the channel bandwidth required to transmit the information-bearing signal. Below, we first derive the spectral characteristics of the class of linearly modulated signals. Then, we consider the nonlinear CPFSK, CPM, and baseband modulated signals with memory.

3.4.1 Power Spectra of Linearly Modulated Signals

Beginning with the form

$$s(t) = \text{Re}\left[v(t)e^{j2\pi f_c t}\right]$$

which relates the bandpass signal $s(t)$ to the equivalent lowpass signal $v(t)$, we may express the autocorrelation function of $s(t)$ as

$$\phi_{ss}(\tau) = \text{Re}\left[\phi_{vv}(\tau)e^{j2\pi f_c \tau}\right] \tag{3.4.1}$$

where $\phi_{vv}(\tau)$ is the autocorrelation function of the equivalent lowpass signal $v(t)$. The Fourier transform of (3.4.1) yields the desired expression for the power density spectrum $\Phi_{ss}(f)$ in the form

$$\Phi_{ss}(f) = \tfrac{1}{2}\left[\Phi_{vv}(f - f_c) + \Phi_{vv}(-f - f_c)\right] \tag{3.4.2}$$

where $\Phi_{vv}(f)$ is the power density spectrum of $v(t)$. It suffices to determine the autocorrelation function and the power density spectrum of the equivalent lowpass signal $v(t)$.

First we consider the digital modulation methods for which $v(t)$ is represented in the general form

$$v(t) = \sum_{n=-\infty}^{\infty} I_n u(t - nT) \tag{3.4.3}$$

where I_n represents the nth real- or complex-valued information symbol and $u(t)$ represents a pulse of arbitrary shape and duration which may be real or complex-valued. Thus (3.4.3) encompasses the modulation methods described in Sec. 3.1.1.

The autocorrelation function of $v(t)$ is

$$\phi_{vv}(t + \tau; t) = \tfrac{1}{2}E[v^*(t)v(t + \tau)]$$

$$= \frac{1}{2} \sum_{n=-\infty}^{\infty} \sum_{m=-\infty}^{\infty} E[I_n^* I_m] u^*(t - nT) u(t + \tau - mT) \quad (3.4.4)$$

We assume that the sequence of information symbols $\{I_n\}$ is wide-sense stationary with mean μ_i and autocorrelation function

$$\phi_{ii}(m) = \tfrac{1}{2}E[I_n^* I_{n+m}] \quad (3.4.5)$$

Hence (3.4.4) can be expressed as

$$\phi_{vv}(t + \tau; t) = \sum_{n=-\infty}^{\infty} \sum_{m=-\infty}^{\infty} \phi_{ii}(m - n) u^*(t - nT) u(t + \tau - mT)$$

$$= \sum_{m=-\infty}^{\infty} \phi_{ii}(m) \sum_{n=-\infty}^{\infty} u^*(t - nT) u(t + \tau - nT - mT) \quad (3.4.6)$$

The second summation in (3.4.6), namely,

$$\sum_{n=-\infty}^{\infty} u^*(t - nT) u(t + \tau - nT - mT)$$

is periodic in the t variable with period T. Consequently $\phi_{vv}(t + \tau; t)$ is also periodic in the t variable with period T. That is,

$$\phi_{vv}(t + T + \tau; t + T) = \phi_{vv}(t + \tau; t) \quad (3.4.7)$$

In addition, the mean value of $v(t)$, which is

$$E[v(t)] = \mu_i \sum_{n=-\infty}^{n} u(t - nT) \quad (3.4.8)$$

is periodic with period T. Therefore $v(t)$ is a stochastic process having a periodic mean and autocorrelation function. Such a process is called a *cyclostationary process* or a *periodically stationary process in the wide sense*, as described in Sec. 1.2.6.

In order to compute the power density spectrum of a cyclostationary process, the dependence of $\phi_{vv}(t + \tau; t)$ on the t variable must be eliminated. This can be accomplished simply by averaging $\phi_{vv}(t + \tau; t)$ over a single period. Thus

$$\bar{\phi}_{vv}(\tau) = \frac{1}{T} \int_{-T/2}^{T/2} \phi_{vv}(t + \tau; t) \, dt$$

$$= \sum_{m=-\infty}^{\infty} \phi_{ii}(m) \sum_{n=-\infty}^{\infty} \frac{1}{T} \int_{-T/2}^{T/2} u^*(t - nT) u(t + \tau - nT - mT) \, dt$$

$$= \sum_{m=-\infty}^{\infty} \phi_{ii}(m) \sum_{n=-\infty}^{\infty} \frac{1}{T} \int_{-T/2-nT}^{T/2-nT} u^*(t) u(t + \tau - mT) \, dt \quad (3.4.9)$$

We interpret the integral in (3.4.9) as the time-autocorrelation function of $u(t)$

and define it as

$$\phi_{uu}(\tau) = \int_{-\infty}^{\infty} u^*(t)u(t+\tau)\,dt \qquad (3.4.10)$$

Consequently (3.4.9) can be expressed as

$$\bar{\phi}_{vv}(\tau) = \frac{1}{T}\sum_{m=-\infty}^{\infty} \phi_{ii}(m)\phi_{uu}(\tau - mT) \qquad (3.4.11)$$

The Fourier transform of the relation in (3.4.11) yields the (average) power density spectrum of $v(t)$ in the form

$$\Phi_{vv}(f) = \frac{1}{T}|U(f)|^2 \Phi_{ii}(f) \qquad (3.4.12)$$

where $U(f)$ is the Fourier transform of $u(t)$, and $\Phi_{ii}(f)$ denotes the power density spectrum of the information sequence, defined as

$$\Phi_{ii}(f) = \sum_{m=-\infty}^{\infty} \phi_{ii}(m)e^{-j2\pi fmT} \qquad (3.4.13)$$

The result in (3.4.12) illustrates the dependence of the power density spectrum of $v(t)$ on the spectral characteristics of the pulse $u(t)$ and the information sequence $\{I_n\}$. That is, the spectral characteristics of $v(t)$ can be controlled by design of the pulse shape $u(t)$ and by design of the correlation characteristics of the information sequence.

Whereas the dependence of $\Phi_{vv}(f)$ on $U(f)$ is easily understood upon observation of (3.4.12), the effect of the correlation properties of the information sequence is more subtle. First of all, we note that for an arbitrary autocorrelation $\phi_{ii}(m)$ the corresponding power density spectrum $\Phi_{ii}(f)$ is periodic in frequency with period $1/T$. In fact, the expression in (3.4.13) relating the spectrum $\Phi_{ii}(f)$ to the autocorrelation $\phi_{ii}(m)$ is in the form of an exponential Fourier series with the $\{\phi_{ii}(m)\}$ as the Fourier coefficients. As a consequence, the autocorrelation sequence $\phi_{ii}(m)$ is given by

$$\phi_{ii}(m) = T\int_{-1/2T}^{1/2T} \Phi_{ii}(f)e^{j2\pi fmT}\,df \qquad (3.4.14)$$

Second, let us consider the case in which the information symbols in the sequence are real and mutually uncorrelated. In this case, the autocorrelation function $\phi_{ii}(m)$ can be expressed as

$$\phi_{ii}(m) = \begin{cases} \sigma_i^2 + \mu_i^2 & m = 0 \\ \mu_i^2 & m \neq 0 \end{cases} \qquad (3.4.15)$$

where σ_i^2 denotes the variance of an information symbol. When (3.4.15) is used to substitute for $\phi_{ii}(m)$ in (3.4.13), we obtain

$$\Phi_{ii}(f) = \sigma_i^2 + \mu_i^2 \sum_{m=-\infty}^{\infty} e^{-j2\pi fmT} \qquad (3.4.16)$$

The summation in (3.4.16) is periodic with period $1/T$. It may be viewed as the exponential Fourier series of a periodic train of impulses with each impulse having an area $1/T$. Therefore (3.4.16) can also be expressed in the form

$$\Phi_{ii}(f) = \sigma_i^2 + \frac{\mu_i^2}{T} \sum_{m=-\infty}^{\infty} \delta\left(f - \frac{m}{T}\right) \tag{3.4.17}$$

Substitution of the expression in (3.4.17) into (3.4.12) yields the desired result for the power density spectrum of $v(t)$ when the sequence of information symbols is uncorrelated. That is,

$$\Phi_{vv}(f) = \frac{\sigma_i^2}{T} |U(f)|^2 + \frac{\mu_i^2}{T^2} \sum_{m=-\infty}^{\infty} \left|U\left(\frac{m}{T}\right)\right|^2 \delta\left(f - \frac{m}{T}\right) \tag{3.4.18}$$

The expression for the power density spectrum in (3.4.18) is purposely separated into two terms to emphasize the two different types of spectral components. The first term is the continuous spectrum, and its shape depends only on the spectral characteristic of the signal pulse $u(t)$. The second term consists of discrete frequency components spaced $1/T$ apart in frequency. Each spectral line has a power that is proportional to $|U(f)|^2$ evaluated at $f = m/T$. We note that the discrete frequency components vanish when the information symbols have zero mean, i.e., $\mu_i = 0$. This condition is usually desirable for the digital modulation techniques under consideration and it is satisfied when the information symbols are equally likely and symmetrically positioned in the complex plane. Thus the system designer can control the spectral characteristics of the digitally modulated signal by proper selection of the characteristics of the information sequence to be transmitted.

Example 3.4.1. To illustrate the spectral shaping resulting from $u(t)$, consider the rectangular pulse shown in Fig. 3.4.1a. The Fourier transform of $u(t)$ is

$$U(f) = AT \frac{\sin \pi f T}{\pi f T} e^{-j\pi f T}$$

Hence

$$|U(f)|^2 = (AT)^2 \left(\frac{\sin \pi f T}{\pi f T}\right)^2 \tag{3.4.19}$$

This spectrum is illustrated in Fig. 3.4.1b. We note that it contains nulls at multiples of $1/T$ in frequency and that it decays inversely as the square of the frequency variable. As a consequence of the spectral nulls in $U(f)$, all but one of the discrete spectral components in (3.4.18) vanish. Thus, upon substitution for $|U(f)|^2$ from (3.4.19), (3.4.18) reduces to

$$\Phi_{vv}(f) = \sigma_i^2 A^2 T \left(\frac{\sin \pi f T}{\pi f T}\right)^2 + A^2 \mu_i^2 \delta(f) \tag{3.4.20}$$

Example 3.4.2. As a second illustration of the spectral shaping resulting from $u(t)$, we consider the raised cosine pulse

$$u(t) = \frac{A}{2}\left[1 + \cos\frac{2\pi}{T}\left(t - \frac{T}{2}\right)\right] \qquad 0 \le t \le T \tag{3.4.21}$$

$u(t)$

A

0 T t

(a)

$|U(f)|^2$

$(AT)^2$

$-\frac{3}{T}$ $-\frac{2}{T}$ $-\frac{1}{T}$ 0 $\frac{1}{T}$ $\frac{2}{T}$ $\frac{3}{T}$ f

(b)

FIGURE 3.4.1
Rectangular pulse and its energy density spectrum $|U(f)|^2$.

This pulse is graphically illustrated in Fig. 3.4.2a. Its Fourier transform is easily derived and it may be expressed in the form

$$U(f) = \frac{AT}{2} \frac{\sin \pi fT}{\pi fT(1 - f^2 T^2)} e^{-j\pi fT} \tag{3.4.22}$$

The square of the magnitude of $U(f)$ is shown in Fig. 3.4.2b. It is interesting to note that the spectrum has nulls at $f = n/T$, $n = \pm 2, \pm 3, \pm 4, \ldots$. Consequently all the discrete spectral components in (3.4.18), except the ones at $f = 0$ and $f = \pm 1/T$ vanish. When compared to the spectrum of the rectangular pulse, the spectrum of the raised cosine pulse has a broader main lobe but the tails decay inversely as f^6.

Example 3.4.3. To illustrate that spectral shaping can also be accomplished by operations performed on the input information sequence, we consider a binary sequence $\{a_n\}$ from which we form the symbols

$$I_n = a_n + a_{n-1} \tag{3.4.23}$$

The $\{a_n\}$ are assumed to be uncorrelated random variables, each having a zero

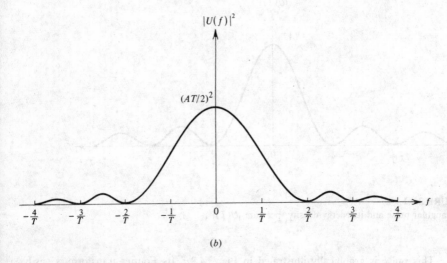

FIGURE 3.4.2
Raised cosine pulse and its energy density spectrum $|U(f)|^2$.

mean and a unit variance. Then the autocorrelation function of the sequence $\{I_n\}$ is

$$\phi_{ii}(m) = E(I_n I_{n+m})$$

$$= \begin{cases} 2 & m = 0 \\ 1 & m = \pm 1 \\ 0 & \text{otherwise} \end{cases} \tag{3.4.24}$$

Hence the power density spectrum of the input sequence is

$$\Phi_{ii}(f) = 2(1 + \cos 2\pi f T)$$

$$= 4\cos^2 \pi f T \tag{3.4.25}$$

and the corresponding power density spectrum for the (lowpass) modulated signal is

$$\Phi_{vv}(f) = \frac{4}{T} |U(f)|^2 \cos^2 \pi f T \tag{3.4.26}$$

3.4.2 Power Spectra of CPFSK and CPM Signals

In this section we derive the power density spectrum for the class of constant amplitude CPM signals that were described in Sec. 3.3.2. We begin by computing the autocorrelation function and its Fourier transform as was done in the case of linearly modulated signals.

The constant amplitude CPM signal is expressed as

$$s(t; \mathbf{I}) = A \cos \left[2\pi f_c t + \phi(t; \mathbf{I}) \right] \tag{3.4.27}$$

where

$$\phi(t; \mathbf{I}) = 2\pi h \sum_{k=-\infty}^{\infty} I_k q(t - kT) \tag{3.4.28}$$

Each symbol in the sequence $\{I_n\}$ can take one of the M values $\{\pm 1, \pm 3, \ldots, \pm(M-1)\}$. These symbols are statistically independent and identically distributed with prior probabilities

$$P_n = P(I_k = n) \qquad n = \pm 1, \pm 3, \ldots, \pm(M-1) \tag{3.4.29}$$

where $\sum P_n = 1$. The pulse $u(t) = q'(t)$ is zero outside of the interval $[0, LT]$, $q(t) = 0$, $t < 0$, and $q(t) = \frac{1}{2}$ for $t > LT$.

The autocorrelation function of the equivalent lowpass signal

$$v(t) = e^{j\phi(t; \mathbf{I})}$$

is

$$\phi_{vv}(t + \tau; t) = E\left\{ \exp\left(j 2\pi h \sum_{k=-\infty}^{\infty} I_k [q(t + \tau - kT) - q(t - kT)] \right) \right\} \tag{3.4.30}$$

First we express the sum in the exponent as a product of exponents. The result is

$$\phi_{vv}(t + \tau; t) = E\left\{ \prod_{k=-\infty}^{\infty} \exp\left(j 2\pi h I_k [q(t + \tau - kT) - q(t - kT)] \right) \right\} \tag{3.4.31}$$

Next, we perform the expectation over the data symbols $\{I_k\}$. Since these symbols are statistically independent, we obtain

$$\phi_{vv}(t + \tau; t)$$

$$= \prod_{k=-\infty}^{\infty} \left\{ \sum_{\substack{n=-(M-1) \\ n \text{ odd}}}^{M-1} P_n \exp\left(j 2\pi h n [q(t + \tau - kT) - q(t - kT)] \right) \right\} \tag{3.4.32}$$

Finally, the average autocorrelation function is

$$\bar{\phi}_{vv}(\tau) = \frac{1}{T} \int_0^T \phi_{vv}(t + \tau; t) \, dt \tag{3.4.33}$$

Although (3.4.32) implies that there are an infinite number of factors in the product, the pulse $u(t) = q'(t) = 0$ for $t < 0$ and $t > LT$, and $q(t) = 0$ for $t < 0$. Consequently, only a finite number of terms in the product have nonzero exponents. Thus, (3.4.32) can be simplified considerably. In addition, if we let $\tau = \xi + mT$ where $0 \le \xi < T$ and $m = 0, 1, \ldots$, the average autocorrelation in (3.4.33) reduces to

$$\bar{\phi}_{vv}(\xi + mT)$$

$$= \frac{1}{T} \int_0^T \prod_{k=1-L}^{m+1} \left\{ \sum_{\substack{n=-(M-1) \\ n \text{ odd}}}^{M-1} P_n \exp\left(j2\pi hn \left[q(t + \xi - (k-m)T) - q(t - kT) \right] \right) \right\}$$

(3.4.34)

Let us focus on $\bar{\phi}_{vv}(\xi + mT)$ for $\xi + mT \ge LT$. In this case, (3.4.34) may be expressed as

$$\bar{\phi}_{vv}(\xi + mT) = \left[\psi(jh) \right]^{m-N} \lambda(\xi) \qquad m \ge L, 0 \le \xi < T \quad (3.4.35)$$

where $\psi(jh)$ is the characteristic function of the random sequence $\{I_n\}$, defined as

$$\psi(jh) = E\left[e^{j\pi h I_n} \right]$$

$$= \sum_{\substack{n=-(M-1) \\ n \text{ odd}}}^{M-1} P_n e^{j\pi hn} \qquad (3.4.36)$$

and $\lambda(\xi)$ is the remaining part of the average autocorrelation function, which may be expressed as

$$\lambda(\xi) = \frac{1}{T} \int_0^T \prod_{k=1-L}^{0} \left\{ \sum_{\substack{n=-(M-1) \\ n \text{ odd}}}^{M-1} P_n \exp\left(j2\pi hn \left[\tfrac{1}{2} - q(t - kT) \right] \right) \right\}$$

$$\times \prod_{k=1-L}^{1} \left\{ \sum_{\substack{n=-(M-1) \\ n \text{ odd}}}^{M-1} P_n \exp\left(j2\pi hn\, q(t + \xi - kT) \right) \right\} dt \qquad m \ge L \quad (3.4.37)$$

Thus, $\bar{\phi}_{vv}(\tau)$ may be separated into a product of $\lambda(\xi)$ and $\psi(jh)$ as indicated in (3.4.35) for $\tau = \xi + mT \ge LT$ and $0 \le \xi < T$. This property is used below.

The Fourier transform of $\bar{\phi}_{vv}(\tau)$ yields the average power density spectrum as

$$\Phi_{vv}(f) = \int_{-\infty}^{\infty} \bar{\phi}_{vv}(\tau) e^{-j2\pi f\tau} \, d\tau$$

$$= 2 \operatorname{Re}\left\{ \int_0^{\infty} \bar{\phi}_{vv}(\tau) e^{-j2\pi f\tau} \, d\tau \right\} \qquad (3.4.38)$$

But

$$\int_0^\infty \bar{\phi}_{vv}(\tau) e^{-j2\pi h\tau} \, d\tau = \int_0^{LT} \bar{\phi}_{vv}(\tau) e^{-j2\pi f\tau} \, d\tau$$

$$+ \int_{LT}^\infty \bar{\phi}_{vv}(\tau) e^{-j2\pi f\tau} \, d\tau \tag{3.4.39}$$

With the aid of (3.4.35) the integral in the range $LT \le \tau < \infty$ may be expressed as

$$\int_{LT}^\infty \bar{\phi}_{vv}(\tau) e^{-2\pi f\tau} \, d\tau = \sum_{m=L}^\infty \int_{mT}^{(m+1)T} \bar{\phi}_{vv}(\tau) e^{-j2\pi f\tau} \, d\tau \tag{3.4.40}$$

Now, let $\tau = \xi + mT$. Then (3.4.40) becomes

$$\int_{LT}^\infty \bar{\phi}_{vv}(\tau) e^{-j2\pi f\tau} \, dt = \sum_{m=L}^\infty \int_0^T \bar{\phi}_{vv}(\xi + mT) e^{-j2\pi f(\xi+mT)} \, d\xi$$

$$= \sum_{m=L}^\infty \int_0^T \lambda(\xi) [\psi(jh)]^{m-L} e^{-j2\pi f(\xi+mT)} \, d\xi$$

$$= \sum_{n=0}^\infty \psi^n(jh) e^{-j2\pi fnT} \int_0^T \lambda(\xi) e^{-j2\pi f(\xi+LT)} \, d\xi \tag{3.4.41}$$

A property of the characteristic function is $|\psi(jh)| \le 1$. For values of h for which $|\psi(jh)| < 1$, the summation in (3.4.41) converges and yields

$$\sum_{n=0}^\infty \psi^n(jh) e^{-j2\pi fnT} = \frac{1}{1 - \psi(jh) e^{-j2\pi f T}} \tag{3.4.42}$$

In this case, (3.4.41) reduces to

$$\int_{LT}^\infty \bar{\phi}_{vv}(\tau) e^{-j2\pi f\tau} \, dt = \frac{1}{1 - \psi(jh) e^{-j2\pi f T}} \int_0^T \bar{\phi}_{vv}(\xi + LT) e^{-j2\pi f(\xi+LT)} \, d\xi \tag{3.4.43}$$

By combining (3.4.38), (3.4.39), and (3.4.43) we obtain the power density spectrum of the CPM signal in the form

$$\Phi_{vv}(f) = 2 \, \mathrm{Re} \left\{ \int_0^{LT} \bar{\phi}_{vv}(\tau) e^{-j2\pi f\tau} \, d\tau \right.$$

$$\left. + \frac{1}{1 - \psi(jh) e^{-j2\pi f T}} \int_{LT}^{(L+1)T} \bar{\phi}_{vv}(\tau) e^{-j2\pi f\tau} \, dt \right\} \tag{3.4.44}$$

This is the desired result when $|\psi(jh)| < 1$. In general the power density spectrum is evaluated numerically from (3.4.44). The average autocorrelation function $\bar{\phi}_{vv}(\tau)$ for the range $0 \le \tau \le (L+1)T$ may be computed numerically from (3.4.34).

For values of h for which $|\psi(jh)| = 1$, e.g., $h = K$, where K is an integer, we can set

$$\psi(jh) = e^{j2\pi\nu} \qquad 0 \le \nu < 1 \tag{3.4.45}$$

Then, the sum in (3.4.41) becomes

$$\sum_{n=0}^{\infty} e^{-j2\pi T(f-\nu/T)n} = \frac{1}{2} + \frac{1}{2T} \sum_{n=-\infty}^{\infty} \delta\left(f - \frac{\nu}{T} - \frac{n}{T}\right) - j\frac{1}{2}\cot \pi T\left(f - \frac{\nu}{T}\right)$$

(3.4.46)

Thus, the power density spectrum now contains impulses located at frequencies

$$f_n = \frac{n+\nu}{T} \qquad 0 \le \nu < 1, n = 0, 1, 2, \ldots \qquad (3.4.47)$$

The result in (3.4.46) can be combined with (3.4.41) and (3.4.39) to obtain the entire power density spectrum, which includes both a continuous spectrum component and a discrete spectrum component.

Let us return to the case for which $|\psi(jh)| < 1$. When the symbols are equally probable, i.e.,

$$P_n = \frac{1}{M} \quad \text{for all } n$$

the characteristic function simplifies to the form

$$\psi(jh) = \frac{1}{M} \sum_{\substack{n=-(M-1) \\ n \text{ odd}}}^{M-1} e^{j\pi hn}$$

$$= \frac{1}{M} \frac{\sin M\pi h}{\sin \pi h} \qquad (3.4.48)$$

Note that in this case $\psi(jh)$ is real. The average autocorrelation function given by (3.4.34) also simplifies in this case to

$$\bar{\phi}_{vv}(\tau) = \frac{1}{T} \int_0^T \prod_{k=1-L}^{[\tau/T]} \frac{1}{M} \frac{\sin 2\pi hM[q(t+\tau-kT) - q(t-kT)]}{\sin 2\pi h[q(t+\tau-kT) - q(t-kT)]} dt \quad (3.4.49)$$

The corresponding expression for the power density spectrum reduces to

$$\Phi_{vv}(f) = 2\left\{ \int_0^{T-} \bar{\phi}_{vv}(\tau) \cos 2\pi f\tau \, d\tau + \frac{1 - \psi(jh)\cos 2\pi fT}{1 + \psi^2(jh) - 2\psi(jh)\cos 2\pi fT} \right.$$

$$\left. \times \int_{LT}^{(L+1)T-} \bar{\phi}_{vv}(\tau)\cos 2\pi f\tau \, d\tau \right\} \quad (3.4.50)$$

Power density spectrum of CPFSK. A closed-form expression for the power density spectrum can be obtained from (3.4.50) when the pulse shape $u(t)$ is rectangular and zero outside the interval $[0, T]$. In this case $q(t)$ is linear for $0 \le t \le T$. The resulting power spectrum may be expressed as

$$\Phi_{vv}(f) = T\left[\frac{1}{M} \sum_{n=1}^{M} A_n^2(f) + \frac{2}{M^2} \sum_{n=1}^{M} \sum_{m=1}^{M} B_{nm}(f)A_n(f)A_m(f) \right] \quad (3.4.51)$$

where

$$A_n(f) = \frac{\sin \pi [fT - (2n - 1 - M)h/2]}{\pi [fT - (2n - 1 - M)h/2]}$$

$$B_{nm}(f) = \frac{\cos (2\pi fT - \alpha_{nm}) - \psi \cos \alpha_{nm}}{1 + \psi^2 - 2\psi \cos 2\pi fT}$$

$$\alpha_{nm} = \pi h(m + n - 1 - M)$$

$$\psi \equiv \psi(jh) = \frac{\sin M\pi h}{M \sin \pi h}$$

(3.4.52)

The power density spectrum of CPFSK for $M = 2$, 4, and 8 is plotted in Figs. 3.4.3, 3.4.4, and 3.4.5 as a function of the normalized frequency fT, with the modulation index $h = 2f_dT$ as a parameter. Note that only one-half of the bandwidth occupancy is shown in these graphs. The origin corresponds to the carrier f_c. The graphs illustrate that the spectrum of CPFSK is relatively smooth and well confined for $h < 1$. As h approaches unity, the spectra become very peaked and, for $h = 1$ when $|\psi| = 1$, we find that impulses occur at M frequencies. When $h > 1$ the spectrum becomes much broader. In communication systems where CPFSK is used, the modulation index is designed to conserve bandwidth, so that $h < 1$.

The special case of binary CPFSK with $h = \frac{1}{2}$ (or $f_d = 1/4T$) and $\psi = 0$ corresponds to MSK. In this case, the spectrum of the signal is

$$\Phi_{vv}(f) = \frac{16A^2T}{\pi^2} \left[\frac{\cos 2\pi fT}{1 - 16f^2T^2} \right]^2$$

(3.4.53)

where the signal amplitude $A = 1$ in (3.4.52). In contrast the spectrum of four-phase offset (quadrature) PSK (OQPSK) with a rectangular pulse $u(t)$ of duration T is

$$\Phi_{vv}(f) = A^2T \left(\frac{\sin \pi fT}{\pi fT} \right)^2$$

(3.4.54)

If we compare these spectral characteristics, we should normalize the frequency variable by the bit rate or the bit interval T_b. Since MSK is binary FSK, it follows that $T = T_b$ in (3.4.53). On the other hand, in OQPSK, $T = 2T_b$ so that (3.4.54) becomes

$$\Phi_{vv}(f) = 2A^2T_b \left(\frac{\sin 2\pi fT_b}{2\pi fT_b} \right)^2$$

(3.4.55)

The spectra of the MSK and OQPSK signals are illustrated in Fig. 3.4.6. Note that the main lobe of MSK is 50 percent wider than that for OQPSK. However, the side lobes in MSK fall off considerably faster. For example, if we compare the bandwidth W that contains 99 percent of the total power, we find that $W = 1.2/T_b$ for MSK and $W \approx 8/T_b$ for OQPSK. Consequently, MSK has a narrower spectral occupancy when viewed in terms of fractional out-of-band

FIGURE 3.4.3
Power density spectrum of binary CPFSK.

FIGURE 3.4.3 (*Continued*).

FIGURE 3.4.4
Power density spectrum of quaternary CPFSK.

Spectral density for four – level CPFSK

$h = 1.05$

$h = 2f_d T$

$h = 1.1$
$h = 1.2$
$h = 1.3$
$h = 1.4$

fT, normalized frequency

(c)

FIGURE 3.4.4 (*Continued*).

power above $fT_b = 1$. Graphs for the fractional out-of-band power for OQPSK and MSK are shown in Fig. 3.4.7. Note that MSK is significantly more bandwidth efficient than QPSK. This efficiency accounts for the popularity of MSK in many digital communications systems.

Even greater bandwidth efficiency than MSK can be achieved by reducing the modulation index. However, the FSK signals will no longer be orthogonal and there will be an increase in the error probability.

Spectral characteristics of CPM. In general, the bandwidth occupancy of CPM depends on the choice of the modulation index h, the pulse shape $u(t)$, and the number of signals M. As we have observed for CPFSK, small values of h result in CPM signals with relatively small bandwidth occupancy, while large values of h result in signals with large bandwidth occupancy. This is also the case for the more general CPM signals.

The use of smooth pulses such as raised cosine pulses of the form

$$u(t) = \begin{cases} \dfrac{1}{2LT}\left(1 - \cos\dfrac{2\pi t}{LT}\right) & 0 \le t \le LT \\ 0 & \text{otherwise} \end{cases} \qquad (3.4.56)$$

where $L = 1$ for full response and $L > 1$ for partial response, result in smaller bandwidth occupancy and, hence, greater bandwidth efficiency than the use of

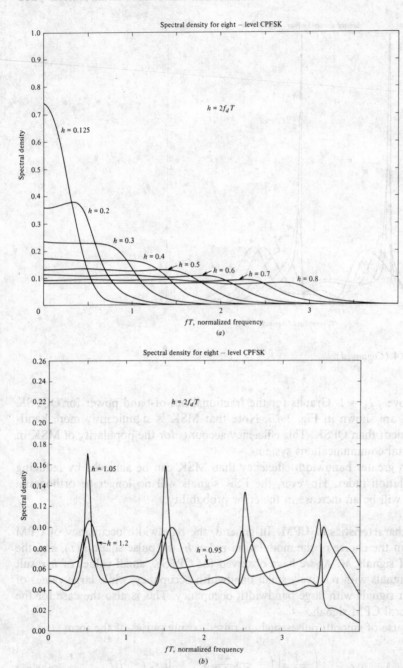

FIGURE 3.4.5
Power density spectrum of octal CPFSK.

FIGURE 3.4.6
Power density spectra of MSK and QPSK. [From Gronemeyer and McBride (1976); © 1976 IEEE; reprinted with permission.]

FIGURE 3.4.7
Fractional out-of-band power (normalized two-sided bandwidth = 2 BT). [From Gronemeyer and McBride (1976); © 1976 IEEE; reprinted with permission.]

FIGURE 3.4.8
Power density spectrum for binary CPM with $h = \frac{1}{2}$ and different pulse shapes. [From Aulin et al. (1981); © 1981 IEEE; reprinted with permission.]

rectangular pulses. For example, Fig. 3.4.8 illustrates the power density spectrum for binary CPM with different partial response raised cosine (LRC) pulses when $h = \frac{1}{2}$. For comparison, the spectrum of binary CPFSK is also shown. Note that as L increases the pulse $u(t)$ becomes smoother and the corresponding spectral occupancy of the signal is reduced.

The effect of varying the modulation index in a CPM signal is illustrated in Fig. 3.4.9 for the case of $M = 4$ and a raised cosine pulse of the form given in (3.4.56) with $L = 3$. Note that these spectral characteristics are similar to the ones illustrated previously for CPFSK, except that these spectra are narrower due to the use of a smoother pulse shape.

Finally, in Fig. 3.4.10 we illustrate the fractional out-of-band power for two-amplitude CPFSK with several different values of h.

3.4.3 Power Spectra of Modulated Signals with Memory

In the last two sections we have determined the spectral characteristics for the class of linearly modulated signals without memory and for the class of angle-modulated signals such as CPFSK and CPM, which are nonlinear and possess memory. In this section we consider the spectral characteristics of linearly modulated signals that have memory that can be modeled by a Markov chain. We have already encountered such signals in Sec. 3.3.3, were we described several types of baseband signals.

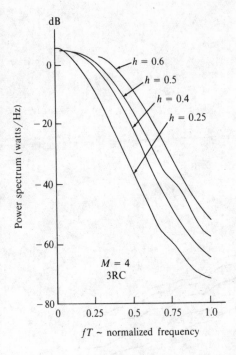

FIGURE 3.4.9
Power density spectrum for $M = 4$ CPM with $3RC$ and different modulation indices. [From Aulin et al. (1981); © 1981 IEEE; reprinted with permission.]

The power density spectrum of a digitally modulated signal that is generated by a Markov chain may be derived by following the basic procedure given in the previous section. Thus, we can determine the autocorrelation function and then evaluate its Fourier transform to obtain the power density spectrum. For signals that are generated by a Markov chain with transition probability matrix **P**, the power density spectrum of the modulated signal may be expressed in the general form [for reference, see Titsworth and Welch (1961)]

$$\Phi(f) = \frac{1}{T^2} \sum_{n=-\infty}^{\infty} \left| \sum_{i=1}^{K} p_i S_i\left(\frac{n}{T}\right) \right|^2 \delta\left(f - \frac{n}{T}\right) + \frac{1}{T} \sum_{i=1}^{K} p_i |S_i(f)|^2$$

$$+ \frac{2}{T} \text{Re}\left[\sum_{i=1}^{K} \sum_{j=1}^{K} p_i S_i^*(f) S_j(f) P_{ij}(f) \right] \qquad (3.4.57)$$

where $S_i(f)$ is the Fourier transform of the signal waveform $s_i(t)$ and $P_{ij}(f)$ is the Fourier transform of the discrete-time sequence $p_{ij}(n)$, defined as

$$P_{ij}(f) = \sum_{n=1}^{\infty} p_{ij}(n) e^{-j2\pi n f T} \qquad (3.4.58)$$

The term $p_{ij}(n)$ denotes the probability that the signal $s_j(t)$ is transmitted n signaling intervals after the transmission of $s_i(t)$. Hence, $\{p_{ij}(n)\}$ are the

FIGURE 3.4.10
Fractional out-of-band power for two-component CPFSK [Mulligan (1988)].

transition probability in the transition probability matrix $\mathbf{P}^n$. Note that $p_{ij}(1) = p_{ij}$.

When there is no memory in the modulation method, the signal waveform transmitted on each signaling interval is independent of the waveforms transmitted in previous signaling intervals. The power density spectrum of the resultant signal may still be expressed in the form of (3.4.57), if the transition probability matrix is replaced by

$$\mathbf{P} = \begin{bmatrix} p_i & p_2 & \cdots & p_K \\ p_1 & p_2 & \cdots & p_K \\ \vdots & & & \\ p_1 & p_2 & \cdots & p_K \end{bmatrix} \tag{3.4.59}$$

and we impose the condition that $\mathbf{P}^n = \mathbf{P}$ for all $n \geq 1$. Under these conditions the expression for the power density spectrum becomes a function of the stationary state probabilities $\{ p_i \}$ only and, hence, it reduces to the simpler form

$$\Phi(f) = \frac{1}{T^2} \sum_{n=-\infty}^{\infty} \left| \sum_{i=1}^{K} p_i S_i \left(\frac{n}{T} \right) \right|^2 \delta \left(f - \frac{n}{T} \right)$$

$$+ \frac{1}{T} \sum_{i=1}^{K} p_i (1 - p_i) \left| S_i(f) \right|^2$$

$$- \frac{2}{T} \sum_{i=1}^{K} \sum_{j=1}^{K} p_i p_j \, \text{Re} \left[S_i(f) S_j^*(f) \right] \qquad (3.4.60)$$

We observe that our previous result for the power density spectrum of memoryless linear modulation given by (3.4.18) may be viewed as a special case of (3.4.60) in which all waveforms are identical except for a set of scale factors that convey the digital information (Prob. 3.22).

We also make the observation that the first term in the expression for the power density spectrum given by either (3.4.57) or (3.4.60) consists of discrete frequency components. This line spectrum vanishes when

$$\sum_{i=1}^{K} p_i S_i \left(\frac{n}{T} \right) = 0 \qquad (3.4.61)$$

The condition in (3.4.61) is usually imposed in the design of practical digital communications systems and is easily satisfied by an appropriate choice of signaling waveforms (Prob. 3.23).

Now, let us determine the power density spectrum of the baseband-modulated signals described in Sec. 3.3.3. First, the NRZ signal is characterized by the two waveforms $s_1(t) = s(t)$ and $s_2(t) = -s(t)$, where $s(t)$ is a rectangular pulse of amplitude A. For $K = 2$, (3.4.60) reduces to

$$\Phi(f) = \frac{(2p - 1)^2}{T^2} \sum_{n=-\infty}^{\infty} \left| S \left(\frac{n}{T} \right) \right|^2 \delta \left(f - \frac{n}{T} \right) + \frac{4p(1 - p)}{T} \left| S(f) \right|^2 \quad (3.4.62)$$

where

$$\left| S(f) \right|^2 = (AT)^2 \left(\frac{\sin \pi f T}{\pi f T} \right)^2 \qquad (3.4.63)$$

Observe that when $p = \frac{1}{2}$, the line spectrum vanishes and $\Phi(f)$ reduces to

$$\Phi(f) = \frac{1}{T} \left| S(f) \right|^2 \qquad (3.4.64)$$

The biphase signal, also called a Manchester-coded signal, is a memoryless binary form of modulation for which $s_1(t) = -s_2(t) = s(t)$, where $s(t)$ is shown in Fig. 3.3.21. Hence, its power density spectrum is given by (3.4.62), where the spectrum of the signal $s(t)$ is

$$\left| S(f) \right|^2 = (AT)^2 \left[\frac{\sin^2 \pi f T/2}{\pi f T/2} \right]^2 \qquad (3.4.65)$$

The NRZI signal is characterized by the transition probability matrix

$$\mathbf{P} = \begin{bmatrix} \frac{1}{2} & \frac{1}{2} \\ \frac{1}{2} & \frac{1}{2} \end{bmatrix} \qquad (3.4.66)$$

Notice that in this case $\mathbf{P}^n = P$ for all $n \geq 1$. Hence, the special form for the power density spectrum given by (3.4.62) applies to this modulation format as well. Consequently, the power density spectrum for the NRZI signal is identical to the spectrum of the NRZ signal.

Delay modulation has a transition probability matrix

$$P = \begin{bmatrix} 0 & \frac{1}{2} & 0 & \frac{1}{2} \\ 0 & 0 & \frac{1}{2} & \frac{1}{2} \\ \frac{1}{2} & \frac{1}{2} & 0 & 0 \\ \frac{1}{2} & 0 & \frac{1}{2} & 0 \end{bmatrix} \qquad (3.4.67)$$

and stationary state probabilities $p_i = \frac{1}{4}$ for $i = 1, 2, 3, 4$. Powers of $\mathbf{P}$ may be obtained by use of the relation

$$\mathbf{P}^4 \rho = \tfrac{1}{4}\rho \qquad (3.4.68)$$

where ρ is the signal correlation matrix with elements

$$\rho_{ij} = \frac{1}{T} \int_0^T s_i(t) s_j(t)\, dt \qquad (3.4.69)$$

and where the four signals $\{s_i(t), \ i = 1, 2, 3, 4\}$ are shown in Fig. 3.3.22. It is easily seen that

$$\rho = \begin{bmatrix} 1 & 0 & 0 & -1 \\ 0 & 1 & -1 & 0 \\ 0 & -1 & 1 & 0 \\ -1 & 0 & 0 & 1 \end{bmatrix} \qquad (3.4.70)$$

Consequently, powers of ρ can be generated from the relation

$$\mathbf{P}^{k+4}\rho = \mathbf{P}^k\rho \qquad k > 1 \qquad (3.4.71)$$

Use of (3.4.67), (3.4.70), and (3.4.71) in (3.4.57) yields the power density spectrum of the Miller code. It may be expressed in the form

$$\Phi(f) = \frac{1}{2\psi^2(17 + 8\cos 8\psi)}[23 - 2\cos\psi - 22\cos 2\psi - 12\cos 3\psi + 5\cos 4\psi$$

$$+ 12\cos 5\psi + 2\cos 6\psi - 8\cos 7\psi + 2\cos 8\psi] \quad (3.4.72)$$

where $\psi = \pi f T$.

The spectra of these four baseband signals are illustrated in Fig. 3.4.11. Observe that the spectra of the NRZ and NRZI signals peak at $f = 0$, whereas the biphase signal has no power at $f = 0$, and peaks in the vicinity of $fT = 0.7$. However, its spectrum is very wide. The Miller code has the narrowest spectrum and a relatively small dc content. Its bandwidth occupancy is approximately

FIGURE 3.4.11
Power spectral density (one-sided) of Miller code (delay modulation), biphase and NRZ/NRZI baseband signals. [From Hecht and Guida (1969); © 1969 IEEE; reprinted with permission.]

one-half of that for the biphase signal. These two characteristics make the Miller code an attractive choice for channels that do not pass dc, such as magnetic recording media.

3.5 BIBLIOGRAPHICAL NOTES AND REFERENCES

The characteristics of signals and systems given in this chapter are very useful in the design of optimum modulation/demodulation and coding/decoding techniques for a variety of channel models. In particular, the digital modulation methods introduced in this chapter are considered in greater detail in Chap. 4, which is concerned with optimum demodulation techniques and their performance in an additive, white gaussian noise channel. A general reference for signal characterization is the book by Franks (1969).

Of particular importance in the design of digital communications systems are the spectral characteristics of the digitally modulated signals, which are presented in this chapter in some depth. Of these modulation techniques, CPM is one of the most important due to its efficient use of bandwidth. For this reason, it has been widely investigated by many researchers and a large number of papers have been published in the technical literature. The most comprehensive treatment of CPM, including its performance and its spectral characteristics, can be found in the book by Anderson et al. (1986). In addition to this text, the tutorial paper by Sundberg (1986) presents the basic concepts and an overview of the performance characteristics of various CPM techniques. This paper also contains over 100 references to published papers on this topic.

There are a large number of references dealing with the spectral characteristics of CPFSK and CPM. As a point of reference we should mention that MSK was invented by Doelz and Heald in 1961. The early work on the power density spectrum of CPFSK and CPM was done by Bennett and Rice (1963), Anderson and Salz (1965), and Bennett and Davey (1965). The book by Lucky et al. (1968) also contains a treatment of the spectral characteristics of CPFSK. Most of the recent work is referenced in the paper by Sundberg (1986). We should also cite the special issue on bandwidth-efficient modulation and coding published by the *IEEE Transactions on Communications* (March 1981), which contains several papers on the spectral characteristics and performance of CPM.

The generalization of MSK to multiple amplitudes was investigated by Weber et al. (1978). The combination of multiple amplitudes with more general CPM was proposed by Mulligan (1988) who investigated its spectral characteristics and its error rate performance in gaussian noise with and without coding.

PROBLEMS

3.1. Prove the following properties of Hilbert transforms:
   (a) If $x(t) = x(-t)$, then $\hat{x}(t) = -\hat{x}(-t)$.
   (b) If $x(t) = -x(-t)$, then $\hat{x}(t) = \hat{x}(-t)$.
   (c) If $x(t) = \cos \omega_0 t$, then $\hat{x}(t) = \sin \omega_0 t$.
   (d) If $x(t) = \sin \omega_0 t$, then $\hat{x}(t) = -\cos \omega_0 t$.
   (e) $\hat{\hat{x}}(t) = -x(t)$.
   (f) $\int_{-\infty}^{\infty} x^2(t)\, dt = \int_{-\infty}^{\infty} \hat{x}^2(t)\, dt$.
   (g) $\int_{-\infty}^{\infty} x(t)\hat{x}(t)\, dt = 0$.

3.2. If $x(t)$ is a stationary random process with autocorrelation function $\phi_{xx}(\tau) = E[x(t)x(t+\tau)]$ and spectral density $\Phi_{xx}(f)$, then show that $\phi_{\hat{x}\hat{x}}(\tau) = \phi_{xx}(\tau)$, $\phi_{x\hat{x}}(\tau) = -\hat{\phi}_{xx}(\tau)$, $\Phi_{\hat{x}\hat{x}}(f) = \Phi_{xx}(f)$.

3.3. Suppose that $n(t)$ is a zero mean stationary narrowband process which is represented by either (3.1.27), (3.1.28), or (3.1.29). The autocorrelation function of the equivalent lowpass process $z(t) = x(t) + jy(t)$ is defined as

$$\phi_{zz}(\tau) = \tfrac{1}{2}E[z^*(t)z(t+\tau)]$$

(a) Show that

$$E[z(t)z(t+\tau)] = 0$$

(b) Suppose $\phi_{zz}(\tau) = N_0\delta(\tau)$, and let

$$V = \int_0^T z(t)\, dt$$

Determine $E(V^2)$ and $E(VV^*) = E(|V|^2)$.

3.4. Determine the autocorrelation function of the stochastic process

$$x(t) = A\sin(2\pi f_c t + \theta)$$

where f_c is a constant and θ is a uniformly distributed phase, i.e.,

$$p(\theta) = \frac{1}{2\pi} \qquad 0 \le \theta \le 2\pi$$

3.5. A lowpass gaussian stochastic process $x(t)$ has a power spectral density

$$\Phi(f) = \begin{cases} N_0 & |f| < B \\ 0 & |f| > B \end{cases}$$

Determine the power spectral density and the autocorrelation function of $y(t) = x^2(t)$.

3.6. Consider an equivalent lowpass digitally modulated signal of the form

$$u(t) = \sum_n [a_n g(t - 2nT) - jb_n g(t - 2nT - T)]$$

where $\{a_n\}$ and $\{b_n\}$ are two sequences of statistically independent binary digits and $g(t)$ is a sinusoidal pulse defined as

$$g(t) = \begin{cases} \sin \dfrac{\pi t}{2T} & 0 < t < 2T \\ 0 & \text{otherwise} \end{cases}$$

This type of signal is viewed as a four-phase PSK signal in which the pulse shape is one-half cycle of a sinusoid. Each of the information sequences $\{a_n\}$ and $\{b_n\}$ is transmitted at a rate of $1/2T$ bits/s and, hence, the combined transmission rate is $1/T$ bits/s. The two sequences are staggered in time by T seconds in transmission. Consequently the signal $u(t)$ is called *staggered four-phase PSK*.

(a) Show that the envelope $|u(t)|$ is a constant, independent of the information a_n on the in-phase component and information b_n on the quadrature component. In other words, the amplitude of the carrier used in transmitting the signal is constant.

(b) Determine the power density spectrum of $u(t)$.

(c) Compare the power density spectrum obtained from (b) with the power density spectrum of the MSK signal. What conclusion can you draw from this comparison?

3.7. Consider a four-phase PSK signal which is represented by the equivalent lowpass signal

$$u(t) = \sum_n I_n g(t - nT)$$

where I_n takes on one of the four possible values $(\pm 1 \pm j)/\sqrt{2}$ with equal probability. The sequence of information symbols $\{I_n\}$ is statistically independent.

(a) Determine and sketch the power density spectrum of $u(t)$ when

$$g(t) = \begin{cases} A & 0 \le t \le T \\ 0 & \text{otherwise} \end{cases}$$

(b) Repeat (a) when

$$g(t) = \begin{cases} A \sin \dfrac{\pi t}{T} & 0 \le t \le T \\ 0 & \text{otherwise} \end{cases}$$

(c) Compare the spectra obtained in (a) and (b) in terms of the 3-dB bandwidth and the bandwidth to the first spectral null.

3.8. The random process $v(t)$ is defined as

$$v(t) = X \cos 2\pi f_c t - Y \sin 2\pi f_c t$$

where X and Y are random variables. Show that $v(t)$ is wide-sense stationary if and only if $E(X) = E(Y) = 0$, $E(X^2) = E(Y^2)$, and $E(XY) = 0$.

216 DIGITAL COMMUNICATIONS

3.9. Carry out the Gram–Schmidt orthogonalization of the signals in Fig. 3.2.1a in the order $s_4(t)$, $s_3(t)$, $s_1(t)$, and, thus, obtain a set of orthonormal functions $\{g_m(t)\}$. Then, determine the vector representation of the signals $\{s_n(t)\}$ by using the orthonormal functions $\{g_m(t)\}$. Also, determine the signal energies.

3.10. Determine the signal space representation of the four signals $s_k(t)$, $k = 1, 2, 3, 4$, shown below, by using as basic functions the orthonormal functions $f_1(t)$ and $f_2(t)$. Plot the signal space diagram and show that this signal set is equivalent to that for a four-phase PSK signal.

FIGURE P3.10

3.11. The power density spectrum of the cyclostationary process

$$v(t) = \sum_{n=-\infty}^{\infty} I_n u(t - nT)$$

was derived in Sec. 3.4.1 by averaging the autocorrelation function $\phi_{vv}(t + \tau, t)$ over the period T of the process and then evaluating the Fourier transform of the average autocorrelation function. An alternative approach is to change the cyclostationary process into a stationary process $v_\Delta(t)$ by adding a random variable Δ, uniformly distributed over $0 \le \Delta < T$, so that

$$v_\Delta(t) = \sum_{n=-\infty}^{\infty} I_n u(t - nT - \Delta)$$

and defining the spectral density of $v(t)$ as the Fourier transform of the autocorrelation function of the stationary process $v_\Delta(t)$. Derive the result in (3.4.1) by evaluating the autocorrelation function of $v_\Delta(t)$ and its Fourier transform.

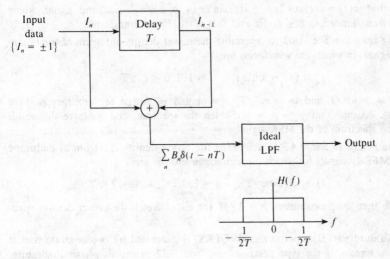

FIGURE P3.12

3.12. A PAM partial response signal (PRS) is generated as shown in Fig. P3.12 by exciting an ideal lowpass filter of bandwidth W by the sequence

$$B_n = I_n + I_{n-1}$$

at a rate $1/T = 2W$ symbols per second. The sequence $\{I_n\}$ consists of binary digits selected independently from the alphabet $\{1, -1\}$ with equal probability. Hence, the filtered signal has the form

$$v(t) = \sum_{n=-\infty}^{\infty} B_n u(t - nT) \qquad T = 1/2W$$

(a) Sketch the signal space diagram for $v(t)$ and determine the probability of occurrence of each symbol.

(b) Determine the autocorrelation and power density spectrum of the three-level sequence $\{B_n\}$.

(c) The signal points of the sequence $\{B_n\}$ form a Markov chain. Sketch this Markov chain and indicate the transition probabilities among the states.

3.13. The lowpass equivalent representation of a PAM signal is

$$u(t) = \sum_n I_n g(t - nT)$$

Suppose $g(t)$ is a rectangular pulse and

$$I_n = a_n - a_{n-2}$$

where $\{a_n\}$ is a sequence of uncorrelated binary-valued $(1, -1)$ random variables which occur with equal probability.

(a) Determine the autocorrelation function of the sequence $\{I_n\}$.

(b) Determine the power density spectrum of $u(t)$.

(c) Repeat (b) if the possible values of the a_n are $(0, 1)$.

3.14. Show that $x(t) = s(t)\cos 2\pi f_c \pm \hat{s}(t)\sin 2\pi f_c t$ is a single-sideband signal, where $s(t)$ is band-limited to $B \leq f_c$ Hz and $\hat{s}(t)$ is its Hilbert transform.

3.15. Use the results in Sec. 3.4.3 to determine the power density spectrum of the binary FSK signals in which the waveforms are

$$s_i(t) = \sin \omega_i t \qquad i = 1, 2, 0 \leq t \leq T$$

where $\omega_1 = n\pi/T$ and $\omega_2 = mT/T$, $n \neq m$ and m and n are arbitrary positive integers. Assume that $p_1 = p_2 = \frac{1}{2}$. Sketch the spectrum and compare this result with the spectrum of the MSK signal.

3.16. Use the results in Sec. 3.4.3 to determine the power density spectrum of multitone FSK (MFSK) signals for which the signals waveforms are

$$s_n(t) = \sin 2\pi nt/T \qquad n = 1, 2, \ldots, M, 0 \leq t \leq T$$

Assume that the probabilities $p_i = 1/M$ for all i. Sketch the power density spectrum.

3.17. A quadrature partial response signal (QPRS) is generated by two separate partial response signals of the type described in Prob. 3.12 placed in phase quadrature. Hence, the QPRS is represented as

$$s(t) = \text{Re}\left[v(t)e^{j\pi f_c t}\right]$$

where

$$v(t) = v_c(t) + jv_s(t)$$

$$= \sum_n B_n u(t - nT) + j\sum_n C_n u(t - nT)$$

and $B_n = I_n + I_{n-1}$ and $C_n = J_n + J_{n-1}$. The sequences $\{B_n\}$ and $\{C_n\}$ are uncorrelated and $I_n = \pm 1$, $J_n = \pm 1$ with equal probability.

    (a) Sketch the signal space diagram for the QPRS signal and determine the probability of occurrence of each symbol.

    (b) Determine the autocorrelation and power density spectrum of $v_c(t)$, $v_s(t)$, and $v(t)$.

    (c) Sketch the Markov chain model and indicate the transition probabilities for the QPRS.

3.18. Determine the autocorrelation functions for the MSK and offset QPSK modulated signals based on the assumption that the information sequences for each of the two signals are uncorrelated and zero mean.

3.19. Sketch the phase-free, the state trellis, and the state diagram for partial response CPM with $h = \frac{1}{2}$ and

$$u(t) = \begin{cases} \dfrac{1}{4T} & 0 \leq t \leq 2T \\ 0 & \text{otherwise} \end{cases}$$

3.20. Determine the number of terminal phase states in the state trellis diagram for (a) a full response binary CPFSK with either $h = \frac{2}{3}$ or $\frac{3}{4}$ and (b) a partial response $L = 3$ binary CPFSK with either $h = \frac{2}{3}$ or $\frac{3}{4}$.

3.21. Show that 16 QAM can be represented as a superposition of two four-phase constant envelope signals where each component is amplified separately before summing, i.e.,

$$s(t) = [A_n \cos \pi f_c t + B_n \sin 2\pi f_c t] + [C_n \cos 2\pi f_c t + D_n \sin 2\pi f_c t]$$

where $\{A_n\}$, $\{B_n\}$, $\{C_n\}$, and $\{D_n\}$ are statistically independent binary sequences with elements from the set $\{+1, -1\}$. Thus, show that the resulting signal is equivalent to

$$s(t) = I_n \cos 2\pi f_c t + Q_n \sin 2\pi f_c t$$

and determine I_n and Q_n in terms of A_n, B_n, C_n, and D_n.

3.22. Use the result in (3.4.60) to derive the expression for the power density spectrum of memoryless linear modulation given by (3.4.18) under the condition that

$$s_k(t) = I_k s(t) \qquad k = 1, 2, \ldots, K$$

where I_k is one of the K possible transmitted symbols that occur with equal probability.

3.23. Show that a necessary and sufficient condition for the absence of the line spectrum component in (3.4.60) is

$$\sum_{i=1}^{K} p_i s_i(t) = 0$$

CHAPTER
4

MODULATION AND DEMODULATION FOR THE ADDITIVE GAUSSIAN NOISE CHANNEL

In Chap. 2, we briefly described the channel encoder and found it convenient to separate it into two parts, the discrete channel encoder, which introduces redundancy in the information sequence, and the digital modulator. Similarly, the channel decoder is separated into a digital demodulator followed by a discrete channel decoder.

This chapter is concerned with the design and performance characteristics of the modulator and demodulator for an additive gaussian noise channel. The design and performance of the discrete channel encoder and decoder are treated in Chap. 5.

As previously stated, the modulator performs the function of mapping the digital sequence into waveforms that are appropriate for the waveform channel. In turn, the demodulator processes the channel-corrupted received waveforms and presents to the decoder a set of real numbers resulting from the waveform processing. In the absence of a discrete channel encoder and decoder, the demodulator also performs the inverse mapping from the set of real numbers to the digital sequence.

The digital information to be transmitted over the channel is assumed to be a sequence of binary digits occurring at a uniform rate of R bits per second. Each binary digit may be transmitted directly by sending either a waveform $s_1(t)$

corresponding to a binary digit 0 or a waveform $s_2(t)$ corresponding to a binary digit 1. This type of transmission is termed *binary signaling*.

As an alternative to transmitting the binary digits directly, the information sequence can be subdivided into groups or blocks, where each block consists of k bits. With k bits per block, there are $2^k = M$ distinct blocks and, hence, we require M different waveforms in order to transmit the k-bit blocks unambiguously. For example, the M waveforms may differ either in amplitude as in pulse amplitude modulation (PAM), in phase as in phase-shift keying (PSK), in frequency as in frequency-shift keying (FSK), or in both amplitude and phase (PAM-PSK). Since the binary digits occur at a rate R, the time interval or signaling interval corresponding to k-bit blocks is $T = k/R$ seconds.

In this chapter we limit our treatment to digital modulation methods that are memoryless. Modulation techniques such as continuous-phase modulation (CPM) are treated in Chap. 6 where we consider communication over band-limited channels. The emphasis of our discussion is on the evaluation of the performance of the communications system that employs different types of signals for transmitting information over an additive, white gaussian noise channel. We begin by describing in more detail the characteristics of the signal waveforms introduced in Chap. 3 for transmission of the information sequence.

4.1 CHARACTERIZATION OF THE SIGNAL WAVEFORMS

In a general M-ary signaling system, the waveforms used to transmit the information will be denoted as $\{s_m(t),\ m = 1, \ldots, M\}$. Usually these waveforms are bandpass and, hence, they are represented as

$$s_m(t) = \text{Re}\left[u_m(t)e^{j2\pi f_c t}\right] \qquad m = 1, 2, \ldots, M, 0 \le t \le T \quad (4.1.1)$$

where $\{u_m(t)\}$ denote the equivalent low-pass waveforms. The M signals are characterized individually by their energy, defined as

$$\mathscr{E}_m = \int_0^T s_m^2(t)\, dt$$

$$= \frac{1}{2} \int_0^T |u_m(t)|^2\, dt \qquad m = 1, 2, \ldots, M \quad (4.1.2)$$

and mutually by their complex-valued cross-correlation coefficients

$$\rho_{jm} = \frac{1}{2\sqrt{\mathscr{E}_m \mathscr{E}_j}} \int_0^T u_m(t) u_j^*(t)\, dt \quad (4.1.3)$$

In terms of the bandpass signal waveforms, we have

$$\text{Re}\left[\rho_{jm}\right] = \frac{1}{\sqrt{\mathscr{E}_m \mathscr{E}_j}} \int_0^T s_m(t) s_j(t)\, dt \quad (4.1.4)$$

The M signal waveforms may also be expressed as linear combinations of a set of orthonormal functions $\{f_n(t)\}$, as previously described in Sec. 3.2. Thus, the waveforms $\{s_m(t)\}$ may be expressed as

$$s_m(t) = \sum_{n=1}^{N} s_{mn} f_n(t) \qquad m = 1, 2, \ldots, M, \, 0 \le t \le T \qquad (4.1.5)$$

where

$$s_{mn} = \int_0^T s_m(t) f_n(t) \, dt \qquad m = 1, 2, \ldots, M, \, n = 1, 2, \ldots, N \qquad (4.1.6)$$

and N is the dimensionality of the signals.

The signal energy $\mathscr{E}_m$ may also be expressed in terms of the signal vector $\mathbf{s}_m = [s_{m1}, s_{m2}, \ldots, s_{mN}]$ as

$$
\begin{aligned}
\mathscr{E}_m &= \int_0^T s_m^2(t) \, dt \\
&= \int_0^T \sum_{n=1}^{N} \sum_{i=1}^{N} s_{mn} s_{mi} f_n(t) f_i(t) \, dt \\
&= \sum_{n=1}^{N} s_{mn}^2 = |\mathbf{s}_m|^2
\end{aligned}
\qquad (4.1.7)
$$

Similarly, the cross-correlation coefficients ρ_{km} may be expressed in terms of the vectors $\mathbf{s}_k = [s_{k1}, s_{k2}, \ldots, s_{kN}]$ and $\mathbf{s}_m = [s_{m1}, s_{m2}, \ldots, s_{mN}]$ as

$$
\begin{aligned}
\mathrm{Re}\,[\rho_{km}] &= \frac{1}{\sqrt{\mathscr{E}_k \mathscr{E}_m}} \int_0^T s_m(t) s_k(t) \, dt \\
&= \frac{1}{\sqrt{\mathscr{E}_k \mathscr{E}_m}} \int_0^T \sum_{n=1}^{N} \sum_{i=1}^{N} s_{mn} s_{ki} f_n(t) f_i(t) \, dt \\
&= \frac{1}{\sqrt{|\mathbf{s}_m|^2 |\mathbf{s}_k|^2}} \sum_{n=1}^{N} s_{mn} s_{kn} \\
&= \frac{\mathbf{s}_m \cdot \mathbf{s}_k}{|\mathbf{s}_m|\,|\mathbf{s}_k|}
\end{aligned}
\qquad (4.1.8)
$$

where $\mathbf{s}_m \cdot \mathbf{s}_k$ denotes the dot product of the two signal vectors.

The analogous expressions for the signal energy and correlation coefficients expressed in terms of the complex-valued, equivalent lowpass signal vectors $\{\mathbf{u}_m\}$ are

$$
\begin{aligned}
\mathscr{E}_m &= \tfrac{1}{2} |\mathbf{u}_m|^2 \qquad m = 1, 2, \ldots, M \\
\rho_{km} &= \frac{\mathbf{u}_m \cdot \mathbf{u}_k^*}{|\mathbf{u}_m|\,|\mathbf{u}_k|}
\end{aligned}
\qquad (4.1.9)
$$

The cross-correlation coefficients between pairs of signal waveforms or signal vectors is one set of parameters for characterizing the similarity or dissimilarity of the signals. Another related parameter is the Euclidean distance

d_{km} between a pair of signals, de ined as

$$d_{km}^2 = |\mathbf{s}_m - \mathbf{s}_k|^2$$

$$= \int_0^T [s_m(t) - s_k(t)]^2 \, dt$$

$$= \mathscr{E}_m + \mathscr{E}_k - 2\sqrt{\mathscr{E}_m \mathscr{E}_k} \, \text{Re}\,(\rho_{km}) \qquad (4.1.10)$$

When $\mathscr{E}_m = \mathscr{E}_k = \mathscr{E}$ for all m and k, (4.1.10) simplifies to

$$d_{km}^2 = 2\mathscr{E}[1 - \text{Re}\,(\rho_{km})] \qquad (4.1.11)$$

Thus, the Euclidean distance is an alternative measure of the similarity or dissimilarity of the set of signal waveforms or the corresponding signal vectors.

Binary antipodal signals. Consider the two signals

$$s_1(t) = -s_2(t) = \sqrt{\frac{2\mathscr{E}}{T}} \cos 2\pi f_c t \qquad 0 \le t \le T \qquad (4.1.12)$$

These are one-dimensional signals for which the equivalent lowpass waveforms are rectangular pulses, i.e.,

$$u_1(t) = -u_2(t) = \sqrt{\frac{2\mathscr{E}}{T}} \qquad 0 \le t \le T \qquad (4.1.13)$$

The corresponding vector representation is $\mathbf{s}_1 = [s_{11}]$ and $\mathbf{s}_2 = [s_{21}]$ where $s_{11} = -s_{21} = \sqrt{\mathscr{E}}$, or equivalently, $\mathbf{u}_1 = [u_{11}]$ and $\mathbf{u}_2 = [u_{21}]$, where $u_{11} = \sqrt{2\mathscr{E}} = -u_{21}$. Figure 4.1.1 illustrates the signal space diagram. The cross-correlation coefficient is

$$\text{Re}\,(\rho_{12}) = \frac{\mathbf{s}_1 \cdot \mathbf{s}_2}{|\mathbf{s}_1|\,|\mathbf{s}_2|} = -1 \qquad (4.1.14)$$

and the corresponding Euclidean distance is

$$d_{12} = \{2\mathscr{E}[1 - \text{Re}\,(\rho_{12})]\}^{1/2} = 2\sqrt{\mathscr{E}} \qquad (4.1.15)$$

Two signals with a cross-correlation coefficient of -1 are called *antipodal*.

FIGURE 4.1.1
Signal-space diagram for binary antipodal signals.

Binary orthogonal signals. Consider the two signals

$$s_1(t) = \sqrt{\frac{2\mathscr{E}}{T}} \cos 2\pi f_c t \qquad 0 \le t \le T$$

$$s_2(t) = \sqrt{\frac{2\mathscr{E}}{T}} \sin 2\pi f_c t \qquad 0 \le t \le T$$

(4.1.16)

where either $f_c = 1/T$ or $f_c \gg 1/T$, so that

$$\mathrm{Re}\,(\rho_{12}) = \frac{1}{\mathscr{E}} \int_0^T s_1(t) s_2(t)\, dt = 0$$

(4.1.17)

Since $\mathrm{Re}\,(\rho_{12}) = 0$, the two signals are *orthogonal*.

The equivalent lowpass waveforms corresponding to $s_1(t)$ and $s_2(t)$ are

$$u_1(t) = \sqrt{\frac{2\mathscr{E}}{T}} \qquad 0 \le t \le T$$

$$u_2(t) = -j\sqrt{\frac{2\mathscr{E}}{T}} \qquad 0 \le t \le T$$

(4.1.18)

Clearly, these signal waveforms can be represented as two-dimensional signal vectors, i.e.,

$$\mathbf{s}_1 = (\sqrt{\mathscr{E}}, 0) \qquad \mathbf{s}_2 = (0, -\sqrt{\mathscr{E}})$$

(4.1.19)

which correspond to the signal space diagram shown in Fig. 4.1.2. Note that $d_{12} = \sqrt{2\mathscr{E}}$.

We observe that the vector representation for the equivalent lowpass signals is $\mathbf{u}_1 = [u_{11}]$ and $\mathbf{u}_2 = [u_{21}]$ where

$$u_{11} = \sqrt{2\mathscr{E}} + j0$$

$$u_{21} = 0 - j\sqrt{2\mathscr{E}}$$

(4.1.20)

FIGURE 4.1.2
Signal-space diagram for binary orthogonal signals.

Hence,

$$\rho_{12} = \frac{\mathbf{u}_1^* \cdot \mathbf{u}_2}{|\mathbf{u}_1| \, |\mathbf{u}_2|} = -j = e^{-j\pi/2} \tag{4.1.21}$$

Multiphase signals. Consider the M-ary PSK signals

$$s_m(t) = \sqrt{\frac{2\mathscr{E}}{T}} \cos\left[2\pi f_c t + \frac{2\pi}{M}(m-1)\right] \qquad m = 1, 2, \ldots, M, 0 \le t \le T$$

$$= \sqrt{\frac{2\mathscr{E}}{T}} \cos\frac{2\pi}{M}(m-1)\cos 2\pi f_c t - \sqrt{\frac{2\mathscr{E}}{T}} \sin\frac{2\pi}{M}(m-1)\sin 2\pi f_c t \tag{4.1.22}$$

The equivalent lowpass signal waveforms are

$$u_m(t) = \sqrt{\frac{2\mathscr{E}}{T}} \, e^{j2\pi(m-1)/M} \qquad m = 1, 2, \ldots, M, 0 \le t \le T \tag{4.1.23}$$

These signal waveforms are represented by the two-dimensional vectors

$$\mathbf{s}_m = \left[\sqrt{\mathscr{E}}\cos\frac{2\pi}{M}(m-1), \sqrt{\mathscr{E}}\sin\frac{2\pi}{M}(m-1)\right] \qquad m = 1, 2, \ldots, M \tag{4.1.24}$$

or in complex-valued form as

$$u_m = \sqrt{2\mathscr{E}} \, e^{j2\pi(m-1)/M}$$

and are described by the signal space diagram shown in Fig. 4.1.3. Clearly, these signals have equal energy. Their complex-valued correlation coefficients are

$$\rho_{km} = \frac{1}{2\mathscr{E}}\int_0^T u_k^*(t)u_m(t)\,dt \qquad k = 1, 2, \ldots, M, \; m = 1, 2, \ldots, M$$

$$= e^{j2\pi(m-k)/M} \tag{4.1.25}$$

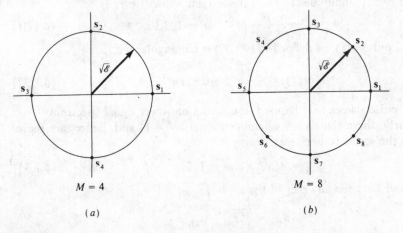

$M = 4$

(a)

$M = 8$

(b)

FIGURE 4.1.3
Signal-space diagrams for $M = 4$ and $M = 8$ PSK signals.

and the real-valued cross-correlation coefficients are

$$\text{Re}\,(\rho_{km}) = \cos\frac{2\pi}{M}(m-k) \tag{4.1.26}$$

Hence, the Euclidean distance between pairs of signals is

$$\begin{aligned}
d_{km} &= \left\{2\mathscr{E}\big(1 - \text{Re}\,(\rho_{km})\big)\right\}^{1/2}\\
&= \left\{2\mathscr{E}\left[1 - \cos\frac{2\pi}{M}(m-k)\right]\right\}^{1/2}
\end{aligned} \tag{4.1.27}$$

The *minimum distance* $d_{\min}$ corresponds to the case in which $|m - k| = 1$, i.e., adjacent signal phases. In this case

$$d_{\min} = \sqrt{2\mathscr{E}\left(1 - \cos\frac{2\pi}{M}\right)} \tag{4.1.28}$$

The special case of four phase signals ($M = 4$) results in the signal vectors

$$\mathbf{s}_1 = (\sqrt{\mathscr{E}},0) \qquad \mathbf{s}_2 = (0,\sqrt{\mathscr{E}}) \qquad \mathbf{s}_3 = (-\sqrt{\mathscr{E}},0) \qquad \mathbf{s}_4 = (0,-\sqrt{\mathscr{E}}) \tag{4.1.29}$$

and the signal space diagram shown in Fig. 4.1.3a. These signal vectors may also be viewed as two pairs of orthogonal signals, i.e., the pair of vectors $(\mathbf{s}_1,\mathbf{s}_3)$ is orthogonal to $(\mathbf{s}_2,\mathbf{s}_4)$. On the other hand, the pair $(\mathbf{s}_1,\mathbf{s}_3)$ is antipodal and so is the pair $(\mathbf{s}_2,\mathbf{s}_4)$. These four signals are called *biorthogonal*. The general class of biorthogonal signals is considered below.

Multiamplitude signals. Consider the M-ary PAM signals

$$\begin{aligned}
s_m(t) &= A_m\sqrt{\frac{2\mathscr{E}}{T}}\,\cos 2\pi f_c t\\
&= A_m\,\text{Re}\left[u(t)e^{j2\pi f_c t}\right] \qquad m = 1,2,\ldots, M, 0 \le t \le T
\end{aligned} \tag{4.1.30}$$

where the signal amplitude A_m takes the discrete values (levels)

$$A_m = 2m - 1 - M \qquad m = 1,2,\ldots, M \tag{4.1.31}$$

The signal pulse $u(t)$, as defined in (4.1.30) is rectangular, i.e.,

$$u(t) = \sqrt{\frac{2\mathscr{E}}{T}} \qquad 0 \le t \le T \tag{4.1.32}$$

but other pulse shapes may be used to obtain a narrower signal spectrum.

Clearly, these signals are one-dimensional ($N = 1$) and, hence, are represented by the scalar components

$$s_{m1} = A_m\sqrt{\mathscr{E}} \qquad m = 1,2,\ldots, M \tag{4.1.33}$$

The distance between any pair of signals is

$$\begin{aligned}
d_{mk} &= \sqrt{(s_{m1} - s_{k1})^2}\\
&= \sqrt{\mathscr{E}}\,|A_m - A_k|
\end{aligned} \tag{4.1.34}$$

FIGURE 4.1.4
Signal-space diagrams for M-ary PAM signals.

Hence, the distance between a pair of adjacent signal points, i.e., the minimum distance, is

$$d_{\min} = 2\sqrt{\mathscr{E}} \qquad (4.1.35)$$

Examples of signal space diagrams for M-ary PAM signals are shown in Fig. 4.1.4.

QAM signals. A quadrature amplitude-modulated (QAM) signal or a quadrature amplitude-shift-keying (QASK) signal is represented as

$$s_m(t) = A_{mc}\sqrt{\frac{2\mathscr{E}}{T}}\,\cos 2\pi f_c t - A_{ms}\sqrt{\frac{2\mathscr{E}}{T}}\,\sin 2\pi f_c t$$

$$= \mathrm{Re}\left[(A_{mc} + jA_{ms})u(t)e^{j2\pi f_c t}\right] \qquad (4.1.36)$$

where A_{mc} and A_{ms} are the information-bearing signal amplitudes of the quadrature carriers and $u(t) = \sqrt{2\mathscr{E}/T}$, $0 \le t \le T$. QAM signals are two-dimensional signals and, hence, they are represented by the vectors

$$\mathbf{s}_m = \left(\sqrt{\mathscr{E}}A_{mc}, \sqrt{\mathscr{E}}A_{ms}\right) \qquad (4.1.37)$$

The distance between any pair of signal vectors is

$$d_{mk} = \sqrt{|\mathbf{s}_m - \mathbf{s}_k|^2}$$

$$= \sqrt{\mathscr{E}\left[(A_{mc} - A_{kc})^2 + (A_{ms} - A_{ks})^2\right]} \qquad k, m = 1, 2, \ldots, M \quad (4.1.38)$$

When the signal amplitudes take the discrete values $\{2m - 1 - M,\ m = 1, 2, \ldots, M\}$, the signal space diagram is rectangular as shown in Fig. 4.1.5. In this case the distance between adjacent points, i.e., the minimum distance, is $d_{\min} = 2\sqrt{\mathscr{E}}$.

M-ary orthogonal signals. In Sec. 3.3.1 we indicated that multidimensional signals can be generated either by subdividing a time interval of duration $N\Delta t$ into N distinct (nonoverlapping) time slots of duration Δt, or by subdividing a

FIGURE 4.1.5
Rectangular QAM signal constellations with $d_{\min} = 2\sqrt{\varepsilon}$.

frequency interval $N\Delta f$ into N distinct frequency slots of width Δf, or by means of a combination of both. To be specific, let us consider the set of M FSK (frequency shift-keyed) signals

$$s_m(t) = \sqrt{\frac{2\mathscr{E}}{T}} \cos\left[2\pi f_c t + 2\pi m \Delta f t\right]$$

$$= \text{Re}\left[u_m(t)e^{j2\pi f_c t}\right] \qquad m = 1, 2, \ldots, M, \, 0 \le t \le T \quad (4.1.39)$$

where the equivalent lowpass waveforms are defined as

$$u_m(t) = \sqrt{\frac{2\mathscr{E}}{T}} \, e^{j2\pi m \Delta f t} \qquad m = 1, 2, \ldots, M, \, 0 \le t \le T \quad (4.1.40)$$

These waveforms are characterized as having equal energy and cross-correlation coefficients

$$\rho_{km} = \frac{\left(\dfrac{2\mathscr{E}}{T}\right)}{2\mathscr{E}} \int_0^T e^{j2\pi(m-k)\Delta f t} \, dt$$

$$= \frac{\sin \pi T(m-k)\,\Delta f}{\pi T(m-k)\,\Delta f} e^{j\pi T(m-k)\Delta f} \quad (4.1.41)$$

(a)

(b)

FIGURE 4.1.6
Cross-correlation coefficient as a function of frequency separation for FSK signals.

The real part of ρ_{km} is

$$\rho_r \equiv \mathrm{Re}\,(\rho_{km}) = \frac{\sin \pi T(m-k)\,\Delta f}{\pi T(m-k)\,\Delta f} \cos \pi T(m-k)\,\Delta f$$

$$= \frac{\sin 2\pi T(m-k)\,\Delta f}{2\pi T(m-k)\,\Delta f} \tag{4.1.42}$$

First, we observe that $\mathrm{Re}\,(\rho_{km}) = 0$ when $\Delta f = 1/2T$ and $m \neq k$. Since $|m - k| = 1$ corresponds to adjacent frequency slots, $\Delta f = 1/2T$ represents the minimum frequency separation between adjacent signals for orthogonality of the M signals. Plots of $\mathrm{Re}\,(\rho_{km})$ versus Δf and $|\rho_{km}|$ versus Δf are shown in Fig. 4.1.6. Note that $|\rho_{km}| = 0$ for multiples of $1/T$ whereas $\mathrm{Re}\,(\rho_{km}) = 0$ for multiples of $1/2T$.

FIGURE 4.1.7
Orthogonal signals for $M = N = 3$.

For the case in which $\Delta f = 1/2T$, the FSK signals are equivalent to the N-dimensional vectors

$$
\begin{aligned}
\mathbf{s}_1 &= (\sqrt{\mathscr{E}}, 0, \ldots, 0) \\
\mathbf{s}_2 &= (0, \sqrt{\mathscr{E}}, 0, \ldots, 0) \\
&\vdots \quad \vdots \\
\mathbf{s}_N &= (0, 0, \ldots, 0, \sqrt{\mathscr{E}})
\end{aligned}
\tag{4.1.43}
$$

where $N = M$. The distance between pairs of signals is

$$
d_{km} = \sqrt{2\mathscr{E}} \quad \text{all } m, k \tag{4.1.44}
$$

which is also the minimum distance. Figure 4.1.7 illustrates the signal space diagram for $M = N = 3$.

Biorthogonal signals. A set of M biorthogonal signals can be constructed from $M/2$ orthogonal signals by simply including the negatives of the orthogonal signals. Thus, we require $N = M/2$ dimensions for the construction of a set of M biorthogonal signals. Figure 4.1.8 illustrates the biorthogonal signals for $M = 4$ and 6.

We note that the correlation between any pair of waveforms is either $\rho_r = -1$ or 0. The corresponding distances are $d = 2\sqrt{\mathscr{E}}$ or $\sqrt{2\mathscr{E}}$, with the latter being the minimum distance.

Simplex signals. Suppose we have a set of M orthogonal waveforms $\{s_m(t)\}$ or, equivalently, their vector representation $\{\mathbf{s}_m\}$. Their geometric mean is

$$
\bar{\mathbf{s}} = \frac{1}{M} \sum_{m=1}^{M} \mathbf{s}_m \tag{4.1.45}
$$

Now, let us construct another set of M signals by subtracting the geometric mean from each of the M orthogonal signals. Thus,

$$
\mathbf{s}'_m = \mathbf{s}_m - \bar{\mathbf{s}} \quad m = 1, 2, \ldots, M \tag{4.1.46}
$$

FIGURE 4.1.8
Signal-space diagrams for $M = 4$ and $M = 6$ biorthogonal signals.

The effect of the subtraction is to translate the origin of the M orthogonal signals to the point $\bar{s}$.

The resulting signal waveforms are called *simplex signals* and have the following properties. First, the energy per waveform is

$$|s'_m| = |s_m - \bar{s}|^2$$

$$= \mathscr{E} - \frac{2}{M}\mathscr{E} + \frac{1}{M}\mathscr{E}$$

$$= \mathscr{E}\left(1 - \frac{1}{M}\right) \qquad (4.1.47)$$

Second, the cross correlation of any pair of signals is

$$\text{Re}(\rho_{mn}) = \frac{s'_m \cdot s'_n}{|s'_m|\,|s'_n|}$$

$$= \frac{-1/M}{(1 - 1/M)} = -\frac{1}{M-1} \qquad (4.1.48)$$

for all m, n. Hence, the set of simplex waveforms is *equally correlated* and requires less energy, by the factor $(1 - 1/M)$, than the set of orthogonal waveforms. Since only the origin was translated, the distance between any pair of signal points is maintained at $d = \sqrt{2\mathscr{E}}$ which is the same as the distance between any pair of orthogonal signals.

Figure 4.1.9 illustrates the simplex signals for $M = 2, 3,$ and 4. Note that the signal dimensionality is $N = M - 1$.

Signal waveforms from binary codes. A set of M signaling waveforms can be generated from a set of M binary code words of the form

$$C_m = [c_{m1}c_{m2} \cdots c_{mN}] \qquad m = 1, 2, \ldots, M \qquad (4.1.49)$$

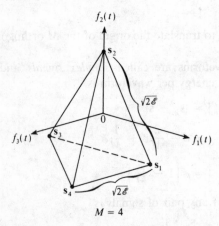

FIGURE 4.1.9
Signal-space diagrams for M-ary simplex signals.

where $c_{mj} = 0$ or 1 for all m and j. Each component of a code word is mapped into an elementary binary PSK waveform as follows:

$$c_{mj} = 1 \Rightarrow s_{mj}(t) = \sqrt{\frac{2\mathscr{E}_c}{T_c}}\, \cos 2\pi f_c t \qquad 0 \le t \le T_c$$

$$c_{mj} = 0 \Rightarrow s_{mj}(t) = -\sqrt{\frac{2\mathscr{E}_c}{T_c}}\, \cos 2\pi f_c t \qquad 0 \le t \le T_c$$

(4.1.50)

where $T_c = T/N$ and $\mathscr{E}_c = \mathscr{E}/N$. Thus, the M code words $\{C_m\}$ are mapped into a set of M waveforms $\{s_m(t)\}$.

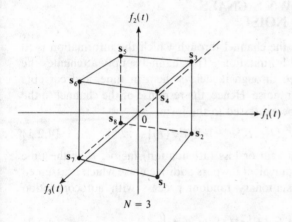

FIGURE 4.1.10
Signal-space diagrams for signals generated from binary codes.

The waveforms can be represented in vector form as

$$s_m = (s_{m1}, s_{m2}, \ldots, s_{mN}) \qquad m = 1, 2, \ldots, M \qquad (4.1.51)$$

where $s_{mj} = \pm \sqrt{\mathscr{E}/N}$ for all m and j. N is called the block length of the code, and it is also the dimension of the M waveforms.

We note that there are 2^N possible waveforms that can be constructed from the 2^N possible binary code words. We may select a subset of $M < 2^N$ signal waveforms for transmission of the information. We also observe that the 2^N possible signal points correspond to the vertices of an N-dimensional hypercube with its center at the origin. Figure 4.1.10 illustrates the signal points in $N = 2$ and 3 dimensions.

Each of the M waveforms has energy $\mathscr{E}$. The cross correlation between any pair of waveforms depends on how we select the M waveforms from the 2^N possible waveforms. This topic is treated in Chap. 5. Clearly, any adjacent signal points have a cross-correlation coefficient

$$\rho_r = \frac{\mathscr{E}\left(1 - \dfrac{1}{N}\right)}{\mathscr{E}} = \frac{N-1}{N} \tag{4.1.52}$$

and a corresponding distance of

$$d = \sqrt{2\mathscr{E}(1 - \rho_r)}$$
$$= \sqrt{2\mathscr{E}/N} \tag{4.1.53}$$

This concludes our discussion on the characterization of the various basic signaling waveforms for digital communication. In the following section we consider the optimum demodulation of these signals and their error probability when used to transmit information over an additive white gaussian noise channel.

4.2 OPTIMUM DEMODULATION FOR COMPLETELY KNOWN SIGNALS IN ADDITIVE GAUSSIAN NOISE

In this chapter we assume that the channel through which the information is to be transmitted has no bandwidth limitations. The channel simply attenuates the signaling waveforms transmitted through it, delays them in time, and corrupts them by the addition of gaussian noise. Hence, the response of the channel to the transmitted signals $s_m(t)$ may be expressed in the form

$$b(t) = \mathrm{Re}\left\{\left[\alpha u_m(t - t_0)e^{-j2\pi f_c t_0} + z(t)\right]e^{j2\pi f_c t}\right\} \tag{4.2.1}$$

where α represents the channel gain or loss (attenuation) factor, t_0 is the time delay, and $z(t)$ represents the equivalent lowpass additive noise which is assumed to be a zero mean gaussian stationary random process with autocorrelation function

$$\phi_{zz}(\tau) = \tfrac{1}{2}E\left[z^*(t)z(t + \tau)\right] \tag{4.2.2}$$

Presently we shall assume that the receiver is synchronized to the transmitter. That is, the delay t_0 is known at the receiver. Later we shall consider methods for synchronizing the receiver to the incoming signal. At this point, it is sufficient to indicate that synchronization is derived from time measurements performed on the received signal. The precision to which one must synchronize in time for purposes of detecting the signal $u_m(t - t_0)$ depends on the duration T of the signaling interval. Usually t_0 must be known to within a small fraction of T, for example, to within ± 5 percent of T for most applications.

In addition to the time delay t_0, the carrier phase $\phi = 2\pi f_c t_0$ must also be considered in processing the signal at the receiver. We note that if t_0 is estimated to within ± 5 percent of T, an error of this magnitude usually results in a large error in the phase ϕ because f_c is usually very large. Furthermore, the oscillator that generates the carrier at the transmitter is not absolutely stable, but may drift in frequency. Consequently even if t_0 can be measured precisely, the uncertainty in f_c results in an uncertainty in the value of the phase ϕ.

In some applications, it is desirable to estimate ϕ to the required precision at the receiver and use the estimate to compensate for the channel phase shift. In other applications, the channel phase shift is ignored in the detection process. We shall consider both cases in this chapter. In this section we assume that the carrier phase ϕ is known perfectly at the receiver. This is called *coherent demodulation*. The estimation of ϕ from the received signal is considered in Sec. 4.5. On the other hand, in Sec. 4.3 the received waveforms are demodulated with no attempt being made to estimate the carrier phase. When the phase is ignored in demodulating the received signal, the processing is called *noncoherent demodulation*.

Since we have assumed that the receiver is synchronized to the transmitter, we may eliminate the dependence of the received signal on t_0. Then the equivalent lowpass signal at the receiver can be expressed in the form

$$r(t) = \alpha e^{-j\phi} u_m(t) + z(t) \qquad 0 \le t \le T \tag{4.2.3}$$

This signal is processed at the receiver as described below. Having observed the received signal in a given signaling interval, the demodulator must decide which one of the M possible signal waveforms was sent. It is desirable that the decision be made with a minimum probability of error.

The probability of a decision error is minimized if the demodulator selects the signal having the largest posterior probability. That is, the demodulator computes the posterior probabilities

$$P[\text{signal } m \text{ was transmitted}|r(t), 0 \le t \le T] \qquad m = 1, 2, \ldots, M \tag{4.2.4}$$

and decides in favor of the signal resulting in the largest posterior probability. This decision criterion is called the *maximum a posteriori probability* (*MAP*) *criterion*.

The posterior probabilities for the M signals can be expressed in terms of the probabilities of the received signal $r(t)$ conditioned on each of the M possible transmitted signals. In order to compute these conditional probabilities, we need a mechanism for representing the received signal $r(t)$ in terms of a set of observable random variables $\{r_n\}$. Such a representation can be accomplished by any one of a number of methods, one of which is the Karhunen–Loéve series expansion. In Appendix 4A this expansion is used in the computation of the posterior probabilities given in (4.2.4) from which we obtain the form for the optimum demodulator.

The computations performed by the optimum demodulator for detection of one of M signals in additive gaussian noise are specified by (4A.19) and (4A.27) of Appendix 4A. When specialized to the case in which the M signals are equally likely and the additive gaussian noise is white, (4A.19) and (4A.27) simplify to the result given in (4A.30). The expression (4A.30) indicates that the optimum demodulator must compute the M decision variables

$$U_m = \text{Re}\left\{ e^{j\phi} \int_0^T r(t) u_m^*(t)\, dt \right\} - \alpha \mathscr{E}_m \qquad m = 1, 2, \ldots, M \tag{4.2.5}$$

and decide in favor of the signal corresponding to the largest decision variable. The first term in (4.2.5) can be generated by one of two methods. One method is to multiply the received signal $r(t)$ by the complex conjugate of $u_m(t)$ for $m = 1, 2, \ldots, M$ and integrate the product over the signaling interval, as illustrated in Fig. 4.2.1. We say that the received signal $r(t)$ is cross-correlated with the complex conjugate of each of the possible transmitted signals and, thus, this demodulator structure is called a *correlation demodulator*. The M outputs of the correlators are then multiplied by $e^{j\phi}$ and the real part of the resulting complex-valued quantities is retained. From these numbers we subtract the corresponding quantities $\alpha\mathcal{E}_m$, $m = 1, 2, \ldots, M$, which represent the bias terms in the decision variables.

A second method for generating the decision variables $\{U_m\}$ is to pass $r(t)$ through a parallel bank M filters, having equivalent lowpass impulse responses $b_m(t) = u_m^*(T - t)$, $m = 1, 2, \ldots, M$, and sampling the outputs of the filters at $t = T$, as illustrated in Fig. 4.2.2. The filters are said to be matched to the M signaling waveforms and the corresponding demodulator is called a *matched filter demodulator*. As in the correlation demodulator, the sampled outputs from the matched filters must be multiplied by $e^{j\phi}$ and the real part of the resulting products is retained. Moreover, the bias term $\{\alpha\mathcal{E}_m\}$ must be subtracted from these numbers to form the decision variables $\{U_m\}$.

Our decision to separate the multiplicative factor $e^{j\phi}$ from $u_m^*(t)$ in the correlation demodulator structure and from $u_m^*(T - t)$ in the matched filter demodulator structure was totally arbitrary. By incorporating the factor $e^{j\phi}$ into $u_m^*(t)$ and $u_m^*(T - t)$, we observe that the two demodulator structures employ the M noiseless channel-perturbed signals in the cross-correlation and matched filtering operations, respectively.

Since the M signal waveforms can be represented by the series expansion

$$s_m(t) = \sum_{n=1}^{N} s_{mn} f_n(t)$$

or, equivalently, as

$$u_m(t) = \sum_{n=1}^{N} u_{mn} g_n(t) \tag{4.2.6}$$

there is an alternative implementation of the optimum demodulator that is equivalent to the cross correlator and matched-filter demodulators illustrated in Figs. 4.2.1 and 4.2.2, respectively. Specifically, the receiver cross correlates the received signal $r(t)$ with each of the N orthonormal waveforms $\{g_n(t)\}$ and samples the output of the N integrators at the end of the signal interval, i.e., at $t = T$. This yields a received signal vector $\mathbf{r}$ with components

$$\mathbf{r} = (r_1 r_2 \cdots r_N) \tag{4.2.7}$$

FIGURE 4.2.1
Cross-correlation-type demodulator.

where

$$r_i = \alpha e^{-j\phi} u_{mi} + z_i \qquad i = 1, 2, \ldots, N \qquad (4.2.8)$$

The terms $\{z_i\}$ represent the additive gaussian noise at the output of the correlators at $t = T$. That is

$$z_i = \int_0^T z(t) g_i^*(t) \, dt \qquad i = 1, 2, \ldots, N \qquad (4.2.9)$$

Since $E[z(t)] = 0$, it follows that $E(z_i) = 0$. Furthermore,

$$
\begin{aligned}
E(z_i z_j^*) &= \int_0^T \int_0^T E[z(t) z^*(\rho)] g_i^*(t) g_j(\tau) \, dt \, d\tau \\
&= 2N_0 \int_0^T g_i^*(t) g_j(t) \, dt \\
&= \begin{cases} 2N_0 & i = j \\ 0 & i \neq j \end{cases}
\end{aligned}
\qquad (4.2.10)
$$

Hence, the noise components are uncorrelated and, because they are gaussian, they are statistically independent.

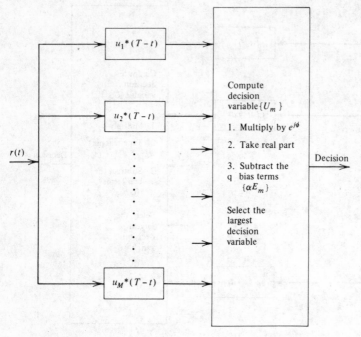

FIGURE 4.2.2
Matched-filter demodulator.

The set of N complex-valued random variables is described statistically by the joint conditional pdf

$$p(\mathbf{r}|\mathbf{u}_m) = \prod_{i=1}^{N} \frac{1}{2\pi\sigma^2} \exp\left\{-\frac{1}{2\sigma^2}\left|r_i - \alpha e^{-j\phi}u_{mi}\right|^2\right\}$$

$$= \left(\frac{1}{2\pi\sigma^2}\right)^N \exp\left\{-\frac{1}{2\sigma_2}\sum_{i=1}^{N}\left|r_i - \alpha e^{-j\phi}u_{mi}\right|^2\right\} \quad (4.2.11)$$

where $\sigma^2 = N_0$.

The posterior probability on which we base the decision is given as

$$P(\mathbf{u}_m \text{ was transmitted}|\mathbf{r}) = \frac{p(\mathbf{r}|\mathbf{u}_m)P(\mathbf{u}_m)}{p(\mathbf{r})} \quad m = 1, 2, \ldots, M \quad (4.2.12)$$

The denominator in (4.2.1) is identical for all signals and, hence, it does not affect the decision. Also, for equally likely signals, $P(\mathbf{u}_m) = 1/M$ for all m. Hence, the selection of the signal that yields the largest posterior probability is equivalent to determining the signal that maximizes $p(\mathbf{r}|\mathbf{u}_m)$. Thus, the MAP criterion reduces to determining

$$\max_m p(\mathbf{r}|\mathbf{u}_m) \quad m = 1, 2, \ldots, M \quad (4.2.13)$$

This is called the *maximum likelihood (ML) criterion*.

From (4.2.11) we observe that the maximum of $p(\mathbf{r}|\mathbf{u}_m)$ is equivalent to determining the signal that maximizes the metric

$$\mu(\mathbf{u}_m) = -\sum_{i=1}^{N} |r_i - \alpha e^{-j\phi} u_{mi}|^2$$

$$= -|\mathbf{r}|^2 + 2\alpha \operatorname{Re}\left[e^{j\phi}\mathbf{r} \cdot \mathbf{u}_m^*\right] - \alpha^2 |\mathbf{u}_m|^2 \qquad m = 1, 2, \ldots, M \quad (4.2.14)$$

Since $|\mathbf{r}|^2$ is common to all metrics and it is independent of $\mathbf{u}_m$, we may simply ignore this term. Furthermore, we may divide (4.2.14) by 2α. Thus, we are left with the metrics, or decision variables,

$$\mu(\mathbf{u}_m) = \operatorname{Re}\left[e^{j\phi}\mathbf{r} \cdot \mathbf{u}_m^*\right] - \alpha \mathscr{E}_m \qquad m = 1, 2, \ldots, M \quad (4.2.15)$$

where $\mathscr{E}_m = |\mathbf{u}_m|^2/2$. Note the similarity and equivalence of the metrics (decision variables) in (4.2.15) to the decision variables given by (4.2.5).

The demodulator implied by (4.2.15) is illustrated in Fig. 4.2.3. After cross correlation with $\{g_n^*(t)\}$ the vector components from the correlators are cross correlated with each of the possible signal vectors $\mathbf{u}_m$, $m = 1, 2, \ldots, M$, and the resulting sum of products are phase shifted by ϕ. The real part of the phase shifted complex numbers is taken and the bias terms $(\alpha \mathscr{E}_m)$ are subtracted from the corresponding terms to yield the decision variables in (4.2.15).

Instead of using cross correlators to obtain the received vector components $\{r_i\}$, we may use a bank of N filters with impulse responses $\{g_n^*(T-t)\}$ that are the matched filters to the orthonormal functions $\{g_n^*(t)\}$. The remaining operations are identical to the ones described in the previous paragraph. The resulting optimum demodulator is shown in Fig. 4.2.4.

When the number of signal waveforms $M \gg N$, as is often the case in practice, the optimum receiver realizations illustrated in Figs. 4.2.3 and 4.2.4 are simpler than the ones in Figs. 4.2.1 and 4.2.2, which employ M correlators or M matched filters, respectively.

An important consideration in the implementation of any of the demodulator structures described above is the fact that one must have knowledge of the channel attenuation factor α and the carrier phase ϕ. As indicated previously, this knowledge is not available *a priori*, but it is obtained from estimates derived from the received signal. Usually, the estimates are formed from observation of the received signal over many signaling intervals. For example, a phase-locked loop is commonly employed for estimating the carrier phase for the purpose of performing coherent demodulation.

Some simplification in the computation of the decision variables given by either (4.2.5) or (4.2.15) is achieved by using equal-energy signaling waveforms. As a consequence of the equal-energy condition, the bias terms in (4.2.5) and (4.2.15) are identical for all M signals and, hence, they can be neglected. In such a case, the attenuation factor α need not be estimated. In QAM and PAM, where the signals have different energies, the estimation of α is required.

When the phase ϕ of the received signal is known perfectly, it can be suppressed from the decision variables in (4.2.5) and (4.2.15). Thus, a further

FIGURE 4.2.3
Cross-correlation-type demodulator.

240

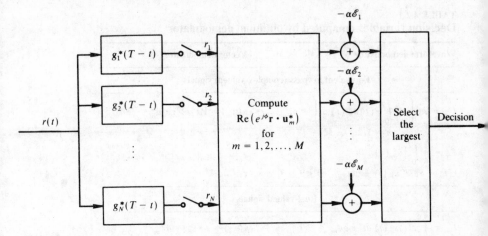

FIGURE 4.2.4
Matched-filter demodulator.

simplification can be made in the computation of the decision variables, especially in (4.2.15), by using the real signal vectors $\{s_m\}$ in place of the complex-valued signals $\{\mathbf{u}_m\}$. The decision variables simplify to

$$\mu(\mathbf{s}_m) = \mathbf{r} \cdot \mathbf{s}_m - \alpha \mathscr{E}_m \qquad m = 1, 2, \ldots, M \qquad (4.2.16)$$

The received vector is expressed as

$$\mathbf{r} = \alpha \mathbf{s}_n + \mathbf{n} \qquad (4.2.17)$$

where $\mathbf{s}_n$ is assumed to be the transmitted signal, $\mathbf{n}$ is the real-valued noise vector, and $\mathbf{r}$ represents the received signal vector. The elements of the noise vector are zero mean and have a variance of $N_0/2$, where N_0 is the power spectral density of the noise.

Table 4.2.1 contains a summary of the decision variables for the waveform demodulator of (4.1.5) and the equivalent vector representations given by (4.2.15) and (4.2.16).

In the remainder of this section we derive the probability of error when the various signal waveforms described in the previous section are used to transmit the information sequence over an additive, white gaussian noise channel.

4.2.1 Probability of Error for Binary Signaling

The two signal waveforms $s_m(t)$, $m = 1, 2$, are generally given as

$$s_m(t) = \operatorname{Re}\left[u_m(t)e^{j2\pi f_c t}\right] \qquad m = 1, 2, 0 \le t \le T \qquad (4.2.18)$$

These waveforms are assumed to have equal energy $\mathscr{E}$ and their equivalent lowpass waveforms $u_m(t)$, $m = 1, 2$, are characterized by the complex-valued correlation coefficient ρ_{12} given by (4.1.3).

TABLE 4.2.1
Decision variables computed by optimum demodulator

Waveform demodulator	Vector demodulator
Equivalent lowpass (complex-valued) signals	
$U_m = \mathrm{Re}\left[e^{j\phi} \int_0^T r(t) u_m^*(t)\, dt \right] - \alpha\mathscr{E}_m$ $\qquad m = 1, 2, \ldots, M$	$\mu(\mathbf{u}_m) = \mathrm{Re}\left[e^{j\phi}\mathbf{r} \cdot \mathbf{u}_m^* \right] - \alpha\mathscr{E}_m$ $\qquad m = 1, 2, \ldots, M$
$\mathscr{E}_m = \dfrac{1}{2} \int_0^T \left|u_m(t)\right|^2 dt$	$\mathscr{E}_m = \tfrac{1}{2}\left|\mathbf{u}_m\right|^2$
$r(t) = \alpha e^{-j\phi} u_m(t) + z(t) \qquad 0 \le t \le T$	$\mathbf{r} = \alpha e^{-j\phi}\mathbf{u}_m + \mathbf{z}$
Real-valued signals	
$U_m = \displaystyle\int_0^T r(t) s_m(t)\, dt - \alpha\mathscr{E}_m$ $\qquad m = 1, 2, \ldots, M$	$\mu(\mathbf{s}_m) = \mathbf{r} \cdot \mathbf{s}_m - \alpha\mathscr{E}_m$ $\qquad m = 1, 2, \ldots, M$
$\mathscr{E}_m = \displaystyle\int_0^T s^2(t)\, dt$	$\mathscr{E}_m = \left|\mathbf{s}_m\right|^2$
$r(t) = \alpha s_m(t) + n(t) \qquad 0 \le t \le T$	$\mathbf{r} = \alpha\mathbf{s}_m + \mathbf{n}$

As demonstrated above, the optimum demodulator forms the decision variables

$$U_m = \mathrm{Re}\left[\int_0^T r(t) u_m^*(t)\, dt \right] \qquad m = 1, 2 \qquad (4.2.19)$$

or, equivalently

$$\mu(\mathbf{u}_m) = \mathrm{Re}\left[e^{j\phi}\mathbf{r} \cdot \mathbf{u}_m^* \right] \qquad m = 1, 2 \qquad (4.2.20)$$

and decides in favor of the signal corresponding to the larger decision variable. Our objective is to determine the performance of the optimum phase-coherent (perfect estimate of the carrier phase ϕ) demodulator, expressed in terms of the average probability of a binary digit error. Let us first evaluate the probability of error for the decision variables given by (4.2.19). Then we shall demonstrate that a similar computation based on the decision variables in (4.2.20) yields the same result.

Suppose the signal $s_1(t)$ is transmitted in the interval $0 \le t \le T$. The equivalent lowpass received signal is

$$r(t) = \alpha e^{-j\phi} u_1(t) + z(t) \qquad 0 \le t \le T \qquad (4.2.21)$$

Substituting (4.2.21) into (4.2.19), we obtain the decision variables

$$\begin{aligned} U_1 &= \mathrm{Re}\left(2\alpha\mathscr{E} + N_1\right) = 2\alpha\mathscr{E} + N_{1r} \\ U_2 &= \mathrm{Re}\left(2\alpha\mathscr{E}\rho + N_2\right) = 2\alpha\mathscr{E}\rho_r + N_{2r} \end{aligned} \qquad (4.2.22)$$

where N_m, $m = 1, 2$, represent the noise components in the decision variables, given by

$$N_m = e^{j\phi} \int_0^T z(t) u_m^*(t)\, dt \qquad (4.2.23)$$

and $N_{mr} = \text{Re}(N_m)$. The noise components N_{mr}, $m = 1, 2$, are jointly gaussian random variables with zero mean.

The probability of error is just the probability that the decision variable U_2 exceeds the decision variable U_1. But

$$P(U_2 > U_1) = P(U_2 - U_1 > 0) = P(U_1 - U_2 < 0) \qquad (4.2.24)$$

For mathematical convenience, we define a new random variable V as

$$V = U_1 - U_2$$

$$= 2\alpha\mathscr{E}(1 - \rho_r) + N_{1r} - N_{2r} \qquad (4.2.25)$$

Since the noise terms N_{1r} and N_{2r} are gaussian, the difference $N_{1r} - N_{2r}$ is also gaussian-distributed and, hence, the random variable V is gaussian-distributed with mean value

$$m_v = E(V) = 2\alpha\mathscr{E}(1 - \rho_r) \qquad (4.2.26)$$

and variance

$$\sigma_v^2 = E\left[(N_{1r} - N_{2r})^2\right]$$

$$= E(N_{1r}^2) - 2E(N_{1r}N_{2r}) + E(N_{2r}^2)$$

$$= 4\mathscr{E}N_0(1 - \rho_r) \qquad (4.2.27)$$

where N_0 is the power spectral density of $z(t)$.

In terms of the random variable V, the probability of error is

$$P(V < 0) = \int_{-\infty}^{0} p(v) \, dv$$

$$= \frac{1}{\sqrt{2\pi}\,\sigma_v} \int_{-\infty}^{0} e^{-(v - m_v)^2 / 2\sigma_v^2} \, dv$$

$$= \tfrac{1}{2} \text{erfc}\left(\sqrt{\frac{\alpha^2\mathscr{E}}{2N_0}(1 - \rho_r)}\right) \qquad (4.2.28)$$

where $\text{erfc}(x)$ is the complementary error function, defined as

$$\text{erfc}(x) = \frac{2}{\sqrt{\pi}} \int_{x}^{\infty} e^{-t^2} \, dt \qquad (4.2.29)$$

By following the same procedure as above, it is easily shown that when $s_2(t)$ is transmitted, the probability that the demodulator will decide in favor of $s_1(t)$ and, thus, make an error, is identical to the expression given in (4.2.28). Therefore the average probability of a binary digit error for equally likely signals is

$$P_2 = \tfrac{1}{2} \text{erfc}\left(\sqrt{\frac{\alpha^2\mathscr{E}}{2N_0}(1 - \rho_r)}\right) \qquad (4.2.30)$$

The computation of the error probability based on the vector representation of signals yields the same result. For equal energy signals the decision variables are

$$\mu(\mathbf{u}_m) = \text{Re}\left(e^{j\phi}\mathbf{r} \cdot \mathbf{u}_m^*\right) \qquad m = 1, 2 \qquad (4.2.31)$$

Let us limit our attention to one- and two-dimensional signals, e.g., antipodal and orthogonal signals. Then, the signal vectors are represented as $\mathbf{s}_1 = (s_{11}, s_{12})$ and $\mathbf{s}_2 = (s_{21}, s_{22})$ or by their equivalent lowpass complex-valued signal vectors $u_1 = \sqrt{2}\,(s_{11} - js_{12})$ and $u_2 = \sqrt{2}\,(s_{21} + js_{22})$.

If we assume that the signal u_1 was transmitted, the received signal is

$$r = \alpha e^{-j\phi} u_1 + z \qquad (4.2.32)$$

where z is the complex-valued gaussian noise component. By substituting for r in (4.2.31) we obtain

$$\begin{aligned}
\mu(u_1) &= 2\alpha\mathcal{E} + N_{1r} \\
\mu(u_2) &= 2\alpha\mathcal{E}(1 - \rho_r) + N_{2r}
\end{aligned} \qquad (4.2.33)$$

where $N_{mr} = \text{Re}(u_m^* z)$, $m = 1, 2$. The statistical characteristics of the noise components are identical to those in (4.2.27). Hence, the probability of error is identical to the result given by (4.2.28).

If the computation of the error probability is based on the real-valued signals $\{\mathbf{s}_m, m = 1, 2\}$, the two decision variables are

$$\mu(\mathbf{s}_m) = \mathbf{r} \cdot \mathbf{s}_m \qquad m = 1, 2 \qquad (4.2.34)$$

Again, let the signal $\mathbf{s}_1$ be transmitted so that $\mathbf{r} = \alpha \mathbf{s}_1 + \mathbf{n}$ where $\mathbf{n}$ is the noise vector. Then

$$\begin{aligned}
\mu(\mathbf{s}_1) &= \alpha\mathcal{E} + \mathbf{s}_1 \cdot \mathbf{n} \\
\mu(\mathbf{s}_2) &= \alpha\mathcal{E}\rho_r + \mathbf{s}_2 \cdot \mathbf{n}
\end{aligned} \qquad (4.2.35)$$

and

$$\begin{aligned}
P_2 &= P\left[\mu(\mathbf{s}_2) > \mu(\mathbf{s}_1)\right] \\
&= P\left[(\mathbf{s}_1 - \mathbf{s}_2) \cdot \mathbf{n} < -\alpha\mathcal{E}(1 - \rho_r)\right]
\end{aligned} \qquad (4.2.36)$$

The gaussian noise variable $y = (\mathbf{s}_1 - \mathbf{s}_2) \cdot \mathbf{n}$ has zero mean and variance

$$\sigma_y^2 \equiv E(y^2) = d_{12}^2 N_0/2 \qquad (4.2.37)$$

Therefore,

$$\begin{aligned}
P_2 &= P\left[y < -\alpha\mathcal{E}(1 - \rho_r)\right] = P\left[y > \frac{\alpha d_{12}^2}{2}\right] \\
&= \frac{1}{2}\,\text{erfc}\left(\sqrt{\frac{\alpha^2 d_{12}^2}{4N_0}}\right) \\
&= \frac{1}{2}\,\text{erfc}\left(\sqrt{\frac{\alpha^2\mathcal{E}}{2N_0}(1 - \rho_r)}\right)
\end{aligned} \qquad (4.2.38)$$

which is identical to the result for the error probability derived above for complex-valued signals. It is interesting to note that the probability of error P_2 given in (4.2.38) is expressed as

$$P_2 = \frac{1}{2} \operatorname{erfc} \left(\sqrt{\frac{\alpha^2 d_{12}^2}{4N_0}} \right) \tag{4.2.39}$$

where d_{12} is the distance of the two signals. Hence, we observe that an increase in the distance between the two signals reduces the probability of error.

(a)

(b)

(c)

FIGURE 4.2.5
(a) Binary PSK signal; (b) equivalent lowpass pulse; (c) block diagram of PSK demodulator.

Observe that the minimum of P_2 with respect to the signal correlation coefficient ρ_r, where $-1 \leq \rho_r \leq 1$, is obtained when $\rho_r = -1$, in which case $s_1(t) = -s_2(t)$. Thus, *the optimum binary signals are antipodal* and their performance on the AWGN channel is

$$P_2 = \frac{1}{2}\,\mathrm{erfc}\left(\sqrt{\frac{\alpha^2 \mathscr{E}}{N_0}}\right)$$

$$= \frac{1}{2}\,\mathrm{erfc}\left(\sqrt{\gamma_b}\right) \tag{4.2.40}$$

$\rho_r = -1$
Antipodal
$P_b = \frac{1}{2}\,\mathrm{erfc}(\sqrt{\gamma_b})$

$\rho_r = 0$
Orthogonal signals
$P_b = \frac{1}{2}\,\mathrm{erfc}(\sqrt{\gamma_b/2})$

P_b, probability of error

γ_b, SNR/bit, dB

FIGURE 4.2.6
Probability of error for binary signals.

where $\gamma_b = \alpha^2 \mathscr{E}/N_0$ is received signal-to-noise ratio (SNR) per information bit.

A special case of antipodal signaling is binary PSK, for which the signaling waveforms are

$$s_1(t) = A \sin 2\pi f_c t \qquad\qquad\qquad\qquad 0 \le t \le T$$
$$s_2(t) = A \sin (2\pi f_c t + \pi) = -A \sin 2\pi f_c t \quad 0 \le t \le T \tag{4.2.41}$$

Figure 4.2.5a illustrates a PSK signal. The corresponding equivalent lowpass signal is $u(t) = jA$, $0 \le t \le T$. That is, $|u(t)|$ is a rectangular pulse as shown in Fig. 4.2.5b. A block diagram of the binary PSK demodulator implemented as a correlator is shown in Fig. 4.2.5c. The received signal, which is either $\alpha A \sin (2\pi f_c t + \phi)$ plus noise or $\alpha A \sin (2\pi f_c t + \pi + \phi)$ plus noise, is multiplied by the $\sin (2\pi f_c t + \hat{\phi})$ and the product is integrated and sampled periodically at the end of each T-second signaling interval. The output of the integrator is reset to zero after sampling. The integrator acts as a lowpass filter, thus rejecting the double frequency components resulting from the multiplication. It is usually called an *integrate-and-dump filter*. The carrier phase ϕ is estimated from the incoming signal, usually by means of a phase-locked loop, described in Sec. 4.5. Ideally, in the absence of noise, $\hat{\phi} = \phi$. The timing signal for sampling the output of the integrator is also derived from the received signal.

A second type of binary signaling commonly used in practice is orthogonal signaling. This corresponds to signals having a cross-correlation coefficient $\rho_r = 0$. Hence the probability of error for phase-coherent detection of binary orthogonal signals is

$$P_2 = \tfrac{1}{2} \operatorname{erfc} \left(\sqrt{\frac{\gamma_b}{2}} \right) \tag{4.2.42}$$

which is 3 dB poorer than the performance of antipodal signaling. Figure 4.2.6 shows the error rate performance of binary antipodal and binary orthogonal signals. Binary orthogonal FSK signals are shown in Fig. 4.2.7.

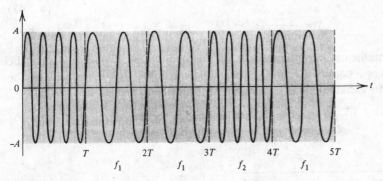

FIGURE 4.2.7
Example of a binary FSK signal.

4.2.2 Probability of Error for M-ary Orthogonal Signals

In this section we evaluate the performance of M-ary orthogonal signals in an AWGN channel. Our starting point is the set of decision variables

$$U_m = \text{Re}\left[e^{j\phi} \int_0^T r(t) u_m^*(t)\, dt \right] \qquad m = 1, 2, \ldots, M \qquad (4.2.43)$$

that are computed in the optimum demodulator for equally likely, equal energy signals.

Let us assume that the transmitted signal is $s_1(t)$ and the equivalent lowpass received signal is

$$r(t) = \alpha e^{-j\phi} u_1(t) + z(t) \qquad (4.2.44)$$

Substitution for $r(t)$ from (4.2.44) into (4.2.43) yields, for the decision variable U_1 corresponding to the output of the filter matched to $u_1(t)$,

$$U_1 = 2\alpha\mathscr{E} + N_{1r} \qquad (4.2.45)$$

Since the signaling waveforms are orthogonal, the other decision variables consist of noise only and they can be expressed as

$$U_m = N_{mr} \qquad m = 2, 3, \ldots, M \qquad (4.2.46)$$

where†

$$N_{mr} = \text{Re}\left[\int_0^T z(t) u_m^*(t)\, dt \right] \qquad m = 1, 2, \ldots, M \qquad (4.2.47)$$

The noise components N_{mr} at the output of the matched filters are zero mean gaussian random variables with variance $\sigma^2 = 2\mathscr{E}N_0$. As a consequence of the orthogonality of the signals, the noise terms $\{N_{mr}\}$ are mutually uncorrelated and, since they are gaussian-distributed, they are mutually statistically independent. Hence

$$p(U_1) = \frac{1}{\sqrt{2\pi}\,\sigma} e^{-(U_1 - 2\alpha\mathscr{E})^2/2\sigma^2}$$

$$(4.2.48)$$

$$p(U_m) = \frac{1}{\sqrt{2\pi}\,\sigma} e^{-U_m^2/2\sigma^2} \qquad m = 2, 3, \ldots, M$$

It is mathematically convenient to derive first the probability that the receiver makes a correct decision. This is the probability that U_1 exceeds all the other decision variables $U_2, U_3, \ldots, U_M$, and it may be expressed as

$$P_c = \int_{-\infty}^{\infty} P(U_2 < U_1, U_3 < U_1, \ldots, U_M < U_1 | U_1) p(U_1)\, dU_1 \qquad (4.2.49)$$

---

† Since the additive noise is white and gaussian, any phase rotation applied to $z(t)$ does not alter the statistics of $z(t)$. Hence the exponential factor $e^{j\phi}$ may be absorbed into $z(t)$.

where $P(U_2 < U_1, U_3 < U_1, \ldots, U_M < U_1|U_1)$ denotes the joint probability that $U_2, U_3, \ldots, U_M$ are all less than U_1, conditioned on U_1. Since the $\{U_m\}$ are statistically independent, this joint probability factors into a product of $M - 1$ marginal probabilities of the form

$$P(U_m < U_1|U_1) = \int_{-\infty}^{U_1} p(U_m)\, dU_m \qquad m = 2, 3, \ldots, M$$

$$= \frac{1}{\sqrt{2\pi}} \int_{-\infty}^{U_1/\sqrt{2\mathscr{E}N_0}} e^{-x^2/2}\, dx \qquad (4.2.50)$$

These probabilities are identical for $m = 2, 3, \ldots, M$ and, hence, the joint probability under consideration is simply the result in (4.2.50) raised to the $(M - 1)$st power. Thus the probability of a correct decision becomes

$$P_c = \int_{-\infty}^{\infty} \left(\frac{1}{\sqrt{2\pi}} \int_{-\infty}^{U_1/\sqrt{2\mathscr{E}N_0}} e^{-x^2/2}\, dx \right)^{M-1} p(U_1)\, dU_1 \qquad (4.2.51)$$

and the probability of a symbol (k-bit character) error is

$$P_M = 1 - P_c \qquad (4.2.52)$$

After substituting for $p(U_1)$ in (4.2.51) and making a change in the variable of integration, we obtain the result

$$P_M = \frac{1}{\sqrt{2\pi}} \int_{-\infty}^{\infty} \left\{ 1 - \left[1 - \tfrac{1}{2}\, \mathrm{erfc}\left(\frac{y}{\sqrt{2}} \right) \right]^{M-1} \right\} e^{-(y - \sqrt{2\gamma})^2/2}\, dy \qquad (4.2.53)$$

where $\gamma = \alpha^2 \mathscr{E}/N_0$ denotes the received SNR for each k-bit symbol. The same expression for the probability of error is obtained when any one of the other $M - 1$ signaling waveforms is transmitted. Since all the signaling waveforms are equally likely, the expression for P_M given in (4.2.53) is the average probability of a symbol error.

In comparing the performance of various digital signaling methods, it is desirable to have the error probability expressed in terms of the SNR per bit, instead of the SNR per symbol. With $M = 2^k$ waveforms, each symbol conveys k bits of information and, hence, $\gamma = k\gamma_b$, where γ_b is the SNR per bit. Thus P_M can be expressed as a function of γ_b.

Sometimes, it is also desirable to convert the probability of a symbol error into an equivalent probability of a binary digit error. For equiprobable orthogonal signals, all symbol errors are equiprobable and occur with probability

$$\frac{P_M}{M - 1} = \frac{P_M}{2^k - 1} \qquad (4.2.54)$$

Furthermore, the are $\binom{k}{n}$ ways in which n bits out of k may be in error. Hence the average number of bit errors per k-bit symbol is

$$\sum_{n=1}^{k} n \binom{k}{n} \frac{P_M}{2^k - 1} = k \frac{2^{k-1}}{2^k - 1} P_M \qquad (4.2.55)$$

FIGURE 4.2.8
Probability of bit error for coherent detection of orthogonal signals.

and the average bit error probability is just the result in (4.2.55) divided by k, the number of bits per symbol. Thus

$$P_b = \frac{2^{k-1}}{2^k - 1} P_M \qquad (4.2.56)$$

The probability of a binary digit error as a function of the SNR per bit is shown in Fig. 4.2.8 for $M = 2, 4, 8, 16, 32,$ and 64. The curves illustrate the advantage of increasing the number of waveforms. That is, by increasing M one can reduce the SNR per bit required to achieve a given probability of error.

Let us investigate the limiting form of P_M as M approaches infinity. To simplify the mathematical development, we first derive a simple upper bound (union bound) on the probability of a symbol error P_M. From the derivation of the probability of error for binary signals, we demonstrated that

$$P_2 = \frac{1}{2} \operatorname{erfc}\left(\sqrt{\frac{\alpha^2 d_{12}^2}{4 N_0}}\right) \tag{4.2.57}$$

where d_{12} is the distance between the two signals. In the case of M-ary orthogonal signals, all signals are equidistant with $d_{km} = \sqrt{2\mathscr{E}}$. Hence, the probability of error in comparing the decision variable U_1, which contains the signal component, with any of the other decision variables $\{U_m, m = 2, 3, \ldots, M\}$ is

$$P_2 = \tfrac{1}{2} \operatorname{erfc}\left(\sqrt{\gamma/2}\right) \tag{4.2.58}$$

where $\gamma = \alpha^2 \mathscr{E}/N_0$ is the NSR per symbol.

Now if we view the receiver for M signals as one that makes $M - 1$ binary decisions between U_1 and the other $M - 1$ matched filter outputs U_m, $m = 2, 3, \ldots, M$, the probability of a symbol error is upper-bounded by the union of the $M - 1$ error events. That is,

$$P_M \le (M - 1) P_2$$
$$\le \frac{M - 1}{2} \operatorname{erfc}\left(\sqrt{\gamma/2}\right) < \frac{M}{2} \operatorname{erfc}\left(\sqrt{\gamma/2}\right) \tag{4.2.59}$$

This bound can be simplified further by upper-bounding the complementary error function as follows:

$$\operatorname{erfc}\left(\sqrt{\gamma/2}\right) < e^{-\gamma/2} \tag{4.2.60}$$

Thus

$$P_M < \frac{M}{2} e^{-\gamma/2} = \tfrac{1}{2} 2^k e^{-\gamma/2}$$
$$< \tfrac{1}{2} \exp\left[-k\left(\frac{\gamma_b}{2} - \ln 2\right)\right] \tag{4.2.61}$$

As $k \to \infty$ or, equivalently, as $M \to \infty$ the probability of error approaches zero exponentially, provided that the SNR per bit satisfies the condition

$$\gamma_b > 2 \ln 2 = 1.39 \qquad (1.42 \text{ dB}) \tag{4.2.62}$$

The simple bounds used above do not produce the smallest lower bound on the SNR per bit γ_b. The problem is that the union bound is very loose when the argument of the complementary error function is small. An alternative approach is to use two different upper bounds, depending on the range of the complementary error function. Beginning with (4.2.53), we observe that

$$\left\{1 - \left[1 - \tfrac{1}{2} \operatorname{erfc}\left(\frac{y}{\sqrt{2}}\right)\right]^{M-1}\right\} \le \frac{M - 1}{2} \operatorname{erfc}\left(\frac{y}{\sqrt{2}}\right) < M e^{-y^2/2} \tag{4.2.63}$$

This is just the union bound used above, which is tight when y is large, i.e., for $y > y_0$, where y_0 depends on M. When y is small, the union bound exceeds unity for large M. Since

$$\left\{1 - \left[1 - \tfrac{1}{2}\,\mathrm{erfc}\left(\frac{y}{\sqrt{2}}\right)\right]^{M-1}\right\} \leq 1 \qquad (4.2.64)$$

for all y, we may use this bound for $y < y_0$ because it is tighter than the union bound. Thus (4.2.53) may be upper-bounded as

$$P_M < \frac{1}{\sqrt{2\pi}} \int_{-\infty}^{y_0} e^{-\left(y-\sqrt{2\gamma}\right)^2/2}\,dy + \frac{M}{\sqrt{2\pi}} \int_{y_0}^{\infty} e^{-y^2/2}e^{-\left(y-\sqrt{2\gamma}\right)^2/2}\,dy \quad (4.2.65)$$

The value of y_0 that minimizes this upper bound is found by differentiating the right-hand side of (4.2.65) and setting the derivative to zero. It is easily verified that the solution is

$$e^{y_0^2/2} = M \qquad (4.2.66)$$

or, equivalently,

$$y_0 = \sqrt{2\ln M} = \sqrt{2\ln 2 \log_2 M}$$
$$= \sqrt{2k \ln 2} \qquad (4.2.67)$$

Having determined y_0, let us now compute simple exponential upper bounds for the integrals in (4.2.65). For the first integral we have

$$\frac{1}{\sqrt{2\pi}} \int_{-\infty}^{y_0} e^{-\left(y-\sqrt{2\gamma}\right)^2/2}\,dy = \frac{1}{\sqrt{\pi}} \int_{-\infty}^{-\left(\sqrt{2\gamma}-y_0\right)/\sqrt{2}} e^{-x^2}\,dx$$

$$= \tfrac{1}{2}\,\mathrm{erfc}\left(\frac{\sqrt{2\gamma}-y_0}{r2}\right) \qquad y_0 \leq \sqrt{2\gamma}$$

$$< e^{-\left(\sqrt{2\gamma}-y_0\right)^2/2} \qquad y_0 \leq \sqrt{2\gamma} \quad (4.2.68)$$

The second integral is upper-bounded as follows:

$$\frac{M}{\sqrt{2\pi}} \int_{y_0}^{\infty} e^{-y^2/2}e^{-\left(y-\sqrt{2\gamma}\right)^2/2}\,dy = \frac{M}{\sqrt{2\pi}}\,e^{-\gamma/2} \int_{y_0-\sqrt{\gamma/2}}^{\infty} e^{-x^2}\,dx$$

$$< \begin{cases} Me^{-\gamma/2} & y_0 \leq \sqrt{\gamma/2} \\ Me^{-\gamma/2}e^{-\left(y_0-\sqrt{\gamma/2}\right)^2} & y_0 \geq \sqrt{\gamma/2} \end{cases} \qquad (4.2.69)$$

Combining the bounds for the two integrals and substituting $e^{y_0^2/2}$ for M, we obtain

$$P_M < \begin{cases} e^{-\left(\sqrt{2\gamma}-y_0\right)^2/2} + e^{(y_0^2-\gamma)/2} & 0 \leq y_0 \leq \sqrt{\gamma/2} \\ e^{-\left(\sqrt{2\gamma}-y_0\right)^2/2} + e^{(y_0^2-\gamma)/2}e^{-\left(y_0-\sqrt{\gamma/2}\right)^2} & \sqrt{\gamma/2} \leq y_0 \leq \sqrt{2\gamma} \end{cases} \qquad (4.2.70)$$

In the range $0 \leq y_0 \leq \sqrt{\gamma/2}$, the bound may be expressed as

$$P_M < e^{(y_0^2-\gamma)/2}\left(1 + e^{-\left(y_0-\sqrt{\gamma/2}\right)^2}\right)$$

$$< 2e^{(y_0^2-\gamma)/2} \qquad 0 \leq y_0 \leq \sqrt{\gamma/2} \qquad (4.2.71)$$

In the range $\sqrt{\gamma/2} \leq y_0 \leq \sqrt{2\gamma}$, the two terms in (4.2.70) are identical. Hence

$$P_M < 2e^{-\left(\sqrt{2\gamma}-y_0\right)^2/2} \qquad \sqrt{\gamma/2} \leq y_0 \leq \sqrt{2\gamma} \qquad (4.2.72)$$

Now we substitute for y_0 and γ. Since $y_0 = \sqrt{2 \ln M} = \sqrt{2k \ln 2}$ and $\gamma = k\gamma_b$, the bounds in (4.2.71) and (4.2.72) may be expressed as

$$P_M < \begin{cases} 2e^{-k(\gamma_b - 2\ln 2)/2} & \ln M \leq \dfrac{\gamma}{4} \\[2mm] 2e^{-k\left(\sqrt{\gamma_b}-\sqrt{\ln 2}\right)^2} & \dfrac{\gamma}{4} \leq \ln M \leq \gamma \end{cases} \qquad (4.2.73)$$

The first upper bound coincides with the union bound presented earlier, but it is loose for large values of M. The second upper bound is better for large values of M. We note that $P_M \to 0$ as $k \to \infty$ ($M \to \infty$) provided that $\gamma_b > \ln 2$. But, $\ln 2$ is the limiting value of the SNR per bit required for reliable transmission when signaling at a rate equal to the capacity of the infinite bandwidth AWGN channel as shown in Sec. 4.2.9. In fact, when the substitutions

$$y_0 = \sqrt{2k \ln 2} = \sqrt{2RT \ln 2}$$
$$\gamma = \frac{TP_{\text{av}}}{N_0} = TC_{\infty} \ln 2 \qquad (4.2.74)$$

are made into the two upper bounds given in (4.2.71) and (4.2.72), where $C_{\infty} = P_{\text{av}}/N_0 \ln 2$ is the capacity of the infinite bandwidth AWGN channel, the result is

$$P_M < \begin{cases} 2 \cdot 2^{-T[(1/2)C_{\infty} - R]} & 0 \leq R \leq \tfrac{1}{4}C_{\infty} \\[2mm] 2 \cdot 2^{-T\left(\sqrt{C_{\infty}}-\sqrt{R}\right)^2} & \tfrac{1}{4}C_{\infty} \leq R \leq C_{\infty} \end{cases} \qquad (4.2.75)$$

Thus we have expressed the bounds in terms of C_{∞} and the bit rate in the channel. The first upper bound is appropriate for rates below $C_{\infty}/4$ while the second upper bound is tighter than the first for rates between $C_{\infty}/4$ and C_{∞}. Clearly, the probability of error can be made arbitrarily small by making $T \to \infty$ ($M \to \infty$ for fixed R), provided that $R < C_{\infty} = P_{\text{av}}/N_0 \ln 2$. Furthermore, we observe that the set of orthogonal waveforms achieves the channel capacity bound as $M \to \infty$, when the rate $R < C_{\infty}$.

As previously indicated, a set of M orthogonal waveforms can be constructed by frequency-shifting the carrier, thus creating frequency orthogonal (FSK) signals. For coherent detection of M-ary FSK, orthogonality among the waveforms is obtained with a frequency separation $\Delta f = 1/2T$ Hz between adjacent frequencies.† Thus, the FSK signals may be represented as

$$s_m(t) = \text{Re}\left[Ae^{j2\pi(f_c + m\Delta f)t}\right] \qquad 0 \leq t \leq T, \ m = 1, 2, \ldots, M \qquad (4.2.76)$$

where $\Delta f = 1/2T$. The bandwidth occupied by each signal is approximately Δf and, hence, the channel bandwidth required for transmission of the M wave-

---

† The FSK waveforms can be generated as CPFSK with a peak frequency deviation $f_d = 1/4T$, or a deviation ratio $h = \frac{1}{2}$.

forms is

$$W = M \Delta f = \frac{M}{2T} = \frac{2^{RT}}{2T} \tag{4.2.77}$$

We observe that, for a fixed rate R, as $M \to \infty$ or equivalently as $T \to \infty$, the channel bandwidth required becomes infinite.

A measure of the bandwidth efficiency of any digital modulation technique is the ratio R/W, where R is the information rate in bits per second and W is the required bandwidth in hertz. For M-ary signaling, $R = k/T$, where T is the symbol duration, and $k = \log_2 M$. In the case of orthogonal FSK signals detected coherently, substitution for $T = k/R$ in (4.2.77) yields the ratio

$$\frac{R}{W} = \frac{2 \log_2 M}{M} \tag{4.2.78}$$

When $M = 2$ and $M = 4$, $R/W = 1$. On the other hand, for $M > 4$ the ratio R/W is less than 1, and as $M \to \infty$, $R/W \to 0$. The ratio W/R is called the *bandwidth expansion factor* and is denoted as B_e.

Another method for constructing M orthogonal waveforms is to make them orthogonal in time. That is, the time interval T can be subdivided into M non-overlapping intervals and a carrier-modulated pulse is transmitted in one of the M time intervals for each k-bit symbol. The channel bandwidth required is approximately the reciprocal of twice the width of the time interval occupied by a pulse and, hence, $W = M/2T$, which is identical to the channel bandwidth required for orthogonal FSK detected coherently.

In general, orthogonal signaling waveforms can be constructed by subdividing the time-frequency plane into nonoverlapping cells and assigning each cell to one of the M waveforms, with the assignment of cells done in a way that satisfies the equal energy constraint.

From the above discussion it is apparent that the penalty paid for increasing the number M of waveforms is an increase in the required channel bandwidth and an increase in the complexity of the demodulator as measured by the number of matched filters or correlators. In return, the advantage of increasing M is a reduction in the SNR per bit required to attain a specified probability of error. In general, orthogonal waveforms are not efficient in terms of their bandwidth utilization. As will be shown in Chap. 5, bandwidth efficiency can be increased by selecting waveforms generated from block codes and convolutional codes.

4.2.3 Probability of Error for M-ary Biorthogonal Signals

As previously indicated in Sec. 4.1, a set of orthogonal waveforms $\{s_m(t)\}$ and their negative counterparts $\{-s_m(t)\}$ constitute a set of biorthogonal waveforms. Hence a set of M biorthogonal waveforms is formed from $M/2$ orthogonal waveforms. As a result, the channel bandwidth required to accommodate the M biorthogonal waveforms is just one-half of that required for a set of M orthogonal waveforms. There is also a reduction in the complexity of the demodulator for the biorthogonal signals relative to that for the orthogonal signals, since the former is implemented with $M/2$ matched filters or cross correlators whereas the latter requires M matched filters or cross correlators.

It is a simple matter to determine an upper (union) bound on the performance of biorthogonal signals when used for digital communication over an AWGN channel. In fact, the union bound derived for the M-ary orthogonal signals is also applicable to biorthogonal signals, although it is somewhat looser due to the fact that the $M/2$ orthogonal signals are separated in distance by $d = \sqrt{2\mathscr{E}}$ whereas the distance between any waveform and its negative is $d = 2\sqrt{\mathscr{E}}$. Hence, the errors are dominated by the signals at the minimum distance of $d = \sqrt{2\mathscr{E}}$. It is less likely (by 3 dB in SNR) that we would mistakenly select a signal when its negative is transmitted. However, instead of relying on the union bound, we evaluate the exact probability of error.

To determine the performance of the optimum demodulator, let us assume that the equivalent lowpass orthogonal waveforms are $u_m(t)$, $m = 1, 2, \ldots, M/2$, and that $u_1(t)$ is the transmitted waveform. As before, the received waveform is expressed as

$$r(t) = \alpha e^{-j\phi} u_1(t) + z(t) \qquad 0 \le t \le T \tag{4.2.79}$$

The sampled outputs of the matched filters or cross correlators yield the decision variables

$$
\begin{aligned}
U_1 &= 2\alpha\mathscr{E} + N_{1r} \\
U_m &= N_{mr} \qquad m = 2, 3, \ldots, M/2
\end{aligned} \tag{4.2.80}
$$

where the noise components $\{N_{mr}\}$, defined as

$$N_{mr} = \mathrm{Re}\left[\int_0^T z(t) u_m^*(t)\, dt\right] \qquad m = 1, 2, \ldots, M/2 \tag{4.2.81}$$

are zero mean, mutually uncorrelated, identically distributed gaussian random variables with variance $\sigma^2 = 2\mathscr{E}N_0$. The decision is made in favor of the signal corresponding to the largest $|U_m|$, while the sign of this term is used to decide whether $u_m(t)$ or $-u_m(t)$ was transmitted. According to this rule, the probability of a correct decision is equal to the probability that $U_1 > 0$ and U_1 exceeds $|U_m|$ for $m = 1, 2, \ldots, M/2$. But

$$
\begin{aligned}
P(|U_m| < U_1 | U_1 > 0) &= \frac{1}{\sqrt{2\pi}\,\sigma} \int_{-U_1}^{U_1} e^{-x^2/2\sigma^2}\, dx \\
&= \frac{1}{\sqrt{2\pi}} \int_{-U_1/\sigma}^{U_1/\sigma} e^{-x^2/2}\, dx
\end{aligned} \tag{4.2.82}
$$

Since the gaussian-distributed decision variables are uncorrelated and, hence, statistically independent, the joint probability that $|U_m| < U_1$ given $U_1 > 0$ for $m = 2, 3, \ldots, M/2$, is simply the result in (4.2.82) raised to the $(M/2 - 1)$st power. Therefore the probability of a correct decision is

$$P_c = \int_0^\infty p(U_1)\, dU_1 \left(\frac{1}{\sqrt{2\pi}} \int_{-U_1/\sigma}^{U_1/\sigma} e^{-x^2/2}\, dx\right)^{M/2-1} \tag{4.3.83}$$

and the probability of error is $P_M = 1 - P_c$. Substitution for $p(U_1)$ into (4.2.83) and a simple change in the variable of integration for the outer integral yields the final result in the form

$$P_M = 1 - \frac{1}{\sqrt{2\pi}} \int_{-\sqrt{2\gamma}}^{\infty} e^{-v^2/2} \, dv \left(\frac{1}{\sqrt{2\pi}} \int_{-(v+\sqrt{2\gamma})}^{v+\sqrt{2\gamma}} e^{-x^2/2} \, dx \right)^{M/2-1} \quad (4.2.84)$$

where, as before, $\gamma = \alpha^2 \mathscr{E}/N_0$ is the received SNR per k-bit symbol. Graphs of P_M as a function of the SNR per bit γ_b for $M = 2, 4, 8, 16, 32$ are shown in Fig. 4.2.9.

FIGURE 4.2.9
Probability of symbol error for biorthogonal signals.

4.2.4   Probability of Error for Simplex Signals

We demonstrated in Sec. 4.1 that simplex signals are a set of M equicorrelated signals with cross-correlation coefficients $\rho_r = -1/(M-1)$. As in the case of the orthogonal signals, the M equicorrelated signals are all equidistant with separation $d_{km} = \sqrt{2\mathscr{E}}$, $k, m = 1, 2, \ldots, M$. However, they achieve this mutual separation with a transmitted signal energy of $\mathscr{E}(M-1)/M$. Therefore, the probability of error for simplex signals is identical to the performance characteristics obtained for orthogonal signals, but this performance is achieved with a savings of

$$10 \log(1 - \rho_r) = 10 \log \frac{M}{M-1} \text{ dB}$$

in SNR.† The error probability may be expressed as

$$P_M = 1 - \frac{1}{\sqrt{2\pi}} \int_{-\infty}^{\infty} \left[\frac{1}{\sqrt{2\pi}} \int_{-\infty}^{v+\sqrt{2\gamma(1-\rho_r)}} e^{-x^2/2}\, dx \right]^{M-1} e^{-v^2/2}\, dv \quad (4.2.85)$$

which follows immediately from (4.2.53). For $M = 2$, the saving in SNR is 3 dB, as we have observed previously, but as M is increased the saving in SNR quickly approaches 0 dB. Hence, for large signaling alphabets, orthogonal waveforms are nearly optimum from the viewpoint of performance.

4.2.5   Probability of Error for M-ary Binary-Coded Signals

In Sec. 4.1 we described the construction of M-ary waveforms from binary codes of block length N, in which the binary digits of a code word are mapped into binary PSK waveforms. The waveforms correspond to the vertices of an N-dimensional hypercube with its center at the origin. The distance between adjacent vertices is $d = \sqrt{2\mathscr{E}/N}$.

Now suppose that we select M vertices from the N^2 possible vertices such that the distance between any pair of vertices in the subset M is

$$d \geq d_{\min} = \sqrt{2\mathscr{E}\delta/N}$$

where δ is a positive integer and $\mathscr{E}/N = \mathscr{E}_c$ is the signal energy per code element. The probability of error may be upper bounded by the union bound, which is

$$P_M \leq \frac{M-1}{2} \operatorname{erfc}\left(\sqrt{\frac{\alpha^2}{4N_0} d_{\min}^2} \right)$$

$$\leq \frac{M-1}{2} \operatorname{erfc}\left(\sqrt{\frac{\alpha^2 \mathscr{E}_c}{2N_0} \delta} \right) \quad (4.2.86)$$

---

† Although a rigorous proof has not been given to date, it is believed that the simplex set is the optimum set of M-ary waveforms for the AWGN channel.

Since each signal waveform conveys k bits of information and each waveform has energy $\mathscr{E}$, it follows that $\mathscr{E} = k\mathscr{E}_b$, where $\mathscr{E}_b$ is the energy per bit. Furthermore, $\mathscr{E}_c = \mathscr{E}/N = k\mathscr{E}_b/N$. Therefore, the union bound may be expressed in terms of the SNR per bit as

$$P_M \leq \frac{M-1}{2} \operatorname{erfc}\left(\sqrt{\gamma_b k\delta/2N}\right) \tag{4.2.87}$$

Thus, we have an upper bound on the probability of error for M-ary signal waveforms obtained from binary block codes. This topic will be discussed in more detail in Chap. 5.

4.2.6   Probability of Error for Multiphase Signals

The signal waveforms considered in Secs. 4.2.2 through 4.2.5 shared the characteristics that, for a fixed information rate R, the channel bandwidth required to transmit the signals increases as the number M of waveforms is increased. In contrast, the multiphase signal waveforms considered in this section and the multiamplitude and QAM signals considered in the next two sections, have the characteristic that the channel bandwidth requirements for a fixed rate actually decrease with an increase in M. As shown below, the penalty in using such bandwidth-efficient waveforms is an increase in the SNR required to achieve a specified level of performance. In short, these bandwidth-efficient signal waveforms are appropriate for channels having a large SNR.

The general representation for a set of M-ary phase signal waveforms is

$$s_m(t) = \operatorname{Re}\left\{u(t)\exp\left[j\left(2\pi f_c t + \frac{2\pi}{M}(m-1) + \lambda\right)\right]\right\}$$

$$m = 1, 2, \ldots, M, \, 0 \leq t \leq T \tag{4.2.88}$$

where λ is an arbitrary initial phase. Its effect is to rotate the signal constellation. The pulse $u(t)$ determines the spectral characteristics of the multiphase signal. If $u(t)$ is a rectangular pulse of the form

$$u(t) = \sqrt{\frac{2\mathscr{E}}{T}} \qquad 0 \leq t \leq T \tag{4.2.89}$$

the signal waveforms may be expressed as

$$s_m(t)\sqrt{\frac{2\mathscr{E}}{T}} A_{mc}\cos 2\pi f_c t - \sqrt{\frac{2\mathscr{E}}{T}} A_{ms}\sin 2\pi f_c t \tag{4.2.90}$$

where

$$A_{cm} = \cos\left[\frac{2\pi}{M}(m-1) + \lambda\right] \qquad m = 1, 2, \ldots, M \tag{4.2.91}$$

$$A_{sm} = \sin\left[\frac{2\pi}{M}(m-1) + \lambda\right] \qquad m = 1, 2, \ldots, M \tag{4.2.92}$$

This signaling method is called phase-shift keying (PSK). Thus, the signal given by (4.2.90) is viewed as two quadrature carriers with amplitudes A_{cm} and

FIGURE 4.2.10
Block diagram of modulator for multiphase PSK.

A_{sm}, which depend on the transmitted phase in each signaling interval. The modulator for generating this PSK signal is shown in block diagram form in Fig. 4.2.10.

The mapping or assignment of k information bits to the $M = 2^k$ possible phases may be done in a number of ways. The preferred assignment is one in which adjacent phases differ by one binary digit as illustrated in Fig. 4.2.11. This mapping is called *Gray encoding*. It is important in the demodulation of the signal because the most likely errors caused by noise involve the erroneous selection of an adjacent phase to the transmitted signal phase. In such a case, only a single bit error occurs in the k-bit sequence.

The general form of the optimum demodulator for detecting one of M signals in an AWGN channel, as derived previously, is one that computes the decision variables

$$U_m = \text{Re}\left(e^{j\phi} \int_0^T r(t)u^*(t) \exp\left\{ -j\left[\frac{2\pi}{M}(m-1) + \lambda \right] \right\} dt \right)$$

$$m = 1, 2, \dots, M \quad (4.2.93)$$

and selects the signal corresponding to the largest decision variable. We observe

FIGURE 4.2.11
PSK signal constellations for $M = 4$ and $M = 8$.

that the exponential factor under the integral in (4.2.93) is independent of the variable of integration and, hence, it can be factored out. As a result, the optimum demodulator can be implemented as a single matched filter or cross correlator which computes the vector

$$V = e^{j\phi} \int_0^T r(t) u^*(t) \, dt \qquad (4.2.94)$$

projects it onto the M unit vectors

$$V_m = \exp\left\{ j \left[\frac{2\pi}{M}(m - 1) + \lambda \right] \right\} \qquad m = 1, 2, \dots, M \qquad (4.2.95)$$

and selects the signal corresponding to the largest value obtained by this projection. Thus

$$U_m = \text{Re}\,(VV_m^*) \qquad m = 1, 2, \dots, M \qquad (4.2.96)$$

Figure 4.2.12 shows a block diagram of a demodulator for recovering the noise-corrupted signal components A_{cm} and A_{sm}, from which the vector V is

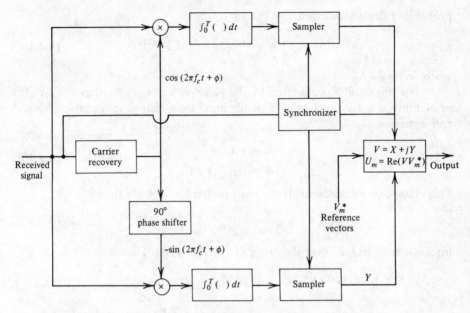

FIGURE 4.2.12
Block diagram of demodulator for PSK signals.

formed. The projection of V onto the unit vectors $V_m^* = a_{cm} - ja_{sm}$ is simply accomplished by the formation of the product $U_m = Xa_{cm} + Ya_{sm}$, $m = 1, 2, \ldots, M$. Equivalently the vector V can be followed by a phase detector which computes the phase of V, denoted by θ, and selects from the set $\{s_m(t)\}$ that signal having a phase closest to θ.

Having described the form of the modulator and demodulator for M-ary PSK, we now evaluate the performance in terms of the probability of error in an AWGN channel. In order to compute the average probability of error, we assume that $s_1(t)$ is transmitted. Then the received waveform is

$$r(t) = \alpha e^{-j\phi} u(t) + z(t) \qquad (4.2.97)$$

Substitution of (4.2.97) into (4.2.94) yields the vector V as

$$V = 2\alpha\mathscr{E} + N \qquad (4.2.98)$$

where the noise component N is a complex-valued gaussian random variable with zero mean and variance $\frac{1}{2}E(|N|^2) = 2\mathscr{E}N_0$. Let $V = X + jY$, where

$$\begin{aligned} X &= 2\alpha\mathscr{E} + \text{Re}(N) \\ Y &= \text{Im}(N) \end{aligned} \qquad (4.2.99)$$

The X and Y components are jointly gaussian random variables, with the joint probability density function

$$p(x, y) = \frac{1}{2\pi\sigma^2} e^{-[(x - 2\alpha\mathscr{E})^2 + y^2]/2\sigma^2} \qquad (4.2.100)$$

where $\sigma^2 = 2\mathscr{E}N_0$.

The phase of V, computed by the phase detector, is $\theta = \tan^{-1} Y/X$. The probability density function of θ is obtained by a change in variables from X and Y to

$$\begin{aligned} R &= \sqrt{X^2 + Y^2} \\ \theta &= \tan^{-1} Y/X \end{aligned} \qquad (4.2.101)$$

This change in variables yields the joint probability density function

$$p(r, \theta) = \frac{r}{2\pi\sigma^2} e^{-(r^2 + 4\alpha^2\mathscr{E}^2 - 4\alpha\mathscr{E}r\cos\theta)/2\sigma^2} \qquad (4.2.102)$$

Integration of $p(r, \theta)$ over the range of r yields $p(\theta)$. That is,

$$\begin{aligned} p(\theta) &= \int_0^\infty p(r, \theta)\, dr \\ &= \frac{1}{2\pi} e^{-\gamma}\left(1 + \sqrt{4\pi\gamma}\,\cos\theta\, e^{\gamma\cos^2\theta}\frac{1}{\sqrt{2\pi}}\int_{-\infty}^{\sqrt{2\gamma}\cos\theta} e^{-x^2/2}\, dx\right) \end{aligned} \qquad (4.2.103)$$

where $\gamma = \alpha^2\mathscr{E}/N_0$ is the SNR per symbol. Figure 4.2.13 illustrates $p(\theta)$ for several values of γ. It is observed that $p(\theta)$ becomes narrower and more peaked about $\theta = 0$ as γ is increased.

A decision error is made if the noise causes the phase to fall outside of the range $-\pi/M \leq \theta \leq \pi/M$. Thus

$$P_M = 1 - \int_{-\pi/M}^{\pi.M} p(\theta)\, d\theta \qquad (4.2.104)$$

In general, the integral of $p(\theta)$ does not reduce to a simple form and must be evaluated numerically, except when $M = 2$ and $M = 4$.

For binary signaling, the PSK waveforms are antipodal and, hence, the probability of error is given by (4.2.40). When $M = 4$, we have in effect two binary PSK signals in phase quadrature as indicated above. With coherent demodulation, there is no cross talk or interference between the signals on the two quadrature carriers and, hence, the bit error probability is identical to (4.2.40). That is,

$$P_2 = \tfrac{1}{2}\,\mathrm{erfc}\left(\sqrt{\gamma_b}\right) \qquad (4.2.105)$$

where γ_b is the SNR per bit. On the other hand, the symbol error probability for $M = 4$ is determined by noting that

$$\begin{aligned} P_c &= (1 - P_2)^2 \\ &= \left[1 - \tfrac{1}{2}\,\mathrm{erfc}\left(\sqrt{\gamma_b}\right)\right]^2 \end{aligned} \qquad (4.2.106)$$

FIGURE 4.2.13
Probability density function $p(\theta)$ for $\gamma = 1, 2, 4, 10$.

where P_c is the probability of a correct decision for the 2-bit symbol. The result in (4.2.106) follows from the statistical independence of the noise on the quadrature carriers. Therefore, the symbol error probability for $M = 4$ is

$$P_4 = 1 - P_c$$

$$= \text{erfc}\left(\sqrt{\gamma_b}\right)\left[1 - \tfrac{1}{4}\,\text{erfc}\left(\sqrt{\gamma_b}\right)\right] \qquad (4.2.107)$$

For $M > 4$, the symbol error probability P_M is obtained by numerically integrating (4.2.104). Figure 4.2.14 illustrates this error probability as a function of the SNR per bit for $M = 2, 4, 8, 16,$ and 32. The graphs clearly illustrate the penalty in SNR per bit as M is increased beyond $M = 4$. For example, at $P_M = 10^{-5}$, the difference between $M = 4$ and $M = 8$ is approximately 4 dB,

FIGURE 4.2.14
Probability of a symbol error for PSK signals.

and the difference between $M = 8$ and $M = 16$ is approximately 5 dB. For large values of M, doubling the number of phases requires an additional 6 dB/bit to achieve the same performance.

An approximation to the error probability for large values of M and $\gamma \gg 1$ (large SNR) may be obtained by first approximating $p(\theta)$. We note that for $\gamma \gg 1$ and $|\theta| < \pi/2$,

$$\frac{1}{\sqrt{2\pi}} \int_{-\infty}^{\sqrt{2\gamma}\,\cos\theta} e^{-x^2/2}\,dx \approx 1 - \frac{e^{-\gamma\cos^2\theta}}{\sqrt{4\pi\gamma}\,\cos\theta}$$

Hence

$$p(\theta) \approx \sqrt{\frac{\gamma}{\pi}} \cos \theta e^{-\gamma \sin^2 \theta} \qquad (4.2.108)$$

Substituting for $p(\theta)$ in (4.2.104) and by performing the change in variable from θ to $u = \sqrt{\gamma} \sin \theta$, we find that

$$P_M \approx 1 - \int_{-\pi/M}^{\pi/M} \sqrt{\gamma/\pi} \cos \theta e^{-\gamma \sin^2 \theta} \, d\theta$$

$$\approx 1 - \left(1 - \frac{2}{\sqrt{\pi}} \int_{\sqrt{\gamma} \sin \pi/M}^{\infty} e^{-u^2} \, du\right)$$

$$\approx \operatorname{erfc}\left(\sqrt{\gamma} \sin \frac{\pi}{M}\right) = \operatorname{erfc}\left(\sqrt{k\gamma_b} \sin \frac{\pi}{M}\right) \qquad (4.2.109)$$

where $k = \log_2 M$. We note that this approximation to the error probability is good for all values of M. For example, when $M = 2$ and $M = 4$ we have $\operatorname{erfc}(\sqrt{\gamma_b})$, which compares favorably with the exact error probabilities given in (4.2.105) and (4.2.107) when $\gamma_b \gg 1$.

The equivalent bit error probability for M-ary PSK is rather tedious to derive due to its dependence on the particular mapping of k-bit symbols into the corresponding signal phases. When a Gray code is used in the mapping, two k-bit symbols corresponding to adjacent signal phases differ in only a single bit. Since the most probable errors due to noise result in the erroneous selection of an adjacent phase to the true phase, most k-bit symbol errors contain only a single bit error. Hence the equivalent bit error probability for M-ary PSK is well approximated as

$$P_b \approx \frac{1}{k} P_M \qquad (4.2.110)$$

Differentially encoded PSK and DPSK. Our treatment of the demodulation of PSK signals assumed that the demodulator had a perfect estimate of the carrier phase available. In practice, however, the carrier phase is extracted from the received signal by performing some nonlinear operation that introduces a phase ambiguity. For example, in binary PSK, the signal is squared in order to remove the modulation and the double-frequency component that is generated is filtered and divided by 2 in frequency in order to extract an estimate of the carrier frequency and phase ϕ. These operations result in a phase ambiguity of 180° in the carrier phase. Similarly, in four phase PSK, the received signal is raised to the fourth power in order to remove the digital modulation and the resulting fourth harmonic of the carrier frequency is filtered and divided by 4 in order to extract the carrier component. These operations yield a carrier frequency component containing the estimate of the carrier phase ϕ, but there is a phase ambiguity of 90° in the phase estimate. Consequently, we do not have an absolute estimate of the carrier phase for demodulation.

The phase ambiguity problem resulting from the estimation of the carrier phase ϕ can be overcome by encoding the information in phase differences between successive signal transmissions as opposed to absolute phase encoding. For example, in binary PSK, the information bit 1 may be transmitted by shifting the phase of the carrier by 180° relative to the previous carrier phase, while the information bit 0 is transmitted by a zero phase shift relative to the phase in the previous signaling interval. In four phase PSK, the relative phase shifts between successive intervals are 0, 90°, 180°, and $-90°$ corresponding to the information bits 00, 01, 11, and 10, respectively. The generalization to $M > 4$ phases is straightforward. The PSK signals resulting from the encoding process are said to be *differentially encoded*. The encoding is performed by a relatively simple logic circuit preceding the modulator.

Demodulation of the differentially encoded PSK signal is performed as described above, assuming an absolute phase reference. Thus, the received signal is demodulated to one of the M possible transmitted phases in each signaling interval. Following the demodulator is a relatively simple phase comparator that compares the phases of the demodulated signal over two consecutive intervals in order to extract the information.

Coherent demodulation of differentially encoded PSK results in a higher probability of error than the error probability derived for absolute phase encoding. With differentially encoded PSK, an error in the demodulated phase of the signal in any given interval will usually result in decoding errors over two consecutive signaling intervals. This is especially the case for error probabilities below 0.1. Therefore, the probability of error in differentially encoded M-ary PSK is approximately twice the probability of error for M-ary PSK with absolute phase encoding. However, this factor of 2 increase in the error probability translates into a relatively small loss in SNR.

The differentially encoded PSK also allows another form of demodulation that does not require the estimation of the carrier phase. Instead, the received signal in any given signaling interval is compared to the phase of the received signal in the previous signaling interval. To elaborate, suppose that the received signal is passed through a filter matched to $u(t)$ and the output of the filter is sampled every T seconds. At the nth signaling interval, this processing yields the vector

$$V_n = 2\alpha\mathscr{E}e^{j(\theta_n-\phi)} + N_n \qquad (4.2.111)$$

where θ_n is the phase angle of the transmitted signal and N_n is the noise component. Similarly the vector from the ·previous signaling interval may be expressed as

$$V_{n-1} = 2\alpha\mathscr{E}e^{j(\theta_{n-1}-\phi)} + N_{n-1}$$

The decision variable is the phase difference between these two complex numbers. Equivalently, we can project V_n onto V_{n-1} and use the phase of the resulting

FIGURE 4.2.15
Block diagram of binary DPSK demodulator.

complex number. That is, the decision variable is the phase of

$$U_n = V_n V_{n-1}^* = 4\alpha^2 \mathscr{E}^2 e^{j(\theta_n - \theta_{n-1})}$$
$$+ 2\alpha \mathscr{E} N_n e^{-j(\theta_{n-1} - \phi)}$$
$$+ 2\alpha \mathscr{E} N_{n-1}^* e^{j(\theta_n - \phi)} + N_n N_{n-1}^* \qquad (4.2.112)$$

which in the absence of noise is $\theta_n - \theta_{n-1}$. Thus the mean value of U_n is independent of the carrier phase. A differentially encoded PSK signal that is demodulated in this manner is called *differential PSK* (DPSK).

In dealing with real bandpass signals, the demodulation of DPSK may be accomplished as illustrated in Fig. 4.2.15. If the pulse $u(t)$ is rectangular, the matched filters in Fig. 4.2.15 may be replaced by integrate-and-dump filters.

In practice, the demodulation of DPSK is often simplified by delaying the received signal by the symbol interval T and using the delayed signal to multiply the received signal. For example, the demodulators for four-phase DPSK with $\lambda = 0$ and $\pi/4$ are shown in Figs. 4.2.16 and 4.2.17, respectively. The two corresponding DPSK signal constellations are shown in Fig. 4.2.18. In these demodulators, the two binary digits are separated by introducing appropriate phase shifts in the delayed signal. Thus, the received signal is projected on the phase-shifted delayed signal. These demodulators are suboptimum, as demonstrated by Park (1978).

Now let us consider the performance of a DPSK demodulator. The derivation of the exact value for the probability of error for M-ary DPSK is extremely difficult except for $M = 2$. The major difficulty is encountered in the determination of the probability density function for the phase of the random variable U_n given in (4.2.112). However, an approximation to the performance of DPSK is easily obtained, as we now demonstrate.

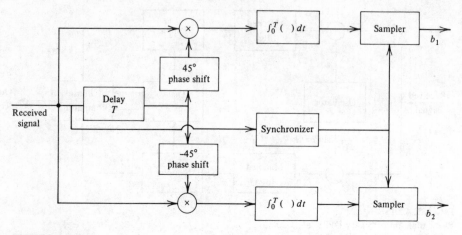

FIGURE 4.2.16
Block diagram of demodulator for four-phase DPSK with $\lambda = 0$.

Without loss of generality, suppose the phase difference $\theta_n - \theta_{n-1} = 0$. Furthermore, the exponential factors $e^{-j(\theta_{n-1} - \phi)}$ and $e^{j(\theta_n - \phi)}$ in (4.2.112) can be absorbed into the gaussian noise components N_n and N_{n-1} without changing their statistical properties. Therefore, U_n can be expressed as

$$U_n = 4\alpha^2 \mathscr{E}^2 + 2\alpha\mathscr{E}(N_n + N_n N_{n-1}^*) + N_n N_{n-1}^* \qquad (4.2.113)$$

The complication in determining the probability density function for the phase is the term $N_n N_{n-1}^*$, which is the product of two complex-valued gaussian random variables. However, at signal-to-noise ratios of practical interest, the term $N_n N_{n-1}^*$

FIGURE 4.2.17
Block diagram of demodulator for four-phase DPSK with $\lambda = \pi/4$.

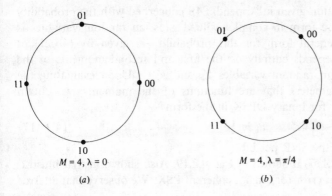

FIGURE 4.2.18
Signal constellations showing relative phase shift between successive intervals for four-phase DPSK.

is small relative to the dominant noise term $2\alpha\mathscr{E}(N_n + N_{n-1}^*)$. If we neglect the term $N_n N_{n-1}^*$ and we also normalize U_n by dividing through by $2\alpha\mathscr{E}$, the new decision variable becomes

$$U_n' = 2\alpha\mathscr{E} + (N_n + N_{n-1}^*)$$
$$= X + jY \qquad (4.2.114)$$

where

$$X = 2\alpha\mathscr{E} + \text{Re}\,(N_n + N_{n-1}^*)$$
$$Y = \text{Im}\,(N_n + N_{n-1}^*)$$

The variables X and Y are uncorrelated gaussian random variables with identical variances $\sigma^2 = 4\mathscr{E}N_0$. The phase of U_n' is

$$\theta_u = \tan^{-1}\frac{Y}{X} \qquad (4.2.115)$$

At this stage we have a problem that is identical to the one that we have solved previously for coherent PSK. The only difference is that the noise variance is now twice as large as in the case of coherent PSK. Thus we conclude that the performance of DPSK is 3 dB poorer than that for coherent PSK. This result is relatively good for $M > 4$, but it is pessimistic for $M = 2$ in the sense that the loss in DPSK relative to coherent PSK is less than 3 dB at large SNR. This is demonstrated below.

In binary DPSK, the two possible transmitted phase differences are zero and π rad. As a consequence, only the real part of the decision variable U_n is needed for recovering the information. Using (4.2.112) with $\theta_n - \theta_{n-1} = 0$ or, equivalently, using (4.2.113), we express the real part of U_n as

$$\text{Re}\,(U_n) = \tfrac{1}{2}(V_n V_{n-1}^* + V_n^* V_{n-1}) \qquad (4.2.116)$$

Since the phase difference between the two successive signaling intervals is zero, an error is made if $\text{Re}\,(U_n) < 0$. The probability that $V_n V_{n-1}^* + V_n^* V_{n-1} < 0$ is a special case of a derivation given in Appendix 4B concerned with the probability that a general quadratic form in complex-valued gaussian random variables is less than zero. The general form for this probability is given by (4B.21) of Appendix 4B and it depends entirely on the first and second moments of the complex-valued gaussian random variables V_n and V_{n-1}. Upon evaluating the moments and the parameters that are functions of the moments, we obtain the probability of error for binary DPSK in the form

$$P_b = \tfrac{1}{2} e^{-\gamma_b} \qquad (4.2.117)$$

where $\gamma_b = \alpha^2 \mathscr{E}_b / N_0$ is the SNR per bit.

The graph of (4.2.117) is shown in Fig. 4.2.19. Also shown in that illustration is the probability of error for binary, coherent PSK. We observe that at error

FIGURE 4.2.19
Probability of error for binary PSK and DPSK.

probabilities of $P_b \leq 10^{-3}$ the difference in SNR between binary PSK and binary DPSK is less than 3 dB. In fact, at $P_b \leq 10^{-5}$ the difference in SNR is less than 1 dB.

The probability of a binary digit error for four-phase DPSK with Gray coding can be expressed in terms of well-known functions, but its derivation is quite involved. We simply state the result at this point and refer the interested reader to Appendix 7A for the details in the derivation. It is expressed in the form

$$P_b = Q(a, b) - \tfrac{1}{2}I_0(ab)\exp\left[-\tfrac{1}{2}(a^2 + b^2)\right] \tag{4.2.118}$$

where $Q(a, b)$ is the Q function defined previously by (1.1.119) and (1.1.120), $I_0(x)$ is the modified Bessel function of order zero, which was defined by (1.1.117), and the parameters a and b are defined as

$$a = \sqrt{2\gamma_b\left(1 - \frac{1}{\sqrt{2}}\right)}$$
$$b = \sqrt{2\gamma_b\left(1 + \frac{1}{\sqrt{2}}\right)} \tag{4.2.119}$$

Figure 4.2.20 illustrates the probability of a binary digit error for two- and four-phase DPSK and coherent PSK signaling obtained from evaluating the exact formulas derived in this section. Since binary DPSK is only slightly inferior to binary PSK at large SNR, and DPSK does not require an elaborate method for estimating the carrier phase, it is often used in digital communications systems. On the other hand, four-phase DPSK is approximately 2.3 dB poorer in performance than four-phase PSK at large SNR. Consequently the choice between these two four-phase systems is not as clear cut. One must weigh the 2.3-dB loss against the reduction in implementation complexity.

Having discussed the performance of multiphase signaling, we now turn our attention to the question of bandwidth efficiency. The channel bandwidth required to transmit a set of multiphase signals is simply the bandwidth of the equivalent lowpass signal pulse $u(t)$, which depends on its detailed characteristics. For our purposes, we assume that $u(t)$ is a pulse of duration T and that its bandwidth W is approximately equal to the reciprocal of T. Thus $W = 1/T$ and since $T = k/R = (\log_2 M)/R$, it follows that

$$W = \frac{R}{\log_2 M} \tag{4.2.120}$$

Therefore, as M is increased, the channel bandwidth required decreases when the information rate R remains fixed. The bandwidth efficiency of multiphase signaling as measured by the ratio R/W in bits per second per hertz is

$$\frac{R}{W} = \log_2 M \tag{4.2.121}$$

FIGURE 4.2.20
Probability of bit error for binary and four-phase PSK and DPSK.

4.2.7 Probability of Error for Multiamplitude Signals

In general, the signal waveforms in M-ary PAM are represented as

$$s_m(t) = A_m s(t)$$

$$= A_m \mathrm{Re}\left[u(t)e^{j2\pi f_c t}\right] \qquad m = 1, 2, \ldots, M, \; 0 \le t \le T \quad (4.2.122)$$

where the signal amplitude A_m takes the discrete values (levels)

$$A_m = 2m - 1 - M \qquad m = 1, 2, \ldots, M \quad (4.2.123)$$

$M = 8$

FIGURE 4.2.21
Signal constellations for $M = 4$ and $M = 8$ PAM.

Note that when $M = 2$, the binary signaling waveforms are antipodal. Figure 4.2.21 illustrates the signal constellations for $M = 4$ and 8, with Gray encoding used in the mapping of the information bits into signal amplitudes.

For this class of signals each waveform has a different energy, namely,

$$\mathscr{E}_m = \int_0^T s_m^2(t)\, dt$$

$$= A_m^2 \frac{1}{2} \int_0^T |u(t)|^2\, dt$$

$$= A_m^2 \mathscr{E} \tag{4.2.124}$$

where $\mathscr{E}$ is the energy of the pulse $u(t)$. Under these conditions, we recall that the optimum demodulator computes the decision variables given in (4.2.5) or (4.2.15), which contain the bias terms $\{a\mathscr{E}_m\}$.

When $u(t)$ is a rectangular pulse, it is defined as

$$u(t) = \sqrt{\frac{2\mathscr{E}}{T}} \qquad 0 \leq t \leq T$$

In this case, the signaling method is usually called amplitude-shift keying (ASK).

The fact that the signaling waveforms have a common waveform $u(t)$ and differ only in amplitude allows us to simplify the demodulator structure considerably, just as we did in the detection of PSK. In particular, we need only employ a single matched filter or cross correlator corresponding to the common waveform $u(t)$, followed by an amplitude detector. The output of the matched filter or cross correlator at the sampling instant is the decision variable

$$U = \text{Re}\left[e^{j\phi} \int_0^T r(t) u^*(t) \, dt \right] \qquad (4.2.125)$$

We observe that estimation of α is unnecessary in generating the decision variable U, but since U is directly proportional to α through its dependence on $r(t)$, knowledge of this parameter is necessary in realizing the optimum amplitude detector, as discussed below.

If $s_m(t)$ is transmitted over an AWGN channel, the received signal, expressed in the equivalent lowpass form, is

$$r(t) = \alpha e^{-j\phi} A_m u(t) + z(t) \qquad (4.2.126)$$

Substitution for $r(t)$ from (4.2.126) into (4.2.125) yields

$$U = 2\alpha\mathscr{E}A_m + \nu$$
$$= \mu_m + \nu \qquad (4.2.127)$$

where, by definition,

$$\mu_m = 2\alpha\mathscr{E}A_m$$
$$\nu = \text{Re}\left[\int_0^T z(t) u^*(t) \, dt \right] \qquad (4.2.128)$$

Thus μ_m is the mean value of U, and ν is a gaussian noise variable having zero mean and variance $\sigma_\nu^2 = 2\mathscr{E}N_0$. Therefore, the probability density function for U is

$$p(U) = \frac{1}{\sqrt{2\pi}\,\sigma_\nu} e^{-(U-\mu_m)^2/2\sigma_\nu^2} \qquad (4.2.129)$$

In the amplitude detector the decision variable U is compared with the M possible received levels and a decision is made in favor of the level nearest U. The factor $2\alpha\mathscr{E}$, which includes the attenuation introduced by the channel, scales the received levels so that, in general, they are different from the transmitted levels. This implies that the amplitude detector must take such a scale factor into account in making its decisions. In practice, such scaling is usually compensated prior to the amplitude detector (and prior to the matched filter or cross correlator) by means of an automatic gain control (AGC) having a time constant that is large relative to the signaling interval. Thus the AGC does not respond to the instantaneous signal amplitude but it does compensate for slow amplitude variations in the received signal caused by the channel. In other words, the AGC may be viewed as an estimator and compensator for the channel attenuation factor α. This means that the signal levels presented to the amplitude detector are invariant to such a scale factor. Figure 4.2.22 illustrates a functional block diagram of a PAM demodulator.

FIGURE 4.2.22
Block diagram of demodulator for PAM.

Let us now determine the probability of error for the amplitude detector. We assume that α is known exactly. On the basis that all levels are equally likely *a priori*, the average probability of error is simply the probability that the noise variable v exceeds in magnitude one-half the distance between levels. However, when either one of the two outside levels $\pm(M-1)$ is transmitted, an error can occur in one direction only. Thus we have

$$P_M = \frac{M-1}{M} P(|U - \mu_k| > 2\alpha\mathscr{E})$$

$$= \frac{M-1}{M} \frac{2}{\sqrt{2\pi}\,\sigma_v} \int_{2\alpha\mathscr{E}}^{\infty} e^{-x^2/2\sigma_v^2}\, dx$$

$$= \frac{M-1}{M} \operatorname{erfc}\left(\sqrt{\frac{\alpha^2\mathscr{E}}{N_0}}\right) \qquad (4.2.130)$$

Since the transmitted energy or power in any signaling interval is a function of the transmitted signal amplitude, it is appropriate to express P_M in terms of the power or energy averaged over the set of M amplitudes. The average transmitted power is

$$P_{\text{av}} = \frac{1}{T} \int_0^T E\left[s_m^2(t)\right] dt$$

$$= \frac{E(A_m^2)}{2T} \int_0^T |u(t)|^2\, dt$$

$$= \frac{E(A_m^2)}{T}\mathscr{E} \qquad (4.2.131)$$

where the expectation is taken over the statistical distribution of the signal amplitudes. But the expected value of A_m^2 is

$$E\left(A_m^2\right) = E\left[(2m - 1 - M)^2\right] = \frac{2}{M} \sum_{m=1}^{M/2} (2m - 1)^2$$

$$= \frac{M^2 - 1}{3} \tag{4.2.132}$$

Substitution of (4.2.132) into (4.2.131) yields

$$P_{\text{av}} = \frac{M^2 - 1}{3} \frac{\mathscr{E}}{T} \tag{4.2.133}$$

FIGURE 4.2.23
Probability of a symbol error for PAM.

and, hence,

$$\mathscr{E} = \frac{3P_{av}T}{M^2 - 1} \qquad (4.2.134)$$

Substitution for $\mathscr{E}$ in the expression for the error probability given in (4.2.130) yields the result

$$P_M = \frac{M-1}{M} \operatorname{erfc}\left(\sqrt{\frac{3}{M^2 - 1}\frac{\alpha^2 P_{av}T}{N_0}}\right) \qquad (4.2.135)$$

If we define an average signal-to-noise per symbol as

$$\gamma_{av} = \frac{\alpha^2 P_{av}T}{N_0} \qquad (4.2.136)$$

then

$$P_M = \frac{M-1}{M} \operatorname{erfc}\left(\sqrt{\frac{3}{M^2 - 1}\gamma_{av}}\right) \qquad (4.2.137)$$

The probability of a symbol error is plotted in Fig. 4.2.23 as a function of the average SNR per bit $\gamma_b = \gamma_{av}/k$. We observe that, just as in the performance of PSK, there is a penalty in increasing the number M of signaling waveforms. In terms of the average SNR per bit, the loss is approximately 4 dB for $M = 4$ and, for large M, the penalty approaches 6 dB for every factor of 2 increase in M.

Finally, we consider the bandwidth requirements for transmitting the PAM waveforms. Two equally bandwidth-efficient methods for transmitting PAM are (1) single-sideband PAM and (2) quadrature PAM, in which the information sequence is split into two parallel sequences that are transmitted via PAM on the quadrature carriers $\cos 2\pi f_c t$ and $\sin 2\pi f_c t$. For both methods the required channel bandwidth W is approximately equal to $1/2T$ and, since $1/T = R/k = R/\log_2 M$ symbols/s, we obtain the result

$$W = \frac{R}{2\log_2 M} \qquad (4.2.138)$$

The bandwidth efficiency for PAM as measured by the ratio R/W in bits per second per hertz is

$$\frac{R}{W} = 2\log_2 M \qquad (4.2.139)$$

When compared with the bandwidth efficiency of PSK given by (4.2.121), we observe that PAM is a factor of 2 better by virtue of the fact that the PAM signal can be transmitted by single sideband or, alternatively, by use of two independent quadrature carriers (double sideband).

4.2.8   Probability of Error for QAM

In Secs. 4.2.6 and 4.2.7 we considered the use of multiple signal phases and multiple signal amplitudes, respectively, for transmitting k information bits per symbol (per waveform) over the AWGN channel. We observed that the bandwidth efficiency R/W of these digital modulations is proportional to $k = \log_2 M$. We also observed that for large values of k, an increase from k to $k + 1$ requires a 6-dB (fourfold) increase in the average transmitted power to maintain the same error rate performance. This rather large penalty can be reduced, as is shown below, by using a combination of multiple phases and multiple amplitudes to transmit the k-bit information symbols.

As we have previously indicated in Sec. 4.1, a quadrature amplitude modulated (QAM) or quadrature amplitude-shift keying (QASK) signal is represented as

$$s_m(t) = \mathrm{Re}\left[(A_{mc} + jA_{ms})u(t)e^{j2\pi f_c t}\right] \qquad 0 \le t \le T$$

$$= A_{mc}\sqrt{\frac{2\mathscr{E}}{T}}\,\cos 2\pi f_c t - A_{ms}\sqrt{\frac{2\mathscr{E}}{T}}\,\sin 2\pi f_c t$$

$$0 \le t \le T, m = 1, 2, \ldots, M \quad (4.2.140)$$

where A_{mc} and A_{ms} are the information-bearing signal amplitudes of the quadrature carriers and $u(t) = \sqrt{2\mathscr{E}/T}$, $0 \le t \le T$. Pulse shapes other than rectangular may be used to reduce the spectral occupancy of the transmitted signal.

QAM signals are two-dimensional signals and, hence, they are represented by the vectors

$$\mathbf{s}_m = \left(\sqrt{\mathscr{E}}A_{mc}, \sqrt{\mathscr{E}}A_{ms}\right) \qquad m = 1, 2, \ldots, M \quad (4.2.141)$$

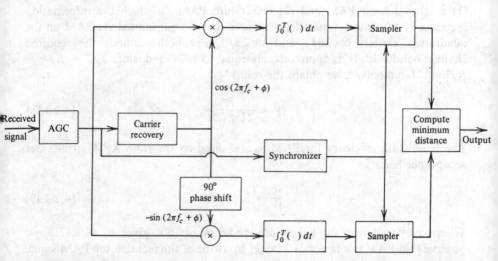

FIGURE 4.2.24
Block diagram of demodulator for QAM.

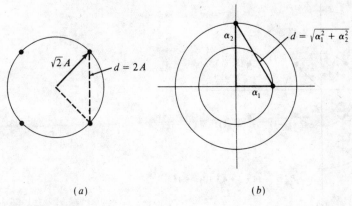

FIGURE 4.2.25
Four-point signal constellations.

or, equivalently, by the complex-valued quantities

$$u_m = \sqrt{2\mathscr{E}}\,(A_{mc} + jA_{ms}) \qquad m = 1, 2, \ldots, M \qquad (4.2.142)$$

The optimum demodulator for QAM computes the decision variables given by either (4.2.5) or (4.2.15) and selects the signal corresponding to the largest decision variable. However, just as in the case of PSK and PAM, the demodulator may be simplified by using a single complex-valued (or two real) correlator or matched filter to $u(t)$, followed by a combined amplitude and phase detector. The need for bias correction by the factors $\{\alpha \mathscr{E}_m\}$ is satisfied by using an automatic gain control (AGC) that tracks and corrects long-term amplitude variations caused by the channel.

The received real signal is demodulated, as shown in Fig. 4.2.24, by multiplying it with the two quadrature carriers appropriately shifted in phase to compensate for the carrier phase shift, integrating over the signaling interval of duration T, and sampling the output every T seconds to obtain the two quadrature components. These components are noise-corrupted estimates of the transmitted vector s_m and may be denoted as $\hat{s}$. The distances between $\hat{s}$ and the M possible transmitted points $\{s_m\}$ are then computed and the decision is made in favor of the signal point that is closest to $\hat{s}$.

Let us now consider some specific signal constellations. We begin with signal sets that have $M = 4$ points. Figure 4.2.25 illustrates two four-point signal sets. The first is a four-phase PSK signal and the second is a QAM signal with two-amplitude levels, labeled α_1 and α_2, and four phases. In order to maintain the same minimum distance in the two sets of signals, the condition that must be satisfied is

$$\alpha_1^2 + \alpha_2^2 = 4A^2$$

Clearly, the solution $\alpha_1 = A$ and $\alpha_2 = \sqrt{3}\,A$ satisfies this condition. The corre-

$$P_{av} = \frac{A^2}{8}(4 \cdot 2 + 4 \cdot 10) = 6A^2$$

(a)

$$P_{av} = \frac{A^2}{8}(4 \cdot 4 + 4 \cdot 8) = 6A^2$$

(b)

FIGURE 4.2.26
Eight-point signal constellations.

sponding four-point QAM constellation requires an average power of

$$P_{av} = \tfrac{1}{2}\big(\alpha_1^2 + \alpha_2^2\big) = 2A^2$$

which is the same as the average power for the $M = 4$ PSK constellation. Hence, for all practical purposes, the error rate performance of the two signal sets is the same. In other words, there is no performance advantage of the two-amplitude QAM signal set over four-phase PSK.

Next, let us consider combined amplitude and phase signals for $M = 8$. In this case, there are many possible signal sets. We shall consider the four signal constellation shown in Fig. 4.2.26, all of which consist of two amplitudes and have a minimum distance between points equal to $2A$. The coordinates (A_{mc}, A_{ms}) for each signal set (normalized by A) are given in the figure. Assuming that the signal points are equally probable, the average transmitted power

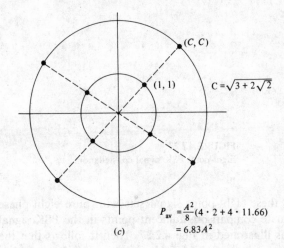

$$P_{av} = \frac{A^2}{8}(4 \cdot 2 + 4 \cdot 11.66)$$
$$= 6.83A^2$$

(c)

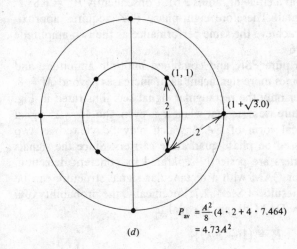

$$P_{av} = \frac{A^2}{8}(4 \cdot 2 + 4 \cdot 7.464)$$
$$= 4.73A^2$$

(d)

FIGURE 4.2.26 (*Continued*).

(normalized) is

$$P_{av} = \frac{1}{M} \sum_{m=1}^{M} \left(A_{mc}^2 + A_{ms}^2 \right)$$

$$= \frac{A^2}{M} \sum_{m=1}^{M} \left(a_m^2 + b_m^2 \right) \qquad (4.2.143)$$

where (a_m, b_m) are the normalized coordinates of the signal points.

The first two signal sets in Fig. 4.2.26 contain points that fall on a rectangular grid. Both have identical average power $P_{av} = 6A^2$. The third signal set requires an average transmitted power of $6.83A^2$, while the fourth requires $4.73A^2$. Therefore the last signal set is approximately 1 dB better than the first two and approximately 1.6 dB better than the third.

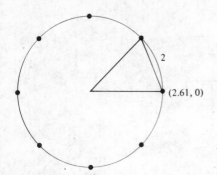

FIGURE 4.2.27
Eight-point PSK signal constellation.

We may also compare these eight-point signal sets with pure eight-phase PSK. To obtain an approximate comparison, adjacent points in the PSK signal are separated by two units as illustrated in Fig. 4.2.27. Then it follows that the points in the PSK signal fall on a circle of radius 2.61. Consequently $P_{av} = 6.83A^2$ for the eight-phase PSK signal. Therefore eight-phase PSK requires approximately 1.6 dB more power to achieve the same performance as the two-amplitude signal set shown in Fig. 4.2.26d.

The difference between pure PSK and combined multiple amplitude and multiple phase signaling becomes more significant as M increases beyond $M = 8$. For $M \geq 16$ we shall consider only the rectangular signal sets illustrated in Fig. 4.1.5. For such a signal structure we note that when k is even, the $M = 2^k$ signal points result in a symmetrical form of QAM which may be viewed as two separate PAM signals impressed on phase-quadrature carriers. Since the signals on the phase-quadrature carriers are perfectly separated by coherent detection, the error rate performance for QAM with a rectangular signal structure can be determined directly from the results of Sec. 4.2.7. Specifically, the probability of a correct decision for the M-ary QAM system is equal to

$$P_c = \left(1 - P_{\sqrt{M}}\right)^2$$

where $P_{\sqrt{M}}$ is the probability of error of a $\sqrt{M}$-ary PAM system, where the average power in each quadrature signal of the PAM is one-half of that in the QAM system. By appropriately modifying (4.2.137), we obtain

$$P_{\sqrt{M}} = \left(1 - \frac{1}{\sqrt{M}}\right)\text{erfc}\left(\sqrt{\frac{3}{M-1}\frac{1}{2}\gamma_{av}}\right)$$

Therefore the probability of a symbol error for the M-ary QAM system is

$$P_M = 1 - \left(1 - P_{\sqrt{M}}\right)^2 = 2\left(1 - \frac{1}{\sqrt{M}}\right)\text{erfc}\left(\sqrt{\frac{3}{2(M-1)}k\gamma_b}\right)$$

$$\times \left[1 - \frac{1}{2}\left(1 - \frac{1}{\sqrt{M}}\right)\text{erfc}\left(\sqrt{\frac{3}{2(M-1)}k\gamma_b}\right)\right] \quad (4.2.144)$$

where γ_{av} is the average SNR per k-bit symbol and γ_b is the average SNR per bit.

We note that this result is exact for $M = 2^k$, when k is even. On the other hand, when k is odd there is no equivalent $\sqrt{M}$-ary PAM signaling system which can be used to equate performance. This is no problem, however, since it is rather easy to determine the error rate for a rectangular signal set. If we employ a rectangular decision region around each point in the signal set, it is relatively straightforward to show that the symbol error probability is tightly upper-bounded as (see Prob. 4.20)

$$
\begin{aligned}
P_M &\leq 1 - \left[1 - \mathrm{erfc}\left(\sqrt{\frac{3\gamma_{av}}{2(M-1)}} \right) \right]^2 \\
&\leq 2\,\mathrm{erfc}\left(\sqrt{\frac{3\gamma_{av}}{2(M-1)}} \right) \left[1 - \frac{1}{2}\mathrm{erfc}\left(\sqrt{\frac{3\gamma_{av}}{2(M-1)}} \right) \right] \\
&\leq 2\,\mathrm{erfc}\left(\sqrt{\frac{3}{2(M-1)} k\gamma_b} \right)
\end{aligned}
\qquad (4.2.145)
$$

for any value of $k \geq 1$.

To see how the performance of M-ary PSK compares with M-ary QAM, consider an M-ary PSK signal set with equally spaced points on a circle of radius r such that the distance between adjacent points is 2, as in the rectangular QAM signal set shown in Fig. 4.1.5. Since the angular distance between adjacent points for the PSK signal set is $2\pi/M$, it is easy to see that the radius of the circle must be

$$
r = \frac{1}{\sin(\pi/M)} \approx \frac{M}{\pi} \qquad M \text{ large}
$$

Hence the average power for M-ary PSK must increase as $(M/\pi)^2$ in order to maintain the same error rate performance as M increases. On the other hand, (4.2.144) indicates that the average power for QAM must increase as $2(M-1)/3$. The ratio

$$
D_M = \frac{3M^2}{2(M-1)\pi^2} \qquad (4.2.146)
$$

indicates the advantage of using QAM instead of PSK. Table 4.2.2 lists this ratio in decibels for several values of M.

TABLE 4.2.2

M	$10\log\left[3M^2 / 2(M-1)\pi^2\right]$, dB
8	1.43
16	4.14
32	7.01
64	9.95

FIGURE 4.2.28
Probability of a symbol error for QAM and PSK.

The error rate performance of QAM given by (4.2.144) is illustrated in Fig. 4.2.28 for $M = 4$, 16, and 64. For purposes of comparison, the performance of coherent PSK for $M = 16$ and $M = 32$ is also illustrated in Fig. 4.2.28. We observe that the difference in performance between QAM and PSK is quite close to the values predicted by the approximation given in (4.2.146). Finally, in Table 4.2.3 we list the factor $10 \log 2(M - 1)/3$, which represents the increase in average power required to maintain a given level of performance for QAM as the number of signal points increases. Note that increasing M by a factor of 2 requires approximately a 3-dB increase in power. Consequently, it costs us 3 dB in power to increase the rate by 1 bit per symbol.

TABLE 4.2.3
QAM constellations

Number of points	Average power in dB
4	3
8	6.7
16	10.0
32	13.2
64	16.2
128	19.2
256	22.3

The rectangular signal sets described above have the distinct advantage of being easily generated as two PAM signals impressed on phase-quadrature carriers. In addition, they are easily demodulated, as illustrated in Fig. 4.2.24, to yield the two quadrature components. As an alternative to computing distances as previously described, each component can be individually detected by comparing it to a set of thresholds. Although many other different types of QAM signal sets have been investigated and reported on in the technical literature few outperform the rectangular signal sets described above, and when they do, their SNR advantage in gaussian noise is relatively small.

Finally, we observe that the bandwidth efficiency of QAM is equivalent to that of PAM, i.e., $R/W = 2\log_2 \sqrt{M} = \log_2 M$.

4.2.9 Comparison of Digital Signaling Methods

The digital signaling techniques presented thus far can be compared in a number of ways. For example, one can compare them on the basis of the SNR required to achieve a specified probability of error. However, such a comparison would not be very meaningful unless it were made on the basis of a fixed data rate or, equivalently, on the basis of a fixed bandwidth. The most compact and meaningful comparison is one that is based on the normalized data rate R/W (bits per second per hertz of bandwidth) versus the SNR per bit required to achieve a given error rate. Table 4.2.4 summarizes the bandwidth efficiency and error

TABLE 4.2.4
Summary of performance and rate of modulation methods

Type of modulation	R/W	P_M
PAM	$2\log_2 M$	(4.2.137) and Fig. 4.2.23
QAM	$\log_2 M$	(4.2.144) and Fig. 4.2.28
PSK	$\log_2 M$	(4.2.104), (4.2.109), and Fig. 4.2.14
Orthogonal	$(2\log_2 M)/M$	(4.2.53) and Fig. 4.2.8

FIGURE 4.2.29
Comparison of several modulation methods at 10^{-5} symbol error probability.

probability formulas for the various types of signals described in this section. A comparison of their performance is illustrated in Fig. 4.2.29 for a symbol error rate $P_M = 10^{-5}$.

The graphs in Fig. 4.2.29 illustrate the fact that PAM, QAM, PSK, and DPSK are bandwidth-efficient digital signaling techniques. That is, for a fixed bandwidth, the rate increases (logarithmically) with an increase in the number of waveforms. The cost for the increase in rate is an increase in the SNR per bit to achieve a specified error rate. Consequently these digital signaling methods are appropriate for channels in which the SNR is large enough to support a normalized rate $R/W \geq 1$. On the other hand, orthogonal signals, as well as biorthogonal and simplex signals, make inefficient use of channel bandwidth in the sense that $R/W \leq 1$. That is, these signaling methods trade bandwidth for a reduction in the SNR per bit required to achieve a given error rate. Therefore they are appropriate for channels in which the SNR is small and there is sufficient bandwidth to allow for $R/W < 1$.

For purposes of comparison, Fig. 4.2.29 also illustrates the capacity of a band-limited AWGN channel, having a bandwidth W and a SNR per bit $\gamma_b = \mathcal{E}_b/N_0$. Referring back to Sec. 2.4, the capacity of such a channel in bits per second is

$$C = W \log_2 \left(1 + \frac{C}{W} \frac{\mathcal{E}_b}{N_0} \right) \tag{4.2.147}$$

Consequently the capacity normalized by the bandwidth W is

$$\frac{C}{W} = \log_2 \left(1 + \frac{C}{W} \frac{\mathcal{E}_b}{N_0} \right) \tag{4.2.148}$$

Hence

$$\frac{\mathcal{E}_b}{N_0} = \frac{2^{C/W} - 1}{C/W} \tag{4.2.149}$$

When $C/W = 1$, $\mathcal{E}_b/N_0 = 1$ (0 dB). As $C/W \to \infty$,

$$\frac{\mathcal{E}_b}{N_0} \approx \frac{2^{C/W}}{C/W}$$

$$\approx \exp\left(\frac{C}{W} \ln 2 - \ln \frac{C}{W} \right)$$

Thus $\mathcal{E}_b/N_0$ increases exponentially as $C/W \to \infty$. On the other hand as $C/W \to 0$,

$$\frac{\mathcal{E}_b}{N_0} = \lim_{C/W \to 0} \frac{2^{C/W} - 1}{C/W} = \ln 2 \tag{4.2.150}$$

which is -1.6 dB.

As we have already shown, the lower limit can be achieved by use of orthogonal signals in the limit as the number of waveforms $M \to \infty$. It can also be obtained by use of either biorthogonal or simplex signals for $M \to \infty$. In the case of orthogonal signals, we showed that the error probability is upper-bounded as

$$P_M < \begin{cases} 2 \cdot 2^{-T[(1/2)C_\infty - R]} & 0 \leq R \leq \frac{1}{4}C_\infty \\ 2 \cdot 2^{-T(\sqrt{C_\infty} - \sqrt{R})^2} & \frac{1}{4}C_\infty \leq R \leq C_\infty \end{cases} \tag{4.2.151}$$

where R is the bit rate and C_∞ is the capacity of the infinite bandwidth white

FIGURE 4.2.30
Channel reliability function for the infinite bandwidth AWGN channel.

gaussian noise channel, defined as

$$C_\infty = \frac{P_{av}}{N_0 \ln 2} \text{ bits/s} \qquad (4.2.152)$$

The exponential factor

$$E(R) = \begin{cases} \frac{1}{2}C_\infty - R & 0 \le R \le \frac{1}{4}C_\infty \\ \left(\sqrt{C_\infty} - \sqrt{R}\right)^2 & \frac{1}{4}C_\infty \le R \le C_\infty \end{cases} \qquad (4.2.153)$$

in (4.2.151) is called the *channel reliability function* for the infinite bandwidth AWGN channel. A plot of $E(R)/C_\infty$ is shown in Fig. 4.2.30. Also shown in this figure is the exponential factor for the union bound on P_M, given by (4.2.61), which may be expressed as

$$P_M \le \frac{1}{2} \cdot 2^{-T[(1/2)C_\infty - R]} \qquad 0 \le R \le \frac{1}{2}C_\infty \qquad (4.2.154)$$

Clearly, the exponential factor in (4.2.154) is not as tight as $E(R)$, due to the looseness of the union bound. In fact, $E(R)$ is known to be exponentially tight for the AWGN channel, which means that we cannot find another reliability function that is greater than $E(R)$ for any rate and still maintain the condition that $P_M \to 0$ as $T \to \infty$. Furthermore, $E(R)$ as given by (4.2.153) can be shown to hold for other signal sets besides orthogonal signals, e.g., simplex, for transmission over the AWGN channel.

Although the error probability can be made arbitrarily small by increasing the number of either orthogonal, biorthogonal, or simplex signals, with $R < C_\infty$, for a relatively modest number of signals, there is a large gap between the actual performance and the best achievable performance given by the channel capacity formula. For example, from Fig. 4.2.29 we observe that a set of $M = 16$ orthogonal signals detected coherently requires a SNR per bit of approximately 7.5 dB to achieve a bit error rate of $P_e = 10^{-5}$. In contrast, the channel capacity formula indicates that for a $C/W = 0.5$, reliable transmission is possible with a SNR of -0.8 dB. This represents a rather large difference of 8.3 dB per bit and serves as a motivation for searching for more efficient signaling waveforms. In Chap. 5 we demonstrate that coded waveforms can reduce this gap considerably.

Similar gaps in performance also exist in the bandwidth-limited region of Fig. 4.2.29, where $R/W > 1$. In this region, however, we must be more clever in how we use coding to improve performance, because we cannot expand the bandwidth as in the power-limited region. The use of coding techniques for bandwidth-efficient transmission is also treated in Chap. 5.

4.3 OPTIMUM DEMODULATION FOR SIGNALS WITH RANDOM PHASE IN ADDITIVE GAUSSIAN NOISE

In this section we are concerned with the form of the optimum demodulator when the carrier phase is unknown at the receiver and no attempt is made to estimate its value. Uncertainty in the phase of the received signal may be attributed to one or both of the following reasons: First, the oscillator that generates the carrier at the transmitter may not possess sufficient phase stability to allow the receiver to estimate the carrier phase from observation of the received signal. Second, the time delay in the propagation of the signal from the transmitter to the receiver may be changing too rapidly and in a random manner, so that the carrier phase of the received signal does not remain fixed long enough to allow for its estimation and tracking at the receiver.

In the absence of knowledge of the carrier phase, we treat this signal parameter as a random variable and assign to it the uniform probability density function (pdf)

$$p(\phi) = \frac{1}{2\pi} \qquad 0 \le \phi \le 2\pi \tag{4.3.1}$$

This is called the *least favorable pdf* for the signal phase. In some sense, the uniform pdf is the one that yields the maximum uncertainty for ϕ. This pdf is used to average over the phase parameter in the computation of the probabilities upon which the demodulator bases its decisions.

Following the notation introduced previously, we assume that there are M signaling waveforms used in the transmission of information, namely,

$$s_m(t) = \text{Re}\left[u_m(t)e^{j2\pi f_c t}\right] \qquad m = 1, 2, \ldots, M \qquad (4.3.2)$$

where $\{u_m(t)\}$ are the equivalent lowpass waveforms. The channel through which the signaling waveforms are transmitted is assumed to introduce an attenuation factor α and a phase shift ϕ_m for each signal. Thus the equivalent lowpass received signal, corrupted by additive noise, is represented as

$$r(t) = \alpha e^{-j\phi_m} u_m(t) + z(t) \qquad (4.3.3)$$

By lifting the restriction that all signals have the same phase shift, we have adopted, in effect, a slightly more general model for the channel, a model which reflects the reasons presented above for the uncertainties in the received signal phase.

In the derivation of the structure for the optimum demodulator, carried out in Appendix 4C, the phases $\{\phi_m\}$ are assumed to be mutually statistically independent and uniformly distributed. The result from this derivation is given by (4C.11) in combination with (4C.6). When specialized to the case in which the additive gaussian noise is white and the M signals are equally likely and have equal energy, (4C.6) and (4C.11) simplify to a demodulator that computes the M decision variables

$$U_m = \left| \int_0^T r(t) u_m^*(t)\, dt \right| \qquad (4.3.4)$$

and selects the signal corresponding to the largest decision variable.

A block diagram of the optimum demodulator is shown in Fig. 4.3.1. It consists of M matched filters with impulse responses $\{u_m^*(T - t)\}$ (or M cross correlators) followed by envelope detectors. Since an envelope detector takes the magnitude of the complex-valued signal, the phase ϕ_m is rendered irrelevant and the detection technique is said to be noncoherent.

As an alternative to an envelope detector, we can use a square-law detector, which in effect yields the decision variables $\{U_m^2\}$, $m = 1, 2, \ldots, M$. When the signaling waveforms are transmitted over a single channel, which is the case under consideration, the performance of the square-law detector is identical to that of the envelope detector. On the other hand, when the same information is transmitted over several parallel channels and combined at the receiver, which is the problem treated in Sec. 4.4, the performance of the square-law detector and combiner is much easier to analyze than the performance of the envelope detector and combiner. For multichannel signaling, the difference in performance between the two types of detection is very small. Hence the performance obtained from the analysis of the square-law detector is a good approximation to the performance achieved with the envelope detector.

Although the analysis of the performance of the optimum noncoherent demodulator is carried out on the basis of the decision variables given by (4.3.4), it is easily seen that the vector formulation of the optimum demodulator yields

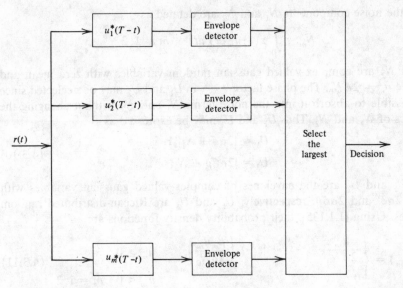

FIGURE 4.3.1
Optimum demodulator for noncoherent detection.

the decision variables

$$\mu(\mathbf{u}_m) = |\mathbf{r} \cdot \mathbf{u}_m| \qquad m = 1, 2, \ldots, M \qquad (4.3.5)$$

which is analogous to (4.3.4).

4.3.1 Noncoherent Detection of Binary Signals in an AWGN Channel

The two signaling waveforms are

$$s_m(t) = \mathrm{Re}\left[u_m(t)e^{j2\pi f_c t}\right] \qquad m = 1, 2 \qquad (4.3.6)$$

The waveforms have equal energy $\mathscr{E}$ and a complex cross-correlation coefficient ρ. The optimum noncoherent demodulator computes the decision variables

$$U_m = \left|\int_0^T r(t)u_m^*(t)\,dt\right| \qquad m = 1, 2 \qquad (4.3.7)$$

where the equivalent lowpass received signal is expressed as

$$r(t) = \alpha e^{-j\phi_m}u_m(t) + z(t) \qquad m = 1, 2 \qquad (4.3.8)$$

In order to derive the probability of error, we assume that the transmitted waveform is $u_1(t)$. Then the decision variables U_1 and U_2 may be expressed as

$$\begin{aligned} U_1 &= |2\alpha\mathscr{E}e^{-j\phi_1} + N_1| \\ U_2 &= |2\alpha\mathscr{E}\rho e^{-j\phi_1} + N_2| \end{aligned} \qquad (4.3.9)$$

where the noise components N_1 and N_2 are defined as

$$N_m = \int_0^T z(t) u_m^*(t)\, dt \qquad m = 1, 2$$

N_1 and N_2 are complex-valued gaussian random variables with zero mean and variance $\sigma_n^2 = 2\mathscr{E}N_0$. The phase factor $e^{-j\phi_1}$ in U_1 and U_2 may be neglected since it is possible to absorb it into the noise terms N_1 and N_2 without changing the statistics of N_1 and N_2. Thus U_1 and U_2 may be expressed as

$$U_1 = |2\alpha\mathscr{E} + N_1|$$
$$U_2 = |2\alpha\mathscr{E}\rho + N_2| \tag{4.3.10}$$

Since U_1 and U_2 are the envelopes of complex-valued gaussian variables with means $2\alpha\mathscr{E}$ and $2\alpha\mathscr{E}\rho$, respectively, U_1 and U_2 are Ricean-distributed random variables. Using (1.1.135), their probability density functions are

$$p(u_m) = \begin{cases} \dfrac{u_m}{2\mathscr{E}N_0} e^{-(u_m^2 + \beta_m^2)/4\mathscr{E}N_0} I_0\left(\dfrac{\beta_m u_m}{2\mathscr{E}N_0} \right) & u_m \geq 0 \\ 0 & u_m < 0 \quad m = 1, 2 \end{cases} \tag{4.3.11}$$

where $\beta_1 = 2\alpha\mathscr{E}$ and $\beta_2 = 2\alpha\mathscr{E}|\rho|$. We note that U_1 and U_2 are statistically dependent as a consequence of the nonorthogonality of the waveforms.

The probability of error is equal to the probability that U_2 exceeds U_1. One approach to computing this probability is to determine first the joint pdf of U_1 and U_2 and, then,

$$P_b = P(U_2 > U_1) = \int_0^\infty \int_{u_1}^\infty p(u_1, u_2)\, du_2\, du_1 \tag{4.3.12}$$

This approach was first used by Helstrom (1955), who determined $p(u_1, u_2)$ and evaluated the double integral in (4.3.12).

An alternative approach is based on the observation that the probability of error may be expressed as

$$P_b = P(U_2 > U_1) = P(U_2^2 > U_1^2) = P(U_1^2 - U_2^2 < 0) \tag{4.3.13}$$

But $U_1^2 - U_2^2 = D$ is a special case of a general quadratic form in complex-valued gaussian random variables, defined in (4B.1) of Appendix 4B. Denoting the characteristic function of D as $\psi(jv)$, we have

$$P_b = P(D < 0) = \int_{-\infty}^0 p(D)\, dD$$

$$= \int_{-\infty}^0 dD\, \frac{1}{2\pi} \int_{-\infty}^\infty \psi(jv) e^{-jvD}\, dv$$

$$= -\frac{1}{2\pi j} \int_{-\infty + j\varepsilon}^{\infty + j\varepsilon} \frac{\psi(jv)}{v}\, dv \tag{4.3.14}$$

where ε is a small positive number inserted in order to move the path of integration away from the singularity at $v = 0$ and which must be positive in order to allow for the interchange in the order of integrations. The integral in (4.3.14) is evaluated in Appendix 4B with the result given by (4B.21). For the special case under consideration, the derivation yields the error probability in the form

$$P_b = Q(a, b) - \tfrac{1}{2}e^{-(a^2+b^2)/2}I_0(ab) \qquad (4.3.15)$$

where

$$a = \sqrt{\frac{\gamma}{2}\left(1 - \sqrt{1 - |\rho|^2}\right)}$$

$$b = \sqrt{\frac{\gamma}{2}\left(1 + \sqrt{1 - |\rho|^2}\right)} \qquad (4.3.16)$$

FIGURE 4.3.2
Probability of error for noncoherent detection.

$Q(a, b)$ is the Q function defined previously in (1.1.120) and $I_0(x)$ is the modified Bessel function of order zero.

The error probability P_b is illustrated in Fig. 4.3.2 for several values of $|\rho|$. P_b is minimized when $\rho = 0$; that is, when the signals are orthogonal. For this case, $a = 0$, $b = \sqrt{\gamma}$, and (4.3.15) simplifies to

$$P_b = Q(0, \sqrt{\gamma}) - \tfrac{1}{2}e^{-\gamma/2} \tag{4.3.17}$$

From the definition of the Q function given in (1.1.120),

$$Q(0, \sqrt{\gamma}) = e^{-\gamma/2} \tag{4.3.18}$$

Substitution for $Q(0, \sqrt{\gamma})$ from (4.3.18) into (4.3.17) yields the probability of error for noncoherent detection of binary orthogonal signals as

$$P_b = \tfrac{1}{2}e^{-\gamma/2} \tag{4.3.19}$$

On the other hand, when $|\rho| = 1$, the error probability in (4.3.15) becomes $P_b = \tfrac{1}{2}$ as expected.

The cross-correlation coefficient ρ of two FSK signals with frequency separation Δf is given by (4.1.41). The magnitude of ρ is

$$|\rho| = \left| \frac{\sin \pi T \Delta f}{\pi T \Delta f} \right| \tag{4.3.20}$$

We observe that $|\rho| = 0$ if $\Delta f = m/T$, where m is an integer. With $m = 1$, twice as much frequency separation is required to obtain orthogonal FSK waveforms for noncoherent detection as for coherent detection.

A block diagram of a binary FSK demodulator is shown in Fig. 4.3.3. This implementation employs two bandpass filters, one tuned to frequency f_1 and the other to f_2. The filter outputs are envelope-detected and sampled to form the two decision variables. Alternatively, the demodulator may be implemented in the form of a correlator using integrate-and-dump filters followed by square-law detection, as shown in Fig. 4.3.4.

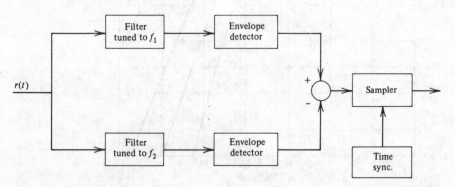

FIGURE 4.3.3
Demodulator for binary FSK.

FIGURE 4.3.4
Alternative form of demodulator for binary FSK.

4.3.2 Noncoherent Detection of M-ary Orthogonal Signals in an AWGN Channel

In this subsection we derive the performance for noncoherent detection of M-ary orthogonal signaling. The M signaling waveforms are assumed to have equal energy and are equally likely to be transmitted. For the AWGN channel, the optimum receiver computes the decision variables

$$U_m = \left| \int_0^T r(t) u_m^*(t) \, dt \right| \qquad m = 1, 2, \ldots, M \tag{4.3.21}$$

and selects the signal corresponding to the largest decision variable.

To derive the probability of error, suppose that the signal $u_1(t)$ is transmitted. Then the equivalent lowpass received signal is

$$r(t) = \alpha e^{-j\phi_1} u_1(t) + z(t) \tag{4.3.22}$$

Since the signals are detected noncoherently, the phase term $e^{-j\phi_1}$ becomes irrelevant and can be dropped from consideration. With this simplification, the decision variables may be expressed as

$$\begin{aligned} U_1 &= |2\alpha\mathscr{E} + N_1| \\ U_m &= |N_m| \qquad m = 2, 3, \ldots, M \end{aligned} \tag{4.3.23}$$

where, by definition

$$N_m = \int_0^T z(t) u_m^*(t) \, dt \qquad m = 1, 2, \ldots, M \tag{4.3.24}$$

The complex-valued gaussian random variables $\{N_m\}$ are mutually uncorrelated and identically distributed with zero mean and variance $\sigma_n^2 = 2\mathscr{E}N_0$. The decision variables $U_1, U_2, \ldots, U_M$ are mutually statistically independent. $U_2, \ldots, U_M$ are described statistically by the Rayleigh probability density function

$$p(u_m) = \frac{u_m}{2\mathscr{E}N_0} e^{-u_m^2/4\mathscr{E}N_0} \qquad m = 2, \ldots, M \tag{4.3.25}$$

while U_1 is described by the Rice probability density function

$$p(u_1) = \frac{u_1}{2\mathscr{E}N_0} \exp\left(-\frac{u_1^2 + 4\alpha^2\mathscr{E}^2}{4\mathscr{E}N_0}\right) I_0\left(\frac{\alpha u_1}{N_0}\right) \tag{4.3.26}$$

The probability of a correct decision is simply the probability that U_1 exceeds all the other $\{U_m\}$. That is,

$$P_c = P(U_2 < U_1, U_3 < U_1, \ldots, U_M < U_1)$$

$$= \int_0^\infty P(U_2 < u_1, U_3 < u_1, \ldots, U_M < u_1 \mid U_1 = u_1) p(u_1) \, du_1 \tag{4.3.27}$$

Since the U_m, $m = 2, 3, \ldots, M$, are statistically independent and identically distributed, the joint probability conditioned on U_1 factors into a product of $M - 1$ identical terms. Thus

$$P_c = \int_0^\infty [P(U_2 < u_1 \mid U_1 = u_1)]^{M-1} p(u_1) \, du_1 \tag{4.3.28}$$

Now

$$P(U_2 < u_1 \mid U_1 = u_1) = \int_0^{u_1} p(u_2) \, du_2$$

$$= 1 - e^{-u_1^2/4\mathscr{E}N_0} \tag{4.3.29}$$

The $(M - 1)$st power of (4.3.29) may be expressed in the form

$$\left(1 - e^{-u_1^2/4\mathscr{E}N_0}\right)^{M-1} = \sum_{n=0}^{M-1} (-1)^n \binom{M-1}{n} e^{-nu_1^2/4\mathscr{E}N_0} \tag{4.3.30}$$

Substitution of this result into (4.3.28) and integration over u_1 yields

$$P_c = \sum_{n=0}^{M-1} (-1)^n \binom{M-1}{n} \frac{1}{n+1} e^{-\gamma n/(n+1)} \tag{4.3.31}$$

where $\gamma = \alpha^2\mathscr{E}/N_0$ is the SNR per symbol. Then the probability of a symbol error, which is $P_M = 1 - P_c$, becomes

$$P_M = \sum_{n=1}^{M-1} (-1)^{n+1} \binom{M-1}{n} \frac{1}{n+1} e^{-nk\gamma_b/(n+1)} \tag{4.3.32}$$

where γ_b is the SNR per bit. Since the signaling waveforms are orthogonal, the equivalent bit error probability can be obtained from P_M by means of the formula given in (4.2.56). This bit error probability is plotted in Fig. 4.3.5 for $M = 2, 4, 8, 16$, and 32. We note that as M increases, the SNR per bit required

FIGURE 4.3.5
Probability of a bit error for noncoherent detection of orthogonal signals.

to achieve a given performance decreases, just as in the case of coherent detection of M signals. The asymptotic behavior for noncoherent detection of M orthogonal signals is also similar to that for coherent detection. Following the derivation given for coherent detection of orthogonal signals in Sec. 4.2.2, it is easy to show that $P_M \to 0$ as $M \to \infty$, provided $\gamma_b > \ln 2$.

FSK signal waveforms are commonly employed for M-ary orthogonal signaling with noncoherent detection. In such a case, orthogonality of the FSK waveforms requires a frequency separation of $\Delta f = 1/T$. Consequently the channel bandwidth required for transmission is

$$W = M \Delta f = \frac{M}{T} \tag{4.3.33}$$

and the corresponding transmission rate R is

$$R = \frac{k}{T} = \frac{1}{T} \log_2 M \qquad \text{bits/s} \qquad (4.3.34)$$

Therefore the bandwidth efficiency is

$$\frac{R}{W} = \frac{\log_2 M}{M} \qquad (4.3.35)$$

which is one-half that obtained for orthogonal M-ary FSK with coherent detection.

4.4 MULTICHANNEL DIGITAL SIGNALING IN AN AWGN CHANNEL

In some applications it is desirable to transmit the same information-bearing signal over several channels. This mode of transmission is used primarily in situations where there is a high probability that one or more of the channels will become unreliable from time to time. For example, radio channels such as ionospheric scatter and tropospheric scatter suffer from signal fading due to multipath, which renders the channels unreliable for short periods of time. As another example, multichannel signaling is sometimes employed in military communication systems as a means for combating jamming.

The effect of fading in multichannel digital transmission is a topic considered in Chap. 7. In this section we confine our attention to multichannel signaling over fixed channels that differ only in attenuation and phase shift. The specific model for the multichannel digital signaling system may be described as follows. The signaling waveforms, in general, are expressed as

$$s_{nm}(t) = \text{Re}\left[u_{nm}(t)e^{j2\pi f_c t}\right] \qquad \begin{array}{l} 0 \leq t \leq T \\ n = 1, 2, \ldots, L \\ m = 1, 2, \ldots, M \end{array} \qquad (4.4.1)$$

where L is the number of channels and M is the number of waveforms. The waveforms are assumed to have equal energy and to be equally probable *a priori*. The waveforms $\{s_{nm}(t)\}$ transmitted over the L channels are scaled by the factors $\{\alpha_n\}$, phase-shifted by $\{\phi_n\}$, and corrupted by additive noise. The equivalent lowpass signals received from the L channels may be expressed as

$$r_n(t) = \alpha_n e^{-j\phi_n} u_{nm}(t) + z_n(t) \qquad \begin{array}{l} 0 \leq t \leq T \\ n = 1, 2, \ldots, L \\ m = 1, 2, \ldots, M \end{array} \qquad (4.4.2)$$

where $\{u_{nm}(t)\}$ are the equivalent lowpass transmitted waveforms and $\{z_n(t)\}$ represent the additive noise processes on the L channels. We assume that $\{z_n(t)\}$ are mutually statistically independent and identically distributed gaussian noise random processes.

We consider two types of processing at the receiver, namely, coherent detection and noncoherent detection. The receiver for coherent detection estimates the channel parameters $\{\alpha_n\}$ and $\{\phi_n\}$ and uses the estimates in computing the decision variables. Suppose we define $g_n = \alpha_n e^{-j\phi_n}$ and let $\hat{g}_n$ be the estimate of g_n. Then the decision variables for coherent detection may be expressed as

$$U_m = \sum_{n=1}^{L} \mathrm{Re}\left[\hat{g}_n^* \int_0^T r_n(t) u_{nm}^*(t)\, dt\right] \qquad m = 1, 2, \ldots, M \qquad (4.4.3)$$

In noncoherent detection no attempt is made to estimate the channel parameters. The demodulator may base its decision either on the sum of the envelopes (envelope detection) or the sum of the squared envelopes (square-law detection) of the matched filter outputs. In general the performance obtained with envelope detection differs little from the performance obtained with square-law detection in AWGN. However, square-law detection of multichannel signaling in AWGN channels is considerably easier to analyze than envelope detection. Therefore we confine our attention to square-law detection of the received signals on the L channels, which produces the decision variables

$$U_m = \sum_{n=1}^{L} \left| \int_0^T r_n(t) u_{nm}^*(t)\, dt \right|^2 \qquad m = 1, 2, \ldots, M \qquad (4.4.4)$$

Let us consider binary signaling first, and assume that $u_{n1}(t)$, $n = 1, 2, \ldots, L$ are the L transmitted waveforms. Then an error is committed if U_2 exceeds U_1 or, equivalently, if the difference $D = U_1 - U_2 < 0$. For noncoherent detection, this difference may be expressed as

$$D = \sum_{n=1}^{L} \left[|X_n|^2 - |Y_n|^2 \right] \qquad (4.4.5)$$

where the variables $\{X_n\}$ and $\{Y_n\}$ are defined as

$$X_n = \int_0^T r_n(t) u_{n1}^*(t)\, dt \qquad n = 1, 2, \ldots, L$$

$$Y_n = \int_0^T r_n(t) u_{n2}^*(t)\, dt \qquad n = 1, 2, \ldots, L \qquad (4.4.6)$$

The $\{X_n\}$ are mutually independent and identically distributed gaussian random variables. The same statement applies to the variables $\{Y_n\}$. However, for any n, X_n and Y_n may be correlated, but we require that X_n and Y_m be independent when $m \neq n$. For coherent detection, the difference $D = U_1 - U_2$ may be expressed as

$$D = \frac{1}{2} \sum_{n=1}^{L} \left(X_n Y_n^* + X_n^* Y_n \right) \qquad (4.4.7)$$

where, by definition,

$$Y_n = \hat{g}_n \qquad n = 1, 2, \ldots, L$$

$$X_n = \int_0^T r_n(t) \left[u_{n1}^*(t) - u_{n2}^*(t) \right] dt \qquad (4.4.8)$$

If the estimates $\{\hat{g}_n\}$ are obtained from observation of the received signal over one or more signaling intervals as described in Appendix 7A, their statistical characteristics are described by the gaussian distribution. Then the $\{Y_n\}$ are characterized as mutually independent and identically distributed gaussian random variables. The same statement applies to the variables $\{X_n\}$. As in noncoherent detection, we allow for correlation between X_n and Y_n, but not between X_m and Y_n for $m \neq n$.

In Appendix 4B we derive the probability that the general quadratic form

$$D = \sum_{n=1}^{L} \left[A |X_n|^2 + B |Y_n|^2 + C X_n Y_n^* + C^* X_n^* Y_n \right] \tag{4.4.9}$$

in complex-valued gaussian random variables is less than zero. This probability, which is given in (4B.21) of Appendix 4B, is the probability of error for binary multichannel signaling in AWGN. A number of special cases are of particular importance.

If the binary signals are antipodal and the estimates of $\{g_n\}$ are perfect, as in coherent PSK, the probability of error takes the simple form

$$P_b = \tfrac{1}{2} \operatorname{erfc}\left(\sqrt{\gamma_b}\right) \tag{4.4.10}$$

where

$$\gamma_b = \frac{\mathscr{E}}{N_0} \sum_{n=1}^{L} |g_n|^2$$

$$= \frac{\mathscr{E}}{N_0} \sum_{n=1}^{L} \alpha_n^2 \tag{4.4.11}$$

is the SNR per bit. If the channels are all identical, $\alpha_n = \alpha$ for all n and, hence,

$$\gamma_b = \frac{L\mathscr{E}}{N_0} \alpha^2 \tag{4.4.12}$$

We observe that $L\mathscr{E}$ is the total transmitted signal energy for the L signals. The interpretation of this result is that the receiver combines the energy from the L channels in an optimum manner. That is, there is no loss in performance in dividing the total transmitted signal energy among the L channels. The same performance is obtained as in the case in which a single waveform having energy $L\mathscr{E}$ is transmitted on one channel. This behavior holds true only if the estimates $\hat{g}_n = g_n$, for all n. If the estimates are not perfect, a loss in performance occurs, the amount of which depends on the quality of the estimates as described in Appendix 7A.

Perfect estimates for $\{g_n\}$ constitute an extreme case. At the other extreme we have binary DPSK signaling. In DPSK, the estimates $\{\hat{g}_n\}$ are simply the (normalized) signal-plus-noise samples at the outputs of the matched filters in the previous signaling interval. This is the poorest estimate that one might consider using in estimating $\{g_n\}$. For binary DPSK the probability of error obtained from (4B.21) is

$$P_b = \frac{1}{2^{2L-1}} e^{-\gamma_b} \sum_{n=0}^{L-1} c_n \gamma_b^n \qquad (4.4.13)$$

where, by definition,

$$c_n = \frac{1}{n!} \sum_{k=0}^{L-1-n} \binom{2L-1}{k} \qquad (4.4.14)$$

and γ_b is the SNR per bit previously defined in (4.4.11) and, for identical channels in (4.4.12). This result can be compared with the single-channel ($L = 1$) error probability. To simplify the comparison, we assume that the L channels have identical attentuation factors. Thus, for the same value of γ_b, the performance of the multichannel system is poorer than that of the single-channel system. That is, splitting the total transmitted energy among L channels results in a loss in performance, the amount of which depends on L.

A loss in performance also occurs in square-law detection of orthogonal signals transmitted over L channels. For binary orthogonal signaling, the expression for the probability of error is identical in form to that for binary DPSK given in (4.4.13), except that γ_b is replaced by $\gamma_b/2$. That is, binary orthogonal signaling with noncoherent detection is 3 dB poorer than binary DPSK. However, the loss in performance due to noncoherently combining the signals received on the L channels is identical to that for binary DPSK.

Figure 4.4.1 illustrates the loss resulting from noncoherent (square-law) combining of the L signals as a function of L. The probability of error is not shown, but it can be easily obtained from the curve of the expression

$$P_b = \tfrac{1}{2} e^{-\gamma_b} \qquad (4.4.15)$$

which is the error probability of binary DPSK shown in Fig. 4.2.19 and then degrading the required SNR per bit γ_b by the noncoherent combining loss corresponding to the value of L.

Now let us return to M-ary orthogonal signaling with square-law detection and combining of the signals on the L channels. The decision variables are given by (4.4.4). Suppose that the signals $u_{n1}(t)$, $n = 1, 2, \ldots, L$, are transmitted over the L AWGN channels. Then

$$U_1 = \sum_{n=1}^{L} |2\mathscr{E}\alpha_n + N_{n1}|^2$$

$$\qquad (4.4.16)$$

$$U_m = \sum_{n=1}^{L} |N_{nm}|^2 \qquad m = 2, 3, \ldots, M$$

where the $\{N_{nm}\}$ are complex-valued zero-mean gaussian random variable with variance $\sigma^2 = \frac{1}{2} E[|N_{nm}|^2] = 2\mathscr{E}N_0$. Hence U_1 is described statistically as a non-central chi-square random variables with $2L$ degrees of freedom and noncentral-

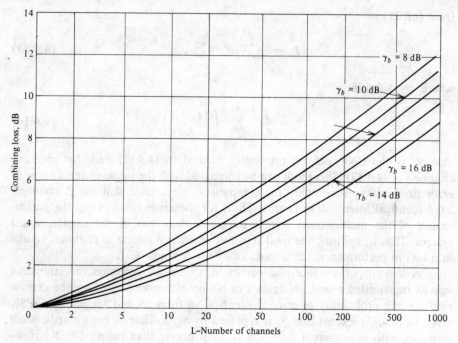

FIGURE 4.4.1
Combining loss in noncoherent detection and combining of binary multichannel signals.

ity parameter

$$s^2 = \sum_{n=1}^{L} (2\mathscr{E}\alpha_n)^2 = 4\mathscr{E}^2 \sum_{n=1}^{L} \alpha_n^2 \qquad (4.4.17)$$

Using (1.1.115), we obtain the pdf of U_1 as

$$p(u_1) = \frac{1}{4\mathscr{E}N_0}\left(\frac{u_1}{s^2}\right)^{(L-1)/2} \exp\left(-\frac{s^2+u_1}{4\mathscr{E}N_0}\right) I_{L-1}\left(\frac{s\sqrt{u_1}}{2\mathscr{E}N_0}\right) \qquad u_1 \geq 0 \quad (4.4.18)$$

On the other hand, the $\{U_m\}$, $m = 2, 3, \ldots, M$, are statistically independent and identically chi-square-distributed random variables, each having $2L$ degrees of freedom. Using (1.1.107), we obtain the pdf for U_m as

$$p(u_m) = \frac{1}{(4\mathscr{E}N_0)^L(L-1)!} u_m^{L-1}e^{-u_m/4\mathscr{E}N_0} \qquad \begin{matrix} u_m \geq 0 \\ m = 2, 3, \ldots, M \end{matrix} \quad (4.4.19)$$

The probability of a symbol error is

$$\begin{aligned}
P_M &= 1 - P_c \\
&= 1 - P(U_2 < U_1, U_3 < U_1, \ldots, U_M < U_1) \\
&= 1 - \int_0^{\infty} \left[P(U_2 < u_1 \mid U_1 = u_1) \right]^{M-1} p(u_1) \, du_1 \quad (4.4.20)
\end{aligned}$$

But

$$P(U_2 < u_1 \mid U_1 = u_1) = 1 - \exp\left(-\frac{u_1}{4\mathscr{E}N_0}\right) \sum_{k=0}^{L-1} \frac{1}{k!}\left(\frac{u_1}{4\mathscr{E}N_0}\right)^k \quad (4.4.21)$$

Hence

$$P_M = 1 - \int_0^\infty \left[1 - e^{-u_1/4\mathscr{E}N_0} \sum_{k=0}^{L-1} \frac{1}{k!}\left(\frac{u_1}{4\mathscr{E}N_0}\right)^k\right]^{M-1} p(u_1)\, du_1$$

$$= 1 - \int_0^\infty \left[1 - e^{-\gamma v} \sum_{k=0}^{L-1} \frac{1}{k!}v^k\right]^{M-1} \gamma v^{L-1} e^{-\gamma(1+v)} I_{L-1}\left(2\gamma\sqrt{v}\right)\, dv \quad (4.4.22)$$

where

$$\gamma = \mathscr{E} \sum_{n=1}^{L} \alpha_n^2 / N_0$$

The integral in (4.4.22) can be evaluated numerically. It is also possible to expand the term $(1 - x)^{M-1}$ in (4.4.22) and carry out the integration term by term. This approach yields an expression for P_M in terms of finite sums, as shown by Blanco (1989).

An alternative approach is to use the union bound

$$P_M < (M - 1)P_2(L) \quad (4.4.23)$$

where $P_2(L)$ is the probability of error in choosing between U_1 and any one of the $(M - 1)$ decision variables $\{U_m\}$, $m = 2, 3, \ldots, M$. From our previous discussion on the performance of binary orthogonal signaling, we have

$$P_2(L) = \frac{1}{2^{2L-1}} e^{-k\gamma_b/2} \sum_{n=0}^{L-1} c_n\left(\frac{k\gamma_b}{2}\right)^n \quad (4.4.24)$$

where c_n is given by (4.4.14). For relatively small values of M, the union bound in (4.4.23) is sufficiently tight for most practical applications.

4.5 CARRIER AND SYMBOL SYNCHRONIZATION

From our discussion of demodulation of the various digitally modulated signals described in this chapter, we have observed that coherent demodulation requires the use of a properly phased carrier at the receiver. The carrier phase must be estimated by the receiver from the incoming signal. In addition, symbol timing must be derived from the received signal for both coherent and noncoherent demodulation in order to synchronously sample the output of the matched filters or correlators at the proper time instant. In this section we consider techniques for both carrier and symbol synchronization that derive the carrier phase and symbol timing from the information-bearing signal.

4.5.1 Carrier Synchronization Techniques

There are two basic approaches for dealing with carrier synchronization at the receiver. One approach is to multiplex, usually in frequency, a special signal, called a pilot signal, that allows the receiver to extract and, thus, to synchronize its local oscillator to the carrier frequency and phase of the received signal. When an unmodulated carrier component is transmitted along with the information-bearing signal, the receiver employs a phase-locked loop (PLL) to acquire and track the carrier component. The PLL is designed to have a narrow bandwidth so that it is not significantly affected by the presence of frequency components from the information-bearing signal.

The signal approach, which appears to be more prevalent in practice, is to derive the carrier synchronization directly from the modulated signal. This approach has the distinct advantage that the total transmitter power is allocated to the transmission of the information-bearing signal. In our treatment of carrier synchronization, we confine our attention to the second approach; hence, we assume that the signal is transmitted via suppressed carrier.

In order to emphasize the importance of extracting an accurate phase estimate, let us consider the effect of a carrier phase error on the demodulation of a double sideband, suppressed carrier (DSB/SC) signal. To be specific, suppose we have an amplitude-modulated signal of the form

$$s(t) = A(t)\cos(2\pi f_c t + \phi) \tag{4.5.1}$$

If we demodulate the signal by multiplying $s(t)$ with the carrier reference

$$c(t) = \cos(2\pi f_c t + \hat{\phi}) \tag{4.5.2}$$

we obtain

$$c(t)s(t) = \tfrac{1}{2}A(t)\cos(\phi - \hat{\phi}) + \tfrac{1}{2}A(t)\cos(4\pi f_c t + \phi + \hat{\phi})$$

The double-frequency component may be removed by passing the product signal $c(t)s(t)$ through a lowpass filter. This filtering yields the information-bearing signal

$$y(t) = \tfrac{1}{2}A(t)\cos(\phi - \hat{\phi}) \tag{4.5.3}$$

Note that the effect of the phase error $\phi - \hat{\phi}$ is to reduce the signal level in voltage by the amount $\cos(\phi - \hat{\phi})$ and in power by the amount $\cos^2(\phi - \hat{\phi})$. Hence, a phase error of $10°$ results in a signal power loss of 0.13 dB and a phase error of $30°$ results in a signal power loss of 1.25 dB in an amplitude-modulated signal.

The effect of carrier phase errors in QAM and multiphase PSK is much more severe. The QAM and M-PSK signals may be represented as

$$s(t) = A(t)\cos(2\pi f_c t + \phi) - B(t)\sin(2\pi f_c t + \phi) \tag{4.5.4}$$

This signal is demodulated by the two quadrature carriers

$$c_c(t) = \cos(2\pi f_c t + \hat{\phi})$$
$$c_s(t) = -\sin(2\pi f_c t + \hat{\phi}) \tag{4.5.5}$$

Multiplication of $s(t)$ with $c_c(t)$ followed by lowpass filtering yields the in-phase component

$$y_I(t) = \tfrac{1}{2}A(t)\cos\left(\phi - \hat{\phi}\right) - \tfrac{1}{2}B(t)\sin\left(\phi - \hat{\phi}\right) \qquad (4.5.6)$$

Similarly, multiplication of $s(t)$ by $c_s(t)$ followed by lowpass filtering yields the quadrature component

$$y_Q(t) = \tfrac{1}{2}B(t)\cos\left(\phi - \hat{\phi}\right) + \tfrac{1}{2}A(t)\sin\left(\phi - \hat{\phi}\right) \qquad (4.5.7)$$

The expressions in (4.5.6) and (4.5.7) clearly indicate that the phase error in the demodulation of QAM and M-PSK signals has a much more severe effect than in the demodulation of a PAM signal. Not only is there a reduction in the power of the desired signal component by the factor $\cos^2\left(\phi - \hat{\phi}\right)$, but there is also cross talk interference from the in-phase and quadrature components. Since the average power levels of $A(t)$ and $B(t)$ are similar, a small phase error causes a large degradation in performance. Hence, the phase accuracy requirements for QAM and multiphase coherent PSK are much higher than DSB/SC PAM.

Squaring loop. Let us first consider phase recovery techniques for DSB/SC PAM signals which are represented by (4.5.1). Note that $E[s(t)] = E[A(t)] = 0$ when the signal levels are symmetric about zero. Consequently, the average value of $s(t)$ does not produce any phase coherent frequency components at any frequency including the carrier. One method for generating a carrier from the received signal is to square the signal and, thus, to generate a frequency component at $2f_c$, which can be used to drive a phase-locked loop (PLL) tuned to $2f_c$. This method is illustrated in the block diagram shown in Fig. 4.5.1. The output of the square-law device is

$$
\begin{aligned}
s^2(t) &= A^2(t)\cos^2\left(2\pi f_c t + \phi\right) \\
&= \tfrac{1}{2}A^2(t) + \tfrac{1}{2}A^2(t)\cos\left(4\pi f_c t + 2\phi\right) \qquad (4.5.8)
\end{aligned}
$$

Since the modulation is a cyclostationary stochastic process, the expected value of $s^2(t)$ is

$$E\left[s^2(t)\right] = \tfrac{1}{2}E\left[A^2(t)\right] + \tfrac{1}{2}E\left[A^2(t)\right]\cos\left(4\pi f_c t + 2\phi\right) \qquad (4.5.9)$$

Hence, there is power at the frequency $2f_c$.

If the output of the square-law device is passed through a bandpass filter tuned to the double-frequency term in (4.5.8), the mean value of the filter is a sinusoid with frequency $2f_c$, phase 2ϕ, and amplitude $E[A^2(t)]H(2f_c)/2$, where $H(2f_c)$ is the gain of the filter at $f = 2f_c$. Thus, the square-law device has produced a periodic component from the input signal $s(t)$. In effect the squaring of $s(t)$ has removed the sign information contained in $A(t)$ and, thus, has resulted in phase coherent frequency components at twice the carrier. The filtered frequency component at $2f_c$ is then used to drive the PLL.

The PLL basically consists of a multiplier, a loop filter, and a voltage-controlled oscillator (VCO). If we assume that the input to the PLL is the sinusoid

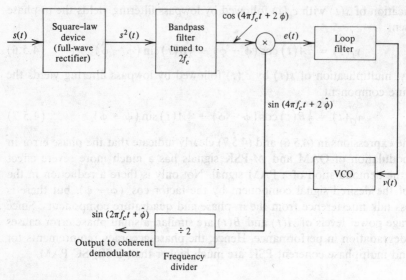

FIGURE 4.5.1
Carrier recovery using a square-law device.

$\cos(4\pi f_c t + 2\phi)$ and the output of the VCO is $\sin(4\pi f_c t + 2\hat{\phi})$, where $\hat{\phi}$ represents the estimate of ϕ, the product of these two signals is

$$e(t) = \cos(4\pi f_c t + 2\phi)\sin(4\pi f_c t + 2\hat{\phi})$$
$$= \tfrac{1}{2}\sin 2(\hat{\phi} + \phi) + \tfrac{1}{2}\sin(8\pi f_c t + 2\phi + 2\hat{\phi}) \qquad (4.5.10)$$

The loop filter is a lowpass filter that responds only to the low-frequency component $\tfrac{1}{2}\sin(\hat{\phi} - \phi)$ and removes the component at $4f_c$. This filter is usually selected to have the relatively simple transfer function

$$G(s) = \frac{1 + \tau_2 s}{1 + \tau_1 s} \qquad (4.5.11)$$

where τ_1 and τ_2 are design parameters ($\tau_1 \gg \tau_2$) that control the bandwidth of the loop. A higher-order filter that contains additional poles may be used if necessary to obtain a better loop response.

The output of the loop filter provides the control voltage $v(t)$ for the VCO. The VCO is basically a sinusoidal signal generator with an instantaneous phase given by

$$4\pi f_c t + 2\hat{\phi}(t) = 4\pi f_c t + K\int_{-\infty}^{t} v(\tau)\,d\tau \qquad (4.5.12)$$

where K is a gain constant in radians per volt-second. Hence,

$$2\hat{\phi} = K\int_{-\infty}^{t} v(\tau)\,d\tau \qquad (4.5.13)$$

FIGURE 4.5.2
Model of phase-locked loop.

By neglecting the double-frequency term resulting from the multiplication of the input signal with the output of the VCO, we may reduce the PLL into the equivalent closed-loop system model shown in Fig. 4.5.2. The sine function of the phase difference $2(\phi - \hat{\phi})$ makes this system nonlinear and, as a consequence, the analysis of its performance in the presence of noise is somewhat involved but, nevertheless, it is mathematically tractable for some simple loop filters.

In normal operation when the loop is tracking the phase of the incoming carrier, the phase error $\phi - \hat{\phi}$ is small and, hence

$$\tfrac{1}{2} \sin 2(\hat{\phi} - \phi) \approx \hat{\phi} - \phi \qquad (4.5.14)$$

With this approximation, the PLL becomes linear and is characterized by the closed-loop transfer function

$$H(s) = \frac{KG(s)/s}{1 + KG(s)/s} \qquad (4.5.15)$$

where the factor of $\tfrac{1}{2}$ has been absorbed into the gain parameter K. By substituting from (4.5.11) for $G(s)$ into (4.5.15) we obtain

$$H(s) = \frac{1 + \tau_2 s}{1 + \left(\tau_2 + \dfrac{1}{K}\right)s + \dfrac{\tau_1}{K}s^2} \qquad (4.5.16)$$

Hence the closed-loop system for the linearized PLL is second order when $G(s)$ is given by (4.5.11). The parameter τ_2 controls the position of the zero, while K and τ_1 are used to control the position of the closed loop system poles. It is customary to express the denominator of $H(s)$ in the standard form

$$D(s) = s^2 + 2\zeta\omega_n s + \omega_n^2 \qquad (4.5.17)$$

where ζ is called the loop damping factor and ω_n is the natural frequency of the loop. In terms of the loop parameters, $\omega_n = \sqrt{K/\tau_1}$, and $\zeta = (\tau_2 + 1/K)/2\omega_n$

the closed loop transfer function becomes

$$H(s) = \frac{(2\zeta\omega_n - \omega_n^2/K)s + \omega_n^2}{s^2 + 2\zeta\omega_n s + \omega_n^2} \tag{4.5.18}$$

The (one-sided) noise equivalent bandwidth (see Prob. 1.22) of the loop is

$$B_{eq} = \frac{\tau_2^2\left(\dfrac{1}{\tau_2^2} + \dfrac{K}{\tau_1}\right)}{4\left(\tau_2 + \dfrac{1}{K}\right)}$$

$$= \frac{1 + (\tau_2\omega_n)^2}{8\zeta\omega_n} \tag{4.5.19}$$

In practice, it is desirable to have a narrow bandwidth PLL in order to minimize the effects of input noise on the estimated phase ϕ.

Effect of noise on phase estimation. In order to evaluate the effects of noise on the estimate of the carrier phase, let us assume that the noise at the input to the PLL is narrowband. For this analysis, we assume that the PLL is tracking a sinusoidal signal of the form

$$s(t) = A_c \cos\left(2\pi f_c t + \phi(t)\right) \tag{4.5.20}$$

which is corrupted by the additive narrowband noise

$$n(t) = x(t)\cos 2\pi f_c t - y(t)\sin 2\pi f_c t \tag{4.5.21}$$

The in-phase and quadrature components of the noise are assumed to be statistically independent, stationary white gaussian noise processes with (two-sided) power spectral density $N_0/2$ W/Hz. By using simple trigonometric identities, the noise term in (4.5.21) can be expressed as

$$n(t) = n_c(t)\cos\left(2\pi f_c t + \phi(t)\right) - n_s(t)\sin\left(2\pi f_c + \phi(t)\right) \tag{4.5.22}$$

where

$$\begin{aligned} n_c(t) &= x(t)\cos\phi(t) + y(t)\sin\phi(t) \\ n_s(t) &= -x(t)\sin\phi(t) + y(t)\cos\phi(t) \end{aligned} \tag{4.5.23}$$

We note that

$$n_c(t) + jn_s(t) = \left[x(t) + jy(t)\right]e^{-j\phi(t)}$$

so that the quadrature components $n_c(t)$ and $n_s(t)$ have exactly the same statistical characteristics as $x(t)$ and $y(t)$.

If $s(t) + n(t)$ is multiplied by the output of the VCO and the double-frequency terms are neglected, the input to the loop filter is the noise-corrupted signal

$$\begin{aligned} e(t) &= A_c \sin\Delta\phi + n_c(t)\sin\Delta\phi - n_s(t)\cos\Delta\phi \\ &= A_c \sin\Delta\phi + n_1(t) \end{aligned} \tag{4.5.24}$$

FIGURE 4.5.3
Equivalent PLL model with additive noise.

where, by definition, $\Delta\phi = \phi - \hat{\phi}$ is the phase error. Thus, we have the equivalent model for the PLL with additive noise as shown in Fig. 4.5.3.

When the power $P_c = A_c^2/2$ of the incoming signal is much larger than the noise power, we may linearize the PLL and, thus, easily determine the effect of the additive noise on the quality of the estimate $\hat{\phi}$. Under these conditions the model for the linearized PLL with additive noise is illustrated in Fig. 4.5.4. Note that the gain parameter A_c may be normalized to unity, provided that the noise terms are scaled by $1/A_c$, i.e., the noise terms become

$$n_2(t) = \frac{n_c(t)}{A_c} \sin \Delta\phi - \frac{n_c(t)}{A_c} \cos \Delta\phi \qquad (4.5.25)$$

Since the noise $n_2(t)$ is additive at the input to the loop, the variance of the phase error $\Delta\phi$, which is also the variance of the VCO output phase, is

$$\sigma_{\hat{\phi}}^2 = \frac{2N_0 B_{eq}}{A_c^2} \qquad (4.5.26)$$

where B_{eq} is the (one-sided) equivalent noise bandwidth of the loop, given in

FIGURE 4.5.4
Linearized PLL model with additive noise.

(4.5.19). Note that $A_c^2/2$ is the power of the input sinusoid and that σ_ϕ^2 is simply the ratio of total noise power within the bandwidth of the PLL divided by the input signal power. Hence,

$$\sigma_\phi^2 = \frac{1}{\gamma_L}$$

where γ_L is defined as the signal-to-noise ratio

$$\text{SNR} \equiv \gamma_L = \frac{A_c^2/2}{N_0 B_{\text{eq}}} \tag{4.5.27}$$

The expression for the variance σ_ϕ^2 of the VCO phase error applies to the case where the SNR is sufficiently high that the linear model for the PLL applies. An exact analysis based on the nonlinear PLL is mathematically tractable when $G(s) = 1$, which results in a first-order loop. In this case the probability density function for the phase error may be derived [see Viterbi (1966)] and has the form

$$p(\Delta\phi) = \frac{\exp(\gamma_L \cos \Delta\phi)}{2\pi I_0(\gamma_L)} \tag{4.5.28}$$

where γ_L is the SNR given by (4.5.27), with B_{eq} being the appropriate noise bandwidth of the first-order loop, and $I_0(\cdot)$ is the modified Bessel function of order zero.

From the expression for $p(\Delta\phi)$ we may obtain the exact value of the variance for the phase error on a first-order PLL. This is plotted in Fig. 4.5.5 as a

FIGURE 4.5.5

Comparison of VCO phase variance for exact and approximate (linear model) first-order PLL. [From *Principles of Coherent Communication* by A. J. Viterbi; © 1966 by McGraw-Hill Book Company. Reprinted with permission of the publisher.]

function of $1/\gamma_L$. Also shown for comparison is the result obtained with the linearized PLL model. Note that the variance for the linear model is close to the exact variance for $\gamma_L > 3$. Hence, the linear model is adequate for practical purposes.

Approximate analyses of the statistical characteristics of the phase error for the nonlinear PLL have also been performed. Of particular importance is the transient behavior of the PLL during initial acquisition. Another important problem is the behavior of the PLL at low SNR. It is known, for example, that when the SNR at the input to the PLL drops below a certain value, there is a rapid deterioration in the performance of the PLL. The loop begins to lose lock and an impulsive-type of noise, characterized as clicks, is generated which degrades the performance of the loop. Results on these topics can be found in the texts by Viterbi (1966), Lindsey (1972), Lindsey and Simon (1973), and Gardner (1979) and in the survey papers by Gupta (1975) and Lindsey and Chie (1981).

Now that we have established the effect of noise on the performance of the PLL, let us return to the problem of carrier synchronization based on the system shown in Fig. 4.5.1. The squaring of the signal is a necessary operation which destroys the modulation and produces a frequency component at $2f_c$. However, the squaring operation leads to a noise enhancement that increases the noise power level at the input to the PLL and results in an increase in the variance of the phase error.

To elaborate on this point, let the input to the squarer be $s(t) + n(t)$, where $s(t)$ is given by (4.5.1) and $n(t)$ represents the bandpass additive gaussian noise process. By squaring $s(t) + n(t)$ we obtain

$$y(t) = s^2(t) + 2s(t)n(t) + n^2(t) \qquad (4.5.29)$$

where $s^2(t)$ is the desired signal component and the other two components are the signal $\times$ noise and noise $\times$ noise terms. By computing the autocorrelation functions and power density spectra of these two noise components, one can easily show that both components have spectral power in the frequency band centered at $2f_c$. Consequently, the bandpass filter with bandwidth B_{bp} centered at $2f_c$, which produces the desired sinusoidal signal component that drives the PLL, also passes noise due to these two terms.

Since the bandwidth of the loop is designed to be significantly smaller than the bandwidth B_{bp} of the bandpass filter, the total noise spectrum at the input to the PLL may be approximated as a constant within the loop bandwidth. This approximation allows us to obtain a simple expression for the variance of the phase error as

$$\sigma_{\hat{\phi}}^2 = \frac{1}{\gamma_L S_L} \qquad (4.5.30)$$

where S_L is called the squaring loss and is given as

$$S_L = \frac{1}{1 + \dfrac{B_{\text{bp}}/2B_{\text{eq}}}{\gamma_L}} \qquad (4.5.31)$$

Since $S_L < 1$, S_L^{-1} represents the increase in the variance of the phase error caused by the added noise (noise × noise terms) that results from the squarer. Note, for example, that when $\gamma_L = B_{bp}/2B_{eq}$, the loss is 3 dB.

Finally we observe that the output of the VCO from the squaring loop must be frequency divided by 2 to generate the phase-locked carrier for signal demodulation. It should be noted that the output of the frequency divider has a phase ambiguity of 180° relative to the phase of the received signal. For this reason, the binary data must be differentially encoded prior to transmission and differentially decoded at the receiver.

Costas loop. A second method for generating a properly phased carrier for a double-sideband suppressed carrier signal is illustrated by the block diagram shown in Fig. 4.5.6. This scheme was developed by Costas (1956) and is called the *Costas loop*. The received signal is multiplied by $\cos(2\pi f_c t + \hat{\phi})$ and $\sin(2\pi f_c t + \hat{\phi})$, which are outputs from the VCO. The two products are

$$y_c(t) = [s(t) + n(t)] \cos(\pi f_c t + \hat{\phi})$$
$$= \tfrac{1}{2}[A(t) + n_c(t)] \cos \Delta\phi + \tfrac{1}{2}n_s(t) \sin \Delta\phi + \text{double frequency terms}$$
$$y_c(t) = [s(t) + n(t)] \sin(2\pi f_c t + \hat{\phi})$$
$$\qquad\qquad (4.5.32)$$
$$= \tfrac{1}{2}[A(t) + n_c(t)] \sin \Delta\phi - \tfrac{1}{2}n_s(t) \cos \Delta\phi + \text{double frequency terms}$$

FIGURE 4.5.6
Block diagram of Costas loop.

where the phase error $\Delta\phi = \hat{\phi} - \phi$. The double-frequency terms are eliminated by the lowpass filters following the multiplications.

An error signal is generated by multiplying the two outputs of the lowpass filters. Thus,

$$e(t) = \tfrac{1}{8}\left\{\left[A(t) + n_c(t)\right]^2 - n_s^2(t)\right\} \sin 2\,\Delta\phi$$

$$-\tfrac{1}{4}n_s(t)\left[A(t) + n_c(t)\right] \cos 2\,\Delta\phi \qquad (4.5.33)$$

This error signal is filtered by the loop filter whose output is the control voltage that drives the VCO.

We note that the error signal into the loop filter consists of the desired term $A^2(t)\sin 2(\hat{\phi} - \phi)$ plus terms that involve signal × noise and noise × noise. These terms are similar to the two noise terms at the input to the PLL for the squaring method. In fact, if the loop filter in the Costas loop is identical to that used in the squaring loop, the two loops are equivalent. Under this condition the probability density function of the phase error and the performance of the two loops are identical.

It is interesting to note that the optimum lowpass filter for rejecting the double-frequency terms in the Costas loop is a filter matched to the signal pulse in the information-bearing signal. If matched filters are employed for the lowpass filters, their outputs could be sampled at the bit rate, at the end of each signal interval, and the discrete-time signal samples could be used to drive the loop. The use of the matched filter results in a smaller noise into the loop.

Finally, we note that, as in the squaring PLL, the output of the VCO contains a phase ambiguity of 180°, necessitating the need for differential encoding of the data prior to transmission and differential decoding at the demodulator.

Decision-feedback loop. A third method for recovering the carrier is based on the concept of decision-directed phase estimation initially described and analyzed by Proakis et al. (1964) and further analyzed by Natali and Walbesser (1969) and others. In this method, decisions from the demodulator, which are usually correct with high probability, are fed back as shown in Fig. 4.5.7.

The received signal is multiplied by the quadrature carriers $c_c(t)$ and $c_s(t)$, as given by (4.5.5), which are derived from the VCO. The product signal

$$s(t)\cos\left(2\pi f_c t + \hat{\phi}\right) = \tfrac{1}{2}\left[A(t) + n_c(t)\right]\cos \Delta\phi$$

$$+ \tfrac{1}{2}n_s(t)\sin \Delta\phi + \text{double frequency terms} \qquad (4.5.34)$$

is used to recover the information carried by $A(t)$. The demodulator makes a decision on the symbol that is received every T seconds. Thus, in the absence of decision errors it reconstructs $A(t)$ free of any noise. This reconstructed signal is used to multiply the product of the second quadrature multiplier, which has been delayed by T seconds to allow the demodulator to reach a decision. Thus, the

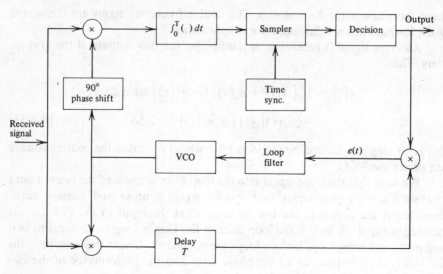

FIGURE 4.5.7
Carrier recovery with a decision-feedback PLL.

input to the loop filter in the absence of decision errors is the error signal

$$e(t) = \tfrac{1}{2}A(t)\{[A(t) + n_c(t)]\sin\Delta\phi - n_s(t)\cos\Delta\}$$

$$+ \text{double frequency terms}$$

$$= \tfrac{1}{2}A^2(t)\sin\Delta\phi + \tfrac{1}{2}A(t)[n_c(t)\sin\Delta\phi - n_s(t)\cos\Delta\phi]$$

$$+ \text{double frequency term} \qquad (4.5.35)$$

The loop filter is lowpass and, hence, it rejects the double frequency term in $e(t)$.

We note that the decision-directed or decision-feedback phase-locked loop (DFPLL) differs from the Costas loop only in the method by which $A(t)$ is rectified for the purpose of removing the modulation. In the Costas loop each of the two quadrature signals used to rectify $A(t)$ is corrupted by noise. In the DFPLL, only one of the signals used to rectify $A(t)$ is corrupted by noise. On the other hand, the squaring loop is similar to the Costas loop in terms of the noise effect on the estimate $\hat{\phi}$. Consequently the DFPLL is superior in performance to both the Costas loop and the squaring loop, provided that the demodulator is operating at error rates below 10^{-2} where an occasional decision error has a negligible effect on $\hat{\phi}$. Quantitative comparisons of the variance of the phase errors in a Costas loop to those in a DFPLL have been made by Lindsey and Simon (1973), and show that the variance of the DFPLL is 4 to 10 times smaller for signal-to-noise ratios per bit above 0 dB.

Carrier-tracking for multiple phase signals. When the digital information is transmitted via M-phase modulation of a carrier, the methods described above

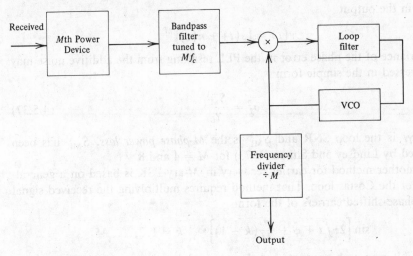

Output

FIGURE 4.5.8
Carrier recovery with Mth power law device for M-ary PSK.

can be generalized to provide the properly phased carrier for demodulation. The received M-phase signal, excluding the additive noise, may be expressed as

$$s(t) = A \cos \left[2\pi f_c t + \phi + \frac{2\pi}{M}(m - 1) \right] \qquad m = 1, 2, \ldots, M \quad (4.5.36)$$

where $2\pi(m-1)/M$ represents the information bearing component of the signal phase. The problem in carrier recovery is to remove the information-bearing component and, thus, to obtain the unmodulated carrier $\cos(2\pi f_c t + \phi)$. One method by which this can be accomplished is illustrated in Fig. 4.5.8, which represents a generalization of the squaring loop. The signal is passed through an Mth power law device, which generates a number of harmonics of f_c. The bandpass filter selects the harmonic $\cos(2\pi M f_c t + M\phi)$ for driving the PLL. The term

$$\frac{2\pi}{M}(m - 1) \cdot M = 2\pi(m - 1) \equiv 0 \qquad (\text{mod } 2\pi) \qquad m = 1, 2, \ldots, M$$

Thus, the information is removed. The VCO output is $\sin(2\pi M f_c t + M\hat{\phi})$, so this output is divided in frequency by M to yield $\sin(2\pi f_c t + \hat{\phi})$, and phase-shifted by $\pi/2$ radians to yield $\cos(2\pi f_c t + \hat{\phi})$. These components are then fed to the demodulator. Although not explicitly shown, there is a phase ambiguity in these reference sinusoids of $360°/M$, which can be overcome by differential encoding of the data at the transmitter and differential decoding after demodulation at the receiver.

Just as in the case of the squaring PLL, the Mth power PLL operates in the presence of noise that has been enhanced by the Mth power law device which results in the output

$$y(t) = [s(t) + n(t)]^M$$

The variance of the phase error in the PLL resulting from the additive noise may be expressed in the simple form

$$\sigma_{\hat{\phi}}^2 = \frac{S_{\mathrm{ML}}^{-1}}{\gamma_L} \tag{4.5.37}$$

where γ_L is the loop SNR and S_{ML}^{-1} is the *M-phase power loss*. S_{ML} has been evaluated by Lindsey and Simon (1973) for $M = 4$ and 8.

Another method for carrier recovery in M-ary PSK is based on a generalization of the Costas loop. That method requires multiplying the received signal by M phase-shifted carriers of the form

$$\sin\left[2\pi f_c t + \hat{\phi} + \frac{\pi}{M}(k-1)\right] \qquad k = 1, 2, \ldots, M$$

lowpass-filtering each product, and then multiplying the outputs of the lowpass filters to generate the error signal. The error signal excites the loop filter, which, in turn, provides the control signal for the VCO. This method is relatively complex to implement and, consequently, has not been generally used in practice.

A simpler method and one that performs relatively well for M-ary PSK is the DFPLL illustrated by the block diagram shown in Fig. 4.5.9. The received signal is demodulated to yield the phase estimate

$$\hat{\theta}_m = \frac{2\pi}{M}(m-1)$$

which, in the absence of a decision error, is the transmitted signal phase. The two outputs of the quadrature multipliers are delayed by the symbol duration T and multiplied by $\cos\theta_m$ and $-\sin\theta_m$ to yield

$$-s(t)\cos\left(2\pi f_c t + \hat{\phi}\right)\sin\theta_m$$

$$= -\tfrac{1}{2}\left[A\cos\theta_m + n_c(t)\right]\sin\theta_m\cos\left(\phi - \hat{\phi}\right)$$

$$+ \tfrac{1}{2}\left[A\sin\theta_m + n_s(t)\right]\sin\theta_m\sin\left(\phi - \hat{\phi}\right) + \text{double frequency terms}$$

$$-s(t)\sin\left(2\pi f_c t + \hat{\phi}\right)\cos\theta_m \tag{4.5.38}$$

$$= \tfrac{1}{2}\left[A\cos\theta_m + n_c(t)\right]\cos\theta_m\sin\left(\phi - \hat{\phi}\right)$$

$$+ \tfrac{1}{2}\left[A\sin\theta_m + n_s(t)\right]\cos\theta_m\cos\left(\phi - \hat{\phi}\right) + \text{double frequency terms}$$

The two signals are added to generate the error signal

$$e(t) = \frac{A}{2}\sin\left(\phi - \hat{\phi}\right) + \frac{1}{2}n_c(t)\sin\left(\phi - \hat{\phi} - \theta_m\right)$$

$$+ \frac{1}{2}n_s(t)\cos\left(\phi - \hat{\phi} - \theta_m\right) + \text{double frequency terms} \tag{4.5.39}$$

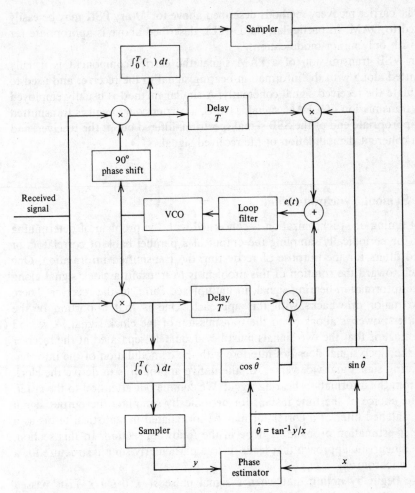

FIGURE 4.5.9
Carrier recovery for M-ary PSK using a decision-feedback PLL.

This error signal is the input to the loop filter that provides the control signal for the VCO.

We observe that the two quadrature noise components in (4.5.39) appear as additive terms. There is no term involving a product of two noise components as in an Mth power law device. Consequently, there is no Mth power loss associated with the decision-feedback PLL.

This M-phase tracking loop also has a phase ambiguity of $360°/M$, necessitating the need for differentially encoding the information sequence prior to transmission and differentially decoding the received sequence after demodulation to recover the information.

The carrier recovery methods described above for M-ary PSK may be easily extended to QAM. In particular, the DFPLL described above is appropriate for QAM with only minor modifications.

In VSB transmission of a PAM signal the carrier component is usually transmitted along with the information-bearing signal to the receiver and used to demodulate the received signal coherently. A similar method is usually employed in SSB transmission of a PAM signal. That is, a carrier component is transmitted at the appropriate end of the SSB signal spectrum, filtered out at the receiver, and used in coherent demodulation of the received signal.

4.5.2 Symbol Synchronization

Symbol timing or synchronization is concerned with the problem of determining a clock for periodically sampling the output of a parallel bank of correlators or matched filters for the purpose of recovering the transmitted information. One approach toward the solution of this problem is to transmit a clock signal along with the information-bearing signal, in multiplexed form, to the receiver. There are two major drawbacks with this approach. One is that a portion of the transmitter power is allocated to the transmission of the clock signal. A second disadvantage is that the two signals must be adequately separated at the receiver so that the clock signal does not interfere with the demodulation of the information-bearing signal and vice versa. An alternative approach is to derive the clock signal from the information-bearing signal. We confine our attention to the latter.

The choice of the time instant for periodically sampling the output signal from a matched filter or a correlator may be formulated in statistical terms as a parameter estimation problem as done in the following section. In this section, we take a heuristic approach that leads us to a particularly simple structure for a symbol synchronizer.

We begin by noting that when a signal pulse $s(t)$, $0 \leq t \leq T$, is passed through a filter matched to it, the output of the filter reaches a maximum at $t = T$. For example, suppose that $s(t)$ is a rectangular pulse as shown in Fig. 4.5.10a. The autocorrelation function of the pulse, which can be obtained by passing the pulse through its matched filter, is the triangular pulse shown in Fig. 4.5.10b. The proper time to sample the output of the matched filter for a maximum output is at $t = T$, i.e., at the peak of the correlation function. Instead, if we sample the output of the matched filter early, say at $t = T - \delta$, or late at $t = T + \delta$, the values of the samples will be smaller (on the average when noise is present) than at the peak value. Since the autocorrelation function is even relative to $t = T$, the values of the correlation function at $t = T - \delta$ and $t = T + \delta$ are equal in magnitude. Hence the difference in the magnitude of these sampled values is zero. When this situation occurs, the proper sampling time is the midpoint between $t = T - \delta$ and $t = T + \delta$, i.e., at $t = T$. This condition forms the basis for a commonly used synchronizer, called an *early-late gate synchronizer*, *described below.*

(a)

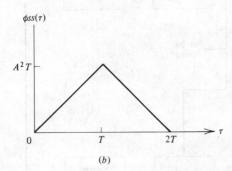

(b)

FIGURE 4.5.10
(a) Rectangular pulse; (b) its autocorrelation function.

The early-late gate synchronizer may be realized by means of matched filters or by means of cross correlators. Let us consider the latter. Suppose that the transmitted pulse is rectangular, as previously shown in Fig. 4.5.10a. The cross correlator for the rectangular pulse is equivalent to an integrator which is sampled and reset to zero at $t = T$. Now suppose that the integration interval is not properly aligned with the arrival of the signal pulse. If the integration begins at $t = -\delta$ and ends at $t = T - \delta$, i.e., the integration begins early, the output of the integrator after T seconds is less (on the average when the noise is present) than the output when the integrator is perfectly synchronized. A similar statement applies if the integrator starts late, say at $t = \delta$, and ends the integration at $t = T + \delta$.

Figure 4.5.11 illustrates an early-late gate synchronizer in block diagram form. The received signal is translated in frequency to lowpass and passed through two correlators, one of which has its reference delayed by $\delta < T$ and the other its reference advanced by δ, relative to the peak of the autocorrelation function. Since the autocorrelation function of any signal pulse is even, the difference (error signal) between the two correlators should be zero. If the timing is off such that the error signals is not zero, the clock should be retarded or advanced, depending on the sign of the error, by an amount that will drive the error to zero. In order to smooth the error signal, it is passed through a lowpass filter. The output of the lowpass filter is a control voltage for a VCC whose

FIGURE 4.5.11
Block diagram of early-late gate synchronizer.

output is used to excite periodically a signal pulse generator that feeds the correlators. If the signal pulses are rectangular, there is no need for a signal generator and the correlators become integrate-and-dump filters.

When the synchronizer is operating in a tracking mode, the correlators, which are offset in time by $+\delta$ and $-\delta$, are affected by adjacent symbols. However, if the information symbols have zero mean, the contribution to the output of the correlators from adjacent symbols averages out to zero in the loop filter.

An equivalent realization of the early-late gate synchronizer that is somewhat easier to implement is shown in Fig. 4.5.12. In this case the clock from the VCC is advanced and retarded by δ and these clock signals are used to sample the outputs of the correlators.

The early-late gate synchronizer is especially appropriate for PSK and PAM, two of the more common forms of digital modulation. For the more general combined amplitude and phase modulation, i.e. QAM, a particularly

FIGURE 4.5.12
Block diagram of early-late gate synchronizer—an alternative form.

effective symbol synchronizer is shown in Fig. 4.5.13. Recall that the demodulation of QAM is accomplished by use of two quadrature carriers $\cos(2\pi f_c t + \hat{\phi})$ and $\sin(2\pi f_c t + \hat{\phi})$, as shown in Fig. 4.2.24. The in-phase and quadrature components resulting from the multiplications are each filtered by a narrow bandpass filter tuned to $1/2T$. The two filter outputs are squared (rectified), summed, and then filtered by a narrowband filter tuned to the clock frequency $1/T$. Thus the resulting sinusoidal waveform is the appropriate clock signal for sampling the outputs of the matched filters to recover the information.

In addition to the symbol synchronizers described above, a number of others have been devised for the various types of digital modulations considered in this chapter. A comprehensive treatment of this important topic is found in a number of books, e.g., Stiffler (1971) and Lindsey (1972).

In the following section we consider the problem of carrier and symbol synchronization as a parameter estimation problem based on the maximum likelihood criterion.

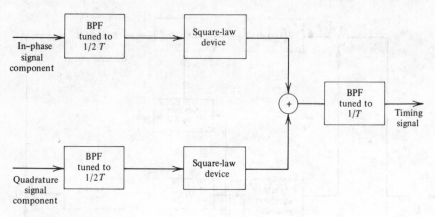

FIGURE 4.5.13
Block diagram of a synchronizer for QAM.

4.5.3 Maximum Likelihood Parameter Estimation

The basis for the foregoing development of carrier and symbol synchronization techniques was rather heuristic. In this section we take a more analytical approach based on statistical parameter estimation theory, in which we treat the carrier phase and the transmission delay as two signal parameters that are to be estimated in the presence of additive noise. The received signal is represented as

$$r(t) = s(t; \phi, \tau) + n(t) \qquad (4.5.40)$$

where $s(t; \phi, \tau)$ is the signal component and $n(t)$ denotes the additive noise. To simplify the notation, we let α denote the parameter vector $\{\phi, \tau\}$ so that the signal component is expressed as $s(t; \alpha)$.

There are basically two criteria that are widely applied to signal parameter estimation, the maximum likelihood (ML) criterion and the maximum *a posteriori* probability (MAP) criterion. In the MAP criterion the signal parameter vector α is modeled as random, and characterized by an *a priori* probability density function $p(\alpha)$. In the maximum likelihood criterion, the signal parameter vector α is treated as deterministic but unknown.

By using an orthonormal expansion on the received signal $r(t)$, such as the Karhunen–Loève expansion described in Appendix 4A, the joint pdf of the random variables $[r_1, r_2, \ldots, r_N] \equiv \mathbf{r}$ in the expansion can be expressed as $p(\mathbf{r}|\alpha)$. Then, the ML estimate of α is the value that maximizes $p(\mathbf{r}|\alpha)$. On the other hand, the MAP estimate is the value of α that maximizes the *a posteriori* probability density function

$$p(\alpha|\mathbf{r}) = \frac{p(\mathbf{r}|\alpha)\,p(\alpha)}{p(\mathbf{r})} \qquad (4.5.41)$$

We note that if there is no prior knowledge of the parameter vector α, we may assume that $p(\alpha)$ is uniform (constant) over the range of values of the parameters. In such a case the value of α that maximizes $p(\alpha|r)$ is identical to the value of α that maximizes $p(r|\alpha)$. Therefore, the MAP and ML estimates are identical.

In our treatment of parameter estimation given below, we view the parameters ϕ and τ as unknown, but deterministic. Hence, we adopt the ML criterion for estimating them.

In the ML estimation of signal parameters we require that the receiver extract the estimate by observing the received signal over a time interval $T_0 \gg T$, which is called the observation interval. Estimates obtained from a single observation interval are sometimes called one shot estimates. In practice, however, the estimation is performed on a continuous basis by using tracking loops (either analog or digital) that continuously update the estimates. Nevertheless, one shot estimates yield insight for tracking loop implementation. In addition, they prove useful in the analysis of the performance of ML estimation and their performance can be related to that obtained with a tracking loop.

Carrier phase estimation. First, let us consider the estimation of the carrier phase ϕ. For simplicity, we assume that τ is known and, in particular, we set $\tau = 0$. If the additive noise is assumed to be a white gaussian random process, it is easily shown (see Appendix 4A and Prob. 4.21) that the probability density function $p(r|\phi)$, in the limit as the number N of signal samples in the observation interval approaches infinity, is proportional to the likelihood function $\Lambda(\phi)$, defined as

$$\Lambda(\phi) = \exp\left\{ -\frac{1}{N_0} \int_{T_0} [r(t) - s(t; \phi)]^2 \, dt \right\}$$

$$= \exp\left\{ -\frac{1}{N_0} \int_{T_0} r^2(t) \, dt + \frac{2}{N_0} \int_{T_0} r(t) s(t; \phi) \, dt \right.$$

$$\left. - \frac{1}{N_0} \int_{T_0} s^2(t; \phi) \, dt \right\}. \qquad (4.5.42)$$

Note that the first term of the exponential factor does not involve the signal parameter ϕ. The third term, which contains the integral of $s^2(t; \phi)$, is a constant equal to the signal energy over the observation interval T_0 for any value of ϕ. Only the second term, which involves the cross correlation of the received signal $r(t)$ with the signal $s(t; \phi)$ depends on the choice of ϕ. Therefore, the likelihood function $\Lambda(\phi)$ may be expressed as

$$\Lambda(\phi) = C \exp\left[\frac{2}{N_0} \int_{T_0} r(t) s(t; \phi) \, dt \right] \qquad (4.5.43)$$

where C is a constant independent of ϕ.

The ML estimate $\hat{\phi}_{ML}$ is the value of ϕ that maximizes $\Lambda(\phi)$ in (4.5.43). Equivalently, the value $\hat{\phi}_{ML}$ also maximizes the logarithm of $\Lambda(\phi)$, i.e., the

log-likelihood function

$$\Lambda_L(\phi) = \frac{2}{N_0} \int_{T_0} r(t) s(t; \phi) \, dt \tag{4.5.44}$$

Note that in defining $\Lambda_L(\phi)$ we have ignored the constant term $\ln C$.

Example 4.5.1. As an example of the optimization to determine the carrier phase, let us consider the transmission of the unmodulated carrier $A \cos(2\pi f_c t)$. The received signal is

$$r(t) = A \cos(2\pi f_c t + \phi) + n(t)$$

where ϕ is the unknown phase. We seek the value ϕ, say $\hat{\phi}_{ML}$, that maximizes

$$\Lambda_L(\phi) = \frac{2A}{N_0} \int_{T_0} r(t) \cos(2\pi f_c t + \phi) \, dt$$

A necessary condition for a maximum is that

$$\frac{d\Lambda_L(\phi)}{d\phi} = 0$$

This condition yields

$$\int_{T_0} r(t) \sin(2\pi f_c t + \hat{\phi}_{ML}) \, dt = 0 \tag{4.5.45}$$

or, equivalently,

$$\hat{\phi}_{ML} = -\tan^{-1} \frac{\int_{T_0} r(t) \sin 2\pi f_c t \, dt}{\int_{T_0} r(t) \cos 2\pi f_c t \, dt} \tag{4.5.46}$$

We observe that the optimality condition given by (4.5.45) implies the use of a PLL to extract the estimate as illustrated in Fig. 4.5.14. The loop filter is an integrator whose bandwidth is proportional to the reciprocal of the integration interval T_0. On the other hand, (4.5.46) implies an implementation that uses quadrature carriers to cross correlate with $r(t)$. Then, $\hat{\phi}_{ML}$ is the arctangent of the ratio of these two correlator outputs, as shown in Fig. 4.5.15. Note that this estimation scheme yields $\hat{\phi}_{ML}$ explicitly.

FIGURE 4.5.14
A PLL for obtaining the ML estimate of the phase of an unmodulated carrier.

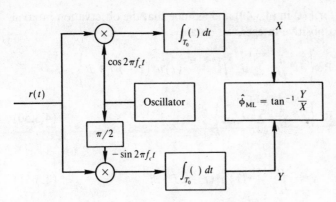

FIGURE 4.5.15
A (one-shot) ML estimate of the phase of an unmodulated carrier.

This example clearly demonstrates that the PLL provides the ML estimate of the phase of an unmodulated carrier.

A problem arises in maximizing either (4.5.43) or (4.5.44) when the signal $s(t; \phi)$ carries the information sequence $\{I_n\}$. In this case we can adopt one of two approaches—either we assume that $\{I_n\}$ is known or we treat $\{I_n\}$ as a random sequence and average over its statistics.

In decision-directed parameter estimation, we assume that the information sequence $\{I_n\}$ over the observation interval has been estimated and, in the absence of demodulation errors, $\tilde{I}_n = I_n$, where $\tilde{I}_n$ denotes the detected value of the information I_n. In this case $s(t; \phi)$ is completely known except for the carrier phase.

To be specific, let us consider the decision-directed phase estimate for the class of linear modulation techniques for which the received *equivalent lowpass signal* may be expressed as

$$r(t) = e^{-j\phi} \sum_n I_n u(t - nT) + z(t)$$

$$= v(t) e^{-j\phi} + z(t) \tag{4.5.47}$$

where $v(t)$ is a known signal if the sequence $\{I_n\}$ is assumed known. The likelihood function and corresponding log-likelihood function for the equivalent lowpass signal are

$$\Lambda(\phi) = C \exp\left\{ \mathrm{Re}\left[\frac{1}{N_0} \int_{t_0} r(t) v^*(t) e^{j\phi} \, dt \right] \right\} \tag{4.5.48}$$

$$\Lambda_L(\phi) = \mathrm{Re}\left\{ \left[\frac{1}{N_0} \int_{T_0} r(t) v^*(t) \, dt \right] e^{j\phi} \right\} \tag{4.5.49}$$

If we substitute for $v(t)$ in (4.5.49) and assume that the observation interval $T_0 = KT$ where K is a positive integer, we obtain

$$\Lambda_L(\phi) = \text{Re}\left\{ e^{j\phi}\frac{1}{N_0}\sum_{n=0}^{K-1} I_n^* \int_{nT}^{(n+1)T} r(t)u^*(t-nT)\,dt \right\}$$

$$= \text{Re}\left\{ e^{j\phi}\frac{1}{N_0}\sum_{n=0}^{K-1} I_n^* y_n \right\} \tag{4.5.50}$$

where, by definition,

$$y_n = \int_{nT}^{(n+1)} r(t)u^*(t-nT)\,dt \tag{4.5.51}$$

Note that y_n is the output of the matched filter in the nth signal interval. The ML estimate of ϕ is easily found from (4.5.50) by differentiatin the log likelihood

$$\Lambda_L(\phi) = \text{Re}\left[\frac{1}{N_0}\sum_{n=0}^{K-1} I_n^* y_n\right]\cos\phi - \text{Im}\left[\frac{1}{N_0}\sum_{n=0}^{K-1} I_n^* y_n\right]\sin\phi$$

with respect to ϕ and setting the derivative to zero. Thus, we obtain

$$\hat{\phi}_{ML} = -\tan^{-1}\frac{\text{Im}\left[\sum_{n=0}^{K-1} I_n^* y_n\right]}{\text{Re}\left[\sum_{n=0}^{K-1} I_n^* y_n\right]} \tag{4.5.52}$$

We call $\hat{\phi}_{ML}$ in (4.5.52) the decision-directed (or decision-feedback) carrier phase estimate. It is easliy shown (Prob. 4.22) that the mean value of $\hat{\phi}_{ML}$ is ϕ, so that the estimate is unbiased. Furthermore, the pdf of $\hat{\phi}_{ML}$ can be obtained (Prob. 4.23) by use of the procedure described in Sec. 4.2.6.

The decision-feedback PLL shown in Fig. 4.5.9 is equivalent to (4.5.52) and, hence, it represents an implementation of the ML estimate for M-PSK signals.

The ML estimate in (4.5.52) is also appropriate for QAM. The ML estimate for offset QPSK is also easily obtained (Prob. 4.24) by maximizing the log likelihood function in (4.5.49), with $v(t)$ given as

$$v(t) = \sum_n I_n u(t-nT) + j\sum_n J_n u\left(t-nT-\frac{T}{2}\right) \tag{4.5.53}$$

where $I_n = \pm 1$ and $J_n = \pm 1$.

Instead of using a decision-directed scheme to obtain the phase estimate, we may treat the data as random variables and simply average $\Lambda(\phi)$ over these random variables prior to maximization. In order to carry out this integration we may use either the actual probability distribution function of the data if it is known or, perhaps, we may assume some probability distribution that might be a reasonable approximation to the true distribution. The following example illustrates the first approach.

Example 4.5.2. Suppose the real signal $s(t)$ carries binary modulation. Then, in a signal interval we have

$$s(t) = A \cos 2\pi f_c t \quad 0 \le t \le T$$

where $A = \pm 1$ with equal probability. Clearly, the pdf of A is given as

$$p(A) = \tfrac{1}{2}\delta(A - 1) + \tfrac{1}{2}\delta(A + 1)$$

Now, the likelihood function $\Lambda(\phi)$ given by (4.5.43) is conditional on a given value of A and must be averaged over the two values. Thus,

$$
\begin{aligned}
\overline{\Lambda}(\phi) &= \int_{-\infty}^{\infty} \Lambda(\phi)\, p(A)\, dA \\
&= \frac{1}{2} \exp\left[\frac{2}{N_0} \int_0^T r(t) \cos(2\pi f_c t + \phi)\, dt \right] \\
&\quad + \frac{1}{2} \exp\left[-\frac{2}{N_0} \int_0^T r(t) \cos(2\pi f_c t + \phi)\, dt \right] \\
&= \cosh\left[\frac{2}{N_0} \int_0^T r(t) \cos(2\pi f_c t + \phi)\, dt \right]
\end{aligned}
$$

and the corresponding log likelihood function is

$$\overline{\Lambda}_L(\phi) = \ln \cosh\left[\frac{2}{N_0} \int_0^T r(t) \cos(2\pi f_c t + \phi)\, dt \right] \qquad (4.5.54)$$

If we differentiate $\overline{\Lambda}_L(\phi)$ and set the derivative to zero, we obtain the ML estimate for the non-decision-directed estimate. Unfortunately, the functional relationship in (4.5.54) is highly nonlinear and, hence, an exact solution is difficult to obtain. On the other hand, approximations are possible. In particular,

$$\ln \cosh x = \begin{cases} \tfrac{1}{2}x^2 & |x| \ll 1 \\ |x| & |x| \gg 1 \end{cases} \qquad (4.5.55)$$

With these approximations, the solution for ϕ becomes tractable.

In this example, we averaged over the two possible values of the information symbol. When the information symbols are M-valued, where M is large, the averaging operation yields highly nonlinear functions of the parameter to be estimated. In such a case, we may simplify the problem by assuming that the information symbols are continuous random variables. For example, we may assume that the symbols are zero mean gaussian. The following example illustrates this approximation and the resulting form for the average likelihood function.

Example 4.5.3. Let us consider the same signal as in Example 4.5.2, but now we assume that the amplitude A is zero mean gaussian with unit variance. Thus,

$$p(A) = \frac{1}{\sqrt{2\pi}} e^{-A^2/2}$$

If we average $\Lambda(\phi)$ over the assumed pdf of A, we obtain the average likelihood $\overline{\Lambda}(\phi)$ in the form

$$\overline{\Lambda}(\phi) = C \exp\left\{ \left[\frac{2}{N_0} \int_0^T r(t) \cos(2\pi f_c t + \phi)\, dt \right]^2 \right\} \qquad (4.5.56)$$

and the corresponding log likelihood as

$$\overline{\Lambda}_L(\phi) = \left[\frac{2}{N_0} \int_0^T r(t) \cos\left(2\pi f_c t + \phi\right) dt \right]^2 \qquad (4.5.57)$$

We can obtain the ML estimate of ϕ by differentiating $\overline{\Lambda}_L(\phi)$ and setting the derivative to zero.

It is interesting to note that the log likelihood function is quadratic under the gaussian assumption and that it is approximately quadratic as indicated in (4.5.55) for small values of the cross correlation of $r(t)$ with $s(t; \phi)$. In other words, if the cross correlation over a single interval is small, the gaussian assumption for the distribution of the information symbols yields a good approximation to the log likelihood function.

In view of these results, we may use the gaussian approximation on all the symbols in the observation interval $T_0 = KT$. Specifically, we assume that the K information symbols are statistically independent and identically distributed. By averaging the likelihood function $\Lambda(\phi)$ over the gaussian pdf for each of the K symbols in the interval $T_0 = KT$, we obtain the result

$$\Lambda(\phi) = C \exp \left\{ \sum_{n=0}^{K-1} \left[\frac{2}{N_0} \int_{nT}^{(n+1)T} r(t) \cos\left(2\pi f_c t + \phi\right) dt \right]^2 \right\} \qquad (4.5.58)$$

If we take the logarithm of (4.5.58), differentiate the resulting log likelihood function, and set the derivative to zero we obtain the condition for the ML estimate as

$$\sum_{n=0}^{K-1} \int_{nT}^{(n+1)T} r(t) \cos\left(2\pi f_c t + \hat{\phi}\right) dt \cdot \int_{nT}^{(n+1)T} r(t) \sin\left(2\pi f_c t + \hat{\phi}\right) dt = 0$$

$$(4.5.59)$$

Although this equation can be manipulated further, its present form suggests the tracking loop configuration illustrated in Fig. 4.5.16. We observe that this loop resembles a Costas loop. We also note that the multiplication of the two signals from the integrators destroys the sign carried by the information symbols. The summer plays the role of the loop filter. In a tracking loop configuration, the summer may be implemented either as a sliding-window digital filter (summer) or as a lowpass digital filter with exponential weighting of the past data.

In view of the similarity and equivalence of the Costas loop to the squaring PLL shown in Fig. 4.5.1, it is apparent that another possible implementation of non-decision-directed ML phase estimation is the squaring loop.

In a similar manner, one can derive non-decision-directed ML phase estimates for QAM and M-PSK. The starting point is to average the likelihood function given by (4.5.48) over the statistical characteristics of the data. Here again, we may use the gaussian approximation (two-dimensional gaussian for complex-valued information symbols) in averaging over the information sequence. This is left as an exercise for the reader. Suffice it to say, the resulting

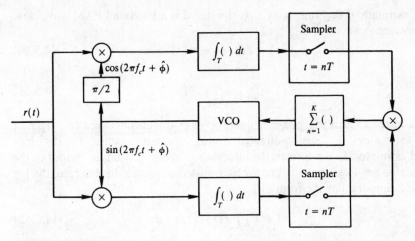

FIGURE 4.5.16
Non-decision-directed PLL for carrier phase estimations of PAM signals.

ML estimates for M-PSK can be realized by use of the M power law PLL shown in Fig. 4.5.8.

We should also mention that carrier phase recovery for CPM signals can be accomplished in a decision-directed manner by use of a PLL. From the optimum demodulator for CPM signals, which is described in Chapter 6, we can generate an error signal that is filtered in a loop filter whose output drives a PLL. A block diagram for this scheme is shown in Fig. 4.5.17.

In the case of rectangular QAM, one may treat the modulation as two independent PAM symbols in phase quadrature, and simply eliminate the sign carried by the information symbols by squaring the signal after demodulation to baseband.

Symbol timing estimation. Now that we have treated the ML estimation of the carrier phase in some detail, let us turn our attention to the problem of obtaining

FIGURE 4.5.17
Carrier phase recovery for CPM signals.

the ML estimate of the time delay τ. If the signal is a baseband PAM waveform, it is represented as

$$r(t) = s(t; \tau) + n(t) \qquad (4.5.60)$$

where

$$s(t; \tau) = \sum_n I_n u(t - nT - \tau) \qquad (4.5.61)$$

As in the case of ML phase estimation we distinguish between two types of timing estimators, a decision-directed timing estimator and a non-decision-directed estimator. In the former, the information symbols from the output of the demodulator are treated as the known transmitted sequence. In this case, the log likelihood function has the form

$$\Lambda_L(\tau) = C_L \int_{T_0} r(t) s(t; \tau)\, dt \qquad (4.5.62)$$

If we substitute (4.5.61) into (4.5.62) we obtain

$$\Lambda_L(\tau) = C_L \sum_n I_n \int_{T_0} r(t) u(t - nT - \tau)\, dt$$

$$= C_L \sum_{n=1} I_n y_n(\tau) \qquad (4.5.63)$$

where $y_n(\tau)$ is defined as

$$y_n(\tau) = \int_{T_0} r(t) u(t - nT - \tau)\, dt \qquad (4.5.64)$$

A necessary condition for τ to be the ML estimate of τ is that

$$\frac{d\Lambda_L(\tau)}{d\tau} = \sum_n I_n \frac{d}{d\tau} \int_{T_0} r(t) u(t - nT - \tau)\, dt$$

$$= \sum_n I_n \frac{d}{d\tau} [y_n(\tau)] = 0 \qquad (4.5.65)$$

The result in (4.5.65) suggests the implementation of the tracking loop shown in Fig. 4.5.18. We should observe that the summation in the loop serves as the loop filter whose bandwidth is controlled by the length of the sliding window in the summation. The output of the loop filter drives the voltage-controlled clock (VCC) which controls the sampling times for the input to the loop.

In the case of non-decision-directed estimation we first average the likelihood ratio $\Lambda(\tau)$ over the pdf of the information symbols to obtain $\overline{\Lambda}(\tau)$ and then differentiate either $\overline{\Lambda}(\tau)$ or $\ln \overline{\Lambda}(\tau) \equiv \overline{\Lambda}_L(\tau)$ to obtain the condition for the maximum likelihood estimate $\hat{\tau}_{ML}$.

In the case of binary (baseband) PAM, where $I_n = \pm 1$ with equal probability, the average over the data yields

$$\overline{\Lambda}_L(\tau) = \sum_n \ln \cosh C y_n(\tau) \qquad (4.5.66)$$

FIGURE 4.5.18
Decision-directed ML estimation of timing for baseband PAM.

just as in the case of the phase estimator. Since $\ln \cosh x \approx x^2/2$ for small x, the square-law approximation

$$\overline{\Lambda}_L(\tau) \approx \tfrac{1}{2} C^2 \sum_n y_n^2(\tau) \tag{4.5.67}$$

is appropriate for low signal-to-noise ratios. For multilevel PAM, we may approximate the statistical characteristics of the information symbols $\{I_n\}$ by the gaussian pdf, with zero mean and unit variance. When we average $\Lambda(\tau)$ over the gaussian pdf, the logarithm of $\overline{\Lambda}(\tau)$ is identical to $\overline{\Lambda}_L(\tau)$ given by (4.5.67). Consequently, the non-decision-directed estimate of τ may be obtained by differentiating (4.5.67). The result is an approximation to the ML estimate of the delay time. The derivative of (4.5.67) is

$$\frac{d}{d\tau} \sum_n y_n^2(\tau) = 2\sum_n y_n(\tau) \frac{dy_n(\tau)}{d\tau} = 0 \tag{4.5.68}$$

where $y_n(\tau)$ is given by (4.5.64).

An implementation of a tracking loop based on the derivative of $\overline{\Lambda}_L(\tau)$ given by (4.5.66) is shown in Fig. 4.5.19. Alternatively, an implementation of a tracking loop based on (4.5.68) is illustrated in Fig. 4.5.20. In both structures we

FIGURE 4.5.19
Non-decision-directed estimation of timing for binary baseband PAM.

FIGURE 4.5.20
Non-decision-directed estimation of timing for baseband PAM.

observe that the summation serves as the loop filter that drives the VCC. It is interesting to note the resemblance of the timing loop in Fig. 4.5.20 to the Costas loop for phase estimation.

The early-late gate synchronizer described in Section 4.5.2 is a non-decision-directed estimator of symbol timing that approximates the maximum likelihood estimator. This assertion can be demonstrated by approximating the derivative of the log likelihood function by the finite difference, i.e.,

$$\frac{d\overline{\Lambda}_L(\tau)}{d\tau} \approx \frac{\overline{\Lambda}_L(\tau + \delta) - \overline{\Lambda}(\tau - \delta)}{2\delta} \qquad (4.5.69)$$

If we substitute for $\overline{\Lambda}_L(\tau)$ from (4.5.67) into (4.5.69) we obtain the approximation for the derivative as

$$\frac{d\overline{\Lambda}_L(\tau)}{d\tau} = \frac{C^2}{4\delta} \sum_n \left[y_n^2(\tau + \delta) - y_n^2(\tau - \delta) \right]$$

$$\approx \frac{C^2}{4\delta} \sum_n \left\{ \left[\int_{T_0} r(t)u(t - nT - \tau - \delta)\, dt \right]^2 \right.$$

$$\left. - \left[\int_{T_0} r(t)u(t - nT - \tau + \delta)\, dt \right]^2 \right\} \qquad (4.5.70)$$

But the mathematical expression in (4.5.70) basically describes the functions performed by the early-late gate symbol synchronizers previously described in Section 4.5.2 and illustrated in Figs. 4.5.11 and 4.5.12.

The techniques described above for ML timing estimation of baseband PAM signals can be extended to carrier modulation signal formats such as QAM and PSK in a straightforward manner, by dealing with the equivalent lowpass form of the signals. Thus, the problem of ML estimation of symbol timing for carrier signals is very similar to the problem formulation for the baseband PAM signal.

Joint estimation of carrier phase and symbol timing. The estimation of the carrier phase and symbol timing may be accomplished separately as described above or jointly. Joint ML estimation of two or more signal parameters yields estimates that are as good and usually better than the estimates obtained from separate optimization of the likelihood function. In other words, the variances of the signal parameters obtained from joint optimization are less than or equal to the variance of parameter estimates obtained from separately optimizing the likelihood function.

Let us consider the joint estimation of the carrier phase and symbol timing. The log likelihood function for these two parameters may be expressed in terms of the equivalent lowpass signals as

$$\Lambda_L(\phi, \tau) = \text{Re}\left\{ \frac{1}{N_0} \int_{T_0} r(t) v^*(t; \phi, \tau)\, dt \right\} \qquad (4.5.71)$$

where $v(t; \phi, \tau)$ is the equivalent lowpass signal which has the general form

$$v(t; \phi, \tau) = e^{-j\phi}\left\{ \sum_n I_n u(t - nT - \tau) + j\sum_n J_n w(t - nT - \tau) \right\} \qquad (4.5.72)$$

where $\{I_n\}$ and $\{J_n\}$ are the two information sequences.

We note that for PAM we may set $J_n = 0$ for all n, and the sequence $\{I_n\}$ is real. For QAM and PSK, we set $I_n = 0$ for all n and the sequence $\{I_n\}$ is complex-valued. For offset QPSK, both sequences $\{I_n\}$ and $\{J_n\}$ are nonzero and $w(t) = u(t - T/2)$.

For decision-directed ML estimation of ϕ and τ, the log likelihood function becomes

$$\Lambda_L(\phi, \tau) = \text{Re}\left\{ \frac{e^{j\phi}}{N_0} \sum_m \left[I_n^* y_n(\tau) + jJ_n^* x_n(\tau) \right] \right\} \qquad (4.5.73)$$

where

$$y_n(\tau) = \int_{T_0} r(t) u^*(t - nT - \tau)\, dt$$

$$x_n(\tau) = \int_{T_0} r(t) w^*(t - nT - \tau)\, dt \qquad (4.5.74)$$

Necessary conditions for the estimates of ϕ and τ to be ML estimates are

$$\frac{\partial \Lambda_L(\phi, \tau)}{\partial \phi} = 0 \qquad \frac{\partial \Lambda_L(\phi, \tau)}{\partial \tau} = 0 \qquad (4.5.75)$$

It is convenient to define

$$A(\tau) + jB(\tau) = \frac{1}{N_0}\sum_n \left[I_n^* y_n(\tau) + jJ_n^* x_n(\tau) \right] \qquad (4.5.76)$$

With this definition, (4.5.73) may be expressed in the simple form

$$\Lambda_L(\phi, \tau) = A(\tau)\cos\phi - jB(\tau)\sin\phi \qquad (4.5.77)$$

Now the conditions in (4.5.75) for the joint ML estimates become

$$\frac{\partial \Lambda(\phi, \tau)}{\partial \phi} = -A(\tau)\sin\phi - B(\tau)\cos\phi = 0 \qquad (4.5.78)$$

$$\frac{\partial \Lambda(\phi, \tau)}{\partial \tau} = \frac{\partial A(\tau)}{\partial \tau}\cos\phi - j\frac{\partial B(\tau)}{\partial \tau}\sin\phi = 0 \qquad (4.5.79)$$

From (4.5.78) we obtain

$$\hat{\phi}_{\mathrm{ML}} = -\tan^{-1}\frac{B(\hat{\tau}_{\mathrm{ML}})}{A(\hat{\tau}_{\mathrm{ML}})} \qquad (4.5.80)$$

The solution to (4.5.79) which incorporates (4.5.80) is

$$\left[A(\tau)\frac{\partial A(\tau)}{\partial \tau} + jB(\tau)\frac{\partial B(\tau)}{\partial \tau} \right]_{\tau=\hat{\tau}_{\mathrm{ML}}} = 0 \qquad (4.5.81)$$

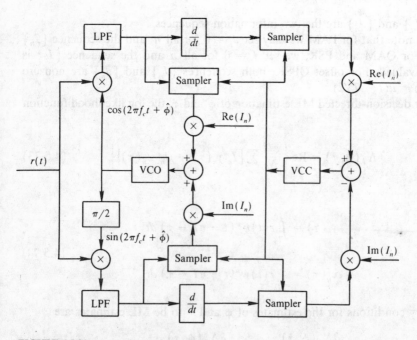

FIGURE 4.5.21
Decision-directed joint tracking loop for carrier phase and symbol timing in QAM and PSK.

The decision-directed tracking loop for QAM (or PSK) obtained from these equations is illustrated in Fig. 4.5.21.

Offset QPSK requires a slightly more complex structure for joint estimation of ϕ and τ. The structure is easily derived from the equations (4.5.76) through (4.5.81).

In addition to the joint estimates given above, it is also possible to derive non-decision-directed estimates of the carrier phase and symbol timing, although we will not pursue this approach.

We should also mention that one can combine the parameter estimation problem with the demodulation of the information sequence $\{I_n\}$. Thus, one can consider the joint maximum likelihood estimation of $\{I_n\}$, the carrier phase ϕ, and the symbol timing parameter τ. Results on these joint estimation problems have appeared in the technical literature, e.g., Kobayashi (1971), Falconer (1976), and Falconer and Salz (1977).

Performance characteristics of ML estimators. The quality of a signal parameter estimate is usually measured in terms of its bias and its variance. In order to define these terms, let us assume that we have a sequence of observations $[x_1, x_2, x_3, \ldots, x_n] = \mathbf{x}$ with pdf $p(\mathbf{x}|\phi)$ from which we extract an estimate of a parameter ϕ. The bias of an estimate, say $\hat{\phi}(\mathbf{x})$, is defined as

$$\text{bias} = E\left[\hat{\phi}(\mathbf{x})\right] - \phi \qquad (4.5.82)$$

where ϕ is the true value of the parameter. When $E[\phi(\mathbf{x})] = \phi$, we say that the estimate is *unbiased*. The variance of the estimate $\phi(x)$ is defined as

$$\sigma_\phi^2 = E\left\{\left[\hat{\phi}(\mathbf{x})\right]^2\right\} - \left\{E\left[\hat{\phi}(\mathbf{x})\right]\right\}^2 \qquad (4.5.83)$$

In general σ_ϕ^2 may be difficult to compute. However, a well known result in parameter estimation [for reference, see Helstrom (1968)] is the Cramér–Rao lower bound on the mean square error defined as

$$E\left\{\left[\hat{\phi}(\mathbf{x}) - \phi\right]^2\right\} \geq \frac{\left[\dfrac{\partial}{\partial \phi} E\left[\hat{\phi}(\mathbf{x})\right]\right]^2}{E\left\{\left[\dfrac{\partial}{\partial \phi} \ln p(\mathbf{x}|\phi)\right]^2\right\}} \qquad (4.5.84)$$

Note that when the estimate is unbiased, the numerator of (4.5.84) is unity and the bound becomes a lower bound on the variance σ_ϕ^2 of the estimate $\hat{\phi}(\mathbf{x})$, i.e.,

$$\sigma_{\hat{\phi}}^2 \geq \frac{1}{E\left\{\left[\dfrac{\partial}{\partial \phi} \ln p(\mathbf{x}|\phi)\right]^2\right\}} \qquad (4.5.85)$$

Since $\ln p(\mathbf{x}|\phi)$ differs from the log likelihood function by a constant factor independent of ϕ, it follows that

$$E\left\{\left[\frac{\partial}{\partial\phi}\ln p(\mathbf{x}|\phi)\right]^2\right\} = E\left\{\left[\frac{\partial}{\partial\phi}\ln \Lambda_L(\phi)\right]^2\right\}$$

$$= -E\left\{\frac{\partial^2}{\partial\phi^2}\ln \Lambda_L(\phi)\right\} \qquad (4.5.86)$$

Therefore, the lower bound on the variance is

$$\sigma_{\hat{\phi}}^2 \geq \frac{1}{E\left\{\left[\frac{\partial}{\partial\phi}\ln \Lambda_L(\phi)\right]^2\right\}} = -\frac{1}{E\left[\frac{\partial^2}{\partial\phi^2}\ln \Lambda_L(\phi)\right]} \qquad (4.5.87)$$

This lower bound is a very useful result. It provides a benchmark for comparing the variance of any practical estimate to the lower bound. Any estimate that is unbiased and whose variance attains the lower bound is called an *efficient estimate*.

In general, efficient estimates are rare. When they exist, they are maximum likelihood estimates. A well known result from parameter estimation theory is that any ML parameter estimate is asymptotically (arbitrarily large number of observations) unbiased and efficient. To a large extent, these desirable properties constitute the importance of ML parameter estimates. It is also known that a ML estimate is asymptotically gaussian distributed [with mean ϕ and variance equal to the lower bound given by (4.5.87)].

In the case of the ML estimates described in this section for the two signal parameters, their variance is generally inversely proportional to the signal-to-noise ratio, or, equivalently, inversely proportional to the signal power multiplied by the observation interval T_0. Furthermore, the variance of the decision-directed estimates, at low error probabilities, are generally lower than the variance of non-decision-directed estimates. In fact, the performance of the ML decision-directed estimates for ϕ and τ attain the lower bound.

The following example is concerned with the evaluation of the Cramér–Rao lower bound on the ML estimate of the carrier phase.

Example 4.5.4. The ML estimate of the phase of an unmodulated carrier was shown in (4.5.45) to satisfy the condition

$$\int_{T_0} r(t)\sin\left(2\pi f_c t + \hat{\phi}_{\mathrm{ML}}\right) dt = 0 \qquad (4.5.88)$$

where

$$r(t) = s(t;\phi) + n(t)$$
$$= A\cos\left(2\pi f_c t + \phi\right) + n(t) \qquad (4.5.89)$$

The condition in (4.5.88) was derived by maximizing the likelihood function

$$\Lambda_L(\phi) = \frac{2}{N_0}\int_{T_0} r(t)s(t;\phi)\, dt \qquad (4.5.90)$$

The variance of $\hat{\phi}_{\text{ML}}$ is lower bounded as

$$\sigma_{\hat{\phi}_{\text{ML}}}^2 \geq \left\{ \frac{2A}{N_0} \int_{T_0} E[r(t)] \cos(2\pi f_c t + \phi) \, dt \right\}^{-1}$$

$$\geq \left\{ \frac{2A}{N_0} \int_{T_0} dt \right\}^{-1} = \frac{N_0}{A^2 T_0}$$

$$\geq \frac{N_0/2T_0}{A^2/2} = \frac{N_0 B_{\text{eq}}}{P_c} \tag{4.5.91}$$

where $P_c = A^2/2$ is the signal power. The factor $1/2T_0$ is simply the equivalent noise bandwidth of the ideal integrator.

From this example, we observe that the variance of the ML phase estimate is lower bounded as

$$\sigma_{\hat{\phi}_{\text{ML}}}^2 \geq \frac{1}{\gamma_L} \tag{4.5.92}$$

where $\gamma_L = P_c/N_0 B_{\text{eq}}$ is the loop SNR. This is also the variance obtained for the phase estimate from a PLL with decision-directed estimation. As we have already observed, non-decision directed estimates do not perform as well due to losses in the nonlinearities required to remove the modulation, e.g., the squaring loss and the Mth power loss.

Similar results can be obtained on the quality of the symbol timing estimates derived above. In addition to their dependence on the SNR, the quality of symbol timing estimates is a function of the signal pulse shape. For example, a

FIGURE 4.5.22
Performance of baseband symbol timing estimate for fixed signal and loop bandwidths. [From "Synchronization Subsystems: Analysis and Design" by L. Franks, 1983. Reprinted with permission of the author.]

FIGURE 4.5.23
Performance of baseband symbol timing estimate for fixed SNR and fixed loop bandwidth. [From "Synchronization Subsystems: Analysis and Design" by L. Franks, 1983. Reprinted with permission of the author.]

pulse shape that is commonly used in practice is one that has a raised cosine spectrum (see Sec. 6.2). For such a pulse, the rms timing error ($\sigma_{\hat{\tau}}$) as a function of SNR is illustrated in Fig. 4.5.22 for both decision-directed and non-decision-directed estimates. Note the significant improvement in performance of the decision-directed estimate compared with the non-decision-directed estimate. Now, if the bandwidth of the pulse is varied, the pulse shape is changed and, hence, the rms value of the timing error also changes. For example, when the bandwidth of the pulse that has a raised cosine spectrum is varied, the rms timing error varies as shown in Fig. 4.5.23. Note that the error decreases as the bandwidth of the pulse increases.

In conclusion, we have presented the ML method for signal parameter estimation and have applied it to the estimation of the carrier phase and symbol timing. We have also related these ML estimates to the estimation techniques presented in Secs. 4.5.1 and 4.5.2, and briefly described their performance characteristics.

4.6 BIBLIOGRAPHICAL NOTES AND REFERENCES

In the derivation of the optimum demodulator for a signal corrupted by AWGN we applied mathematical techniques that were originally used in deriving optimum receiver structures for radar signals. For example, the matched filter was first proposed by North (1943) for use in radar detection, and it is sometimes called the North filter. Derivation of the optimum demodulator based on the use of the Karhunen–Loève expansion can be found in the classical texts by Davenport and Root (1958), Helstrom (1960, 1968), and Van Trees (1968). Its use in radar detection theory is described in the paper by Kelly et al. (1960). These detection methods are based on the hypothesis testing methods developed by statisticians, e.g., Neyman and Pearson (1933) and Wald (1947).

The geometric approach to signal design and detection, which was presented in the context of digital modulation and which has its roots in Shannon's original work, is conceptually appealing and is now widely used since its introduction in the text by Wozencraft and Jacobs (1965).

Design and analysis of signal constellations for the AWGN channel has received considerable attention in the technical literature. Of particular significance is the performance analysis of two-dimensional (QAM) signal constellations that has been treated in the papers of Cahn (1960), Hancock and Lucky (1960), Campopiano and Glazer (1962), Lucky and Hancock (1962), Salz et al. (1971), Simon and Smith (1973), Thomas et al. (1974), and Foschini et al. (1974). Signal design based on multidimensional signal constellations has been described and analyzed in the paper by Gersho and Lawrence (1984).

Carrier recovery and timing synchronization are two topics that have been thoroughly investigated over the past three decades. The Costas loop was invented in 1956 and the decision-directed phase estimation methods were described in 1964. The work on decision-directed estimation was motivated by earlier work of Price (1962a, b). Comprehensive treatments of phase-locked loops first appeared in the books by Viterbi (1966) and Gardner (1966). Books that cover carrier phase recovery and time synchronization techniques were written by Stiffler (1971), Lindsey (1972), and Lindsey and Simon (1973).

A number of tutorial papers have appeared in *IEEE* journals on the PLL and on time synchronization. We cite, for example, the paper by Gupta (1975), which treats both analog and digital implementations of PLL's, and the paper by Lindsey and Chie (1981), which is devoted to the analysis of digital PLL's. In addition, the tutorial paper by Franks (1980) describes both carrier phase and symbol synchronization methods, including methods based on the maximum likelihood estimation criterion. The paper by Franks is contained in a special issue of the *IEEE Transactions on Communications* (August 1980) devoted to synchronization.

Application of the maximum likelihood criterion to parameter estimation was first described in the context of radar parameter estimation (range and range rate). Subsequently, this optimal criterion was applied to carrier phase and symbol timing estimation as well as to joint parameter estimation with data symbols. Papers on these topics have been published by several researchers, including Falconer (1976), Mengali (1977), Falconer and Salz (1977), and Meyers and Franks (1980).

The Cramér–Rao lower bound on the variance of a parameter estimate is derived and evaluated in a number of standard texts on detection and estimation theory, such as Helstrom (1968) and Van Trees (1968). It is also described in several books on mathematical statistics, such as the book by Cramér (1946).

APPENDIX 4A
OPTIMUM DEMODULATOR FOR A COMPLETELY KNOWN
SIGNAL IN ADDITIVE GAUSSIAN NOISE

In this appendix we derive the structure of the optimum demodulator for detecting one out of M possible transmitted signals with a minimum probability of error. The mathematical model for the received signal is

$$r(t) = \alpha e^{-j\phi} u_m(t) + z(t) \qquad \begin{matrix} 0 \le t \le T \\ m = 1, 2, \ldots, M \end{matrix} \qquad (4A.1)$$

where α denotes the channel attenuation, ϕ denotes the carrier phase, and $z(t)$ denotes the additive gaussian noise process that corrupts the signal.

The optimum demodulator computes the set of M posterior probabilities indicated in (4.2.4) and decides in favor of the signal corresponding to the largest posterior probability. In order to compute the posterior probabilities, we need a method for representing the signal $r(t)$ in terms of a set of observable random variables $\{r_n\}$. One possible set consists of a set of samples of $r(t)$ taken on the interval $(0, T)$. The major difficulty in dealing with the samples of $r(t)$ is that, in general, they are correlated. The representation employed here, the Karhunen–Loève series expansion, has the desirable property that the random variables $\{r_n\}$ are uncorrelated. Since the $\{r_n\}$ are gaussian, they are also statistically independent.

First we present the Karhunen–Loève series expansion and its important properties. Toward this end, let $z(t)$, $0 \le t \le T$, denote a complex-valued, possibly nonstationary, random process with zero mean and autocorrelation function

$$\phi_{zz}(t, \tau) = \tfrac{1}{2} E\left[z(t) z^*(\tau)\right] \qquad (4A.2)$$

We expand $z(t)$ in a series of the form

$$z(t) = \lim_{N \to \infty} \sum_{n=1}^{N} z_n f_n(t) \qquad (4A.3)$$

where $\{z_n\}$ are the coefficients in the expansion and $\{f_n(t)\}$ are a set of orthonormal functions on the interval $(0, T)$. That is,

$$\int_0^T f_m(t) f_n^*(t)\, dt = \delta_{mn} \qquad (4A.4)$$

where δ_{mn} is the Kronecker delta, defined as

$$\delta_{mn} = \begin{cases} 1 & m = n \\ 0 & m \ne n \end{cases} \qquad (4A.5)$$

The orthonormality condition implies that the coefficients may be expressed as

$$z_n = \int_0^T z(t) f_n^*(t)\, dt \qquad (4A.6)$$

If we impose the condition that the coefficients $\{z_n\}$ be mutually uncorrelated, i.e., $E[z_k z_j^*] = 0$ for $k \neq j$, then the set of orthonormal functions $\{f_n(t)\}$ are the eigenfunctions of the integral equation

$$\int_0^T \phi_{zz}(t, \tau) f_n(\tau)\, d\tau = \lambda_n f_n(t) \tag{4A.7}$$

where λ_n is the eigenvalue corresponding to the eigenfunction $f_n(t)$. The proof of this statement is as follows. We write (4A.3), in the limit as $N \to \infty$, as

$$z(t) = \sum_{m=1}^{\infty} z_m f_m(t) \tag{4A.8}$$

and, when we multiply both sides of this equation by z_n^* and take the expected value, we obtain

$$E[z(t)z_n^*] = \sum_{m=1}^{\infty} E(z_m z_n^*) f_m(t)$$

$$= E(|z_n|^2) f_n(t) \tag{4A.9}$$

On the other hand, if we conjugate (4A.6), multiply both sides of it by $z(t)$, and take the expected value, we obtain

$$E[z(t)z_n^*] = \int_0^T E[z(t)z^*(\tau)] f_n(\tau)\, d\tau$$

$$= 2\int_0^T \phi_{zz}(t, \tau) f_n(\tau)\, d\tau \tag{4A.10}$$

By combining the results in (4A.9) and (4A.10), we obtain the integral equation

$$\int_0^T \phi_{zz}(t, \tau) f_n(\tau)\, d\tau = \tfrac{1}{2} E(|z_n|^2) f_n(t) \tag{4A.11}$$

which is identical to (4A.7) with $\lambda_n = \tfrac{1}{2} E(|z_n|^2)$.

The converse statement to the above can also be easily established. That is, if the orthonormal functions $\{f_n(t)\}$ satisfy the integral equation in (4A.7), then

$$E[z_m z_n^*] = 2\lambda_n \delta_{mn} \tag{4A.12}$$

The eigenfunctions $\{f_n(t)\}$ can also be used to represent the signals $\{u_m(t)\}$ in the form of a series. That is,

$$u_m(t) = \lim_{N \to \infty} \sum_{n=1}^{N} u_{mn} f_n(t) \tag{4A.13}$$

where

$$u_{mn} = \int_0^T u_m(t) f_n^*(t)\, dt \tag{4A.14}$$

Consequently the received signal $r(t)$ can be represented as

$$r(t) = \lim_{N \to \infty} \sum_{n=1}^{N} r_n f_n(t) \tag{4A.15}$$

where the coefficents $\{r_n\}$ are given as

$$r_n = \alpha e^{-j\phi} u_{mn} + z_n \qquad \begin{array}{l} m = 1, 2, \ldots, M \\ n = 1, 2, \ldots \end{array} \qquad (4A.16)$$

Thus the random process $r(t)$ is represented by the coefficients $\{r_n\}$ in the expansion (4A.15).

Given the coefficients $\{r_n\}$, we can use Bayes' theorem to express the posterior probabilities for the M signals, given in (4.2.1), as

$$P(m|\mathbf{r}) = \frac{p(\mathbf{r}|m)P(m)}{p(\mathbf{r})} \qquad m = 1, \ldots, M \qquad (4A.17)$$

where $P(m|\mathbf{r})$ is the posterior probability that $u_m(t)$ was transmitted given $\mathbf{r} = (r_1, r_2, \ldots)$, $P(m)$ denotes the prior probability for the mth signal, and $p(\mathbf{r})$ denotes the probability density function of $\mathbf{r}$, which can be expressed as

$$p(\mathbf{r}) = \sum_{k=1}^{M} p(\mathbf{r}|k)P(k) \qquad (4A.18)$$

Substitution of (4A.18) into (4A.17) and division of both numerator and denominator by $p(\mathbf{r}|m)$ yields the result

$$P(m|\mathbf{r}) = \frac{P(m)}{\displaystyle\sum_{k=1}^{M} \Lambda_{km}(\mathbf{r})P(k)} \qquad (4A.19)$$

where

$$\Lambda_{km}(\mathbf{r}) = \frac{p(\mathbf{r}|k)}{p(\mathbf{r}|m)} \qquad (4A.20)$$

denotes the likelihood ratio of the signals (k, m). It is apparent from (4A.19) that the likelihood ratios $\Lambda_{km}(\mathbf{r})$, $k, m = 1, 2, \ldots, M$, are sufficient statistics for the computation of the posterior probabilities. Consequently we focus our attention on the computation of $\Lambda_{km}(\mathbf{r})$.

Since the $\{r_n\}$ given by (4A.16) are complex-valued statistically independent gaussian random variables with mean $\alpha e^{-j\phi} u_{mn}$ and variance $\frac{1}{2}E(|z_n|^2) = \lambda_n$, the joint probability density function for $\mathbf{r}_N = (r_1, r_2, \ldots, r_N)$ can be expressed as

$$p(\mathbf{r}_N|m) = \left(\prod_{n=1}^{N} 2\pi\lambda_n \right)^{-1} \exp\left(-\sum_{n=1}^{N} |r_n - \alpha e^{-j\phi} u_{mn}|^2 / 2\lambda_n \right)$$

$$m = 1, 2, \ldots, M \quad (4A.21)$$

Using (4A.21) to form the likelihood ratio given in (4A.20), we obtain, after some cancelation of like terms,

$$\Lambda_{km}(\mathbf{r}_N) = \exp\left\{ \text{Re}\left[\alpha e^{j\phi} \sum_{n=1}^{N} r_n(u_{kn}^* - u_{mn}^*)/\lambda_n \right] + \alpha^2 \sum_{n=1}^{N} \frac{|u_{mn}|^2 - |u_{kn}|^2}{2\lambda_n} \right\}$$

$$(4A.22)$$

We focus our attention on the terms in the exponent of (4A.22), in the limit as $N \to \infty$. First we define a set of functions $\{q_m(t)\}$ as

$$q_m(t) = \lim_{N \to \infty} \sum_{n=1}^{N} \frac{u_{mn}}{\lambda_n} f_n(t) \qquad m = 1, 2, \ldots, M \qquad (4A.23)$$

Then, by use of (4A.15) and (4A.23), it is easily established that the relations

$$\lim_{N \to \infty} \sum_{n=1}^{N} \frac{r_n u_{mn}^*}{\lambda_n} = \int_0^T r(t) q_m^*(t) \, dt \qquad m = 1, 2, \ldots, M \qquad (4A.24)$$

and

$$\lim_{N \to \infty} \sum_{n=1}^{N} \frac{|u_{mn}|^2}{\lambda_n} = \int_0^T u_m(t) q_m^*(t) \, dt \qquad m = 1, 2, \ldots, M \qquad (4A.25)$$

must hold. In addition, we note that the functions $\{q_m(t)\}$ can be expressed as the solutions of the integral equation

$$\int_0^T \phi_{zz}(t, \tau) q_m(\tau) \, d\tau = u_m(t) \qquad m = 1, 2, \ldots, M \qquad (4A.26)$$

That (4A.26) is true can be easily established by substituting (4A.23) into (4A.26) and using (4A.7) to obtain $u_m(t)$ in the series form given by (4A.13).

As a consequence of the relations in (4A.24) and (4A.25), the likelihood ratio in (4A.22), in the limit as $N \to \infty$, becomes

$$\lim_{N \to \infty} \Lambda_{km}(\mathbf{r}_N) = \Lambda_{km}(\mathbf{r}) = \exp \left(\text{Re} \left\{ \alpha e^{j\phi} \int_0^T r(t) [q_k^*(t) - q_m^*(t)] \, dt \right\} \right.$$

$$\left. + \frac{\alpha^2}{2} \left[\int_0^T u_m(t) q_m^*(t) \, dt - \int_0^T u_k(t) q_k^*(t) \, dt \right] \right) \qquad (4A.27)$$

The terms of the form

$$\text{Re} \left[\alpha e^{j\phi} \int_0^T r(t) q_m^*(t) \, dt \right] \qquad m = 1, 2, \ldots, M$$

that appear in the likelihood ratio can be generated by passing the received signal $r(t)$ through a parallel bank of M filters having impulse responses $q_m^*(T - t)$, $m = 1, 2, \ldots, M$, sampling their outputs at time $t = T$, multiplying each output by $\alpha e^{j\phi}$, and, finally, taking the real part of each complex number. Alternatively the terms given above can be generated by passing $r(t)$ through a parallel bank of M cross correlators (multipliers followed by T-second integrators), sampling their outputs at $t = T$, multiplying each output by $\alpha e^{j\phi}$, and, finally, taking the real part of each complex number.

Included in the likelihood ratio are also terms of the form

$$\int_0^T u_m(t) q_m^*(t) \, dt \qquad m = 1, 2, \ldots, M$$

which represent fixed bias terms that are independent of the received signal $r(t)$.

A significant simplification in the computation of the posterior probabilities given in (4A.19) occurs when all the M signals are equally likely to be transmitted *a priori*. That is, the prior probabilities $P(m)$, $m = 1, 2, \ldots, M$, are all equal to $1/M$. Then it can be easily shown that the decision rule based on the computation of the posterior probabilities is equivalent to computing the M real numbers

$$Q_m = \text{Re}\left[\alpha e^{j\phi} \int_0^T r(t) q_m^*(t)\, dt\right]$$

$$-\frac{\alpha^2}{2} \int_0^T u_m(t) q_m^*(t)\, dt \qquad m = 1, 2, \ldots, M \qquad (4A.28)$$

and deciding in favor of the signal corresponding to the largest of the $\{Q_m\}$.

A further simplification occurs when, in addition to equally likely signals, we have additive white (infinite bandwidth) gaussian noise. In this case $\phi_{zz}(t, \tau) = N_0 \delta(t - \tau)$ and, hence, from (4A.26) we obtain $q_m(t) = u_m(t)/N_0$. Consequently (4A.28) becomes

$$Q_m = \text{Re}\left[\frac{\alpha}{N_0} e^{j\phi} \int_0^T r(t) u_m^*(t)\, dt\right] - \frac{\alpha^2}{N_0} \mathcal{E}_m \qquad m = 1, 2, \ldots, M \qquad (4A.29)$$

where $\mathcal{E}_m$ is the energy in the mth transmitted signal. Since the factor α/N_0 is common to all M terms, we may divide every term by this factor. The result is then the set of M decision variables

$$U_m = \text{Re}\left[e^{j\phi} \int_0^T r(t) u_m^*(t)\, dt\right] - \alpha \mathcal{E}_m \qquad m = 1, 2, \ldots, M \qquad (4A.30)$$

which are further discussed in Sec. 4.2.

APPENDIX 4B
THE PROBABILITY OF ERROR FOR
MULTICHANNEL RECEPTION OF BINARY SIGNALS

In multichannel communications systems that employ binary signaling for transmitting information over the AWGN channel, the decision variable can be expressed as a special case of the general quadratic form

$$D = \sum_{k=1}^{L} \left(A|X_k|^2 + B|Y_k|^2 + C X_k Y_k^* + C^* X_k^* Y_k\right) \qquad (4B.1)$$

in complex-valued gaussian random variables. A, B, and C are constants; X_k

and Y_k are a pair of correlated complex-valued gaussian random variables. For the channels considered, the L pairs $\{X_k, Y_k\}$ are mutually statistically independent and identically distributed.

The probability of error is the probability that D is less than zero. This probability is evaluated below.

The computation begins with the characteristic function, denoted by $\psi_D(jv)$, of the general quadratic form. The probability that D is less than zero, denoted here as the probability of error P_b, is

$$P_b = P(D < 0) = \int_{-\infty}^{0} p(D)\, dD \qquad (4\text{B}.2)$$

where $p(D)$, the probability density function of D, is related to $\psi_D(jv)$ by the Fourier transform, i.e.,

$$p(D) = \frac{1}{2\pi} \int_{-\infty}^{0} \psi_D(jv) e^{-jvD}\, dv$$

Hence

$$P_b = \int_{-\infty}^{0} dD \, \frac{1}{2\pi} \int_{-\infty}^{\infty} \psi_D(jv) e^{-jvD}\, dv \qquad (4\text{B}.3)$$

Let us interchange the order of integrations and carry out first the integration with respect to D. The result is

$$P_b = \frac{-1}{2\pi j} \int_{-\infty+j\varepsilon}^{\infty+j\varepsilon} \frac{\psi_D(jv)}{v}\, dv \qquad (4\text{B}.4)$$

where a small positive number ε has been inserted in order to move the path of integration away from the singularity at $v = 0$ and which must be positive in order to allow for the interchange in the order of integrations.

Since D is the sum of statistically independent random variables, the characteristic function of D factors into a product of L characteristic functions, with each function corresponding to the individual random variables d_k, where

$$d_k = A|X_k|^2 + B|Y_k|^2 + CX_k Y_k^* + C^* X_k^* Y_k$$

The characteristic function of d_k is

$$\phi_{d_k}(jv) = \frac{v_1 v_2}{(v + jv_1)(v - jv_2)} \exp\left[\frac{v_1 v_2(-v^2 \alpha_{1k} + jv\alpha_{2k})}{(v + jv_1)(v - jv_2)} \right] \qquad (4\text{B}.5)$$

where the parameters v_1, v_2, α_{1k}, and α_{2k} depend on the means $\overline{X}_k$ and $\overline{Y}_k$ and the second (central) moments $\mu_{xx}, \mu_{yy}, \mu_{xy}$ of the complex-valued gaussian variables X_k and Y_k through the following definitions ($|C|^2 - AB > 0$):

$$v_1 = \sqrt{w^2 + \frac{1}{\left(\mu_{xx}\mu_{yy} + |\mu_{xy}|^2\right)\left(|C|^2 - AB\right)}} - w$$

$$v_2 = \sqrt{w^2 + \frac{1}{\left(\mu_{xx}\mu_{yy} - |\mu_{xy}|^2\right)\left(|C|^2 - AB\right)}} + w$$

$$w = \frac{A\mu_{xx} + B\mu_{yy} + C\mu_{xy}^* + C\mu_{xy}^*}{2\left(\mu_{xx}\mu_{yy} + |\mu_{xy}|^2\right)\left(|C|^2 - AB\right)} \tag{4B.6}$$

$$\alpha_{1k} = \left(|C|^2 - AB\right)\left(|\bar{X}_k|^2\mu_{yy} + |\bar{Y}_k|^2\mu_{xx} - \bar{X}_k^*\bar{Y}_k\mu_{xy} - \bar{X}_k\bar{Y}_k^*\mu_{xy}^*\right)$$

$$\alpha_{2k} = A|\bar{X}_k|^2 + B|\bar{Y}_k|^2 + C\bar{X}_k^*\bar{Y}_k + C^*\bar{X}_k\bar{Y}_k^*$$

Now as a result of the independence of the random variables d_k, the characteristic function of D is

$$\psi_D(jv) = \prod_{k=1}^{L} \psi_{d_k}(jv)$$

$$\psi_D(jv) = \frac{(v_1 v_2)^L}{(v + jv_1)^L (v - jv_2)^L} \exp\left[\frac{v_1 v_2 (jv\alpha_2 - v^2\alpha_1)}{(v + jv_1)(v - jv_2)}\right] \tag{4B.7}$$

where

$$\alpha_1 = \sum_{k=1}^{L} \alpha_{1k}$$

$$\alpha_2 = \sum_{k=1}^{L} \alpha_{2k} \tag{4B.8}$$

The result of (4B.7) is substituted for $\psi_D(jv)$ in (4B.4) and we obtain

$$P_b = -\frac{(v_1 v_2)^L}{2\pi j} \int_{-\infty + j\varepsilon}^{\infty + j\varepsilon} \frac{dv}{v(v + jv_1)^L (v - jv_2)^L} \exp\left[\frac{v_1 v_2 (jv\alpha_2 - v^2\alpha_1)}{(v + jv_1)(v - jv_2)}\right] \tag{4B.9}$$

This integral is evaluated as follows.

The first step is to express the exponential function in the form

$$\exp\left(-A_1 + \frac{jA_2}{v + jv_1} - \frac{jA_3}{v - jv_2}\right)$$

where one can easily verify that the constants A_1, A_2, and A_3, are given as

$$A_1 = \alpha_1 v_1 v_2$$

$$A_2 = \frac{v_1^2 v_2}{v_1 + v_2}(\alpha_1 v_1 + \alpha_2) \tag{4B.10}$$

$$A_3 = \frac{v_1 v_2^2}{v_1 + v_2}(\alpha_1 v_2 - \alpha_2)$$

Second, a conformal transformation is made from the v plane onto the p plane via the change in variable

$$p = -\frac{v_1}{v_2}\frac{v - jv_2}{v + jv_1} \tag{4B.11}$$

In the p plane the integral given by (4B.9) becomes

$$P_b = \frac{\exp\left[v_1 v_2(-2\alpha_1 v_1 v_2 + \alpha_2 v_1 - \alpha_2 v_2)/(v_1 + v_2)^2\right]}{(1 + v_2/v_1)^{2L-1}}\frac{1}{2\pi j}\int_\Gamma f(p)\,dp \tag{4B.12}$$

where

$$f(p) = \frac{[1 + (v_2/v_1)p]^{2L-1}}{p^L(1-p)}\exp\left(\frac{A_2(v_2/v_1)}{v_1 + v_2}p + \frac{A_3(v_1/v_2)}{v_1 + v_2}\frac{1}{p}\right) \tag{4B.13}$$

and Γ is a circular contour of radius less than unity that encloses the origin.

The third step is to evaluate the integral

$$\frac{1}{2\pi j}\int_\Gamma f(p)\,dp = \frac{1}{2\pi j}\int_\Gamma \frac{(1 + (v_2/v_1)p)^{2L-1}}{p^L(1-p)}$$

$$\times \exp\left(\frac{A_2(v_2/v_1)}{v_1 + v_2}p + \frac{A_3(v_1/v_2)}{v_1 + v_2}\frac{1}{p}\right)dp \tag{4B.14}$$

In order to facilitate subsequent manipulations, the constants $a \geq 0$ and $b \geq 0$ are introduced and defined as follows:

$$\frac{a^2}{2} = \frac{A_3(v_1/v_2)}{v_1 + v_2}$$

$$\frac{b^2}{2} = \frac{A_2(v_2/v_1)}{v_1 + v_2} \tag{4B.15}$$

Let us also expand the function $[1 + (v_2/v_1)p]^{2L-1}$ in the binomial series. We obtain, as a result,

$$\frac{1}{2\pi j}\int_\Gamma f(p)\,dp = \sum_{k=0}^{2L-1}\binom{2L-1}{k}\left(\frac{v_2}{v_1}\right)^k$$

$$\times \frac{1}{2\pi j}\int_\Gamma \frac{p^k}{p^L(1-p)}\exp\left(\frac{a^2}{2}\frac{1}{p} + \frac{b^2}{2}p\right)dp \tag{4B.16}$$

The contour integral given in (4B.16) is one representation of the Bessel function. It can be solved by making use of the relations

$$
I_n(ab) = \begin{cases} \dfrac{1}{2\pi j}\left(\dfrac{a}{b}\right)^n \displaystyle\int_\Gamma \dfrac{1}{p^{n+1}}\exp\left(\dfrac{a^2}{2}\dfrac{1}{p}+\dfrac{b^2}{2}p\right)dp \\[4mm] \dfrac{1}{2\pi j}\left(\dfrac{b}{a}\right)^n \displaystyle\int_\Gamma p^{n-1}\exp\left(\dfrac{a^2}{2}\dfrac{1}{p}+\dfrac{b^2}{2}p\right)dp \end{cases}
$$

where $I_n(x)$ is the nth-order modified Bessel function of the first kind and the series representation of the Q function in terms of Bessel functions, i.e.,

$$
Q(a,b) = \exp\left(-\frac{a^2+b^2}{2}\right)\sum_{n=0}^{\infty}\left(\frac{a}{b}\right)^n I_n(ab)
$$

First consider the case $0 \le k \le L-2$ in (4B.16). For this case, the resulting contour integral can be written in the form†

$$
\frac{1}{2\pi j}\int_\Gamma \frac{1}{p^{L-k}(1-p)}\exp\left(\frac{a^2}{2}\frac{1}{p}+\frac{b^2}{2}p\right)dp
$$

$$
= Q(a,b)\exp\left(\frac{a^2+b^2}{2}\right)+\sum_{n=1}^{L-1-k}\left(\frac{b}{a}\right)^n I_n(ab) \qquad (4B.17)
$$

Next consider the term $k = L-1$. The resulting contour integral can be expressed in terms of the Q function as follows:

$$
\frac{1}{2\pi j}\int_\Gamma \frac{1}{p(1-p)}\exp\left(\frac{a^2}{2}\frac{1}{p}+\frac{b^2}{2}p\right)dp = Q(a,b)\exp\left(\frac{a^2+b^2}{2}\right) \qquad (4B.18)
$$

Finally, consider the case $L \le k \le 2L-1$. We have

$$
\frac{1}{2\pi j}\int_\Gamma \frac{p^{k-L}}{1-p}\exp\left(\frac{a^2}{2}\frac{1}{p}+\frac{b^2}{2}p\right)dp
$$

$$
= \sum_{n=0}^{\infty}\frac{1}{2\pi j}\int_\Gamma p^{k-L+n}\exp\left(\frac{a^2}{2}\frac{1}{p}+\frac{b^2}{2}p\right)dp
$$

$$
= \sum_{n=k+1-L}^{\infty}\left(\frac{a}{b}\right)^n I_n(ab) = Q(a,b)\exp\left(\frac{a^2+b^2}{2}\right)-\sum_{n=0}^{k-L}\left(\frac{a}{b}\right)^n I_n(ab)
$$

$$
(4B.19)
$$

---

†This contour integral is related to the generalized Q function, defined as

$$
Q_m(a,b) = \int_b^{\infty} x(x/a)^{m-1}\exp\left(-\frac{x^2+a^2}{2}\right)I_{m-1}(ax)\,dx
$$

in the following manner:

$$
Q_m(a,b)\exp\left(\frac{a^2+b^2}{2}\right) = \frac{1}{2\pi j}\int_\Gamma \frac{1}{p^m(1-p)}\exp\left(\frac{a^2}{2}\frac{1}{p}+\frac{b^2}{2}p\right)dp
$$

Collecting the terms that are indicated on the right-hand side of (4B.16) and using the results given in (4B.17) through (4B.19), the following expression for the contour integral is obtained after some algebra:

$$
\frac{1}{2\pi j} \int_\Gamma f(p)\, dp = \left(1 + \frac{v_2}{v_1}\right)^{2L-1} \left[\exp\left(\frac{a^2 + b^2}{2}\right) Q(a,b) - I_0(ab)\right]
$$
$$
+ I_0(ab) \sum_{k=0}^{L-1} \binom{2L-1}{k}\left(\frac{v_2}{v_1}\right)^k + \sum_{n=1}^{L-1} I_n(ab)
$$
$$
\times \left\{ \sum_{k=0}^{L-1-n} \binom{2L-1}{k}\left[\left(\frac{b}{a}\right)^n \left(\frac{v_2}{v_1}\right)^k - \left(\frac{a}{b}\right)^n \left(\frac{v_2}{v_1}\right)^{2L-1-k}\right]\right\}
$$

$$(4B.20)$$

Equation (4B.20) in conjunction with (4B.12) gives the result for the probability of error. A further simplification results when one uses the following identity, which can be proved easily:

$$
\exp\left[\frac{v_1 v_2}{(v_1 + v_2)^2}(-2\alpha_1 v_1 v_2 + \alpha_2 v_1 - \alpha_2 v_2)\right] = \exp\left(-\frac{a^2 + b^2}{2}\right)
$$

Therefore it follows that

$$
P_b = Q(a,b) - I_0(ab)\exp\left(-\frac{a^2 + b^2}{2}\right)
$$
$$
+ \frac{I_0(ab)\exp\left[-(a^2 + b^2)/2\right]}{(1 + v_2/v_1)^{2L-1}} \sum_{k=0}^{L-1} \binom{2L-1}{k}\left(\frac{v_2}{v_1}\right)^k
$$
$$
+ \frac{\exp\left[-(a^2 + b^2)/2\right]}{(1 + v_2/v_1)^{2L-1}} \sum_{n=1}^{L-1} I_n(ab) \qquad (4B.21)
$$
$$
\times \left\{ \sum_{k=0}^{L-1-n} \binom{2L-1}{k}\left[\left(\frac{b}{a}\right)^n \left(\frac{v_2}{v_1}\right)^k - \left(\frac{a}{b}\right)^n \left(\frac{v_2}{v_1}\right)^{2L-1-k}\right]\right\} \qquad L > 1
$$

$$
P_b = Q(a,b) - \frac{v_2/v_1}{1 + v_2/v_1}I_0(ab)\exp\left(-\frac{a^2 + b^2}{2}\right) \qquad L = 1
$$

This is the desired expression for the probability of error. It is now a simple matter to relate the parameters a and b to the moments of the pairs $\{X_k, Y_k\}$. Substituting for A_2 and A_3 from (4B.10) into (4B.15), we obtain

$$
a = \left[\frac{2v_1^2 v_2(\alpha_1 v_2 - \alpha_2)}{(v_1 + v_2)^2}\right]^{1/2}
$$
$$
b = \left[\frac{2v_1 v_2^2(\alpha_1 v_1 + \alpha_2)}{(v_1 + v_2)^2}\right]^{1/2}
$$

$$(4B.22)$$

Since v_1, v_2, α_1, and α_2 have been given in (4B.6) and (4B.8) directly in terms of the moments of the pairs X_k and Y_k, our task is completed.

APPENDIX 4C
OPTIMUM DEMODULATOR FOR SIGNALS WITH RANDOM PHASE IN ADDITIVE GAUSSIAN NOISE

As in Appendix 4A, we assume that there are M signal waveforms of the form

$$s_n(t) = \text{Re}\left[u_m(t)e^{j2\pi f_c t}\right] \quad \begin{array}{l} m = 1, 2, \ldots, M \\ 0 \leq t \leq T \end{array} \quad (4C.1)$$

where $\{u_m(t)\}$ represent the equivalent lowpass signal waveforms. The equivalent lowpass received signal plus noise can be represented as

$$r(t) = \alpha e^{-j\phi_m}u_m(t) + z(t) \quad (4C.2)$$

where α represents the attenuation in the channel, $\{\phi_m\}$ represent the carrier or channel phase shifts for the M signals, and $z(t)$ is the additive zero mean stationary gaussian noise with autocorrelation function $\phi_{zz}(\tau)$.

In the following discussion the phases $\{\phi_m\}$ are assumed to be unknown at the receiver. For purposes of this derivation, the phases $\{\phi_m\}$ are assumed to be mutually statistically independent and uniformly distributed, i.e.,

$$p(\phi_m) = \frac{1}{2\pi} \quad \begin{array}{l} 0 \leq \phi_m < 2\pi \\ m = 1, 2, \ldots, M \end{array} \quad (4C.3)$$

The Karhunen–Loève expansion for $r(t)$ is

$$r(t) = \lim_{N \to \infty} \sum_{n=1}^{N} r_n f_n(t) \quad (4C.4)$$

where the coefficients $\{r_n\}$ are given as

$$r_n = \alpha e^{-j\phi_m}u_{mn} + z_n \quad \begin{array}{l} m = 1, 2, \ldots, M \\ n = 1, 2, \ldots \end{array} \quad (4C.5)$$

Given the observable coefficients $[r_n]$, we compute the posterior probabilities

$$P(m|\mathbf{r}) = \frac{p(\mathbf{r}|m)P(m)}{p(\mathbf{r})}$$

$$= \frac{P(m)}{\displaystyle\sum_{k=1}^{M} \Lambda_{km}(\mathbf{r})P(k)} \quad (4C.6)$$

where $P(k)$ are the prior probabilities and $\Lambda_{km}(\mathbf{r})$ denotes the likelihood ratio of signals (k, m), defined as

$$\Lambda_{km}(\mathbf{r}) = \frac{p(\mathbf{r}|k)}{p(\mathbf{r}|m)} = \frac{\int_0^{2\pi} p(\mathbf{r}|k, \phi_k)p(\phi_k)\,d\phi_k}{\int_0^{2\pi} p(\mathbf{r}|m, \phi_m)p(\phi_m)\,d\phi_m} \quad (4C.7)$$

The joint probability density function for $\mathbf{r}_N = (r_1, r_2, \ldots, r_N)$ can be expressed as

$$p(\mathbf{r}_N|m, \phi_m) = \left(\prod_{n=1}^{N} 2\pi\lambda_n\right)^{-1} \exp\left(-\sum_{n=1}^{N} |r_n - \alpha e^{-j\phi_m}u_{mn}|^2/2\lambda_n\right)$$

$$m = 1, 2, \ldots, M \quad (4C.8)$$

Then

$$p(\mathbf{r}_N|m) = \int_0^{2\pi} p(\mathbf{r}_N|m, \phi_m) p(\phi_m) \, d\phi_m$$

$$= \left(\prod_{n=1}^{N} 2\pi\lambda_n \right)^{-1} \exp\left[-\sum_{n=1}^{N} \left(|r_n|^2 + \alpha^2 |u_{mn}|^2 \right) / 2\lambda_n \right]$$

$$\times I_0\left(\alpha \left| \sum_{n=1}^{N} r_n u_{mn}^* / \lambda_n \right| \right) \qquad m = 1, 2, \ldots, M \qquad (4C.9)$$

where $I_0(x)$ is the modified Bessel function of order zero. Using (4C.9) we form the likelihood ratio

$$\Lambda_{km}(\mathbf{r}_N) = \frac{p(\mathbf{r}_N|k)}{p(\mathbf{r}_N|m)}$$

$$= \exp\left[-\frac{\alpha^2}{2} \sum_{n=1}^{N} \left(|u_{kn}|^2 - |u_{mn}|^2 \right) / \lambda_n \right] \frac{I_0\left(\alpha \left| \sum_{n=1}^{N} r_n u_{kn}^* / \lambda_n \right| \right)}{I_0\left(\alpha \left| \sum_{n=1}^{N} r_n u_{mn}^* / \lambda_n \right| \right)} \qquad (4C.10)$$

Now we take the limit of $\Lambda_{km}(\mathbf{r}_N)$ as $N \to \infty$ and we obtain

$$\lim_{N \to \infty} \Lambda_{km}(\mathbf{r}_N) = \Lambda_{km}(\mathbf{r})$$

$$= \exp\left\{ -\frac{\alpha^2}{2} \left[\int_0^T u_k(t) q_k^*(t) \, dt - \int_0^T u_m(t) q_m^*(t) \, dt \right] \right\}$$

$$\times \frac{I_0\left[\alpha \left| \int_0^T r(t) q_k^*(t) \, dt \right| \right]}{I_0\left[\alpha \left| \int_0^T r(t) q_m^*(t) \, dt \right| \right]} \qquad (4C.11)$$

where $q_m(t)$ has been defined previously in (4A.23).

Our interpretation of the processing implied by (4C.11) is as follows. The received signal $r(t)$ is passed through a parallel bank of M filters with impulse responses $\{q_m^*(T - t)\}$ which are sampled at $t = T$. Equivalently, a bank of M cross correlators can be employed in place of the matched filters. Following each matched filter or cros correlator is an envelope detector having an output that is equal to the magnitude of the complex-valued signal from the matched filter or cross correlator. The outputs from the envelope detectors form the arguments of the Bessel function $I_0(x)$ that enter into the computation of the likelihood ratio. The terms in the exponent of (4C.11) represent fixed bias terms that are independent of the received signal $r(t)$.

Let us now consider some special cases. First, when the prior probabilities are equal, the demodulator that bases its decision on the maximum of the posterior probabilities $P(m|\mathbf{r})$ is equivalent to one that bases its decision on the maximum of the $p(\mathbf{r}|m)$. That is, the MAP receiver is equivalent to the maximum likelihood receiver. But this implies that the receiver computes the M real

numbers

$$Q_m = \exp\left[-\frac{\alpha^2}{2}\int_0^T u_m(t)q_m^*(t)\,dt\right]I_0\left[\alpha\left|\int_0^T r(t)q_m^*(t)\,dt\right|\right]$$

$$m = 1, 2, \ldots, M \quad (4C.12)$$

and decides in favor of the signal corresponding to the largest of the $\{Q_m\}$.

A significant simplification in the processing occurs when, in addition to equally likely signals, we have equal energy signals and white gaussian noise. Then all the bias terms are identical and, hence, they may be neglected. Furthermore, as a consequence of the monotonicity of the Bessel function $I_0(x)$, only the argument of $I_0(x)$ need be computed. Thus, under these conditions, the demodulator bases its decision on the M decision variables

$$U_m = \left|\int_0^T r(t)u_m^*(t)\,dt\right| \qquad m = 1, 2, \ldots, M \qquad (4C.13)$$

The performance of this demodulator is discussed in Sec. 4.3.

PROBLEMS

4.1. The Schwarz inequality is

$$\left|\int_a^b x(t)y(t)\,dt\right|^2 \le \int_a^b |x(t)|^2\,dt\int_a^b |y(t)|^2\,dt$$

Equality holds if $x(t) = Ay^*(t)$, where A is an arbitrary constant.

Figure P4.1 illustrates a linear filter with input $u(t) + z(t)$, impulse response $c(t)$, and output $v(t)$, where

$$v(t) = \int_0^T [u(\tau) + z(\tau)]c(t-\tau)\,d\tau$$

At $t = T$, we have

$$v(T) = \int_0^T u(t)c(T-\tau)\,d\tau + \int_0^T z(\tau)c(T-\tau)\,d\tau$$

$$= w(T) + N(T)$$

The noise process $z(t)$ is zero mean and has an autocorrelation function $\phi_{zz}(\tau) = N_0\delta(\tau)$. The SNR at the output of the filter is

$$\text{SNR} = \frac{|w(T)|^2}{E[|N(T)|^2]}$$

FIGURE P4.1

Use the Schwarz inequality to show that the impulse response of the filter that maximizes the SNR is $c(t) = u^*(T - t)$.

4.2. Let $z(t) = x(t) + jy(t)$ be a complex-valued zero mean white gaussian noise process with autocorrelation function $\phi_{zz}(\tau) = N_0\delta(\tau)$. Let $u_m(t), m = 1, 2, \ldots, M$, be a set of M orthogonal equivalent lowpass waveforms defined on the interval $0 \leq t \leq T$. Define

$$N_{mr} = \text{Re}\left[\int_0^T z(t) u_m^*(t) \, dt\right] \qquad m = 1, 2, \ldots, M$$

(a) Determine the variance of N_{mr}.

(b) Show that $E(N_{mn}N_{kr}) = 0$ for $k \neq m$.

4.3. The two equivalent lowpass signals shown in Fig. P4.3 are used to transmit a binary sequence over an additive white gaussian noise channel. The received signal can be expressed as

$$r(t) = \alpha u_i(t) e^{-j\phi} + z(t) \qquad \begin{matrix} 0 \leq t \leq T \\ i = 1, 2 \end{matrix}$$

where $z(t)$ is a zero mean gaussian noise process with autocorrelation function

$$\phi_{zz}(\tau) = \tfrac{1}{2}E[z^*(t)z(t + \tau)] = N_0\delta(\tau)$$

(a) Determine the transmitted energy in $u_1(t)$ and $u_2(t)$ and the cross-correlation coefficient ρ_{12}.

(b) Suppose the receiver is implemented by means of coherent detection using two matched filters, one matched to $u_1(t)$ and the other to $u_2(t)$. Sketch the equivalent lowpass impulse responses of the matched filters.

(c) Sketch the noise-free response of the two matched filters when the transmitted signal is $u_2(t)$. Assume $\alpha = 1$ and $\phi = 0$.

(d) Suppose the receiver is implemented by means of two cross correlators (multipliers followed by integrators) in parallel. Sketch the output of each integrator as a function of time for the interval $0 \leq t \leq T$ when the transmitted signal is $u_2(t)$. Assume $\alpha = 1$ and $\phi = 0$.

(e) Compare the sketches in (c) and (d). Are they the same? Explain briefly.

(f) From your knowledge of the signal characteristics, give the probability of error for this binary communications system.

FIGURE P4.3

4.4. This problem deals with the characteristics of a DPSK signal.

(a) Suppose we wish to transmit the binary sequence given below by binary DPSK. Let $s(t) = A\cos(2\pi f_c t + \theta)$ represent the transmitted signal in any signaling interval of duration T. Give the phase of the transmitted signal for the following data sequence. Begin with $\theta = 0$ for the phase of the first bit to be transmitted.

$$\text{Data:} \quad 1 \quad 1 \quad 0 \quad 1 \quad 0 \quad 0 \quad 0 \quad 1 \quad 0 \quad 1 \quad 0 \quad 1 \quad 1 \quad 0$$

(b) If the data sequence is uncorrelated, determine and sketch the power density spectrum of the signal transmitted by DPSK.

4.5. In the detection of binary PSK, the decision variable is

$$U = \text{Re}\left[c^{j\hat{\phi}} \int_0^T r(t) u^*(t)\, dt \right] \begin{smallmatrix} \text{"1"} \\ \gtrless \\ \text{"0"} \end{smallmatrix} 0$$

where $\hat{\phi}$ is an estimate of the phase shift imparted on the transmitted signal by the channel. Suppose the received signal is

$$r(t) = \alpha e^{-j\phi} u(t) + z(t)$$

where $z(t)$ is zero mean white gaussian noise with spectral density N_0. Derive the probability of error as a function of the phase error $\phi - \hat{\phi}$. How much of a phase error can be tolerated if the loss in SNR is not to exceed 3 dB?

4.6. In two-phase DPSK, the received signal in one signaling interval is used as a phase reference for the received signal in the following signaling interval. The decision variable is

$$D = \text{Re}\left(V_m V_{m-1}^* \right) \begin{smallmatrix} \text{"1"} \\ \gtrless \\ \text{"0"} \end{smallmatrix} 0$$

where

$$V_k = 2\alpha\mathscr{E} e^{j(\theta_k - \phi)} + N_k$$

represents the complex-valued output of the filter matched to the transmitted signal $u(t)$. N_k is a complex-valued gaussian variable having zero mean and statistically independent components.

(a) If we write $V_k = X_k + jY_k$, show that D is equivalent to

$$d = \left(\frac{X_m + X_{m-1}}{2} \right)^2 + \left(\frac{Y_m + Y_{m-1}}{2} \right)^2 - \left(\frac{X_m - X_{m-1}}{2} \right)^2 - \left(\frac{Y_m - Y_{m-1}}{2} \right)^2$$

(b) For mathematical convenience, suppose that $\theta_k = \theta_{k-1}$. Show that the random variables U_1, U_2, U_3, and U_4 are statistically independent gaussian variables, where $U_1 = (X_m + X_{m-1})/2$, $U_2 = (Y_m + Y_{m-1})/2$, $U_3 = (X_m - X_{m-1})/2$, and $U_4 = (Y_m - Y_{m-1})/2$.

(c) Define the random variables $W_1 = U_1^2 + U_2^2$ and $W_2 = U_3^2 + U_4^2$. Then

$$D = W_1 - W_2 \begin{smallmatrix} \text{"1"} \\ \gtrless \\ \text{"0"} \end{smallmatrix} 0$$

Determine the probability density functions for W_1 and W_2.

(d) Determine the probability of error P_b, where

$$P_b = P(D < 0) = P(W_1 - W_2 < 0) = \int_0^\infty P(W_2 > w_1 | w_1)\, p(w_1)\, dw_1$$

4.7. Suppose that we have a complex-valued gaussian random variable $z = x + jy$, where (x, y) are statistically independent variables with zero mean and variance $E(x^2) = E(y^2) = \sigma^2$. Let

$$r = z + m \qquad \text{where } m = m_r + jm_i$$

and define r as

$$r = a + jb$$

Clearly, $a = x + m_r$, and $b = y + m_i$. Determine the following probability density functions:

(a) $p(a, b)$
(b) $p(u, \phi)$, where $u = \sqrt{a^2 + b^2}$ and $\phi = \tan^{-1} b/a$
(c) $p(u)$

Note: In (b) it is convenient to define $\theta = \tan^{-1} m_i/m_r$ so that

$$m_r = \sqrt{m_r^2 + m_i^2} \cos \theta \qquad \text{and} \qquad m_i = \sqrt{m_r^2 + m_i^2} \sin \theta.$$

Furthermore, you must use the relation

$$\frac{1}{2\pi} \int_0^{2\pi} e^{\alpha \cos (\phi - \theta)} \, d\phi = I_0(\alpha) = \sum_{n=0}^{\infty} \frac{\alpha^{2n}}{2^{2n}(n!)^2}$$

where $I_0(\alpha)$ is the modified Bessel function of order zero.

4.8. A ternary communication system transmits one of three signals, $u(t)$, 0, or $-u(t)$, every T seconds. The received signal is either $r(t) = ae^{-j\phi}u(t) + z(t)$, $r(t) = z(t)$, or $r(t) = -ae^{-j\phi}u(t) + z(t)$, where $z(t)$ is white gaussian noise with $E[z(t)] = 0$ and $\phi_{zz}(\tau) = \frac{1}{2}E[z(t)z^*(\tau)] = N_0\delta(t - \tau)$. The optimum receiver computes the quantity

$$U = \text{Re}\left[e^{j\phi} \int_0^T r(t)u^*(t) \, dt \right]$$

and compares U with a threshold A and a threshold $-A$. If $U > A$, the decision is made that $u(t)$ was sent. If $U < -A$, the decision is made in favor of $-u(t)$. If $-A < U < A$, the decision is made in favor of 0.

(a) Determine the three conditional probabilities of error: P_e given that $u(t)$ was sent, P_e given that $-u(t)$ was sent, and P_e given that 0 was sent.
(b) Determine the average probability of error P_e as a function of the threshold A, assuming that the three symbols are equally probable a priori.
(c) Determine the value of A that minimizes P_e.

4.9. The two equivalent lowpass signals shown in Fig. P4.9 are used to transmit a binary information sequence. The transmitted signals, which are equally probable, are

FIGURE P4.9

corrupted by additive zero mean white gaussian noise having an equivalent lowpass representation $z(t)$ with an autocorrelation function

$$\phi_{zz}(\tau) = \tfrac{1}{2}E[z^*(t)z(t+\tau)]$$
$$= N_0\delta(\tau)$$

(a) What is the transmitted signal energy?

(b) What is the probability of a binary digit error if coherent detection is employed at the receiver?

(c) What is the probability of a binary digit error if noncoherent detection is employed at the receiver?

4.10. In Sec. 4.2.1 it was shown that the minimum frequency separation for orthogonality of binary FSK signals with coherent detection is $\Delta f = 1/2T$. However, a lower error probability is possible with coherent detection of FSK if Δf is increased beyond $1/2T$. Show that the optimum value of Δf is $0.715/T$ and determine the probability of error for this value of Δf.

4.11. The equivalent lowpass waveforms for three signal sets are shown in Fig. P4.11. Each set may be used to transmit one of four equally probable messages over an

FIGURE P4.11

additive white gaussian noise channel. The equivalent lowpass noise $z(t)$ has zero mean and autocorrelation function $\phi_{zz}(\tau) = N_0\delta(\tau)$.

(a) Classify the signal waveforms in set I, set II, and set III. In other words, state the category or class to which each signal set belongs.

(b) What is the *average* transmitted energy for each signal set?

(c) For signal set I, specify the average probability of error if the signals are detected coherently.

(d) For signal set II, give a union bound on the probability of a symbol error if the detection is performed (i) coherently and (ii) noncoherently.

(e) Is it possible to use noncoherent detection on signal set III? Explain.

(f) Which signal set or signal sets would you select if you wished to achieve a bit rate to bandwidth (R/W) ratio of at least 2 *Briefly* explain your answer.

4.12. Let $z(t)$, $0 \le t \le T$, be a complex-valued zero mean stationary gaussian random process with autocorrelation function $\phi_{zz}(\tau)$. A Fourier series expansion of $z(t)$ is

$$z(t) = \lim_{N \to \infty} \sum_{n=-N}^{N} z_n e^{jn\omega_0 t} \qquad \omega_0 = \frac{2\pi}{T}$$

Show that the random variables $\{z_n\}$ in the series expansion are correlated.

4.13. Prove that if the set of orthonormal functions $\{f_n(t)\}$ satisfy the integral equation

$$\int_0^T \phi_{zz}(t, \tau) f_n(\tau)\, d\tau = \lambda_n f_n(t)$$

the coefficients $\{z_n\}$ in the Karhunen–Loève expansion of the stationary stochastic process $z(t)$, i.e.,

$$z(t) = \lim_{N \to \infty} \sum_{n=1}^{N} z_n f_n(t)$$

must satisfy the condition $E(z_n z_m^*) = 0$ for $m \ne n$.

4.14. Consider the Karhunen–Loève expansions

$$q(t) = \lim_{N \to \infty} \sum_{n=1}^{N} \frac{u_n}{\lambda_n} f_n(t)$$

$$u(t) \lim_{N \to \infty} \sum_{n=1}^{N} u_n f_n(t)$$

where the functions $\{f_n(t)\}$ are orthonormal and satisfy the integral equation

$$\int_0^T \phi_{zz}(t - \tau) f_n(\tau)\, d\tau = \lambda_n f_n(t)$$

Show that $q(t)$ and $u(t)$ are related through the integral equation

$$\int_0^T \phi_{zz}(t - \tau) q(\tau)\, d\tau = u(t)$$

4.15. For a given $u(t)$ and autocorrelation function $\phi_{zz}(\tau)$, the integral equation for $q(t)$ in Prob. 4.14 determines the impulse response of a filter matched to $u(t)$ and the characteristics of the power density spectrum of the additive noise. This may be easily seen by considering the integral equation

$$\int_{-\infty}^{\infty} \phi_{zz}(t - \tau) q(\tau)\, d\tau = u(t)$$

over the time interval $-\infty < t < \infty$.

White gaussian
noise

$Q^*(f) = G(f) V(f)$

FIGURE P4.15

(a) By computing the Fourier transform of both sides of the integral equation, show that

$$Q(f) = \frac{U(f)}{\Phi_{zz}(f)}$$

(b) The filter matched to $u(t)$ for colored (nonwhite) gaussian noise has the transfer function

$$Q^*(f) = \frac{U^*(f)}{\Phi_{zz}(f)}$$

This matched filter can be realized as a cascade of a noise-whitening filter with transfer function $G(F)$ and a matched filter for the white noise signal, as shown in the block diagram in Fig. P4.15. Determine $G(f)$ and $V(f)$.

4.16. The two equivalent lowpass signals shown in Fig. P4.16 are used to transmit a binary sequence. The equivalent lowpass impulse response of the channel is $h(t) = 4\delta(t) - 2\delta(t - T)$. To avoid pulse overlap between successive transmissions, the transmission rate in bits per second is selected to be $R = 1/2T$. The transmitted signals are equally probable and are corrupted by additive zero mean white gaussian noise having an equivalent lowpass representation $z(t)$ with an autocorrelation function

$$\phi_{zz}(\tau) = \tfrac{1}{2}E[z^*(t)z(t + \tau)] = N_0\delta(\tau)$$

(a) Sketch the two possible equivalent lowpass noise-free received waveforms.
(b) Specify the optimum receiver and sketch the equivalent lowpass impulse responses of all filters used in the optimum receiver. Assume coherent detection of the signals.

FIGURE P4.16

(c) Determine the average probability of error for *coherent detection* of the signals. Show your work in arriving at your answer.

(d) What is the error probability if *noncoherent detection* is used?

4.17. Consider a quaternary ($M = 4$) communication system that transmits, every T seconds, one of four equally probable signals: $u_1(t)$, $-u_1(t)$, $u_2(t)$, $-u_2(t)$. The signals $u_1(t)$ and $u_2(t)$ are orthogonal with equal energy. The additive noise is white gaussian with zero mean and autocorrelation function $\phi_{zz}(\tau) = N_0\delta(\tau)$. The demodulator consists of two filters matched to $u_1(t)$ and $u_2(t)$ and their outputs at the sampling instant are U_1 and U_2. The demodulator bases its decision on the following rule:

$$U_1 > |U_2| \Rightarrow u_1(t) \qquad U_1 < -|U_2| \Rightarrow -u_1(t)$$
$$U_2 > |U_1| \Rightarrow u_2(t) \qquad U_2 < -|U_1| \Rightarrow -u_2(t)$$

Since the signal set is biorthogonal, the error probability is given by (4.2.87). Express this error probability in terms of a single integral and, thus, show that the symbol error probability for a biorthogonal signal set with $M = 4$ is identical to that for four-phase PSK. *Hint*: A change in variables from U_1 and U_2 to $W_1 = U_1 + U_2$ and $W_2 = U_1 - U_2$ simplifies the problem.

4.18. The input $s(t)$ to a bandpass filter is

$$s(t) = \text{Re}\left[u(t)e^{j2\pi f_c t}\right]$$

where $u(t)$ is a rectangular pulse as shown in Fig. P4.18(a).

(a) Determine the output $y(t)$ of a bandpass filter for all $t \geq 0$ if the impulse response of the filter is

$$g(t) = \text{Re}\left[2h(t)e^{j2\pi f_c t}\right]$$

where $h(t)$ is an exponential as shown in Fig. P4.18(b).

(b) Sketch the *equivalent lowpass output* of the filter.

(c) When would you sample the output of the filter if you wished to have the maximum output at the sampling instant? What is the value of the maximum output?

(d) Suppose that in addition to the input signal $s(t)$ there is additive white gaussian noise

$$n(t) = \text{Re}\left[z(t)e^{j2\pi f_c t}\right]$$

(a) (b)

FIGURE P4.18

where $\phi_{zz}(\tau) = N_0\delta(\tau)$. At the sampling instant determined in part (c) the signal sample is corrupted by an additive gaussian noise term. Determine its mean and variance.

(e) What is the signal-to-noise ratio γ of the sampled output?

(f) Determine the signal-to-noise ratio when $h(t)$ is the matched filter to $u(t)$ and compare this result with the value of γ obtained in part (e).

4.19. $X_1, X_2, \ldots, X_N$ are a set of N statistically independent and identically distributed real gaussian random variables with moments $E(X_i) = m$ and $\text{var}(X_i) = \sigma^2$.

(a) Define

$$U = \sum_{n=1}^{N} X_n$$

Evaluate the SNR of U, which is defined as

$$(\text{SNR})_U = \frac{[E(U)]^2}{2\sigma_U^2}$$

where σ_U^2 is the variance of U.

(b) Define

$$V = \sum_{n=1}^{N} X_n^2$$

Evaluate the SNR of V, which is defined as

$$(\text{SNR})_V = \frac{[E(V)]^2}{2\sigma_V^2}$$

where σ_V^2 is the variance of V.

(c) Plot $(\text{SNR})_U$ and $(\text{SNR})_V$ versus m^2/σ^2 on the same graph and, thus, compare the SNR's graphically.

(d) What does the result in (c) imply regarding coherent detection and combining versus square-law detection and combining of multichannel signals?

4.20. Derive the upper bound on the symbol error probability for QAM, given by (4.2.145).

4.21. The signal

$$r(t) = s(t; \theta) = n(t) \qquad 0 \le t \le T$$

is used to estimate the parameter θ. The additive noise $n(t)$ is zero mean gaussian noise with spectral density $\Phi_{nn}(f)$. Use the Karhunen–Loève expansion to determine the pdf $p(\mathbf{r}|\theta)$ where $\mathbf{r} = [r_1, r_2, \ldots, r_N]$ and take the limit as $N \to \infty$. Show that for white gaussian noise, $\lim_{N \to \infty} p(\mathbf{r}|\theta)$ is proportional to the likelihood function $\Lambda(\theta)$ given by (4.5.42).

4.22. Show that the mean value of the ML estimate in (4.5.52) is ϕ, i.e., that the estimate is unbiased.

4.23. Determine the pdf of the ML phase estimate in (4.5.52).

4.24. Determine the ML phase estimate for offset QPSK.

4.25. By averaging $p(\mathbf{r}|\phi)$ over the uniformly distributed phase, derive the decision metrics

$$\mu(\mathbf{u}_m) = |\mathbf{r} \cdot \mathbf{u}_m| \qquad m = 1, 2, \ldots, M$$

for the optimum noncoherent demodulator of M equal energy signals.

4.26. Use the approximation given in (4.2.108) for the pdf of the signal phase at high SNR to obtain the approximate variance of the ML phase estimate given by (4.5.52).

4.27. Determine the signal and noise components at the input to a fourth power ($M = 4$) PLL that is used to generate the carrier phase for demodulation of QPSK. By ignoring all noise components except the ones that are linear in the noise $n(t)$, determine the variance of the phase estimate at the output of the PLL.

CHAPTER
5

EFFICIENT SIGNALING WITH CODED WAVEFORMS

In Chap. 4 we considered the problem of digital signaling by means of $M = 2^k$ signal waveforms, where each waveform conveys k bits of information. We observed that some modulation methods provide better performance than others. In particular, we demonstrated that orthogonal signaling waveforms allowed us to make the probability of error arbitrarily small by letting the number of waveforms $M \to \infty$, provided that the SNR per bit $\gamma_b \geq -1.6$ dB. Thus, we can operate at the capacity of the additive, white gaussian noise channel in the limit as the bandwidth expansion factor $B_e = W/R \to \infty$. This is a heavy price to pay because B_e grows exponentially with the block length k. Such inefficient use of channel bandwidth is highly undesirable.

In this chapter we consider signal waveforms generated from either binary or nonbinary sequences. The resulting waveforms are generally characterized by a bandwidth expansion factor that grows only linearly with k. Consequently, coded waveforms offer the potential for greater bandwidth efficiency than orthogonal M-ary waveforms. We shall demonstrate that coded waveforms offer performance advantages not only in power-limited applications where $R/W < 1$ but also in bandwidth-limited systems where $R/W > 1$.

5.1 MODEL OF DIGITAL COMMUNICATIONS SYSTEM EMPLOYING CODING

The block diagram shown in Fig. 5.1.1 illustrates the basic elements of a digital communications system employing coded waveforms. The input to the encoder is assumed to be a sequence of binary digits (bits) occurring at a rate R bits/s. Two types of encoding of the information sequence are considered in this chapter. The first is block encoding, by which we mean that blocks of k information bits are encoded into corresponding blocks of n bits ($n > k$). Each block of n bits from the encoder constitutes a code word contained in a set of $M = 2^k$ possible code words. The code rate, defined as the ratio k/n and denoted by R_c, is a measure of the amount of redundancy introduced by the encoder. Thus the bit rate at the output of the block encoder is R/R_c.

The second type of encoding to be considered is linear convolutional encoding of the information sequence. The convolutional encoder may be viewed as a linear finite-state shift register with an output sequence consisting of a selected set of linear combinations of the input sequence. The number of output bits from the shift register for each input bit is a measure of the redundancy in the code and the reciprocal of this quantity is defined as the code rate R_c. Thus the bit rate at the output of the convolutional encoder is also R/R_c.

The binary digits from the encoder are fed into a modulator which maps each bit into an elementary signal waveform. Binary PSK or FSK are commonly used signaling waveforms for transmitting the bits in the coded sequence. The channel over which the waveforms are transmitted is assumed to corrupt the signal by the addition of white gaussian noise. The resulting received signal is processed first by the demodulator and then by the decoder.

The demodulator may be viewed as the matched filter to the signal waveform corresponding to each transmitted bit. Its output, sampled at the rate R/R_c, may or may not be quantized. At one extreme, the demodulator can be used to make firm decisions on whether each coded bit is a 0 or a 1. Thus the output is quantized to two levels, denoted as 0 and 1. We say that the demodulator has made a "hard decision" on each bit. The sequence of detected bits from the demodulator is fed into the decoder, which takes the detected bits and

FIGURE 5.1.1
Model of digital communications system with channel encoding and decoding.

attempts to recover the information sequence. Since the decoder operates on the hard decisions made by the demodulator, the decoding process is termed *hard-decision decoding*. At the other extreme, the analog (unquantized) output from the demodulator can be fed to the decoder. The decoder, in turn, makes use of the additional information contained in the unquantized samples to recover the information sequence with a higher reliability than that achieved with hard decisions. We refer to the resulting decoding as *soft-decision decoding*. The same term is used to describe decoding with quantized outputs from the demodulator, where the number of quantization levels exceeds two.

When the output of the demodulator is quantized to Q levels, the combination of modulator, channel, and demodulator is reduced to an equivalent discrete-time, discrete-amplitude channel with an input sequence selected from the alphabet $\{0, 1\}$ and an output sequence of symbols taken from the alphabet $\{0\,1, 2, \ldots, Q - 1\}$. As a consequence of the assumption that the additive gaussian noise in the channel is white, the equivalent discrete-time, discrete-amplitude channel is a discrete memoryless channel (DMC) which is completely characterized by a set of transition probabilities $P(k|j)$ where $j = 0, 1$ and $k = 0, 1, \ldots, Q - 1$.

When $Q = 2$ (hard decisions), the channel output is binary and the transition probabilities $P(0|0)$ and $P(1|1)$ denote the probabilities for correct reception, conditional on a 0 and a 1 having been transmitted, respectively. On the other hand, the probabilities $P(1|0)$ and $P(0|1)$ represent the two error probabilities corresponding to the events "a 0 is transmitted and is demodulated into a 1" and "a 1 is transmitted and is demodulated into a 0." Typical modulation and demodulation techniques such as binary PSK and binary FSK result in the symmetry conditions $P(0|0) = P(1|1) = 1 - p$ and $P(0|1) = P(1|0) = p$. The resulting binary-input, binary-output DMC, previously discussed in Chap. 2, is called a *binary symmetric channel* (BSC) with crossover probability p.

We may also consider the more general coding problem is which the input to the modulator consists of symbols selected from the (nonbinary) alphabet $\{0, 1, 2, \ldots, q - 1\}$. Such a situation arises when the encoder is nonbinary, having an input that consists of blocks of k bits which in turn form k-bit symbols that can be represented by the integers $\{0, 1, 2, \ldots, 2^k - 1\}$. The output sequence of the nonbinary encoder also consists of k-bit symbols. These k-bit symbols can be transmitted directly by employing $q = 2^k$ signaling waveforms at the modulator so that each k-bit symbol is mapped into one of the waveforms. Then, if the demodulator is quantized to Q levels where $Q \geq q$, the AWGN channel becomes a DMC characterized by the transition probabilities $P(i|j)$, $j = 0, 1, \ldots, q - 1$ and $i = 0, 1, \ldots, Q - 1$.

In this chapter, we shall consider the performance achieved by coding for the AWGN channel with hard-decision decoding and soft-decision decoding. The emphasis will be on linear binary codes. The performance characteristics that we obtain will depend on a number of code parameters such as the rate of the code, the number of code words in the code, and the distance properties of the code.

The design of coded modulation for efficient transmission of information may be divided into two basic approaches. One is the algebraic approach, which is primarily concerned with the design of coding and decoding techniques for specific classes of codes, such as cyclic block codes, convolutional codes, etc. The second is the probabilistic approach, which is concerned with the analysis of the performance of a general class of coded signals. This approach yields bounds on the probability of error that can be attained for communication over a channel having some specified characteristic.

In this section we adopt the probabilistic approach to coded modulation. The algebraic approach based on block codes and on convolutional codes is treated in Secs. 5.2 and 5.3.

5.1.1 Random Coding Based on M-ary Binary Coded Signals

Let us consider a set of M coded signal waveforms constructed from a set of n-dimensional binary code words of the form

$$\mathbf{C}_i = [c_{i1}c_{i2} \cdots c_{in}] \qquad i = 1, 2, \ldots, M \tag{5.1.1}$$

where $c_{ij} = 0$ or 1. Each bit in the code word is mapped into a binary PSK waveform, so that the signal waveform corresponding to the code word $\mathbf{C}_i$ may be expressed as

$$s_i(t) = \sum_{j=1}^{n} s_{ij} f_j(t) \qquad i = 1, 2, \ldots, M \tag{5.1.2}$$

where

$$s_{ij} = \begin{cases} \sqrt{\mathscr{E}_c} & \text{when } c_{ij} = 1 \\ -\sqrt{\mathscr{E}_c} & \text{when } c_{ij} = 0 \end{cases} \tag{5.1.3}$$

and $\mathscr{E}_c$ is the energy per code bit. Thus, the waveforms $s_i(t)$ are equivalent to the n-dimensional vectors

$$\mathbf{s}_i = (s_{i1}s_{i2} \cdots s_{in}) \qquad i = 1, 2, \ldots, M \tag{5.1.4}$$

which correspond to the vertices of a hypercube in n-dimensional space.

Now, suppose that the information rate into the encoder is R bits per second and we encode blocks of k bits at a time into one of the M waveforms. Hence, $k = RT$ and $M = 2^k = 2^{RT}$ signals are required. It is convenient to define a parameter D as

$$D = \frac{n}{T} \text{ dimensions/s} \tag{5.1.5}$$

Thus, $n = DT$ is the dimensionality of the signal space.

The hypercube has $2^n = 2^{DT}$ vertices of which $M = 2^{RT}$ may be used to transmit the information. If we impose the condition that $D > R$, the fraction of

FIGURE 5.1.2
An ensemble at 2^{nM} communications system. Each system employs a different set of M signal from the set of 2^{nM} possible choices.

the vertices that we use as signal points is

$$F = \frac{2^k}{2^n} = \frac{2^{RT}}{2^{DT}} = 2^{-(D-R)T} \qquad (5.1.6)$$

Clearly, if $D > R$, we have $F \to 0$ as $T \to \infty$.

The question that we wish to pose is the following. Can we choose a subset $M = 2^{RT}$ vertices out of the $2^n = 2^{DT}$ available vertices such that the probability of error $P_e \to 0$ as $T \to \infty$ or, equivalently, as $n \to \infty$? Since the fraction F of vertices used approaches zero as $T \to \infty$, it should be possible to select M signal waveforms having a minimum distance that increases as $T \to \infty$ and, thus, $P_e \to 0$.

Instead of attempting to find a single set of M coded waveforms for which we compute the error probability, let us consider the ensemble of $(2^n)^M$ distinct ways in which we can select M vertices from the 2^n available vertices of the hypercube. Associated with each of the 2^{nM} selections there is a communication system, consisting of a modulator, a channel, and a demodulator which is optimum for the selected set of M waveforms. Thus, there are 2^{nM} communication systems, one for each choice of the M coded waveforms, as illustrated in Fig. 5.1.2. Each communication system is characterized by its probability of error.

Suppose that our choice of M coded waveforms is based on random selection from the set of 2^{nM} possible sets of codes. Thus, the random selection of the mth code, denoted as $\{\mathbf{s}_i\}_m$, occurs with probability

$$P(\{\mathbf{s}_i\}_m) = \frac{1}{2^{nM}} \qquad (5.1.7)$$

and the corresponding conditional probability of error for this choice of coded signals is $P_e(\{\mathbf{s}_i\}_m)$. Then, the average probability of error over the ensemble of codes is

$$\overline{P}_e = \sum_{m=1}^{2^{nM}} P_e(\{\mathbf{s}_i\}_m) P(\{\mathbf{s}_i\}_m)$$

$$= \frac{1}{2^{nM}} \sum_{m=1}^{2^{nM}} P_e(\{\mathbf{s}_i\}_m) \qquad (5.1.8)$$

where the overbar on P_e denotes an average over the ensemble of codes.

It is clear that some choices of codes will result in a large probability of error. For example, the code that assigns all M k-bit sequences to the same vertex of the hypercube will result in a large probability of error. In such a case, $P_e(\{\mathbf{s}_i\}_m) > \overline{P}_e$. However, there will also be choices of codes for which $P_e(\{\mathbf{s}_i\}_m) < \overline{P}_e$. Consequently, if we obtain an upper bound on $\overline{P}_e$, this bound will also hold for those codes for which $P_e(\{\mathbf{s}_i\}_m) < \overline{P}_e$. Furthermore, if $\overline{P}_e \to 0$ as $T \to \infty$, then we conclude that for these codes $P(\{\mathbf{s}_i\}_m) \to 0$ as $T \to \infty$.

In order to determine an upper bound on $\overline{P}_e$, we consider the transmission of a k-bit message $\mathbf{X}_k \equiv [x_1 x_2 x_3 \cdots x_k]$, where $x_j = 0$ or 1 for $j = 1, 2, \ldots, k$. The conditional probability of error averaged over the ensemble of codes is

$$\overline{P_e(\mathbf{X}_k)} = \sum_{\text{all codes}} P_e(\mathbf{X}_k, \{\mathbf{s}_i\}_m) P(\{\mathbf{s}_i\}_m) \qquad (5.1.9)$$

where $P_e(\mathbf{X}_k, \{\mathbf{s}_i\}_m)$ is the conditional probability of error for a given k-bit message $\mathbf{X}_k$, which is transmitted by use of the code $\{\mathbf{s}_i\}_m$. For the mth code, the probability of error $P_e(\mathbf{X}_k, \{\mathbf{s}_i\}_m)$ is upper bounded as

$$P_e(\mathbf{X}_k, \{\mathbf{s}_i\}_m) \le \sum_{\substack{l=1 \\ l \ne k}}^{M} P_{2m}(\mathbf{s}_l, \mathbf{s}_k) \qquad (5.1.10)$$

where $P_{2m}(\mathbf{s}_l, \mathbf{s}_k)$ is the probability of error for a binary communications system that employs the signal vectors $\mathbf{s}_l$ and $\mathbf{s}_k$ to communicate one of two equally likely k-bit messages. Hence,

$$\overline{P_e(\mathbf{X}_k)} \le \sum_{\text{all codes}} P_e(\{\mathbf{s}_i\}_m) \sum_{\substack{l=1 \\ l \ne k}}^{M} P_{2m}(\mathbf{s}_l, \mathbf{s}_k) \qquad (5.1.11)$$

If we interchange the order of the summations in (5.1.11) we obtain

$$\overline{P_e(X_k)} \le \sum_{\substack{l=1 \\ l \ne k}}^{M} \left[\sum_{\text{all codes}} P_e(\{\mathbf{s}_i\}_m) P_{2m}(\mathbf{s}_l, \mathbf{s}_k) \right]$$

$$\le \sum_{\substack{l=1 \\ l \ne k}}^{M} \overline{P_2(\mathbf{s}_l, \mathbf{s}_k)} \qquad (5.1.12)$$

where $\overline{P_2(\mathbf{s}_l, \mathbf{s}_k)}$ represents the ensemble average of $P_{2m}(\mathbf{s}_l, \mathbf{s}_k)$ over the 2^{nM} codes or the 2^{nM} communications systems.

For the additive white gaussian noise channel, the binary error probability $P_{2m}(\mathbf{s}_l, \mathbf{s}_k)$ is†

$$P_{2m}(\mathbf{s}_l, \mathbf{s}_k) = \frac{1}{2} \operatorname{erfc}\left(\sqrt{\frac{d_{lk}^2}{4N_0}} \right) \qquad (5.1.13)$$

where $d_{lk}^2 = |\mathbf{s}_l - \mathbf{s}_k|^2$. If $\mathbf{s}_l$ and $\mathbf{s}_k$ differ in d coordinates,

$$d_{lk}^2 = |\mathbf{s}_l - \mathbf{s}_k|^2 = \sum_{j=1}^{n} (s_{lj} - s_{kj})^2 = d\left(2\sqrt{\mathscr{E}_c}\right)^2 = 4d\mathscr{E}_c \qquad (5.1.14)$$

Hence,

$$P_{2m}(\mathbf{s}_l, \mathbf{s}_k) = \frac{1}{2} \operatorname{erfc}\left(\sqrt{\frac{d\mathscr{E}_c}{N_0}} \right) \qquad (5.1.15)$$

Now, we can average $P_{2m}(\mathbf{s}_l, \mathbf{s}_k)$ over the ensemble of codes. Since all the codes are equally probable, the signal vector $\mathbf{s}_l$ is equally likely to be any of the 2^n possible vertices of the hypercube and it is statistically independent of the signal vector $\mathbf{s}_k$. Therefore, $P(s_{li} = s_{ki}) = \frac{1}{2}$ and $P(s_{li} \neq s_{ki}) = \frac{1}{2}$, independently for all $i = 1, 2, \ldots, n$. Consequently, the probability that $\mathbf{s}_l$ and $\mathbf{s}_k$ differ in d positions is simply

$$P(d) = \binom{n}{d} \left(\frac{1}{2}\right)^n \qquad (5.1.16)$$

Hence, the expected value of $P_{2m}(\mathbf{s}_l, \mathbf{s}_k)$ over the ensemble of codes may be expressed as

$$\overline{P_2(\mathbf{s}_l, \mathbf{s}_k)} = \sum_{d=0}^{n} P(d) \left[\frac{1}{2} \operatorname{erfc}\left(\sqrt{\frac{d\mathscr{E}_c}{N_0}} \right) \right]$$

$$= \frac{1}{2^{n+1}} \sum_{d=0}^{n} \binom{n}{d} \operatorname{erfc}\left(\sqrt{\frac{d\mathscr{E}_c}{N_0}} \right) \qquad (5.1.17)$$

The result in (5.1.17) can be simplified if we upper-bound the complementary error function as

$$\operatorname{erfc}\left(\sqrt{\frac{d\mathscr{E}_c}{N_0}} \right) < e^{-d\mathscr{E}_c/N_0}$$

Thus,

$$\overline{P_2(\mathbf{s}_l, \mathbf{s}_k)} \leq \frac{1}{2^{n+1}} \sum_{d=0}^{n} \binom{n}{d} e^{-d\mathscr{E}_c/N_0}$$

$$\leq \frac{1}{2^{n+1}} \left(1 + e^{-\mathscr{E}_c/N_0}\right)^n$$

$$\leq \frac{1}{2} \left(\frac{1 + e^{-\mathscr{E}_c/N_0}}{2} \right)^n \qquad (5.1.18)$$

---

† Throughout this chapter, the channel gain (attenuation) factor is normalized to $\alpha = 1$.

We observe that the right-hand side of (5.1.18) is independent of the indices l and k. Hence, when we substitute the bound in (5.1.18) into (5.1.12) we obtain

$$\overline{P_e(\mathbf{X}_k)} \leq \sum_{\substack{l=1 \\ l \neq k}}^{M} \overline{P_2(\mathbf{s}_l, \mathbf{s}_k)} = \frac{M-1}{2}\left(\frac{1 + e^{-\mathscr{E}_c/N_0}}{2}\right)^n$$

$$< \frac{M}{2}\left(\frac{1 + e^{-\mathscr{E}_c/N_0}}{2}\right)^n$$

Finally, the unconditional average error probability $\overline{P}_e$ is obtained by averaging $\overline{P_e(X_k)}$ over all possible k-bit information sequences. Thus,

$$\overline{P}_e = \sum_k \overline{P_e(\mathbf{X}_k)}\, P(\mathbf{X}_k) < \frac{M}{2}\left(\frac{1 + e^{-\mathscr{E}_c/N_0}}{2}\right)^n \sum_k P(\mathbf{X}_k)$$

$$< \frac{M}{2}\left(\frac{1 + e^{-\mathscr{E}_c/N_0}}{2}\right)^n \tag{5.1.19}$$

This result can be expressed in a more convenient form by first defining a parameter R_0, which is called the *cutoff rate* and has units of bits per dimension, as

$$R_0 = \log_2 \frac{2}{1 + e^{-\mathscr{E}_c/N_0}}$$

$$= 1 - \log_2\left(1 + e^{-\mathscr{E}_c/N_0}\right) \quad \text{antipodal signaling} \tag{5.1.20}$$

Then, (5.1.19) becomes

$$\overline{P}_e < \frac{M}{2} 2^{-nR_0} = \frac{2^{RT}}{2} 2^{-nR_0} \tag{5.1.21}$$

Since $n = DT$, (5.1.21) may be expressed as

$$\overline{P}_e < \tfrac{1}{2} 2^{-T(DR_0 - R)} \tag{5.1.22}$$

The parameter R_0 is plotted as a function of $\mathscr{E}_c/N_0$ in Fig. 5.1.3. We observe that $0 \leq R_0 \leq 1$. Consequently, $P_e \to 0$ as $T \to \infty$ provided that the information rate $R < DR_0$.

Alternatively, (5.1.21) may be expressed as

$$\overline{P}_e < \tfrac{1}{2} 2^{-n(R_0 - R/D)} \tag{5.1.23}$$

The ratio R/D also has units of bits per dimension and may be defined as

$$R_c = \frac{R}{D} = \frac{R}{n/T} = \frac{RT}{n} = \frac{k}{n} \tag{5.1.24}$$

Hence, R_c is the code rate and

$$\overline{P}_e < \tfrac{1}{2} 2^{-n(R_0 - R_c)} \tag{5.1.25}$$

We conclude that when $R_c < R_0$, the average probability of error $P_e \to 0$ as the code block length $n \to \infty$. Since the average value of the probability error can be made arbitrarily small as $n \to \infty$, it follows that there exist codes in the ensemble of 2^{nM} codes that have a probability of error no larger than $\overline{P}_e$.

From the derivation of the average error probability given above, we conclude that good codes exist. Although we do not normally select codes at random, it is interesting to consider the question of whether or not a randomly selected code is likely to be a good code. In fact, we can easily show that there are many good codes in the ensemble. First, we note that $\bar{P}_e$ is an ensemble average of error probabilities over all codes and that all these probabilities are obviously positive quantities. If a code is selected at random, the probability that its error probability $P_e > \alpha \bar{P}_e$ is less than $1/\alpha$. Consequently, no more than 10 percent of the codes have an error probability that exceeds $10\bar{P}_e$ and no more than 1 percent of the codes have an error probability that exceeds $100\bar{P}_e$.

We should emphasize that codes with error probabilities exceeding $\bar{P}_e$ are not necessarily poor codes. For example, suppose that an average error rate of $\bar{P}_e < 10^{-10}$ can be attained by using codes with dimensionality n_0 when $R_0 > R_c$. Then, if we select a code with error probability $1000\bar{P}_e = 10^{-7}$, we may compensate for this reduction in error probability by increasing n from n_0 to $n = 10n_0/7$. Thus, by a modest increase in dimensionality, we have a code with $\bar{P}_e < 10^{-10}$. In summary, good codes are abundant and, hence, they are easily found even by random selection.

It is also interesting to express the average error probability in (5.1.25) in terms of the SNR per bit γ_b. To accomplish this, we express the energy per signal waveform as

$$\mathscr{E} = n\mathscr{E}_c = k\mathscr{E}_b \tag{5.1.26}$$

Hence, $n = k\mathscr{E}_b/\mathscr{E}_c$. We also note that $R_c\mathscr{E}_b/\mathscr{E}_c = 1$. Therefore, (5.1.25) may be expressed as

$$\bar{P}_e < \tfrac{1}{2}2^{-k(\gamma_b/\gamma_0 - 1)} \tag{5.1.27}$$

FIGURE 5.1.4
Lower bound on SNR per bit γ_b for binary antipodal signals.

where γ_0 is a normalized SNR parameter, defined as

$$\gamma_0 = \frac{R_c}{R_0}\gamma_b$$
$$= \frac{R_c\gamma_b}{1 - \log_2\left(1 + e^{-R_c\gamma_b}\right)} \qquad (5.1.28)$$

Now, we note that $\overline{P}_e \to 0$ as $k \to 0$, provided that the SNR per bit $\gamma_b > \gamma_0$.

The parameter γ_0 is plotted in Fig. 5.1.4 as a function of $R_c\gamma_b$. Note that as $R_c\gamma_b \to 0$, $\gamma_0 \to 2\ln 2$. Consequently, the error probability for M-ary binary coded signals is equivalent to the error probability obtained from the union bound for M-ary orthogonal signals, provided that the signal dimensionality is sufficiently large so that $\gamma_0 \approx 2\ln 2$.

The dimensionality parameter D that we introduced in (5.1.5) is proportional to the channel bandwidth required to transmit the signals. Recall from the sampling theorem that a signal of bandwidth W may be represented by samples taken at a rate of $2W$ samples per second. Thus, in the time interval of length T there are $n = 2WT$ samples or, equivalently, n degrees of freedom (dimensions). Consequently, D may be equated with $2W$.

Finally, we note that the binary coded signals considered in this section are appropriate when the SNR per dimension is small, e.g., $\mathscr{E}_c/N_0 < 10$. However, when $\mathscr{E}_c/N_0 > 10$, R_0 saturates at 1 bit per dimension. Since the code rate is

FIGURE 5.1.5
Signal alphabet consisting of four amplitude levels.

restricted to be less than R_0, binary coded signals become inefficient at $\mathscr{E}_c/N_0 >$ 10. In such a case, we may use nonbinary-coded signals to achieve an increase in the number of bits per dimension. For example, multiple amplitude coded signal sets can be constructed from nonbinary codes by mapping each code element into one of q possible amplitude levels (as in PAM or ASK). Such codes are considered below.

5.1.2 Random Coding Based on M-ary Multiamplitude Signals

Instead of constructing binary-coded signals, suppose we employ nonbinary codes with code words of the form given by (5.1.1), where the code elements c_{ij} are selected from the set $\{0, 1, \ldots, q-1\}$. Each code element is mapped into one of q possible amplitude levels. Thus, we construct signals corresponding to n-dimensional vectors $\{\mathbf{s}_i\}$ as in (5.1.4), where the components $\{s_{ij}\}$ are selected from a multiamplitude set of q possible values. Now, we have q^n possible signals from which we select $M = 2^{RT}$ signals to transmit k-bit blocks of information. The q amplitudes corresponding to the code elements $\{0, 1, \ldots, q-1\}$ may be denoted as $\{a_1, a_2, \ldots, a_q\}$ and they are assumed to be selected according to some specified probabilities $\{p_i\}$. The amplitude levels are assumed to be equally spaced over the interval $[-\sqrt{\mathscr{E}_c}, \sqrt{\mathscr{E}_c}]$. For example, Fig. 5.1.5 illustrates the amplitude values for $q = 4$. In general, adjacent amplitude levels are separated by $2\sqrt{\mathscr{E}_c}/(q-1)$. This assignment guarantees not only that each component s_{ij} is peak energy limited to $\sqrt{\mathscr{E}_c}$, but, also, each code word is constrained in average energy to satisfy the condition

$$|\mathbf{s}_i|^2 < n\mathscr{E}_c \tag{5.1.29}$$

By repeating the derivation given above for random selection of codes in an AWGN channel, we find [for reference, see Wozencraft and Jacobs (1965)] that the average probability of error is upper bounded as

$$\bar{P}_e < M2^{-nR_0} = 2^{RT}2^{-nR_0} = 2^{-n(R_0 - R/D)} \tag{5.1.30}$$

where R_0 is defined as

$$R_0 = -\log_2\left(\sum_{l=1}^{q}\sum_{m=1}^{q} p_l p_m e^{-d_{lm}^2/4N_0}\right) \tag{5.1.31}$$

and

$$d_{lm} = |a_l - a_m| \qquad l, m = 1, 2, \ldots, q \tag{5.1.32}$$

In the special case where all the amplitude levels are equally likely, $p_l = p_m = 1/q$ and (5.1.31) reduces to

$$R_0 = -\log_2 \left(\frac{1}{q^2} \sum_{l=1}^{q} \sum_{m=1}^{q} e^{-d_{lm}^2/4N_0} \right) \qquad (5.1.33)$$

For example, where $q = 2$ and $a_1 = -\sqrt{\mathscr{E}_c}$, $a_2 = \sqrt{\mathscr{E}_c}$, we have $d_{11} = d_{22} = 0$, $d_{12} = d_{21} = 2\sqrt{\mathscr{E}_c}$ and, hence

$$R_0 = \log_2 \frac{2}{1 + e^{-\mathscr{E}_c/N_0}} \qquad q = 2$$

which agrees with our previous result. When $q = 4$, $a_1 = -\sqrt{\mathscr{E}_c}$, $a_2 = -\sqrt{\mathscr{E}_c}/3$, $a_3 = \sqrt{\mathscr{E}_c}/3$, and $a_4 = \sqrt{\mathscr{E}_c}$, we have $d_{mm} = 0$ for $m = 1, 2, 3, 4$, $d_{12} = d_{23} = d_{34} =$

FIGURE 5.1.6
Cutoff rate R_0 for equally spaced q-level amplitude modulation with equal probabilities $p_c = 1/q$. (From *Principles of Communication Engineering* by J. M. Wozencraft and I. M. Jacobs, © 1965 by John Wiley and Sons, Inc. Reprinted with permission of the publisher.)

$d_{21} = d_{32} = d_{43} = 2\sqrt{\mathscr{E}_c}/3$, $d_{13} = d_{31} = d_{24} = d_{42} = 4\sqrt{\mathscr{E}_c}/3$, and $d_{14} = d_{41} = \sqrt{\mathscr{E}_c}$. Hence,

$$R_0 = \log_2 \frac{8}{2 + 3e^{-\mathscr{E}_c/9N_0} + 2e^{-4\mathscr{E}_c/9N_0} + e^{-\mathscr{E}_c/N_0}} \qquad q = 4 \quad (5.1.34)$$

Clearly, R_0 now saturates at 2 bits per dimensions as $\mathscr{E}_c/N_0$ increases.

The graphs of R_0 as a function of $\mathscr{E}_c/N_0$ for equally spaced and equally likely amplitude levels are shown in Fig. 5.1.6 for $q = 2, 3, 4, 8, 16, 32,$ and 64. Note that the saturation level now occurs at $\log_2 q$ bits per dimension. Consequently, for high SNR, $P_e \to 0$ as $n \to \infty$, provided that $R < DR_0 = 2WR_0$ bits per second.

If we remove the peak energy constraint on each of the elements, but retain the average energy constraint per code word as given by (5.1.29), it is possible to obtain a larger upper bound on the number of bits per dimension. For this case, the result obtained by Shannon (1959b) is

$$R_0^* = \frac{1}{2}\left[1 + \frac{\mathscr{E}_c}{N_0} - \sqrt{1 + \left(\frac{\mathscr{E}_c}{N_0}\right)^2}\right]\log_2 e$$

$$+ \frac{1}{2}\log_2\left[\frac{1}{2}\left(1 + \sqrt{1 + \left(\frac{\mathscr{E}_c}{N_0}\right)^2}\right)\right] \qquad (5.1.35)$$

The graph of R_0^* as a function of the SNR per dimension $\mathscr{E}_c/N_0$ is also shown in Fig. 5.1.6. It is clear that our selection of the equally spaced, equally likely amplitude levels that result in R_0 is suboptimum. However, these coded signals are easily generated and implemented in practice. This is an important advantage that justifies their use.

5.1.3 Comparison of R_0^* with the Capacity of the AWGN Channel

The channel capacity of the band-limited additive, white gaussian noise channel with an average power constraint on the input signal was derived in Sec. 2.4, and is given as

$$C = W \log_2\left(1 + \frac{P_{av}}{WN_0}\right) \text{ bits/s} \qquad (5.1.36)$$

where P_{av} is the average power of the input signal and W is the channel bandwidth. It is interesting to express the capacity of this channel in terms of bits per dimension and the average power in terms of energy per dimension. With $D = 2W$ and

$$\mathscr{E}_c = \frac{\mathscr{E}}{n} = \frac{P_{av}T}{n}$$

FIGURE 5.1.7
Comparison of cutoff rate R_0^* with the channel capacity for an AWGN channel.

we have

$$P_{av} = \frac{n}{T} \mathcal{E}_c = D \mathcal{E}_c \tag{5.1.37}$$

By defining $C_n = C/2W = C/D$ and substituting for W and P_{av}, (5.1.36) may be expressed as

$$C_n = \frac{1}{2} \log_2 \left(1 + 2\frac{\mathcal{E}_c}{N_0}\right)$$

$$= \frac{1}{2} \log_2 \left(1 + 2R_c \gamma_b\right) \text{ bits/dimension} \tag{5.1.38}$$

This expression for the normalized capacity may be compared with R_0^*, as shown in Fig. 5.1.7. Since C_n is the ultimate upper limit on the transmission rate R/D, $R_o^* < C_n$ as expected. We also observe that for small values of $\mathcal{E}_c/N_0$, the difference between R_0^* and C_n is approximately 3 dB. Therefore, the use of randomly selected, optimum average power limited, multiamplitude signals yields a rate function R_0^* that is within 3 dB of the channel capacity. More elaborate bounding techniques are required to show that the probability of error can be made arbitrarily small when $R < DC_n = 2WC_n = C$.

5.1.4 Channel Reliability Functions

In Sec. 4.2.2 we derived exponential bounds on the error probability for M-ary orthogonal signals on an infinite bandwidth AWGN channel. We demonstrated that P_M is upper bounded as

$$P_M < 2 \cdot 2^{-TE(R)} \qquad (5.1.39)$$

where $E(R)$ was called the channel reliability function. Recall that for M-ary orthogonal signals transmitted on an infinite bandwidth AWGN channel, the channel reliability function is

$$E(R) = \begin{cases} \frac{1}{2}C_\infty - R & 0 \le R \le \frac{1}{4}C_\infty \\ \left(\sqrt{C_\infty} - \sqrt{R}\right)^2 & \frac{1}{4}C_\infty \le R \le C_\infty \end{cases} \qquad (5.1.40)$$

In this case, we note that $E(0) = C_\infty/2$ and that as the rate $R \to C_\infty$, $E(R) \to 0$.

The bound given by (5.1.39) and (5.1.40) has been shown by Gallager (1965) to be *exponentially tight*. This means that there does not exist another reliability function, say $E_1(R)$, which satisfies the condition $E_1(R) > E(R)$ for any R. Consequently, the error probability is bounded from above and below as

$$K_l 2^{-TE(R)} \le P_e \le K_u 2^{-TE(R)} \qquad (5.1.41)$$

where the constants have only a weak dependence on T, i.e., they vary slowly with T.

Since orthogonal signals provide essentially the same performance as the optimum simplex signals for large M, the lower bound in (5.1.41) applies for any signal set. Hence, the reliability function $E(R)$ given by (5.1.40) determines the exponential characteristics of the error probability for digital signaling over the infinite bandwidth AWGN channel.

By use of random coding, we also demonstrated that the average probability of error is upper bounded as

$$\bar{P}_e \le \frac{1}{2} 2^{-nE_r(R)} \qquad (5.1.42)$$

where the exponential factor is given as

$$E_r(R) = R_0 - \frac{R}{D} \qquad (5.1.43)$$

Note that $E_r(0) = R_0$ and $E_r(R_0 D) = 0$. We recall that the upper bound in (5.1.41) has resulted from union bounding the error events.

In general, the exponential behavior of the error probability obtained from the union bound is not as tight as we would like. Through more elaborate bounding techniques, it is possible to derive both lower and upper bounds on the error probability for a variety of different channel models. These bounds have the general form

$$K_l e^{-nE_l(R)} \leq P_e \leq K_u e^{-nE_u(R)} \tag{5.1.44}$$

where the reliability functions satisfy the condition $E_l(R) \geq E_u(R)$. These functions depend on the channel models. Their general characteristic behavior is described in depth by Gallager (1968).

5.2 LINEAR BLOCK CODES

A block code consists of a set of fixed-length vectors called *code words*. The length of a code word is the number of elements in the vector and is denoted by n. The elements of a code word are selected from an alphabet of q elements. When the alphabet consists of two elements, 0 and 1, the code is a binary code and the elements of any code word are called bits. When the elements of a code word are selected from an alphabet having q elements ($q > 2$), the code is nonbinary. It is interesting to note that when q is a power of 2, i.e., $q = 2^b$ where b is a positive integer, each q-ary element has an equivalent binary representation consisting of b bits and, thus, a nonbinary code of block length N can be mapped into a binary code of block length $n = bN$.

There are 2^n possible code words in a binary block code of length n. From these 2^n code words we may select $M = 2^k$ code words ($k < n$) to form a code. Thus a block of k information bits is mapped into a code word of length n selected from the set of $M = 2^k$ code words. We refer to the resulting block code as an (n, k) code, and the ratio $k/n \equiv R_c$ is defined to be the rate of the code. More generally, in a code having q elements, there are q^n possible code words. A subset of $M = 2^k$ code words may be selected to transmit k-bit blocks of information.

Besides the code rate parameter R_c, an important parameter of a code word is its weight, which is simply the number of nonzero elements that it contains. In general, each code word has its own weight. The set of all weights in a code constitutes the weight distribution of the code. When all the M code words have equal weight, the code is called a *fixed-weight code* or a *constant-weight code*.

The encoding and decoding functions involve the arithmetic operations of addition and multiplication performed on code words. These arithmetic operations are performed according to the conventions of the algebraic field which has as its elements the symbols contained in the alphabet. For example, the symbols in a binary alphabet are 0 and 1; hence the field has two elements. In general, a field F consists of a set of elements that has two arithmetic operations defined on its elements, namely, addition and multiplication, which satisfy the following properties (axioms):

Addition.

1. The set F is closed under addition, i.e., if $a, b \in F$, then $a + b \in F$.
2. Addition is associative, i.e., if a, b, and c are elements of F then $a + (b + c) = (a + b) + c$

3. Addition is commutative, i.e., $a + b = b + a$

4. The set contains an element called *zero* that satisfies the condition $a + 0 = a$

5. Every element in the set has its own negative element. Hence, if b is an element, its negative is denoted as $(-b)$. The subtraction of two elements, such as $a - b$, is defined as $a + (-b)$.

Multiplication.

1. The set F is closed under multiplication, i.e., if $a, b \in F$, then $ab \in F$.

2. Multiplication is associative, i.e., $a(bc) = (ab)c$

3. Multiplication is commutative, i.e., $ab = ba$

4. Multiplication is distributive over addition, i.e., $(a + b)c = ac + bc$

5. Every element of F, except zero, has an inverse. Hence, if $b \in F$ ($b \neq 0$), its inverse is defined as b^{-1} and $bb^{-1} = 1$. The division of two elements, such as $a \div b$, is defined as ab^{-1}.

We are very familiar with the field of real numbers and the field of complex numbers. These fields have an infinite number of elements. However, as indicated above, codes are constructed from fields with a finite number of elements. A finite field with q elements is generally called a *Galois field* and denoted as $GF(q)$.

Every field must have a *zero* element and a *one* element. Hence, the simplest field is GF(2). In general, when q is a prime, we can construct the finite field $GF(q)$ consisting of the elements $\{0, 1, \ldots, q - 1\}$. The addition and multiplication operations on the elements of $GF(q)$ are defined modulo-q and denoted as (mod-q). For example, the addition and multiplication tables for GF(2) are

+	0	1		·	0	1
0	0	1		0	0	0
1	1	0		1	0	1

which are operations (mod-2). Similarly, the field GF(5) is a set consisting of the elements $\{0, 1, 2, 3, 4\}$. The addition and multiplication tables for GF(5) are

+	0	1	2	3	4		·	0	1	2	3	4
0	0	1	2	3	4		0	0	0	0	0	0
1	1	2	3	4	0		1	0	1	2	3	4
2	2	3	4	0	1		2	0	2	4	1	3
3	3	4	0	1	2		3	0	3	1	4	2
4	4	0	1	2	3		4	0	4	3	2	1

In general, the finite field $GF(q)$ can be constructed only if q is a prime or a power of a prime. When q is a prime, multiplication and addition are based on modulo-q arithmetic as illustrated above. If $q = p^m$ where p is a prime and m is any positive integer, it is possible to extend the field $GF(p)$ to the field $GF(p^m)$.

This is called the *extension field* of GF(p). Multiplication and addition of the elements in the extension field are based on modulo-p arithmetic.

With this brief introduction on the arithmetic operations that may be performed on the elements of code words, let us now consider some basic characteristics of block codes.

Suppose $\mathbf{C}_i$ and $\mathbf{C}_j$ are any two code words in an (n, k) block code. A measure of the difference between the code words in the number of corresponding elements or positions in which they differ. This measure is called the *Hamming distance* between the two code words and it is denoted as d_{ij}. Clearly d_{ij} for $i \neq j$ satisfies the condition $0 < d_{ij} \leq n$. The smallest value of the set $\{d_{ij}\}$ for the M code words is called the *minimum distance* of the code and is denoted as $d_{\min}$. Since the Hamming distance is a measure of the separation between pairs of code words, it is intimately related to the cross-correlation coefficient between corresponding pairs of waveforms generated from the code words. The relationship is discussed in Sec. 5.2.4.

Besides characterizing a code as being binary or nonbinary, one can also describe it as either linear or nonlinear. Suppose $\mathbf{C}_i$ and $\mathbf{C}_j$ are two code words in an (n, k) block code and let α_1 and α_2 be any two elements selected from the alphabet. Then the code is said to be linear if and only if $\alpha_1 \mathbf{C}_i + \alpha_2 \mathbf{C}_j$ is also a code word. Implied by this definition is that a linear code must contain the all-zero code word. Consequently a constant-weight code is nonlinear.

Suppose we have a binary linear block code and let $\mathbf{C}_i$, $i = 1, 2, \ldots, M$, denote the M code words. For convenience, let $\mathbf{C}_1$ denote the all-zero code word, i.e., $\mathbf{C}_1 = [0 \quad 0 \quad \cdots \quad 0]$ and let w_r denote the weight of the rth code word. It follows that w_r is the Hamming distance between the code words $\mathbf{C}_r$ and $\mathbf{C}_1$. Thus the distance $d_{1r} = w_r$. In general, the distance d_{ij} between any pair of code words $\mathbf{C}_i$ and $\mathbf{C}_j$ is simply equal to the weight of the code word formed by taking the difference between $\mathbf{C}_i$ and $\mathbf{C}_j$. Since the code is linear, the difference (equivalent to taking the modulo-2 sum for a binary code) between $\mathbf{C}_i$ and $\mathbf{C}_j$ is also a code word having a weight included in the set $\{w_r\}$. Hence the weight distribution of a linear code completely characterizes the distance properties of the code. The minimum distance of the code is, therefore,

$$d_{\min} = \min_{r, r \neq 1} \{w_r\} \qquad (5.2.1)$$

A number of elementary concepts from linear algebra are particularly useful in dealing with linear block codes. Specifically, the set of all n-tuples (vectors with n elements) form a vector space S. If we select a set of $k < n$ linearly independent vectors from S and from these construct the set of all linear combinations of these vectors, the resulting set forms a subspace of S, say S_c of dimension k. Any set of k linearly independent vectors in the subspace S_c constitutes a basis. Now consider the set of vectors in S which are orthogonal to every vector in a basis for S_c (and hence orthogonal to all vectors in S_c). This set of vectors is also a subspace of S and is called the *null space* of S_c. If the dimension of S_c is k, the dimension of the null space is $n - k$.

Expressed in terms appropriate for binary block codes, the vector space S consists of the 2^n binary valued n-tuples. The linear (n, k) code is a set of 2^k n-tuples called *code words* which forms a subspace S_c over the field of two elements. Since there are 2^k code words in S_c, a basis for S_c has k code words. That is, k linearly independent code words are required to construct 2^k linear combinations, thus generating the entire code. The null space of S_c is another linear code which consists of 2^{n-k} code words of block length n and $n - k$ information bits. Its dimension is $n - k$. In Sec. 5.2.1 we consider these relationships in greater detail.

5.2.1 The Generator Matrix and the Parity Check Matrix

Let $x_{m1}, x_{m2}, \ldots, x_{mk}$ denote the k information bits encoded into the code word C_m. Throughout this chapter we follow the established convention in coding of representing code words as row vectors. Thus the vector of k information bits into the encoder is denoted as

$$\underline{X}_m = [x_{m1} \quad x_{m2} \quad \cdots \quad x_{mk}]$$

and the output of the encoder is the vector

$$C_m = [c_{m1} \quad c_{m2} \quad \cdots \quad c_{mn}]$$

The encoding operation performed in a linear binary block encoder can be represented by a set of n equations of the form

$$c_{mj} = x_{m1}g_{1j} + x_{m2}g_{2j} + \cdots + x_{mk}g_{kj} \qquad j = 1, 2, \ldots, n \qquad (5.2.2)$$

where $g_{ij} = 0$ or 1 and $x_{mi}g_{ij}$ represents the product of x_{mi} and g_{ij}. The linear equations in (5.2.2) may also be represented in matrix form as

$$C_m = X_m G \qquad (5.2.3)$$

where G, called the *generator matrix* of the code, is

$$G = \begin{bmatrix} \leftarrow g_1 \rightarrow \\ \leftarrow g_2 \rightarrow \\ \vdots \\ \leftarrow g_k \rightarrow \end{bmatrix} = \begin{bmatrix} g_{11} & g_{12} & \cdots & g_{1n} \\ g_{21} & g_{22} & \cdots & g_{2n} \\ \cdots\cdots\cdots\cdots\cdots\cdots \\ g_{k1} & g_{k2} & \cdots & g_{kn} \end{bmatrix} \qquad (5.2.4)$$

Note that any code word is simply a linear combination of the vectors $\{g_i\}$ of G, i.e.,

$$C_m = x_{m1}g_1 + x_{m2}g_2 + \cdots + x_{mk}g_k \qquad (5.2.5)$$

Since the linear (n, k) code with 2^k code words is a subspace of dimension k, the row vectors $\{g_i\}$ of the generator matrix G must be linearly independent, i.e., they must span the subspace of k dimensions. In other words, the $\{g_i\}$ must be a basis for the (n, k) code. We note that the set of basis vectors is not unique and, hence, G is not unique. We also note that since the subspace has dimension k, the rank of G is k.

Any generator matrix of an (n, k) code can be reduced by row operations (and column permutations) to the "systematic form,"

$$\mathbf{G} = [\mathbf{I}_k \vdots \mathbf{P}] = \begin{bmatrix} 1 & 0 & 0 & \cdots & 0 & p_{11} & p_{12} & \cdots & p_{1n-k} \\ 0 & 1 & 0 & \cdots & 0 & p_{21} & p_{22} & \cdots & p_{2n-k} \\ \cdots & \cdots & \cdots & \cdots & \cdots & \cdots & \cdots & \cdots & \cdots \\ 0 & 0 & 0 & \cdots & 1 & p_{k1} & p_{k2} & \cdots & p_{kn-k} \end{bmatrix} \quad (5.2.6)$$

where $\mathbf{I}_k$ is a $k \times k$ identity matrix and $\mathbf{P}$ is a $k \times (n - k)$ matrix that determines the $n - k$ redundant bits or parity check bits. We note that a generator matrix of the systematic form generates a linear block code in which the first k bits of each code word are identical to the information bits to be transmitted and the remaining $n - k$ bits of each code word are linear combinations of the k information bits. These $n - k$ redundant bits are called *parity check bits*. The resulting (n, k) code is called a *systematic code*.

An (n, k) code generated by a generator matrix that is not in the systematic form of (5.2.6) is called *nonsystematic*. However, such a generator matrix is equivalent to a generator matrix of the systematic form in the sense that one can be obtained from the other by elementary row operations and column permutations. The two (n, k) linear codes generated by the two equivalent generator matrices are said to be equivalent, and one can be obtained from the other by a permutation of the places of every element. Thus every linear (n, k) code is equivalent to a linear systematic (n, k) code.

Example 5.2.1. Consider a $(7, 4)$ code with generator matrix

$$\mathbf{G} = \begin{bmatrix} 1 & 0 & 0 & 0 & 1 & 0 & 1 \\ 0 & 1 & 0 & 0 & 1 & 1 & 1 \\ 0 & 0 & 1 & 0 & 1 & 1 & 0 \\ 0 & 0 & 0 & 1 & 0 & 1 & 1 \end{bmatrix} = [\mathbf{I}_4 \vdots \mathbf{P}] \quad (5.2.7)$$

A typical code word may be expressed as

$$\mathbf{C}_m = [x_{m1} x_{m2} x_{m3} x_{m4} c_{m5} c_{m6} c_{m7}]$$

where the $\{x_{mj}\}$ represent the 4 information bits and the $\{c_{mj}\}$ represent the 3 parity check bits given by

$$\begin{aligned} c_{m5} &= x_{m1} + x_{m2} + x_{m3} \\ c_{m6} &= x_{m2} + x_{m3} + x_{m4} \\ c_{m7} &= x_{m1} + x_{m2} + x_{m4} \end{aligned} \quad (5.2.8)$$

A linear systematic (n, k) binary block encoder may be implemented by using a k-bit shift register and $(n - k)$ modulo-2 adders tied to the appropriate stages of the shift register. The $(n - k)$ adders generate the parity check bits which are subsequently stored temporarily in a second shift register of length $n - k$. The k-bit block of information bits is shifted into the k-bit shift register and the $n - k$ parity check bits are computed. Then the k information bits followed by the $n - k$ parity check bits are shifted out of the two shift registers

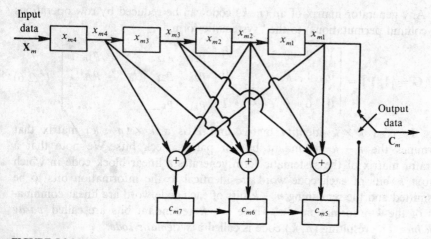

FIGURE 5.2.1
A linear shift register for generating a $(7, 4)$ binary code.

and fed to the modulator. This encoding is illustrated in Fig. 5.2.1, for the $(7, 4)$ code of Example 5.2.1.

Associated with any linear (n, k) code is the dual code of dimension $(n - k)$. The dual code is a linear $(n, n - k)$ code with 2^{n-k} code vectors which is the null space of the (n, k) code. The generator matrix for the dual code, denoted by $\mathbf{H}$, consists of $(n - k)$ linearly independent code vectors selected from the null space. Any code word $\mathbf{C}_m$ of the (n, k) code is orthogonal to any code word in the dual code. Hence any code word of the (n, k) code is orthogonal to every row of the matrix $\mathbf{H}$, i.e.,

$$\mathbf{C}_m \mathbf{H}' = \mathbf{0} \qquad (5.2.9)$$

where $\mathbf{0}$ denotes an all zero row vector with $n - k$ elements, and $\mathbf{C}_m$ is a code word of the (n, k) code. Since (5.2.9) holds for every code word of the (n, k) code, it follows that

$$\mathbf{G} \mathbf{H}' = \mathbf{0} \qquad (5.2.10)$$

where $\mathbf{0}$ is now a $k \times (n - k)$ matrix with all-zero elements.

Now suppose that the linear (n, k) code is systematic and its generator matrix $\mathbf{G}$ is given by the systematic form in (5.2.6). Then, since $\mathbf{G} \mathbf{H}' = \mathbf{0}$, it follows that

$$\mathbf{H} = [-\mathbf{P}' \vdots \mathbf{I}_{n-k}] \qquad (5.2.11)$$

The negative sign in (5.2.11) may be dropped when dealing with binary codes, since modulo-2 subtraction is identical to modulo-2 addition.

Example 5.2.2. For the systematic $(7, 4)$ code generated by matrix $\mathbf{G}$ given by (5.2.7), we have, according to (5.2.11), the matrix $\mathbf{H}$ in the form

$$\mathbf{H} = \begin{bmatrix} 1 & 1 & 1 & 0 & | & 1 & 0 & 0 \\ 0 & 1 & 1 & 1 & | & 0 & 1 & 0 \\ 1 & 1 & 0 & 1 & | & 0 & 0 & 1 \end{bmatrix} \qquad (5.2.12)$$

Now, the product $\mathbf{C}_m \mathbf{H}'$ yields the three equations

$$\begin{aligned} x_{m1} + x_{m2} + x_{m3} + c_{m5} &= 0 \\ x_{m2} + x_{m3} + x_{m4} + c_{m6} &= 0 \\ x_{m1} + x_{m2} + x_{m4} + c_{m7} &= 0 \end{aligned} \qquad (5.2.13)$$

Thus we observe that the product $\mathbf{C}_m \mathbf{H}'$ is equivalent to adding the parity check bits to the corresponding linear combinations of the information bits used to compute c_{mj}, $j = 5, 6, 7$. That is, the equations in (5.2.13) are equivalent to the equations in (5.2.8). The matrix $\mathbf{H}$ may be used by the decoder to check that a received code word $\mathbf{Y}$ satisfies the condition given in (5.2.13), i.e., $\mathbf{YH}' = \mathbf{0}$. In so doing, the decoder checks the received parity check bits with the corresponding linear combination of the bits y_1, y_2, y_3, y_4 which formed the parity check bits at the transmitter. It is therefore appropriate to call $\mathbf{H}$ the *parity check matrix* associated with the (n, k) code.

We make the following observation regarding the relation of the minimum distance of a code to its parity check matrix $\mathbf{H}$. The product $\mathbf{C}_m \mathbf{H}'$ with $\mathbf{C}_m \neq \mathbf{0}$ represents a linear combination of the n columns of $\mathbf{H}'$. Since $\mathbf{C}_m \mathbf{H}' = \mathbf{0}$, the column vectors of $\mathbf{H}$ are linearly dependent. Suppose $\mathbf{C}_j$ denotes the minimum weight code word of a linear (n, k) code. It must satisfy the condition $\mathbf{C}_j \mathbf{H}' = \mathbf{0}$. Since the minimum weight is equal to the minimum distance, it follows that $d_{\min}$ of the columns of $\mathbf{H}$ are linearly dependent. Alternatively, we may say that no more than $d_{\min} - 1$ columns of $\mathbf{H}$ are linearly independent. Since the rank of $\mathbf{H}$ is at most $n - k$, then $n - k \geq d_{\min} - 1$. Therefore $d_{\min}$ is upper-bounded as

$$d_{\min} \leq n - k + 1 \qquad (5.2.14)$$

Given a linear binary (n, k) code with minimum distance $d_{\min}$, we can construct a linear binary $(n + 1, k)$ code by appending one additional parity check bit to each code word. The check bit is usually selected to be a check bit on all the bits in the code word. Thus the added check bit is a 0 if the original code word has an even number of 1s and it is a 1 if the code word has an odd number of 1s. Consequently, if the minimum weight and, hence, the minimum distance of the code is odd, the added parity check bit increases the minimum distance by 1. We call the $(n + 1, k)$ code an *extended code*. Its parity check matrix is

$$\mathbf{H}_e = \left[\begin{array}{ccccc|c} & & & & & 0 \\ & & \mathbf{H} & & & 0 \\ & & & & & \vdots \\ & & & & & 0 \\ \hline 1 & 1 & 1 & \cdots & 1 & 1 \end{array} \right] \qquad (5.2.15)$$

where $\mathbf{H}$ is the parity check matrix of the original code.

A systematic (n, k) code can also be shortened by setting a number of the information bits to zero. That is, a linear (n, k) code which consists of k information bits and $n - k$ check bits can be shortened into a $(n - l, k - l)$ linear code by setting the first l bits to zero. These l bits are not transmitted. The $n - k$ check bits are computed in the usual manner, as in the original code. Since

$$\mathbf{C}_m = \mathbf{X}_m \mathbf{G}$$

the effect of setting the first l bits of $\mathbf{X}_m$ to 0 is equivalent to reducing the number of rows of $\mathbf{G}$ by removing the first l rows. Equivalently, since

$$\mathbf{C}_m \mathbf{H}' = \mathbf{0}$$

we may remove the first l columns of $\mathbf{H}$. The shortened $(n - l, k - l)$ code consists of 2^{k-l} code words. The minimum distance of these 2^{k-l} code words is at least as large as the minimum distance of the original (n, k) code.

5.2.2 Some Specific Linear Block Codes

In this subsection we shall briefly describe three types of linear block codes that are frequently encountered in practice and list their important parameters.

Hamming codes. There are both binary and nonbinary Hamming codes. We limit our discussion to the properties of binary Hamming codes. This is a class of codes with the property that

$$(n, k) = (2^m - 1, 2^m - 1 - m) \tag{5.2.16}$$

where m is any positive integer. For example, if $m = 3$, we have a $(7, 4)$ code.

The parity check matrix $\mathbf{H}$ of a Hamming code has a special property that allows us to describe the code rather easily. Recall that the parity check matrix of an (n, k) code has $n - k$ rows and n columns. For the binary (n, k) Hamming code, the $n = 2^m - 1$ columns consist of all possible binary vectors with $n - k = m$ elements, except the all-zero vector. For example, the $(7, 4)$ code considered in Examples 5.2.1 and 5.2.2 is a Hamming code. Its parity check matrix consists of the seven column vectors (001), (010), (011), (100), (101), (110), (111).

If we desire to generate a systematic Hamming code, the parity check matrix $\mathbf{H}$ can be easily arranged in the systematic form (5.2.11). Then the corresponding generator matrix $\mathbf{G}$ can be obtained from (5.2.11).

We make the observation that no two columns of $\mathbf{H}$ are linearly dependent, for otherwise the two columns would be identical. However, for $m > 1$, it is possible to find three columns of $\mathbf{H}$ which add to zero. Consequently $d_{\min} = 3$ for an (n, k) Hamming code.

By adding an overall parity bit, a Hamming (n, k) code can be modified to yield an $(n + 1, k)$ code with $d_{\min} = 4$. On the other hand, an (n, k) Hamming code may be shortened to $(n - l, k - l)$ by removing l rows of its generator matrix $\mathbf{G}$ or, equivalently, by removing l columns of its parity check matrix $\mathbf{H}$.

The weight distribution for the class of Hamming (n, k) codes is known and is expressed in compact form by the weight enumerating polynomial

$$
\begin{aligned}
A(z) &= \sum_{i=0}^{n} A_i z^i \\
&= \frac{1}{n+1} \left[(1+z)^n + n(1+z)^{(n-1)/2}(1-z)^{(n+1)/2} \right] \quad (5.2.17)
\end{aligned}
$$

where A_i is the number of code words of weight i.

Hadamard codes. A Hadamard code is obtained by selecting as code words the rows of a Hadamard matrix. A Hadamard matrix $\mathbf{M}_n$ is an $n \times n$ matrix (n an even integer) of 1s and 0s with the property that any row differs from any other row in exactly $n/2$ positions.† One row of the matrix contains all zeros. The other rows contain $n/2$ zeros and $n/2$ ones.

For $n = 2$, the Hadamard matrix is

$$
\mathbf{M}_2 = \begin{bmatrix} 0 & 0 \\ 0 & 1 \end{bmatrix} \quad (5.2.18)
$$

Furthermore, from $\mathbf{M}_n$, we can generate the Hadamard matrix $\mathbf{M}_{2n}$ according to the relation

$$
\mathbf{M}_{2n} = \begin{bmatrix} \mathbf{M}_n & \mathbf{M}_n \\ \mathbf{M}_n & \overline{\mathbf{M}}_n \end{bmatrix} \quad (5.2.19)
$$

where $\overline{\mathbf{M}}_n$ denotes the complement (0s replaced by 1s and vice versa) of $\mathbf{M}_n$. Thus, by substituting (5.2.18) into (5.2.19), we obtain

$$
\mathbf{M}_4 = \begin{bmatrix} 0 & 0 & 0 & 0 \\ 0 & 1 & 0 & 1 \\ 0 & 0 & 1 & 1 \\ 0 & 1 & 1 & 0 \end{bmatrix} \quad (5.2.20)
$$

The complement of $\mathbf{M}_4$ is

$$
\overline{\mathbf{M}}_4 = \begin{bmatrix} 1 & 1 & 1 & 1 \\ 1 & 0 & 1 & 0 \\ 1 & 1 & 0 & 0 \\ 1 & 0 & 0 & 1 \end{bmatrix} \quad (5.2.21)
$$

Now the rows of $\mathbf{M}_4$ and $\overline{\mathbf{M}}_4$ form a linear binary code of block length $n = 4$ having $2n = 8$ code words. The minimum distance of the code is $d_{\min} = n/2 = 2$.

By repeated application of (5.2.19) we can generate Hadamard codes with block length $n = 2^m$, $k = \log_2 2n = \log_2 2^{m+1} = m + 1$, and $d_{\min} = n/2 = 2^{m-1}$, where m is a positive integer. In addition to the important special case

---

† Sometimes the elements of the Hadamard matrix are denoted as $+1$ and -1. Then the rows of the Hadamard matrix are all mutually orthogonal. We also note that the $M = 2^k$ signal waveforms, constructed from Hadamard code words by mapping each bit in a code word into a binary PSK signal, are orthogonal.

TABLE 5.2.1
Weight distribution of Golay (23, 12) and extended Golay (24, 12) codes†

Weight	Number of code words	
	(23, 12) Code	(24, 12) Code
0	0	1
7	253	0
8	506	759
11	1288	0
12	1288	2576
15	506	0
16	253	759
23	1	0
24	0	1

† Peterson and Weldon [1972].

where $n = 2^m$, Hadamard codes of other block lengths are possible, but the codes are not linear.

Golay code. The Golay code is a binary linear (23, 12) code with $d_{min} = 7$. The extended Golay code obtained by adding an overall parity to the (23, 12) is a binary linear (24, 12) code with $d_{min} = 8$. Table 5.2.1 lists the weight distribution of the code words in the Golay (23, 12) and the extended Golay (24, 12) codes. We discuss the generation of the Golay code in Sec. 5.2.3.

5.2.3 Cyclic Codes

Cyclic codes are a subset of the class of linear codes which satisfy the following cyclic shift property: If $\mathbf{C} = [c_{n-1}, c_{n-2}, \ldots, c_1, c_0]$ is a code word of a cyclic code, then $[c_{n-2}, c_{n-3}, \ldots, c_0, c_{n-1}]$, obtained by a cyclic shift of the elements of $\mathbf{C}$, is also a code word. That is, all cyclic shifts of $\mathbf{C}$ are code words. As a consequence of the cyclic property, the codes possess a considerable amount of structure which can be exploited in the encoding and decoding operations. A number of efficient encoding and hard-decision decoding algorithms have been devised for cyclic codes which make it possible to implement long block codes with a large number of code words in practical communications systems. A description of specific algorithms is beyond the scope of this book. Our primary objective is to briefly describe a number of characteristics of cyclic codes.

In dealing with cyclic codes it is convenient to associate with a code word $\mathbf{C} = [c_{n-1}, c_{n-2}, \ldots, c_1, c_0]$ a polynomial $C(p)$ of degree $\leq n - 1$, defined as

$$C(p) = c_{n-1}p^{n-1} + c_{n-2}p^{n-2} + \cdots + c_1 p + c_0 \qquad (5.2.22)$$

For a binary code, each of the coefficients of the polynomial is either zero or one.

Now suppose we form the polynomial

$$pC(p) = c_{n-1}p^n + c_{n-2}p^{n-1} + \cdots + c_1 p^2 + c_0 p$$

This polynomial cannot represent a code word, since its degree may be equal to n (when $c_{n-1} = 1$). However, if we divide $pC(p)$ by $p^n + 1$, we obtain

$$\frac{pC(p)}{p^n + 1} = c_{n-1} + \frac{C_1(p)}{p^n + 1} \tag{5.2.23}$$

where

$$C_1(p) = c_{n-2}p^{n-1} + c_{n-3}p^{n-2} + \cdots + c_0 p + c_{n-1}$$

We note that the polynomial $C_1(p)$ represents the code word $\mathbf{C}_1 = [c_{n-2}, \ldots, c_0, c_{n-1}]$, which is just the code word $\mathbf{C}$ shifted cyclicly by one position. Since $C_1(p)$ is the remainder obtained by dividing $pC(p)$ by $p^n + 1$, we say that

$$C_1(p) = pC(p) \bmod (p^n + 1) \tag{5.2.24}$$

In a similar manner, if $C(p)$ represents a code word in a cyclic code, then $p^i C(p) \bmod (p^n + 1)$ is also a code word of the cyclic code. Thus we may write

$$p^i C(p) = Q(p)(p^n + 1) + C_i(p) \tag{5.2.25}$$

where the remainder polynomial $C_i(p)$ represents a code word of the cyclic code and $Q(p)$ is the quotient.

We can generate a cyclic code by using a *generator polynomial* $g(p)$ of degree $n - k$. The generator polynomial of an (n, k) cyclic code is a factor of $p^n + 1$ and has the general form

$$g(p) = p^{n-k} + g_{n-k-1}p^{n-k-1} + \cdots + g_1 p + 1 \tag{5.2.26}$$

We also define a *message polynomial* $X(p)$ as

$$X(p) = x_{k-1}p^{k-1} + x_{k-2}p^{k-2} + \cdots + x_1 p + x_0 \tag{5.2.27}$$

where $[x_{k-1}, x_{k-2}, \ldots, x_1, x_0]$ represent the k information bits. Clearly, the product $X(p)g(p)$ is a polynomial of degree less than or equal to $n - 1$ which may represent a code word. We note that there are 2^k polynomials $\{X_i(p)\}$ and, hence, there are 2^k possible code words that can be formed from a given $g(p)$.

Suppose we denote these code words as

$$C_m(p) = X_m(p)g(p) \qquad m = 1, 2, \ldots, 2^k \tag{5.2.28}$$

To show that the code words in (5.2.28) satisfy the cyclic property, consider any code word $C(p)$ in (5.2.28). A cyclic shift of $C(p)$ produces

$$C_1(p) = pC(p) + c_{n-1}(p^n + 1) \tag{5.2.29}$$

and, since $g(p)$ divides both $p^n + 1$ and $C(p)$, it also divides $C_1(p)$, i.e., $C_1(p)$ can be represented as

$$C_1(p) = X_1(p)g(p)$$

TABLE 5.2.2
(7, 4) Cyclic code

Generator polynomial: $g_1(p) = p^3 + p^2 + 1$

Information bits				Code words						
p^3	p^2	p^1	p^0	p^6	p^5	p^4	p^3	p^2	p^1	p^0
0	0	0	0	0	0	0	0	0	0	0
0	0	0	1	0	0	0	1	1	0	1
0	0	1	0	0	0	1	1	0	1	0
0	0	1	1	0	0	1	0	1	1	1
0	1	0	0	0	1	1	0	1	0	0
0	1	0	1	0	1	1	1	0	0	1
0	1	1	0	0	1	0	1	1	1	0
0	1	1	1	0	1	0	0	0	1	1
1	0	0	0	1	1	0	1	0	0	0
1	0	0	1	1	1	0	0	1	0	1
1	0	1	0	1	1	1	0	0	1	0
1	0	1	1	1	1	1	1	1	1	1
1	1	0	0	1	0	1	1	1	0	0
1	1	0	1	1	0	1	0	0	0	1
1	1	1	0	1	0	0	0	1	1	0
1	1	1	1	1	0	0	1	0	1	1

Therefore a cyclic shift of any code word $C(p)$ generated by (5.2.28) yields another code word.

From the above, we see that code words possessing the cyclic property can be generated by multiplying the 2^k message polynomials with a unique polynomial $g(p)$, called the generator polynomial of the (n, k) cyclic code, which divides $p^n + 1$ and has degree $n - k$. The cyclic code generated in this manner is a subspace S_c of the vector space S. The dimension of S_c is k.

Example 5.2.3. Consider a code with the block length $n = 7$. The polynomial $p^7 + 1$ has the following factors:

$$p^7 + 1 = (p + 1)(p^3 + p^2 + 1)(p^3 + p + 1) \qquad (5.2.30)$$

To generate a (7, 4) cyclic code, we may take as a generator polynomial one of the following two polynomials:

$$g_1(p) = p^3 + p^2 + 1$$
$$g_2(p) = p^3 + p + 1 \qquad (5.2.31)$$

The codes generated by $g_1(p)$ and $g_2(p)$ are equivalent. The code words in the (7, 4) code generated by $g_1(p) = p^3 + p^2 + 1$ are given in Table 5.2.2.

In general the polynomial $p^n + 1$ may be factored as

$$p^n + 1 = g(p)h(p)$$

where $g(p)$ denotes the generator polynomial for the (n, k) cyclic code and

TABLE 5.2.3
(7, 3) Dual code

Generator polynomial $p^4 h_1(p^{-1}) = p^4 + p^2 + p + 1$

Information bits			Code words						
p^2	p^1	p^0	p^6	p^5	p^4	p^3	p^2	p^1	p^0
0	0	0	0	0	0	0	0	0	0
0	0	1	0	0	1	0	1	1	1
0	1	0	0	1	0	1	1	1	0
0	1	1	0	1	1	1	0	0	1
1	0	0	1	0	1	1	1	0	0
1	0	1	1	0	0	1	0	1	1
1	1	0	1	1	1	0	0	1	0
1	1	1	1	1	0	0	1	0	1

$h(p)$ denotes the *parity polynomial* that has degree k. The latter may be used to generate the dual code.

For this purpose, we define the *reciprocal polynomial* of $h(p)$ as

$$p^k h(p^{-1}) = p^k(p^{-k} + h_{k-1}p^{-k+1} + h_{k-2}p^{-k+2} + \cdots + h_1 p^{-1} + 1)$$
$$= 1 + h_{k-1}p + h_{k-2}p^2 + \cdots + h_1 p^{k-1} + p^k \qquad (5.2.32)$$

Clearly, the reciprocal polynomial is also a factor of $p^n + 1$. Hence, $p^k h(p^{-1})$ is the generator polynomial of an $(n, n - k)$ cyclic code. This cyclic code is the dual code to the (n, k) code generated from $g(p)$. Thus, the $(n, n - k)$ dual code constitutes the null space of the (n, k) cyclic code.

Example 5.2.4. Let us consider the dual code to the $(7,4)$ cyclic code generated in Example 5.2.3. This dual code is a $(7,3)$ cyclic code associated with the parity polynomial

$$h_1(p) = (p + 1)(p^3 + p + 1)$$
$$= p^4 + p^3 + p^2 + 1 \qquad (5.2.33)$$

The reciprocal polynomial is

$$p^4 h_1(p^{-1}) = 1 + p + p^2 + p^4$$

This polynomial generates the $(7,3)$ dual code given in Table 5.2.3. The reader can verify that the code words in the $(7,3)$ dual code are orthogonal to the code words in the $(7,4)$ cyclic code of Example 5.2.3. Note that neither the $(7,4)$ nor the $(7,3)$ codes are systematic.

It is desirable to show how a generator matrix can be obtained from the generator polynomial of a cyclic (n, k) code. As previously indicated, the generator matrix for an (n, k) code can be constructed from any set of k linearly independent code words. Hence, given the generator polynomial $g(p)$, an easily generated set of k linearly independent code words is the code words corresponding to the set of k linearly independent polynomials

$$p^{k-1}g(p), p^{k-2}g(p), \ldots, pg(p), g(p)$$

Since any polynomial of degree less than or equal to $n - 1$ and divisible by $g(p)$ can be expressed as a linear combination of this set of polynomials, the set forms a basis of dimension k. Consequently the code words associated with these polynomials form a basis of dimension k for the (n, k) cyclic code.

Example 5.2.5. The four rows of the generator matrix for the $(7,4)$ cyclic code with generator polynomial $g_1(p) = p^3 + p^2 + 1$ are obtained from the polynomials

$$p^i g_1(p) = p^{3+i} + p^{2+i} + p^i \qquad i = 3, 2, 1, 0$$

It is easy to see that the generator matrix is

$$G_1 = \begin{bmatrix} 1 & 1 & 0 & 1 & 0 & 0 & 0 \\ 0 & 1 & 1 & 0 & 1 & 0 & 0 \\ 0 & 0 & 1 & 1 & 0 & 1 & 0 \\ 0 & 0 & 0 & 1 & 1 & 0 & 1 \end{bmatrix} \qquad (5.2.34)$$

Similarly, the generator matrix for the $(7,4)$ cyclic code generated by the polynomial $g_2(p) = p^3 + p + 1$ is

$$G_2 = \begin{bmatrix} 1 & 0 & 1 & 1 & 0 & 0 & 0 \\ 0 & 1 & 0 & 1 & 1 & 0 & 0 \\ 0 & 0 & 1 & 0 & 1 & 1 & 0 \\ 0 & 0 & 0 & 1 & 0 & 1 & 1 \end{bmatrix} \qquad (5.2.35)$$

The parity check matrices corresponding to G_1 and G_2 can be constructed in the same manner by using the respective reciprocal polynomials (Prob. 5.8).

Note that the generator matrix obtained by this construction is not in systematic form. We can construct the generator matrix of a cyclic code in the systematic form $G = [I_k \vdots P]$, from the generator polynomial as follows. First, we observe that the lth row of G corresponds to a polynomial of the form $p^{n-l} + R_l(p)$, $l = 1, 2, \ldots, k$, where $R_l(p)$ is a polynomial of degree less than $n - k$. This form can be obtained by dividing p^{n-l} by $g(p)$. Thus, we have

$$\frac{p^{n-l}}{g(p)} = Q_l(p) + \frac{R_l(p)}{g(p)} \qquad l = 1, 2, \ldots, k$$

or, equivalently,

$$p^{n-l} = Q_l(p)g(p) + R_l(p) \qquad i = 1, 2, \ldots, k \qquad (5.2.36)$$

where $Q_l(p)$ is the quotient. But $p^{n-1} + R_l(p)$ is a code word of the cyclic code since $p^{n-l} + R_l(p) = Q_l(p)g(p)$. Therefore the desired polynomial corresponding to the lth row of $\mathbf{G}$ is $p^{n-l} + R_l(p)$.

Example 5.2.6. For the $(7,4)$ cyclic code with generator polynomial $g_2(p) = p^3 + p + 1$, previously discussed in Example 5.2.5, we have

$$p^6 = (p^3 + p + 1)g(p) + p^2 + 1$$

$$p^5 = (p^2 + 1)g(p) + p^2 + p + 1$$

$$p^4 = pg(p) + p^2 + p$$

$$p^3 = g(p) + p + 1$$

Hence the generator matrix of the code in systematic form is

$$\mathbf{G}_2 = \begin{bmatrix} 1 & 0 & 0 & 0 & 1 & 0 & 1 \\ 0 & 1 & 0 & 0 & 1 & 1 & 1 \\ 0 & 0 & 1 & 0 & 1 & 1 & 0 \\ 0 & 0 & 0 & 1 & 0 & 1 & 1 \end{bmatrix} \tag{5.2.37}$$

and the corresponding parity check matrix is

$$\mathbf{H}_2 = \begin{bmatrix} 1 & 1 & 1 & 0 & 1 & 0 & 0 \\ 0 & 1 & 1 & 1 & 0 & 1 & 0 \\ 1 & 1 & 0 & 1 & 0 & 0 & 1 \end{bmatrix} \tag{5.2.38}$$

It is left as an exercise for the reader to demonstrate that the generator matrix $\mathbf{G}_2$ given by (5.2.35) and the systematic form given by (5.2.37) generate the same set of code words (Prob. 5.2).

The method for constructing the generator matrix $\mathbf{G}$ in systematic form according to (5.2.36) also implies that a systematic code can be generated directly from the generator polynomial $g(p)$. Suppose that we multiply the message polynomial $X(p)$ by p^{n-k}. Thus, we obtain

$$p^{n-k}X(p) = x_{k-1}p^{n-1} + x_{k-2}p^{n-2} + \cdots + x_1 p^{n-k+1} + x_0 p^{n-k}$$

In a systematic code, this polynomial represents the first k bits in the code word $C(p)$. To this polynomial we must add a polynomial of degree less than $n - k$ representing the parity check bits. Now, if $p^{n-k}X(p)$ is divided by $g(p)$ the result is

$$\frac{p^{n-k}X(p)}{g(p)} = Q(p) + \frac{r(p)}{g(p)}$$

or, equivalently,

$$p^{n-k}X(p) = Q(p)g(p) + r(p) \tag{5.2.39}$$

where $r(p)$ has degree less than $n - k$. Clearly, $Q(p)g(p)$ is a code word of the cyclic code. Hence, by adding (modulo-2) $r(p)$ to both sides of (5.2.39) we obtain the desired systematic code.

To summarize, the systematic code may be generated by (1) multiplying the message polynomial $X(p)$ by p^{n-k}, (2) dividing $p^{n-k}X(p)$ by $g(p)$ to obtain the remainder $r(p)$, and (3) adding $r(p)$ to $p^{n-k}X(p)$. Below we demonstrate how these computations can be performed by using shift registers with feedback.

Since $p^n + 1 = g(p)h(p)$ or, equivalently, $g(p)h(p) = 0 \bmod(p^n + 1)$, we say that the polynomials $g(p)$ and $h(p)$ are *orthogonal*. Furthermore, the polynomials $p^i g(p)$ and $p^j h(p)$ are also orthogonal for all i and j. However, the vectors corresponding to the polynomials $g(p)$ and $h(p)$ are orthogonal only if the ordered elements of one of these vectors are reversed. The same statement applies to the vectors corresponding to $p^i g(p)$ and $p^j h(p)$. In fact, if the parity polynomial $h(p)$ is used as a generator for the $(n, n-k)$ dual code, the set of code words obtained is simply the same code words generated by the reciprocal polynomial except that the code vectors are reversed. This implies that the generator matrix for the dual code obtained from the reciprocal polynomial $p^k h(p^{-1})$ can also be obtained indirectly from $h(p)$. Since the parity check matrix $\mathbf{H}$ for the (n, k) cyclic code is the generator matrix for the dual code, it follows that $\mathbf{H}$ can also be obtained from $h(p)$. The following example illustrates these relationships.

Example 5.2.7. The dual code to the $(7, 4)$ cyclic code generated by $g_1(p) = p^3 + p^2 + 1$ is the $(7, 3)$ dual code that is generated by the reciprocal polynomial $p^4 h_1(p^{-1}) = p^4 + p^2 + p + 1$. However, we may also use $h_1(p)$ to obtain the generator matrix for the dual code. Then, the matrix corresponding to the polynomials $p^i h_1(p)$, $i = 2, 1, 0$, is

$$
G_{h1} = \begin{bmatrix} 1 & 1 & 1 & 0 & 1 & 0 & 0 \\ 0 & 1 & 1 & 1 & 0 & 1 & 0 \\ 0 & 0 & 1 & 1 & 1 & 0 & 1 \end{bmatrix}
$$

The generator matrix for the $(7, 3)$ dual code, which is the parity check matrix for the $(7, 4)$ cyclic code, consists of the rows of $\mathbf{G}_{h1}$ taken in reverse order. Thus,

$$
H_1 = \begin{bmatrix} 0 & 0 & 1 & 0 & 1 & 1 & 1 \\ 0 & 1 & 0 & 1 & 1 & 1 & 0 \\ 1 & 0 & 1 & 1 & 1 & 0 & 0 \end{bmatrix}
$$

The reader may verify that $G_1 H_1' = 0$.

We note that the column vectors of H_1 consist of all seven binary vectors of length 3, except the all-zero vector. But this is just the description of the parity check matrix for a $(7, 4)$ Hamming code. Therefore the $(7, 4)$ cyclic code is equivalent to the $(7, 4)$ Hamming code discussed previously in Examples 5.2.1 and 5.2.2.

Encoders for cyclic codes. The encoding operations for generating a cyclic code may be performed by a linear feedback shift register based on the use of either the generator polynomial or the parity polynomial. First, let us consider the use of $g(p)$.

As indicated above, the generation of a systematic cyclic code involves three steps, namely multiplying the message polynomial $X(p)$ by p^{n-k}, dividing

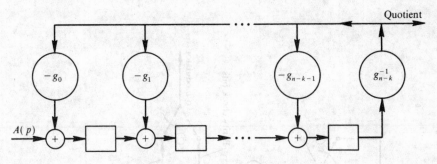

FIGURE 5.2.2
A feedback shift register for dividing the polynomial $A(p)$ by $g(p)$.

the product by $g(p)$, and, finally, adding the remainder to $p^{n-k}X(p)$. Of these three steps, only the division is nontrivial.

The division of the polynomial $A(p) = p^{n-k}X(p)$ of degree $n-1$ by the polynomial

$$g(p) = g_{n-k}p^{n-k} + g_{n-k-1}p^{n-k} + \cdots + g_1 p + g_0$$

may be accomplished by the $(n-k)$ stage feedback shift register illustrated in Fig. 5.2.2. Initially, the shift register contains all zeros. The coefficients of $A(p)$ are clocked into the shift register one (bit) coefficient at a time, beginning with the higher-order coefficients, i.e., with a_{n-1}, followed by a_{n-2}, and so on. After the kth shift the first nonzero output of the quotient is $q_1 = g_{n-k}a_n$. Subsequent outputs are generated as illustrated in Fig. 5.2.2. For each output coefficient in the quotient we must subtract the polynomial $g(p)$ multiplied by that coefficient, as in ordinary long division. This subtraction is performed by means of the feedback part of the shift register. Thus, the feedback shift register in Fig. 5.2.2 performs division of two polynomials.

In our case, $g_{n-k} = g_0 = 1$ and for binary codes the arithmetic operations are performed in modulo-2 arithmetic. Consequently, the subtraction operations reduce to modulo-2 addition. Furthermore, we are only interested in generating the parity check bits for each code word, since the code is systematic. Consequently, the encoder for the cyclic code takes the form illustrated in Fig. 5.2.3. The first k bits at the output of the encoder are simply the k information bits. These k bits are also clocked simultaneously into the shift register, since the switch 1 is in the closed position. Note that the polynomial multiplication of p^{n-k} with $X(p)$ is not performed explicitly. After the k information bits are all clocked into the encoder, the positions of the two switches are reversed. At this time the contents of the shift register are simply the $n-k$ parity check bits, which correspond to the coefficients of the remainder polynomial. These $(n-k)$ bits are clocked out one at a time and sent to the modulator.

FIGURE 5.2.3
Encoding of a cyclic code by use of the generator polynomial $g(p)$.

FIGURE 5.2.4
The encoder for the $(7,4)$ cyclic code with generator polynomial $g(p) = p^3 + p + 1$.

Example 5.2.8. The shift register for encoding the $(7,4)$ cyclic code with generator polynomial $g(p) = p^3 + p + 1$ is illustrated in Fig. 5.2.4. Suppose the input message bits are 0110. The contents of the shift register are as follows.

Input	Shift	Shift register contents		
	0	0	0	0
0	1	0	0	0
1	2	1	1	0
1	3	1	0	1
0	4	1	0	0

Hence, the three parity check bits are 100, which correspond to the code bits $c_5 = 0$, $c_6 = 0$, and $c_7 = 1$.

Instead of using the generator polynomial, we may implement the encoder for the cyclic code by making use of the parity polynomial

$$h(p) = p^k + h_{k-1}p^{k-1} + \cdots + h_1 p + 1$$

The encoder is shown in Fig. 5.2.5. Initially, the k information bits are shifted into the shift register and simultaneously fed to the modulator. After all k information bits are in the shift register, the switch is thrown into position 2 and the shift register is clocked $n - k$ times to generate the $n - k$ parity check bits as illustrated in Fig. 5.2.5.

Example 5.2.9. The parity polynomial for the $(7,4)$ cyclic code generated by $g(p) = p^3 + p + 1$ is $h(p) = p^4 + p^2 + p + 1$. The encoder for this code based on the parity polynomial is illustrated in Fig. 5.2.6. If the input to the encoder is the message bits 0110, the parity check bits are $c_5 = 0$, $c_6 = 0$, and $c_7 = 1$, as is easily verified.

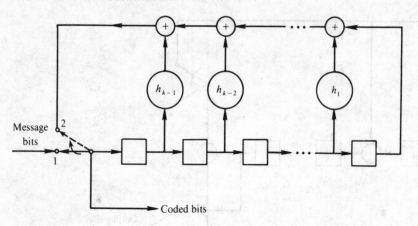

FIGURE 5.2.5
Encoder for an (n, k) cyclic code based on parity polynomial $h(p)$.

It should be noted that the encoder based on the generator polynomial is simpler when $n - k < k$ ($k > n/2$), i.e., for high rate codes ($R_c > 1/2$), while the encoder based on the parity polynomial is simpler when $k < n - k$ ($k < n/2$), which corresponds to low rate codes ($R_c < 1/2$).

Cyclic Hamming codes. The class of cyclic codes include the Hamming codes, which have a block length $n = 2^m - 1$ and $n - k = m$ parity check bits, where m is any positive integer. The cyclic Hamming codes are equivalent to the Hamming codes described in Sec. 5.2.2.

Cyclic (23, 12) Golay code. The linear (23, 12) Golay code previously described in Sec. 5.2.2 can be generated as a cyclic code by means of the generator polynomial

$$g(p) = p^{11} + p^9 + p^7 + p^6 + p^5 + p + 1 \tag{5.2.40}$$

The code words have a minimum distance $d_{\min} = 7$.

Maximum-length shift-register codes. Maximum-length shift-register codes are a class of cyclic codes with

$$(n, k) = (2^m - 1, m) \tag{5.2.41}$$

FIGURE 5.2.6
The encoder for the (7, 4) cyclic code based on the parity polynomial $h(p) = p^4 + p^2 + p + 1$.

FIGURE 5.2.7
Three-stage ($m = 3$) shift register with feedback.

where m is a positive integer. The code words are usually generated by means of an m-stage digital shift register with feedback, based on the parity polynomial. For each code word to be transmitted, the m information bits are loaded into the shift register and the switch is thrown from position ① to position ②. The contents of the shift register are shifted to the left one bit at a time for a total of $2^m - 1$ shifts. This operation generates a systematic code with the desired output length $n = 2^m - 1$. For example, the code words generated by the $m = 3$ stage shift register in Fig. 5.2.7 are listed in Table 5.2.4.

We note that, with the exception of the all-zero code word, all the code words generated by the shift register are different cyclic shifts of a single code word. The reason for this structure is easily seen from the state diagram of the shift register, which is illustrated in Fig. 5.2.8 for $m = 3$. When the shift register is loaded initially and shifted $2^m - 1$ times, it will cycle through all possible $2^m - 1$ states. Hence the shift register is back to its original state in $2^m - 1$ shifts. Consequently the output sequence is periodic with length $n = 2^m - 1$. Since there are $2^m - 1$ possible states, this length corresponds to the largest possible period. This explains why the $2^m - 1$ code words are different cyclic shifts of a single code word.

Maximal-length shift-register codes exist for any positive value of m. Table 5.2.5 lists the stages connected to the modulo-2 adder which result in a maximum-length shift register for $2 \leq m \leq 34$.

TABLE 5.2.4
Maximum-length shift-register code for $m = 3$

Information bits	Code words
0 0 0	0 0 0 0 0 0 0
0 0 1	0 0 1 1 1 0 1
0 1 0	0 1 0 0 1 1 1
0 1 1	0 1 1 1 0 1 0
1 0 0	1 0 0 1 1 1 0
1 0 1	1 0 1 0 0 1 1
1 1 0	1 1 0 1 0 0 1
1 1 1	1 1 1 0 1 0 0

FIGURE 5.2.8
The seven states for the $m = 3$ maximal length shift register.

TABLE 5.2.5
Shift-register connections for generating maximum-length sequences†

m	Stages connected to modulo-2 adder	m	Stages connected to modulo-2 adder	m	Stages connected to modulo-2 adder
2	1, 2	13	1, 10, 11, 13	24	1, 18, 23, 24
3	1, 3	14	1, 5, 9, 14	25	1, 23
4	1, 4	15	1, 15	26	1, 21, 25, 26
5	1, 4	16	1, 5, 14, 16	27	1, 23, 26, 27
6	1, 6	17	1, 15	28	1, 26
7	1, 7	18	1, 12	29	1, 28
8	1, 5, 6, 7	19	1, 15, 18, 19	30	1, 8, 29, 30
9	1, 6	20	1, 18	31	1, 29
10	1, 8	21	1, 20	32	1, 11, 31, 32
11	1, 10	22	1, 22	33	1, 21
12	1, 7, 9, 12	23	1, 19	34	1, 8, 33, 34

† Forney [1970].

Another characteristic of the code words in a maximum-length shift-register code is that each code word, with the exception of the all-zero code word, contains 2^{m-1} ones and $2^{m-1} - 1$ zeros. Hence all these code words have identical weights, namely, $w = 2^{m-1}$. Since the code is linear, this weight is also the minimum distance of the code, i.e.,

$$d_{min} = 2^{m-1}$$

Finally, we note that the (7, 3) maximum-length shift-register code shown in Table 5.2.4 is identical to the (7, 3) code given in Table 5.2.3, which is the dual of the (7, 4) Hamming code given in Table 5.2.2. This is not a coincidence. The maximum-length shift-register codes are the dual codes of the cyclic Hamming $(2^m - 1, 2^m - 1 - m)$ codes.

The shift register for generating the maximum-length code may also be used to generate a periodic binary sequence with period $n = 2^m - 1$. The binary periodic sequence exhibits a periodic autocorrelation $\phi(m)$ with values $\phi(m) = n$ for $m = 0, \pm n, \pm 2n, \ldots$, and $\phi(m) = -1$ for all other shifts as described in Sec. 8.2.3. This impulse-like autocorrelation implies that the power spectrum is nearly white and, hence, the sequence resembles white noise. As a consequence, maximum-length sequences are called pseudo-noise (PN) sequences and find use in the scrambling of data and in the generation of spread spectrum signals.

Bose-Chaudhuri-Hocquenghem (BCH) codes. BCH codes are a large class of cyclic codes that include both binary and nonbinary alphabets. Binary BCH codes may be constructed with parameters

$$n = 2^m - 1$$
$$n - k \le mt$$
$$d_{min} = 2t + 1$$

(5.2.42)

where m ($m \geq 3$) and t are arbitrary positive integers. Hence, this class of binary codes provides the communications system designer with a large selection of block lengths and code rates. Nonbinary BCH codes include the powerful Reed–Solomon codes that are described later.

The generator polynomials for BCH codes can be constructed from factors of ($p^{2^m-1} + 1$). Table 5.2.6 lists the coefficients of generator polynomials for BCH codes of block lengths $7 \leq n \leq 255$, corresponding to $3 \leq m \leq 8$. The coefficients are given in octal form, with the left-most digit corresponding to the highest degree term of the generator polynomial. Thus, the coefficients of the generator polynomial for the $(15, 5)$ code are 2467, which in binary form is $10, 100, 110, 111$. Consequently, the generator polynomial is $g(p) = p^{10} + p^8 + p^5 + p^4 + p^2 + p + 1$.

A more extensive list of generator polynomials for BCH codes is given by Peterson and Weldon (1972), who tabulate the polynomial factors of ($p^{2^m-1} + 1$) for $m \leq 34$.

5.2.4 Optimum Soft-Decision Decoding of Linear Block Codes

In this subsection we derive the performance of linear binary block codes on an AWGN channel when optimum (unquantized) soft-decision decoding is employed at the receiver. The bits of a code word may be transmitted by any one of the binary signaling methods described in Chap. 4. For our purposes we consider binary (or quaternary) coherent PSK, which is the most efficient method, and binary orthogonal FSK either with coherent detection or noncoherent detection.

Let $\mathcal{E}$ denote the transmitted signal energy per code word and let $\mathcal{E}_c$ denote the signal energy required to transmit a single element (bit) in the code word. Since there are n bits per code word, $\mathcal{E} = n\mathcal{E}_c$, and since each code word conveys k bits of information, the energy per information bit is

$$\mathcal{E}_b = \frac{\mathcal{E}}{k} = \frac{n}{k}\mathcal{E}_c = \mathcal{E}_c/R_c \qquad (5.2.43)$$

The code words are assumed to be equally likely *a priori* with prior probability $1/M$.

Suppose the bits of a code word are transmitted by binary PSK. Thus each code word results in one of M signaling waveforms. From Chap. 4 we know that the optimum receiver, in the sense of minimizing the average probability of a code word error, for the AWGN channel can be realized as a parallel bank of M filters matched to the M possible transmitted waveforms. The outputs of the M matched filters at the end of each signaling interval, which encompasses the transmission of n bits in the code word, are compared and the code word corresponding to the largest matched filter output is selected. Alternatively, M cross correlators can be employed. In either case, the receiver implementation can be simplified. That is, an equivalent optimum receiver can be realized by use of a single filter (or cross correlator) matched to the binary PSK waveform used to

TABLE 5.2.6
**Coefficients of generator polynomials (in octal form) for BCH codes
of lengths $7 \leq n \leq 255$†**

n	k	t	g(p)
7	4	1	13
15	11	1	23
	7	2	721
	5	3	2467
31	26	1	45
	21	2	3551
	16	3	107657
	11	5	5423325
	6	7	313365047
63	57	1	103
	51	2	12471
	45	3	1701317
	39	4	166623567
	36	5	1033500423
	30	6	157464165547
	24	7	17323260404441
	18	10	1363026512351725
	16	11	6331141367235453
	10	13	472622305527250155
	7	15	5231045543503271737
127	120	1	211
	113	2	41567
	106	3	11554743
	99	4	3447023271
	92	5	624730022327
	85	6	130704476322273
	78	7	26230002166130115
	71	9	6255010713253127753
	64	10	1206534025570773100045
	57	11	335265252505705053517721
	50	13	54446512523314012421501421
	43	14	17721772213651227521220574343
	36	15	31460746665220750447645745721735
	29	21	403114461367670603667530141176155
	22	23	123376070404722522435445626637647043
	15	27	22057042445604554770523013762217604353
	8	31	7047264052751030651476224271567733130217
255	247	1	435
	239	2	267543
	231	3	156720665
	223	4	75626641375
	215	5	23157564726421
	207	6	16176560567636227
	199	7	7633031270420722341
	191	8	2663470176115333714567
	187	9	52755313540001322236351
	179	10	22624710717340432416300455

TABLE 5.2.6 (*Continued*).

n	k	t	g(p)
	171	11	15416214212342356077061630637
	163	12	75004155100756025515747245 14601
	155	13	37575130054076650157225 06464677633
	147	14	1642130173537165525304165305441011711
	139	15	4614017320601755615707227302474535 67445
	131	18	2157133314715101512612502774421420241 65471
	123	19	1206140522420660037172103265161412262 72506267
	115	21	6052666557210024726363640460027635255 6313472737
	107	22	2220577232206625631241730023534742017 6574750154441
	99	23	1065666725347317422274141620157433225 2411076432303431
	91	25	6750265030327444172723631724732511075 550762720724344561
	87	26	1101367634147432364352316343071720462 06722545273311721317
	79	27	6670003563765750002027034420736617462 1015326711766541342355
	71	29	2402471052064432151555417211233116320 544425036255764322 1706035
	63	30	1075447505516354432531521735770700366 6111726455267613656702543301
	55	31	7315425203501100133015275306032054325 4143267550105570444260354 73617
	47	42	2533542017062646563033041377406233175 12333414544604500506602455 2543173
	45	43	1520205605523416113110134637642370156 36700244707623730332021570250 51541
	37	45	5136330255067007414177447245437530420 73570617432343234764435473740304 4003
	29	47	3025715536673071465527064012361377115 34224232420117411406025475741040356 5037
	21	55	1256215257060332656001773153607612103 22734140565307454252115312161446651 3473725
	13	59	4641732005052564544426573714250066004 33067744547656140317467721357026134 460500547
	9	63	1572602521747246320103104325535513461 41623672120440745451127661155477055 61677516057

† From paper by Stenbit (1964), © 1964 IEEE; reprinted with permission.

transmit each bit in the code word, followed by a decoder which forms the M decision variables corresponding to the M code words.

To be specific, let y_j, $j = 1, 2, \ldots, n$, represent the n sampled outputs of the matched filter for any particular code word. Since the signaling is binary coherent PSK, the output y_j may be expressed either as

$$y_j = 2\mathscr{E}_c + v_j \tag{5.2.44}$$

when the jth bit of a code word is a 1 or as

$$y_j = -2\mathscr{E}_c + v_j \tag{5.2.45}$$

when the jth bit is a 0. The variables $\{v_j\}$ represent additive white gaussian noise at the sampling instants. Each v_j has a zero mean and a variance of $2\mathscr{E}_c N_0$. From knowledge of the M possible transmitted code words and upon reception of the $\{y_j\}$, the optimum decoder forms the M decision variables

$$U_i = \sum_{j=1}^{n} (2c_{ij} - 1) y_j \qquad i = 1, 2, \ldots, M \tag{5.2.46}$$

where c_{ij} denotes the bit in the jth position of the ith code word. Thus, if $c_{ij} = 1$, the weighting factor $2c_{ij} - 1 = 1$, and if $c_{ij} = 0$, the weighting factor $2c_{ij} - 1 = -1$. In this manner, the weighting $2c_{ij} - 1$ aligns the signal components in $\{y_j\}$ such that the decision variable corresponding to the actual transmitted code word will have a mean value $2\mathscr{E}_c n$ while the other $M - 1$ decision variables will have smaller mean values.

Although the computations involved in forming the decision variables for soft-decision decoding according to (5.2.46) are relatively simple, it may still be impractical to compute (5.2.46) for all the possible code words when the number of code words is large, e.g., $M > 2^{10}$. In such a case it is still possible to implement soft-decision decoding using algorithms which employ techniques for discarding improbable code words without computing their entire decision variables as given by (5.2.46). Several different types of soft-decision decoding algorithms have been described in the technical literature. The interested reader is referred to the papers by Forney [1966], Weldon [1971], Chase [1972], Wainberg and Wolf [1973], Wolf [1978], and Matis and Modestino [1982].

In determining the probability of error for a linear block code, we note that when such a code is employed on a binary-input, output-symmetric channel such as the AWGN channel with optimum soft-decision decoding, the error probability for the transmission of the mth code word is the same for all m. Hence we assume for simplicity that the all-zero code word $\mathbf{C}_1$ is transmitted. For correct decoding of $\mathbf{C}_1$, the decision variable U_1 must exceed all the other $M - 1$ decision variables $U_2, U_3, \ldots, U_M$. All the decision variables are gaussian-distributed. The mean value of U_1 is $2\mathscr{E}_c n$ while the mean value of U_m, $m = 2, 3, \ldots, M$, is $2\mathscr{E}_c n(1 - 2w_m/n)$. The variance for each decision variable is $2\mathscr{E}_c N_0 n$. The derivation of the exact expression for the probability of correct decoding or, equivalently, the probability of a code word error is complicated by the correlations among the M decision variables. The cross-correlation coefficients between

$\mathbf{C}_1$ and the other $M - 1$ code words are

$$\rho_m = 1 - 2w_m/n \qquad m = 2, \ldots, M \qquad (5.2.47)$$

where w_m denotes the weight of the mth code word.

Instead of attempting to derive the exact error probability, we resort to a union bound. The probability that U_m exceeds U_1, given previously in (4.2.30), is

$$P_2(m) = \tfrac{1}{2} \operatorname{erfc} \left[\sqrt{\frac{\mathscr{E}}{2N_0} (1 - \rho_m)} \right] \qquad (5.2.48)$$

where $\mathscr{E} = k\mathscr{E}_b$ is the transmitted energy per waveform. Substitution for ρ_m from (5.2.47) and for $\mathscr{E}$ yields

$$P_2(m) = \tfrac{1}{2} \operatorname{erfc} \left(\sqrt{\frac{\mathscr{E}_b}{N_0} R_c w_m} \right)$$

$$= \tfrac{1}{2} \operatorname{erfc} \left(\sqrt{\gamma_b R_c w_m} \right) \qquad (5.2.49)$$

where γ_b is the SNR per bit and R_c is the code rate. Then the average probability of a code word error is bounded from above by the sum of the binary error events given by (5.2.49). Thus

$$P_M \leq \sum_{m=2}^{M} P_2(m)$$

$$\leq \sum_{m=2}^{M} \tfrac{1}{2} \operatorname{erfc} \left(\sqrt{\gamma_b R_c w_m} \right) \qquad (5.2.50)$$

The computation of the probability of error for soft-decision decoding according to (5.2.50) requires knowledge of the weight distribution of the code. Weight distributions of many codes are given in a number of texts on coding theory, e.g., Berlekamp [1968] and MacWilliams and Sloane [1977].

A somewhat looser bound is obtained by noting that

$$\operatorname{erfc} \left(\sqrt{\gamma_b R_c w_m} \right) \leq \operatorname{erfc} \left(\sqrt{\gamma_b R_c d_{\min}} \right) < \exp \left(-\gamma_b R_c d_{\min} \right) \qquad (5.2.51)$$

Consequently

$$P_M \leq \frac{M-1}{2} \operatorname{erfc} \left(\sqrt{\gamma_b R_c d_{\min}} \right) < \tfrac{1}{2} \exp \left(-\gamma_b R_c d_{\min} + k \ln 2 \right) \qquad (5.2.52)$$

This bound is particularly useful since it does not require knowledge of the weight distribution of the code. When the upper bound in (5.2.52) is compared with the performance of an uncoded binary PSK system, which is upper-bounded as $\tfrac{1}{2} \exp(-\gamma_b)$, we find that coding yields a gain of approximately $10 \log(R_c d_{\min} - k \ln 2/\gamma_b)$ dB. We may call this the *coding gain*. We note that its value depends on the code parameters and also on the SNR per bit γ_b.

The expression for the probability of error in Sec. 4.2.4 for equicorrelated waveforms gives us yet a third approximation to the error probabilities for coded waveforms. We know that the maximum cross-correlation coefficient between a pair of coded waveforms is

$$\rho_{\max} = 1 - \frac{2}{n} d_{\min} \qquad (5.2.53)$$

If we assume as a worst case that all the M code words have a cross-correlation coefficient equal to $\rho_{\max}$, then with $\rho_r = \rho_{\max}$ gives the code word error probability. Since some code words are separated by more than the minimum distance, the error probability evaluated for $\rho_r = \rho_{\max}$ is actually an upper bound. Thus

$$P_M \le 1 - \frac{1}{\sqrt{2\pi}} \int_{-\infty}^{\infty} e^{-v^2/2} \left(\frac{1}{\sqrt{2\pi}} \int_{-\infty}^{v+\sqrt{4\gamma_b R_c d_{\min}}} e^{-x^2/2} \, dx \right)^{M-1} dv \quad (5.2.54)$$

The bounds on the performance of linear block codes given above are in terms of the block error or code word error probability. The evaluation of the equivalent bit error probability P_b is much more complicated. In general, when a block error is made, some of the k information bits in the block will be correct and some will be in error. For orthogonal waveforms the conversion factor that multiplies P_M to yield P_b is $2^{k-1}/2^k - 1$. This factor is unity for $k = 1$ and approaches $\frac{1}{2}$ as k increases, which is equivalent to assuming that, on the average, half of the k bits will be in error when a block error occurs. The conversion factor for coded waveforms depends in a complicated way on the distance properties of the code, but is certainly no worse than assuming that, on the average, half of the k bits will be in error when a block error occurs. Consequently $P_b \le P_M/2$.

The bounds on performance given by (5.2.50), (5.2.52), and (5.2.54) also apply to the case in which a pair of bits of a code word are transmitted by quaternary PSK, since quaternary PSK may be viewed as being equivalent to two independent binary PSK waveforms transmitted in phase quadrature. Furthermore, the bounds in (5.2.52) and (5.2.54), which depend only on the minimum distance of the code, apply as well to nonlinear binary block codes.

If binary orthogonal FSK is used to transmit each bit of a code word on the AWGN channel, the optimum receiver can be realized by means of two matched filters, one matched to the frequency corresponding to a transmission of a 0, and the other filter matched to the frequency corresponding to a transmission of a 1, followed by a decoder which forms the M decision variables corresponding to the M possible code words. The detection at the receiver may be coherent or noncoherent. In either case, let y_{0j} and y_{1j} denote the input samples to the combiner. The decision variables formed by the decoder may be expressed as

$$U_i = \sum_{j=1}^{n} \left[c_{ij} y_{1j} + (1 - c_{ij}) y_{0j} \right] \qquad i = 1, 2, \ldots, M \qquad (5.2.55)$$

where c_{ij} represents the jth bit in the ith code word. The code word corresponding to the largest of the $\{U_i\}$ is selected as the transmitted code word.

If the detection of the binary FSK waveforms is coherent, the random variables $\{y_{0j}\}$ and $\{y_{1j}\}$ are gaussian and, hence, the decision variables $\{U_i\}$

are also gaussian. In this case bounds on the performance of the code are easily obtained. To be specific, suppose that the all-zero code word $\mathbf{C}_1$ is transmitted. Then

$$
\begin{aligned}
y_{0j} &= 2\mathcal{E}_c + v_{0j} & j = 1, 2, \ldots, n \\
y_{1j} &= v_{1j} & j = 1, 2, \ldots, n
\end{aligned}
\tag{5.2.56}
$$

where the $\{v_{ij}\}$, $i = 0, 1$, $j = 1, 2, \ldots, n$, are mutually statistically independent gaussian random variables with zero mean and a variance equal to $2\mathcal{E}_c N_0$. Consequently U_1 is gaussian with a mean of $2\mathcal{E} n$ and a variance of $2\mathcal{E}_c N_0 n$. On the other hand, the decision variable U_m, corresponding to the code word having weight w_m, is gaussian with a mean of $2\mathcal{E}_c n(1 - w_m/n)$ and a variance of $2\mathcal{E}_c N_0 n$. Since the decision variables are correlated, we again resort to a union bound. The correlation coefficients are given as

$$
\rho_m = 1 - w_m/n
\tag{5.2.57}
$$

Using the binary error probability given in (5.2.48) with $\rho_m = 1 - w_m/n$, we obtain the probability that U_m exceeds U_1 in the form

$$
P_2(m) = \tfrac{1}{2} \operatorname{erfc}\left(\sqrt{\frac{\gamma_b}{2} R_c w_m} \right)
\tag{5.2.58}
$$

Comparison of this result with that given in (5.2.49) for coherent PSK reveals that coherent PSK requires 3 dB less SNR to achieve the same performance. This is not surprising in view of the fact that uncoded PSK is 3 dB better than binary orthogonal FSK with coherent detection. Hence the advantage of PSK over FSK is maintained in the coded waveforms. We conclude, then, that the bounds given in (5.2.50), (5.2.52), and (5.2.54) apply to coded waveforms transmitted by binary orthogonal coherent FSK with γ_b replaced by $\gamma_b/2$.

If square-law detection of the binary orthogonal FSK signal is employed at the receiver, the performance is further degraded by the noncoherent combining loss. Suppose again that the all-zero code word is transmitted. Then the decision variables are given by (5.2.55) where the input variables to the decoder are now

$$
\begin{aligned}
y_{0j} &= |2\mathcal{E}_c + N_{0j}|^2 & j = 1, 2, \ldots, n \\
y_{1j} &= |N_{1j}|^2 & j = 1, 2, \ldots, n
\end{aligned}
\tag{5.2.59}
$$

where $\{N_{0j}\}$ and $\{N_{1j}\}$ represent complex-valued mutually statistically independent gaussian random variables with zero mean and a variance equal to $2\mathcal{E}_c N_0$. The decision variable U_1 is given as

$$
U_1 = \sum_{j=1}^{n} y_{0j}
\tag{5.2.60}
$$

while the decision variable corresponding to the code word having weight w_m is statistically equivalent to a decision variable of a code word in which $c_{mj} = 1$ for $1 \leq j \leq w_m$ and $c_{mj} = 0$ for $w_m + 1 \leq j \leq n$. Hence U_m may be expressed as

$$
U_m = \sum_{j=1}^{w_m} y_{1j} + \sum_{j=w_m+1}^{n} y_{0j}
\tag{5.2.61}
$$

The difference between U_1 and U_m is

$$U_1 - U_m = \sum_{j=1}^{w_m} (y_{0j} - y_{1j}) \tag{5.2.62}$$

and the probability of error is simply the probability that $U_1 - U_m < 0$. But this difference is a special case of the general quadratic form in complex-valued gaussian random variables described previously in Sec. 4.4. The expression for the probability of error in deciding between U_1 and U_m is

$$P_2(m) = \frac{1}{2^{2w_m - 1}} \exp\left(-\frac{\gamma_b}{2} R_c w_m\right) \sum_{i=0}^{w_m - 1} K_i \left(\frac{\gamma_b}{2} R_c w_m\right)^i \tag{5.2.63}$$

where, by definition,

$$K_i = \frac{1}{i!} \sum_{r=0}^{w_m - 1 - i} \binom{2w_m - 1}{r} \tag{5.2.64}$$

The union bound obtained by summing $P_2(m)$ over $2 \le m \le M$ provides us with an upper bound on the probability of a code word error.

As an alternative we may use the minimum distance instead of the weight distribution to obtain the looser upper bound

$$P_M \le \frac{M-1}{2^{2d_{\min} - 1}} \exp\left(-\frac{\gamma_b}{2} R_c d_{\min}\right) \sum_{i=0}^{d_{\min} - 1} K_i \left(\frac{\gamma_b}{2} R_c d_{\min}\right)^i \tag{5.2.65}$$

A measure of the noncoherent combining loss inherent in the square-law detection and combining of the n elementary binary FSK waveforms in a code word can be obtained from Fig. 4.4.1 where $d_{\min}$ is used in place of L. The loss obtained is relative to the case in which the n elementary binary FSK waveforms are first detected coherently and combined as in (5.2.55) and then the sums are square-law-detected or envelope-detected to yield the M decision variables. The binary error probability for the latter case is

$$P_2(m) = \tfrac{1}{2} \exp\left(-\frac{\gamma_b}{2} R_c w_m\right) \tag{5.2.66}$$

and, hence,

$$P_M \le \sum_{m=2}^{M} P_2(m)$$

If $d_{\min}$ is used instead of the weight distribution, the union bound for the code word error probability in the latter case is

$$P_M \le \frac{M-1}{2} \exp\left(-\frac{\gamma_b}{2} R_c d_{\min}\right) \tag{5.2.67}$$

The channel bandwidth required to transmit the coded waveforms can be determined as follows. If binary PSK is used to transmit each bit in a code word, the required bandwidth is approximately equal to the reciprocal of the time interval devoted to the transmission of each bit. For an information rate of R bits/s, the time available to transmit k information bits and $n - k$ redundant (parity) bits (n total bits) is $T = k/R$. Hence

$$W = \frac{1}{T/n} = \frac{n}{k/R} = \frac{R}{R_c} \tag{5.2.68}$$

Therefore the bandwidth expansion factor B_e for the coded waveforms is

$$B_e = \frac{W}{R}$$

$$= \frac{n}{k} = \frac{1}{R_c} \tag{5.2.69}$$

On the other hand, if binary FSK with noncoherent detection is employed for transmitting the bits in a code word, $W \approx 2n/T$ and, hence, the bandwidth expansion factor increases by approximately a factor of 2 relative to binary PSK. In any case, B_e increases inversely with the code rate or, equivalently, it increases linearly with the block size n.

We are now in a position to compare the performance characteristics and bandwidth requirements of coded signaling waveforms with orthogonal signaling waveforms. A comparison of the expression for P_M given in (4.2.53) for orthogonal waveforms and in (5.2.54) for coded waveforms with coherent PSK indicates that the coded waveforms result in a loss of at most $10 \log n/2d_{\min}$ dB relative to orthogonal waveforms having the same number of waveforms. On the other hand, if we compensate for the loss in SNR due to coding by increasing the number of code words so that coded transmission requires $M_c = 2^{k_c}$ waveforms and orthogonal signaling requires $M_o = 2^{k_o}$ waveforms, then [from the union bounds in (4.2.61) and (5.2.52)] the performance obtained with the two sets of signaling waveforms at high SNR is about equal if

$$k_o = 2R_c d_{\min} \tag{5.2.70}$$

Under this condition the bandwidth expansion factor for orthogonal signaling can be expressed as

$$B_{eo} = \frac{M_0}{2 \log_2 M_o} = \frac{2^{k_o}}{2k_o} = \frac{2^{2R_c d_{\min}}}{4R_c d_{\min}} \tag{5.2.71}$$

while for coded signaling waveforms we have $B_{ec} = 1/R_c$. The ratio of B_{eo} given in (5.2.71) to B_{ec}, which is

$$\frac{B_{eo}}{B_{ec}} = \frac{2^{2R_c d_{\min}}}{4d_{\min}} \tag{5.2.72}$$

provides a measure of the relative bandwidth between orthogonal signaling and signaling with coded coherent PSK waveforms.

For example, suppose we use a $(63, 33)$ binary cyclic code which has a minimum distance $d_{\min} = 12$. The bandwidth ratio for orthogonal signaling relative to this code, given by (5.2.72), is 127. This is indicative of the bandwidth efficiency obtained through coding relative to orthogonal signaling.

5.2.5 Hard-Decision Decoding

The bounds given in Sec. 5.2.4 on the performance of coded signaling waveforms on the AWGN channel are based on the premise that the samples from the matched filter or cross correlator are not quantized. Although this processing yields the best performance, the basic limitation is the computational burden of forming M decision variables and comparing these to obtain the largest. The amount of computation becomes excessive when the number M of code words is large.

To reduce the computational burden, the analog samples can be quantized and the decoding operations are then performed digitally. In this subsection we consider the extreme situation in which each sample corresponding to a single bit of a code word is quantized to two levels, zero and one. That is, a (hard) decision is made as to whether each transmitted bit in a code word is a 0 or a 1. The resulting discrete-time channel which consists of the modulator, the AWGN channel, and the demodulator constitutes a BSC with crossover probability p. If coherent PSK is employed in transmitting and receiving the bits in each code word, then

$$p = \tfrac{1}{2} \operatorname{erfc}\left(\sqrt{\frac{\mathscr{E}_c}{N_0}}\right)$$

$$= \tfrac{1}{2} \operatorname{erfc}\left(\sqrt{\gamma_b R_c}\right) \tag{5.2.73}$$

On the other hand, if FSK is used to transmit the bits in each code word,

$$p = \tfrac{1}{2} \operatorname{erfc}\left(\sqrt{\tfrac{1}{2}\gamma_b R_c}\right) \tag{5.2.74}$$

for coherent detection and

$$p = \tfrac{1}{2} \exp\left(-\tfrac{1}{2}\gamma_b R_c\right) \tag{5.2.75}$$

for noncoherent detection.

Minimum-distance (maximum-likelihood) decoding. The n bits from the demodulator corresponding to a received code word are passed to the decoder, which compares the received code word with the M possible transmitted code words and decides in favor of the code word that is closest in Hamming distance (number of bit positions in which two code words differ) to the received code word. This minimum distance decoding rule is optimum in the sense that it results in a minimum probability of a code word error for the binary symmetric channel.

A conceptually simple, albeit computationally inefficient, method for decoding is to first add (modulo-2) the received code word vector to all the M possible transmitted code words $\mathbf{C}_i$ to obtain the error vectors $\mathbf{e}_i$. Hence $\mathbf{e}_i$ represents the error event that must have occurred on the channel in order to transform the code word $\mathbf{C}_i$ into the particular received code word. The number of errors in transforming $\mathbf{C}_i$ into the received code word is just equal to the number of 1s in

$\mathbf{e}_i$. Thus, if we simply compute the weight of each of the M error vectors $\{\mathbf{e}_i\}$ and decide in favor of the code word that results in the smallest weight error vector, we have, in effect, a realization of the minimum distance decoding rule.

A more efficient method for hard-decision decoding makes use of the parity check matrix $\mathbf{H}$. To elaborate, suppose that $\mathbf{C}_m$ is the transmitted code word and $\mathbf{Y}$ is the received code word at the output of the demodulator. In general, $\mathbf{Y}$ may be expressed as

$$\mathbf{Y} = \mathbf{C}_m + \mathbf{e}$$

where $\mathbf{e}$ denotes an arbitrary binary error vector. The product $\mathbf{YH'}$ yields

$$\begin{aligned} \mathbf{YH'} &= (\mathbf{C}_m + \mathbf{e})\mathbf{H'} \\ &= \mathbf{C}_m\mathbf{H'} + \mathbf{e}\mathbf{H'} \\ &= \mathbf{e}\mathbf{H'} = \mathbf{S} \end{aligned} \tag{5.2.76}$$

where the $(n-k)$-dimensional vector $\mathbf{S}$ is called the *syndrome of the error pattern*. In other words, the vector $\mathbf{S}$ has components that are zero for all parity check equations that are satisfied and nonzero for all parity check equations that are not satisfied. Thus $\mathbf{S}$ contains the pattern of failures in the parity checks.

We emphasize that the syndrome $\mathbf{S}$ is a characteristic of the error pattern and not of the transmitted code word. Furthermore, we observe that there are 2^n possible error patterns and only 2^{n-k} syndromes. Consequently different error patterns result in the same syndrome.

Suppose we construct a decoding table in which we list all the 2^k possible valid code words in the first row, beginning with the all-zero code word in the first (left-most) column. This all-zero code word also represents the all-zero error pattern. We fill in the first column by listing first all $(n-1)$ error patterns $\{\mathbf{e}_i\}$ of weight 1. If $n < 2^{n-k}$, we may then list all double error patterns, then all triple error patterns, etc., until we have a total of 2^{n-k} entries in the first column. Thus, the number of rows that we can have is 2^{n-k}, which is equal to the number of syndromes. Next, we add each error pattern in the first column to the corresponding code words. Thus, we fill in the remainder of the $n \times (n-k)$ table as follows:

$$\begin{array}{ccccc} \mathbf{C}_1 & \mathbf{C}_2 & \mathbf{C}_3 & \cdots & \mathbf{C}_{2^k} \\ \mathbf{e}_2 & \mathbf{C}_2 + \mathbf{e}_2 & \mathbf{C}_3 + \mathbf{e}_2 & \cdots & \mathbf{C}_{2^k} + \mathbf{e}_2 \\ \mathbf{e}_3 & \mathbf{C}_2 + \mathbf{e}_3 & \mathbf{C}_3 + \mathbf{C}_3 + \mathbf{e}_3 & \cdots & \mathbf{C}_{2^k} + \mathbf{e}_3 \\ \vdots & \vdots & \vdots & & \vdots \\ \mathbf{e}_{2^{n-k}} & \mathbf{C}_2 + \mathbf{e}_{2^{n-k}} & \mathbf{C}_3 + \mathbf{e}_{2^{n-k}} & \cdots & \mathbf{C}_{2^k} + \mathbf{e}_{2^{n-k}} \end{array}$$

This table is called a *standard array*. Each row including the first consists of k possible received code words that would result from the corresponding error pattern in the first column. Each row is called a *coset* and the first (left-most) code word (or error pattern) is called a *coset leader*. Therefore, a coset consist of all the possible received code words resulting from a particular error pattern (coset leader).

TABLE 5.2.7
Standard array for the (5, 2) code

	Code Words		
0 0 0 0 0	0 1 0 1 1	1 0 1 0 1	1 1 1 1 0
0 0 0 0 1	0 1 0 1 0	1 0 1 0 0	1 1 1 1 1
0 0 0 1 0	0 1 0 0 1	1 0 1 1 1	1 1 1 0 0
0 0 1 0 0	0 1 1 1 1	1 0 0 0 1	1 1 0 1 0
0 1 0 0 0	0 0 0 1 1	1 1 1 0 1	1 0 1 1 0
1 0 0 0 0	1 1 0 1 1	0 0 1 0 1	0 1 1 1 0
1 1 0 0 0	1 0 0 1 1	0 1 1 0 1	0 0 1 1 0
1 0 0 1 0	1 1 0 0 1	0 0 1 1 1	0 1 1 0 0

Example 5.2.10. Let us construct the standard array for the $(5, 2)$, systematic code with generator matrix given by

$$G = \begin{bmatrix} 1 & 0 & 1 & 0 & 1 \\ 0 & 1 & 0 & 1 & 1 \end{bmatrix}$$

This code has a minimum distance $d_{min} = 3$. The standard array is given in Table 5.2.7. Note that in this code, the coset leaders consist of the all-zero error pattern, five error patterns of weight 1, and two error patterns of weight 2. Although many more double error patterns exist, there is only room for two to complete the table. These were selected such that their corresponding syndromes are distinct from those of the single error patterns.

Now, suppose that e_i is a coset leader and that C_m was the transmitted code word. Then, the error pattern e_i would result in the received code word

$$Y = C_m + e_i$$

The syndrome is

$$S = (C_m + e_i)H' = C_m H' + e_i H' = e_i H'$$

Clearly, all received code words in the same coset have the same syndrome, since the latter depends only on the error pattern. Furthermore, each coset has a different syndrome. Having established this characteristic of the standard array, we may simply construct a syndrome decoding table in which we list the 2^{n-k} syndromes and the corresponding 2^{n-k} coset leaders that represent the minimum weight error patterns. Then, given a received code vector Y, we compute the syndrome

$$S = YH'$$

For the computed S we find the corresponding (most likely) error vector, say $\hat{e}_m$. This error vector is added to Y to yield the decoded word

$$\hat{C}_m = Y \oplus \hat{e}_m$$

Example 5.2.11. Consider the $(5, 2)$ code with the standard array given in Table 5.2.7. The syndromes versus the most likely error patterns are given in Table 5.2.8.

TABLE 5.2.8
Syndrome table for the (5, 2) code

Syndrome	Error pattern
0 0 0	0 0 0 0 0
0 0 1	0 0 0 0 1
0 1 0	0 0 0 1 0
1 0 0	0 0 1 0 0
0 1 1	0 1 0 0 0
1 0 1	1 0 0 0 0
1 1 0	1 1 0 0 0
1 1 1	1 0 0 1 0

Now suppose the actual error vector on the channel is

$$\mathbf{e} = \begin{bmatrix} 1 & 0 & 1 & 0 & 0 \end{bmatrix}$$

The syndrome computed for the error is $\mathbf{S} = \begin{bmatrix} 0 & 0 & 1 \end{bmatrix}$. Hence, the error determined from the table is $\hat{\mathbf{e}} = \begin{bmatrix} 0 & 0 & 0 & 0 & 1 \end{bmatrix}$. When $\hat{\mathbf{e}}$ is added to $\mathbf{Y}$ the result is a decoding error. In other words, the (5, 2) code corrects all single errors and only two doubles, namely $\begin{bmatrix} 1 & 1 & 0 & 0 & 0 \end{bmatrix}$ and $\begin{bmatrix} 1 & 0 & 0 & 1 & 0 \end{bmatrix}$.

Syndrome decoding of cyclic codes. As described above, hard-decision decoding of a linear block code may be accomplished by first computing the syndrome $\mathbf{S} = \mathbf{YH}'$, then using a table lookup to find the most probable error pattern $\mathbf{e}$ corresponding to the computed syndrome $\mathbf{S}$, and, finally, adding the error pattern $\mathbf{e}$ to the received vector $\mathbf{Y}$ to obtain the most probable code word $\hat{C}_m$. When the code is cyclic, the syndrome computation may be performed by a shift register similar in form to that used for encoding.

To elaborate, let us consider a systematic cyclic code and let us represent the received code vector $\mathbf{Y}$ by the polynomial $Y(p)$. In general, $\mathbf{Y} = \mathbf{C} + \mathbf{e}$ where $\mathbf{C}$ is the transmitted code word and $\mathbf{e}$ is the error vector. Hence, we have

$$Y(p) = C(p) + e(p)$$
$$= X(p)g(p) + e(p) \tag{5.2.77}$$

Now, suppose we divide $Y(p)$ by the generator polynomial $g(p)$. This division will yield

$$\frac{Y(p)}{g(p)} = Q(p) + \frac{R(p)}{g(p)}$$

or, equivalently,

$$Y(p) = Q(p)g(p) + R(p) \tag{5.2.78}$$

The remainder $R(p)$ is a polynomial of degree less than or equal to $n - k - 1$. If we combine (5.2.77) with (5.2.78) we obtain

$$e(p) = [X(p) + Q(p)]g(p) + R(p) \tag{5.2.79}$$

Shift	Register contents
0	000
1	100
2	010
3	001
4	010
5	101
6	100
7	110

FIGURE 5.2.10
Syndrome computation for the $(7, 4)$ cyclic code with generator polynomial $g(p) = p^3 + p + 1$ and received vector $\mathbf{Y} = [1001101]$.

Error detection and error correction capability. It is clear from the discussion above that when the syndrome consists of all zeros, the received code word is one of the 2^k possible transmitted code words. Since the minimum separation between a pair of code words is d_{min}, it is possible for an error pattern of weight d_{min} to transform one of these 2^k code words in the code into another code word. When this happens we have an *undetected error*. On the other hand, if the actual number of errors is less than d_{min}, the syndrome will have a nonzero weight. When this occurs, we have detected the presence of one or more errors on the channel. Clearly, the (n, k) block code is capable of *detecting* $d_{min} - 1$ errors. Error detection may be used in conjunction with an automatic repeat-request (ARQ) scheme for retransmission of the code word.

The *error correction capability* of a code also depends on the minimum distance. However, the number of correctable error patterns is limited by the number of possible syndromes or coset leaders in the standard array. To determine the error correction capability of an (n, k) code, it is convenient to view the 2^k code words as points in n-dimensional space. If each code word is viewed as the center of a sphere of radius (Hamming distance) t, the largest value that t may have without intersection (or tangency) of any pair of the 2^k spheres is $t = [(d_{min} - 1)/2]$ where $[x]$ denote the largest integer contained in x. Within each sphere lie all the possible received code words of distance less than or equal to t from the valid code word. Consequently, any received code vector that falls within a sphere is decoded into the valid code word at the center of the sphere. This implies that an (n, k) code with minimum distance d_{min} is capable of correcting $t = [(d_{min} - 1)/2]$ errors. Figure 5.2.11 is a two-dimensional representation of the code words and the spheres.

As described above, a code may be used to detect $d_{min} - 1$ errors or to correct $t = [(d_{min} - 1)/2]$ errors. Clearly, to correct t errors implies that we have

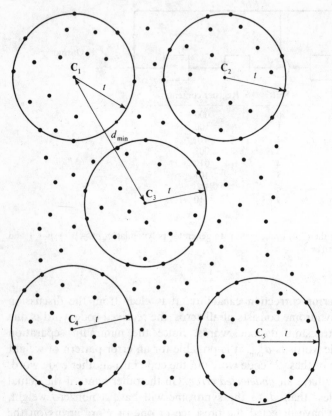

FIGURE 5.2.11
A representation of code words as centers of spheres of radius $t = [(d_{min} - 1)/2]$.

detected t errors. However, it is also possible to detect more than t errors if we compromise in the error correction capability of the code. For example, a code with $d_{min} = 7$ can correct $t = 3$ errors. If we wish to detect four errors, we can do so by reducing the radius of the sphere around each code word from 3 to 2. Thus, patterns with four errors are detectable but only patterns of two errors are correctable. In other words, when only two errors occur these are corrected, and when three or four errors occur the receiver may ask for a retransmission. If more than four errors occur they will go undetected if the code word falls within a sphere of radius 2. Similarly, for $d_{min} = 7$, five errors can be detected and one error corrected. In general, a code with minimum distance d_{min} can detect e_c errors and correct e_i errors, where

$$e_d + e_c \leq d_{min} - 1$$

Probability of error based on error correction. We conclude this section with the derivation of the probability of error for hard-decision decoding of linear binary block codes based on error correction only.

From the above discussion it is clear that the optimum decoder for a binary symmetric channel will decode correctly if (but not necessarily only if) the number of errors in a code word is less than half the minimum distance d_{min} of the code. That is, any number of errors up to

$$t = \left[\frac{d_{min} - 1}{2} \right]$$

are always correctable. Since the binary symmetric channel is memoryless, the bit errors occur independently. Hence the probability of m errors in a block of n bits is

$$P(m, n) = \binom{n}{m} p^m (1 - p)^{n-m} \qquad (5.2.81)$$

and, therefore, the probability of a code word error is upper-bounded by the expression

$$P_M \leq \sum_{m=t+1}^{n} P(m, n) \qquad (5.2.82)$$

Equality holds in (5.2.82) if the linear block code is a perfect code. In order to describe the basic characteristics of a perfect code, suppose we place a sphere of radius t around each of the possible transmitted code words. Each sphere around a code word contains the set of all code words of Hamming distance less than or equal to t from the code word. Now the number of code words in a sphere of radius $t = [(d_{min} - 1)/2]$ is

$$1 + \binom{n}{1} + \binom{n}{2} + \cdots + \binom{n}{t} = \sum_{i=0}^{t} \binom{n}{i}$$

Since there are $M = 2^k$ possible transmitted code words, there are 2^k nonoverlapping spheres each having a radius t. The total number of code words enclosed in the 2^k spheres cannot exceed the 2^n possible received code words. Thus a t error correcting code must satisfy the inequality

$$2^k \sum_{i=0}^{t} \binom{n}{i} \leq 2^n$$

or, equivalently,

$$2^{n-k} \geq \sum_{i=0}^{t} \binom{n}{i} \qquad (5.2.83)$$

A perfect code has the property that all spheres of Hamming distance $t = [(d_{min} - 1)/2]$ around the $M = 2^k$ possible transmitted code words are disjoint and every received code word falls in one of the spheres. Thus every received code word is, at most, a distance t from one of the possible transmitted code words and (5.2.83) holds with equality. For such a code, all error patterns of weight less than or equal to t are corrected by the optimum (minimum distance) decoder. On the other hand, any error pattern of weight $t + 1$ or greater cannot be corrected. Consequently the expression for the error probability given in

(5.2.82) holds with equality. The Golay $(23, 12)$ code, having $d_{min} = 7$ and $t = 3$, is a perfect code. The Hamming codes, which have the parameters $n = 2^{n-k} - 1$, $d_{min} = 3$, and $t = 1$, are also perfect codes. These two nontrivial codes and the trivial code consisting of two code words of odd length n and $d_{min} = n$ are the only perfect binary block codes. These codes are optimum on the BSC in the sense that they result in a minimum error probability among all codes having the same block length and the same number of information bits.

The optimality property defined above also holds for quasiperfect codes. A quasiperfect code is characterized by the property that all spheres of Hamming radius t around the M possible transmitted code words are disjoint and every received code word is at most at distance $t + 1$ from one of the possible transmitted code words. For such a code, all error patterns of weight less than or equal to t and some error patterns of weight $t + 1$ are correctable, but any error pattern of weight $t + 2$ or greater leads to incorrect decoding of the code word. Clearly (5.2.82) is an upper bound on the error probability and

$$P_M \geq \sum_{m=t+2}^{n} P(m, n) \qquad (5.2.84)$$

is a lower bound.

A more precise measure of the performance for quasiperfect codes can be obtained by making use of the inequality in (5.2.83). That is, the total number of code words outside the 2^k spheres of radius t is

$$N_{t+1} = 2^n - 2^k \sum_{i=0}^{t} \binom{n}{i}$$

If these code words are equally subdivided into 2^k sets and each set is associated with one of the 2^k spheres, then each sphere is enlarged by the addition of

$$\beta_{t+1} = 2^{n-k} - \sum_{i=0}^{t} \binom{n}{i} \qquad (5.2.85)$$

code words having distance $t + 1$ from the transmitted code word. Consequently, of the $\binom{n}{t+1}$ error patterns of distance $t + 1$ from each code word we can correct β_{t+1} error patterns. Thus the error probability for decoding the quasiperfect code may be expressed as

$$P_M = \sum_{m=t+2}^{n} P(m, n) + \left[\binom{n}{t+1} - \beta_{t+1}\right] p^{t+1}(1 - p)^{n-t-1} \qquad (5.2.86)$$

There are many known quasiperfect codes, although they do not exist for all choices of n and k. Since such codes are optimum for the binary symmetric channel, any (n, k) linear block code must have an error probability that is at least as large as (5.2.86). Consequently (5.2.86) is a lower bound on the probability of error for any (n, k) linear block code, where t is the largest integer such that $\beta_{t+1} \geq 0$.

Another pair of upper and lower bounds is obtained by considering two code words which differ by the minimum distance. First we note that P_M cannot be less than the probability of erroneously decoding the transmitted code word as its nearest neighbor which is at distance $d_{\min}$ from the transmitted code word. That is,

$$P_M \geq \sum_{m=[d_{\min}/2]+1}^{d_{\min}} \binom{d_{\min}}{m} p^m (1-p)^{d_{\min}-m} \qquad (5.2.87)$$

On the other hand, P_M cannot be greater than $M-1$ times the probability of erroneously decoding the transmitted code word as its nearest neighbor, which is at distance $d_{\min}$ from the transmitted code word. That is a union bound, which is expressed as

$$P_M \leq (M-1) \sum_{m=[d_{\min}/2]+1}^{d_{\min}} \binom{d_{\min}}{m} p^m (1-p)^{d_{\min}-m} \qquad (5.2.88)$$

When M is large, the lower bound in (5.2.87) and the upper bound in (5.2.88) are very loose.

A tight upper bound on P_M can be obtained by applying the Chernoff bound presented earlier in Sec. 1.1.6. We assume again that the all-zero code was transmitted. In comparing the received code word to the all-zero code word and to a code word of weight w_m, the probability of a decoding error, obtained from the Chernoff bound (Prob. 5.20), is upper-bounded by the expression

$$P_2(w_m) \leq [4p(1-p)]^{w_m/2} \qquad (5.2.89)$$

The union of these binary decisions yields the upper bound

$$P_M \leq \sum_{m=2}^{M} [4p(1-p)]^{w_m/2} \qquad (5.2.90)$$

A simpler version of (5.2.90) is obtained if we employ $d_{\min}$ in place of the weight distribution. That is,

$$P_M \leq (M-1)[4p(1-p)]^{d_{\min}/2} \qquad (5.2.91)$$

Of course (5.2.90) is a tighter upper bound that (5.2.91).

In Sec. 5.2.6, we compare the various bounds given above for a specific code, namely, the Golay (23,12) code. In addition, we compare the error rate performance of hard-decision decoding and soft-decision decoding.

5.2.6 Comparison of Performance between Hard-Decision and Soft-Decision Decoding

It is both interesting and instructive to compare the bounds on the error rate performance of linear block codes for soft-decision decoding and hard-decision decoding on an AWGN channel. For illustrative purposes we shall use the Golay (23,12) code, which has the relatively simple weight distribution given in Table 5.2.1. As stated previously, this code has a minimum distance $d_{\min} = 7$.

First we compute and compare the bounds on the error probability for hard-decision decoding. Since the Golay $(23, 12)$ code is a perfect code, the exact error probability for hard-decision decoding is

$$P_M = \sum_{m=4}^{23} \binom{23}{m} p^m (1 - p)^{23 - m}$$

$$= 1 - \sum_{m=0}^{3} \binom{23}{m} p^m (1 - p)^{23 - m} \qquad (5.2.92)$$

where p is the probability of a binary digit error for the binary symmetric channel. Binary (or four-phase) coherent PSK is assumed to be the modulation/demodulation technique for the transmission and reception of the binary digits contained in each code word. Thus the appropriate expression for p is given by (5.2.73). In addition to the exact error probability given by (5.2.92), we have the lower bound given by (5.2.87) and the three upper bounds given by (5.2.88), (5.2.90), and (5.2.91).

Numerical results obtained from these bounds are compared with the exact error probability in Fig. 5.2.12. We observe that the lower bound is very loose. At $P_M = 10^{-5}$, the lower bound is off by approximately 2 dB from the exact error probability. At $P_M = 10^{-2}$, the difference increases to approximately 4 dB. Of the three upper bounds, the one given by (5.2.88) is the tightest; it differs by less than 1 dB from the exact error probability at $P_M = 10^{-5}$. The Chernoff bound in (5.2.90) which employs the weight distribution is also relatively tight. Finally, the Chernoff bound which employs only the minimum distance of the code is the poorest of the three. At $P_M = 10^{-5}$, it differs from the exact error probability by approximately 2 dB. All three upper bounds are very loose for error rates above $P_M = 10^{-2}$.

It is also interesting to compare the performance between soft-decision decoding and hard-decision decoding. For this comparison, we use the upper bounds on the error probability for soft-decision decoding given by (5.2.50) and (5.2.52) and the exact error probability for hard-decision decoding given by (5.2.92). Figure 5.2.13 illustrates these performance characteristics. We observe that the two bounds for soft-decision decoding differ by approximately 0.5 dB at $P_M = 10^{-6}$ and by approximately 1 dB at $P_M = 10^{-2}$. We also observe that the difference in performance between hard-decision and soft-decision decoding is approximately 2 dB in the range $10^{-2} < P_M < 10^{-6}$. In the range $P_M > 10^{-2}$, the curve of the error probability for hard-decision decoding crosses the curves for the bounds. This behavior indicates that the bounds for soft-decision decoding are loose when $P_M > 10^{-2}$.

The 2-dB difference between hard-decision and soft-decision decoding is a characteristic that applies not only to the Golay code, but is a fundamental result that applies in general to coded digital communications over the AWGN channel. This result is derived below by computing the capacity of the AWGN channel with hard-decision and soft-decision decoding.

FIGURE 5.2.12
Comparison of bounds with exact error probability for hard-decision decoding of Golay (23, 12) code.

The channel capacity of the BSC in bits per code symbol, derived earlier in Sec. 2.4, is

$$C = 1 + p \log_2 p + (1 - p) \log_2 (1 - p) \qquad (5.2.93)$$

where the probability of a bit error for binary, coherent PSK on an AWGN channel is given by (5.2.73). Suppose we use (5.2.73) for p, let $C = R_c$ in (5.2.93), and then determine the value of γ_b that satisfies this equation. The result is shown in Fig. 5.2.14 as a graph of R_c versus γ_b. For example, suppose that we are interested in using a code with rate $R_c = \frac{1}{2}$. For this code rate we note that the

FIGURE 5.2.13
Comparison of soft-decision decoding with hard-decision decoding for the Golay (23, 12) code.

minimum SNR per bit required to achieve capacity with hard-decision decoding is approximately 1.8 dB.

What is the limit on the minimum SNR as the code rate approaches zero? For small values of R_c, the probability p can be approximated as

$$p \simeq \frac{1}{2} - \sqrt{\frac{\gamma_b R_c}{\pi}} \qquad (5.2.94)$$

When the expression for p is substituted in (5.2.93) and the logarithms in (5.2.93) are approximated by

$$\log_2 (1 + x) \simeq \left(x - \frac{x^2}{2}\right) \bigg/ \ln 2$$

FIGURE 5.2.14
Code rate as a function of the minimum SNR per bit for soft-decision and hard-decision decoding.

the channel capacity formula reduces to

$$C = \frac{2}{\pi \ln 2} \gamma_b R_c \qquad (5.2.95)$$

Now we set $C = R_c$. Thus, in the limit as R_c approaches zero, we obtain the result

$$\gamma_b = \frac{\pi}{2} \ln 2 \qquad (0.37 \, \text{dB}) \qquad (5.2.96)$$

The capacity of the binary-input AWGN channel with soft-decision decoding can be computed in a similar manner. The expression for the capacity in bits per code symbol, derived earlier in Sec. 2.4, is

$$C = \frac{1}{2} \sum_{k=0}^{1} \int_{-\infty}^{\infty} p(y|k) \log_2 \frac{p(y|k)}{p(y)} \, dy \qquad (5.2.97)$$

where $p(y|k)$, $k = 0, 1$, denote the probability density functions of the demodulator output conditioned on the transmitted bit being a 0 and a 1, respectively. For the AWGN channel we have

$$p(y|k) = \frac{1}{\sqrt{2\pi} \sigma} e^{-(y-m_k)^2/2\sigma^2} \qquad k = 0, 1 \qquad (5.2.98)$$

where $m_0 = -2\mathscr{E}_c$, $m_1 = 2\mathscr{E}_c$, $\sigma^2 = 2\mathscr{E}_c N_0$, and $\mathscr{E}_c = R_c \mathscr{E}_b$. The unconditional probability density $p(y)$ is simply one-half of the sum of $p(y|1)$ and $p(y|0)$. As R_c approaches zero, the expression in (5.2.97) for the channel capacity can be approximated as (see Prob. 5.21)

$$C \approx \frac{\gamma_b R_c}{\ln 2} \qquad (5.2.99)$$

Again, we set $C = R_c$. Thus, as $R_c \rightarrow 0$, the minimum SNR per bit to achieve capacity is

$$\gamma_b = \ln 2 \qquad (-1.6\,\text{dB}) \tag{5.2.100}$$

By using (5.2.98) in (5.2.97) and setting $C = R_c$, a numerical solution can be obtained for code rates in the range $0 \le R_c \le 1$. The result of this solution is also shown in Fig. 5.2.14.

From the above, we observe that in the limit as R_c approaches zero, the difference in SNR γ_b between hard-decision and soft-decision decoding is $\pi/2$, which is approximately 2 dB. On the other hand, as R_c increases toward unity, the difference in γ_b between these two decoding techniques decreases. For example, at $R_c = 0.8$ the difference is about 1.5 dB.

The curves in Fig. 5.2.14 provide more information than just the difference in performance between soft- and hard-decision decoding. These curves also specify the minimum SNR per bit that is required for a given code rate. For example, a code rate of $R_c = 0.8$ can provide arbitrarily small error probability at an SNR per bit of 2 dB, when soft-decision decoding is used. By comparison, an uncoded binary PSK requires 9.6 dB to achieve an error probability of 10^{-5}. Hence, a 7.6-dB gain is possible by employing a rate $R_c = 4/5$ code. Unfortunately, to achieve such a large coding gain usually implies the use of an extremely long block length code which leads to a very complex receiver. Nevertheless, the curves in Fig. 5.2.14 provide a benchmark for comparing the coding gains achieved by practically implementable codes with the ultimate limits for either soft- or hard-decision decoding.

Instead of comparing the difference between hard- and soft-decision decoding based on the channel capacity relations, we may perform similar comparisons based on the random coding rate parameters. In Sec. 5.1 we demonstrated that the ensemble average probability of error for randomly selected binary code words is upper-bounded as

$$\overline{P}_e < 2^{-n(R_0 - R_c)} \tag{5.2.101}$$

where $R_c = k/n$ is the code rate and the cutoff rate R_0 represents the upper bound on R_c such that $\overline{P}_e \rightarrow 0$ as $n \rightarrow \infty$. For unquantized (soft-decision) decoding R_0 is given as

$$R_0 = \log_2 \frac{2}{1 + e^{-\mathscr{E}_c/N_0}} \tag{5.2.102}$$

where $\mathscr{E}_c/N_0 = R_c\gamma_b$ is the SNR per dimension. This result was derived in Sec. 5.1.

On the other hand, if the output of the demodulator is quantized to Q levels prior to decoding, the Chernoff bound may be used to upper-bound the ensemble average binary error probability $P_2(\mathbf{s}_l, \mathbf{s}_m)$ defined in Sec. 5.1. The result of this derivation is the same upper bound on $\overline{P}_e$ given in (5.2.101) but

FIGURE 5.2.15
Comparison of R_0 (soft-decision decoding) with R_Q (hard-decision decoding) as a function of the SNR per dimension.

with R_0 replaced by R_Q, where

$$R_Q = \max_{\{p_j\}} \left\{ -\log_2 \sum_{i=0}^{Q-1} \left[\sum_{j=0}^{1} p_j \sqrt{P(i|j)} \right]^2 \right\} \qquad (5.2.103)$$

In (5.2.103), $\{p_j\}$ are the prior probabilities of the two signals at the input to the channel and $\{P(i|j)\}$ denote the transition probabilities of the channel. For example, in the case of a binary symmetric channel we have $p_1 = p_0 = \frac{1}{2}$, $P(0|0) = P(1|1) = 1 - p$, and $P(0|0) = p$. Hence,

$$R_Q = \log_2 \frac{2}{1 + \sqrt{4p(1-p)}} \qquad Q = 2 \qquad (5.2.104)$$

where

$$p = \frac{1}{2} \operatorname{erfc}\left(\sqrt{\gamma_b R_c}\right) \qquad (5.2.105)$$

A plot of R_Q versus $10 \log \mathscr{E}_c/N_0$ is illustrated in Fig. 5.2.15 for $Q = 2$ and $Q = \infty$ (soft-decision decoding). We note that the difference in decoder performance between unquantized soft-decision decoding and hard-decision decoding is approximately 2 dB. In fact, it is easily demonstrated again that as $\mathscr{E}_c/N_0 \to 0$, the loss in performance due to hard-decision decoding is $10 \log_{10} \pi/2 \approx 2$ dB, which is the same decibel difference that was obtained in our comparison of the channel capacity relations. We mention that about 1 dB of this loss can be recovered by quantizing the output of the demodulator to three levels instead of two (see Prob. 5.14). Additional improvements are possible by quantizing the output into more than three levels. This topic is discussed further in Sec. 5.2.9.

5.2.7 Bounds on Minimum Distance of Linear Block Codes

The expressions for the probability of error derived in this chapter for soft-decision and hard-decision decoding of linear binary block codes clearly indicate the importance that the minimum distance parameter plays in the performance of the code. If we consider soft-decision decoding, for example, the upper bound on the error probability given by (5.2.52) indicates that, for a given code rate $R_c = k/n$, the probability of error in an AWGN channel decreases exponentially with $d_{\min}$. When this bound is used in conjunction with the lower bound on $d_{\min}$ given below, we obtain an upper bound on P_M that can be achieved by many known codes. Similarly, we may use the upper bound given by (5.2.82) for the probability of error for hard-decision decoding in conjunction with the lower bound on $d_{\min}$ to obtain an upper bound on the error probability for linear binary block codes on the binary symmetric channel.

On the other hand, an upper bound on $d_{\min}$ can be used to determine a lower bound on the probability of error achieved by the best code. For example, suppose that hard-decision decoding is employed. In this case, we have the two lower bounds on P_M given by (5.2.86) and (5.2.87), with the former being the tighter of the two. When either one of these two bounds is used in conjunction with an upper bound on $d_{\min}$, the result is a lower bound on P_M for the best (n, k) code. Thus upper and lower bounds on $d_{\min}$ are important in assessing the capabilities of codes.

A simple upper bound on the minimum distance of an (n, k) binary or non-binary linear block code was given in (5.2.14) as $d_{\min} \leq n - k + 1$. It is convenient to normalize this expression by the block size n. That is,

$$\frac{d_{\min}}{n} \leq (1 - R_c) + \frac{1}{n} \qquad (5.2.106)$$

where R_c is the code rate. For large n, the factor $1/n$ can be neglected.

If a code has the largest possible distance, i.e., $d_{\min} = n - k + 1$, it is called a *maximum-distance-separable code*. Except for the trivial repetition-type codes, there are no binary maximum-separable codes. In fact, the upper bound in (5.2.106) is extremely loose for binary codes. On the other hand, nonbinary codes with $d_{\min} = n - k + 1$ do exist. For example, the Reed–Solomon codes, which are a subclass of BCH codes, are maximum-distance-separable.

In addition to the upper bound given above, there are several relatively tight bounds on the minimum distance of linear block codes. We shall briefly describe four important bounds, three of which are upper bounds and the other a lower bound. The derivations of these bounds are lengthy and are not of particular interest in our subsequent discussion. The interested reader may refer to Chap. 4 in the book by Peterson and Weldon [1972] for those derivations.

One upper bound on the minimum distance can be obtained from the inequality in (5.2.83). By taking the logarithm of both sides of (5.2.83) and dividing by n, we obtain

$$1 - R_c \geq \frac{1}{n} \log_2 \left[\sum_{i=0}^{t} \binom{n}{i} \right] \tag{5.2.107}$$

Since the error-correcting capability of the code, measured by t, is related to the minimum distance, the above relation is an upper bound on the minimum distance. It is called the *Hamming upper bound*.

The asymptotic form of (5.2.107) is obtained by letting $n \to \infty$. Now, for any n, let t_0 be the largest integer t for which (5.2.107) holds. Then, it can be shown [see Peterson and Weldon (1972)] that as $n \to \infty$, the ratio t/n for any (n,k) block code cannot exceed t_0/n, where t_0/n satisfies the equation

$$1 - R_c = H\left(\frac{t_0}{n}\right) \tag{5.2.108}$$

and $H(x)$ is the binary entropy function defined previously by (2.2.10).

The generalization of the Hamming bound to nonbinary codes is simply

$$1 - R_c \geq \frac{1}{n} \log_q \left[\sum_{i=0}^{t} \binom{n}{i}(q-1)^i \right] \tag{5.2.109}$$

Another upper bound, developed by Plotkin (1960), may be stated as follows. The number of check digits required to achieve a minimum distance $d_{\min}$ in an (n, k) linear block code satisfies the inequality

$$n - k \geq \frac{q d_{\min} - 1}{q - 1} - 1 - \log_q d_{\min} \tag{5.2.110}$$

where q is the alphabet size. When the code is binary, (5.2.110) may be expressed as

$$\frac{d_{\min}}{n}\left(1 - \frac{1}{2d_{\min}} \log_2 d_{\min}\right) \leq \frac{1}{2}\left(1 - R_c + \frac{2}{n}\right)$$

In the limit as $n \to \infty$ with $d_{\min}/n \leq 1/2$, (5.2.110) reduces to

$$\frac{d_{\min}}{n} \leq \frac{1}{2}(1 - R_c) \tag{5.2.111}$$

Finally, there is another tight upper bound on the minimum distance obtained by Elias [see Berlekamp (1968)]. It may be expressed in its asymptotic form as

$$\frac{d_{\min}}{n} \leq 2A(1 - A) \tag{5.2.112}$$

where the parameter A is related to the code rate through the equation

$$R_c = 1 + A \log_2 A + (1 - A) \log_2 (1 - A) \qquad 0 \leq A \leq \tfrac{1}{2} \tag{5.2.113}$$

Lower bounds on the minimum distance of (n, k) block codes also exist. In particular, binary block codes exist that have a normalized minimum distance that asymptotically satisfies the inequality

$$\frac{d_{min}}{n} \geq \alpha \qquad (5.2.114)$$

where α is related to the code rate through the equation

$$R_c = 1 - H(\alpha)$$
$$= 1 + \alpha \log_2 \alpha + (1 - \alpha) \log_2 \alpha \qquad 0 \leq \alpha \leq \tfrac{1}{2} \qquad (5.2.115)$$

This lower bound is a special case of a lower bound developed by Gilbert (1952) and Varsharmov (1957), that applies to nonbinary and binary block codes.

The asymptotic bounds given above are plotted in Fig. 5.2.16 for binary codes. Also plotted in the figure for purposes of comparison are curves on the minimum distance as a function of code rate of BCH codes of block lengths $n = 31$ and 63. We observe that for $n = 31$ and 63 the normalized minimum distance falls well above the Varsharmov–Gilbert lower bound. As the block length n increases, the efficiency of the BCH codes diminishes. For example, when $n = 1023$, the curve for the normalized minimum distance falls close to the Varsharmov–Gilbert bound. As n increases beyond $n = 1023$, the normalized minimum distance of the BCH codes continues to decrease and falls below the Varsharmov–Gilbert bound. That is, d_{min}/n approaches zero as n tends to infinity. Consequently the BCH codes, which are the most important class of cyclic codes, are not very efficient at large block lengths.

FIGURE 5.2.16
Upper and lower bounds on normalized minimum distance as a function of code rate.

5.2.8 Nonbinary Block Codes and Concatenated Block Codes

A nonbinary block code consists of a set of fixed-length code words in which the elements of the code words are selected from an alphabet of q symbols, denoted as $\{0, 1, 2, \ldots, q - 1\}$. Usually, $q = 2^k$ so that k information bits are mapped into one of the q symbols. The length of the nonbinary code word is denoted as N and the number of information symbols encoded into a block of N symbols is denoted as K. The minimum distance of the nonbinary code is denoted as $D_{\min}$. A systematic (N, K) block code consists of K information symbols and $N - K$ parity check symbols.

Among the various types of nonbinary linear block codes, the Reed–Solomon codes are some of the most important for practical applications. As indicated previously, they are a subset of the BCH codes, which in turn are a class of cyclic codes. These codes are described by the parameters

$$
\begin{aligned}
N &= q - 1 = 2^k - 1 \qquad K = 8 \;\Rightarrow\; N = 255 \\
K &= 1, 2, 3, \ldots, N - 1 \qquad\qquad\qquad (5.2.116) \\
D_{\min} &= N - K + 1
\end{aligned}
$$

$$
R_c = \frac{K}{N}
$$

Such a code is guaranteed to correct up to

$$
\begin{aligned}
t &= \left[\frac{D_{\min} - 1}{2} \right] \\
&= \left[\frac{N - K}{2} \right] \qquad\qquad (5.2.117)
\end{aligned}
$$

symbol errors. Of course these codes may be extended or shortened in the manner described previously for binary block codes.

The weight distribution $\{A_i\}$ of the class of Reed–Solomon codes is known. The coefficients in the weight enumerating polynomial are given as

$$
A_i = \binom{N}{i}(q - 1) \sum_{j=0}^{i-D} (-1)^j \binom{i-1}{j} q^{i-j-D} \qquad i \geq D_{\min} \quad (5.2.118)
$$

where $D \equiv D_{\min}$ and $q = 2^k$.

One reason for the importance of the Reed–Solomon codes is their good distance properties. A second reason for their importance is the existence of efficient hard-decision decoding algorithms which make it possible to implement relatively long codes in many practical applications where coding is desirable.

A nonbinary code is particularly matched to an M-ary modulation technique for transmitting the 2^k possible symbols. Specifically, M-ary orthogonal signaling, e.g., M-ary FSK, is frequently used. Each of the 2^k symbols in the q-ary alphabet is mapped to one of the $M = 2^k$ orthogonal signals. Thus the transmission of a code word is accomplished by transmitting N orthogonal signals, where each signal is selected from the set of $M = 2^k$ possible signals.

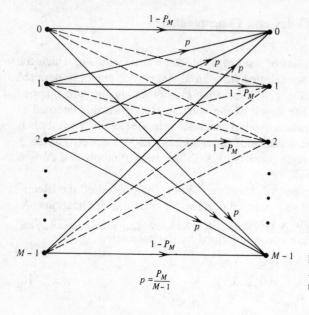

$$p = \frac{P_M}{M-1}$$

FIGURE 5.2.17
M-ary input, M-ary output, symmetric memoryless channel.

The optimum demodulator for such a signal corrupted by AWGN consists of M matched filters (or cross correlators) whose outputs are passed to the decoder, either in the form of soft decisions or in the form of hard decisions. If hard decisions are made by the demodulator, the symbol error probability P_M and the code parameters are sufficient to characterize the performance of the decoder. In fact, the modulator, the AWGN channel, and the demodulator form an equivalent discrete (M-ary) input, discrete (M-ary) output, symmetric memoryless channel characterized by the transition probabilities $P_c = 1 - P_M$ and $P_M/(M-1)$. This channel model, which is illustrated in Fig. 5.2.17, is a generalization of the BSC.

The performance of the hard-decision decoder may be characterized by the following upper bound on the code word error probability:

$$P_e \le \sum_{i=t+1}^{N} \binom{N}{i} P_M^i (1 - P_M)^{N-i} \qquad (5.2.119)$$

where t is the number of errors guaranteed to be corrected by the code.

When a code word error is made, the corresponding symbol error probability is

$$P_{es} = \frac{1}{N} \sum_{i=t+1}^{N} i \binom{N}{i} P_M^i (1 - P_M)^{N-i} \qquad (5.2.120)$$

Furthermore, if the symbols are converted to binary digits, the bit error probability corresponding to (5.2.120) is

$$P_{eb} = \frac{2^{k-1}}{2^k - 1} P_{es} \qquad (5.2.121)$$

Example 5.2.13. Let us evaluate the performance of an $N = 2^5 - 1 = 31$ Reed–Solomon code with $D_{min} = 3, 5, 9,$ and 17. The corresponding values of K are 29, 27, 23, and 15. The modulation is $M = 32$ orthogonal FSK with noncoherent detection at the receiver.

The probability of a symbol error is given by (4.3.32) and may be expressed as

$$P_M = \frac{1}{M} e^{-\gamma} \sum_{n=2}^{M} (-1)^n \binom{M}{n} e^{\gamma/n} \qquad (5.2.122)$$

where γ is the SNR per code symbol. By using (5.2.122) in (5.2.120) and combining the result with (5.2.121) we obtain the bit error probability. The results of these computations are plotted in Fig. 5.2.18. Note that the more powerful codes (large D_{min}) give poorer performance at low SNR per bit than the weaker codes. On the other hand, at high SNR the more powerful codes give better performance. Hence, there are crossovers among the various codes, as illustrated for example in Fig. 5.2.18 for the $t = 1$ and $t = 8$ codes. Crossovers also occur among the $t = 1, 2,$ and

FIGURE 5.2.18
Performance of several $n = 31$, t error correcting Reed–Solomon codes with 32-ary FSK modulation on an AWGN channel (noncoherent demodulation).

4 codes at smaller values of SNR per bit. Similarly, the curves for $t = 4$ and 8 and for $t = 8$ and 2 cross in the region of high SNR. This is the characteristic behavior for noncoherent detection of the coded waveforms.

If the demodulator does not make a hard decision on each symbol, but instead, passes the unquantized matched filter outputs to the decoder, soft-decision decoding can be performed. This decoding involves the formation of $q^K = 2^{kK}$ decision variables, where each decision variable corresponds to one of the q^K code words and consists of a sum of N matched filter outputs corresponding to the N code symbols. The matched filter outputs may be added coherently, or they may be envelope-detected and then added, or they may be square-law detected and then added. If coherent detection is used and the channel noise is AWGN, the computation of the probability of error is a straightforward extension of the binary case considered in Sec. 5.2.4. On the other hand, when envelope detection or square-law detection and noncoherent combining are used to form the decision variables, the computation of the decoder performance is considerably more complicated.

Concatenated block codes. A concatenated code consists of two separate codes which are combined to form a larger code. Usually one code is selected to be nonbinary and the other is binary. The two codes are concatenated as illustrated in Fig. 5.2.19. The nonbinary (N, K) code forms the outer code and the binary code forms the inner code. Code words are formed by subdividing a block of kK information bits into K groups, called *symbols*, where each symbol consists of k bits. The K k-bit symbols are encoded into N k-bit symbols by the outer encoder, as is usually done with a nonbinary code. The inner encoder takes each k-bit symbol and encodes it into a binary block code of length n. Thus we obtain a concatenated block code having a block length of Nn bits and containing kK information bits. That is, we have created an equivalent (Nn, Kk) long binary code. The bits in each code word are transmitted over the channel by means of PSK or, perhaps, by FSK.

We also indicate that the minimum distance of the concatenated code is $d_{\min}D_{\min}$, where $D_{\min}$ is the minimum distance of the outer code and $d_{\min}$ is the

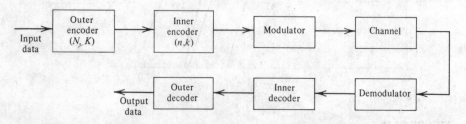

FIGURE 5.2.19
Block diagram of a communications system employing a concatenated code.

minimum distance of the inner code. Furthermore, the rate of the concatenated code is Kk/Nn, which is equal to the product of the two code rates.

A hard-decision decoder for a concatenated code is conveniently separated into an inner decoder and an outer decoder. The inner decoder takes the hard decisions on each group of n bits, corresponding to a code word of the inner code, and makes a decision on the k information bits based on maximum-likelihood (minimum-distance) decoding. These k bits represent one symbol of the outer code. When a block of N k-bit symbols are received from the inner decoder, the outer decoder makes a hard decision on the K k-bit symbols based on maximum-likelihood decoding.

Soft-decision decoding is also a possible alternative with a concatenated code. Usually the soft-decision decoding is performed on the inner code, if it is selected to have relatively few code words, i.e., if 2^k is not too large. The outer code is usually decoded by means of hard-decision decoding, especially if the block length is long and there are many code words. On the other hand, there may be a significant gain in performance when soft-decision decoding is used on both the outer and inner codes, to justify the additional decoding complexity. This is the case in digital communications over fading channels, as we will demonstrate in Chap. 7.

We conclude this subsection with the following example.

Example 5.2.14. Suppose that the $(7, 4)$ Hamming code described previously in Examples 5.2.1 and 5.2.2 is used as the inner code in a concatenated code in which the outer code is a Reed–Solomon code. Since $k = 4$, we select the length of the Reed–Solomon code to be $N = 2^4 - 1 = 15$. The number of information symbols K per outer code word may be selected over the range $1 \le K \le 14$ in order to achieve a desired code rate.

5.2.9 Communications System Design Based on the Cutoff Rate

In the foregoing discussion we characterized coding and modulation performance in terms of the error probability, which is certainly a meaningful criterion for system design. However, in many cases, the computation of the error probability is extremely difficult, especially if nonlinear operations such as signal quantization are performed in processing the signal at the receiver, or if the additive noise is nongaussian.

Instead of attempting to compute the exact probability of error for specific codes, we may use the ensemble average probability of error for randomly selected code words. The channel is assumed to have q input symbols $\{0, 1, \ldots, q - 1\}$, Q output symbols $\{0, 1, \ldots, Q - 1\}$, and is characterized by the transition probabilities $P(i|j)$, where $j = 0, 1, \ldots, q - 1$ and $i = 0, 1, \ldots, Q - 1$, with $Q \ge q$. The input symbols occur with probabilities $\{p_j\}$ and are assumed to be statistically independent. In addition, the noise on the channel is assumed to be statistically independent in time, so that there is no dependence among successive received symbols. Under these conditions, the ensemble average probability of error for randomly selected code words may be derived by applying the Chernoff bound [for reference, see Viterbi and Omura (1979)].

The general result that is obtained for the discrete memoryless channel is

$$\overline{P}_e < 2^{-n(R_Q - R/D)} \tag{5.2.123}$$

where n is the block length of the code, R is the information rate in bits per second, D is the number of dimensions per second, and R_Q is the cutoff rate for a quantizer with Q levels, defined as

$$R_Q = \max_{\{p_j\}} \left\{ -\log_2 \sum_{i=0}^{Q-1} \left[\sum_{j=0}^{q-1} p_j \sqrt{P(i|j)} \right]^2 \right\} \tag{5.2.124}$$

From the viewpoint of code design, the combination of modulator, waveform channel, and demodulator constitutes a discrete-time channel with q inputs and Q outputs. The transition probabilities $\{P(i|j)\}$ depend on the channel noise characteristics, the number of quantization levels, and the type of quantizer, e.g., uniform or nonuniform. For example, in the binary-input AWGN channel, the output of the correlator at the sampling instant may be expressed as

$$p(y|j) = \frac{1}{\sqrt{2\pi}\sigma} e^{-(y-m_j)^2/2\sigma^2} \qquad j = 0,1 \tag{5.2.125}$$

where $m_0 = -2\mathscr{E}_c$, $m_1 = 2\mathscr{E}_c$, and $\sigma^2 = 2\mathscr{E}_c N_0$. These two pdf's are shown in Fig. 5.2.20. Also illustrated in the figure is a quantization scheme which subdivides the real line into five regions. From such a subdivision we may compute the transition probabilities and optimally select the thresholds that subdivide the regions in a way that maximizes R_Q for any given Q. Thus,

$$P(i|j) = \int_{r_i} p(y|j) \, dy \tag{5.2.126}$$

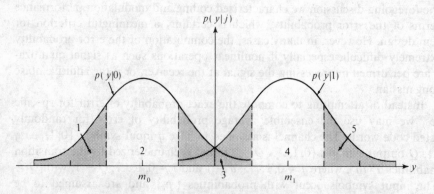

FIGURE 5.2.20
Example of quantization of the demodulator output into five levels.

where the integral of $p(y|j)$ is evaluated over the region r_i that corresponds to the transition probability $P(i|j)$.

The value of the rate R_Q in the limit as $Q \to \infty$ yields the cutoff rate for the unquantized decoder. It is relatively straightforward to show that as $Q \to \infty$, the first summation (sum from $i = 0$ to $Q - 1$) in (5.2.124) becomes an integral and the transition probabilities are replaced by the corresponding pdf's. Thus, when the channel consists of q discrete inputs and one continuous output y, which represents the unquantized output from a matched filter or a cross correlator in a system that employs either PSK or a multiamplitude (ASK) modulation, the cutoff rate is given by

$$R_0 = \max_{\{p_j\}} \left\{ -\log_2 \int_{-\infty}^{\infty} dy \left[\sum_{j=0}^{q-1} p_j \sqrt{p(y|j)} \right]^2 \right\} \qquad (5.2.127)$$

where p_j, $0 \le j \le q - 1$, is the probability of transmitting the jth symbol and $p(y|j)$ is the conditional probability density function of the output y from the matched filter or cross correlator when the jth signal is transmitted. This is the desired expression for unquantized (soft-decision) decoding.

We observe that when the input signal is binary PSK with $p_0 = p_1 = 1/2$ and the noise is additive, white, and gaussian, (5.2.127) reduces to the familiar result given previously in (5.2.102) and (5.1.20).

The general expressions in (5.2.127) and (5.2.124) allow us to compare the performance of various receiver implementations based on a different number of quantization levels.

Example 5.2.15. Let us compare the performance of a binary PSK input signal in an AWGN channel when the receiver quantizes the output to $Q = 2$, 4, and 8 levels. To simplify the optimization problem for the quantization of the signal at the output of the demodulator, the quantization levels are placed at $0, \pm\tau_h, \pm 2\tau_h, \ldots,$ $\pm (2^{b-1} - 1)\tau_h$, where τ_h is the *quantizer step-size parameter* which is to be selected, and b is the number of bits of the quantizer. A good strategy for the selection of τ_h is to choose it to minimize the SNR per bit γ_b that is required for operation at a code rate R_0. This implies that the step-size parameter must be optimized for every SNR, which in a practical implementation of the receiver means that the SNR must be measured. Fortunately, τ_h does not exhibit high sensitivity to small changes in SNR, so that it is possible to optimize τ_h for one SNR and obtain good performance for a wide range of SNR's about this nominal value by using a fixed τ_h.

Based on this approach, the expression for R_Q given by (5.2.124) was evaluated for $b = 1$ (hard-decision decoding), 2, and 3 bits, corresponding to $Q = 2$, 4, and 8 levels of quantization. The results are plotted in Fig. 5.2.21. The value of R_0 for unquantized soft-decision decoding, obtained by evaluating (5.2.127) is also shown in Fig. 5.2.21. We observe that two-bit quantization with $\tau_h = 1.0$ gains about 1.4 dB over hard-decision decoding and three-bit quantization with $\tau_h = 0.5$ yields an additional 0.4-dB improvement. Thus, with a three-bit quantizer we are within 0.2 dB of the unquantized soft-decision decoding limit. Clearly, there is little to be gained in increasing the precision any further.

FIGURE 5.2.21
Effect of quantization on the performance of a coded communications system operating at a rate $R = R_0$ or $R = R_Q$, with binary PSK modulation on an AWGN channel.

When a nonbinary code is used in conjunction with M-ary ($M = q$) signaling, the received signal at the output of the M matched filters may be represented by the vector $\mathbf{y} = (y_1, y_2, \ldots, y_M)$. The cutoff rate for this M-input, M-output (unquantized) channel is

$$R_0 = \max_{\{p_j\}} \left\{ -\log_2 \sum_{j=0}^{M-1} \sum_{i=0}^{M-1} p_j p_i \int_{-\infty}^{\infty} \sqrt{p(\mathbf{y}|j) p(\mathbf{y}|i)} \, d\mathbf{y} \right\} \quad (5.2.128)$$

where $p(\mathbf{y}|j)$ is the conditional probability density function of the output vector $\mathbf{y}$ from the demodulator given that the jth signal was transmitted. Note that (5.2.128) is similar in form to (5.2.127) except that we now have an M-fold integral to perform because there are M outputs from the demodulator.

Let us assume that the M signals are orthogonal so that the M-outputs conditioned on a particular input signal are statistically independent. As a consequence,

$$p(\mathbf{y}|j) = p_{s+n}(y_j) \prod_{\substack{i=0 \\ i \neq j}}^{M-1} p_n(y_i) \quad (5.2.129)$$

where $p_{s+n}(y_j)$ is the pdf of the matched filter output corresponding to the transmitted signal and $\{p_n(y_i)\}$ corresponds to the noise-only outputs from the other $M - 1$ matched filters. When (5.2.129) is incorporated into (5.2.128) we obtain

$$R_0 = \max_{\{p_j\}} \left\{ -\log_2 \left[\sum_{j=0}^{M-1} p_j^2 + \sum_{\substack{j=0 \\ l \neq j}}^{M-1} \sum_{l=0}^{M-1} p_l p_j \left(\int_{-\infty}^{\infty} dy \sqrt{p_{s+n}(y) p_n(y)} \right)^2 \right] \right\}$$

$$(5.2.130)$$

The maximization of R_0 over the set of input probabilities yields $p_j = 1/M$ for $1 \leq j \leq M$. Consequently, (5.2.130) reduces to

$$R_0 = \log_2 \frac{M}{1 + (M-1) \left(\int_{-\infty}^{\infty} \sqrt{p_{s+n}(y) p_n(y)} \, dy \right)^2}$$

$$= \log_2 M - \log_2 \left[1 + (M-1) \left(\int_{-\infty}^{\infty} \sqrt{p_{s+n}(y) p_n(y)} \, dy \right)^2 \right] \quad (5.2.131)$$

This is the desired result for the cutoff rate of an M-ary input, M-ary vector output unquantized channel.

For a phase coherent detection of the M-ary orthogonal signals the appropriate pdf's are

$$P_{s+n}(y) = \frac{1}{\sqrt{2\pi}\,\sigma} e^{-(y-m)^2/2\sigma^2}$$

$$p_n(y) = \frac{1}{\sqrt{2\pi}\,\sigma} e^{-y^2/2\sigma^2}$$

$$(5.2.132)$$

where $m = 2\mathscr{E}$ and $\sigma^2 = 2\mathscr{E}N_0$. Substitution of these relations into (5.2.131) and evaluating the integral yields

$$R_0 = \log_2 \frac{M}{1 + (M-1)e^{-\mathscr{E}/2N_0}}$$

$$= \log_2 \frac{M}{1 + (M-1)e^{-R_w \gamma_b/2}} \quad (5.2.133)$$

where $\mathscr{E}$ is the received energy per waveform, R_w is the information rate in bits per waveform, and $\gamma_b = \mathscr{E}_b/N_0$ is the SNR per bit.

We should emphasize that the rate parameter R_w has imbedded in it the code rate R_c. For example, if $M = 2$ and the code is binary, $R_w = R_c$. More generally, if the code is binary and $M = 2^\nu$, then each M-ary waveform conveys $R_w = \nu R_c$ bits of information. It is also interesting to note that if the code is binary and $M = 2$, (5.2.133) reduces to

$$R_0 = \log_2 \frac{2}{1 + e^{-R_c \gamma_b/2}} \qquad M = 2 \text{ orthogonal signals} \quad (5.2.134)$$

which is 3 dB worse than the cutoff rate for antipodal signals. If we set $R_w = R_0$

FIGURE 5.2.22
SNR per bit required to operate at a rate R_0 with M-ary orthogonal signals detected coherently in an AWGN channel.

in (5.2.133) and solve for γ_b we obtain

$$\gamma_b = \frac{2}{R_0} \ln \frac{M-1}{M \cdot 2^{-R_0} - 1} \tag{5.2.135}$$

The graphs for R_0 versus γ_b for several values of M are illustrated in Fig. 5.2.22. Note that the curve for any value of M saturates at $R_0 = \log_2 M$.

It is also interesting to consider the limiting form of (5.2.133) in the limit as $M \to \infty$. This yields

$$\lim_{M \to \infty} R_0 = \frac{\mathscr{E}}{2N_0 \ln 2} \text{ bits/waveform} \tag{5.2.136}$$

Since $\mathscr{E} = P_{av}T$, where T is the time interval per waveform, it follows that

$$\lim_{M \to \infty} \frac{R_0}{T} = \frac{P_{av}}{2N_0 \ln 2} = \frac{1}{2} C_\infty \tag{5.2.137}$$

Hence, in the limit as $M \to \infty$, the cutoff rate is one-half of the capacity for the infinite bandwidth AWGN channel. Alternatively, the substitution of $\mathscr{E} = R_0 \mathscr{E}_b$ in (5.2.136) yields $\gamma_b = 2 \ln 2$ (1.4 dB), which is the minimum SNR required to operate at R_0 (as $M \to \infty$). Hence, signaling at a rate R_0 requires 3 dB more power than the Shannon limit.

The value of R_0 given in (5.2.133) is based on the use of M-ary orthogonal signals, which are clearly suboptimal when M is small. If we attempt to maximize R_0 by selecting the best set of M waveforms, we should not be surprised to find that the simplex set of waveforms is optimum. In fact, R_0 for these optimum waveforms is simply given as

$$R_0 = \log_2 \frac{M}{1 + (M-1)e^{-M\mathscr{E}/2(M-1)N_0}} \tag{5.2.138}$$

If we compare this expression with (5.2.133), we observe that R_0 in (5.2.138) simply reflects the fact that the simplex set is more energy efficient by the factor $M/(M-1)$.

In the case of noncoherent detection, the probability density functions corresponding to signal-plus-noise and noise alone may be expressed as

$$p_{s+n}(y) = ye^{-(y^2+a^2)/2}I_0(ay) \qquad y \geq 0 \tag{5.2.139}$$
$$p_n(y) = ye^{-y^2/2} \qquad y \geq 0$$

where, by definition, $a = \sqrt{2\mathscr{E}/N_0}$. The computation of R_0 given by (5.2.131) does not yield a closed form solution. Instead, the integral in (5.2.131) must be evaluated numerically. Results for this case have been given by Jordan (1966) and Bucher (1980). For example, the (normalized) cutoff rate R_0 for M-ary orthogonal signals with noncoherent detection is shown in Fig. 5.2.23 for $M = 2, 4, 8, 16$. For purposes of comparison we also plot the cutoff rate for hard-decision

FIGURE 5.2.23
SNR per bit required to operate at a rate R_0 with M-ary orthogonal signals detected noncoherently in an AWGN channel.

decoding ($Q = 2$) of the M-ary symbols. In this case, we have

$$R_Q = \log_2 \frac{M}{\left[\sqrt{P_M} + \sqrt{(M - 1)(1 - P_M)}\right]^2} \qquad Q = 2 \qquad (5.2.140)$$

where P_M is the probability of a symbol error. For a relatively broad range of rates, the difference between soft- and hard-decision decoding is approximately 2 dB.

The most striking characteristic of the performance curves in Fig. 5.2.23 is that there is an optimum code rate for any given M. Unlike the case of coherent detection where the SNR per bit decreases monotonically with a decrease in code rate, the SNR per bit for noncoherent detection reaches a minimum in the vicinity of a normalized rate of 0.5, and increases for both high and low rates. The minimum is rather broad, so there is really a range of rates from 0.2 to 0.9 where the SNR per bit is within 1 dB of the minimum. This characteristic behavior in the performance with noncoherent detection is attributed to the nonlinear characteristic of the demodulator.

5.2.10 Interleaving of Coded Data for Channels with Burst Errors

Most of the well known codes that have been devised for increasing the reliability in the transmission of information are effective when the errors caused by the channel are statistically independent. This is the case for the AWGN channel. However, there are channels that exhibit bursty error characteristics. One example is the class of channels characterized by multipath and fading, which is described in detail in Chap. 7. Signal fading due to time-variant multipath propagation often causes the signal to fall below the noise level, thus resulting in a large number of errors. A second example is the class of magnetic recording channels (tape or disk) in which defects in the recording media result in clusters of errors. Such error clusters are not usually corrected by codes that are optimally designed for statistically independent errors.

Considerable work has been done on the construction of codes that are capable of correcting burst errors. Probably the best known burst error correcting codes are the subclass of cyclic codes called Fire codes, named after P. Fire (1959) who discovered them. Another class of cyclic codes for burst error correction were subsequently discovered by Burton (1969).

A *burst* of errors of length b is defined as a sequence of b-bit errors, the first and last of which are 1's. The *burst error correction capability* of a code is defined as one less than the length of the shortest uncorrectable burst. It is relatively easy to show that a systematic (n, k) code, which has $n - k$ parity check bits, can correct bursts of length $b \leq [(n - k)/2]$.

An effective method for dealing with burst error channels is to interleave the coded data in such a way that the bursty channel is transformed into a channel having independent errors. Thus, a code designed for independent channel errors (short bursts) is used.

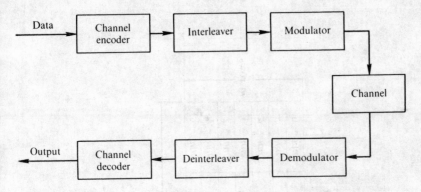

FIGURE 5.2.24
Block diagram of system employing interleaving for burst-error channel.

A block diagram of a system that employs interleaving is shown in Fig. 5.2.24. The encoded data are reordered by the interleaver and transmitted over the channel. At the receiver, after (either hard or soft decisions) demodulation, the deinterleaver puts the data in proper sequence and passes it to the decoder. As a result of the interleaving/deinterleaving, error bursts are spread out in time so that errors within a code word appear to be independent.

The interleaver can take one of two forms—a block structure or a convolutional structure. A block *interleaver* formats the encoded data in a rectangular array of m rows and n columns. Usually, each row of the array constitutes a code word of length n. An *interleaver of degree m* consists of m rows (m code words) as illustrated in Fig. 5.2.25. The bits are read out column-wise and transmitted over the channel. At the receiver, the deinterleaver stores the data in the same rectangular array format, but it is read out row-wise, one code word at a time. As a result of this reordering of the data during transmission, a burst of errors of length $l = mb$ is broken up into m bursts of length b. Thus, an (n, k) code that can handle burst errors of length $b \le [(n - k)/2]$ can be combined with an interleaver of degree m to create an interleaved (mn, mk) block code that can handle bursts of length mb.

A *convolutional interleaver* can be used in place of a block interleaver in much the same way. Convolutional interleavers are better matched for use with the class of convolutional codes that is described in the following section. Convolutional interleaver structures have been described in the literature by Ramsey (1970) and Forney (1971).

5.3 CONVOLUTIONAL CODES

A convolutional code is generated by passing the information sequence to be transmitted through a linear finite-state shift register. In general, the shift register consists of L (k-bit) stages and n linear algebraic function generators, as shown

FIGURE 5.2.25
A block interleaver for coded data.

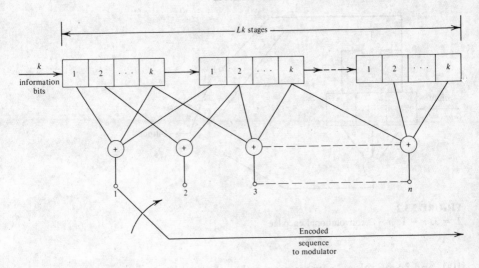

FIGURE 5.3.1
Convolutional encoder.

in Fig. 5.3.1. The input data to the encoder, which is assumed to be binary, is shifted into and along the shift register k bits at a time. The number of output bits for each k-bit input sequence is n bits. Consequently the code rate is defined as $R_c = k/n$, consistent with the definition of the code rate for a block code. The parameter L is called the *constraint length* of the convolutional code.†

One method for describing a convolutional code is to give its generator matrix, just as we did for block codes. In general, the generator matrix for a convolutional code is semi-infinite since the input sequence is semi-infinite in length. As an alternative to specifying the generator matrix, we will use a functionally equivalent representation in which we specify a set of n vectors, one vector for each of the n modulo-2 adders. Each vector has Lk dimensions and contains the connections of the encoder to that modulo-2 adder. A 1 in the ith position of the vector indicates that the corresponding stage in the shift register is connected to the modulo-2 adder and a 0 in a given position indicates that no connection exists between that stage and the modulo-2 adder.

To be specific, let us consider the binary convolutional encoder with constraint length $L = 3$, $k = 1$, and $n = 3$, which is shown in Fig. 5.3.2. Initially, the shift register is assumed to be in the all-zero state. Suppose the first input bit is a 1. Then the output sequence of 3 bits is 111. Suppose the second bit is a 0. The output sequence will then be 001. If the third bit is a 1, the output will be

---

†In many cases, the constraint length of the code is given in bits rather than k-bit bytes. Hence the shift register may be called a *K-stage shift register*, where $K = Lk$. Furthermore, K may not be a multiple of k, in general.

FIGURE 5.3.2
$L = 3$, $k = 1$, $n = 3$ convolutional encoder.

100, and so on. Now, suppose we number the function generators that generate each three-bit output sequence as 1, 2, and 3, from top to bottom, and similarly number each corresponding function generator. Then, since only the first stage is connected to the first function generator (no modulo-2 adder is needed), the generator is

$$\mathbf{g}_1 = [100]$$

The second function generator is connected to stages 1 and 3. Hence

$$\mathbf{g}_2 = [101]$$

Finally,

$$\mathbf{g}_3 = [111]$$

The generators for this code are more conveniently given in octal form as (4, 5, 7). We conclude that, when $k = 1$, we require n generators, each of dimension L to specify the encoder.

For a rate k/n binary convolutional code with $k > 1$ and constraint length L, the n generators are Lk-dimensional vectors, as stated above. The following example illustrates the case in which $k = 2$ and $n = 3$.

Example 5.3.1. Consider the rate 2/3 convolutional encoder illustrated in Fig. 5.3.3. In this encoder, 2 bits at a time are shifted into it and 3 output bits are generated. The generators are

$$\mathbf{g}_1 = [1011]$$
$$\mathbf{g}_2 = [1101]$$
$$\mathbf{g}_3 = [1010]$$

In octal form, these generators are (13, 15, 12).

There are three alternative methods that are often used to describe a convolutional code. These are the tree diagram, the trellis diagram, and the state

FIGURE 5.3.3
$L = 2$, $k = 2$, $n = 3$ convolutional encoder.

diagram. For example, the tree diagram for the convolutional encoder shown in Fig. 5.3.2 is illustrated in Fig. 5.3.4. Assuming that the encoder is in the all-zero state initially, the diagram shows that, if the first input bit is a 0, the output sequence is 000 and, if the first bit is a 1, the output sequence is 111. Now, if the first input bit is a 1 and the second bit is a 0, the second set of 3 output bits is 001. Continuing through the tree, we see that if the third bit is a 0 the output is 011, while if the third bit is a 1 the output is 100. Given that a particular sequence has taken us to a particular node in the tree, the branching rule is to follow the upper branch if the next input bit is a 0 and the lower branch if the bit is a 1. Thus we trace a particular path through the tree that is determined by the input sequence.

Close observation of the tree that is generated by the convolutional encoder shown in Fig. 5.3.2 reveals that the structure repeats itself after the third stage. This behavior is consistent with the fact that the constraint length $L = 3$. That is, the 3-bit output sequence at each stage is determined by the input bit and the 2 previous input bits, i.e., the 2 bits contained in the first two stages of the shift register. The bit in the last stage of the shift register is shifted out at the right and does not affect the output. Thus we may say that the 3-bit output sequence for each input bit is determined by the input bit and the four possible states of the shift register, denoted as $a = 00$, $b = 01$, $c = 10$, $d = 11$. If we label each node in the tree to correspond to the four possible states in the shift register, we find that at the third stage there are two nodes with the label a, two with the label b, two with the label c, and two with the label d. Now we observe that all branches emanating from two nodes having the same label (same state) are identical in the sense that they generate identical output sequences. This means that the two nodes having the same label can be merged. If we do this to the tree shown in

FIGURE 5.3.4
Tree diagram for rate $1/3$, $L = 3$ convolutional code.

Fig. 5.3.4, we obtain another diagram, which is more compact, called a *trellis*. For example, the trellis diagram for the convolutional encoder of Fig. 5.3.2 is shown in Fig. 5.3.5. In drawing this diagram, we use the convention that a solid line denotes the output generated by the input bit 0 and a dotted line denotes the output generated by the input bit 1. In the example being considered, we observe that, after the initial transient, the trellis contains four nodes at each stage, corresponding to the four states of the shift register, a, b, c, d. After the second stage, each node in the trellis has two incoming paths and two outgoing paths. Of the two outgoing paths, one corresponds to the input bit 0 and the other corresponds to the path followed if the input bit is a 1.

Since the output of the encoder is determined by the input and the state of the encoder, an even more compact diagram than the trellis is the state diagram. The state diagram is simply a graph of the possible states of the encoder and the possible transitions from one state to another. For example, the state diagram for the encoder shown in Fig. 5.3.2 is illustrated in Fig. 5.3.6. This diagram shows that the possible transitions are $a \xrightarrow{0} a$, $a \xrightarrow{1} c$, $b \xrightarrow{0} a$, $b \xrightarrow{1} c$, $c \xrightarrow{0} b$, $c \xrightarrow{1} d$, $d \xrightarrow{0} b$, and $d \xrightarrow{1} d$, where $\alpha \xrightarrow{1} \beta$ denotes the transition from state α to state β

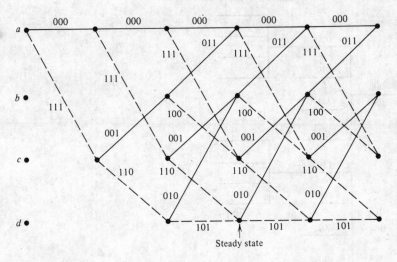

FIGURE 5.3.5
Trellis diagram for rate 1/3, $L = 3$ convolutional code.

FIGURE 5.3.6
State diagram for rate 1/3, $L = 3$ convolutional code.

when the input bit is a 1. The 3 bits shown next to each branch in the state diagram represent the output bits. A dotted line in the graph indicates that the input bit is a 1, while the solid line indicates that the input bit is a 0.

Example 5.3.2. Let us consider the $k = 2$, rate $\frac{2}{3}$ convolutional code described in Example 5.3.1 and shown in Fig. 5.3.3. The first 2 input bits may be 00, 01, 10, or 11. The corresponding output bits are 000, 010, 111, 101. When the next pair of input bits enter the encoder, the first pair is shifted to the second stage. The corresponding output bits depend on the pair of bits shifted into the second stage and the new pair of input bits. Hence the tree diagram for this code, shown in Fig. 5.3.7, has four branches per node corresponding to the four possible pairs of input

FIGURE 5.3.7
Tree diagram for $L = 2$, $k = 2$, $n = 3$ convolutional code.

symbols. Since the constraint length of the code is $L = 2$, the tree begins to repeat after the second stage. As illustrated in Fig. 5.3.7, all the branches emanating from nodes labeled a (state a) yield identical outputs. By merging the nodes having identical labels, we obtain the trellis, which is shown in Fig. 5.3.8. Finally, the state diagram for this code is shown in Fig. 5.3.9.

To generalize, we state that a rate k/n, constraint length L, convolutional code is characterized by 2^k branches emanating from each node of the tree diagram. The trellis and the state diagrams each have $2^{k(L-1)}$ possible states. There are 2^k branches entering each state and 2^k branches leaving each state (in the trellis and tree, this is true after the initial transient).

FIGURE 5.3.8
Trellis diagram for $L = 2$, $k = 2$, $n = 3$ convolutional code.

The three types of diagrams described above are also used to represent nonbinary convolutional codes. When the number of symbols in the code alphabet is $q = 2^k$, $k > 1$, the resulting nonbinary code may also be represented as an equivalent binary code. The following example considers a convolutional code of this type.

Example 5.3.3. Let us consider the convolutional code generated by the encoder shown in Fig. 5.3.10. This code may be described as a binary convolutional code with parameters $L = 2$, $k = 2$, $n = 4$, $R_c = \frac{1}{2}$, and having the generators

$$\mathbf{g}_1 = [1010] \qquad \mathbf{g}_2 = [0101] \qquad \mathbf{g}_3 = [1110] \qquad \mathbf{g}_4 = [1001]$$

Except for the difference in rate, this code is similar in form to the rate 2/3, $k = 2$ convolutional code considered in Example 5.3.1.

Alternatively, the code generated by the encoder in Fig. 5.3.10 may be described as a nonbinary ($q = 4$) code with one quaternary symbol as an input and two quaternary symbols as an output. In fact, if the output of the encoder is treated by the modulator and demodulator as q-ary ($q = 4$) symbols which are transmitted over the channel by means of some M-ary ($M = 4$) modulation technique, the code is appropriately viewed as nonbinary.

In any case, the tree, the trellis, and the state diagrams are independent of how we view the code. That is, this particular code is characterized by a tree with

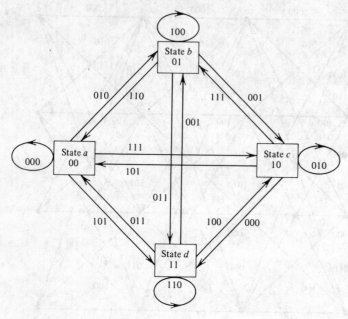

FIGURE 5.3.9
State diagram for $L = 2$, $k = 2$, $n = 3$ convolutional code.

FIGURE 5.3.10
$L = 2$, $k = 2$, $n = 4$ convolutional encoder.

four branches emanating from each node, or a trellis with four possible states and four branches entering and leaving each state or, equivalently, by a state diagram having the same parameters as the trellis.

5.3.1 The Transfer Function of a Convolutional Code

The distance properties and the error rate performance of a convolutional code can be obtained from its state diagram. Since a convolutional code is linear, the set of Hamming distances of the code sequences generated up to some stage in the tree, from the all-zero code sequence, is the same as the set of distances of the code sequences with respect to any other code sequence. Consequently we assume without loss of generality that the all-zero code sequence is the input to the encoder.

The state diagram shown in Fig. 5.3.6 will be used to demonstrate the method for obtaining the distance properties of a convolutional code. First we label the branches of the state diagram as either $D^0 = 1$, D^1, D^2, or D^3, where the exponent of D denotes the Hamming distance of the sequence of output bits corresponding to each branch from the sequence of output bits corresponding to the all-zero branch. The self-loop at node a can be eliminated since it contributes nothing to the distance properties of a code sequence relative to the all-zero code sequence. Furthermore node a is split into two nodes, one of which represents the input and the other of which represents the output of the state diagram. Figure 5.3.11 illustrates the resulting diagram. We use this diagram, which now consists of five nodes because node a was split into two, to write the four state equations

$$X_c = D^3 X_a + D X_b$$
$$X_b = D X_c + D X_d$$
$$X_d = D^2 X_c + D^2 X_d$$
$$X_e = D^2 X_b$$

(5.3.1)

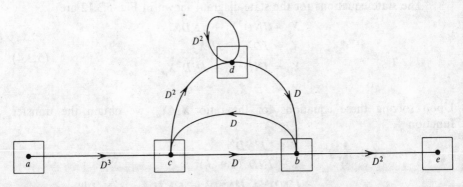

FIGURE 5.3.11
State diagram for rate $1/3$, $L = 3$ convolutional code.

The transfer function for the code is defined as $T(D) = X_e/X_a$. By solving the state equations given above, we obtain

$$T(D) = \frac{D^6}{1 - 2D^2}$$
$$= D^6 + 2D^8 + 4D^{10} + 8D^{12} + \cdots$$
$$= \sum_{d=6}^{\infty} a_d D^d \tag{5.3.2}$$

where, by definition,

$$a_d = \begin{cases} 2^{(d-6)/2} & d \text{ even} \\ 0 & d \text{ odd} \end{cases} \tag{5.3.3}$$

The transfer function for this code indicates that there is a single path of Hamming distance $d = 6$ from the all-zero path that merges with the all-zero path at a given node. From the state diagram shown in Fig. 5.3.6 or the trellis diagram shown in Fig. 5.3.5, it is observed that the $d = 6$ path is $a\ c\ b\ e$. There is no other path from node a to node e having a distance $d = 6$. The second term in (5.3.2) indicates that there are two paths from node a to node e having a distance $d = 8$. Again, from the state diagram or the trellis we observe that these paths are $a\ c\ d\ b\ e$ and $a\ c\ b\ c\ b\ e$. The third term in (5.3.2) indicates that there are four paths of distance $d = 10$, and so forth. Thus the transfer function gives us the distance properties of the convolutional code. The minimum distance of the code is called the *minimum free distance* and denoted as d_{free}. In our example, $d_{\text{free}} = 6$.

The transfer function can be used to provide more detailed information than just the distance of the various paths. Suppose we introduce a factor N into all branch transitions caused by the input bit 1. Thus, as each branch is traversed, the cumulative exponent on N increases by one only if that branch transition is due to an input bit 1. Furthermore, we introduce a factor of J into each branch of the state diagram so that the exponent of J will serve as a counting variable to indicate the number of branches in any given path from node a to node e. For the rate $1/3$ convolutional code in our example, the state diagram that incorporates the additional factors of J and N is shown in Fig. 5.3.12.

The state equations for the state diagram shown in Fig. 5.3.12 are

$$\begin{aligned} X_c &= JND^3 X_a + JND X_b \\ X_b &= JD X_c + JD X_d \\ X_d &= JND^2 X_c + JND^2 X_d \\ X_e &= JD^2 X_b \end{aligned} \tag{5.3.4}$$

Upon solving these equations for the ratio X_e/X_a, we obtain the transfer function

$$T(D, N, J) = \frac{J^3 N D^6}{1 - JND^2(1 + J)}$$
$$= J^3 N D^6 + J^4 N^2 D^8 + J^5 N^2 D^8 + J^5 N^3 D^{10}$$
$$+ 2J^6 N^3 D^{10} + J^7 N^3 D^{10} + \cdots \tag{5.3.5}$$

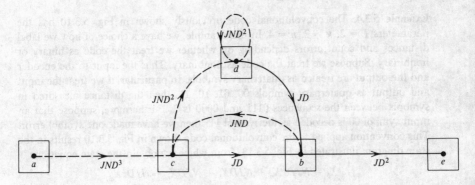

FIGURE 5.3.12
State diagram for rate $1/3$, $L = 3$ convolutional code.

This form for the transfer function gives the properties of all the paths in the convolutional code. That is, the first term in the expansion of $T(D, N, J)$ indicates that the distance $d = 6$ path is of length 3 and of the 3 information bits, one is a 1. The second and third terms in the expansion of $T(D, N, J)$ indicate that of the two $d = 8$ terms, one is of length 4 and the second has length 5. Two of the 4 information bits in the path having length 4 and two of the 5 information bits in the path having length 5 are 1s. Thus the exponent of the factor J indicates the length of the path that merges with the all-zero path for the first time, the exponent of the factor N indicates the number of 1s in the information sequence for that path, and the exponent of D indicates the distance of the sequence of encoded bits for that path from the all-zero sequence.

The factor J is particularly important if we are transmitting a sequence of finite duration, say m bits. In such a case the convolutional code is truncated after m nodes or m branches. This implies that the transfer function for the truncated code is obtained by truncating $T(D, N, J)$ at the term J^m. On the other hand, if we are transmitting an extremely long sequence, i.e., essentially an infinite-length sequence, we may wish to suppress the dependence of $T(D, N, J)$ on the parameter J. This is easily accomplished by setting $J = 1$. Hence, for the example given above, we have

$$T(D, N, 1) \equiv T(D, N) = \frac{ND^6}{1 - 2ND^2}$$

$$= ND^6 + 2N^2D^8 + 4N^3D^{10} + \cdots$$

$$= \sum_{d=6}^{\infty} a_d N^{(d-4)/2} D^d \qquad (5.3.6)$$

where the coefficients $\{a_d\}$ are defined by (5.3.3).

The procedure outlined above for determining the transfer function of a binary convolutional code is easily extended to nonbinary codes. In the following example, we determine the transfer function of the nonbinary convolutional code previously introduced in Example 5.3.3.

Example 5.3.4. The convolutional code previously shown in Fig. 5.3.10 has the parameters $L = 2$, $k = 2$, $n = 4$. In this example, we have a choice of how we label distances and count errors depending on whether we treat the code as binary or nonbinary. Suppose we treat the code as nonbinary. Thus the input to the encoder and the output are treated as quaternary symbols. In particular, if we treat the input and output as quaternary symbols 00, 01, 10, 11, then the distance measured in symbols between the sequences 0111 and 0000 is 2. Furthermore, suppose that an input symbol 00 is decoded as the symbol 11; then we have made one symbol error. This convention applied to the convolutional code shown in Fig. 5.3.10 results in the state diagram illustrated in Fig. 5.3.13, from which we obtain the state equations

$$X_b = NJD^2X_a + NJDX_b + NJDX_c + NJD^2X_d$$
$$X_c = NJD^2X_a + NJD^2X_b + NJDX_c + NJDX_d$$
$$X_d = NJD^2X_a + NJDX_b + NJD^2X_c + NJDX_d \tag{5.3.7}$$
$$X_e = JD^2(X_b + X_c + X_d)$$

Solution of these equations leads to the transfer function

$$T(D, N, J) = \frac{3NJ^2D^4}{1 - 2NJD - NJD^2} \tag{5.3.8}$$

This expression for the transfer function is particularly appropriate when the quaternary symbols at the output of the encoder are mapped into a corresponding set of quaternary waveforms $s_m(t)$, $m = 1, 2, 3, 4$, e.g., four orthogonal waveforms. Thus there is a one-to-one correspondence between code symbols and signal waveforms.

Alternatively, for example, the output of the encoder may be transmitted as a sequence of binary digits by means of binary PSK. In such a case, it is appropriate to measure distance in terms of bits. When this convention is employed, the state diagram is labeled as shown in Fig. 5.3.14. Solution of the state equations obtained from this state diagram yields a transfer function that is different from the one given in (5.3.8).

Some convolutional codes exhibit a characteristic behavior which is called *catastrophic error propagation*. When a code that has this characteristic is used on a binary symmetric channel, it is possible for a finite number of channel errors to cause an infinite number of decoding errors. Such a code can be identified from its state diagram. It will contain a zero distance path (a path with multiplier $D^0 = 1$) from some nonzero state back to the same state. This means that one can loop around this zero distance path an infinite number of times without increasing the distance relative to the all-zero path. But, if this self-loop corresponds to the transmission of a 1, the decoder will make an infinite number of errors. Since such codes are easily recognized, they are easily avoided in practice.

5.3.2 Optimum Decoding of Convolutional Codes—The Viterbi Algorithm

In the decoding of a block code for a memoryless channel, we computed the distances (Hamming distance for hard-decision decoding and Euclidean distance for soft-decision decoding) between the received code word and the 2^k possible

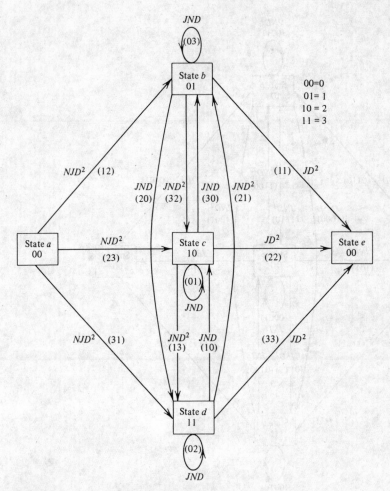

FIGURE 5.3.13
State diagram for $L = 2$, $k = 2$, rate $1/2$ nonbinary code.

transmitted code words. Then we selected the code word that was closest in distance to the received code word. This decision rule, which requires the computation of 2^k distances, or *metrics*, as they are called in this section, is optimum in the sense that it results in a minimum probability of error for the binary symmetric channel with $p < \frac{1}{2}$ and the additive white gaussian noise channel.

Unlike a block code which has a fixed length n, a convolutional code has no well-defined block size. Of course, a convolutional code can be truncated periodically at some stage, thus forcing it to have a fixed length. When this is done, a number of 0 bits are appended to the last information bit for the purpose of clearing the shift register of all information bits. Since the 0s appended carry no information, the efficiency of the code is reduced below the code rate R_c. In order

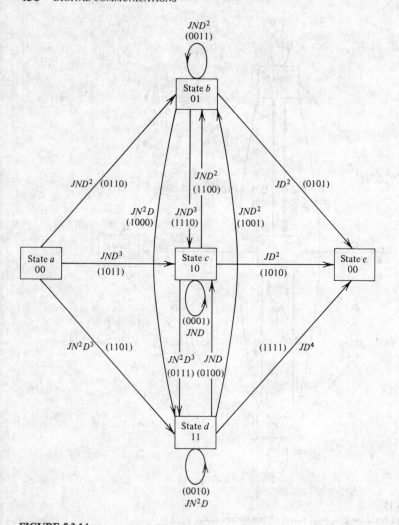

FIGURE 5.3.14
State diagram for $L = 2$, $k = 2$, rate $1/2$ convolutional code with output treated as a binary sequence.

to maintain the code rate as close to R_c as possible, when a convolutional code is truncated periodically, the period is selected to be several times the constraint length. Hence the resulting truncated convolutional code is equivalent to a long block code. Even if the convolutional code is truncated periodically in this manner, it is unnecessary to compute metrics for all possible branches in the tree in order to achieve optimum decoding. That this is true can be shown by referring to the trellis shown in Fig. 5.3.5 for the convolutional code shown in Fig. 5.3.2.

Consider the two paths in the trellis that begin at the initial state a and remerge at state a after three state transitions (three branches), corresponding to the two information sequences 000 and 100 and the transmitted sequences 000 000 000 and 111 001 011, respectively. We denote the transmitted bits by $\{c_{jm}, j = 1, 2, 3; m = 1, 2, 3\}$ where the index j indicates the jth branch and the index m indicates the mth bit in that branch. Correspondingly, we define $\{y_{jm}, j = 1, 2, 3; m = 1, 2, 3\}$ as the output of the demodulator. For example, if the demodulator performs hard-decision decoding, its output for each transmitted bit is either 0 or 1. On the other hand, if soft-decision decoding is employed and the coded sequence is transmitted by binary coherent PSK, the demodulator output is

$$y_{jm} = 2\mathcal{E}_c(2c_{jm} - 1) + v_{jm} \tag{5.3.9}$$

where v_{jm} represents the additive noise and $\mathcal{E}_c$ is the transmitted signal energy for each code bit.

A metric is defined for the jth branch of the rth path through the trellis as the logarithm of the joint probability of the sequence $\{y_{jm}, m = 1, 2, 3\}$ conditioned on the transmitted sequence $\{c_{jm}^{(r)}, m = 1, 2, 3\}$ for the rth path. That is,

$$\mu_j^{(r)} = \log P\left(\mathbf{Y}_j | \mathbf{C}_j^{(r)}\right) \qquad j = 1, 2, 3, \ldots \tag{5.3.10}$$

Furthermore, a metric for the rth path consisting of B branches through the trellis is defined as

$$U^{(r)} = \sum_{j=1}^{B} \mu_j^{(r)} \tag{5.3.11}$$

The criterion for deciding between two paths through the trellis is to select the one having the larger metric. This rule maximizes the probability of a correct decision or, equivalently, it minimizes the probability of error for the sequence of information bits. For example, suppose that hard-decision decoding is performed by the demodulator which yields the received sequence $\{101\ 000\ 100\}$. Let $r = 0$ denote the three-branch all-zero path and let $r = 1$ denote the second three-branch path that begins in the initial state a and remerges with the all-zero path at state a after three transitions. The metrics for these two paths are

$$\begin{aligned} U^{(0)} &= 6\log(1 - p) + 3\log p \\ U^{(1)} &= 4\log(1 - p) + 5\log p \end{aligned} \tag{5.3.12}$$

where p is the probability of a bit error. Assuming that $p < \frac{1}{2}$, we find that the metric $U^{(0)}$ is larger than the metric $U^{(1)}$. This result is consistent with the observation that the all-zero path is at Hamming distance $d = 3$ from the received sequence while the $r = 1$ path is at Hamming distance $d = 5$ from the received path. Thus the Hamming distance is an equivalent metric for hard-decision decoding.

Similarly, suppose that soft-decision decoding is employed and the channel adds white gaussian noise to the signal. Then the demodulator output is de-

458 DIGITAL COMMUNICATIONS

scribed statistically by the probability density function

$$p\left(y_{jm}|c_{jm}^{(r)}\right) = \frac{1}{\sqrt{2\pi}\,\sigma}e^{-[y_{jm}-2\mathscr{E}_c(2c_{jm}^{(r)}-1)]^2/2\sigma^2} \tag{5.3.13}$$

where $\sigma^2 = 2\mathscr{E}_c N_0$ is the variance of the additive gaussian noise. If we neglect the terms that are common to all branch metrics, the branch metric for the jth branch of the rth path may be expressed as

$$\mu_j^{(r)} = \sum_{m=1}^{n} y_{jm}\left(2c_{jm}^{(r)} - 1\right) \tag{5.3.14}$$

where, in our example, $n = 3$. Thus the metrics for the two paths under consideration are

$$U^{(0)} = \sum_{j=1}^{3}\sum_{m=1}^{3} y_{jm}\left(2c_{jm}^{(0)} - 1\right)$$

$$U^{(1)} = \sum_{j=1}^{3}\sum_{m=1}^{3} y_{jm}\left(2c_{jm}^{(1)} - 1\right) \tag{5.3.15}$$

Having defined the branch metrics and path metrics computed by the decoder, we now consider the question of optimum decoding of the convolutionally encoded information sequence. We consider the two paths described above, which merge at state a after three transitions. We note that any particular path through the trellis that stems from this node will add identical terms to the path metrics $U^{(0)}$ and $U^{(1)}$. Consequently, if $U^{(0)} > U^{(1)}$ at the merged node a after three transitions, $U^{(0)}$ will continue to be larger than $U^{(1)}$ for any path that stems from node a. This means that the path corresponding to $U^{(1)}$ can be discarded from further consideration. The path corresponding to the metric $U^{(0)}$ is called the *survivor*. Similarly, one of the two paths that merge at state b can be eliminated on the basis of the two corresponding metrics. This procedure is repeated at state c and state d. As a result, after the first three transitions, there are four surviving paths, one terminating at each state, and a corresponding metric for each survivor.

With the reception of the signal elements $\{y_{4m}, m = 1,2,3\}$, corresponding to the transmission of the fourth information bit, we form the four pairs of metrics, one pair for the two paths that merge at each state. For example, at the fourth stage of the trellis, one of the two paths that merge at state a stems from state a while the second stems from state b. Consequently the metric for the first path that merges at state a is equal to the metric from the previous stage plus an increment to that metric which is proportional to the logarithm of the probability of receiving the sequence $\{y_{4m}, m = 1,2,3\}$ conditioned on the transmitted sequence 000. The metric for the second path that merges at state a, which stems from state b, is equal to the metric for the survivor at state b plus an increment to that metric which is proportional to the probability of receiving the sequence $\{y_{4m}, m = 1,2,3\}$ conditioned on the transmitted sequence 011. The two metrics for the paths that merge at state a are then compared, and the path with the smaller metric is discarded. This procedure is repeated again at the other three nodes. Thus there are four paths that survive at each stage of the trellis and there are four metrics, one for each of the surviving paths. The decoding algorithm we have just described was developed to Viterbi (1967). It is optimum in the sense that it minimizes the probability of error in decoding the entire sequence of information bits.

In general, when a binary convolutional code with $k = 1$ and constraint length L is decoded by means of the Viterbi algorithm, there are 2^{L-1} states. Hence there are 2^{L-1} surviving paths at each stage and 2^{L-1} metrics, one for each surviving path. Furthermore, a binary convolutional code in which k bits at a time are shifted into the encoder that consists of L (k-bit) shift-register stages generates a trellis that has $2^{k(L-1)}$ states. Consequently the decoding of such a code by means of the Viterbi algorithm requires keeping track of $2^{k(L-1)}$ surviving paths and $2^{k(L-1)}$ metrics. At each stage of the trellis there are 2^k paths that merge at each node. Since each path that converges at a common node requires the computation of a metric, there are 2^k metrics computed for each node. Of the 2^k paths that merge at each node, only one survives, and this is the most-probable (minimum-distance) path. Thus the number of computations in decoding performed at each stage increases exponentially with k and L. The exponential increase in computational burden limits the use of the Viterbi algorithm to relatively small values L and k.

The decoding delay in decoding a long information sequence that has been convolutionally encoded is usually too long for most practical applications. Moreover, the memory required to store the entire length of surviving sequences is large and expensive. A solution to this problem is to modify the Viterbi algorithm in a way which results in a fixed decoding delay without significantly affecting the optimal performance of the algorithm. The modification is to retain at any given time t only the most recent δ decoded information bits (symbols) in each surviving sequence. As each new information bit (symbol) is received, a final decision is made on the bit (symbol) received δ branches back in the trellis, by comparing the metrics in the surviving sequences and deciding in favor of the bit in the sequence having the largest metric. If δ is chosen sufficiently large, all surviving sequences will contain the identical decoded bit (symbol) δ branches back in time. That is, with high probability, all surviving sequences at time t stem from the same node at $t - \delta$. It has been found experimentally (computer simulation) that a delay $\delta \geq 5L$ results in a negligible degradation in the performance relative to the optimum Viterbi algorithm.

5.3.3 Probability of Error for Soft-Decision Decoding

In Sec. 5.3.2 we described the Viterbi algorithm, which is the optimum decoding algorithm for a convolutionally encoded sequence transmitted over a memoryless channel. The topic of this subsection is the error rate performance of the Viterbi algorithm on an additive white gaussian noise channel with soft-decision decoding.

In deriving the probability of error for convolutional codes, the linearity property for this class of codes is employed to simplify the derivation. That is, we assume that the all-zero sequence is transmitted and we determine the probability of error in deciding in favor of another sequence. The coded binary digits for the jth branch of the convolutional code, denoted as $\{c_{jm}, m = 1, 2, \ldots, n\}$ and defined in Sec. 5.3.2, are assumed to be transmitted by binary PSK (or four-phase PSK) and detected coherently at the demodulator. The output of the demodulator, which is the input to the Viterbi decoder, is the sequence $\{y_{jm}, m = 1, 2, \ldots, n; j = 1, 2, \ldots\}$ where y_{jm} is defined in (5.3.9).

The Viterbi soft-decision decoder forms the branch metrics defined by (5.3.14) and from these computes the path metrics

$$U^{(r)} = \sum_{j=1}^{B} \mu_j^{(r)} = \sum_{j=1}^{B} \sum_{m=1}^{n} y_{jm}\left(2c_{jm}^{(r)} - 1\right) \tag{5.3.16}$$

where r denotes any one of the competing paths at each node and B is the number of branches (information symbols) in a path. For example, the all-zero path, denoted as $r = 0$, has a path metric

$$U^{(0)} = \sum_{j=1}^{B} \sum_{m=1}^{n} \left(-2\mathscr{E} + \nu_{jm}\right)(-1)$$

$$= 2\mathscr{E}_c Bn + \sum_{j=1}^{B} \sum_{m=1}^{n} \nu_{jm} \tag{5.3.17}$$

Since the convolutional code does not necessarily have a fixed length, we derive its performance from the probability of error for sequences that merge with the all-zero sequence for the first time at a given node in the trellis. In particular, we define the first-event error probability as the probability that another path which merges with the all-zero path at node B has a metric that exceeds the metric of the all-zero path for the first time. Suppose the incorrect path, call it $r = 1$, that merges with the all-zero path differs from the all-zero path in d bits, i.e., there are d 1s in the path $r = 1$ and the rest are 0s. The probability of error in the pairwise comparison of the metrics $U^{(0)}$ and $U^{(1)}$ is

$$P_2(d) = P(U^{(1)} \geq U^{(0)}) = P(U^{(1)} - U^{(0)} \geq 0)$$

$$P_2(d) = P\left[2\sum_{j=1}^{B} \sum_{m=1}^{n} y_{jm}\left(c_{jm}^{(1)} - c_{jm}^{(0)}\right) \geq 0\right] \tag{5.3.18}$$

Since the coded bits in the two paths are identical except in the d positions, (5.3.18) can be written in the simpler form

$$P_2(d) = P\left(\sum_{l=1}^{d} y_l' > 0\right) \tag{5.3.19}$$

where the index l runs over the set of d bits in which the two paths differ and the set $\{y_l'\}$ represents the input to the decoder for these d bits.

The $\{y_l'\}$ are independent and identically distributed gaussian random variables with mean $-2\mathscr{E}_c$, and variance $2\mathscr{E}_c N_0$. Consequently the probability of error in the pairwise comparison of these two paths that differ in d bits is

$$P_2(d) = \tfrac{1}{2}\operatorname{erfc}\left(\sqrt{\frac{\mathscr{E}_c}{N_0}d}\right)$$

$$= \tfrac{1}{2}\operatorname{erfc}\left(\sqrt{\gamma_b R_c d}\right) \qquad (5.3.20)$$

where $\gamma_b = \mathscr{E}_b/N_0$ is the received SNR per bit and R_c is the code rate.

Although we have derived the first-event error probability for a path of distance d from the all-zero path, there are many possible paths with different distances that merge with the all-zero path at a given node B. In fact, the transfer function $T(D)$ provides a complete description of all the possible paths that merge with the all-zero path at node B and their distances. Thus we can sum the error probability in (5.3.20) over all possible path distances. Upon performing this summation, we obtain an upper bound on the first-event error probability in the form

$$P_e \le \sum_{d=d_{\text{free}}}^{\infty} a_d P_2(d)$$

$$\le \frac{1}{2}\sum_{d=d_{\text{free}}}^{\infty} a_d \operatorname{erfc}\left(\sqrt{\gamma_b R_c d}\right) \qquad (5.3.21)$$

where a_d denotes the number of paths of distance d from the all-zero path which merge with the all-zero path for the first time.

There are two reasons why (5.3.21) is an upper bound on the first-event error probability. One is that the events which result in the error probabilities $\{P_2(d)\}$ are not disjoint. This can be seen from observation of the trellis. Second, by summing over all possible $d \ge d_{\text{free}}$ we have implicitly assumed that the convolutional code has infinite length. If the code is truncated periodically after B nodes, the upper bound in (5.3.21) can be improved by summing the error events for $d_{\text{free}} \le d \le B$. This refinement has some merit in determining the performance of short convolutional codes, but the effect on performance is negligible when B is large.

The upper bound in (5.3.21) can be expressed in a slightly different form if the complementary error function is upper-bounded by an exponential. That is,

$$\operatorname{erfc}\left(\sqrt{\gamma_b R_c d}\right) \le e^{-\gamma_b R_c d} = D^d\big|_{D=e^{-\gamma_b R_c}} \qquad (5.3.22)$$

If we use (5.3.22) in (5.3.21), the upper bound on the first-event error probability can be expressed as

$$P_e < \tfrac{1}{2}T(D)\big|_{D=e^{-\gamma_b R_c}} \qquad (5.3.23)$$

Although the first-event error probability provides a measure of the performance of a convolutional code, a more useful measure of performance is the bit error probability. This probability can be upper-bounded by the procedure used in bounding the first-event error probability. Specifically, we know that when an incorrect path is selected, the information bits in which the selected path differs from the correct path will be decoded incorrectly. We also know that the exponents in the factor N contained in the transfer function $T(D, N)$ indicate the number of information bit errors (number of 1s) in selecting an incorrect path that merges with the all-zero path at some node B. If we multiply the pairwise error probability $P_2(d)$ by the number of incorrectly decoded information bits for the incorrect path at the node where they merge, we obtain the bit error rate for that path. The average bit error probability is upper-bounded by multiplying each pairwise error probability $P_2(d)$ by the corresponding number of incorrectly decoded information bits, for each possible incorrect path that merges with the correct path at the Bth node, and summing over all d.

The appropriate multiplication factors corresponding to the number of information bit errors for each incorrectly selected path may be obtained by differentiating $T(D, N)$ with respect to N. In general, $T(D, N)$ can be expressed as

$$T(D, N) = \sum_{d=d_{\text{free}}}^{\infty} a_d D^d N^{f(d)} \tag{5.3.24}$$

where $f(d)$ denotes the exponent of N as a function of d. Taking the derivative of $T(D, N)$ with respect to N and setting $N = 1$, we obtain

$$\left. \frac{dT(D, N)}{dN} \right|_{N=1} = \sum_{d=d_{\text{free}}}^{\infty} a_d f(d) D^d$$

$$= \sum_{d=d_{\text{free}}}^{\infty} \beta_d D^d \tag{5.3.25}$$

where $\beta_d = a_d f(d)$. Thus the bit error probability for $k = 1$ is upper-bounded by

$$P_b < \sum_{d=d_{\text{free}}}^{\infty} \beta_d P_2(d)$$

$$< \frac{1}{2} \sum_{d=d_{\text{free}}}^{\infty} \beta_d \, \text{erfc}\left(\sqrt{\gamma_b R_c d}\right) \tag{5.3.26}$$

If the complementary error function is upper-bounded by an exponential as indicated in (5.3.22), then (5.3.26) can be expressed in the simple form

$$P_b < \frac{1}{2} \sum_{d=d_{\text{free}}}^{\infty} \beta_d D^d \bigg|_{D=e^{-\gamma_b R_c}}$$

$$< \frac{1}{2} \left. \frac{dT(D, N)}{dN} \right|_{N=1, \, D=e^{-\gamma_b R_c}} \tag{5.3.27}$$

If $k > 1$, the equivalent bit error probability is obtained by dividing (5.3.26) and (5.3.27) by k.

The expressions for the probability of error given above are based on the assumption that the code bits are transmitted by binary coherent PSK. The results also hold for four-phase coherent PSK, since this modulation/demodulation technique is equivalent to two independent (phase-quadrature) binary PSK systems. Other modulation and demodulation techniques such as coherent and noncoherent binary FSK can be accommodated by recomputing the pairwise error probability $P_2(d)$. That is, a change in the modulation and demodulation technique used to transmit the coded information sequence affects only the computation of $P_2(d)$. Otherwise the derivation for P_b remains the same.

Although the above derivation of the error probability for Viterbi decoding of a convolutional code applies to binary convolutional codes, it is relatively easy to generalize it to nonbinary convolutional codes in which each nonbinary symbol is mapped into a distinct waveform. In particular, the coefficients $\{\beta_d\}$ in the expansion of the derivative of $T(D, N)$, given by (5.3.25), represent the number of symbol errors in two paths separated in distance (measured in terms of symbols) by d symbols. Again, we denote the probability of error in a pairwise comparison of two paths that are separated in distance by d as $P_2(d)$. Then the symbol error probability, for a k-bit symbol, is upper-bounded by

$$P_M \le \sum_{d=d_{\text{free}}}^{\infty} \beta_d P_2(d)$$

The symbol error probability can be converted into an equivalent bit error probability. For example, if 2^k orthogonal waveforms are used to transmit the k-bit symbols, the equivalent bit error probability is P_M multiplied by the factor $(2^{k-1})/(2^k - 1)$, as shown in Chap. 4.

5.3.4 Probability of Error for Hard-Decision Decoding

We now consider the performance achieved by the Viterbi decoding algorithm on a binary symmetric channel. For hard-decision decoding of the convolutional code, the metrics in the Viterbi algorithm are the Hamming distances between the received sequence and the $2^{k(L-1)}$ surviving sequences at each node of the trellis.

As in our treatment of soft-decision decoding, we begin by determining the first-event error probability. The all-zero path is assumed to be transmitted. Suppose that the path being compared with the all-zero path at some node B has distance d from the all-zero path. If d is odd, the all-zero path will be correctly selected if the number of errors in the received sequence is less than $(d + 1)/2$; otherwise the incorrect path will be selected. Consequently the probability of selecting the incorrect path is

$$P_2(d) = \sum_{k=(d+1)/2}^{d} \binom{d}{k} p^k (1-p)^{d-k} \qquad (5.3.28)$$

where p is the probability of a bit error for the binary symmetric channel. If d is even, the incorrect path is selected when the number of errors exceeds $d/2$. If the number of errors equals $d/2$, there is a tie between the metrics in the two paths, which may be resolved by randomly selecting one of the paths; thus an error occurs half the time. Consequently the probability of selecting the incorrect path is

$$P_2(d) = \sum_{k=d/2+1}^{d} \binom{d}{k} p^k (1-p)^{d-k} + \frac{1}{2} \binom{d}{d/2} p^{d/2} (1-p)^{d/2} \quad (5.3.29)$$

As indicated in Sec.5.3.3, there are many possible paths with different distances that merge with the all-zero path at a given node. Therefore there is no simple exact expression for the first-event error probability. However, we can overbound this error probability by the sum of the pairwise error probabilities $P_2(d)$ over all possible paths which merge with the all-zero path at the given node. Thus we obtain the union bound

$$P_e < \sum_{d=d_{\text{free}}}^{\infty} a_d P_2(d) \quad (5.3.30)$$

where the coefficients $\{a_d\}$ represent the number of paths corresponding to the set of distances $\{d\}$. These coefficients are the coefficients in the expansion of the transfer function $T(D)$ or $T(D, N)$.

Instead of using the expressions for $P_2(d)$ given in (5.3.28) and (5.3.29), we can use the upper bound

$$P_2(d) < [4p(1-p)]^{d/2} \quad (5.3.31)$$

which was given in Sec. 5.2.5. Use of this bound in (5.3.30) yields a looser upper bound on the first-event error probability, in the form

$$P_e < \sum_{d=d_{\text{free}}}^{\infty} a_d [4p(1-p)]^{d/2}$$
$$< T(D)|_{D=\sqrt{4p(1-p)}} \quad (5.3.32)$$

Let us now determine the probability of a bit error. As in the case of soft-decision decoding, we make use of the fact that the exponents in the factors of N that appear in the transfer function $T(D, N)$ indicate the number of nonzero information bits which are in error when an incorrect path is selected over the all-zero path. By differentiating $T(D, N)$ with respect to N and setting $N = 1$, the exponents of N become multiplication factors of the corresponding error-event probabilities $P_2(d)$. Thus we obtain the expression for the upper bound on the bit error probability, in the form

$$P_b < \sum_{d=d_{\text{free}}}^{\infty} \beta_d P_2(d) \quad (5.3.33)$$

where the $\{\beta_d\}$ are the coefficients in the expansion of the derivative of $T(D, N)$, evaluated at $N = 1$. For $P_2(d)$ we may use either the expressions given in (5.3.28) and (5.3.29) or the upper bound in (5.3.31). If the latter is used, the upper bound on P_b can be expressed as

$$P_b < \frac{dT(D, N)}{dN}\bigg|_{N=1, \, D=\sqrt{4p(1-p)}} \qquad (5.3.34)$$

When $k > 1$, the results given in (5.3.33) and (5.3.34) for P_b should be divided by k.

A comparison of the error probability for the rate $1/3$, $L = 3$ convolutional code with soft-decision decoding and hard-decision decoding is made in Fig. 5.3.15. We note that the Chernoff upper bound given by (5.3.34) is less than

FIGURE 5.3.15
Comparison of soft-decision and hard-decision decoding for $L = 3$, $k = 1$, $n = 3$ convolutional code.

1 dB above the tighter upper bound given by (5.3.33) in conjunction with (5.3.28) and (5.3.29). The advantage of the Chernoff bound is its computational simplicity. In comparing the performance between soft-decision and hard-decision decoding, we note that the difference obtained from the upper bounds is approximately 2.5 dB for $10^{-6} \leq P_b \leq 10^{-2}$.

Finally, we should mention that the ensemble average error rate performance of a convolutional code on a discrete memoryless channel, just as in the case of a block code, can be expressed in terms of the cutoff rate parameter R_0 as [for the derivation, see Viterbi and Omura (1979)]

$$\bar{P}_b < \frac{(q-1)q^{-LR_0/R_c}}{\left[1 - q^{-(R_0 - R_c)/R_c}\right]^2} \qquad R_c \leq R_0$$

where q is the number of channel input symbols, L is the constraint length of the code, R_c is the code rate, and R_0 is the cutoff rate previously defined in Secs. 5.1 and 5.2. Therefore, conclusions reached by computing R_0 for various channel conditions apply to both block codes and convolutional codes.

5.3.5 Distance Properties of Binary Convolutional Codes

In this subsection we shall tabulate the minimum free distance and the generators for several binary, short-constraint-length convolutional codes for several code rates. These binary codes are optimal in the sense that, for a given rate, and a given constraint length, they have the largest possible d_{free}. The generators and the corresponding values of d_{free} tabulated below have been obtained by Odenwalder (1970), Larsen (1973), Paaske (1974), and Daut et al. (1982) by use of computer search methods.

TABLE 5.3.1
Rate 1/2 maximum free distance codes†

Constraint length L	Generators in octal		d_{free}	Upper bound on d_{free}
3	5	7	5	5
4	15	17	6	6
5	23	35	7	8
6	53	75	8	8
7	133	171	10	10
8	247	371	10	11
9	561	753	12	12
10	1167	1545	12	13
11	2335	3661	14	14
12	4335	5723	15	15
13	10533	17661	16	16
14	21675	27123	16	17

† Odenwalder (1970) and Larsen (1973).

TABLE 5.3.2
Rate 1/3 maximum free distance codes†

Constraint length L	Generators in octal			d_{free}	Upper bound on d_{free}
3	5	7	7	8	8
4	13	15	17	10	10
5	25	33	37	12	12
6	47	53	75	13	13
7	133	145	175	15	15
8	225	331	367	16	16
9	557	663	711	18	18
10	1117	1365	1633	20	20
11	2353	2671	3175	22	22
12	4767	5723	6265	24	24
13	10533	10675	17661	24	24
14	21645	35661	37133	26	26

† Odenwalder (1970) and Larsen (1973).

Heller (1968) has derived a relatively simple upper bound on the minimum free distance of a rate $1/n$ convolutional code. It is given as

$$d_{\text{free}} \leq \min_{l \geq 1} \left[\frac{2^{l-1}}{2^l - 1} (L + l - 1) n \right] \qquad (5.3.35)$$

where $[x]$ denotes the largest integer contained in x. For purposes of comparison, this upper bound is also given in the tables for the rate $1/n$ codes. For rate k/n convolutional codes, Daut et al. (1982) has given a modification to Heller's

TABLE 5.3.3
Rate 1/4 maximum free distance codes†

Constraint length L	Generators in octal				d_{free}	Upper bound on d_{free}
3	5	7	7	7	10	10
4	13	15	15	17	13	15
5	25	27	33	37	16	16
6	53	67	71	75	18	18
7	135	135	147	163	20	20
8	235	275	313	357	22	22
9	463	535	733	745	24	24
10	1117	1365	1633	1653	27	27
11	2387	2353	2671	3175	29	29
12	4767	5723	6265	7455	32	32
13	11145	12477	15537	16727	33	33
14	21113	23175	35527	35537	36	36

† Larsen (1973).

TABLE 5.3.4
Rate 1/5 maximum free distance codes†

Constraint length L	Generators in octal					d_{free}	Upper bound on d_{free}
3	7	7	7	5	5	13	13
4	17	17	13	15	15	16	16
5	37	27	33	25	35	20	20
6	75	71	73	65	57	22	22
7	175	131	135	135	147	25	25
8	257	233	323	271	357	28	28

† Daut et al. (1982).

TABLE 5.3.5
Rate 1/6 maximum free distance codes†

Constraint length L	Generators in octal			d_{free}	Upper bound on d_{free}
3	7	7	7	16	16
	7	5	5		
4	17	17	13	20	20
	13	15	15		
5	37	35	27	24	24
	33	25	35		
6	73	75	55	27	27
	65	47	57		
7	173	151	135	30	30
	135	163	137		
8	253	375	331	34	34
	235	313	357		

† Daut et al. (1982).

bound. The values obtained from this upper bound for k/n codes are also tabulated.

Tables 5.3.1 through 5.3.7 list the parameters of rate $1/n$ convolutional codes for $n = 2, 3, \ldots, 8$. Tables 5.3.8 through 5.3.11 list the parameters of several rate k/n convolutional codes for $k \leq 4$ and $n \leq 8$.

5.3.6 Nonbinary Dual-k Codes and Concatenated Codes

Our treatment of convolutional codes thus far has been focused primarily on binary codes. Binary codes are particularly suitable for channels in which binary or quaternary PSK modulation and coherent demodulation is possible. However,

TABLE 5.3.6
Rate 1/7 maximum free distance codes†

Constraint length L	Generators in octal				d_{free}	Upper bound on d_{free}
3	7	7	7	7	18	18
	5	5	5			
4	17	17	13	13	23	23
	13	15	15			
5	35	27	25	27	28	28
	33	35	37			
6	53	75	65	75	32	32
	47	67	57			
7	165	145	173	135	36	36
	135	147	137			
8	275	253	375	331	40	40
	235	313	357			

† Daut et al. (1982).

TABLE 5.3.7
Rate 1/8 maximum free distance codes†

Constraint length L	Generators in octal				d_{free}	Upper bound on d_{free}
3	7	7	5	5	21	21
	5	7	7	7		
4	17	17	13	13	26	26
	13	15	15	17		
5	37	33	25	25	32	32
	35	33	27	37		
6	57	73	51	65	36	36
	75	47	67	57		
7	153	111	165	173	40	40
	135	135	147	137		
8	275	275	253	371	45	45
	331	235	313	357		

† Daut et al. (1982).

there are many applications in which PSK modulation and coherent demodulation is not suitable or possible. In such cases, other modulation techniques, e.g., M-ary FSK, are employed in conjunction with noncoherent demodulation. Nonbinary codes are particularly matched to M-ary signals that are demodulated noncoherently.

In this subsection we describe a class of nonbinary convolutional codes, called *dual-k codes*, which are easily decoded by means of the Viterbi algorithm

TABLE 5.3.8
Rate 2/3 maximum free distance codes†

Constraint length L	Generators in octal			d_{free}	Upper bound on d_{free}
2	17	06	15	3	4
3	27	75	72	5	6
4	236	155	337	7	7

† Daut et al. (1982).

TABLE 5.3.9
Rate $k/5$ maximum free distance codes†

Rate	Constraint length L	Generators in octal					d_{free}	Upper bound on d_{free}
2/5	2	17	07	11	12	04	6	6
	3	27	71	52	65	57	10	10
	4	247	366	171	266	373	12	12
3/5	2	35	23	75	61	47	5	5
4/5	2	237	274	156	255	337	3	4

† Daut et al. [1982].

TABLE 5.3.10
Rate $k/7$ maximum free distance codes†

Rate	Constraint length L	Generators in octal				d_{free}	Upper bound on d_{free}
2/7	2	05 15	06 13	12 17	15	9	9
	3	33 25	55 53	72 75	47	14	14
	4	312 171	125 266	247 373	366	18	18
3/7	2	45 57	21 43	36 71	62	8	8
4/7	2	130 156	067 255	237 337	274	6	7

† Daut et al. (1982).

TABLE 5.3.11
Rates 3/4 and 3/8 maximum free distance codes†

Rate	Constraint length L	Generators in octal				d_{free}	Upper bound on d_{free}
3/4	2	13	25	61	47	4	4
3/8	2	15	42	23	61	8	8
		51	36	75	47		

† Daut et al. (1982).

using either soft-decision or hard-decision decoding. They are also suitable either as an outer code or as an inner code in a concatenated code, as will also be described below.

A dual-k rate 1/2 convolutional encoder may be represented as shown in Fig. 5.3.16. It consists of two ($L = 2$) k-bit shift-register stages and $n = 2k$ function generators. Its output is two k-bit symbols. We note that the code considered previously in Example 5.3.3 is a dual-2 convolutional code.

The $2k$ function generators for the dual-k codes have been given by Viterbi and Jacobs (1975). These may be expressed in the form

$$
\begin{bmatrix} \leftarrow \mathbf{g}_1 \rightarrow \\ \leftarrow \mathbf{g}_2 \rightarrow \\ \vdots \\ \leftarrow \mathbf{g}_k \rightarrow \end{bmatrix} =
\begin{bmatrix}
1 & 0 & 0 & \cdots & & 0 & 1 & 0 & 0 & \cdots & & 0 \\
0 & 1 & 0 & \cdots & & 0 & 0 & 1 & 0 & \cdots & & 0 \\
0 & 0 & 0 & \cdots & 0 & 1 & 0 & 0 & & \cdots & 0 & 1
\end{bmatrix} = [\mathbf{I}_k \vdots \mathbf{I}_k]
$$

$$
\begin{bmatrix} \leftarrow \mathbf{g}_{k+1} \rightarrow \\ \leftarrow \mathbf{g}_{k+2} \rightarrow \\ \vdots \\ \leftarrow \mathbf{g}_{2k} \rightarrow \end{bmatrix} =
\begin{bmatrix}
1 & 1 & 0 & 0 & & \cdots & & 0 & 1 & 0 & 0 & & \cdots & & 0 \\
0 & 0 & 1 & 0 & & \cdots & & 0 & 0 & 1 & 0 & & \cdots & & 0 \\
0 & 0 & 0 & 1 & 0 & \cdots & & 0 & 0 & 0 & 1 & 0 & \cdots & & 0 \\
\hline
0 & 0 & 0 & & & \cdots & 0 & 1 & & & & & & \\
1 & 0 & 0 & & & \cdots & 0 & 0 & 0 & 0 & & & \cdots & & 0 & 1
\end{bmatrix}
$$

$$
=
\begin{bmatrix}
1 & 1 & 0 & 0 & & \cdots & & 0 & \\
0 & 0 & 1 & 0 & & \cdots & & 0 & \\
0 & 0 & 0 & 1 & 0 & \cdots & & 0 & \\
\hline
0 & 0 & 0 & & & \cdots & 0 & 1 & & \mathbf{I}_k \\
1 & 0 & 0 & & & \cdots & 0 & 0 &
\end{bmatrix}
$$

$$(5.3.36)$$

where $\mathbf{I}_k$ denotes the $k \times k$ identity matrix.

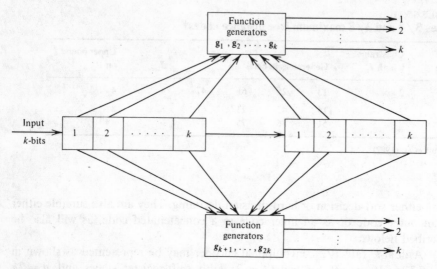

FIGURE 5.3.16
Encoder for rate $1/2$ dual-k codes.

The general form for the transfer function of a rate $1/2$ dual-k code has been derived by Odenwalder (1976). It is expressed as

$$T(D, N, J) = \frac{(2^k - 1) D^4 J^2 N}{1 - NJ \left[2D + (2^k - 3) D^2 \right]}$$

$$= \sum_{i=4}^{\infty} a_i D^i N^{f(i)} J^{h(i)} \qquad (5.3.37)$$

where D represents the Hamming distance for the q-ary ($q = 2^k$) symbols, the $f(i)$ exponent on N represents the number of information symbol errors produced in selecting a branch in the tree or trellis other than a corresponding branch on the all-zero path, and the $h(i)$ exponent on J is equal to the number of branches in a given path. Note that the minimum free distance is $d_{\text{free}} = 4$ symbols ($4k$ bits).

Lower-rate dual-k convolutional codes can be generated in a number of ways, the simplest of which is to repeat each symbol generated by the rate $1/2$ code r times, where $r = 1, 2, \ldots, m$ ($r = 1$ corresponds to each symbol appearing once). If each symbol in any particular branch of the tree or trellis or state diagram is repeated r times, the effect is to increase the distance parameter from D to D^r. Consequently the transfer function for a rate $1/2r$ dual-k code is

$$T(D, N, J) = \frac{(2^k - 1) D^{4r} J^2 N}{1 - NJ \left[2D^r + (2^k - 3) D^{2r} \right]} \qquad (5.3.38)$$

In the transmission of long information sequences, the path length parameter J in the transfer function may be suppressed by setting $J = 1$. The resulting transfer function $T(D, N)$ may be differentiated with respect to N, and N is set to unity. This yields

$$\frac{dT(D, N)}{dN}\bigg|_{N=1} = \frac{(2^k - 1)D^{4r}}{\left[1 - 2D^r - (2^k - 3)D^{2r}\right]^2}$$

$$= \sum_{i=4r}^{\infty} \beta_i D^i \qquad (5.3.39)$$

where β_i represents the number of symbol errors associated with a path having distance D^i from the all-zero path, as described previously in Sec. 5.3.3. The expression in (5.3.39) may be used to evaluate the error probability for dual-k codes under various channel conditions.

Performance of dual-k codes with M-ary modulation. Suppose that a dual-k code is used in conjunction with M-ary orthogonal signaling at the modulator, where $M = 2^k$. Each symbol from the encoder is mapped into one of the M possible orthogonal waveforms. The channel is assumed to add white gaussian noise. The demodulator consists of M matched filters.

If the decoder performs hard-decision decoding, the performance of the code is determined by the symbol error probability P_M. This error probability has been computed in Chap. 4 for both coherent and noncoherent detection. From P_M we can determine $P_2(d)$ according to (5.3.28) or (5.3.29), which is the probability of error in a pairwise comparison of the all-zero path with a path that differs in d symbols. The probability of a bit error is upper-bounded as

$$P_b < \frac{2^{k-1}}{2^k - 1} \sum_{d=4r}^{\infty} \beta_d P_2(d) \qquad (5.3.40)$$

The factor $2^{k-1}/2^k - 1$ is used to convert the symbol error probability to the bit error probability.

Instead of hard-decision decoding, suppose the decoder performs soft-decision decoding using the output of a demodulator which employs a square-law detector. The expression for the bit error probability given by (5.3.40) still applies, but now $P_2(d)$ is given as

$$P_2(d) = \frac{1}{2^{2d-1}} \exp\left(-\frac{\gamma_b}{2} R_c d\right) \sum_{i=0}^{d-1} K_i \left(\frac{\gamma_b}{2} R_c d\right)^i \qquad (5.3.41)$$

where

$$K_i = \frac{1}{i!} \sum_{l=0}^{d-1-i} \binom{2d-1}{l}$$

and $R_c = 1/2r$ is the code rate. This expression follows from the result given previously in (5.2.63).

Concatenated codes. In Sec. 5.2.8 we considered the concatenation of two block codes to form a long block code. Now that we have described convolutional codes, we broaden our viewpoint and consider the concatenation of a block code with a convolutional code or the concatenation of two convolutional codes.

As described previously, the outer code is usually chosen to be nonbinary with each symbol selected from an alphabet of $q = 2^k$ symbols. This code may be a block code, such as a Reed–Solomon code, or a convolutional code, such as a dual-k code. The inner code may be either binary or nonbinary and either a block or a convolutional code. For example, a Reed–Solomon code may be selected as the outer code and a dual-k code may be selected as the inner code. In such a concatenation scheme, the number of symbols in the outer code q equals 2^k, so that each symbol of the outer code maps into a k-bit symbol of the dual-k code. M-ary orthogonal signals may be used to transmit the symbols.

The decoding of such concatenated codes may also take a variety of different forms. If the inner code is a convolutional code having a short constraint length, the Viterbi algorithm provides an efficient means for decoding, using either soft-decision or hard-decision decoding.

If the inner code is a block code, and the decoder for this code performs soft-decision decoding, the outer decoder may also perform soft-decision decoding using as inputs the metrics corresponding to each word of the inner code. On the other hand, the inner decoder may make a hard decision after receipt of the code word and feed the hard decisions to the outer decoder. Then the outer decoder must perform hard-decision decoding.

The following example describes a concatenated code in which the outer code is a convolutional code and the inner code is a block code.

Example 5.3.5. Suppose we construct a concatenated code by selecting a dual-k code as the outer code and a Hadamard code as the inner code. To be specific, we select a rate $1/2$ dual-5 code and a Hadamard $(16, 5)$ inner code. The dual-5 rate $1/2$ code has a minimum free distance $D_{\text{free}} = 4$ and the Hadamard code has a minimum distance $d_{\min} = 8$. Hence the concatenated code has an effective minimum distance of 32. Since there are 32 code words in the Hadamard code and 32 possible symbols in the outer code, in effect, each symbol from the outer code is mapped into one of the 32 Hadamard code words.

The probability of a symbol error in decoding the inner code may be determined from the results of the performance of block codes given in Secs. 5.2.4 and 5.2.5, for soft-decision and hard-decision decoding, respectively. First, suppose that hard-decision decoding is performed in the inner decoder with the probability of a code word (symbol of outer code) error denoted as P_{32}, since $M = 32$. Then the performance of the outer code and, hence, the performance of the concatenated code is obtained by using this error probability in conjunction with the transfer function for the dual-5 code given by (5.3.37).

On the other hand, if soft-decision decoding is used on both the outer and the inner codes, the soft-decision metrics from each received Hadamard code word is passed to the Viterbi algorithm, which computes the accumulated metrics for the competing paths through the trellis. We shall give numerical results on the performance of concatenated codes of his type in our discussion of coding for Rayleigh fading channels.

5.3.7 Other Decoding Algorithms
for Convolutional Codes

The Viterbi algorithm described in Sec. 5.3.2 is the optimum decoding algorithm (in the sense of maximum-likelihood decoding of the entire sequence) for convolutional codes. However, it requires the computation of 2^{kL} metrics at each node of the trellis and the storage of $2^{k(L-1)}$ metrics and $2^{k(L-1)}$ surviving sequences, each of which may be about $5kL$ bits long. The computational burden and the storage required to implement the Viterbi algorithm make it impractical for convolutional codes with a large constraint length.

Prior to the discovery of the optimum algorithm by Viterbi, a number of other algorithms had been proposed for decoding convolutional codes. The earliest was the sequential decoding algorithm originally proposed by Wozencraft (1957, 1961), and subsequently modified by Fano (1963).

The Fano sequential decoding algorithm searches for the most probable path through the tree or trellis by examining one path at a time. The increment added to the metric along each branch is proportional to the probability of the received signal for that branch, just as in Viterbi decoding, with the exception that an additional negative constant is added to each branch metric. The value of this constant is selected such that the metric for the correct path will increase on the average, while the metric for any incorrect path will decrease on the average. By comparing the metric of a candidate path with a moving (increasing) threshold, Fano's algorithm detects and discards incorrect paths.

To be more specific, let us consider a memoryless channel. The metric for the rth path through the tree or trellis from the first branch to branch B may be expressed as

$$U^{(r)} = \sum_{j=1}^{B} \sum_{m=1}^{n} \mu_{jm}^{(r)} \qquad (5.3.42)$$

where

$$\mu_{jm}^{(r)} = \log_2 \frac{p\left(y_{mn}|c_{jm}^{(r)}\right)}{p(y_{jm})} - K \qquad (5.3.43)$$

In (5.3.43), y_{jm} is the demodulator output sequence, $p(y_{jm}|c_{jm}^{(r)})$ denotes the pdf of y_{jm} conditional on the code bit $c_{jm}^{(r)}$ for the mth bit of the jth branch of the rth path, and K is a positive constant. K is selected as indicated above so that the incorrect paths will have a decreasing metric while the correct path will have an increasing metric on the average. Note that the term $p(y_{jm})$ in the denominator is independent of the code sequence and, hence, may be subsumed in the constant factor.

The metric given by (5.3.43) is generally applicable for either hard- or soft-decision decoding. However, it can be considerably simplified when hard-decision decoding is employed. Specifically, if we have a BSC with transition (error) probability p, the metric for each received bit, consistent with the form in

(5.3.43), is given as

$$\mu_{jm}^{(r)} = \begin{cases} \log_2 2(1-p) - R_c & \text{if } \tilde{y}_{jm} = c_{jm}^{(r)} \\ \log 2p - R_c & \text{if } \tilde{y}_{jm} \neq c_{jm}^{(r)} \end{cases} \tag{5.3.44}$$

where $\tilde{y}_{jm}$ is the hard-decision output from the demodulator and $c_{jm}^{(r)}$ is the mth code bit in the jth branch of the rth path in the tree and R_c is the code rate. Note that this metric requires some (approximate) knowledge of the error probability.

Example 5.3.6. Suppose we have a rate $R_c = 1/3$ binary convolutional code for transmitting information over a BSC with $p = 0.1$. By evaluating (5.3.44) we find that

$$\mu_{jm}^{(r)} = \begin{cases} 0.52 & \text{if } \tilde{y}_{jm} = c_{jm}^{(r)} \\ -2.65 & \text{if } \tilde{y}_{jm} \neq c_{jm}^{(r)} \end{cases} \tag{5.3.45}$$

To simplify the computations, the metric in (5.3.45) may be normalized. It is well approximated as

$$\mu_{jm}^{(r)} = \begin{cases} 1 & \text{if } \tilde{y}_{jm} = c_{jm}^{(r)} \\ -5 & \text{if } \tilde{y}_{jm} \neq c_{jm}^{(r)} \end{cases} \tag{5.3.46}$$

Since the code rate is $1/3$, there are three output bits from the encoder for each input bit. Hence, the branch metric consistent with (5.3.46) is

$$\mu_j^{(r)} = 3 - 6d$$

or, equivalently, is

$$\mu_j^{(r)} = 1 - 2d \tag{5.3.47}$$

where d is the Hamming distance of the three received bits from the three branch bits. Thus, the metric $\mu_j^{(r)}$ is simply related to the Hamming distance of the received bits to the code bits in the jth branch of the rth path.

Initially the decoder may be forced to start on the correct path by the transmission of a few known bits of data. Then it proceeds forward from node to node, taking the most probable (largest metric) branch at each node and increasing the threshold such that the threshold is never more than some preselected value, say τ, below the metric. Now suppose that the additive noise (for soft-decision decoding) or demodulation errors resulting from noise on the channel (for hard-decision decoding) cause the decoder to take an incorrect path because it appears more probable than the correct path. This is illustrated in Fig. 5.3.17. Since the metrics of an incorrect path decrease on the average, the metric will fall below the current threshold, say τ_0. When this occurs, the decoder backs up and takes alternative paths through the tree or trellis, in order of decreasing branch metrics, in an attempt to find another path that exceeds the threshold τ_0. If it is successful in finding an alternative path, it continues along that path, always selecting the most probable branch at each node. On the other hand, if

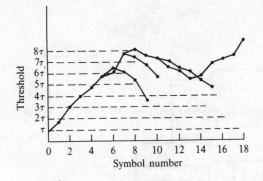

FIGURE 5.3.17
An example of the path search in sequential decoding. [From paper by Jordan (1966), © 1966 IEEE. Reprinted with permission.]

no path exists that exceeds the threshold τ_0, the threshold is reduced by an amount τ and the original path is retraced. If the original path does not stay above the new threshold, the decoder resumes its backward search for other paths. This procedure is repeated, with the threshold reduced by τ for each repetition until the decoder finds a path that remains above the adjusted threshold. A simplified flow diagram of Fano's algorithm is shown in Fig. 5.3.18.

The sequential decoding algorithm requires a buffer memory in the decoder to store incoming demodulated data during periods when the decoder is searching for alternate paths. When a search terminates, the decoder must be capable of processing demodulated bits sufficiently fast to empty the buffer prior to commencing a new search. Occasionally, during extremely long searches, the buffer may overflow. This causes loss of data, a condition which can be remedied by retransmission of the lost information. In this regard we should mention that the cutoff rate R_0 has a special meaning in sequential decoding. It is the rate above which the average number of decoding operations per decoded digit becomes infinite, and it is denoted as the *computational cutoff rate* R_{comp}. In practice, sequential decoders usually operate at rates near R_0.

The Fano sequential decoding algorithm has been successfully implemented in several communications systems. Its error rate performance is comparable to that of Viterbi decoding. However, in comparison with Viterbi decoding, sequential decoding has a significantly larger decoding delay. On the positive side, sequential decoding requires less storage than Viterbi decoding and, hence, it appears attractive for convolutional codes with a large constraint length. The issues of computational complexity and storage requirements for sequential decoding are interesting and have been thoroughly investigated. For an analysis of these topics and other characteristics of the Fano algorithm, the interested reader may refer to Gallager (1968), Wozencraft and Jacobs (1965), Savage (1966), and Forney (1974).

Another type of sequential decoding algorithm, called a *stack algorithm*, has been proposed independently by Jelinek (1969) and Zigangirov (1966). In contrast to the Viterbi algorithm, which keeps track of $2^{(L-1)k}$ paths and corresponding metrics, the stack sequential decoding algorithm deals with fewer

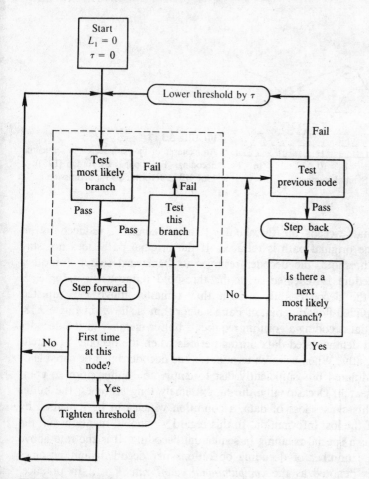

FIGURE 5.3.18
A simplified flow diagram of Fano's algorithm. [From paper by Jordan (1966), © 1966 IEEE. Reprinted with permission.]

paths and their corresponding metrics. In a stack algorithm the more probable paths are ordered according to their metrics, with the path at the top of the stack having the largest metric. At each step of the algorithm, only the path at the top of the stack is extended by one branch. This yields 2^k successors and their corresponding metrics. These 2^k successors along with the other paths are then reordered according to the values of the metrics and all paths with metrics that fall below some preselected amount from the metric of the top path may be discarded. Then the process of extending the path with the largest metric is repeated. Figure 5.3.19 illustrates the first few steps in a stack algorithm.

It is apparent that when none of the 2^k extensions of the path with the largest metric remains at the top of the stack, the next step in the search involves

Rate $\frac{1}{3}$ code

Branch metric = $1-2d$

d = Hamming distance

Received sequence: 101 111 110 011

Stack with accumulated path metrics

Step a	Step b	Step c	Step d	Step e	Step f
-1	-2	-3	-2	-1	-2
-3	-3	-3	-3	-3	-3
	-4	-4	-4	-4	-4
		-5	-5	-5	-4
			-8	-7	-5
				-8	-7
					-8

FIGURE 5.3.19

An example of the stack algorithm for decoding a rate 1/3 convolutional code.

the extension of another path which has climbed to the top of the stack. It follows that the algorithm does not necessarily advance by one branch through the trellis in every iteration. Consequently some amount of storage must be provided for newly received signals and previously received signals in order to allow the algorithm to extend the search along one of the shorter paths, when such a path reaches the top of the stack.

In a comparison of the stack algorithm with the Viterbi algorithm, the stack algorithm requires fewer metric computations, but this computational savings is offset to a large extent by the computations involved in reordering the stack after every iteration. In comparison with the Fano algorithm, the stack algorithm is computationally simpler since there is no retracing over the same path as is done in the Fano algorithm. On the other hand, the stack algorithm requires more storage than the Fano algorithm.

A third alternative to the optimum Viterbi decoder is a method called *feedback decoding* [see Heller (1975)], which has been applied to decoding for a BSC (hard-decision decoding). In feedback decoding, the decoder makes a hard decision on the information bit at stage j based on metrics computed from stage j to stage $j + m$, where m is a preselected positive integer. Thus the decision on the information bit is either 0 or 1 depending on whether the minimum Hamming distance path which begins at stage j and ends at stage $j + m$ contains a 0 or 1 in the branch emanating from stage j. Once a decision is made on the information bit at stage j, only that part of the tree which stems from the bit selected at stage j is kept (half the paths emanating from node j) and the remaining part is discarded. This is the feedback feature of the decoder.

The next step is to extend the part of the tree that has survived to stage $j + 1 + m$ and consider the paths from stage $j + 1$ to $j + 1 + m$ in deciding on the bit at stage $j + 1$. Thus this procedure is repeated at every stage. The parameter m is simply the number of stages in the tree that the decoder looks ahead before making a hard decision. Since a large value of m results in a large amount of storage, it is desirable to select m as small as possible. On the other hand, m must be sufficiently large to avoid a severe degradation in performance. To balance these two conflicting requirements, m is usually selected in the range $L \leq m \leq 2L$ where L is the constraint length. Note that this decoding delay is significantly smaller than the decoding delay in a Viterbi decoder, which is usually about $5L$.

Example 5.3.7. Let us consider the use of a feedback decoder for the rate $1/3$ convolutional code shown in Fig. 5.3.2. Figure 5.3.20 illustrates the tree diagram and the operation of the feedback decoder for $m = 2$. That is, in decoding the bit at branch j, the decoder considers the paths at branches j, $j + 1$, and $j + 2$. Beginning with the first branch, the decoder computes eight metrics (Hamming distances) and decides that the bit for the first branch is 0 if the minimum disance path is contained in the upper part of the tree, and 1 if the minimum distance path is contained in the lower part of the tree. In this example, the received sequence for the first three branches is assumed to be 101111110, so that the minimum distance path is in the upper part of the tree. Hence the first output bit is 0.

    The next step is to extend the upper part of the tree (the part of the tree that has survived) by one branch, and to compute the eight metrics for branches 2, 3, and 4. For the assumed received sequence 111110011, the minimum-distance path is contained in the lower part of the section of the tree that survived from the first step. Hence the second output bit is 1. The third step is to extend this lower part of the tree and to repeat the procedure described for the first two steps.

Instead of computing metrics as described above, a feedback decoder for the BSC may be efficiently implemented by computing the syndrome from the

Step 1 : Upper-tree metrics : 7,6,3,2* ; lower-tree metrics : 5,4,3,4 →0
Step 2 : Upper-tree metrics : 7,6,5,6 ; lower-tree metrics : 3,6,1*, 2 →1

FIGURE 5.3.20
An example of feedback decoding for a rate 1/3 convolutional code.

TABLE 5.3.12
Upper bounds on coding gain for soft-decision decoding of some convolution codes

Rate 1 / 2 codes			Rate 1 / 3 codes		
Constraint length L	d_{free}	Upper bound (dB)	Constraint length L	d_{free}	Upper bound (dB)
3	5	3.98	3	8	4.26
4	6	4.77	4	10	5.23
5	7	5.44	5	12	6.02
6	8	6.02	6	13	6.37
7	10	6.99	7	15	6.99
8	10	6.99	8	16	7.27
9	12	7.78	9	18	7.78
10	12	7.78	10	20	8.24

received sequence and using a table lookup method for correcting errors. This method is similar to the one described for decoding block codes. For some convolutional codes, the feedback decoder simplifies to a form called a *majority logic decoder* or a *threshold decoder* [see Massey (1963) and Heller (1975)].

5.3.8 Practical Considerations in the Application of Convolutional Codes

Convolutional codes are widely used in many practical applications of communications system design. Viterbi decoding is predominantly used for short constraint lengths ($L \leq 10$) while sequential decoding is used for long constraint length codes, where the complexity of Viterbi decoding becomes prohibitive. The choice of constraint length is dictated by the desired coding gain.

From the error probability results for soft-decision decoding given by (5.3.26) it is apparent that the coding gain achieved by a convolutional code over an uncoded binary PSK or QPSK system is

$$\text{coding gain} \leq \log_{10} \left(R_c d_{\text{free}} \right)$$

We also know that the minimum free distance d_{free} can be increased either by decreasing the code rate or by increasing the constraint length, or both. Table 5.3.12 provides a list of upper bounds on the coding gain for several convolutional codes. For purposes of comparison, Table 5.3.13, taken from the paper of Jacobs (1974), lists the actual coding gains and the upper bounds for several short constraint length convolutional codes with Viterbi decoding. We should note that the coding gain increases toward the asymptotic limit as the SNR per bit increases.

These results are based on soft-decision Viterbi decoding. If hard-decision decoding is used, the coding gains are reduced by approximately 2 dB for the AWGN channel.

TABLE 5.3.13
Coding gain (dB) for soft-decision Viterbi decoding†

P_b	$\mathscr{E}_b / N_0$ uncoded (dB)	$R_c = 1/3$		$R_c = 1/2$			$R_c = 2/3$		$R_c = 3/4$	
		$L = 7$	$L = 8$	$L = 5$	$L = 6$	$L = 7$	$L = 6$	$L = 8$	$L = 6$	$L = 9$
10^{-3}	6.8	4.2	4.4	3.3	3.5	3.8	2.9	3.1	2.6	2.6
10^{-5}	9.6	5.7	5.9	4.3	4.6	5.1	4.2	4.6	3.6	4.2
10^{-7}	11.3	6.2	6.5	4.9	5.3	5.8	4.7	5.2	3.9	4.8

† From Jacobs (1974); © 1974 IEEE; reprinted with permission.

Larger coding gains than the ones listed in the above tables are achieved by employing long constraint lengths convolutional codes, e.g., $L = 50$, and decoding such codes by sequential decoding. Invariably, sequential decoders are implemented for hard-decision decoding to reduce complexity. Figure 5.3.21 illustrates the error rate performance of several constraint length $L = 7$ convolutional codes for rates $1/2$ and $1/3$ and for sequential decoding (with hard decisions) of a rate $1/2$ and a rate $1/3$ constraint length $L = 41$ convolutional codes. Note that the $L = 41$ codes achieve an error rate of 10^{-6} at 2.5 and 3 dB, which are within 4 to 4.5 dB of the channel capacity limit, i.e., in the vicinity of the cutoff rate limit. However, the rate $1/2$ and rate $1/3$, $L = 7$ codes with soft-decision Viterbi decoding operate at about 5 and 4.4 dB at 10^{-6}, respectively. These short constraint length codes achieve a coding gain of about 6 dB at 10^{-6}, while the long constraint codes gain about 7.5 to 8 dB.

Two important issues in the implementation of Viterbi decoding are (1) the effect of path memory truncation, which is a desirable feature that ensures a fixed decoding delay, and (2) the degree of quantization of the input signal to the Viterbi decoder. As a rule of thumb, we stated that path memory truncation to about five constraint lengths has been found to result in negligible performance loss. Figure 5.3.22 illustrates the performance obtained by simulation for a rate $1/2$, constraint lengths $L = 3$, 5, and 7 codes with memory path length of 32 bits. In addition to path memory truncation, the computations were performed with eight-level (three bits) quantized input signals from the demodulator. The broken line curves are performance results obtained from the upper bound in the bit error rate given by (5.3.26). Note that the simulation results are close to the theoretical upper bounds, which indicate that the degradation due to path memory truncation and quantization of the input signal has a minor effect on performance (0.20 to 0.30 dB).

Figure 5.3.23 illustrates the bit error rate performance obtained via simulation for hard-decision decoding of convolutional codes with $L = 3$ to 8. Note that with the $L = 8$ code, an error rate of 10^{-5} requires about 6 dB, which represents a coding gain of nearly 4 dB relative to uncoded QPSK.

The effect of input signal quantization is further illustrated in Fig. 5.3.24 for a rate $1/2$, $L = 5$ code. Note that three-bit quantization (eight levels) is about

FIGURE 5.3.21
Performance of rate $1/2$ and rate $1/3$ Viterbi and sequential decoding. [From paper by Omura and Levitt (1982). © 1982 IEEE; reprinted with permission.]

2 dB better than hard-decision decoding, which is the ultimate limit between soft-decision decoding and hard-decision decoding on the AWGN channel. The combined effect of signal quantization and path memory trunction for the rate $1/2$, $L = 5$ code with 8-, 16-, and 32-bit path memories and either one- or three-bit quantization is shown in Fig. 5.3.25. It is apparent from these results that a path memory as short as three constraint lengths does not seriously degrade performance.

When the signal from the demodulator is quantized to more than two levels, another problem that must be considered is the spacing between quantization

FIGURE 5.3.22

Bit error probability for rate $1/2$ Viterbi decoding with eight-level quantized inputs to the decoder and 32-bit path memory. [From paper by Heller and Jacobs (1971). © 1971 IEEE. Reprinted with permission.]

levels. Figure 5.3.26 illustrates the simulation results for an eight-level uniform quantizer as a function of the quantizer threshold spacing. We observe that there is an optimum spacing between thresholds (approximately equal to 0.5). However, the optimum is sufficiently broad (0.4 to 0.7) so that once it is set, there is little degradation resulting from variations in the AGC level of the order of ±20%.

Finally, we should point out some important results in the performance degradation due to carrier phase variations. Figure 5.3.27 illustrates the performance of a rate $1/2$, $L = 7$ code with eight-level quantization and a carrier

FIGURE 5.3.23

Performance of rate $1/2$ codes with hard-decision Viterbi decoding and 32-bit path memory truncation. [From paper by Heller and Jacobs (1971). © 1971 IEEE. Reprinted with permission.]

FIGURE 5.3.24
Performance of rate $1/2$, $L = 5$ code with eight-, four-, and two-level quantization at the input to the Viterbi decoder. Path truncation length = 32 bits. [From paper by Heller and Jacobs (1971). © 1971 IEEE. Reprinted with permission.]

FIGURE 5.3.25
Performance of rate $1/2$, $L = 5$ code with 32-, 16-, and 8-bit path memory truncation and eight- and two-level quantization. [From paper by Heller and Jacobs (1971). © 1971 IEEE. Reprinted with permission.]

FIGURE 5.3.26
Error rate performance of rate $1/2$, $L = 5$ Viterbi decoder for $\mathscr{E}_b/N_0 = 3.5$ dB and eight-level quantization as a function of quantizer threshold level spacing for equally spaced thresholds. [From paper by Heller and Jacobs (1971). © 1971 IEEE. Reprinted with permission.]

phase tracking loop SNR γ_L. Recall that in a PLL, the phase error has a variance that is inversely proportional to γ_L. The results in Fig. 5.3.27 indicate that the degradation is large when the loop SNR is small ($\gamma_L < 12$ dB) and causes the error rate performance to bottom out at relatively high error rate.

5.4 CODED MODULATION FOR BANDWIDTH-CONSTRAINED CHANNELS

In the treatment of block and convolutional codes in Secs. 5.2 and 5.3, respectively, performance improvement was achieved by expanding the bandwidth of the transmitted signal by an amount equal to the reciprocal of the code rate. Recall for example that the improvement in performance achieved by an (n, k) binary block code with soft-decision decoding is approximately $10 \log_{10} (R_c d_{\min} - k \ln 2/\gamma_b)$ compared with uncoded binary or quaternary PSK. For example, when $\gamma_b = 10$ the $(24, 12)$ extended Golay code gives a coding gain of 5 dB. This coding gain is achieved at a cost of doubling the bandwidth of the transmitted signal and, of course, at the additional cost in receiver implementation complexity. Thus, coding provides an effective method for trading bandwidth and implementation complexity against transmitter power. This situation applies to digital communications systems that are designed to operate in the power-limited region where $R/W < 1$.

In this section we consider the use of coded signals for bandwidth-constrained channels. For such channels, the digital communications system is designed to use bandwidth-efficient multilevel/phase modulation, such as ASK, PSK, DPSK, or QAM and operates in the region where $R/W > 1$. When coding is applied to the bandwidth-constrained channel, a performance gain is desired without expanding the signal bandwidth. This goal can be achieved by increasing the number of signals over the corresponding uncoded system to compensate for the redundancy introduced by the code.

FIGURE 5.3.27
Performance of a rate $1/2$, $L = 7$ code with Viterbi decoding and eight-level quantization as a function of the carrier phase tracking loop SNR γ_L. [From paper by Heller and Jacobs (1971). © 1971 IEEE. Reprinted with permission.]

For example, suppose that a system employing uncoded four-phase PSK modulation achieves an $R/W = 2$ bits per second per hertz at an error probability of 10^{-6}. For this error rate the SNR per bit is $\gamma_b = 10.5$ dB. We may try to reduce the SNR per bit by use of coded signals, but this must be done without expanding the bandwidth. If we choose a rate $R_c = 2/3$ code, it must be accompanied by an increase in the number of signal points from four (two bits per symbol) to eight (three bits per symbol). Thus the rate $2/3$ code used in conjunction with eight-phase PSK, for example, yields the same data throughput as uncoded four-phase PSK. However, we recall that an increase in the number of signal phases from four to eight requires an additional 4 dB approximately in signal power to maintain the same error rate. Hence, if coding is to provide a benefit, the performance gain of the rate 2/3 code must overcome this 4-dB penalty.

If the modulation is treated as a separate operation independent of the encoding, the use of very powerful codes (large constraint length convolutional codes or large block length block codes) is required to offset the loss and provide some significant coding gain. On the other hand, if the modulation is an integral part of the encoding process and is designed in conjunction with the code to increase the minimum Euclidean distance between pairs of coded signals, the loss from the expansion of the signal set is easily overcome and a significant coding gain is achieved with relatively simple codes. The key to this integrated modulation and coding approach is to devise an effective method for mapping the coded bits into signal points such that the minimum Euclidean distance is maximized. Such a method was developed by Ungerboeck (1982), based on the principle of *mapping by set partitioning*. We describe this principle by means of two examples.

Example 5.4.1 An 8-PSK signal constellation. Let us partition the eight-phase signal constellation shown in Fig. 5.4.1 into subsets of increasing minimum Euclidean distance. In the eight-phase signal set the signal points are located on a circle of radius $\sqrt{\mathscr{E}}$ and have a minimum distance separation of

$$d_0 = 2\sqrt{\mathscr{E}} \, \sin \frac{\pi}{8} = \sqrt{(2 - \sqrt{2})\mathscr{E}} = 0.765\sqrt{\mathscr{E}}$$

In the first partitioning the eight points are subdivided into two subsets of four points each, such that the minimum distance between points increases to $d_1 = \sqrt{2\mathscr{E}}$.

In the second level of partitioning, each of the two subsets is subdivided into two subsets of two points, such that the minimum distance increases to $d_2 = 2\sqrt{\mathscr{E}}$. This results in four subsets of two points each. Finally, the last stage of partitioning leads to eight subsets, where each subset contains a single point. Note that each level of partitioning increases the minimum Euclidean distance between signal points. The results of these three stages of partitioning are illustrated in Fig. 5.4.1. The way in which the coded bits are mapped into the partitioned signal points is described below.

Example 5.4.2 A 16-QAM signal constellation. The 16-point rectangular signal constellation shown in Fig. 5.4.2 is first divided into two subsets by assigning alternate points to each subset as illustrated in the figure. Thus, the distance between points is increased from $2\sqrt{\mathscr{E}}$ to $2\sqrt{2\mathscr{E}}$ by the first partitioning. Further partitioning of the two subsets leads to greater separation in Euclidean distance between signal points as illustrated in Fig. 5.4.2. It is interesting to note that for the rectangular signal constellations, each level of partitioning increases the minimum Euclidean distance by $\sqrt{2}$, i.e., $d_{i+1}/d_i = \sqrt{2}$ for all i.

In these two examples, the partitioning was carried out to the limit where each subset contains only a single point. In general, this may not be necessary. For example, the 16-point QAM signal constellation may be partitioned only twice to yield four subsets of four points each. Similarly, the eight-phase PSK signal constellation can be partitioned twice to yield four subsets of two points each.

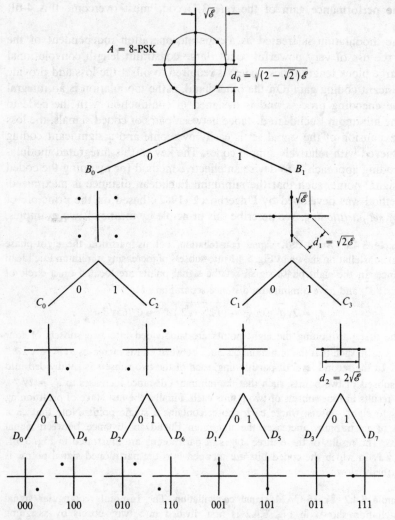

FIGURE 5.4.1
Set partitioning of an 8-PSK signal set.

The degree to which the signal is partitioned depends on the characteristics of the code. In general, the encoding process is performed as illustrated in Fig. 5.4.3. A block of m information bits is separated into two groups of lengths k_1 and k_2. The k_1 bits are encoded into n bits while the k_2 bits are left uncoded. Then, the n bits from the encoder are used to select one of the 2^n possible subsets in the partitioned signal set while the k_2 bits are used to select one of the 2^{k_2} signal points in each subset. When $k_2 = 0$, all m information bits are encoded.

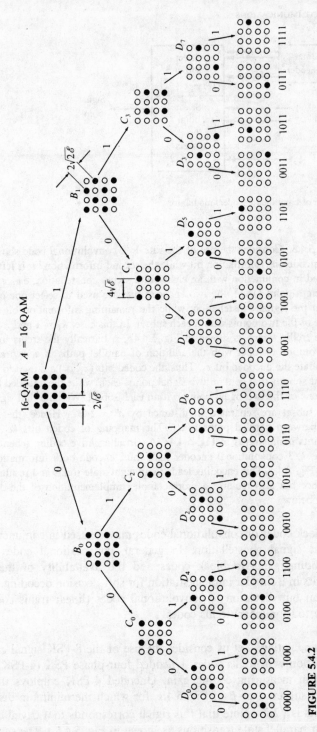

FIGURE 5.4.2
Set partitioning of 16-QAM signal.

491

FIGURE 5.4.3
General structure of combined encoder/modulator.

Example 5.4.3. Consider the use of the rate $1/2$ convolutional code shown in Fig. 5.4.4 to encode one information bit while the second information bit is left uncoded. When used in conjunction with an eight-point signal constellation, e.g., eight-phase PSK or eight-point QAM, the two encoded bits are used to select one of the four subsets in the signal constellation, while the remaining information bit is used to select one of the two points within each subset. In this case, $k_1 = 1$ and $k_2 = 1$. The four-state trellis, which is shown in Fig. 5.4.4b, is basically the trellis for the rate $1/2$ convolution encoder with the addition of parallel paths in each transition to accommodate the uncoded bit c_3. Thus, the coded bits (c_1, c_2) are used to select one of the four subsets that contain two signal points each, while the uncoded bit is used to select one of the two signal points within each subset. We note that signal points within a subset are separated in distance by $d_2 = 2\sqrt{\mathscr{E}}$. Hence, the Euclidean distance between parallel paths is d_2. The mapping of coded bits (c_1, c_2, c_3) to signal points is illustrated in Fig. 5.4.4c. As an alternative coding scheme we may use a rate $2/3$ convolutional encoder and, thus, encode both information bits as shown in Fig. 5.4.5. This encoding leads to an eight-state trellis and results in better performance but also requires a more complex implementation of the decoder as described below.

Either block codes or convolutional codes may be used in conjunction with the partitioned signal constellation. In general, convolutional codes provide comparable coding gains to block codes and the availability of the Viterbi algorithm results in a simpler implementation for soft-decision decoding. For this reason, we limit our dicussion to convolutional codes (linear trellis codes) and more generally to (nonlinear) trellis codes.

Trellis-coded modulation. Let us consider the use of the 8-PSK signal constellation in conjunction with trellis codes. Uncoded four-phase PSK (4-PSK) is used as a reference in measuring coding gain. Uncoded 4-PSK employs the signal points in either subset A or B of Fig. 5.4.1, for which the minimum distance of the signal points is $\sqrt{2\mathscr{E}}$. Note that this signal corresponds to a trivial one-state trellis with four parallel state transitions as shown in Fig. 5.4.6a. The subsets D_0, D_2, D_4, and D_6 are used as the signal points for the purpose of illustration.

(a) Encoder

$(c_1, c_2, c_3) = 000$

(b) Four-state trellis

(c) Mapping of
coded bits (c_1, c_2, c_3)
to signal points

FIGURE 5.4.4
Four-state trellis-coded 8-PSK modulation.

For the coded 8-PSK modulation, we may use the four-state trellis shown in Fig. 5.4.6b. Note that each branch in the trellis corresponds to one of the four subsets C_0, C_1, C_2, C_3. For the eight-point constellation, each of the subsets C_0, C_1, C_2, C_3, contains two signal points. Hence, the state transition C_0 contains the two signal points corresponding to the bits (000, 100) or (0, 4) in octal representation. Similarly, C_2 contains the two signal points corresponding to (010, 110) or to (2, 6) in octal, C_1 contains the points corresponding to (001, 101) or (1, 5) in octal, and C_3 contains the points corresponding to (011, 111) or (3, 7) in octal. Thus, each transition in the four-state trellis contains two parallel paths as shown in more detail in Fig. 5.4.6c. Note that any two signal paths that diverge from one state and remerge at the same state after more than one transition have a squared Euclidean distance of $d_0^2 + 2d_1^2 = d_0^2 + d_2^2$ between them. For example, the signal paths 0, 0, 0 and 2, 1, 2 are separated by $d_0^2 + d_2^2 =$

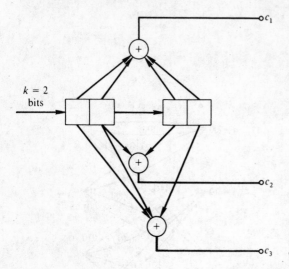

FIGURE 5.4.5
Rate 2/3 convolutional encoder for encoding both information bits.

(a) One-state trellis

(b) Four-state trellis

(c) Four-state trellis

FIGURE 5.4.6
Uncoded 4-PSK and trellis-coded 8-PSK modulation.

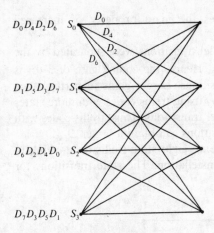

$D_0 D_4 D_2 D_6 \quad S_0$

D_0
D_4
D_2
D_6

$D_1 D_5 D_3 D_7 \quad S_1$

$D_6 D_2 D_4 D_0 \quad S_2$

$D_7 D_3 D_5 D_1 \quad S_3$

FIGURE 5.4.7
An alternative four-state trellis code.

$[(0.765)^2 + 4]\mathscr{E} = 4.585\mathscr{E}$. On the other hand, the squared Euclidean distance between parallel transitions is $d_2^2 = 4\mathscr{E}$. Hence, the minimum Euclidean distance separation between paths that diverge from any state and remerge at the same state in the four-state trellis is $d_2 = 2\sqrt{\mathscr{E}}$. This minimum distance in the trellis code is called the *free Euclidean distance* and denoted as D_{fed}.

In the four-state trellis of Fig. 5.4.6b, $D_{\text{fed}} = 2\sqrt{\mathscr{E}}$. When compared with the Euclidean distance $d_0 = \sqrt{2\mathscr{E}}$ for the uncoded 4-PSK modulation, we observe that the four-state trellis code gives a coding gain of 3 dB.

We should emphasize that the four-state trellis code illustrated in Fig. 5.4.6b is optimum in the sense that it provides the largest free Euclidean distance. Clearly, many other four-state trellis codes can be constructed, including the one shown in Fig. 5.4.7, which consists of four distinct transitions from each state to all other states. However, neither this code nor any of the other possible 4-state trellis codes gives a larger D_{fed}.

The construction of the optimum four-state trellis code for the eight-point constellation was performed on the basis of the following heuristic rules:

(a) Parallel transitions (when they occur) are assigned to signal points separated by the maximum Euclidean distance, e.g., $d_2 = 2\sqrt{\mathscr{E}}$ for 8-PSK in the four subsets C_0, C_1, C_2, C_3.

(b) The transition originating from and merging into any state is assigned the subsets (C_0, C_2) or (C_1, C_3), which have a maximum distance $d_1 = \sqrt{2\mathscr{E}}$.

(c) The signal points should occur with equal frequency.

We note that rules (a) and (b) guarantee that the Euclidean distance associated with single and multiple paths that diverge from any state and remerge in that state exceeds the Euclidean distance of uncoded 4-PSK. Rule (c) guarantees that the trellis code will have a regular structure.

We should indicate that the specific mapping of coded bits into signal points, as illustrated in Fig. 5.4.1 where the eight signal points are represented in an equivalent binary form, is not important. Other mappings can be devised by permuting subsets in a way that preserves the main property of increased minimum distance among the subsets.

In the four-state trellis code, the parallel transitions were separated by the Euclidean distance $2\sqrt{\mathscr{E}}$, which is also D_{fed}. Hence, the coding gain of 3 dB is limited by the distance of the parallel transitions. Larger gains in performance relative to uncoded 4-PSK can be achieved by using trellis codes with more states which allow for the elimination of the parallel transitions. Thus, trellis codes with eight or more states would use distinct transitions to obtain a larger D_{fed}.

For example, in Fig. 5.4.8 we illustrate an eight-state trellis code owing to Ungerboeck (1982) for the 8-PSK signal constellation. The state transitions for

Eight-state trellis

FIGURE 5.4.8
Eight-state trellis code for coded 8-PSK modulation.

maximizing the free Euclidean distance were determined from application of the three basic rules given above. In this case, we note that the minimum squared Euclidean distance is

$$D_{\text{fed}}^2 = d_0^2 + 2d_1^2 = 4.585\mathscr{E}$$

which, when compared with $d_0^2 = 2\mathscr{E}$ for uncoded 4-PSK, represents a gain of 3.6 dB. Ungerboeck (1982, 1987) has also found rate $2/3$ trellis codes with 16, 32, 64, 128, and 256 states that achieve coding gains ranging from 4 to 5.75 dB for 8-PSK modulation.

The basic principle of set partitioning is easily extended to larger PSK signal constellations that yield greater bandwidth efficiency. For example, 3 bits per second per hertz can be achieved with either uncoded 8-PSK or with trellis-coded 16-PSK modulation. Ungerboeck (1987) has devised trellis codes and has evaluated the coding gains achieved by simple rate $1/2$ and rate $2/3$ convolutional codes for the 16-PSK signal constellations. These results are summarized below.

Soft-decision Viterbi decoding for trellis-coded modulation is accomplished in two steps. Since each branch in the trellis corresponds to a signal subset, the first step in decoding is to determine the best signal point within each subset, i.e., the point in each subset that is closest in distance to the received point. We may call this *subset decoding*. In the second step, the signal point selected from each subset and its squared distance metric are used for the corresponding branch in the Viterbi algorithm to determine the signal path through the code trellis that has the minimum sum of squared distances from the sequence of received (noisy channel output) signals.

The error rate performance of the trellis-coded signals in the presence of additive gaussian noise can be evaluated by following the procedure described in Sec. 5.3 for convolutional codes. Recall that this procedure involves the computation of the probability of error for all different error events and summing these error event probabilities to obtain a union bound on the first-event error probability. We note, however, that at high SNR the first-event error probability is dominated by the leading term which has the minimum distance D_{fed}. Consequently, at high SNR the first-event error probability is well approximated as

$$P_e \approx N_{\text{fed}} \left[\frac{1}{2} \operatorname{erfc} \sqrt{\frac{D_{\text{fed}}^2}{4N_0}} \right] \qquad (5.4.1)$$

where N_{fed} denotes the number of signal sequences with distance D_{fed} that diverge at any state and remerge at that state after one or more transitions.

In computing the coding gain achieved by trellis-coded modulation, we usually focus on the gain achieved by increasing D_{fed} and neglect the effect of N_{fed}. However, trellis codes with a large number of states may result in a large N_{fed} that cannot be ignored in assessing the overall coding gain.

In addition to the trellis-coded PSK modulations described above, powerful trellis codes have also been developed for ASK and QAM signal constellations.

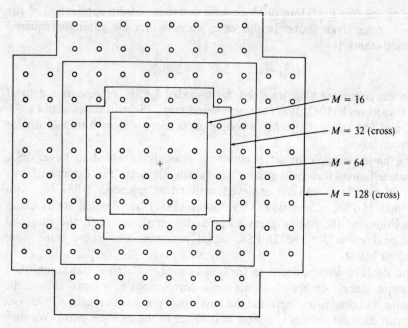

FIGURE 5.4.9
Rectangular two-dimensional (QAM) signal constellations.

Of particular practical importance is the class of trellis-coded two-dimensional rectangular signal constellations. Figure 5.4.9 illustrates these signal constellations for M-QAM where $M = 16$, 32, 64, and 128. The $M = 32$ and 128 constellations have a cross pattern and are sometimes called *cross constellations*. The underlying rectangular grid containing the signal points in M-QAM is called a *lattice of type* Z_2 (the subscript indicates the dimensionality of the space). When set partitioning is applied to this class of signal constellations, the minimum Euclidean distance between successive partitions is $d_{i+1}/d_i = \sqrt{2}$ for all i, as previously observed in Example 5.4.2.

Figure 5.4.10 illustrates an eight-state trellis code that can be used with any of the M-QAM rectangular signal constellations for which $M = 2^k$, where $k = 4, 5, 6, \ldots$, etc. With the eight-state trellis we associate eight signal subsets, so that any of the M-QAM signals sets for $M \geq 16$ are suitable. For $M = 2^{m+1}$, two input bits ($k_1 = 2$) are encoded into $n = 3$ ($n = k_1 + 1$) bits that are used to select one of the eight states corresponding to the eight subsets. The additional $k_2 = m - k_1$ input bits are used to select signal points within a subset and result in parallel transitions in the eight-state trellis. Hence, 16-QAM involves two parallel transitions in each branch of the trellis. More generally, the choice of an $M = 2^{m+1}$-point QAM signal constellation implies that the eight-state trellis contains 2^{m-2} parallel transitions in each branch.

$D_0 D_2 D_4 D_6 \quad S_0$

D_0
D_2
D_4
D_6

$D_1 D_7 D_5 D_3 \quad S_1$

$D_6 D_4 D_2 D_0 \quad S_2$

$D_7 D_1 D_3 D_5 \quad S_3$

$D_4 D_6 D_0 D_2 \quad S_4$

$D_5 D_3 D_1 D_7 \quad S_5$

$D_2 D_0 D_6 D_4 \quad S_6$

D_3
D_5
D_7

$D_3 D_5 D_7 D_1 \quad S_7$

D_1

FIGURE 5.4.10
Eight-state trellis for rectangular QAM signal constellations.

The assignment of signal subsets to transitions is based on the same set of basic (heuristic) rules described above for the 8-PSK signal constellation. Thus, the four (branches) transitions originating from or leading to the same state are assigned either the subsets D_0, D_2, D_4, D_6 or D_1, D_3, D_5, D_7. Parallel transitions are assigned signal points contained within the corresponding subsets. This eight-state trellis code provides a coding gain of 4 dB. The Euclidean distance of parallel transitions exceeds the free Euclidean distance and, hence, the code performance is not limited by parallel transitions.

Larger size trellis codes for M-QAM provide even larger codings gains. For example, trellis codes with 2^ν states for an $M = 2^{m+1}$ QAM signal constellation can be constructed by convolutionally encoding k_1 input bits into $k_1 + 1$ output bits. Thus, a rate $R_c = k_1/(k_1 + 1)$ convolutional code is employed for this purpose. Usually, the choice of $k_1 = 2$ provides a significant fraction of the total

TABLE 5.4.1
Coding gains for trellis-coded ASK signals†

Number of states	k_1	Code rate $\dfrac{k_1}{k_1 + 1}$	$m = 1$ coding gain (dB) of 4-ASK vs. uncoded 2-ASK	$m = 2$ coding gain (dB) of 8-ASK vs. uncoded 4-ASK	$m \to \infty$ asymptotic coding gain (dB)	$m \to \infty$ N_{fed}
4	1	1/2	2.55	3.31	3.52	4
8	1	1/2	3.01	3.77	3.97	4
16	1	1/2	3.42	4.18	4.39	8
32	1	1/2	4.15	4.91	5.11	12
64	1	1/2	4.47	5.23	5.44	36
128	1	1/2	5.05	5.81	6.02	66

† Ungerboeck (1987).

coding gain that is achievable. The additional $k_2 = m - k_1$ input bits are uncoded and are transmitted in each signal interval by selecting signal points within a subset.

The following tables given in the paper by Ungerboeck (1987) provide a summary of coding gains achievable with trellis-coded modulation. Table 5.4.1 summarizes the coding gains achieved for trellis-coded (one-dimensional) ASK modulation with rate 1/2 trellis codes. Note that the coding gain with a 128-state trellis code is 5.8 dB for octal ASK, which is close to the channel cutoff rate R_0 and less than 4 dB from the channel capacity limit for error rates in the range of 10^{-6} to 10^{-8}. We should also observe that the number of paths N_{fed} with free Euclidean distance D_{fed} becomes large with an increase in the number of states.

Table 5.4.2 lists the coding gain for trellis-coded 16-PSK. Again, we observe that the coding gain for eight or more trellis stages exceeds 4 dB, relative to uncoded 8-PSK. A simple rate 1/2 code yields 5.33-dB gain with a 128-stage trellis.

TABLE 5.4.2
Coding gains for trellis-coded 16-PSK modulation†

Number of states	k_1	Code rate $\dfrac{k_1}{k_1 + 1}$	$m = 3$ coding gain (dB) of 16-PSK vs. uncoded 8-PSK	$m \to \infty$ N_{fed}
4	1	1/2	3.54	4
8	1	1/2	4.01	4
16	1	1/2	4.44	8
32	1	1/2	5.13	8
64	1	1/2	5.33	2
128	1	1/2	5.33	2
256	2	2/3	5.51	8

† Ungerboeck (1987).

TABLE 5.4.3
Coding gains for trellis-coded QAM modulation†

Number of states	k_1	Code Rate $\dfrac{k_1}{k_1 + 1}$	$m = 3$ gain (dB) of 16-QAM vs. uncoded 8-QAM	$m = 4$ gain (dB) of 32-QAM vs. uncoded 16-QAM	$m = 5$ gain (dB) of 64-QAM vs. uncoded 32-QAM	$m = \infty$ asymptotic coding gain (dB)	N_{fed}
4	1	1/2	3.01	3.01	2.80	3.01	4
8	2	2/3	3.98	3.98	3.77	3.98	16
16	2	2/3	4.77	4.77	4.56	4.77	56
32	2	2/3	4.77	4.77	4.56	4.77	16
64	2	2/3	5.44	5.44	4.23	5.44	56
128	2	2/3	6.02	6.02	5.81	6.02	344
256	2	2/3	6.02	6.02	5.81	6.02	44

† Ungerboeck (1987).

Table 5.4.3 contains the coding gains obtained with trellis-coded QAM signals. Relatively simple rate 2/3 trellis codes yield a gain of 6 dB with 128 trellis stages for $m = 3$ and 4.

The results in these tables clearly illustrate the significant coding gains that are achievable with relatively simple trellis codes. A 6-dB coding gain is close to the cutoff rate R_0 for the signal sets under consideration. Additional gains that would lead to transmission in the vicinity of the channel capacity bound are difficult to attain without a significant increase in coding/decoding complexity.

Since the channel capacity provides the ultimate limit on code performance, we should emphasize that continued partitioning of large signal sets quickly leads to signal point separation within any subset that exceeds the free Euclidean distance of the code. In such cases, parallel transitions are no longer the limiting factor on D_{fed}. Usually, a partition to eight subsets is sufficient to obtain a coding gain of 5 to 6 dB with simple rate 1/2 or rate 2/3 trellis codes with either 64 or 128 trellis stages, as indicated in Tables 5.4.1 through 5.4.3.

Convolutional encoders for the linear trellis codes listed in Tables 5.4.1 through 5.4.3 for the M-ASK, M-PSK, and M-QAM signal constellations are given in the papers by Ungerboeck (1982, 1987). The encoders may be realized either with feedback or without feedback. For example, Fig. 5.4.11 illustrates three feedback-free convolutional encoders corresponding to 4-, 8-, and 16-state trellis codes for 8-PSK and 16-QAM signal constellations. Equivalent realizations of these trellis codes based on systematic convolutional encoders with feedback are shown in Fig. 5.4.12. Usually, the systematic convolutional encoders are preferred in practical application.

A potential problem with linear trellis codes is that the modulated signal sets are not usually invariant to phase rotations. This poses a problem in practical

(a) 4-state encoder

(b) 8-state encoder

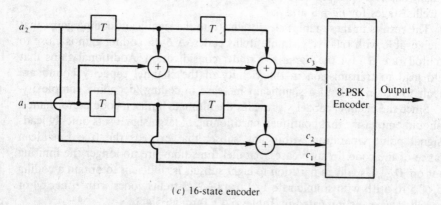

(c) 16-state encoder

FIGURE 5.4.11
Minimal feedback-free convolutional encoders for 8-PSK and 16-QAM signals. [From paper by Ungerboeck (1982). © 1982 IEEE. Reprinted with permission.]

applications where differential encoding is usually employed to avoid phase ambiguities when a receiver must recover the carrier phase after a temporary loss of signal. The problem of phase invariance and differential encoding/decoding was solved by Wei (1984a, b) who devised linear and nonlinear trellis codes that are rotationally invariant to either 180° or 90° phase rotations, respectively. For example, Fig. 5.4.13 illustrates a nonlinear eight-state convolutional encoder for a

FIGURE 5.4.12
Equivalent realizations of systematic convolutional encoders with feedback for 8-PSK and 16-QAM. [From paper by Ungerboeck (1982). © 1982 IEEE. Reprinted with permission.]

32-QAM rectangular signal constellation that is invariant to 90° phase rotations. This trellis code has been adopted as an international standard for 9600 and 14400 bits/s (high speed) telephone line modems.

Trellis-coded modulation schemes have also been developed for multidimensional signals. In practical systems, multidimensional signals are transmitted as a sequence of either one-dimensional (ASK) or two-dimensional (QAM) signals. Trellis codes based on 4-, 8-, and 16-dimensional signal constellations have been constructed and some of these codes have been implemented in commercially available modems. A potential advantage of trellis-coded multidimensional signals is that we can use smaller constituent two-dimensional signal constellations that allow for a tradeoff between coding gain and implementation

FIGURE 5.4.13
Eight-state nonlinear convolutional encoder for 32-QAM signal set which exhibits invariance to 90° phase rotations.

(a) Encoder

(b) 32-point QAM (cross) signal

Binary sequence below signal point: $c_5 c_4 c_3 c_2 c_1$

FIGURE 5.4.13 (Continued).

complexity. The papers by Wei (1987), Ungerboeck (1987), Gersho and Lawrence (1984), and Forney et al. (1984) treat multidimensional signal constellations for trellis-coded modulation.

Finally, we should mention that a new design technique for trellis-coded modulation based on lattices and cosets of a sublattice has recently been described by Calderbank and Sloane (1987) and Forney (1988). This method for constructing trellis codes provides an alternative to the set partitioning method described above. However, the two methods are closely related. In this alternative method, a block of k_1 bits is fed to a convolutional encoder. Each block of k_1 input bits produces an output symbol that is a coset of the sublattice Λ' which is a subset of the chosen lattice. A second block of k_2 input bits is used to select one of the points in the coset at the output of the convolutional encoder. It is apparent that the cosets of the sublattice are akin to the subsets in set partitioning and the elements of the cosets are akin to the signal poir ts within a subset. This new method has led to the discovery of new powerful trellis codes involving larger signal constellations, many of which are listed in the paper by Calderbank and Sloane (1987).

5.5 BIBLIOGRAPHICAL NOTES AND REFERENCES

The pioneering work on coding and coded waveforms for digital communications was done by Shannon (1948a, b), Hamming (1950), and Golay (1949). These works were rapidly followed with papers on code performance by Gilbert (1952), new codes by Muller (1954) and Reed (1954), and coding techniques for noisy channels by Elias (1954, 1955) and Slepian (1956).

During the period 1960–1970 there were a number of significant contributions in the development of coding theory and decoding algorithms. In particular, we cite the papers by Reed and Solomon (1960) on Reed–Solomon codes, the papers by Hocquenghem (1959) and Bose and Ray-Chaudhuri (1960a, b) on BCH codes, and the Ph.D. dissertation of Forney (1966a) on concatenated codes. These works were followed by the papers of Goppa (1970, 1971) on the construction of a new class of linear cyclic codes, now called Goppa codes [also, see Berlekamp (1973)], and the paper of Justesen (1972) on a constructive technique for asymptotically good codes. During this period, work on decoding algorithms was primarily focused on BCH codes. The first decoding algorithm for binary BCH codes was developed by Peterson (1960). A number of refinements and generalizations by Chien (1964), Forney (1965), Massey (1965), and Berlekamp (1968) led to the development of a computationally efficient algorithm for BCH codes which is described in detail by Lin and Costello (1983).

In parallel with these developments on block codes are the developments in convolutional codes which were invented by Elias (1955). The major problem in convolutional coding was decoding. Wozencraft and Reiffen (1961) described a sequential decoding algorithm for convolutional codes. This algorithm was later modified and refined by Fano (1963) and it is now called the Fano algorithm. Subsequently, the stack algorithm was devised by Ziganzirov (1966) and Jelinek (1969) and the Viterbi algorithm was devised by Viterbi (1967). The optimality and the relatively modest complexity for small constraint lengths have served to make Viterbi algorithm the most popular in decoding of convolutional codes with $L \leq 10$.

One of the most important contributions in coding during the 1970s was the work of Ungerboeck and Csajka (1976) on coding for bandwidth-constrained channels. In this paper it was demonstrated that a significant coding gain can be achieved through the introduction of redundancy in a bandwidth-constrained channel and trellis codes were described for achieving coding gains of 3 to 4 dB. This work has generated much interest among researchers and has led to a large number of publications over the past 10 years. A number of references can be found in the papers by Ungerboeck (1982, 1987a, b) and Forney et al. (1984). Additional papers on coded modulation for bandwidth-constrained channels may also be found in the Special Issue on Voiceband Telephone Data Transmission, *IEEE Journal on Selected Areas in Communication* (September 1984).

The use of the cutoff rate parameter as a design criterion was proposed and developed in a paper by Wozencraft and Kennedy (1966) and in the book by Wozencraft and Jacobs (1965). It was used by Jordan (1966) in the design of coded waveforms for M-ary orthogonal signals with coherent and noncoherent detection. Following these pioneering works, the cutoff rate has been widely used as a design criterion for coded signals in a variety of different channel conditions.

In addition to the references given above on coding, decoding, and coded signal design, we should include the collection of papers published by the IEEE Press entitled *Key Papers in the Development of Coding Theory*, edited by Berlekamp (1972). This book contains important papers that were published in the first 25-year history of coding theory. We should also cite the Special Issue in Error-Correcting Codes, published by the *IEEE Transactions on Communications* in October 1971.

APPENDIX 5A
COMMUNICATION LINK BUDGET ANALYSIS

In the design of radio communications systems that transmit over line-of-sight microwave channels and satellite channels, the system designer must specify the size of the transmit and receive antennas, the transmitted power, and the SNR required to achieve a given level of performance at some desired data rate. The system design procedure is relatively straightforward and is outlined below.

FIGURE 5A.1-1

Let us begin with a transmit antenna that radiates isotropically in free space at a power level of P_T watts as shown in Fig. 5A.1. The power density at a distance d from the antenna is $P_T/4\pi d^2$ w/m². If the transmit antenna has some directivity in a particular direction, the power density in that direction is increased by a factor called the antenna gain and denoted by G_T. In such a case the power density at a distance d is $P_T G_T/4\pi d^2$ w/m². The product $P_T G_T$ is usually called the *effective radiated power* (ERP or EIRP), which is basically the radiated power relative to an isotropic antenna for which $G_T = 1$.

A receiving antenna pointed in the direction of the radiated power gathers a portion of the power that is proportional to its cross sectional area. Hence, the received power extracted by the antenna may be expressed as

$$P_R = \frac{P_T G_T A_R}{4\pi d^2} \tag{5A.1}$$

where A_R is the *effective area of the antenna*. From electromagnetic field theory, we obtain the basic relationship between the gain of an antenna and its effective area as

$$A_R = \frac{G_R \lambda^2}{4\pi} \text{ m}^2 \tag{5A.2}$$

where $\lambda = c/f$ is the wavelength of the transmitted signal, c is the speed of light $(3 \times 10^8 \text{ m/s})$, and f is the frequency of the transmitted signal.

If we substitute (5A.2) for A_R into (5A.1) we obtain the expression for the received power in the form

$$P_R = \frac{P_T G_T G_R}{(4\pi d/\lambda)^2} \tag{5A.3}$$

The factor

$$L_s = \left(\frac{\lambda}{4\pi d} \right)^2 \tag{5A.4}$$

is called the *free-space path loss*. If other losses, such as atmospheric losses, are encountered in the transmission of the signal, they may be accounted for by introducing an additional loss factor, say L_a. Therefore, the received power may be written in general as

$$P_R = P_T G_T G_R L_s L_a \qquad (5A.5)$$

As indicated above, the important characteristics of an antenna are its gain and its effective area. These quantities are generally dependent on the wavelength of the radiated power and the physical dimensions of the antenna. For example, a parabolic (dish) antenna of diameter D has an effective area

$$A_R = \frac{\pi D^2}{4}\eta \qquad (5A.6)$$

where $\pi D^2/4$ is the physical area and η is the *illumination efficiency factor* which falls in the range $0.5 \le \eta \le 0.6$. Hence, the antenna gain for a parabolic antenna of diameter D is

$$G_R = \eta\left(\frac{\pi D}{\lambda}\right)^2 \quad \text{parabolic antenna} \qquad (5A.7)$$

As a second example, a horn antenna of physical area A has an efficiency factor of 0.8, an effective area of $A_R = 0.8A$, and an antenna gain of

$$G_R = \frac{10A}{\lambda^2} \quad \text{horn antenna} \qquad (5A.8)$$

Another parameter that is related to the gain (directivity) of an antenna is its beamwidth, which we denote as Θ_B and which is illustrated graphically in Fig. 5A.2. Usually, the beamwidth is measured as the -3-dB width of the antenna pattern. For example, the -3-dB beamwidth of a parabolic antenna is approximately

$$\Theta_B = 70(\lambda/D) \text{ deg} \qquad (5A.9)$$

so that G_T is inversely proportional to Θ_B^2. That is, a decrease of the beamwidth by a factor of 2, which is obtained by doubling the diameter D, increases the antenna gain by a factor of 4 (6 dB).

Based on the general relationship for the received signal power given by (5A.5), the system designer can compute P_R from a specification of the antenna gains and the distance between the transmitter and the receiver. Such computations are usually done on a power basis, so that

$$(P_R)_{dB} = (P_T)_{dB} + (G_T)_{dB} + (G_R)_{dB} + (L_s)_{dB} + (L_a)_{dB} \qquad (5A.10)$$

Example 5A.1. Suppose we have a satellite in geosynchronous orbit (36,000 km above the earth's surface) which radiates 100 w of power, i.e., 20 dB above 1 w (20 dBw). The transmit antenna has a gain of 17 dB, so that the ERP = 37 dBw. Also, suppose that the earth station employs a 3-m parabolic antenna and that the

(*a*) Beamwidth of antenna

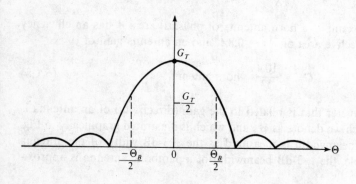

(*b*) Antenna pattern

FIGURE 5A.1-2

downlink is operating at a frequency of 4 GHz. The efficiency factor is $\eta = 0.5$. By substituting these numbers into (5A.7) we obtain the value of the antenna gain as 39 dB. The free-space path loss is

$$L_s = 195.6 \text{ dB}$$

No other losses are assumed. Therefore, the received signal power is

$$(P_R)_{\text{dB}} = 20 + 17 + 39 - 195.6$$

$$= -119.6 \text{ dBw}$$

or, equivalently,

$$P_R = 1.1 \times 10^{-12} \text{ w}$$

To complete the link budget computation we must also consider the effect of the additive noise at the receiver front end. Thermal noise that arises at the receiver front end has a relatively flat power density spectrum up to about 10^{12} Hz, and is given as

$$N_0 = \mathcal{K} T_0 \text{ w/Hz} \qquad (5A.11)$$

where $\mathcal{K}$ is Boltzmann's constant $(1.38 \times 10^{-23} \text{ w-s/deg})$ and T_0 is the noise temperature in degrees Kelvin. Therefore, the total noise power in the signal bandwidth W is $N_0 W$.

The performance of the digital communications system is specified by the $\mathcal{E}_b / N_0$ required to keep the error rate performance below some given value. Since

$$\frac{\mathcal{E}_b}{N_0} = \frac{T_b P_R}{N_0} = \frac{1}{R} \frac{P_R}{N_0} \qquad (5A.12)$$

it follows that

$$\frac{P_R}{N_0} = R \left(\frac{\mathcal{E}_b}{N_0} \right)_{\text{req}} \qquad (5A.13)$$

where $(\mathcal{E}_b / N_0)_{\text{req}}$ is the required SNR per bit. Hence, if we have P_R / N_0 and the required SNR per bit, we can determine the maximum data rate that is possible.

Example 5A.2. For the link considered in Example 5A.1, the received signal power is

$$P_R = 1.1 \times 10^{-12} \text{ w} \quad (-119.6 \text{ dBw})$$

Now, suppose the receiver front end has a noise temperature of 300K, which is typical for receiver in the 4-GHz range. Then

$$N_0 = 4.1 \times 10^{-21} \text{ w/Hz}$$

or, equivalently, -203.9 dBw/Hz. Therefore,

$$\frac{P_R}{N_0} = -119.6 + 203.9 = 84.3 \text{ dB-Hz}$$

If the required SNR per bit is 10 dB, then, from (5A.13), we have the available rate as

$$R_{dB} = 84.3 - 10$$

$$= 74.3 \text{ dB} \quad (\text{with respect to } 1 \text{ bit/s})$$

This corresponds to a rate of 26.9 megabits per second, which is equivalent to about 420 PCM channels, each operating at 64,000 bits per second.

It is a good idea to introduce some safety margin, which we will call the *link margin* M_{dB} in the above computations for the capacity of the communication link. Typically, this may be selected as $M_{dB} = 6$ dB. Then, the link budget computation for the link capacity may be expressed in the simple form

$$R_{dB} = \left(\frac{P_R}{N_0}\right)_{dB\text{-}Hz} - \left(\frac{\mathscr{E}_b}{N_0}\right)_{req} - M_{dB}$$

$$= (P_T)_{dBw} + (G_T)_{dB} + (G_R)_{dB}$$

$$+ (L_a)_{dB} + (L_s)_{dB} - \left(\frac{\mathscr{E}_b}{N_0}\right)_{req} - M_{dB} \qquad (5A.14)$$

PROBLEMS

5.1. The generator matrix for a linear binary code is

$$\mathbf{G} = \begin{bmatrix} 0 & 0 & 1 & 1 & 1 & 0 & 1 \\ 0 & 1 & 0 & 0 & 1 & 1 & 1 \\ 1 & 0 & 0 & 1 & 1 & 1 & 0 \end{bmatrix}$$

(a) Express $\mathbf{G}$ in the systematic [I|P] form.
(b) Determine the parity check matrix $\mathbf{H}$ for the code.
(c) Construct the table of syndromes for the code.
(d) Determine the minimum distance of the code.
(e) Demonstrate that the code word corresponding to the information sequence 101 is orthogonal to $\mathbf{H}$.

5.2. List the code words generated by the matrices given in (5.2.35) and (5.2.37) and, thus, demonstrate that these matrices generate the same set of code words.

5.3. The weight distribution of Hamming codes is known [1]. Expressed as a polynomial in powers of x, the weight distribution for the binary Hamming codes of block length n is

$$A(x) = \sum_{i=0}^{n} A_i x^i$$

$$= \frac{1}{n+1} \left[(1+x)^n + n(1+x)^{(n-1)/2}(1-x)^{(n+1)/2} \right]$$

where A_i is the number of code words of weight i. Use this formula to determine the weight distribution of the (7, 4) Hamming code and check your result with the list of code words given in Table 5.2.2.

5.4. The polynomial

$$g(p) = p^4 + p + 1$$

is the generator for the (15, 11) Hamming binary code.
(*a*) Determine a generator matrix **G** for this code in systematic form.
(*b*) Determine the generator polynomial for the dual code.

5.5. For the (7, 4) cyclic Hamming code with generator polynomial $g(p) = p^3 + p^2 + 1$, construct an (8, 4) extended Hamming code and list all the code words. What is $d_{\min}$ for the extended code?

5.6. An (8, 4) linear block code is constructed by shortening a (15, 11) Hamming code generated by the generator polynomial $g(p) = p^4 + p + 1$.
(*a*) Construct the code words of the (8, 4) code and list them.
(*b*) What is the minimum distance of the (8, 4) code?

5.7. The polynomial $p^{15} + 1$ when factored yields

$$p^{15} + 1 = (p^4 + p^3 + 1)(p^4 + p^3 + p^2 + p + 1)$$

$$\times (p^4 + p + 1)(p^2 + p + 1)(p + 1)$$

(*a*) Construct a systematic (15, 5) code using the generator polynomial

$$g(p) = (p^4 + p^3 + p^2 + p + 1)(p^4 + p + 1)(p^2 + p + 1)$$

(*a*) What is the minimum distance of the code?
(*c*) How many random errors per code word can be corrected?
(*d*) How many errors can be detected by this code?
(*e*) List the code words of a (15, 2) code constructed from the generator polynomial

$$g(p) = (p^{15} + 1)/(p^2 + p + 1)$$

and determine the minimum distance.

5.8. Construct the parity check matrices H_1 and H_2 corresponding to the generator matrices G_1 and G_2 given by (5.2.34) and (5.2.35), respectively.

5.9. Construct an extended $(8,4)$ code from the $(7,4)$ Hamming code by specifying the generator matrix and the parity check matrix.

5.10. A systematic $(6,3)$ code has the generator matrix

$$G = \begin{bmatrix} 1 & 0 & 0 & 1 & 1 & 0 \\ 0 & 1 & 0 & 0 & 1 & 1 \\ 0 & 0 & 1 & 1 & 0 & 1 \end{bmatrix}$$

Construct the standard array and determine the correctable error patterns and their corresponding syndromes.

5.11. Construct the standard array for the $(7,3)$ code with generator matrix

$$G = \begin{bmatrix} 1 & 0 & 0 & 1 & 0 & 1 & 1 \\ 0 & 1 & 0 & 1 & 1 & 1 & 0 \\ 0 & 0 & 1 & 0 & 1 & 1 & 1 \end{bmatrix}$$

and determine the correctable error patterns and their corresponding syndromes.

5.12. Determine the correctable error patterns (of least weight) and their syndromes for the systematic $(7,4)$ cyclic Hamming code.

5.13. Prove that if the sum of two error patterns $\mathbf{e}_1$ and $\mathbf{e}_2$ is a valid code word $\mathbf{C}_j$, then each pattern has the same syndrome.

5.14. Consider the binary-input, ternary-output channel with transition probabilities shown in Fig. P5.14 where e denotes an erasure. For the AWGN channel, α and p

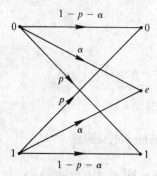

FIGURE P5.14

are defined as

$$\alpha = \frac{1}{\sqrt{\pi N_0}} \int_{-\beta}^{\beta} e^{-(x + \sqrt{\mathscr{E}_c})^2 / N_0} \, dx$$

$$p = \frac{1}{\sqrt{\pi N_0}} \int_{\beta}^{\infty} e^{-(x + \sqrt{\mathscr{E}_c})^2 / N_0} \, dx$$

(*a*) Determine R_Q for $Q = 3$ as a function of the probabilities α and p.

(*b*) The rate parameter R_Q depends on the choice of the threshold β, through the probabilities α and p. For any $\mathscr{E}_c / N_0$, the value of β that maximizes R_Q can be determined by trial and error. For example, it can be shown that for $\mathscr{E}_c/N_0$ below 0 dB, $\beta_{opt} \approx 0.65\sqrt{N_0}/2$; for $1 \le \mathscr{E}_c/N_0 \le 10$, β_{opt} varies approximately linearly between $0.65\sqrt{N_0}/2$ and $1.0\sqrt{N_0}/2$. By using $\beta = 0.65\sqrt{N_0}/2$ for the entire range of $\mathscr{E}_c/N_0$, plot R_Q versus $\mathscr{E}_c/N_0$ and compare this result with R_Q $(Q = \infty)$.

5.15. Prove that any two n-tuples in the same row of a standard array add to produce a valid code word.

5.16. Beginning with a $(15, 7)$ BCH code, construct a shortened $(12, 4)$ code. Give the generator matrix for the shortened code.

5.17. In Sec. 5.2.2 it was indicated that when an (n, k) Hadamard code is mapped into waveforms by means of binary PSK, the corresponding $M = 2^k$ waveforms are orthogonal. Determine the bandwidth expansion factor for the M orthogonal waveforms and compare this with the bandwidth requirements of orthogonal FSK detected coherently.

5.18. Show that the signaling waveforms generated from a maximum-length shift-register code by mapping each bit in a code word into a binary PSK signal are equicorrelated with correlation coefficient $\rho_r = -1/(M - 1)$, i.e., the M waveforms form a simplex set.

5.19. Compute the error probability obtained with a $(7, 4)$ Hamming code on an AWGN channel, both for hard-decision and soft-decision decoding. Use the formulas given in (5.2.50), (5.2.52), (5.2.79), (5.2.90), and (5.2.91).

5.20. Use the results in Sec. 1.1.6 to obtain the Chernoff bound for hard-decision decoding given by (5.2.89) and (5.2.90). Assume that the all-zero code word is transmitted and determine an upper bound on the probability that code word C_m, having weight w_m, is selected. This occurs if $w_m/2$ or more bits are in error. To apply the Chernoff bound, define a sequence of w_m random variables as

$$X_i = \begin{cases} 1 & \text{with probability } p \\ -1 & \text{with probability } 1 - p \end{cases}$$

where $i = 1, 2, \ldots, w_m$, and p is the probability of error. For the BSC the $\{X_i\}$ are statistically independent.

5.21. The channel capacity for a binary-input AWGN channel is given as [see (2.4.18)]

$$C = \max_{P(x)} \sum_x P(x) \int_{-\infty}^{\infty} p(y|x) \log \frac{p(y|x)}{p(y)} \, dy$$

where C is maximized when $P(1) = P(0) = \frac{1}{2}$. When the bits in a code word are transmitted by binary PSK, the pdf's for the output are

$$p(y|0) = \frac{1}{\sqrt{2\pi}\,\sigma} e^{-(y+m^2)/2\sigma^2}$$

$$p(y|1) = \frac{1}{\sqrt{2\pi}\,\sigma} e^{-(y-m)^2/\sigma^2}$$

where $m = 2\mathscr{E}_c$, $\sigma^2 = 2\mathscr{E}_c N_0$, $\mathscr{E}_c$ is energy per element of the code word, $\mathscr{E}_b = \mathscr{E}_c / R_c$, and R_c is the code rate. Determine the minimum SNR per bit to achieve capacity. *Hint*: As $R_c \to 0$,

$$\ln \frac{p(y|x)}{p(y)} = \ln \frac{2}{1 + e^{\alpha y}} \approx -\tfrac{1}{2}\alpha y$$

where α is a function of the SNR and R_c.

5.22. Figure P5.22 depicts a rate $1/2$, constraint length $L = 2$, convolutional code.
  (a) Sketch the tree diagram, the trellis diagram, and the state diagram.
  (b) Solve for the transfer function $T(D, N, J)$ and, from this, specify the minimum free distance.

FIGURE P5.22

5.23. A rate $1/2$, $L = 3$, binary convolutional encoder is shown in Fig. P5.23.
  (a) Draw the tree diagram, the trellis diagram, and the state diagram.
  (b) Determine the transfer function $T(D, N, J)$ and, from this, specify the minimum free distance.

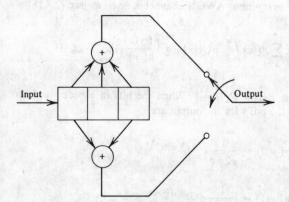

FIGURE P5.23

5.24. Sketch the convolutional encoders for the following codes:

    (*a*) Rate $1/2$, $L = 5$, maximum free distance code (Table 5.3.1).
    (*b*) Rate $1/3$, $L = 5$, maximum free distance code (Table 5.3.2).
    (*c*) Rate $2/3$, $L = 2$, maximum free distance code (Table 5.3.8).

5.25. Draw the state diagram for the rate $2/3$, $L = 2$, convolutional code indicated in Prob. 5.16*c* and, for each transition, show the output sequence and the distance of the output sequence from the all-zero sequence.

5.26. Consider the $L = 3$, rate $1/2$, convolutional code shown in Fig. P5.15. Suppose the code is used on a binary symmetric channel and the received sequence for the first eight branches is 0001100000001001. Trace the decisions on a trellis diagram and label the survivors' Hamming distance metric at each node level. If a tie occurs in the metrics required for a decision, always choose the upper path (arbitrary choice).

5.27. Use the transfer function derived in Prob. 5.15 for the $R_c = 1/2$, $L = 3$, convolutional code to compute the probability of a bit error for an AWGN channel with (*a*) hard-decision and (*b*) soft-decision decoding. Compare the performance by plotting the results of the computation on the same graph.

5.28. Use the generators given by (5.3.36) to obtain the encoder for a dual-3, rate $1/2$ convolutional code. Determine the state diagram and derive the transfer function $T(D, N, J)$.

5.29. Draw the state diagram for the convolutional code generated by the encoder shown in Fig. P5.29 and, thus, determine if the code is catastrophic or noncatastrophic. Also give an example of a rate $1/2$, $L = 4$, convolutional encoder that exhibits catastrophic error propagation.

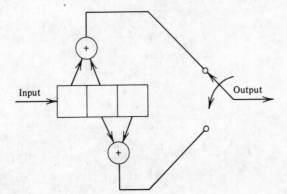

FIGURE P5.29

5.30. A trellis coded signal is formed as shown in Fig. P5.30 by encoding one bit by use of a rate $1/2$ convolutional code, while three additional information bits are left

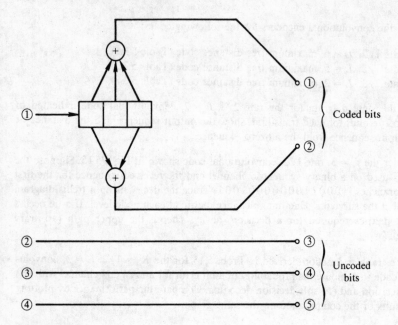

FIGURE P5.30

uncoded. Perform the set partitioning of a 32-QAM (cross) constellation and indicate the subsets in the partition. By how much is the distance between adjacent signal points increased as a result of partitioning?

CHAPTER 6

DIGITAL SIGNALING OVER A BANDWIDTH CONSTRAINED LINEAR FILTER CHANNEL

In Chaps. 4 and 5 we considered the transmission of digital information over an additive gaussian noise channel. Such a channel altered the transmitted signals only to the extent that it introduced an attenuation factor and a phase shift, in addition to the additive gaussian noise. In effect, the channel bandwidth was unrestricted. When the channel is band-limited, its effect on the transmitted signal can still be modeled by a simple attenuation and a phase shift, provided the signaling waveforms are relatively narrower in bandwidth than the channel. That is, for sufficiently narrowband signals, the channel appears to have a constant gain (attenuation) and a constant delay for all frequency components contained in the transmitted signal. In such a case, however, only a portion of the available channel bandwidth is being used. A signal design based on use of only a portion of the available channel bandwidth is generally viewed as an inefficient one.

In this chapter we consider the problem of efficient modulation and demodulation of digital information transmitted over a channel that is band-limited to W Hz. The channel is modeled as a linear filter having an equivalent

519

low-pass impulse response $c(t)$ and a frequency response $C(f)$ that is zero for $|f| > W$.

The first topic that we shall discuss is the design of the signal pulse $u(t)$, in a linearly modulated signal represented as

$$\sum_{n=0}^{\infty} I_n u(t - nT)$$

which efficiently utilizes a band-limited channel. We shall see that when the channel is ideal for $|f| \leq W$, a signal pulse $u(t)$ can be designed which allows us to transmit at symbol rates that are comparable to or exceeding the channel bandwidth W. On the other hand, when the channel is not ideal, signal transmission at a symbol rate equal to or exceeding W results in intersymbol interference among a number of adjacent received symbols.

The second topic treated in this chapter is the structure of the demodulator for processing the received signal which is corrupted by intersymbol interference and additive gaussian noise. A number of different demodulator structures are described and their performances are evaluated.

In practical digital communications systems designed to transmit at high speed over band-limited channels, the frequency response $C(f)$ of the channel is not known with sufficient precision to design a fixed (time-invariant) demodulator for mitigating the intersymbol interference. Consequently a part of this demodulator is made adaptive. Specifically, the filter or signal processing algorithm for handling the intersymbol interference at the demodulator contains a number of parameters which are adaptively adjusted on the basis of measurements of the channel characteristics. Such adaptive filters and adaptive signal processing algorithms are called *adaptive equalizers* and *adaptive equalization algorithms*, respectively. A major part of this chapter is devoted to the discussion of algorithms for adjusting the parameters of adaptive equalizers and to the convergence properties of the algorithms.

6.1 CHARACTERIZATION OF BAND-LIMITED CHANNELS

Of the various channels available for digital communications, telephone channels are by far the most widely used. Such channels are characterized as "band-limited linear filters." This is certainly the proper characterization when frequency division multiplexing (FDM) is used as a means for establishing channels in the telephone network. Recent additions to the telephone network employ pulse code modulation (PCM) for digitizing and encoding the analog signal and time-division multiplexing (TDM) for establishing multiple channels. Nevertheless, filtering is still used on the analog signal prior to sampling and encoding. Consequently, even though the present telephone network employs a mixture of FDM and TDM for transmission, the linear filter model for telephone channels is still appropriate.

For our purposes, a band-limited channel such as a telephone channel will be characterized as a linear filter having an equivalent lowpass frequency response characteristic $C(f)$. Its equivalent lowpass impulse response is denoted as $c(t)$. Then, if a signal of the form

$$s(t) = \text{Re}\left[v(t)e^{j2\pi f_c t}\right] \tag{6.1.1}$$

is transmitted over a bandpass telephone channel, the equivalent lowpass received signal is

$$r(t) = \int_{-\infty}^{\infty} v(\tau)c(t - \tau)\, d\tau + z(t) \tag{6.1.2}$$

where the integral represents the convolution of $c(t)$ with $v(t)$, and $z(t)$ denotes the additive noise. Alternatively, the signal term can be represented in the frequency domain as $V(f)C(f)$, where $V(f)$ is the Fourier transform of $v(t)$.

If the channel is band-limited to W Hz, then $C(f) = 0$ for $|f| > W$. As a consequence, any frequency components in $V(f)$ above $|f| = W$ will not be passed by the channel. For this reason, we limit the bandwidth of the transmitted signal to W Hz also.

Within the bandwidth of the channel, we may express the frequency response $C(f)$ as

$$C(f) = |C(f)|e^{j\theta(f)} \tag{6.1.3}$$

where $|C(f)|$ is the amplitude response characteristic and $\theta(f)$ is the phase response characteristic. Furthermore, the envelope delay characteristic is defined as

$$\tau(f) = -\frac{1}{2\pi}\frac{d\theta(f)}{df} \tag{6.1.4}$$

A channel is said to be "nondistorting" or "ideal" if the amplitude response $|C(f)|$ is constant for all $|f| \le W$ and $\theta(f)$ is a linear function of frequency, i.e., $\tau(f)$ is a constant for all $|f| \le W$. On the other hand, if $|C(f)|$ is not constant for all $|f| \le W$, we say that the channel "distorts the transmitted signal $V(f)$ in amplitude," and, if $\tau(f)$ is not constant for all $|f| \le W$, we say that the channel "distorts the signal $V(f)$ in delay."

As a result of the amplitude and delay distortion caused by the nonideal channel frequency response characteristic $C(f)$, a succession of pulses transmitted through the channel at rates comparable to the bandwidth W are smeared to the point that they are no longer distinguishable as well-defined pulses at the receiving terminal. Instead, they overlap and, thus, we have intersymbol interference. As an example of the effect of delay distortion on a transmitted pulse, Fig. 6.1.1a illustrates a band-limited pulse having zeros periodically spaced in time at points labeled $\pm T$, $\pm 2T$, etc. If information is conveyed by the pulse amplitude, as in PAM for example, then one can transmit a sequence of pulses each of which has a peak at the periodic zeros of the other pulses. However, transmission of the pulse through a channel modeled as having a linear envelope delay characteristic $\tau(f)$ [quadratic phase $\theta(f)$] results in the received pulse shown in Fig. 6.1.1.b having zero crossings that are no longer periodically spaced. Consequently a

(a)

(b)

(c)

FIGURE 6.1.1
Effect of channel distortion: (a) channel input, (b) channel output, (c) equalizer output.

sequence of successive pulses would be smeared into one another and the peaks of the pulses would no longer be distinguishable. Thus the channel delay distortion results in intersymbol interference. As will be discussed in this chapter, it is possible to compensate for the nonideal frequency response characteristic of the channel by use of a filter or equalizer at the demodulator. Figure 6.1.1*c* illustrates the output of a linear equalizer which compensates for the linear distortion in the channel.

The extent of the intersymbol interference on a telephone channel can be appreciated by observing a frequency response characteristic of the channel. Figure 6.1.2 illustrates the measured average amplitude and delay as a function of

FIGURE 6.1.2
Average amplitude and delay characteristics of medium-range telephone channel.

FIGURE 6.1.3
Impulse response of average channel with amplitude and delay shown in Fig. 6.1.2.

frequency for a medium-range (180 to 725-mi) telephone channel of the switched telecommunications network as given by Duffy and Tratcher (1971). We observe that the usable band of the channel extends from about 300 Hz to about 3000 Hz. The corresponding impulse response of this average channel is shown in Fig. 6.1.3. Its duration is about 10 ms. In comparison, the transmitted symbol rates on such a channel may be of the order of 2500 pulses or symbols per second. Hence intersymbol interference might extend over 20 to 30 symbols.

In addition to linear distortion, signals transmitted through telephone channels are subject to other impairments, specifically nonlinear distortion, frequency offset, phase jitter, impulse noise, and thermal noise.

Nonlinear distortion in telephone channels arises from nonlinearities in amplifiers and compandors used in the telephone system. This type of distortion is usually small and it is very difficult to correct.

A small *frequency offset*, usually less than 5 Hz, results from the use of carrier equipment in the telephone channel. Such an offset cannot be tolerated in high-speed digital transmission systems which use synchronous phase-coherent demodulation. The offset is usually compensated for by the carrier recovery loop in the demodulator.

Phase jitter is basically a low-index frequency modulation of the transmitted signal with the low frequency harmonics of the power line frequency (50 or 60 Hz). Phase jitter poses a serious problem in digital transmission of high rates. However, it can be tracked and compensated for, to some extent, at the demodulator.

Impulse noise is an additive disturbance. It arises primarily from the switching equipment in the telephone system. *Thermal* (gaussian) *noise* is also present at levels of 20 to 30 dB below the signal.

TABLE 6.1.1
Specifications for voice-bandwidth telephone channels with C-type conditioning

Characteristic	3002 Channel	3002 / C1 Channel	3002 / C2 Channel	3002 / C4 Channel
Loss at 1000 Hz	16 dB ± 1 dB	16 dB ± 1 dB	16 dB ± 1 dB	16 dB ± 1 dB
Maximum variation of loss				
(a) Short term	± 3 dB	± 3 dB	± 3 dB	± 3 dB
(b) Long term	± 4 dB	± 4 dB	± 4 dB	± 4 dB
Frequency response relative to 1000 Hz	300–3000 Hz: −3 to +12 dB 500–2500 Hz: −2 to +8 dB	3000–2700 Hz: −2 to +6dB 1000–2400 Hz: −1 to +3 dB 2700–3000 Hz: −3 to +12 dB	300–3000 Hz: −2 to +6 dB 500–2800 Hz: −1 to +3 dB	300–3200 Hz: −2 to +6 dB 500–3000 Hz: −2 to +3 dB
Envelope delay distortion	800–2600 Hz: ≤ 1.75 ms	1000–2400 Hz: ≤ 1 ms 800–2600 Hz: ≤ 1.75 ms	1000–2600 Hz: ≤ 0.5 ms 600–2600 Hz: ≤ 1.5 ms 500–2800 Hz: ≤ 3.0 ms	1000–2600 Hz: ≤ 0.3 ms 800–2000 Hz: ≤ 0.5 ms 600–3000 Hz: ≤ 1.5 ms 500–3000 Hz: ≤ 3.0 ms
Frequency error	± 5 Hz	± 5 Hz	± 5 Hz	± 5 Hz
Peak-to-peak phase jitter	10°	10°	10°	10°

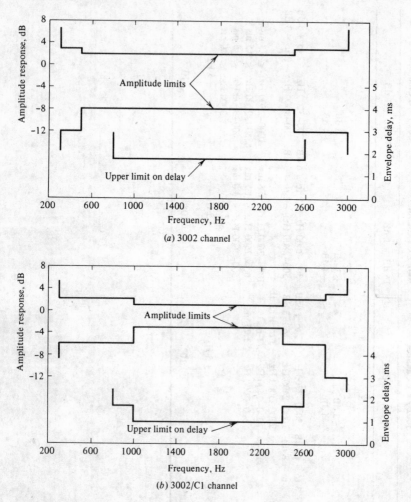

FIGURE 6.1.4
Amplitude and delay specifications for voice-bandwidth telephone channels with C-type conditioning.

Voice-bandwidth telephone channels that can be leased from telephone companies are categorized by the specifications which they meet. A number of these specifications are listed in Table 6.1.1 for types 3002 and 3002 with C1, C2, and C4 conditioning. It should be noted that the type 3002 channel is unconditioned. Types 3002-C1, 3002-C2, and 3002-C4 conditioned circuits employ analog equalizers to improve the frequency response characteristics of the channel. Figure 6.1.4a-d illustrates the amplitude and delay characteristics for these four channels.

The degree to which one must be concerned with these channel impairments depends on the transmission rate over the channel and the modulation technique. For rates below 1800 bits/s ($R/W < 1$), one can choose a modulation technique, for example FSK, which is relatively insensitive to the amount of

FIGURE 6.1.4 (*Continued*).

distortion encountered on typical telephone channels from all the sources listed above. For rates between 1800 and 2400 bits/s ($R/W \approx 1$), a more bandwidth-efficient modulation technique such as four-phase PSK is usually employed. At these rates some form of compromise equalization is often employed to compensate for the average amplitude and delay distortion in the channel. In addition, the carrier recovery method is designed to compensate for the frequency offset. The other channel impairments are not that serious in their effects on the error rate performance at these rates. At transmission rates above 2400 bits/s ($R/W > 1$), bandwidth-efficient coded modulation techniques such as trellis-coded QAM, PAM, and PSK are employed. For such rates, special attention must be paid to linear distortion, frequency offset, and phase jitter. Linear distortion is usually compensated for by means of an adaptive equalizer. Phase jitter is handled by a combination of signal design and some type of phase compensation at the demodulator. At rates above 9600 bits/s, special attention must be paid not only to linear distortion, phase jitter, and frequency offset, but also to the other channel impairments mentioned above.

Unfortunately, a channel model that encompasses all the impairments listed above becomes difficult to analyze. For mathematical tractability the channel model that is adopted in this chapter is a linear filter that introduces amplitude and delay distortion and adds gaussian noise.

Besides the telephone channels, there are other physical channels that exhibit some form of time dispersion, and thus, introduce intersymbol interference. Radio channels such as shortwave ionospheric propagation (HF) and tropospheric scatter are two examples of time-dispersive channels. In these channels, time dispersion and, hence, intersymbol interference is the result of multiple propagation paths with different path delays. The number of paths and the relative time delays among the paths vary with time and, for this reason, these radio channels are usually called "time-variant multipath channels." The time-variant multipath conditions give rise to a wide variety of frequency response characteristics. Consequently the frequency response characterization that is used for telephone channels is inappropriate for time-variant multipath channels. Instead, these radio channels are characterized statistically, as explained in more detail in Chap. 7, in terms of the scattering function which, in brief, is a two-dimensional representation of the average received signal power as a function of relative time delay and Doppler frequency.

For illustrative purposes, a scattering function measured on a medium-range (150-mi) tropospheric scatter channel is shown in Fig. 6.1.5. The total time duration (multipath spread) of the channel response is approximately $0.7\ \mu s$ on the average and the spread between "half-power points" in Doppler frequency is a little less than 1 Hz on the strongest path and somewhat larger on the other paths. Typically, if one is transmitting at a rate of 10^7 symbols/s over such a channel, the multipath spread of $0.7\ \mu s$ will result in intersymbol interference that spans about seven symbols.

Although in this chapter we deal almost exclusively with the linear time-invariant filter model for a band-limited channel, the adaptive equalization techniques that are presented for combating intersymbol interference are also applicable to time-variant multipath channels, under the condition that the time variations in the channel are relatively slow in comparison to the total channel bandwidth or, equivalently, to the symbol transmission rate over the channel.

6.2 SIGNAL DESIGN FOR BAND-LIMITED CHANNELS

It was shown in Chap. 3 that the equivalent lowpass transmitted signal for several different types of digital modulation techniques has the common form

$$\sum_{n=0}^{\infty} I_n u(t - nT) \tag{6.2.1}$$

FIGURE 6.1.5
Scattering function of a medium-range tropospheric scatter channel.

where $\{I_n\}$ represents the discrete information-bearing sequence of symbols and $u(t)$ a pulse which, for purposes of this discussion, is assumed to have a band-limited frequency response characteristic $U(f)$, i.e., $U(f) = 0$ for $|f| > W$. This signal is transmitted over a channel having a frequency response $C(f)$, also limited to $|f| \leq W$. Consequently the received signal can be represented as

$$r(t) = \sum_{n=0}^{\infty} I_n h(t - nT) + z(t) \qquad (6.2.2)$$

where

$$h(t) = \int_{-\infty}^{\infty} u(t)c(t - \tau)\, d\tau \qquad (6.2.3)$$

and $z(t)$ represents the additive white Gaussian noise.

Let us suppose that the received signal is passed first through a filter and then sampled at a rate $1/T$ samples/s. We shall show in a subsequent section that an optimum filter from the point of view of signal detection is a filter matched to the received pulse. That is, the frequency response of the receiving filter is $H^*(f)$. We denote the output of the receiving filter as

$$y(t) = \sum_{n=0}^{\infty} I_n x(t - nT) + v(t) \qquad (6.2.4)$$

where $x(t)$ is the pulse representing the response of the receiving filter to the input pulse $h(t)$ and $v(t)$ is the response of the receiving filter to the noise $z(t)$.

Now, if $y(t)$ is sampled at times $t = kT + \tau_0$, $k = 0, 1, \ldots$, we have

$$y(kT + \tau_0) \equiv y_k = \sum_{n=0}^{\infty} I_n x(kT - nT + \tau_0) + v(kT + \tau_0) \quad (6.2.5)$$

$$y_k = \sum_{n=0}^{\infty} I_n x_{k-n} + v_k \qquad k = 0, 1, \ldots \quad (6.2.6)$$

where τ_0 is the transmission delay through the channel. The sample values can be expressed as

$$y_k = x_0 \left(I_k + \frac{1}{x_0} \sum_{\substack{n=0 \\ n \neq k}}^{\infty} I_n x_{k-n} \right) + v_k \qquad k = 0, 1, \ldots \quad (6.2.7)$$

We regard x_0 as an arbitrary scale factor which we arbitrarily set to unity for convenience. Then

$$y_k = I_k + \sum_{\substack{n=0 \\ n \neq k}}^{\infty} I_n x_{k-n} + v_k \quad (6.2.8)$$

The term I_k represents the desired information symbol at the kth sampling instant, the term

$$\sum_{\substack{n=0 \\ n \neq k}}^{\infty} I_n x_{k-n}$$

represents the intersymbol interference (ISI) and v_k is the additive gaussian noise variable at the kth sampling instant.

The amount of intersymbol interference and noise in a digital communications system can be viewed on an oscilloscope. For ASK signals, we can display the received signal $y(t)$ on the vertical input with the horizontal sweep rate set at $1/T$. The resulting oscilloscope display is called an *eye pattern* because of its resemblance to the human eye. For example, Fig. 6.2.1 illustrates the eye patterns for binary and four-level ASK modulation. The effect of ISI is to cause the eye to close, thereby reducing the margin for additive noise to cause errors. Figure 6.2.2 graphically illustrates the effect of intersymbol interference in reducing the opening of a binary eye. Note that intersymbol interference distorts the position of the zero crossings and causes a reduction in the eye opening. Thus, it causes the system to be more sensitive to a synchronization error.

For PSK and QAM it is customary to display the "eye pattern" as a two-dimensional scatter diagram illustrating the sampled values $\{y_k\}$ that represent the decision variables at the sampling instants. Figure 6.2.3 illustrates such an eye pattern for an 8-PSK signal. In the absence of intersymbol interference and noise, the superimposed signals at the sampling instants would result in eight

BINARY   **QUATERNARY**

FIGURE 6.2.1
Examples of eye patterns for binary and quaternary amplitude shift keying (or PAM).

distinct points corresponding to the eight transmitted signal phases. Intersymbol interference and noise result in a deviation of the received samples $\{y_k\}$ from the desired 8-PSK signal. The larger the intersymbol interference and noise, the larger the scattering of the received signal samples relative to the transmitted signal points.

Below, we consider the problem of signal design under the condition that there is no intersymbol interference at the sampling instants.

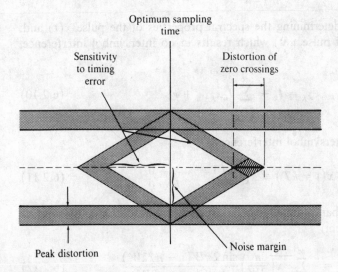

FIGURE 6.2.2
Effect of intersymbol interference on eye opening.

Transmitted eight-phase
signal

Received signal
samples at the output of demodulator

(a)

(b)

FIGURE 6.2.3
Two-dimensional digital "eye patterns."

6.2.1 Design of Band-Limited Signals for No Intersymbol Interference—The Nyquist Criterion

For the discussion in this section and in Sec. 6.2.2, we assume that the band-limited channel has ideal frequency response characteristics, i.e., $C(f) = 1$ for $|f| \leq W$. Then the pulse $x(t)$ has a spectral characteristic $X(f) = |U(f)|^2$ where

$$x(t) = \int_{-W}^{W} X(f) e^{j2\pi ft} \, df \qquad (6.2.9)$$

We are interested in determining the spectral properties of the pulse $x(t)$ and, hence, the transmitted pulse $u(t)$ which results in no intersymbol interference. Since

$$y_k = I_k + \sum_{\substack{n=0 \\ n \neq k}}^{\infty} I_n x_{k-n} + \nu_k \qquad (6.2.10)$$

the condition of no intersymbol interference is

$$x(t = kT) \equiv x_k = \begin{cases} 1 & k = 0 \\ 0 & k \neq 0 \end{cases} \qquad (6.2.11)$$

Since $x(t)$ is a band-limited signal, it can be represented in the infinite series (sampling theorem)

$$x(t) = \sum_{n=-\infty}^{\infty} x\left(\frac{n}{2W}\right) \frac{\sin 2\pi W(t - n/2W)}{2\pi W(t - n/2W)} \qquad (6.2.12)$$

where

$$x\left(\frac{n}{2W}\right) = \int_{-W}^{W} X(f) e^{j2\pi f n/2W} \, df \tag{6.2.13}$$

Now suppose that the transmission for the symbols $\{I_k\}$ is chosen to be $2W$ symbols/s. This particular choice for the symbol rate is called the *Nyquist rate*. If we let $T = 1/2W$, the series expansion for $x(t)$ given in (6.2.12) becomes

$$x(t) = \sum_{n=-\infty}^{\infty} x(nT) \frac{\sin \pi(t - nT)/T}{\pi(t - nT)/T} \tag{6.2.14}$$

To avoid intersymbol interference, the coefficients $x(nT)$ must all be zero except the one at $n = 0$. Consequently the pulse $x(t)$ that gives no intersymbol interference is

$$x(t) = \frac{\sin \pi t/T}{\pi t/T} \tag{6.2.15}$$

and its spectrum is an ideal rectangular characteristic

$$X(f) = \begin{cases} T & |f| \le \dfrac{1}{2T} \\ 0 & |f| > \dfrac{1}{2T} \end{cases} \tag{6.2.16}$$

The graphs of $x(t)$ and $X(f)$ are illustrated in Fig. 6.2.4.

There are two basic problems associated with this pulse shape. One is the problem of realizing a pulse having the rectangular spectral characteristic $X(f)$ given above. That is, $X(f)$ is not physically realizable. The other problem is concerned with the fact that the tails in $x(t)$ decay as $1/t$. Consequently a mistiming error in sampling results in an infinite series of intersymbol interference components. Such a series is not absolutely summable and, hence, the sum of the resulting interference does not converge.

In order to avoid these problems, let us restrict the transmission rate to $1/T < 2W$ symbols/s and attempt to determine if pulses exist that satisfy the condition of no intersymbol interference.

Now

$$x(t) = \int_{-W}^{W} X(f) e^{j2\pi f t} \, df \tag{6.2.17}$$

We wish to sample $x(t)$ at a rate $1/T$ so that

$$x(kT) = \int_{-W}^{W} X(f) e^{j2\pi f kT} \, df \qquad k = 0, \pm 1, \pm 2, \dots \tag{6.2.18}$$

Since $W > 1/2T$, $x(kT)$ can be expressed as

$$x(kT) = \sum_{n=-N}^{N} \int_{(2n-1)/2T}^{(2n+1)/2T} X(f) e^{j2\pi f kT} \, df$$

$$= \sum_{n=-N}^{N} \int_{-1/2T}^{1/2T} X\left(f + \frac{n}{T}\right) e^{j2\pi f kT} \, df \tag{6.2.19}$$

FIGURE 6.2.4
Band-limited signal for no intersymbol interference.

where N is the integer $[2TW]$. Let us define

$$X_{eq}(f) = \sum_{n=-N}^{N} X\left(f + \frac{n}{T}\right), \qquad |f| \le \frac{1}{2T} \qquad (6.2.20)$$

Then, by making use of the relation given by (6.2.20) in (6.2.19), we obtain

$$x(kT) = \int_{-1/2T}^{1/2T} X_{eq}(f) e^{j2\pi fkT} \, df \qquad (6.2.21)$$

Now the condition for no intersymbol interference, i.e., $x(kT) = 0$ for $k \ne 0$, $x(0) = 1$, requires that

$$X_{eq}(f) = \begin{cases} T & |f| \le 1/2T \\ 0 & |f| > 1/2T \end{cases} \qquad (6.2.22)$$

Figure 6.2.5 shows two examples of spectra that have bandwidths $W > 1/2T$ and $X_{eq}(f)$ that results in no intersymbol interference. It is apparent that when $X(f)$ has odd symmetry about the frequency $1/2T$, then $X_{eq}(f)$ has the rectangular spectral characteristic as shown in Fig. 6.2.5. Therefore it is possible to design band-limited pulses that give no intersymbol interference on an ideal channel of bandwidth W provided that the symbol rate does not exceed $2W$. This

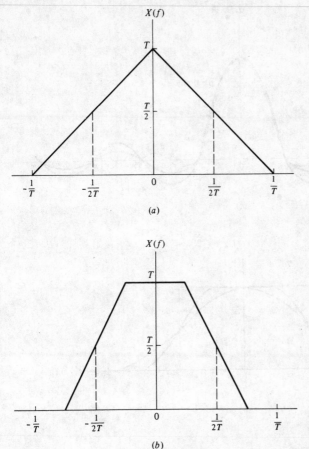

FIGURE 6.2.5
Two examples of band-limited pulse spectra which result in no intersymbol interference.

is the Nyquist criterion for signal design with no intersymbol interference. Practically, if one selects the rate so that

$$W < \frac{1}{T} < 2W$$

it is possible to design a variety of pulses that have good spectral characteristics and are free of intersymbol interference on an ideal channel.

A pulse that has found wide use in digital transmission on band-limited channels such as telephone lines is one that has the raised cosine spectral characteristic,

$$X(f) = \begin{cases} T & 0 \le |f| \le (1 - \beta)/2T \\ \dfrac{T}{2}\left[1 - \sin \pi T\left(f - \dfrac{1}{2T}\right)\Big/ \beta\right] & (1 - \beta)/2T \le |f| \le (1 + \beta)/2T \end{cases}$$

$$(6.2.23)$$

FIGURE 6.2.6
Pulses having a raised cosine spectrum.

where β is called the *rolloff parameter*. The pulse $x(t)$ having this spectrum is

$$x(t) = \frac{\sin \pi t/T}{\pi t/T} \frac{\cos \beta \pi t/T}{1 - 4\beta^2 t^2/T^2} \qquad (6.2.24)$$

The tails of the pulse decay as $1/t^3$. Hence a mistiming error in sampling leads to a series of intersymbol interference components that converge. Figure 6.2.6 illustrates the spectral characteristic $X(f)$ and the pulse $x(t)$ for several values of β.

6.2.2 Design of Band-Limited Signals with Controlled Intersymbol Interference (Partial-Response Signals)

A band-limited signal $x(t)$ can be represented, in general, as a weighted linear superposition of $\sin 2\pi Wt/2\pi Wt$ pulses spaced at intervals $1/2W$ apart, as previously stated in (6.2.12). W is the bandwidth of the baseband signal $x(t)$ and

$x(n/2W)$ is the set of weighting coefficients. The frequency characteristic of $x(t)$ can be expressed in the form

$$X(f) = \begin{cases} \dfrac{1}{2W} \displaystyle\sum_{n=-\infty}^{\infty} x\left(\dfrac{n}{2W}\right) e^{-jn\pi f/W} & |f| \le W \\ 0 & |f| > W \end{cases} \tag{6.2.25}$$

As demonstrated in Sec. 6.2.1, an ideal channel having a bandwidth W can be used to transmit digital information at a symbol rate of $2W$ symbols/s without intersymbol interference at the sampling instants. This rate is achieved by a modulator and a demodulator that have an overall $\sin 2\pi Wt/2\pi Wt$ impulse response. However, as indicated previously, such an impulse response is physically unrealizable. By reducing the symbol rate to less than $2W$ symbols/s, we showed that physically realizable pulses exist which satisfy the constraint of no intersymbol interference at the sampling instants.

In this section we impose the condition that the symbol rate be $2W$ symbols/s but we remove the constraint that there be no intersymbol interference at the sampling instants. Thus we obtain a class of physically realizable pulses that are called *partial-response signals*.

We confine the following discussion to PAM signaling. In describing the characteristics of the signal pulses, we neglect the presence of the additive noise. Hence the PAM signal at the output of the receiving filter can be expressed in the form

$$y(t) = \sum_{n=0}^{\infty} A_n x\left(t - \frac{n}{2W}\right) \tag{6.2.26}$$

where the symbol rate is $2W$ symbols/s. For the ideal but physically unrealizable $\sin 2\pi Wt/2\pi Wt$ pulse,

$$x\left(\frac{n}{2W}\right) = \begin{cases} 1 & n = 0 \\ 0 & \text{otherwise} \end{cases} \tag{6.2.27}$$

and the sequence $\{A_n\}$ is recovered at the receiver by sampling the received signal once every T seconds at times $t = nT = n/2W$, $n = 0, 1, 2, \ldots$.

To circumvent the difficulty with the $\sin 2\pi Wt/2\pi Wt$ pulse shape, suppose we return to the general form of the band-limited signal and allow for at least two of the weighting coefficients $x(n/2W)$ to be nonzero. In so doing we encounter intersymbol interference at the sampling instants when transmission occurs at the Nyquist rate ($2W$ pulses/s). However, the intersymbol interference is deterministic or "controlled" and, hence, it can be taken into account at the receiver, as will be discussed below.

One special case is the duobinary signal pulse specified in the time domain by

$$x\left(\frac{n}{2W}\right) = \begin{cases} 1 & n = 0, 1 \\ 0 & \text{otherwise} \end{cases} \tag{6.2.28}$$

FIGURE 6.2.7
Time domain and frequency domain characteristics of a duobinary signal.

and in the frequency domain as

$$X(f) = \begin{cases} \dfrac{1}{2W}(1 + e^{-j\pi f/W}) = \dfrac{1}{W}e^{-j\pi f/2W}\cos\dfrac{\pi f}{2W} & |f| \le W \\ 0 & |f| > W \end{cases} \quad (6.2.29)$$

The pulse $x(t)$ and the magnitude of the frequency characteristic $X(f)$ are shown in Fig. 6.2.7. Note that the frequency characteristic of the duobinary pulse does not possess sharp discontinuities and, as a result, it can be more easily approximated by a physically realizable filter.

A second special case, called a *modified duobinary signal*, is specified in the time domain by

$$x\left(\frac{n}{2W}\right) = \begin{cases} 1 & n = -1 \\ -1 & n = 1 \\ 0 & \text{otherwise} \end{cases} \quad (6.2.30)$$

and in the frequency domain as

$$X(f) = \begin{cases} \dfrac{1}{2W}(e^{j\pi f/W} - e^{-j\pi f/W}) = \dfrac{j}{W}\sin\dfrac{\pi f}{W} & |f| \le W \\ 0 & |f| > W \end{cases} \quad (6.2.31)$$

The pulse $x(t)$ and the magnitude of the frequency characteristic $X(f)$ for the modified duobinary signal are shown in Fig. 6.2.8. Again, note the smoothness of

FIGURE 6.2.8
Time domain and frequency domain characteristics of a modified duobinary signal.

the function $X(f)$. Also note that $X(0) = 0$, a characteristic that makes this signal especially suitable for SSB transmission on a carrier due to the ease with which the two sidebands can be separated by filtering.

One can generate other interesting pulse shapes at will as shown by Kretzmer (1966) and Lucky et al. (1968). However, as the number of nonzero weighting coefficients is increased, the problem of unraveling the "controlled" intersymbol interference becomes more cumbersome and impractical. Moreover, the recovery of the data sequence becomes more complex as the number of nonzero coefficients $x(n/2W)$ is increased beyond two.

In the following discussion we consider the problem of detecting the partial-response duobinary and modified duobinary signals. First consider the case of a duobinary pulse used in conjunction with binary PAM. The received signal at the sampling instant, in the absence of noise, is

$$B_n = A_n + A_{n-1} \qquad n = 1, 2, \dots \qquad (6.2.32)$$

where $\{A_n\}$ is the set of amplitudes of the transmitted pulses with each A_n taking on the value either $+1$ or -1. Thus we observe that B_n has three possible levels, -2, $+2$, and 0, where the level $B_n = -2$ corresponds to the ordered pair

(A_n, A_{n-1}) taking on value $(-1, -1)$, the level $B_n = +2$ corresponds to $(1, 1)$, and the level $B_n = 0$ corresponds to either $(-1, 1)$ or $(1, -1)$. If A_{n-1} is the decoded symbol from the $(n-1)$st signaling interval, its effect on B_n, the received signal in the nth signaling interval, can be eliminated by subtraction, thus allowing A_n to be decoded. This process can be repeated sequentially.

The disadvantage with this procedure is that errors arising from noise tend to propagate. For example, if A_{n-1} is in error, then its effect on B_n is not eliminated but, in fact, is reinforced by the incorrect subtraction. Hence A_n is also likely to be in error.

Error propagation can be avoided by precoding the data at the transmitter instead of eliminating the intersymbol interference by subtraction at the receiver. The precoding that accomplishes the desired result is as follows. Let $\{D_n\}$ denote the data sequence consisting of 1s and 0s that is to be transmitted. From the sequence $\{D_n\}$ a new sequence $\{P_n\}$ is generated:

$$P_n = D_n \ominus P_{n-1} \qquad n = 1, 2, \ldots \qquad (6.2.33)$$

where the symbol $\ominus$ denotes modulo-2 subtraction.† Then the transmitted pulse in the nth signaling interval is $-x(t)$ if $P_n = 0$ and $+x(t)$ if $P_n = 1$. In other words, the transmitted signal amplitude A_n that modulates the duobinary pulse $x(t)$ is obtained from the sequence $\{P_n\}$ using the following relation, which converts binary $(0, 1)$ signals to bipolar (± 1) signals:

$$A_n = 2P_n - 1 \qquad (6.2.34)$$

The received sequence $\{B_n\}$, for the precoded transmitted symbols, is given again by the relation $B_n = A_n + A_{n-1}$. If we substitute for A_n into B_n, we have

$$B_n = 2(P_n + P_{n-1} - 1) \qquad (6.2.35)$$

Since $D_n = P_n \oplus P_{n-1}$ (mod 2), it follows that the data sequence D_n is obtained from B_n by using the relation

$$D_n = \frac{B_n}{2} + 1 \qquad (\text{mod } 2) \qquad (6.2.36)$$

An example to illustrate the operation described above is given in Table 6.2.1.

In the presence of additive noise the received signal, now having the form $B_n + v_n$, is compared with the two thresholds set at -1 and $+1$. The data sequence D_n is obtained according to the rule

$$D_n = \begin{cases} 1 & \text{if } -1 < B_n + v_n < 1 \\ 0 & \text{if } |B_n + v_n| \geq 1 \end{cases} \qquad (6.2.37)$$

One can easily show that the precoding operation performed in duobinary signaling is identical to the precoding of the data that is performed in binary

---

†Although this is identical to modulo-2 addition, it is convenient to view the precoding operation for duobinary in terms of modulo-2 subtraction.

TABLE 6.2.1
Binary signaling with duobinary pulses

Data sequence D_n:		1	1	1	0	1	0	0	1	0	0	0	1	1	0	1
Precoded sequence P_n:	0	1	0	1	1	0	0	0	1	1	1	1	0	1	1	0
Transmitted sequence A_n:	−1	1	−1	1	1	−1	−1	−1	1	1	1	1	−1	1	1	−1
Received sequence B_n:		0	0	0	2	0	−2	−2	0	2	2	2	0	0	2	0
Decoded sequence D_n:		1	1	1	0	1	0	0	1	0	0	0	1	1	0	1

differential phase-shift keying (DPSK). Again, let $\{D_n\}$ be the data sequence. In DPSK, when $D_k = 1$, the phase of the signal transmitted in the kth signaling interval is shifted by π rad relative to the phase of the signal in the preceding signaling interval; when $D_k = 0$, the phase of the signal remains unchanged relative to the phase of the signal in the preceding signaling interval. When this rule is applied to the information sequence in Table 6.2.1, one obtains directly the sequence $\{A_n\}$ of the transmitted binary digits.

Let us now consider the case of binary PAM signaling employing the modified duobinary pulse at the basic pulse shape. Again, let $\{A_n\}$ denote the sequence of transmitted pulse amplitudes. The received signal will take on three possible levels. In the absence of noise this signal can be represented as

$$B_n = A_n - A_{n-2} \qquad n = 2, 3, \dots \qquad (6.2.38)$$

In order to avoid error propagation, the information sequence is precoded as is done for the duobinary signal. Thus the data sequence $\{D_n\}$ is transformed into a new sequence $\{P_n\}$ where

$$P_n = D_n \oplus P_{n-2} \qquad n = 2, 3, \dots \qquad (6.2.39)$$

where the symbol $\oplus$ denotes modulo-2 addition. The transmitted amplitude A_n is obtained from the sequence $\{P_n\}$ by the relation $A_n = 2P_n - 1$. The rule for decoding the received sequence $\{B_n\}$ is obtained by substituting for A_n into B_n. Thus

$$B_n = 2(P_n - P_{n-2}) \qquad (6.2.40)$$

Since $D_n = P_n \ominus P_{n-2}$ (mod 2), it follows that the data sequence D_n is obtained from B_n by the decoding rule

$$D_n = \frac{B_n}{2} \pmod 2 \qquad (6.2.41)$$

When additive noise is present, the received signal at the sampling instant can be expressed as $B_n + \nu_n$. For this case the decision variable $B_n + \nu_n$ is compared with the thresholds -1 and $+1$ and the data sequence is obtained according to the rule

$$D_n = \begin{cases} 0 & \text{if } -1 < B_n + v_n < 1 \\ 1 & \text{if } |B_n + v_n| \geq 1 \end{cases} \quad (6.2.42)$$

The transition from binary signaling to multilevel PAM signaling using the partial-response pulses is relatively simple. Our discussion will be confined to duobinary and modified duobinary pulses. When these two pulses are used for signaling, the transmission of one of M equally spaced levels at the Nyquist rate gives rise at each sampling instant, in the absence of noise, to one of $2M - 1$ equally spaced signal levels. The $(2M - 1)$-level signal is mapped by the receiver back into an M-level signal.

In order to avoid error propagation at the receiver, the data is precoded before transmission. Let $\{D_n\}$ denote the input data sequence which consists of elements from the alphabet of the M numbers $0, 1, 2, \ldots, M - 1$, and let $\{P_n\}$ denote the output of the precoder. The relationship between $\{P_n\}$ and $\{D_n\}$ is

$$P_n = D_n \ominus P_{n-1} \quad (\text{mod } M) \quad (6.2.43)$$

for duobinary and

$$P_n = D_n \oplus P_{n-2} \quad (\text{mod } M) \quad (6.2.44)$$

for modified duobinary. The precoded symbol P_n is converted into the transmitted amplitude level A_n according to the relation

$$A_n = 2P_n - (M - 1) \quad (6.2.45)$$

In the absence of noise, the received signal at the sampling instants, for the duobinary pulse, can be represented as

$$B_n = A_n + A_{n-1} \quad n = 1, 2, \ldots \quad (6.2.46)$$

The rule for converting the received sequence $\{B_n\}$ into the data sequence $\{D_n\}$ is obtained as follows:

$$B_n = 2[P_n + P_{n-1} - (M - 1)] \quad (6.2.47)$$

But

$$D_n = P_n + P_{n+1} \quad (\text{mod } M) \quad (6.2.48)$$

Hence

$$D_n = \frac{B_n}{2} + (M - 1) \quad (\text{mod } M) \quad (6.2.49)$$

In the presence of noise the received signal-plus-noise is quantized to the nearest of the acceptable signal levels and the rule given above is used on the quantized values to obtain the data sequence $\{D_n\}$.

For the modified duobinary pulse the received signal at the sampling instants, in the absence of noise, is represented as

$$B_n = A_n - A_{n-2} \quad n = 2, 3, \ldots \quad (6.2.50)$$

To obtain the rule for converting the received sequence $\{B_n\}$ into the data

TABLE 6.2.2
Four-level signal transmission with duobinary pulses

Data sequence D_n:		0	0	1	3	1	2	0	3	3	2	0	1	0
Precoded sequence P_n:	0	0	0	1	2	3	3	1	2	1	1	3	2	2
Transmitted sequence A_n:	−3	−3	−3	−1	1	3	3	−1	1	−1	−1	3	1	1
Received sequence B_n:		−6	−6	−4	0	4	6		2	0	0	−2	2	4	2
Decoded sequence D_n:		0	0	1	3	1	2	0	3	3	2	0	1	0

sequence $\{D_n\}$, we substitute for A_n. Thus

$$B_n = 2(P_n - P_{n-2}) \tag{6.2.51}$$

But

$$D_n = P_n - P_{n-2} \quad (\text{mod } M) \tag{6.2.52}$$

Hence

$$D_n = \frac{B_n}{2} \quad (\text{mod } M) \tag{6.2.53}$$

Two examples of the precoding and decoding operations, one for duobinary and the other for modified duobinary, are illustrated in Tables 6.2.2 and 6.2.3 for the case of four-level signaling.

In the discussion above it was assumed that the data sequence $\{D_n\}$ consists of elements chosen from the set $0, 1, 2, \ldots, M-1$. In the usual situation the data source is binary. When this is the case, any group of k binary digits can be mapped uniquely into one of the M numbers $0, 1, 2, \ldots, M-1$, where $M = 2^k$. However, in order to ensure that an error in an adjacent level of the received signal results in only a single binary digit error, the binary source output

TABLE 6.2.3
Four-level signal transmission with modified duobinary pulses

Data sequence D_n:			0	1	3	1	2	0	3	3	2	0	1	0
Precoded sequence P_n:	0	0	0	1	3	2	1	2	0	1	2	1	3	1
Transmitted sequence A_n:	−3	−3	−3	−1	3	1	−1	1	−3	−1	1	−1	3	−1
Received sequence B_n:			0	2	6	2	−4	0	−2	−2	4	0	2	0
Decoded sequence D_n:			0	1	3	1	2	0	3	3	2	0	1	0

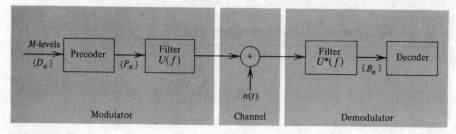

FIGURE 6.2.9
Block diagram of the modulator and demodulator for partial-response signals.

should be coded using the Gray code. The precoding operation is then performed on the Gray-coded data stream.

Probability of error for symbol-by-symbol demodulation. Let us now evaluate the performance in terms of the probability of error for digital PAM signaling with duobinary and modified duobinary pulses transmitted over a white gaussian noise channel. The model of the communications system is shown in Fig. 6.2.9.

At the transmitter, the M-level sequence $\{D_n\}$ is precoded as described earlier in this section. The precoder output $\{P_n\}$ is scaled as described previously to yield the sequence of amplitudes $\{A_n\}$. The filter with transfer function $G(f)$ is excited at the Nyquist rate by impulses having as areas the sequence of amplitudes $\{A_n\}$. The partial-response function $X(f)$ is divided equally between modulator and demodulator for the purpose of suppressing the additive noise in the channel. Hence the filter at the demodulator is matched to the transmitted pulse, and the cascade of the modulator and demodulator filters results in the frequency characteristic $|U(f)|^2 = |X(f)|$, where $X(f)$ is the frequency characteristic of the partial-response pulse. The matched filter output is fed to the decoder, which decodes the received levels modulo-M to recover the data sequence $\{D_n\}$.

The channel corrupts the signal transmitted through it by the addition of white gaussian noise with zero mean value and spectral density $\Phi_{zz}(f) = N_0$.

The input symbols are assumed to be equally probable. Then it is easily shown that in the absence of noise, the output levels have a triangular probability distribution, i.e.,

$$P(B_n = 2ld) = \frac{M - |l|}{M^2} \qquad l = -(M-1), -(M-2), \ldots, 0, 1, 2, \ldots, M-1$$

$$(6.2.54)$$

where B_n denotes the received level in the nth signaling interval and $2d$ is the distance between any two adjacent received levels.

By assuming that a symbol error is committed whenever the additive noise takes the signal of a given level into another level, a tight upper bound on the average probability of a symbol error, denoted by P_M, is easily computed for the case of equally likely data symbols. The observed real-valued decision variable, call it y, is gaussian with mean value $E(y) = B_n = 2ld$ and variance σ^2, where

$$\sigma^2 = \frac{N_0}{2} \int_{-W}^{W} |U(f)|^2 \, df$$

$$= \frac{N_0}{2} \int_{-W}^{W} |X(f)| \, df \qquad (6.2.55)$$

Evaluation of (6.2.55) for the duobinary and modified duobinary pulses yields the variance $\sigma^2 = 2N_0/\pi$. Hence the bound on P_M can be expressed in the form

$$P_m \leq \sum_{1=-(M-2)}^{M-2} P(|y - 2ld| > d | B_n = 2ld) \cdot P(B_n = 2ld)$$

$$+ 2P[y + 2(M-1)d > d | B_n = -2(M-1)d] \cdot P[B_n = -2(M-1)d]$$

$$= P(|y| > d | B_n = 0)\left\{ 2 \sum_{l=0}^{M-1} P(B_n + 2ld) - P(B_n = 0) \right.$$

$$\left. - P[B_n = -2(M-1)d] \right\}$$

$$= \left(1 - \frac{1}{M^2}\right) P(|y| > d | B_n = 0)$$

But

$$P(|y| > d | B_n = 0) = \frac{2}{\sqrt{2\pi}\,\sigma} \int_d^\infty e^{-x^2/2\sigma^2} \, dx$$

Therefore the average probability of a symbol error is

$$P_M \leq \left(1 - \frac{1}{M^2}\right) \text{erfc}\left(\sqrt{\frac{\pi d^2}{4N_0}}\right) \qquad (6.2.56)$$

P_M can also be expressed in terms of the average power transmitted into the channel. For a signal with M levels, equally spaced about zero, the average power at the output of the transmitting filter is

$$P_{av} = \frac{\overline{l^2}}{T} \int_{-W}^{W} |U(f)|^2 \, df$$

$$= \frac{\overline{l^2}}{T} \int_{-W}^{W} |X(f) \, df$$

$$= \frac{4\overline{l^2}}{\pi T} \qquad (6.2.57)$$

where $\overline{l^2}$ is the mean-square value of the transmitted levels and T is the duration of a signaling interval. For the case in which the transmitted levels are equally likely and adjacent levels are separated by a distance $2d$, the value of $\overline{l^2}$ is $d^2(M^2 - 1)/3$. Thus

$$P_{av} = \frac{4d^2(M^2 - 1)}{3\pi T} \qquad (6.2.58)$$

Solving (6.2.58) for d^2, we obtain

$$d^2 = \frac{3\pi}{4(M^2 - 1)} TP_{av} \qquad (6.2.59)$$

When we substitute for d^2 in (6.2.56), we obtain P_M in terms of the average transmitted power P_{av}. That is,

$$P_M \leq \left(1 - \frac{1}{M^2}\right) \text{erfc}\left[\sqrt{\frac{3}{M^2 - 1}\left(\frac{\pi}{4}\right)^2 \gamma_{av}}\right] \qquad (6.2.60)$$

where $\gamma_{av} = TP_{av}/N_0$ is the average SNR per symbol.

The expression given above for the probability of error for multilevel PAM signaling using duobinary and modified duobinary pulses can be compared with the error probability for multilevel PAM signaling using either the ideal, but physically unrealizable, $(\sin 2\pi Wt)/2\pi Wt$ band-limited pulse, or a pulse having a raised cosine spectrum that is transmitted at a rate less than $2W$ symbol/s. For these cases, the expression for P_M derived previously in Sec. 4.2.7 applies. That is,

$$P_M = \left(1 - \frac{1}{M}\right) \text{erfc}\left(\sqrt{\frac{3}{M^2 - 1}\gamma_{av}}\right) \qquad (6.2.61)$$

Therefore, for large values of M, the use of partial-response duobinary and modified duobinary pulses results in a loss of $(\pi/4)^2$ or 2.1 dB in signal-to-noise ratio relative to the best achievable performance. This loss in SNR is due, for the most part, to the fact that the symbol-by-symbol detection technique described above is sub-optimum. This loss can be completely recovered, as will be shown in Sec. 6.7, by means of maximum-likelihood sequence estimation using the Viterbi algorithm, which is the optimum detection technique for dealing with the controlled intersymbol interference.

Alternative characterization of partial-response signals. We conclude this subsection by presenting another interpretation of a partial-response signal. Suppose that the partial-response signal is generated, as shown in Fig. 6.2.10, by passing the discrete-time sequence of amplitudes. $\{A_n\}$ through a discrete-time filter with coefficients $x_n \equiv x(n/2W)$, $n = 0, 1, \ldots, N - 1$, and using the output sequence $\{B_n\}$ from this filter to excite periodically with an input $B_n\delta(t - nT)$ an analog filter having an impulse response $\sin 2\pi Wt/2\pi Wt$. The resulting output signal is identical to the partial-response signal given by (6.2.26).

Since

$$B_n = \sum_{k=0}^{N-1} x_k A_{n-k} \qquad (6.2.62)$$

the sequence of symbols $\{B_n\}$ is correlated as a consequence of the filtering

FIGURE 6.2.10
An alternative method for generating a partial-response signal.

performed on the sequence $\{A_n\}$. In fact, the autocorrelation function of the sequence $\{B_n\}$ is

$$\phi(m) = E(B_n B_{n+m})$$

$$= \sum_{k=0}^{N-1} \sum_{l=0}^{N-1} x_k x_l E(A_{n-k} A_{n+m-l}) \qquad (6.2.63)$$

When the input sequence is zero mean and white,

$$E(A_{n-k} A_{n+m-l}) = \delta_{m+k-l} \qquad (6.2.64)$$

where we have used the normalization $E(A_n^2) = 1$. Substitution of (6.2.64) into (6.2.63) yields the desired autocorrelation function for $\{B_n\}$ in the form

$$\phi(m) = \sum_{k=0}^{N-1-|m|} x_k x_{k+|m|} \qquad m = 0, \pm 1, \ldots, \pm(N-1) \qquad (6.2.65)$$

The corresponding power density spectrum is

$$\Phi(f) = \sum_{m=-(N-1)}^{N-1} \phi(m) e^{-j2\pi fmT}$$

$$= \left| \sum_{m=0}^{N-1} x_m e^{-j2\pi fmT} \right|^2 \qquad (6.2.66)$$

where $T = 1/2W$ and $|f| \le 1/2T = W$.

From (6.2.66) we observe that the discrete-time filter serves to shape the spectrum of the transmitted signal. The analog filter, however, does no spectral shaping at all. For example, in the duobinary pulse the coefficients of the discrete-time filter are $x_0 = x_1 = 1$, and $x_n = 0$ otherwise. The resulting power density spectrum is

$$\Phi(f) = |1 + e^{-j2\pi fT}|^2$$

$$= 4\cos^2 \pi fT = 4\cos^2 \frac{\pi f}{2W} \qquad |f| \leq W \qquad (6.2.67)$$

But this power density spectrum is identical to that obtained by exciting an analog filter which has the frequency response

$$X(f) = 2\cos \frac{\pi f}{2W} \qquad |f| \leq W \qquad (6.2.68)$$

by a sequence of uncorrelated symbols $\{A_n\}$.

The major difference in these two formulations is that in one the spectral shaping is obtained by designing the pulse shape $x(t)$, while in the second the spectral shaping is achieved by the correlation properties of the $\{B_n\}$. Consequently the design of a partial-response signal pulse is equivalent to the problem of selecting the coefficients in the discrete-time filter shown in Fig. 6.2.10 to achieve a desirable correlation function or power density spectrum.

6.3. OPTIMUM DEMODULATOR FOR INTERSYMBOL INTERFERENCE AND ADDITIVE GAUSSIAN NOISE

In this section we derive the structure of the optimum demodulator for digital transmission through a nonideal band-limited channel with additive gaussian noise. We begin with the transmitted signal given by (6.2.1). The received signal, in turn, is

$$r(t) = \sum_n I_n h(t - nT) + z(t) \qquad (6.3.1)$$

where $h(t)$ represents the response of the channel to the input signal pulse $u(t)$ and $z(t)$ represents the additive white gaussian noise.

First we demonstrate that the optimum demodulator can be realized as a filter matched to $h(t)$, followed by a sampler operating at the symbol rate $1/T$ and a subsequent processing algorithm for estimating the information sequence $\{I_n\}$ from the sample values. Consequently the samples at the output of the matched filter are sufficient for the estimation of the sequence $\{I_n\}$. The Karhunen-Loève expansion provides the means for deriving the optimum (maximum-likelihood) demodulator.

Maximum-likelihood demodulator. Following the procedure described in Appendix 4A, we expand the received signal $r(t)$ in the series

$$r(t) = \lim_{N \to \infty} \sum_{k=1}^{N} r_k f_k(t) \qquad (6.3.2)$$

where the $\{f_k(t)\}$ are the set of orthogonal eigenfunctions of the integral equation in the Karhunen–Loève expansion method and the $\{r_k\}$ are the observable random variables obtained by projecting $r(t)$ onto the set of orthonormal functions. It is easily shown that

$$r_k = \sum_n I_n h_{kn} + z_k \qquad k = 1, 2, \dots \tag{6.3.3}$$

where h_{kn} is the value obtained from projecting $h(t - nT)$ onto $f_k(t)$ and z_k is the value obtained by projecting $z(t)$ onto $f_k(t)$. The sequence $\{z_k\}$ is gaussian with zero mean and covariance $\frac{1}{2}E(z_k^* z_m) = \lambda_k \delta_{km}$ where the $\{\lambda_m\}$ are the eigenvalues of the integral equation.

The joint probability density function of the random variables $r_1, r_2, \dots, r_N \equiv \mathbf{r}_N$ conditioned on the transmitted sequence $I_1, I_2, \dots, I_p \equiv \mathbf{I}_p$ is

$$p(\mathbf{r}_N | \mathbf{I}_p) = \left(\prod_{k=1}^{N} 2\pi\lambda_k \right)^{-1} \exp\left(-\frac{1}{2} \sum_{k=1}^{N} \left| r_k - \sum_n I_n h_{kn} \right|^2 \Big/ \lambda_k \right) \tag{6.3.4}$$

In the limit as the number N of observable random variables approaches infinity, the logarithm of $p(\mathbf{r}_N | \mathbf{I}_p)$ is proportional to the quantity $J_0(\mathbf{I}_p)$, defined as

$$J_0(\mathbf{I}_p) = -\int_{-\infty}^{\infty} \left| r(t) - \sum_n I_n h(t - nT) \right|^2 dt$$

$$= -\int_{-\infty}^{\infty} |r(t)|^2 \, dt + 2\,\mathrm{Re} \sum_n \left[I_n^* \int_{-\infty}^{\infty} r(t) h^*(t - nT) \, dt \right]$$

$$- \sum_n \sum_m I_n^* I_m \int_{-\infty}^{\infty} h^*(t - nT) h(t - mT) \, dt \tag{6.3.5}$$

The maximum-likelihood estimates of the symbols $I_1, I_2, \dots, I_p$ are those that maximize this quantity. We note, however, that the integral of $|r(t)|^2$ is independent of the information sequence $\{I_n\}$ and, hence, it may be discarded. The other integral involving $r(t)$ gives rise to the variables

$$y_n \equiv y(nT) = \int_{-\infty}^{\infty} r(t) h^*(t - nT) \, dt \tag{6.3.6}$$

These variables can be generated by passing $r(t)$ through a filter matched to $h(t)$ and sampling the output at the symbol rate $1/T$. The samples $\{y_n\}$ form a set of sufficient statistics for the computation of $J_0(\mathbf{I}_p)$ or, equivalently, of the quantity

$$J(\mathbf{I}_p) = 2\,\mathrm{Re} \left(\sum_n I_n^* y_n \right) - \sum_n \sum_m I_n^* I_m x_{n-m} \tag{6.3.7}$$

where, by definition, $x(t)$ is the response of the matched filter to $h(t)$ and

$$x_n \equiv x(nT) = \int_{-\infty}^{\infty} h^*(t) h(t + nT) \, dt \tag{6.3.8}$$

Hence $x(t)$ represents the output of a filter having an impulse response $h^*(-t)$

and an excitation $h(t)$. In other words, $x(t)$ represents the autocorrelation function of $h(t)$. Consequently $\{x_n\}$ represents the samples of the autocorrelation function of $h(t)$, taken periodically at $1/T$. We are not particularly concerned with the noncausal characteristic of the filter matched to $h(t)$ since in practice we can introduce a sufficiently large delay to ensure causality of the matched filter.

The brute force method for determining the particular sequence $I_1, I_2, I_3, \ldots, I_p$ that maximizes (6.3.7) is the exhaustive method. That is, we evaluate $J(\mathbf{I}_p)$ for all possible sequences and select the sequence that yields the largest $J(\mathbf{I}_p)$. This signal processing method has been termed *maximum-likelihood sequence estimation* (MLSE). A more efficient computational method for implementing MLSE is the Viterbi algorithm, which has already been described for decoding of convolutional codes and which will be described again in this chapter in the context of dealing with intersymbol interference.

In addition to the optimum MLSE method specified by (6.3.7), there are also a number of other algorithms for processing the sampled matched filter outputs. By design, these other algorithms employ symbol-by-symbol estimation of the information sequence $\{I_n\}$, whereas in MLSE the estimation is performed on the entire sequence. In order to describe these algorithms, we deal with the signal given by (6.3.6). If we substitute for $r(t)$ in (6.3.6) using the expression given in (6.3.1), we obtain

$$y_k = \sum_n I_n x_{k-n} + \nu_k \tag{6.3.9}$$

where ν_k denotes the additive noise sequence of the output of the matched filter, i.e.,

$$\nu_k = \int_{-\infty}^{\infty} z(t) h^*(t - kT) \, dt \tag{6.3.10}$$

By expressing y_k in the form

$$y_k = x_0 \left(I_k + \frac{1}{x_0} \sum_{n \neq k} I_n x_{k-n} \right) + \nu_k \tag{6.3.11}$$

we observe that I_k is the desired symbol at the kth sampling instant, the term

$$\frac{1}{x_0} \sum_{n \neq k} I_n x_{k-n}$$

represents the intersymbol interference, and ν_k is the additive noise component. Since the intersymbol interference term depends on the particular transmitted sequence, it is a random variable at any sampling instant. In high-speed transmission on telephone channels and on some radio channels this term is often sufficiently large to mask the desired signal component I_k. Under these circumstances, a decision about the symbol I_k cannot be made with a small error probability if that decision is based entirely on the single received variable y_k. Consequently even the symbol-by-symbol estimation (equalization) methods that

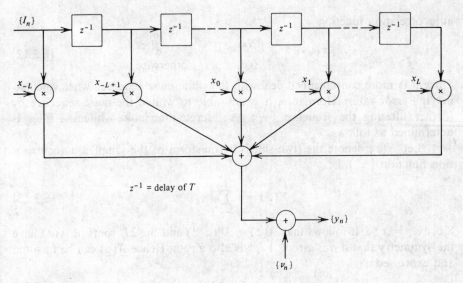

FIGURE 6.3.1
Equivalent discrete-time model of channel with intersymbol interference.

are described below are based on processing several samples of the received signal $\{y_k\}$ at a time in order to estimate a single symbol. Before we begin our discussion of these techniques, it is desirable, for mathematical convenience, to introduce and equivalent discrete-time model for the intersymbol interference.

A discrete-time model for intersymbol interference. Since the transmitter sends discrete-time symbols at a rate $1/T$ symbols/s and the sampled output of the matched filter at the receiver is also a discrete-time signal with samples occurring at a rate $1/T$ per second, it follows that the cascade of the analog filter at the transmitter with impulse response $u(t)$, the channel with impulse response $c(t)$, the matched filter at the receiver with impulse response $h^*(-t)$, and the sampler can be represented by an equivalent discrete-time tranversal filter having tap gain coefficients $\{x_k\}$. We assume that $x_k = 0$ for $|k| > L$, where L is some arbitrary positive integer. In spite of the restriction that the channel is band-limited, this assumption holds in all practical communications systems. Consequently we have an equivalent discrete-time transversal filter that spans a time interval of $2LT$ seconds. Its input is the sequence of information symbols $\{I_k\}$ at the output of the matched filter. That is, the set of noise variables $\{v_k\}$ and its output is the discrete-time sequence $\{y_k\}$ given by (6.3.9). The equivalent discrete-time model is shown in Fig. 6.3.1.

The major difficulty with this discrete-time model occurs in the evaluation of performance of the various equalization or estimation techniques that are discussed in the following sections. The difficulty is caused by the correlations in the noise sequence $\{v_k\}$ at the output of the matched filter. That is, the set of noise variables $\{v_k\}$ is a gaussian-distributed sequence with zero mean and autocorrelation function

$$\tfrac{1}{2}E(v_k^* v_j) = \begin{cases} N_0 x_{k-j} & |k - j| \le L \\ 0 & \text{otherwise} \end{cases} \qquad (6.3.12)$$

Since it is more convenient to deal with the white-noise sequence when calculating the error rate performance, it is desirable to whiten the noise sequence by further filtering the sequence $\{y_k\}$. A discrete-time noise-whitening filter is determined as follows.

Let $X(z)$ denote the (two-sided) z transform of the sampled autocorrelation function $\{x_k\}$, i.e.,

$$X(z) = \sum_{k=-L}^{L} x_k z^{-k} \qquad (6.3.13)$$

Since $x_k = x_{-k}^*$, it follows that $X(z) = X^*(z^{-1})$ and the $2L$ roots of $X(z)$ have the symmetry that if ρ is a root, $1/\rho^*$ is also a root. Hence $X(z)$ can be factored and expressed as

$$X(z) = F(z)F^*(z^{-1}) \qquad (6.3.14)$$

where $F(z)$ is a polynomial of degree L having the roots $\rho_1, \rho_2, \ldots, \rho_L$ and $F^*(z^{-1})$ is a polynomial of degree L having the roots $1/\rho_1^*, 1/\rho_2^*, \ldots, 1/\rho_L^*$. Then an appropriate noise-whitening filter has a z transform $1/F^*(z^{-1})$. Since there are 2^L possible choices for the roots of $F^*(z^{-1})$, each choice resulting in a filter characteristic that is identical in magnitude but different in phase from other choices of the roots, we propose to choose the unique $F^*(z^{-1})$ having minimum phase, i.e., the polynomial having all its roots inside the unit circle. Thus, when all the roots of $F^*(z^{-1})$ are inside the unit circle $1/F^*(z^{-1})$ is a physically realizable, stable, recursive discrete-time filter.† Consequently passage of the sequence $\{y_k\}$ through the digital filter $1/F^*(z^{-1})$ results in an output sequence $\{v_k\}$ which can be expressed as

$$v_k = \sum_{n=0}^{L} f_n I_{k-n} + \eta_k \qquad (6.3.15)$$

where $\{\eta_k\}$ is a white gaussian noise sequence and $\{f_k\}$ is a set of tap coefficients of an equivalent discrete-time transversal filter having a transfer function $F(z)$.

In summary, the cascade of the transmitting filter $u(t)$, the channel $c(t)$, the matched filter $h^*(-t)$, the sampler, and the discrete-time noise-whitening filter $1/F^*(z^{-1})$ can be represented as an equivalent discrete-time transversal filter having the set $\{f_k\}$ as its tap coefficients. The additive noise sequence $\{\eta_k\}$ corrupting the output of the discrete-time transversal filter is a white gaussian noise sequence having zero mean and variance N_0. Figure 6.3.2 illustrates the

---

† By removing the stability condition, we can also show $F^*(z^{-1})$ to have roots on the unit circle.

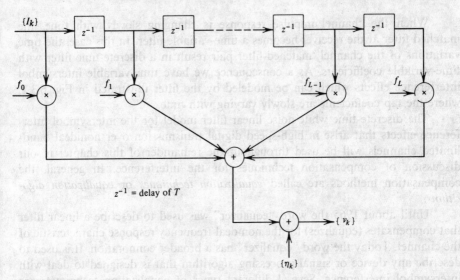

z^{-1} = delay of T

FIGURE 6.3.2
Equivalent discrete-time model of intersymbol interference-channel with WGN.

model of the equivalent discrete-time system with white noise. We refer to this
model as the *equivalent discrete-time white-noise filter model*.

For example, suppose that the transmitter signal pulse $u(t)$ has duration T
and unit energy and the received signal pulse is $h(t) = u(t) + au(t - T)$. Then
the sampled autocorrelation function is given as

$$x_k = \begin{cases} a^* & k = -1 \\ 1 + |a|^2 & k = 0 \\ a & k = 1 \end{cases} \qquad (6.3.16)$$

The z transform of x_k is

$$X(z) = \sum_{k=-1}^{1} x_k z^{-k}$$

$$= a^* z + (1 + |a|^2) + a z^{-1}$$

$$= (az^{-1} + 1)(a^* z + 1)$$

Under the assumption that $|a| > 1$, one chooses $F(z) = az^{-1} + 1$ so that the
equivalent transversal filter consists of two taps having tap gain coefficients
$f_0 = 1$, $f_1 = a$. We note that the correlation sequence $\{x_k\}$ may be expressed in
terms of the $\{f\}$ as

$$x_k = \sum_{n=0}^{L-k} f_n^* f_{n+k} \qquad k = 0, 1, 2, \ldots, L \qquad (6.3.17)$$

When the channel impulse response is changing slowly with time, the matched filter at the receiver becomes a time-variable filter. In this case, the time variations of the channel/matched-filter pair result in a discrete-time filter with time-variable coefficients. As a consequence we have time-variable intersymbol interference effects which can be modeled by the filter illustrated in Fig. 6.3.2, where the tap coefficients are slowly varying with time.

The discrete-time white-noise linear filter model for the intersymbol interference effects that arise in high-speed digital transmission over nonideal band-limited channels will be used throughout the remainder of this chapter in our discussion of compensation techniques for the interference. In general, the compensation methods are called *equalization techniques* or *equalization algorithms*.

Until about 1965 the word "equalizer" was used to describe a linear filter that compensates (equalizes) for the nonideal frequency response characteristic of the channel. Today the word "equalizer" has a broader connotation. It is used to describe any device or signal processing algorithm that is designed to deal with intersymbol interference. Several different types of equalization techniques are described in Sec. 6.4. We begin our discussion with the simplest, namely, a linear equalizer.

6.4 LINEAR EQUALIZATION

The linear filter most often used for equalization is the transversal filter shown in Fig. 6.4.1. Its input is the sequence $\{v_k\}$ given in (6.3.15) and its output is the estimate of the information sequence $\{I_k\}$. The estimate of the kth symbol may be expressed as

$$\hat{I}_k = \sum_{j=-K}^{K} c_j v_{k-j} \qquad (6.4.1)$$

where $\{c_j\}$ are the $(2K+1)$ tap weight coefficients of the filter. The estimate $\hat{I}_k$ is quantized to the nearest (in distance) information symbol to form the decision $\tilde{I}_k$. If $\tilde{I}_k$ is not identical to the transmitted information symbol I_k, an error has been made.

Considerable research has been performed on the criterion for optimizing the filter coefficients $\{c_k\}$. Since the most meaningful measure of performance for a digital communications system is the average probability of error, it is desirable to choose the coefficients to minimize this performance index. However, the probability of error is a highly nonlinear function of $\{c_j\}$. Consequently, the probability of error as a performance index for optimizing the tap weight coefficients of the equalizer is impractical.

Two criteria have found widespread use in optimizing the equalizer coefficients $\{c_j\}$. One is the peak distortion criterion and the other is the mean-square-error criterion.

FIGURE 6.4.1
Linear transversal filter.

6.4.1 Peak Distortion Criterion and the Zero-Facing Algorithm

The peak distortion is simply defined as the worst-case intersymbol interference at the output of the equalizer. The minimization of this performance index is called the *peak distortion criterion*. First we consider the minimization of the peak distortion assuming that the equalizer has an infinite number of taps. Then we will discuss the case in which the transversal equalizer spans a finite time duration.

We observe that the cascade of the discrete-time linear filter model having an impulse response $\{f_n\}$ and an equalizer having an impulse response $\{c_n\}$ can be represented by a single equivalent filter having the impulse response

$$q_n = \sum_{j=-\infty}^{\infty} c_j f_{n-j} \tag{6.4.2}$$

That is, $\{q_n\}$ is simply the convolution of $\{c_n\}$ and $\{f_n\}$. The equalizer is assumed to have an infinite number of taps. Its output at the kth sampling instant can be expressed in the form

$$\hat{I}_k = q_0 I_k + \sum_{n \neq k} I_n q_{k-n} + \sum_{j=-\infty}^{\infty} c_j \eta_{j-j} \tag{6.4.3}$$

The first term in (6.4.3) represents a scaled version of the desired symbol. For convenience, we normalize q_0 to unity. The second term is the intersymbol interference. The peak value of this interference, which is called the *peak distortion*, is

$$D = \sum_{\substack{n=-\infty \\ n \neq 0}}^{\infty} |q_n|$$

$$= \sum_{\substack{n=-\infty \\ n \neq 0}}^{\infty} \left| \sum_{j=-\infty}^{\infty} c_j f_{n-j} \right| \qquad (6.4.4)$$

Thus D is a function of the equalizer tap weights.

With an equalizer having an infinite number of taps, it is possible to select the tap weights so that $D = 0$, i.e., $q_n = 0$ for all n except $n = 0$. That is, the intersymbol interference can be completely eliminated. The values of the tap weights for accomplishing this goal are determined from the condition

$$q_n = \sum_{j=-\infty}^{\infty} c_j f_{n-j} = \begin{cases} 1 & n = 0 \\ 0 & n \neq 0 \end{cases} \qquad (6.4.5)$$

By taking the z transform of (6.4.5), we obtain

$$Q(z) = C(z)F(z) = 1 \qquad (6.4.6)$$

or, simply,

$$C(z) = \frac{1}{F(z)} \qquad (6.4.7)$$

where $C(z)$ denotes the z transform of the $\{c_j\}$. We note that the equalizer, with transfer function $C(z)$, is simply the inverse filter to the linear filter model $F(z)$. In other words, complete elimination of the intersymbol interference requires the use of an inverse filter to $F(z)$. We call such a filter a *zero-forcing filter*. Figure 6.4.2 illustrates in block diagram the equivalent discrete-time channel and equalizer.

The cascade of the noise-whitening filter having the transfer function $1/F^*(z^{-1})$ and the zero-forcing equalizer having the transfer function $1/F(z)$ results in an equivalent zero-forcing equalizer having the transfer function

$$C'(z) = \frac{1}{F(z)F^*(z^{-1})} = \frac{1}{X(z)} \qquad (6.4.8)$$

FIGURE 6.4.2
Block diagram of channel with zero-forcing equalizer.

FIGURE 6.4.3
Block diagram of channel with equivalent zero-forcing equalizer.

as shown in Fig. 6.4.3. This combined filter has as its input the sequence $\{y_k\}$ of samples from the matched filter, given by (6.3.9). Its output consists of the desired symbols corrupted only by additive zero mean gaussian noise. The impulse response of the combined filter is

$$c'_k = \frac{1}{2\pi J} \oint C'(z) z^{k-1} \, dz$$

$$= \frac{1}{2\pi j} \oint \frac{z^{k-1}}{X(z)} \, dz \tag{6.4.9}$$

where the integration is performed on a closed contour that lies within the region of convergence of $C'(z)$. Since $X(z)$ is a polynomial with $2L$ roots ($\rho_1, \rho_2, \ldots, \rho_L$, $1/\rho_1^*, 1/\rho_2^*, \ldots, 1/\rho_L^*$), it follows that $C'(z)$ must converge in an annular region in the z plane that includes the unit circle ($z = e^{j\theta}$). Consequently the closed contour in the integral can be the unit circle.

The performance of the infinite-tap equalizer that completely eliminates the intersymbol interference can be expressed in terms of the signal-to-noise ratio (SNR) at its output. For mathematical convenience we normalize the received signal power to unity. This implies that $q_0 = 1$ and that the expected value of $|I_k|^2$ is also unity. Then the SNR is simply the reciprocal of the noise variance σ_n^2 at the output of the equalizer.

The value of σ_n^2 can be simply determined by observing that the noise sequence $\{\nu_k\}$ at the input to the equivalent zero-forcing equalizer $C'(z)$ has a zero mean and a power spectral density

$$\Phi_{\nu\nu}(\omega) = N_0 X(e^{j\omega T}) \qquad |\omega| \le \frac{\pi}{T} \tag{6.4.10}$$

where $X(e^{j\omega T})$ is obtained from $X(z)$ by the substitution $z = e^{j\omega T}$. Since $C'(z) = 1/X(z)$, it follows that the noise sequence at the output of the equalizer has a power spectral density

$$\Phi_{nn}(\omega) = \frac{N_0}{X(e^{j\omega T})} \qquad |\omega| \le \frac{\pi}{T} \tag{6.4.11}$$

Consequently the variance of the noise variable at the output of the equalizer is

$$\sigma_n^2 = \frac{T}{2\pi} \int_{-\pi/T}^{\pi/T} \Phi_{nn}(\omega) \, d\omega$$

$$= \frac{TN_0}{2\pi} \int_{-\pi/T}^{\pi/T} \frac{d\omega}{X(e^{j\omega T})} \tag{6.4.12}$$

and the SNR for the zero-forcing equalizer is

$$\gamma_\infty = \frac{1}{\sigma_n^2}$$

$$= \left(\frac{TN_0}{2\pi} \int_{-\pi/T}^{\pi/T} \frac{d\omega}{X(e^{j\omega T})} \right)^{-1} \tag{6.4.13}$$

where the subscript on γ indicates that the equalizer has an infinite number of taps.

The spectral characteristics $X(e^{j\omega T})$ corresponding to the Fourier transform of the sampled sequence $\{x_n\}$ has an interesting relationship to the analog filter $H(\omega)$ used at the receiver. Since

$$x_k = \int_{-\infty}^{\infty} h^*(t) h(t + kT) \, dt$$

use of Parseval's theorem yields

$$x_k = \frac{1}{2\pi} \int_{-\infty}^{\infty} |H(\omega)|^2 e^{j\omega kT} \, d\omega \tag{6.4.14}$$

where $H(\omega)$ is the Fourier transform of $h(t)$. But the integral in (6.4.14) can be expressed in the form

$$x_k = \frac{1}{2\pi} \int_{-\pi/T}^{\pi/T} \left[\sum_{n=-\infty}^{\infty} \left| H\left(\omega + \frac{2\pi n}{T} \right) \right|^2 \right] e^{j\omega kT} \, d\omega \tag{6.4.15}$$

Now the Fourier transform of $\{x_k\}$ is

$$X(e^{j\omega T}) = \sum_{k=-\infty}^{\infty} x_k e^{-j\omega kT} \tag{6.4.16}$$

and the inverse transform yields

$$x_k = \frac{T}{2\pi} \int_{-\pi/T}^{\pi/T} X(e^{j\omega T}) e^{j\omega kT} \, d\omega \tag{6.4.17}$$

From a comparison of (6.4.15) and (6.4.17), we obtain the desired relationship between $X(e^{j\omega T})$ and $H(\omega)$. That is

$$X(e^{j\omega T}) = \frac{1}{T} \sum_{n=-\infty}^{\infty} \left| H\left(\omega + \frac{2\pi n}{T} \right) \right|^2 \qquad |\omega| \le \frac{\pi}{T} \tag{6.4.18}$$

where the right-hand side of (6.4.18) is called the *folded spectrum* of $|H(\omega)|^2$. We also observe that $|H(\omega)|^2 = X(\omega)$, where $X(\omega)$ is the Fourier transform of the waveform $x(t)$ and $x(t)$ is the response of the matched filter to the input $h(t)$. Therefore the right-hand side of (6.4.18) can also be expressed in terms of $X(\omega)$.

Substitution for $X(e^{j\omega T})$ in (6.4.13) using the result in (6.4.18) yields the desired expression for the SNR in the form

$$\gamma_\infty = \left[\frac{T^2 N_0}{2\pi} \int_{-\pi/T}^{\pi/T} \frac{d\omega}{\displaystyle\sum_{n=-\infty}^{\infty} \left| H\left(\omega + \frac{2\pi n}{T} \right) \right|^2} \right]^{-1} \qquad (6.4.19)$$

We observe that if the folded spectral characteristic of $H(\omega)$ possesses any zeros, the integrand becomes infinite and the SNR goes to zero. In other words, the performance of the equalizer is poor whenever the folded spectral characteristic possesses nulls or takes on small values. This behavior occurs primarily because the equalizer, in eliminating the intersymbol interference, enhances the additive noise. For example, if the channel contains a spectral null in its frequency response, the linear zero-forcing equalizer attempts to compensate for this by introducing an infinite gain at that frequency. But this compensates for the channel distortion at the expense of enhancing the additive noise. On the other hand, an ideal channel coupled with an appropriate signal design that results in no intersymbol interference will have a folded spectrum that satisfies the condition

$$\sum_{n=-\infty}^{\infty} \left| H\left(\omega + \frac{2\pi n}{T} \right) \right|^2 = T \qquad |\omega| \leq \frac{\pi}{T} \qquad (6.4.20)$$

In this case the SNR achieves its maximum value, namely,

$$\gamma_\infty = \frac{1}{N_0} \qquad (6.4.21)$$

Let us now turn our attention to an equalizer having $(2K + 1)$ taps. Since $c_j = 0$ for $|j| > K$, the convolution of $\{f_n\}$ with $\{c_n\}$ is zero outside the range $-K \leq n \leq K + L - 1$. That is, $q_n = 0$ for $n < -K$ and $n > K + L - 1$. With q_0 normalized to unity, the peak distortion is

$$D = \sum_{\substack{n=-K \\ n \neq 0}}^{K+L-1} |q_n|$$

$$= \sum_{\substack{n=-K \\ n \neq 0}}^{K+L-1} \left| \sum_j c_j f_{n-j} \right| \qquad (6.4.22)$$

Although the equalizer has $(2K + 1)$ adjustable parameters, there are $2K + L$ nonzero values in the response $\{q_n\}$. Therefore it is generally impossible to eliminate completely the intersymbol interference at the output of the equalizer.

There is always some residual interference when the optimum coefficients are used. The problem is to minimize D with respect to the coefficients $\{c_j\}$.

The peak distortion given by (6.4.22) has been shown by Lucky (1965) to be a convex function of the coefficients $\{c_j\}$. That is, it possesses a global minimum and no relative minima. Its minimization can be carried out numerically using, for example, the method of steepest descent. Little more can be said for the general solution to this minimization problem. However, for one special but important case, the solution for the minimization of D is known. This is the case in which the distortion at the input to the equalizer, defined as

$$D_0 = \frac{1}{|f_0|} \sum_{n=1}^{L} |f_n| \qquad (6.4.23)$$

is less than unity. This condition is equivalent to having the eye open prior to equalization. That is, the intersymbol interference is not severe enough to close the eye. Under this condition, the peak distortion D is minimized by selecting the equalizer coefficients to force $q_n = 0$ for $1 \le |n| \le K$ and $q_0 = 1$. In other words, the general solution to the minimization of D, when $D_0 < 1$, is the zero-forcing solution for $\{q_n\}$ in the range $1 \le |n| \le K$. However, the values of $\{q_n\}$ for $K + 1 \le n \le K + L - 1$ are nonzero, in general. These nonzero values constitute the residual intersymbol interference at the output of the equalizer.

The zero-forcing solution when $D_0 < 1$ can be obtained in practice with the following steepest-descent recursive algorithm for adjusting the equalizer coefficients $\{c_j\}$:

$$c_j^{(k+1)} = c_j^{(k)} + \Delta \varepsilon_k I_{k-j}^* \qquad J = -K, \ldots, -1, 0, 1, \ldots, K \qquad (6.4.24)$$

where $c_j^{(k)}$ is the value of the jth coefficient at time $t = kT$, $\varepsilon_k = I_k - \hat{I}_k$ is the error signal at time $t = kT$, and Δ is a scale factor that controls the rate of adjustment, as will be explained later in this section. This is the zero-forcing algorithm. In effect, by cross-correlating the error sequence $\{\varepsilon_k\}$ with the desired sequence $\{I_k\}$ we attempt to force the cross correlations $\varepsilon_k I_{k-j}^*$, $j = -K, \ldots, K$, to zero. In other words, the coefficients are optimally adjusted when the error is orthogonal to the sequence $\{I_k\}$. The demonstration that this leads to the desired solution is quite simple. We have

$$E(\varepsilon_k I_{k-j}^*) = E\left[\left(I_k - \hat{I}_k\right) I_{k-j}^*\right]$$

$$= E\left(I_k I_{k-j}^*\right) - E\left(\hat{I}_k I_{k-j}^*\right) \qquad j = -K, \ldots, K \qquad (6.4.25)$$

We assume that the information symbols are uncorrelated, i.e., $E(I_k I_j^*) = \delta_{kj}$, and that the information sequence $\{I_k\}$ is uncorrelated with the additive noise sequence $\{\eta_k\}$. For $\hat{I}_k$ we use the expression given in (6.4.3). Then, after taking the expected values in (6.4.25) we obtain

$$E(\varepsilon_k I_{k-j}^*) = \delta_{j0} - q_j \qquad j = -K, \ldots, K \qquad (6.4.26)$$

Therefore the conditions

$$E(\varepsilon_k I_{k-j}^*) = 0 \qquad j = -K, \ldots, K \qquad (6.4.27)$$

FIGURE 6.4.4
An adaptive zero-forcing equalizer.

are fulfilled when $q_0 = 1$ and $q_n = 0$ for $1 \le |n| \le K$. Figure 6.4.4 illustrates the equalizer with tap adjustments based on the zero-forcing algorithm.

Two points should be emphasized about the zero-forcing algorithm. One is that the algorithm is optimum in the sense of minimizing D, only if $D_0 < 1$. The zero-forcing solution is not optimum if $D_0 > 1$. The second point is that the peak distortion criterion and, hence, the zero-forcing algorithm neglects the effects of additive noise. These limitations do not exist if the mean-square-error criterion is used in the adjustment of the equalizer coefficients, as discussed below.

6.4.2 Mean-Square-Error (MSE) Criterion and the LMS Algorithm

In the MSE criterion the tap weight coefficients $\{c_j\}$ of the equalizer are adjusted to minimize the mean-square value of the error

$$\varepsilon_k = I_k - \hat{I}_k \qquad (6.4.28)$$

where I_k is the information symbol transmitted in the kth signaling interval and $\hat{I}_k$ is the estimate of that symbol at the output of the equalizer, defined previously in (6.4.1). When the information symbols $\{I_k\}$ are complex-valued, the performance index for the MSE criterion, denoted by J, is defined as

$$J = E|\varepsilon_k|^2$$

$$= E|I_k - \hat{I}_k|^2 \tag{6.4.29}$$

On the other hand, when the information symbols are real-valued, the performance index is simply the square of the real part of ε_k. In either case, J is a quadratic function of the equalizer coefficients $\{c_j\}$. In the following discussion, we consider the minimization of the complex-valued form given in (6.4.29).

First we shall derive the tap weight coefficients that minimize J when the equalizer has an infinite number of taps. In this case the estimate $\hat{I}_k$ is expressed as

$$\hat{I}_k = \sum_{j=-\infty}^{\infty} c_j v_{k-j} \tag{6.4.30}$$

Substitution of (6.4.30) into the expression for J given in (6.4.29) and expansion of the result yields a quadratic function of the coefficients $\{c_j\}$. This function can be easily minimized with respect to the $\{c_j\}$ to yield a set (infinite in number) of linear equations for the $\{c_j\}$. Alternatively, the set of linear equations can be obtained by invoking the orthogonality principle in mean-square estimation. That is, we select the coefficients $\{c_j\}$ to render the error ε_k orthogonal to the signal sequence $\{v_{k-l}^*\}$ for $-\infty < l < \infty$. Thus

$$E(\varepsilon_k v_{k-l}^*) = 0 \qquad -\infty < l < \infty \tag{6.4.31}$$

Substitution for ε_k in (6.4.31) yields

$$E\left[\left(I_k - \sum_{j=-\infty}^{\infty} c_j v_{k-j}\right) v_{k-l}^*\right] = 0$$

or, equivalently,

$$\sum_{j=-\infty}^{\infty} c_j E(v_{k-j} v_{k-l}^*) = E(I_k v_{k-l}^*) \qquad -\infty < l < \infty \tag{6.4.32}$$

To evaluate the moments in (6.4.32), we use the expression for v_k given in (6.3.15). Thus we obtain

$$E(v_{k-j} v_{k-l}^*) = \sum_{n=0}^{L} f_n^* f_{n+l-j} + N_0 \delta_{lj}$$

$$= \begin{cases} x_{l-j} + N_0 \delta_{lj} & |l-j| \le L \\ 0 & \text{otherwise} \end{cases} \tag{6.4.33}$$

and

$$E(I_k v_{k-l}^*) = \begin{cases} f_{-l}^* & -L \le l \le 0 \\ 0 & \text{otherwise} \end{cases} \qquad (6.4.34)$$

Now, if we substitute (6.4.33) and (6.4.34) into (6.4.32) and take the z transform of both sides of the resulting equation, we obtain

$$C(z)\left[F(z)F^*(z^{-1}) + N_0\right] = F^*(z^{-1}) \qquad (6.4.35)$$

Therefore the transfer function of the equalizer based on the MSE criterion is

$$C(z) = \frac{F^*(z^{-1})}{F(z)F^*(z^{-1}) + N_0} \qquad (6.4.36)$$

When the noise whitening filter is incorporated into $C(z)$, we obtain an equivalent equalizer having the transfer function

$$C'(z) = \frac{1}{F(z)F^*(z^{-1}) + N_0}$$
$$= \frac{1}{X(z) + N_0} \qquad (6.4.37)$$

We observe that the only difference between this expression for $C'(z)$ and the one based on the peak distortion criterion is the noise spectral density factor N_0 that appears in (6.4.37). When N_0 is very small in comparison with the signal, the coefficients that minimize the peak distortion D are approximately equal to the coefficients that minimize the MSE performance index J. That is, in the limit as $N_0 \to 0$, the two criteria yield the same solution for the tap weights. Consequently, when $N_0 = 0$, the minimization of the MSE results in complete elimination of the intersymbol interference. On the other hand, that is not the case when $N_0 \ne 0$. In general, when $N_0 \ne 0$, there is both residual intersymbol interference and additive noise at the output of the equalizer.

A measure of the residual intersymbol interference and additive noise is obtained by evaluating the minimum value of J, denoted by $J_{\min}$, when the transfer function $C(z)$ of the equalizer is given by (6.4.36). Since $J = E|\varepsilon_k|^2 = E(\varepsilon_k I_k^*) - E(\varepsilon_k \hat{I}_k^*)$, and since $E(\varepsilon_k \hat{I}_k^*) = 0$ by virtue of the orgothonality conditions given in (6.4.31), it follows that

$$
\begin{aligned}
J_{\min} &= E(\varepsilon_k I_k^*) \\
&= E|I_k|^2 - \sum_{j=-\infty}^{\infty} c_j E(v_{k-j} I_k^*) \\
&= 1 - \sum_{j=-\infty}^{\infty} c_j f_{-j} \qquad (6.4.38)
\end{aligned}
$$

This particular form for $J_{\min}$ is not very informative. More insight on the performance of the equalizer as a function of the channel characteristics is obtained when the summation in (6.4.38) is transformed into the frequency domain. This can be accomplished by first noting that the summation in (6.4.38) is the convolution of $\{c_j\}$ with $\{f_j\}$, evaluated at a shift of zero. Thus, if $\{b_k\}$ denotes the convolution of these two sequences, the summation in (6.4.38) is simply equal to b_0. Since the z transform of the sequence $\{b_k\}$ is

$$
\begin{aligned}
B(z) &= C(z)F(z) \\
&= \frac{F(z)F^*(z^{-1})}{F(z)F^*(z^{-1}) + N_0} \\
&= \frac{X(z)}{X(z) + N_0}
\end{aligned}
\tag{6.4.39}
$$

the term b_0 is

$$
\begin{aligned}
b_0 &= \frac{1}{2\pi j} \oint \frac{B(z)}{z} \, dz \\
&= \frac{1}{2\pi j} \oint \frac{X(z)}{z[X(z) + N_0]} \, dz
\end{aligned}
\tag{6.4.40}
$$

The contour integral in (6.4.40) can be transformed into an equivalent line integral by the change in variable $z = e^{j\omega T}$. The result of this change in variable is

$$
b_0 = \frac{T}{2\pi} \int_{-\pi/T}^{\pi/T} \frac{X(e^{j\omega T})}{X(e^{j\omega T}) + N_0}
\tag{6.4.41}
$$

Finally, substitution of the result in (6.4.41) for the summation in (6.4.38) yields the desired expression for the minimum MSE in the form

$$
\begin{aligned}
J_{\min} &= 1 - \frac{T}{2\pi} \int_{-\pi/T}^{\pi/T} \frac{X(e^{j\omega T})}{X(e^{j\omega T}) + N_0} \, d\omega \\
&= \frac{T}{2\pi} \int_{-\pi/T}^{\pi/T} \frac{N_0}{X(e^{j\omega T}) + N_0} \, d\omega \\
&= \frac{T}{2\pi} \int_{-\pi/T}^{\pi/T} \frac{N_0}{\frac{1}{T} \sum_{n=-\infty}^{\infty} \left| H\left(\omega + \frac{2\pi n}{T}\right) \right|^2 + N_0} \, d\omega
\end{aligned}
\tag{6.4.42}
$$

In the absence of intersymbol interference, $X(e^{j\omega T}) = 1$ and, hence,

$$
J_{\min} = \frac{N_0}{1 + N_0}
\tag{6.4.43}
$$

We observe that $0 \le J_{\min} \le 1$. Furthermore, the relationship between the output SNR γ_∞ and $J_{\min}$ must be

$$
\gamma_\infty = \frac{1 - J_{\min}}{J_{\min}}
\tag{6.4.44}
$$

More importantly, this relation between γ_∞ and $J_{\min}$ also holds when there is residual intersymbol interference in addition to the noise.

Let us now turn our attention to the case in which the transversal equalizer spans a finite time duration. The output of the equalizer in the k th signaling interval is

$$\hat{I}_k = \sum_{j=-K}^{K} c_j v_{k-j} \qquad (6.4.45)$$

The MSE for the equalizer having $2K + 1$ taps, denoted as $J(K)$, is

$$J(K) = E|I_k - \hat{I}_k|^2$$

$$= E\left| I_k - \sum_{j=-K}^{K} c_j v_{k-j} \right|^2 \qquad (6.4.46)$$

Minimization of $J(K)$ with respect to the tap weights $\{c_j\}$ or, equivalently, forcing the error $\varepsilon_k = I_k - \hat{I}_k$ to be orthogonal to the signal samples $v_{j-l}^*, |l| \leq K$, yields the following set of simultaneous equations:

$$\sum_{j=-K}^{K} c_j \Gamma_{lj} = \xi_l \qquad l = -K, \ldots, -1, 0, 1, \ldots, K \qquad (6.4.47)$$

where

$$\Gamma_{lj} = \begin{cases} x_{l-j} + N_0 \delta_{lj} & |l-j| \leq L \\ 0 & \text{otherwise} \end{cases} \qquad (6.4.48)$$

and

$$\xi_l = \begin{cases} f_{-l}^* & -L \leq l \leq 0 \\ 0 & \text{otherwise} \end{cases} \qquad (6.4.49)$$

It is convenient to express the set of linear equations in matrix form. Thus

$$\Gamma C = \xi \qquad (6.4.50)$$

where C denotes the column vector of $(2K + 1)$ tap weight coefficients, Γ denotes the $(2K + 1) \times (2K + 1)$ Hermitian covariance matrix with elements Γ_{ij}, and ξ is a $(2K + 1)$-dimensional column vector with elements ξ_l. The solution of (6.4.50) is

$$C_{\text{opt}} = \Gamma^{-1} \xi \qquad (6.4.51)$$

The optimum tap weight coefficients given by (6.4.51) minimize the performance index $J(K)$, with the result that the minimum value of $J(K)$ is

$$J_{\min}(K) = 1 - \sum_{j=-K}^{K} c_j f_{-j}$$

$$= 1 - \xi'^* \Gamma^{-1} \xi \qquad (6.4.52)$$

where ξ' represents the transpose of the column vector ξ.

The solution for C_{opt} involves inverting the covariance matrix Γ. Alternatively, an iterative procedure may be used to determine C_{opt}. Probably the simplest iterative procedure is the method of steepest descent, in which one begins by choosing arbitrarily the vector $\mathbf{C}$, say $\mathbf{C}_0$. This initial choice of coefficients corresponds to some point on the quadratic MSE surface in the $(2K + 1)$-dimensional space of coefficients. The gradient vector $\mathbf{G}_0$, having the $(2K + 1)$ gradient components $\frac{1}{2}\partial J/\partial c_{0k}, k = -K, \ldots, -1, 0, 1, \ldots, K$, is then computed at this point on the MSE surface and each tap weight is changed in the direction opposite to its corresponding gradient component. The change in the jth tap weight is proportional to the size of the jth gradient component. Thus succeeding values of the coefficient vector $\mathbf{C}$ are obtained according to the relation

$$\mathbf{C}_{k+1} = \mathbf{C}_k - \Delta \mathbf{G}_k \qquad k = 0, 1, 2, \ldots \qquad (6.4.53)$$

where the gradient vector $\mathbf{G}_k$ is

$$\mathbf{G}_k = \frac{dJ}{2d\mathbf{C}_k} = \Gamma \mathbf{C}_k - \boldsymbol{\xi} = -E(\varepsilon_k \mathbf{V}_k^*) \qquad (6.4.54)$$

The vector $\mathbf{C}_k$ represents the set of coefficients at the kth iteration, $\varepsilon_k = I_k - \hat{I}_k$ is the error signal at the kth iteration, $\mathbf{V}_k$ is the vector of received signal samples that make up the estimate $\hat{I}_k$, i.e., $\mathbf{V}_k = (v_{k+K}, \ldots, v_k, \ldots, v_{k-K})$, and Δ is a positive number chosen small enough to ensure convergence of the iterative procedure. If the minimum MSE is reached for some $k = k_0$, then $\mathbf{G}_k = \mathbf{0}$ so that no further change occurs in the tap weights. In general, $J_{\min}(K)$ cannot be attained for a finite value of k_0 with the steepest-descent method. It can, however, be approached as closely as desired for some finite value of k_0.

The basic difficulty with the method of steepest descent for determining the optimum tap weights is the lack of knowledge of the gradient vector $\mathbf{G}_k$ which depends on both the covariance matrix Γ and the vector $\boldsymbol{\xi}$ of cross correlations. In turn, these quantities depend on the coefficients $\{f_k\}$ of the equivalent discrete-time channel model and on the covariances of the information sequence and the additive noise, all of which may be unknown at the receiver in general. To overcome the difficulty, estimates of the gradient vector may be used. That is, the algorithm for adjusting the tap weight coefficients may be expressed in the form

$$\hat{\mathbf{C}}_{k+1} = \hat{\mathbf{C}}_k - \Delta \hat{\mathbf{G}}_k \qquad (6.4.55)$$

where $\hat{\mathbf{G}}_k$ denotes an estimate of the gradient vector $\mathbf{G}_k$ and $\hat{\mathbf{C}}_k$ denotes the estimate of the vector of coefficients.

From (6.4.54) we note that $\mathbf{G}_k$ is the negative of the expected value of the $\varepsilon_k \mathbf{V}_k^*$. Consequently an estimate of $\mathbf{G}_k$ is

$$\hat{\mathbf{G}}_k = -\varepsilon_k \mathbf{V}_k^* \qquad (6.4.56)$$

Since $E(\hat{\mathbf{G}}_k) = \mathbf{G}_k$, the estimate $\hat{\mathbf{G}}_k$ is an unbiased estimate of the true gradient vector $\mathbf{G}_k$. Incorporation of (6.4.56) into (6.4.55) yields the algorithm

$$\hat{\mathbf{C}}_{k+1} = \hat{\mathbf{C}}_k + \Delta \varepsilon_k \mathbf{V}_k^* \qquad (6.4.57)$$

FIGURE 6.4.5
Linear adaptive equalizer based on MSE criterion.

This is the basic LMS (least-mean-square) algorithm for recursively adjusting the tap weight coefficients of the equalizer first proposed by Widrow and Hoff (1960). It is illustrated in the equalizer shown in Fig. 6.4.5.

The basic algorithm given by (6.4.57) and some of the possible variations of it have been incorporated into many commercial adaptive equalizers which are used in high speed modems. Three variations of the basic algorithm are obtained by using only sign information contained in the error signal ε_k and/or in the components of $\mathbf{V}_k$. Hence the three possible variations are

$$c_{(k+1)j} = c_{kj} + \Delta \operatorname{sgn}(\varepsilon_k) v^*_{k-j} \qquad j = -K, \ldots, -1, 0, 1, \ldots, K \qquad (6.4.58)$$

$$c_{(k+1)j} = c_{kj} + \Delta \varepsilon_k \operatorname{sgn}(v^*_{k-j}) \qquad j = -K, \ldots, -1, 0, 1, \ldots, K \qquad (6.4.59)$$

$$c_{(k+1)j} = c_{kj} + \Delta \operatorname{sgn}(\varepsilon_k) \operatorname{sgn}(v^*_{k-j}) \qquad j = -K, \ldots, -1, 0, 1, \ldots, K \qquad (6.4.60)$$

where $\operatorname{sgn}(x)$ is defined as

$$\operatorname{sgn}(x) = \begin{cases} 1+j & \operatorname{Re}(x) > 0, \operatorname{Im}(x) > 0 \\ 1-j & \operatorname{Re}(x) > 0, \operatorname{Im}(x) < 0 \\ -1+j & \operatorname{Re}(x) < 0, \operatorname{Im}(x) > 0 \\ -1-j & \operatorname{Re}(x) < 0, \operatorname{Im}(x) < 0 \end{cases} \qquad (6.4.61)$$

Clearly the algorithm in (6.4.60) is the most easily implemented, but it gives the slowest rate of convergence to the others.

Several other variations of the LMS algorithm are obtained by averaging or filtering the gradient vectors over several iterations prior to making adjustments of the equalizer coefficients. For example, the average over N gradient vectors is

$$\overline{\hat{\mathbf{G}}}_{mN} = -\frac{1}{N} \sum_{n=0}^{N-1} \varepsilon_{mN+n} \mathbf{V}_{mN+n}^* \qquad (6.4.62)$$

and the corresponding recursive equation for updating the equalizer coefficients once every N iterations is

$$\hat{\mathbf{C}}_{(k+1)N} = \hat{\mathbf{C}}_{kN} - \Delta \overline{\hat{\mathbf{G}}}_{kN} \qquad (6.4.63)$$

In effect, the averaging operation performed in (6.4.62) reduces the noise in the estimate of the gradient vector as shown by Gardner (1984).

An alternative approach is to filter the noisy gradient vectors by a lowpass filter and use the output of the filter as an estimate of the gradient vector. For example, a simple lowpass filter for the noisy gradients yields as an output

$$\overline{\hat{\mathbf{G}}}_k = w\overline{\hat{\mathbf{G}}}_{k-1} + (1 - w)\hat{\mathbf{G}}_k \qquad \overline{\mathbf{G}}(0) = \hat{\mathbf{G}}(0) \qquad (6.4.64)$$

where the choice of $0 \leq w < 1$ determines the bandwidth of the lowpass filter. When w is close to unity, the filter bandwidth is small and the effective averaging is performed over many gradient vectors. On the other hand, when w is small, the lowpass filter has a large bandwidth and, hence, it provides little averaging of the gradient vectors. With the filtered gradient vectors given by (6.4.64) in place of $\mathbf{G}_k$, we obtain the filtered gradient LMS algorithm given by

$$\hat{\mathbf{C}}_{k+1} = \hat{\mathbf{C}}_k - \Delta \overline{\hat{\mathbf{G}}}_k \qquad (6.4.65)$$

In the above discussion it was assumed that the receiver has knowledge of the transmitted information sequence in forming the error signal between the desired symbol and its estimate. Such knowledge can be made available during a short training period in which a signal with a known information sequence is transmitted to the receiver for initially adjusting the tap weights. The length of this sequence must be at least as long as the length of the equalizer so that the spectrum of the transmitted signal adequately covers the bandwidth of the channel being equalized.

In practice the training sequence is often selected to be a periodic pseudorandom sequence, such as a maximum length shift-register sequence whose period N is equal to the length of the equalizer ($N = 2K + 1$). In this case, the gradient is usually averaged over the length of the sequence as indicated in (6.4.62) and the equalizer is adjusted once a period according to (6.4.63). A practical scheme for continuous adjustment of the tap weights may be either a decision-directed mode of operation in which decisions on the information symbols are assumed to be correct and used in place of I_k in forming the error signal ε_k, or a known pseudo-random-probe sequence may be inserted in the information-bearing signal either additively or by interleaving in time and the tap weights adjusted by comparing the received probe symbols with the known transmitted probe symbols. In the decision-directed mode of operation, the error signal becomes $\varepsilon_k = \tilde{I}_k - \hat{I}_k$ where $\tilde{I}_k$ is the decision of the receiver based on the estimate $\hat{I}_k$. As long as the receiver is operating at low error rates, an occasional error will have a negligible effect on the convergence of the algorithm.

If the channel response changes, this change is reflected in the coefficients $\{f_k\}$ of the equivalent discrete-time channel model. It is also reflected in the error signal ε_k since it depends on $\{f_k\}$. Hence the tap weights will be changed according to (6.4.57) to reflect the change in the channel. A similar change in the tap weights occurs if the statistics of the noise or the information sequence change. Thus the equalizer is adaptive.

Examples of performance characteristics for the MSE equalizer

Example 6.4.1. First we consider an equivalent discrete-time channel model consisting of two components f_0 and f_1, which are normalized to $|f|^2 + |f_1|^2 = 1$. Then

$$F(z) = f_0 + f_1 z^{-1} \qquad (6.4.66)$$

and

$$X(z) = f_0 f_1^* z + 1 + f_0^* f_1 z^{-1} \qquad (6.4.67)$$

The corresponding frequency response is

$$X(e^{j\omega T}) = f_0 f_1^* e^{j\omega T} + 1 + f_0^* f_1 e^{-j\omega T}$$
$$= 1 + 2|f_0| |f_1| \cos(\omega T + \theta) \qquad (6.4.68)$$

where θ is the angle of $f_0 f_1^*$. We note that this channel characteristic possesses a null at $\omega = \pi/T$ when $f_0 = f_1 = 1/\sqrt{2}$.

A linear equalizer with an infinite number of taps, adjusted on the basis of the MSE criterion, will have the minimum MSE given by (6.4.42). Evaluation of the integral in (6.4.42) for the $X(e^{j\omega T})$ given in (6.4.64) yields the result

$$J_{\min} = \frac{N_0}{\sqrt{N_0^2 + 2N_0\left(|f_0|^2 + |f_1|^2\right) + \left(|f_0|^2 - |f_1|^2\right)^2}}$$

$$= \frac{N_0}{\sqrt{N_0^2 + 2N_0 + \left(|f_0|^2 - |f_1|^2\right)^2}} \qquad (6.4.69)$$

Let us consider the special case in which $f_0 = f_1 = 1/\sqrt{2}$. The minimum MSE is $J_{\min} = N_0/\sqrt{N_0^2 + 2N_0}$ and the corresponding output SNR is

$$\gamma_\infty = \sqrt{1 + \frac{2}{N_0}} - 1$$

$$\approx \left(\frac{2}{N_0}\right)^{1/2} \qquad N_0 \ll 1 \qquad (6.4.70)$$

This result should be compared with the output SNR of $1/N_0$ obtained in the case of no intersymbol interference. A significant loss in SNR occurs from this channel.

Example 6.4.2. As a second example we consider an exponentially decaying characteristic of the form

$$f_k = \sqrt{1 - a^2}\, a^k \qquad k = 0, 1, \ldots$$

where $a < 1$. The Fourier transform of this sequence is

$$X(e^{j\omega T}) = \frac{1 - a^2}{1 + a^2 - 2a \cos \omega T} \tag{6.4.71}$$

which is a function that contains a minimum at $\omega = \pi/T$.

The output SNR for this channel is

$$\gamma_\infty = \frac{1}{\left(\sqrt{1 + 2N_0 \dfrac{1 + a^2}{1 - a^2} + N_o^2} - 1 \right)}$$

$$\approx \frac{1 - a^2}{(1 + a^2) N_0} \qquad N_0 \ll 1 \tag{6.4.72}$$

Therefore the loss in SNR due to the presence of the interference is

$$10 \log_{10} \left[(1 - a^2)/(1 + a^2) \right]$$

As a further illustration of the performance limitations of a linear equalizer in the presence of severe intersymbol interference, we show in Fig. 6.4.6 the probability of error for binary (antipodal) signaling, as measured by Monte Carlo simulation, for the three discrete-time channel characteristic shown in Fig. 6.4.7. For purposes of comparison, the performance obtained for a channel with no intersymbol interference is also illustrated in Fig. 6.4.6. The equivalent discrete-time channel shown in Fig. 6.4.7a is typical of the response of a data-quality telephone channel. In contrast, the equivalent discrete-time channel characteristics shown in Fig. 6.4.7b and c result in severe intersymbol interference. The spectral characteristics $|X(e^{j\omega})|$ for the three channels, illustrated in Fig. 6.4.8, clearly show that the channel in Fig. 6.4.7c has the worst spectral characteristic. Hence the performance of the linear equalizer for this channel is the poorest of the three cases. Next in performance is the channel shown in Fig. 6.4.7b and, finally, the best performance is obtained with the channel shown in Fig. 6.4.7a. In fact, the error rate of the latter is within 3 dB of the error rate achieved with no interference.

One conclusion reached from the results on output SNR γ_∞ and the limited probability of error results illustrated in Fig. 6.4.6 is that a linear equalizer yields good performance on channels such as telephone lines, where the spectral characteristics of the channels are well behaved and do not exhibit spectral nulls. On the other hand, multipath radio channels frequently possess nulls in their time-variant spectral characteristics. Therefore a linear equalizer is inadequate as a compensator for the intersymbol interference encountered on these channels.

FIGURE 6.4.6
Error rate performance of linear MSE equalizer.

In a practical implementation the linear adaptive equalizer shown in Fig. 6.4.5 can be realized either at lowpass (baseband) or at bandpass. For example, Fig. 6.4.9 illustrates the demodulation of QAM (or multiphase PSK) by first translating the signal to baseband and equalizing the baseband signal with an equalizer having complex-valued coefficients. In effect, the complex equalizer with complex-valued (in-phase and quadrature components) input is equivalent to four parallel equalizers with real-valued tap coefficients as shown in Fig. 6.4.10.

As an alternative, we may equalize the signal at passband. This is accomplished as shown in Fig. 6.4.11 for a two-dimensional signal constellation such as QAM and PSK. The received signal is filtered and, in parallel, it is passed through a Hilbert transformer, called a phase splitting filter. Thus, we have the equivalent of in-phase and quadrature components at passband which are fed to a passband complex equalizer. Following the equalization, the signal is down

FIGURE 6.4.7
Three discrete-time channel characteristics.

converted to baseband and detected. The error signal generated for the purpose of adjusting the equalizer coefficients is formed at baseband and frequency translated to passband as illustrated in Fig. 6.4.11.

Linear feedback equalizer. The transversal equalizer with a finite number of taps is an all-zero filter. A small improvement in performance can be achieved by

FIGURE 6.4.8*a*
Amplitude spectrum for the channel shown in Fig. 6.4.7*a*.

FIGURE 6.4.8*b*
Amplitude spectrum for the channel shown in Fig. 6.4.7*b*.

FIGURE 6.4.8c
Amplitude spectrum for the channel shown in Fig. 6.4.7c.

using a linear equalizer that has both poles and zeros. In such an equalizer, the estimate of the symbol at the kth signaling interval, $\tilde{I}_k$, is a weighted linear combination of the input signal sequence v_{k-n}, $n = -K, \ldots, -1, 0$, and of the previous outputs $\hat{I}_{k-1}, \hat{I}_{k-2}, \ldots, \hat{I}_{k-K}$. Thus

$$\hat{I}_k = \sum_{j=-K}^{0} c_j v_{k-j} + \sum_{j=1}^{K} c_j \hat{I}_{k-j} \qquad (6.4.73)$$

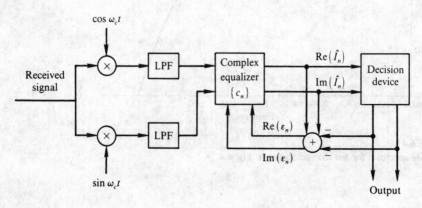

FIGURE 6.4.9
QAM signal demodulation.

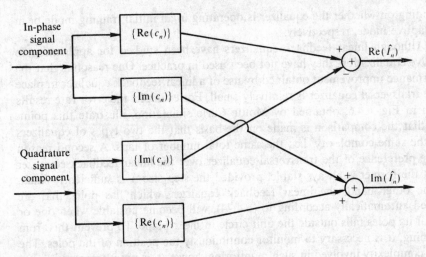

FIGURE 6.4.10
Complex-valued baseband equalizer for QAM signals.

The filter coefficients may be adjusted recursively by forcing the error to be orthogonal to the data in the estimate. That is

$$\hat{\mathbf{C}}_{k+1} = \hat{\mathbf{C}}_k + \Delta \varepsilon_k \hat{V}_k^*$$ (6.4.74)

where the vector $\mathbf{C}_k$ consists of the filter coefficients $c_{-K}, \ldots, c_{-1}, c_0, c, \ldots, c_K$, the vector $\hat{\mathbf{V}}_k$ consists of the data $v_{k+K}, \ldots, v_k, \hat{I}_{k-1}, \ldots, \hat{I}_{k-K}$, and the error is either

$$\varepsilon_k = I_k - \hat{I}_k$$

or

$$\varepsilon_k = \tilde{I}_k - \hat{I}_k$$

FIGURE 6.4.11
QAM or PSK signal equalization at passband.

depending on whether the equalizer is operating in an initial training mode or in an adaptive mode, respectively.

Although linear feedback equalizers have been studied for application to adaptive equalization, they have not been used in practice. One reason is that the performance improvement obtained by use of a linear feedback equalizer in place of the transversal equalizer is relatively small. For example, the error rate results shown in Fig. 6.4.6, obtained by Monte Carlo simulation, illustrate this point. Note that the comparison is made on the basis that the two types of equalizers have the same complexity, i.e., the same total number of taps. A second reason for the preference of the transversal equalizer over the linear feedback equalizer is that the former is always stable, provided the step size Δ is sufficiently small. On the other hand, the linear feedback equalizer, which has poles that are adjusted automatically according to (6.4.74), will become unstable when one or more of its poles falls outside the unit circle in the z plane. To prevent this from happening, it is necessary to monitor continuously the position of the poles. The added complexity involved in such monitoring, however, is not warranted in view of the small improvement in performance.

The limitations of the linear equalizer in coping with severe intersymbol interference have motivated a considerable amount of research on more effective equalization methods. In Secs. 6.5, 6.6, and 6.7 we present nonlinear equalization techniques for dealing with intersymbol interference. As we shall see, the nonlinear techniques provide adequate compensation for severe intersymbol interference.

Convergence properties of the LMS algorithm. The convergence properties of the LMS algorithm given by (6.4.57) are governed by the step-size parameter Δ. We shall now consider the choice of the parameter Δ to ensure convergence of the steepest-descent algorithm in (6.4.53) which employs the exact value of the gradient.

From (6.4.53) and (6.4.54) we have

$$\mathbf{C}_{k+1} = \mathbf{C}_k - \Delta\mathbf{G}_k$$
$$= (\mathbf{I} - \Delta\Gamma)\mathbf{C}_k + \Delta\xi \tag{6.4.75}$$

where $\mathbf{I}$ is the identity matrix, Γ is the autocorrelation matrix of the received signal, $\mathbf{C}_k$ is the $(2K + 1)$-dimensional vector of equalizer tap gains, and ξ is the vector of cross correlations given by (6.4.49). The recursive relation in (6.4.75) can be represented as a closed-loop control system as shown in Fig. 6.4.12. Unfortunately, the set of $(2K + 1)$ first-order difference equations in (6.4.75) are coupled through the autocorrelation matrix Γ. In order to solve these equations and, thus, to establish the convergence properties of the recursive algorithm, it is mathematically convenient to decouple the equations by performing a linear transformation. The appropriate transformation is obtained by noting that the matrix Γ is Hermitian and, hence, can be represented as

$$\Gamma = \mathbf{U}\Lambda\mathbf{U}'^* \tag{6.4.76}$$

FIGURE 6.4.12
Closed-loop control system representation of recursive equation in (6.4.75).

where U is the normalized model matrix of Γ and Λ is a diagonal matrix with diagonal elements equal to the eigenvalues of Γ.

When (6.4.76) is substituted into (6.4.75) and if we define the transformed (orthogonalized) vectors $C_k^o = U'^* C_k$ and $\xi^o = U'^* \xi$, we obtain

$$C_{k+1}^o = (I - \Delta\Lambda)C_k^o + \Delta\xi^o \qquad (6.4.77)$$

This set of first-order difference equations is now decoupled. Their convergence is determined from the homogeneous equation

$$C_{k+1}^o = (I - \Delta\Lambda)C_k^o \qquad (6.4.78)$$

We see that the recursive relation will converge provided that all the poles lie inside the unit circle, i.e.,

$$|1 - \Delta\lambda_k| < 1 \qquad k = -K, \ldots, -1, 0, 1, \ldots, K \qquad (6.4.79)$$

where $\{\lambda_k\}$ is the set of $(2K + 1)$ (possibly nondistinct) eigenvalues of Γ. Since Γ is an autocorrelation matrix, it is positive-definite and, hence, $\lambda_k > 0$ for all k. Consequently convergence of the recursive relation in (6.4.77) is ensured if Δ satisfies the inequality

$$0 < \Delta < \frac{2}{\lambda_{max}} \qquad (6.4.80)$$

where λ_{max} is the largest eigenvalue of Γ.

Since the largest eigenvalue of a positive-definite matrix is less than the sum of all the eigenvalues of the matrix and, furthermore, since the sum of the eigenvalues of a matrix is equal to its trace, we have the following simple upper bound on λ_{max}:

$$\lambda_{max} < \sum_{k=-K}^{K} \lambda_k = \text{trace } \Gamma = (2K + 1)\Gamma_{kk}$$

$$= (2K + 1)(x_0 + N_0) \qquad (6.4.81)$$

From (6.4.78) and (6.4.79) we observe that rapid convergence occurs when $|1 - \Delta\lambda_k|$ is small, i.e., when the pole positions are far from the unit circle. But we cannot achieve this desirable condition and still satisfy (6.4.80) if there is a large difference between the largest and smallest eigenvalues of Γ. In other words, even if we select Δ to be near the upper bound given in (6.4.80), the convergence rate of the recursive MSE algorithm is determined by the smallest eigenvalue $\lambda_{\min}$. Consequently the ratio $\lambda_{\max}/\lambda_{\min}$ ultimately determines the convergence rate. If $\lambda_{\max}/\lambda_{\min}$ is small, Δ can be selected so as to achieve rapid convergence. However, if the ratio $\lambda_{\max}/\lambda_{\min}$ is large, as is the case when the channel frequency response has deep spectral nulls, the convergence rate of the algorithm will be slow.

Excess MSE due to noisy gradient estimates. The recursive algorithm in (6.4.57) for adjusting the coefficients of the linear equalizer employs unbiased noisy estimates of the gradient vector. The noise in these estimates causes random fluctuations in the coefficients about their optimal values and, thus, leads to an increase in the MSE at the output of the equalizer. That is, the final MSE is $J_{\min} + J_\Delta$, where J_Δ is the variance of the measurement noise. The term J_Δ due to the estimation noise has been termed "excess mean-square error" by Widrow (1966).

The total MSE at the output of the equalizer for any set of coefficients $\mathbf{C}$ can be expressed as

$$J = J_{\min} + \left(\mathbf{C} - \mathbf{C}_{\text{opt}}\right)'^{*}\Gamma\left(\mathbf{C} - \mathbf{C}_{\text{opt}}\right) \qquad (6.4.82)$$

where $\mathbf{C}_{\text{opt}}$ represents the optimum coefficients defined by (6.4.51). This expression for the MSE can be simplified by performing the linear orthogonal transformation used above to establish convergence. The result of this transformation applied to (6.4.82) is

$$J = J_{\min} + \sum_{k=-K}^{K} \lambda_k E|c_k^o - c_{k\,\text{opt}}^o|^2 \qquad (6.4.83)$$

where the $\{c_k^o\}$ are the set of transformed equalizer coefficients. The excess MSE is the expected value of the second term in (6.4.83), i.e.,

$$J_\Delta = \sum_{k=-K}^{K} \lambda_k E|c_k^o - c_{k\,\text{opt}}^o|^2 \qquad (6.4.84)$$

To determine J_Δ, we assume that the mean values of the equalizer coefficients $\mathbf{C}_k$ have converged to their optimum values $\mathbf{C}_{\text{opt}}$. Then, the term $\Delta\varepsilon_k\mathbf{V}_k^*$ in the LMS algorithm given by (6.4.57) is a zero mean noise vector. Its covariance is

$$\text{cov}\left[\Delta\varepsilon_k\mathbf{V}_k^*\right] = \Delta^2 E\left[|\varepsilon_k|^2\mathbf{V}_k^*\mathbf{V}_k'\right] \qquad (6.4.85)$$

To a first approximation, we assume that $|\varepsilon_k|^2$ is uncorrelated with the signal vector. Although this assumption is not strictly true, it simplifies the derivation and yields useful results. [The reader may refer to the papers by Mazo (1979), Jones et al. (1982), and Gardner (1984) for further discussion on this assumption.] Then,

$$\text{cov}\left[\Delta\varepsilon_k \mathbf{V}_k^*\right] = \Delta^2 E\left[|\varepsilon_k|^2\right] E\left[\mathbf{V}_k^* \mathbf{V}_k'\right]$$

$$= \Delta^2 J_{\min}\Gamma \tag{6.4.86}$$

For the orthogonalized coefficient vector $\mathbf{C}_k^o$ with additive noise, we have the equation

$$\mathbf{C}_{k+1}^o = (\mathbf{I} - \Delta\Lambda)\mathbf{C}_k^o + \Delta\xi^o + \mathbf{w}_k^o \tag{6.4.87}$$

where $\mathbf{w}_k^o$ is the additive noise vector, which is related to the noise $\Delta\varepsilon_k \mathbf{V}_k^*$ through the transformation

$$\mathbf{w}_k^o = \mathbf{U}^*\left[\Delta\varepsilon_k \mathbf{V}_k^*\right] = \Delta\varepsilon_k \mathbf{U}^* \mathbf{V}_k^* \tag{6.4.88}$$

It is easily seen that the covariance matrix of this noise vector is

$$\text{cov}\left(\mathbf{w}_k^o\right) = \Delta^2 J_{\min}\mathbf{U}^*\Gamma\mathbf{U}$$

$$= \Delta^2 J_{\min}\Lambda \tag{6.4.89}$$

Therefore the components of $\mathbf{w}_k^o$ are uncorrelated and each component has the variance $\sigma_k^2 = \Delta^2\lambda_k J_{\min}$.

Since the noise components of $\mathbf{w}_k^o$ are uncorrelated, we may consider the set of uncoupled difference equations in (6.4.87) separately. Each first-order difference equation represents a filter with impulse response

$$h_k(n) = (1 - \Delta\lambda_k)^n \tag{6.4.90}$$

If we assume that the noise sequence that excites this filter is white with variance $\sigma_k^2 = \Delta^2\lambda_k J_{\min}$, then the variance of the noise at the output of this filter is

$$E|c_k^o - c_{k\,\text{opt}}^o|^2 = \sigma_k^2 \sum_{n=0}^{\infty} h_k^2(n) = \frac{\Delta^2\lambda_k J_{\min}}{1 - (1 - \Delta\lambda_k)^2} \tag{6.4.91}$$

When we substitute the result in (6.4.91) into (6.4.84) we obtain the expression for the excess mean square error as

$$E|c^o - c_{k\,\text{opt}}^o|^2 = \frac{\Delta^2 J_{\min}\lambda_k}{1 - (1 - \Delta\lambda_k)^2} \tag{6.4.92}$$

Hence the excess MSE is

$$J_\Delta = \Delta^2 J_{\min} \sum_{k=-K}^{K} \frac{\lambda_k^2}{1 - (1 - \Delta\lambda_k)^2} \tag{6.4.93}$$

The expression in (6.4.93) can be simplified when Δ is selected such that $\Delta\lambda_k \ll 1$ for all k. Then

$$J_\Delta \approx \tfrac{1}{2}\Delta J_{min} \sum_{k=-K}^{K} \lambda_k$$

$$\approx \tfrac{1}{2}\Delta J_{min} \text{ trace } \Gamma$$

$$\approx \frac{\Delta(2K+1)J_{min}(x_0 + N_0)}{2} \tag{6.4.94}$$

Note that $(x_0 + N_0)$ represents the received signal plus noise power.

It is desirable to have $J_\Delta < J_{min}$. That is, Δ should be selected such that

$$\frac{J_\Delta}{J_{min}} \approx \frac{\Delta(2K+1)(x_0 + N_0)}{2} < 1$$

or, equivalently,

$$\Delta < \frac{2}{(2K+1)(x_0 + N_0)} \tag{6.4.95}$$

For example, if Δ is selected as

$$\Delta = \frac{0.2}{(2K+1)(x_0 + N_0)} \tag{6.4.96}$$

the degradation in the output SNR of the equalizer due to the excess MSE is less than 1 dB.

The analysis given above on the excess mean square error is based on the assumption that the mean value of the equalizer coefficients has converged to their optimum values $\mathbf{C}_{opt}$. Under this condition, the step size Δ should satisfy the bound in (6.4.95). On the other hand, we have determined that convergence of the mean coefficient vector requires that $\Delta < 2/\lambda_{max}$. While a choice of Δ near the upper bound $2/\lambda_{max}$ may lead to initial convergence of the deterministic (known) steepest-descent gradient algorithm, such a large value of Δ will usually result in instability of the LMS stochastic gradient algorithm.

The initial convergence or transient behavior of the LMS algorithm has been investigated by several researchers. Their results clearly indicate that the step size must be reduced in direct proportion to the length of the equalizer as specified by (6.4.95). Hence, the upper bound given by (6.4.95) is also necessary to ensure the initial convergence of the LMS algorithm. The papers by Gitlin and Weinstein (1979) and Ungerboeck (1972) contain an analysis of the transient behavior and the convergence properties of the LMS algorithm.

The following example serves to reinforce the important points made above regarding the initial convergence of the LMS algorithm.

Example 6.4.3. The LMS algorithm was used to adaptively equalize a communication channel for which the autocorrelation matrix Γ has an eigenvalue spread of

FIGURE 6.4.13
Initial convergence characteristics of the LMS algorithm with different step sizes. (From *Digital Signal Processing* by J. G. Proakis and D. G. Manolakis, 1988, Macmillan Publishing Company. Reprinted with permission of the publisher.)

$\lambda_{\max}/\lambda_{\min} = 11$. The number of taps selected for the equalizer was $2K + 1 = 11$. The input signal plus noise power $(x_0 + N_0)$ was normalized to unity. Hence, the upper bound on Δ given by (6.4.95) is 0.18. Figure 6.4.13 illustrates the initial convergence characteristics of the LMS algorithm for $\Delta = 0.045, 0.09$, and 0.115, by averaging the (estimated) MSE in 200 simulations. We observe that by selecting $\Delta = 0.09$ (one-half of the upper bound) we obtain relatively fast initial convergence. If we divide Δ by a factor of 2 to $\Delta = 0.045$, the convergence rate is reduced but the excess mean square error is also reduced so that the LMS algorithm performs better in steady state (in a time-invariant signal environment). Finally, we note that a choice of $\Delta = 0.115$, which is still far below the upper bound, causes large undesirable fluctuations in the output MSE of the algorithm.

In a digital implementation of the LMS algorithm, the choice of the step-size parameter becomes even more critical. In an attempt to reduce the excess mean square error, it is possible to reduce the step-size parameter to the point where the total mean square error actually increases. This condition occurs when the estimated gradient components of the vector $\varepsilon_k \mathbf{V}_k^*$ after multiplication by the small step-size parameter Δ are smaller than one-half of the least significant bit in the fixed-point representation of the equalizer coefficients. In such a case, adaptation ceases. Consequently, it is important for the step size to be large enough to bring the equalizer coefficients in the vicinity of $\mathbf{C}_{\text{opt}}$. If it is desired to decrease the step size significantly, it is necessary to increase the precision in the equalizer coefficients. Typically, 16 bits of precision may be used for the coefficients, with about 10 to 12 of the most significant bits used for arithmetic operations in the equalization of the data. The remaining least significant bits are required to provide the necessary precision for the adaptation process. Thus the scaled, estimated gradient components $\Delta \varepsilon_k \mathbf{V}_k^*$ usually affect only the least significant bits in any one iteration. In effect, the added precision also allows for the noise to be averaged out, since many incremental changes in

Mean square error

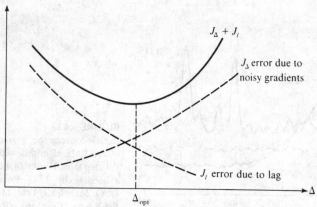

FIGURE 6.4.14
Excess mean-square error J_Δ and lag error J_l as a function of the step. (From *Digital Signal Processing*, by J. G. Proakis and D. G. Manolakis, 1988, Macmillan Publishing Company. Reprinted with permission of the publisher.)

the least significant bits are required before any change occurs in the upper more significant bits used in arithmetic operations for equalizing the data. For an analysis of roundoff errors in a digital implementation of the LMS algorithm, the reader is referred to the papers by Gitlin and Weinstein (1979), Gitlin et al. (1982), and Caraiscos and Liu (1984).

As a final point, we should indicate that the LMS algorithm is appropriate for tracking slowly time-variant signal statistics. In such a case, the minimum MSE and the optimum coefficient vector will be time-variant. In other words, $J_{\min}(n)$ is a function of time and the $(2K + 1)$ dimensional error surface is moving with the time index n. The LMS algorithm attempts to follow the moving minimum $J_{\min}(n)$ in the $(2K + 1)$-dimensional space, but it is always lagging behind due to its use of (estimated) gradient vectors. As a consequence, the LMS algorithm incurs another form of error, called the *lag error*, whose mean square value decreases with an increase in the step size Δ. The total MSE error can now be expressed as

$$J_{\text{total}} = J_{\min}(n) + J_\Delta + J_l$$

where J_l denotes the mean square error due to the lag.

In any given nonstationary adaptive equalization problem, if we plot the errors J_Δ and J_l as a function of Δ, we expect these errors to behave as illustrated in Fig. 6.4.14. We observe that J_Δ increases with an increase in Δ while J_l decreases with an increase in Δ. The total error will exhibit a minimum, which will determine the optimum choice of the step-size parameter.

When the statistical time variations of the signal occur rapidly, the lag error will dominate the performance of the adaptive equalizer. In such a case,

$J_l > J_{\min} + J_\Delta$, even when the largest possible value of Δ is used. When this condition occurs, the LMS algorithm is inappropriate for the application and one must rely on the more complex recursive least-squares algorithms described in Sec. 6.8 to obtain faster convergence and tracking.

Probability of error performance of linear MSE equalizer. Thus far we have discussed the performance of the linear equalizer in terms of the minimum achievable MSE $J_{\min}$ and the output SNR γ, which is related to $J_{\min}$ through the formula in (6.4.44). Unfortunately there is no simple relationship between these quantities and the probability of error. The reason is that the linear MSE equalizer contains some residual intersymbol interference at its output. This situation is unlike that of the infinitely long zero-forcing equalizer for which there is no residual interference, but only gaussian noise. The residual interference at the output of the MSE equalizer is not well characterized as an additional gaussian noise term and, hence, the output SNR does not translate easily into an equivalent error probability.

One approach to computing the error probability is a brute force method that yields an exact result. To illustrate this method, let us consider a PAM signal in which the information symbols are selected from the set of values $2n - M - 1$, $n = 1, 2, \ldots, M$, with equal probability. Now consider the decision on the symbol I_n. The estimate of I_n is

$$\hat{I}_n = q_0 I_n + \sum_{k \neq n} I_k q_{n-k} + \sum_{j=-K}^{K} c_j \eta_{n-j} \tag{6.4.97}$$

where $\{q_n\}$ represent the convolution of the impulse response of the equalizer and equivalent channel, i.e.,

$$q_n = \sum_{k=-K}^{K} c_k f_{n-k} \tag{6.4.98}$$

and the input signal to the equalizer is

$$v_k = \sum_{j=0}^{L} f_j I_{k-j} + \eta_k \tag{6.4.99}$$

The first term in the right-hand side of (6.4.97) is the desired symbol, the middle term is the intersymbol interference, and the last term is the gaussian noise. The variance of the noise is

$$\sigma_n^2 = N_0 \sum_{j=-K}^{K} c_j^2 \tag{6.4.100}$$

For an equalizer with $(2K + 1)$ taps and a channel response that spans $L + 1$ symbols, the number of symbols involved in the intersymbol interference is $2K + L$.

Define

$$D = \sum_{k \neq n} I_k q_{n-k} \tag{6.4.101}$$

For a particular sequence of $2K + L$ information symbols, say the sequence $\mathbf{I}_J$, the intersymbol interference term $D \equiv D_J$ is fixed. The probability of error for a fixed D_J is

$$P_M(D_J) = 2 \frac{(M - 1)}{M} P(N + D_J > q_0)$$

$$= \frac{(M - 1)}{M} \operatorname{erfc} \left[\sqrt{\frac{(q_0 - D_J)^2}{2\sigma_n^2}} \right] \qquad (6.4.102)$$

where N denotes the additive noise term. The average probability of error is obtained by averaging $P_M(D_J)$ over all possible sequences $\mathbf{I}_J$. That is,

$$P_M = \sum_{\mathbf{I}_J} P_M(D_J) P(\mathbf{I}_J)$$

$$= \frac{M - 1}{M} \sum_{\mathbf{I}_J} \operatorname{erfc} \left[\sqrt{\frac{(q_0 - D_J)^2}{2\sigma_n^2}} \right] P(\mathbf{I}_J) \qquad (6.4.103)$$

When all the sequences are equally likely,

$$P(\mathbf{I}_J) = \frac{1}{M^{2K+L}} \qquad (6.4.104)$$

The conditional error probability terms $P_M(D_J)$ are dominated by the sequence that yields the largest value of D_J. This occurs when $I_n = \pm(M - 1)$ and the signs of the information symbols match the signs of the corresponding $\{q_n\}$. Then

$$D_J^* = (M - 1) \sum_{k \neq 0} |q_k|$$

and

$$P_M(D_J^*) = \frac{(M - 1)}{M} \operatorname{erfc} \left[\sqrt{\frac{q_0^2}{2\sigma_n^2} \left(1 - \frac{M - 1}{q_0} \sum_{k \neq 0} |q_k| \right)^2} \right] \qquad (6.4.105)$$

Thus an upper bound on the average probability of error for equally likely symbol sequences is

$$P_M \leq P_M(D_J^*) \qquad (6.4.106)$$

If the exact error probability in (6.4.103) proves to be too cumbersome and too time consuming to evaluate because of the large number of terms in the sum and if the upper bound is too loose, one can resort to one of a number of different approximate methods that have been devised which are known to yield tight bounds on P_M. A discussion of these different approaches would take us too far afield. The interested reader is referred to the papers by Saltzberg (1968), Lugannani (1969), Ho and Yeh (1970), Shimbo and Celebiler (1971), Glave (1972), Yao (1972), and Yao and Tobin (1976).

6.4.3 Fractionally Spaced Equalizers

In the linear equalizer structures that we have described in the previous section, the equalizer taps are spaced at the symbol rate, i.e., at the reciprocal of the signaling rate $1/T$. This tap spacing is optimum if the equalizer is preceded by a filter matched to the channel distorted transmitted pulse. When the channel characteristics are unknown, the receiver filter is usually matched to the transmitted signal pulse and the sampling time is optimized for this suboptimum filter. In general, this approach leads to an equalizer performance that is very sensitive to the choice of sampling time.

The limitations of the symbol rate equalizer are most easily evident in the frequency domain. From (6.2.5), the spectrum of the signal at the input to the equalizer may be expressed as

$$Y_T(f) = \sum_n X\left(f - \frac{n}{T}\right) e^{j2\pi(f - n/T)\tau_0} \qquad (6.4.107)$$

where $Y_T(f)$ is the folded or aliased spectrum, where the folding frequency is $1/2T$. Note that the received signal spectrum is dependent on the choice of the sampling delay τ_0. The signal spectrum at the output of the equalizer is $C_T(f)Y_T(f)$ where

$$C_T(f) = \sum_{k=-K}^{K} c_k e^{-j2\pi fkT} \qquad (6.4.108)$$

It is clear from these relationships that the symbol rate equalizer can only compensate for the frequency response characteristics of the aliased received signal. It cannot compensate for the channel distortion inherent in $X(f)e^{j2\pi f\tau_0}$.

In contrast to the symbol rate equalizer, a fractionally spaced equalizer (FSE) is based on sampling the incoming signal at least as fast as the Nyquist rate. For example, if the transmitted signal consists of pulses having a raised cosine spectrum with a rolloff factor β, its spectrum extends to $F_{\max} = (1 + \beta)/2T$. This signal can be sampled at the receiver at a rate

$$2F_{\max} = \frac{1 + \beta}{T} \qquad (6.4.109)$$

and then passed through an equalizer with tap spacing of $T/(1 + \beta)$. For example, if $\beta = 1$, we would have a $T/2$-spaced equalizer. If $\beta = 0.5$, we would have a $2T/3$-spaced equalizer, and so forth. In general, then, a digitally implemented fractionally spaced equalizer has tap spacing of MT/N where M and N are integers and $N > M$. Usually, a $T/2$-spaced equalizer is used in many applications.

Since the frequency response of the FSE is

$$C_T(f) = \sum_{k=-K}^{K} c_k e^{-j2\pi fkT} \qquad (6.4.110)$$

where $T' = MT/N$, it follows that $C_{T'}(f)$ can equalize the received signal spectrum beyond the Nyquist frequency $f = 1/2T$ to $f = (1 + \beta)/T = N/MT$. The equalized spectrum is

$$C_{T'}Y_{T'}(f) = C_{T'}(f)\sum_n X\left(f - \frac{n}{T'}\right)e^{j2\pi(f-n/T')\tau_0}$$

$$= C_{T'}(f)\sum_n X\left(f - \frac{nN}{MT}\right)e^{j2\pi(f-nN/MT)\tau_0} \qquad (6.4.111)$$

Since $X(f) = 0$ for $|f| > N/MT$, (6.4.111) may be expressed as

$$C_{T'}(f)Y_{T'}(f) = C_{T'}(f)X(f)e^{j2\pi f\tau_0} \qquad |f| \le \frac{1}{2T'} \qquad (6.4.112)$$

Thus, we observe that the FSE compensates for the channel distortion in the received signal before the aliasing effects due to symbol rate sampling. In other words, $C_{T'}(f)$ can compensate for any arbitrary timing phase.

The FSE output is sampled at the symbol rate $1/T$ and has the spectrum

$$\sum_k C_T\left(f - \frac{k}{T}\right)X\left(f - \frac{k}{T}\right)e^{-j2\pi(f-k/T)\tau_0} \qquad (6.4.113)$$

In effect, the optimum FSE is equivalent to the optimum linear receiver consisting of the matched filter followed by a symbol rate equalizer.

Let us now consider the adjustment of the tap coefficients in the FSE. The input to the FSE may be expressed as

$$y\left(\frac{kMT}{N}\right) = \sum_n I_n x\left(\frac{kMT}{N} - nT\right) + \nu\left(\frac{kMT}{N}\right) \qquad (6.4.114)$$

In each symbol interval, the FSE produces an output of the form

$$\hat{I}_k = \sum_{n=-K}^{K} c_j y\left(kT - \frac{nMT}{N}\right) \qquad (6.4.115)$$

where the coefficients of the equalizer are selected to minimize the MSE. This optimization leads to a set of linear equations for the equalizer coefficients which have the optimum solution

$$\mathbf{C}_{\text{opt}} = \mathbf{A}^{-1}\boldsymbol{\alpha} \qquad (6.4.116)$$

where $\mathbf{A}$ is the covariance matrix of the input data and $\boldsymbol{\alpha}$ is the vector of cross correlations. These equations are identical in form to those for the symbol rate equalizer, but there are some subtle differences. One is that $\mathbf{A}$ is Hermitian, but not Toeplitz. In addition, $\mathbf{A}$ exhibits periodicities that are inherent in a cyclostationary process as shown by Qureshi (1985). As a result of the fractional spacing, some of the eigenvalues of $\mathbf{A}$ are nearly zero. Attempts have been made by Long et al. (1988a, b) to exploit this property in the coefficient adjustment by use of singular value decomposition (SVD), with a goal toward reducing the excess mean square error which is proportional to the number of nontrivial eigenvalues of $\mathbf{A}$.

As indicated by (6.4.115) the FSE has a tap spacing of MT/N, but the output of the equalizer is computed at the symbol rate to yield estimates $\{\hat{I}_k\}$ of the transmitted symbols $\{I_k\}$. The possibility of adjusting the equalizer at the sampling rate has been investigated by Ling (1987) with a goal aimed toward increasing the speed of convergence of the equalizer. In this case, it is necessary to perform intersymbol interpolation at the receiver in order to generate a desired signal from which the error signal is formed. Ling found that the baseband complex-valued FSE does not yield faster convergence for sample rate adjustment compared with symbol rate adjustment.

Simulation results demonstrating the effectiveness of the FSE over a symbol rate equalizer have been given in the papers by Qureshi and Forney (1977) and Gitlin and Weinstein (1981). We cite two examples from these papers. First, Fig. 6.4.15 illustrates the performance of the symbol rate equalizer and a $T/2$-FSE for a channel with high-end amplitude distortion, whose characteristics are also shown in this figure. The symbol-spaced equalizer was preceded with a filter matched to the transmitted pulse which had a (square-root) raised cosine spectrum with a 20 percent rolloff ($\beta = 0.2$). The FSE did not have any filter preceding it. The symbol rate was 2400 symbols per second and the modulation was QAM. The received SNR was 30 dB. Both equalizers had 31 taps; hence, the $T/2$-FSE spanned one-half of the time interval of the symbol rate equalizer. Nevertheless, the FSE outperformed the symbol rate equalizer when the latter was optimized at the best sampling time. Furthermore, the FSE did not exhibit any sensitivity to timing phase as illustrated in Fig. 6.4.15.

Similar results were obtained by Gitlin and Weinstein. For a channel with poor envelope delay characteristics, the SNR performance of the symbol rate equalizer and a $T/2$-FSE are illustrated in Fig. 6.4.16. In this case, both equalizers had the same time span. The T-spaced equalizer had 24 taps while the FSE had 48 taps. The symbol rate was 2400 symbols per second and the data rate was 9600 bits per second with 16-QAM modulation. The signal pulse had a raised cosine spectrum with $\beta = 0.12$. Note again that the FSE outperformed the T-spaced equalizer by several decibels even when the latter was adjusted for optimum sampling. The results in these two papers clearly demonstrate the superior performance achieved with a fractionally spaced equalizer.

6.4.4 Self-Recovering (Blind) Equalization

In the conventional zero-forcing or minimum MSE equalizers described above, we assumed that a known training sequence is transmitted to the receiver for the purpose of initially adjusting the equalizer coefficients. However, there are some applications, such as multipoint communication networks, where it is desirable for the receiver to synchronize to the received signal and to adjust the equalizer without having a known training sequence available. Equalization techniques based on initial adjustment of the coefficients without the benefit of a training sequence are said to be *self-recovering* or *blind*.

(a) Channel with high-end amplitude distortion (HA)

(b) Equalizer performance

FIGURE 6.4.15
T and $T/2$ equalizer performance as a function of timing phase for 2400 symbols per second. [From paper by Qureshi and Forney (1977). © 1977 IEEE. Reprinted with permission.]

FIGURE 6.4.16
Performance of T and $T/2$ equalizers as a function of timing phase for 2400 symbols per second 16-QAM on a channel with poor envelope delay. [From paper by Gitlin and Weinstein (1981). Reprinted with permission from Bell System Technical Journal. © 1981 AT & T.]

Several researchers have investigated blind equalization algorithms. Sato (1975) considered multilevel PAM signaling and a blind equalization algorithm that ignored the amplitude information in the signal. The algorithm used only the sign of the signal in generating the error signal for adjusting the equalizer coefficients. In effect, the amplitude information in the received signal was treated as a random noise added to a binary signal.

Blind equalization for two-dimensional signal constellations has been considered by Godard (1980), Benveniste and Goursat (1984), Picchi and Prati (1987), and Foschini (1985).

Godard considered the problem of combined equalization and carrier phase recovery and tracking. The carrier phase tracking is performed at baseband, following the equalizer as shown in Fig. 6.4.17. Based on this structure we may express the equalizer output as

$$\hat{I}_k = \sum_{n=-K}^{K} c_n v_{k-n} \qquad (6.4.117)$$

and the input to the decision device as $\hat{I}_k \exp(-j\hat{\phi}_k)$, where $\hat{\phi}_k$ is the carrier phase estimate in the kth symbol interval.

If the desired symbol were known, we could form the error signal

$$\varepsilon_k = I_k - \hat{I}_k e^{-j\hat{\phi}_k} \qquad (6.4.118)$$

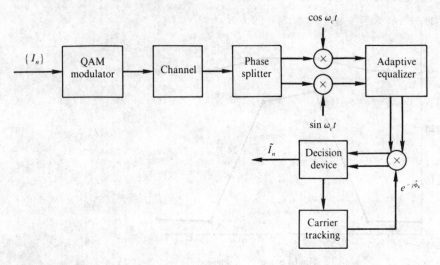

FIGURE 6.4.17
Godard scheme for combined adaptive (blind) equalization and carrier phase tracking.

and minimize the MSE with respect to $\hat{\phi}_k$ and $\{c_n\}$, i.e.,

$$\min_{\hat{\phi}_k, \mathbf{C}} E\left[|I_k - \hat{I}_k e^{-j\hat{\phi}_k}|^2\right] \qquad (6.4.119)$$

This criterion leads us to use the LMS algorithm for recursively estimating $\mathbf{C}$ and ϕ_k. The LMS algorithm based on knowledge of the transmitted sequence is

$$\hat{\mathbf{C}}_{k+1} = \hat{\mathbf{C}}_k + \Delta_c \left(I_k - \hat{I}_k e^{-j\hat{\phi}_k}\right) V_k^* e^{j\hat{\phi}_k} \qquad (6.4.120)$$

$$\hat{\phi}_{k+1} = \hat{\phi}_k + \Delta_\phi \operatorname{Im}\left[I_k \hat{I}_k^* e^{j\hat{\phi}_k}\right] \qquad (6.4.121)$$

where Δ_c and Δ_ϕ are the step-size parameters for the two recursive equations. Note that these recursive equations are coupled together. Unfortunately, these equations will not converge, in general, when the desired symbol sequence $\{I_k\}$ is unknown. In other words, a bootstrapping procedure by which the output decision of the detector is assumed to be correct and used in place of I_k in (6.4.120) and (6.4.121) will not lead to convergence of the algorithm to $\mathbf{C}_{\text{opt}}$ and ϕ_{opt}.

The approach proposed by Godard is to use a criterion that depends on the amount of intersymbol interference at the output of the equalizer but one that is independent of the signal constellation and the carrier phase. For example, a cost function that is independent of carrier phase and has the property that its minimum leads to a small MSE is

$$G^{(p)} = E\left[|\hat{I}_k|^p - |I_k|^p\right]^2 \qquad (6.4.122)$$

where p is a positive real integer. Minimization of $G^{(p)}$ with respect to the equalizer coefficients results in the equalization of the signal amplitude only.

Based on this observation, Godard selected a more general cost function, called the *dispersion of order p*, defined as

$$D^{(p)} = E\left[|\hat{I}_k|^p - R_p\right]^2 \tag{6.4.123}$$

where R_p is a positive real constant. As in the case of $G^{(p)}$, we observe that $D^{(p)}$ is independent of the carrier phase.

The minimization of $D^{(p)}$ with respect to the equalizer coefficients can be performed recursively according to the steepest descent algorithm

$$\mathbf{C}_{k+1} = \mathbf{C}_k - \delta_p \left[\frac{dD^{(p)}}{d\mathbf{C}_k}\right] \tag{6.4.124}$$

where δ_p is the step-size parameter. By differentiating $D^{(p)}$ and dropping the expectation operation we obtain the following LMS-type algorithm for adjusting the equalizer coefficients:

$$\hat{\mathbf{C}}_{k+1} = \hat{\mathbf{C}}_k + \Delta_p \mathbf{V}_p^* \hat{I}_k |\hat{I}_k|^{p-2} \left(R_p - |\hat{I}_k|^p\right) \tag{6.4.125}$$

where Δ_p is the step-size parameter and the optimum choice of R_p is

$$R_p = \frac{E\left[|I_k|^{2p}\right]}{E\left[|I_k|^p\right]} \tag{6.4.126}$$

As expected the recursion in (6.4.125) for $\hat{\mathbf{C}}_k$ does not require knowledge of the carrier phase. Carrier phase tracking may be carried out in a decision-directed mode according to (6.4.121).

Of particular importance is the case $p = 2$, which leads to the relatively simple algorithm

$$\hat{\mathbf{C}}_{k+1} = \hat{\mathbf{C}}_k + \Delta V_k^* \hat{I}_k \left(R_2 - |\hat{I}_k|^2\right)$$
$$\hat{\phi}_{k+1} = \hat{\phi}_k + \Delta_\phi \,\mathrm{Im}\left[\tilde{I}_k \hat{I}_k^* e^{j\hat{\phi}_k}\right] \tag{6.4.127}$$

where $\tilde{I}_k$ is the output decision based on $\hat{I}_k$, and

$$R_2 = \frac{E\left[|I_k|^4\right]}{E\left[|I_k|^2\right]} \tag{6.4.128}$$

Convergence of the algorithm given in (6.4.127) was demonstrated in the paper by Godard (1980). Initially, the equalizer coefficients were set to zero except for the center (reference) tap which was set according to the condition

$$|c_0|^2 > \frac{E|I_k|^4}{2|x_0|^2 \left[E|I_k|^2\right]^2} \tag{6.4.129}$$

which is sufficient, but not necessary, for convergence of the algorithm. Simulation results performed by Godard on simulated telephone channels with typical frequency response characteristics and transmission rates of 7200 to 12,000 bits per second indicate that the algorithm in (6.4.127) performs well and leads to convergence in 5000 to 20,000 iterations depending on the signal constellation. Initially, the eye pattern was closed prior to equalization. The number of iterations required for convergence is about an order of magnitude greater than the number of iterations required to equalize the channels with a known training sequence. No apparent difficulties were encountered in using the decision-directed phase estimation algorithm in (6.4.127) from the beginning of the equalizer adjustment process.

Benveniste and Goursat (1984) derived two algorithms, for quadrature carrier QAM systems, which differ from Godard's algorithm in the manner in which the error signal is generated. In brief, their first algorithm may be expressed as

$$\hat{\mathbf{C}}_{k+1} = \hat{\mathbf{C}}_k + \Delta_c \varepsilon_{k1} \mathbf{V}_k^* e^{j\hat{\phi}_k}$$
$$\hat{\phi}_{k+1} = \hat{\phi}_k + \Delta_\phi \operatorname{Im}\left(\hat{I}_k I_{ks}^*\right)$$
$$\hat{I}_k = \mathbf{V}_k' \mathbf{C}_k e^{-j\hat{\phi}_k} \qquad (6.4.130)$$
$$I_{ks} = \alpha\left[\operatorname{sgn}\left(\operatorname{Re}\hat{I}_k\right) + j\operatorname{sgn}\left(\operatorname{Im}\hat{I}_k\right)\right]$$
$$\varepsilon_{k1} = I_{ks} - \hat{I}_k$$

where α is a positive scale factor. The second algorithm may be expressed as

$$\hat{\mathbf{C}}_{k+1} = \hat{\mathbf{C}}_k + \Delta_c \varepsilon_{k2} \mathbf{V}_k^* e^{j\hat{\phi}_k}$$
$$\hat{\phi}_{k+1} = \hat{\phi}_k + \Delta_\phi \operatorname{Im}\left(\hat{I}_k \varepsilon_{k2}^*\right)$$
$$\varepsilon_k = \tilde{I}_k - \hat{I}_k \qquad (6.4.131)$$
$$\varepsilon_{k2} = \beta_1 \varepsilon_k + \beta_2 |\varepsilon_k| \varepsilon_{k1}$$

where ε_k is the conventional error and ε_{k1} is given in (6.4.130). The weighting constants β_1 and β_2 are selected by the designer and allow us to provide for smooth switching to the decision-directed mode by letting $\beta_2 \to 0$ and $\beta_1 \to 1$. The functional block diagram for the second algorithm given by (6.4.131) is shown in Fig. 6.4.18. The block diagram for the first algorithm is the same except that the conventional error ε_k is not used. These algorithms were shown to provide faster convergence than the Godard algorithm. For example, convergence was obtained on simulated typical telephone channels in about 2000 iterations.

Picchi and Prati (1987) proposed a decision-directed algorithm that represents a departure from the ones proposed by Godard (1980) and Benveniste and Goursat (1984). The Picchi and Prati algorithm employs the same decision-directed error signal as in the conventional tracking LMS algorithm given in (6.4.120) and (6.4.121) but it is modified by use of a "stop-and-go" (binary valued) flag that inhibits adaptation of the equalizer coefficients and carrier phase when the reliability of the decision is low (output error signal is very small), and allows adaptation when the reliability of the decisions is high (output error signal is large). This modified stop-and-go decision-directed algorithm results in a convergence speed that is comparable to that obtained with the Benveniste–Goursat algorithm.

Modifications to the Godard algorithm have been proposed by Foschini (1985), who applied blind equalization to terrestrial radio systems that employ QAM signals. Foschini demonstrated that his modified algorithms can be used to

FIGURE 6.4.18
Block diagram of Benveniste–Goursat blind equalization algorithm of (6.4.131).

recover equalization without a training sequence following a signal fade due to multipath propagation.

In conclusion, the results presented above indicate that self-recovering equalization and carrier phase are possible in digital communication systems when a training signal is not provided for initial adjustment of the adaptive equalizer.

6.5   DECISION-FEEDBACK EQUALIZATION

The decision-feedback equalizer, depicted in Fig. 6.5.1, consists of two filters, a feedforward filter and a feedback filter. As shown, both have taps spaced at the symbol interval T. The input to the feedforward section is the received signal sequence $\{v_k\}$. In this respect, the feedforward filter is identical to the linear transversal equalizer described in Sec. 6.4. The feedback filter has as its input the sequence of decisions on previously detected symbols. Functionally the feedback filter is used to remove that part of the intersymbol interference from the present estimate caused by previously detected symbols as described by Austin (1967), Monsen (1971), George et al. (1971), and Proakis (1975).

From the description given above, it follows that the equalizer output can be expressed as

$$\hat{I}_k = \sum_{j=-K_1}^{0} c_j v_{k-j} + \sum_{j=1}^{K_2} c_j \tilde{I}_{k-j} \qquad (6.5.1)$$

where $\hat{I}_k$ is an estimate of the kth information symbol, $\{c_j\}$ are the tap coefficients of the filter, and $\{\tilde{I}_{k-1}, \ldots, \tilde{I}_{k-K_2}\}$ are previously detected symbols. The equalizer is assumed to have $(K_1 + 1)$ taps in its feedforward section and K_2

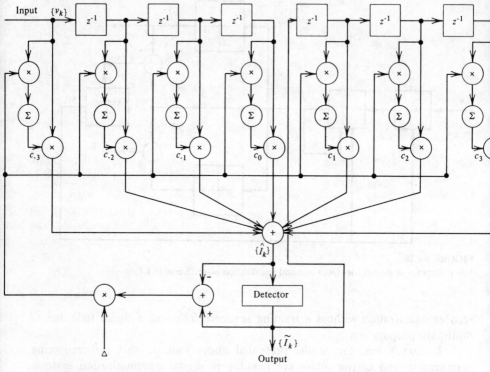

FIGURE 6.5.1
Decision-feedback equalizer.

taps in its feedback section. It should be observed that this nonlinear equalizer differs from the linear equalizer with feedback discussed in Sec. 6.4 only in that the feedback filter contains previously detected symbols $\{\tilde{I}_k\}$ whereas in the linear equalizer the feedback section contains the estimates $\{\hat{I}_k\}$.

Both the peak distortion criterion and the MSE criterion result in a mathematically tractable optimization of the equalizer coefficients as can be concluded from the papers by George et al. (1971), Price (1972), Salz (1973), and Proakis (1975). Since the MSE criterion is more prevalent in practice, we focus our attention on it. Based on the assumption that previously detected symbols in the feedback filter are correct, the minimization of MSE

$$J(K_1, K_2) = E|I_k - \hat{I}_k|^2 \qquad (6.5.2)$$

leads to the following set of linear equations for the coefficients of the feedforward filter:

$$\sum_{j=-K_1}^{0} \psi_{lj} c_j = f_{-l}^* \qquad l = -K_1, \ldots, -1, 0 \qquad (6.5.3)$$

where

$$\psi_{lj} = \sum_{m=0}^{-l} f_m^* f_{m+l-j} + N_0 \delta_{lj} \qquad l, j = -K_1, \ldots, -1, 0 \qquad (6.5.4)$$

The coefficients of the feedback filter of the equalizer are given in terms of the coefficients of the feedforward section by the following expression:

$$c_k = - \sum_{j=-K_1}^{0} c_j f_{k-j} \qquad k = 1, 2, \ldots, K_2 \qquad (6.5.5)$$

The values of the feedback coefficients result in complete elimination of intersymbol interference from previously detected symbols, provided that previous decisions are correct and that $K_2 \geq L$ (see Prob. 6.8).

In an adaptive mode, the filter coefficients $\{c_j\}$ are adjusted recursively in order for the filter to follow time variations in the channel response. For this purpose, an error signal is formed by taking the difference between the detected symbol $\tilde{I}_k$ and the estimate $\hat{I}_k$, i.e., $\varepsilon_k = \tilde{I}_k - \hat{I}_k$. This error signal is scaled by a (step-size) factor Δ and the resulting product is used to adjust the coefficients according to the steepest-descent algorithm

$$\hat{\mathbf{C}}_{k+1} = \hat{\mathbf{C}}_k + \Delta \varepsilon_k \tilde{\mathbf{V}}_k^* \qquad (6.5.6)$$

where $\hat{\mathbf{C}}_k$ is the vector of tap gain coefficients in the kth signaling interval and the vector $\tilde{\mathbf{V}}_k$ has the components $(v_{k+K_1}, \ldots, v_k, \tilde{I}_{k-1}, \ldots, \tilde{I}_{k-K_2})$.

We now turn our attention to the performance achieved with decision-feedback equalization. The exact evaluation of the performance is complicated to some extent by occasional incorrect decisions made by the detector, which then propagate down the feedback section. In the absence of decision errors, the minimum MSE is given as

$$J_{\min}(K_1) = 1 - \sum_{j=-K_1}^{0} c_j f_{-j} \qquad (6.5.7)$$

By going to the limit ($K_1 \to \infty$) of an infinite number of taps in the feedforward filter, we obtain the smallest achievable MSE, denoted as $J_{\min}$. With some effort $J_{\min}$ can be expressed in terms of the spectral characteristics of the channel and additive noise, as shown by Salz (1973). This more desirable form for $J_{\min}$ is

$$J_{\min} = \exp \left\{ \frac{T}{2\pi} \int_{-\pi/T}^{\pi/T} \ln \left[\frac{N_0}{X(e^{j\omega T}) + N_0} \right] d\omega \right\} \qquad (6.5.8)$$

The corresponding output SNR is

$$\gamma_\infty = \frac{1 - J_{\min}}{J_{\min}}$$

$$= -1 + \exp \left\{ \frac{T}{2\pi} \int_{-\pi/T}^{\pi/T} \ln \left[\frac{N_0 + X(e^{j\omega T})}{N_0} \right] d\omega \right\} \qquad (6.5.9)$$

We observe again that in the absence of intersymbol interference, $X(e^{j\omega T}) = 1$ and, hence, $J_{min} = N_0/(1 + N_0)$. The corresponding output SNR is $\gamma_\infty = 1/N_0$.

Example 6.5.1. It is interesting to compare the value of J_{min} for the decision-feedback equalizer with the value of J_{min} obtained with the linear MSE equalizer. For example, let us consider the discrete-time equivalent channel consisting of two paths f_0 and f_1. The minimum MSE for this channel is

$$
\begin{aligned}
J_{min} &= \exp\left\{ \frac{T}{2\pi} \int_{-\pi/T}^{\pi/T} \ln\left[\frac{N_0}{1 + N_0 + 2|f_0||f_1|\cos(\omega T + \theta)} \right] d\omega \right\} \\
&= N_0 \exp\left\{ -\frac{1}{2\pi} \int_{-\pi}^{\pi} \ln[1 + N_0 + 2|f_0||f_1|\cos\omega]\, d\omega \right\} \\
&= \frac{2N_0}{1 + N_0 + \sqrt{(1 + N_0)^2 - 4|f_0 f_1|^2}}
\end{aligned}
\tag{6.5.10}
$$

We note that J_{min} is minimized when $|f_0| = |f_1| = 1/\sqrt{2}$. Then

$$
\begin{aligned}
J_{min} &= \frac{2N_0}{1 + N_0 + \sqrt{(1 + N_0)^2 - 1}} \\
&\approx 2N_0 \qquad N_0 \ll 1
\end{aligned}
\tag{6.5.11}
$$

The corresponding output SNR is

$$
\gamma_\infty \approx \frac{1}{2N_0} \qquad N_0 \ll 1
\tag{6.5.12}
$$

Therefore there is a 3-dB degradation in output SNR due to the presence of intersymbol interference. In comparison, the performance loss for the linear equalizer is very severe. Its output SNR as given by (6.4.70) is $\gamma_\infty \approx (2/N_0)^{1/2}$ for $N_0 \ll 1$.

Example 6.5.2. Consider the exponentially decaying channel characteristic of the form

$$
f_k = (1 - a^2)^{1/2} a^k \qquad k = 0, 1, 2, \ldots
\tag{6.5.13}
$$

where $a < 1$. The output SNR of the decision-feedback equalizer is

$$
\begin{aligned}
\gamma_\infty &= -1 + \exp\left\{ \frac{1}{2\pi} \int_{-\pi}^{\pi} \ln\left[\frac{1 + a^2 + (1 - a^2)/N_0 - 2a\cos\omega}{1 + a^2 - 2a\cos\omega} \right] d\omega \right\} \\
&= -1 + \frac{1}{2N_0}\Big\{ 1 - a^2 + N_0(1 + a^2) \\
&\qquad\qquad + \sqrt{[1 - a^2 + N_0(1 + a^2)]^2 - 4a^2 N_0^2} \Big\} \\
&\approx \frac{(1 - a^2)[1 + N_0(1 + a^2)/(1 - a^2)] - N_0}{N_0} \\
&\approx \frac{1 - a^2}{N_0} \qquad N_0 \ll 1
\end{aligned}
\tag{6.5.14}
$$

Thus the loss in SNR is $10\log_{10}(1 - a^2)$ dB. In comparison, the linear equalizer has a loss of $10\log_{10}[(1 - a^2)/(1 + a^2)]$ dB.

FIGURE 6.5.2
Performance of decision-feedback equalizer with and without error propagation.

These results illustrate the superiority of the decision-feedback equalizer over the linear equalizer when the effect of decision errors on performance is neglected. It is apparent that a considerable gain in performance can be achieved relative to the linear equalizer by the inclusion of the decision-feedback section which eliminates the intersymbol interference from previously detected symbols.

One method of assessing the effect of decision errors on the error rate performance of the decision-feedback equalizer is Monte Carlo simulation on a digital computer. For purposes of illustration we offer the following results for binary PAM signaling through the equivalent discrete-time channel models shown in Fig. 6.4.7b and c.

The results of the simulation are displayed in Fig. 6.5.2. First of all, a comparison of these results with those presented in Fig. 6.4.6 leads us to conclude that the decision-feedback equalizer yields a significant improvement in performance relative to the linear equalizer having the same number of taps. Second, these results indicate that there is still a significant degradation in performance of the decision-feedback equalizer due to the residual intersymbol interference, especially on channels with severe distortion such as the one shown in Fig. 6.4.7c. Finally, the performance loss due to incorrect decisions being fed back is 2 dB, approximately, for the channel responses under consideration. Additional results on the probability of error for a decision-feedback equalizer with error propagation may be found in the paper by Duttweiler et al. (1974).

The structure of the DFE that is analyzed above employs a T-spaced filter for the feedforward section. The optimality of such a structure is based on the assumption that the analog filter preceding the DFE is matched to the channel-corrupted pulse response and its output is sampled at the optimum time instant. In practice, the channel response is not known *a priori* so it is not possible to design an ideal matched filter. In view of this difficulty, it is customary in practical applications to use a fractionally spaced feedforward filter. Of course, the feedback filter tap spacing remains at T. The use of the FSE for the feedforward filter eliminates the system sensitivity to a timing error.

6.5.1 Predictive Decision-Feedback Equalizer

Belfiore and Park (1979) proposed another DFE structure that is equivalent to the one shown in Fig. 6.5.1 under the condition that the feedforward filter has an infinite number of taps. This structure consists of a FSE as a feedforward filter and a linear predictor as a feedback filter, as shown in the configuration given in Fig. 6.5.3. Let us briefly consider the performance characteristics of this equalizer.

First of all, the noise at the output of the infinite length feedforward filter has the power spectrum

$$\frac{N_0 X(e^{j\omega T})}{|N_0 + X(e^{j\omega T})|^2} \qquad |\omega| \le \frac{\pi}{T} \tag{6.5.15}$$

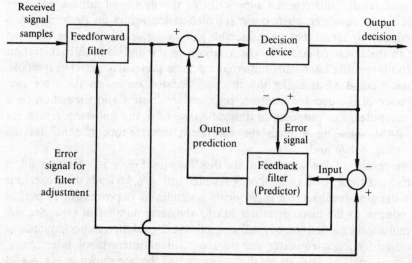

FIGURE 6.5.3
Block diagram of predictive DFE.

The residual intersymbol interference has the power density spectrum

$$\left| 1 - \frac{X(e^{j\omega T})}{N_0 + X(e^{j\omega T})} \right|^2 = \frac{N_0^2}{|N_0 + X(e^{j\omega T})|^2} \qquad |\omega| \le \frac{\pi}{T} \qquad (6.5.16)$$

The sum of these two spectra represents the power density spectrum of the total noise and intersymbol interference at the output of the feedforward filter. Thus, on adding (6.5.15) and (6.5.16) we obtain

$$E(\omega) = \frac{N_0}{N_0 + X(e^{j\omega T})} \qquad |\omega| \le \pi/T \qquad (6.5.17)$$

As we have observed previously, if $X(e^{j\omega T}) = 1$, the channel is ideal and, hence, it is not possible to reduce the MSE any further. On the other hand, if there is channel distortion, the power in the error sequence at the output of the feedforward filter can be reduced by means of linear prediction based on past values of the error sequence.

If $B(\omega)$ represents the frequency response of the infinite length feedback predictor, i.e.,

$$B(\omega) = \sum_{n=1}^{\infty} b_n e^{-j\omega nT} \qquad (6.5.18)$$

then the error at the output of the predictor is

$$E(\omega) - E(\omega)B(\omega) = E(\omega)[1 - B(\omega)] \qquad (6.5.19)$$

The minimization of the mean square value of this error, i.e.,

$$J = \frac{1}{2\pi} \int_{-\pi/T}^{\pi/T} |1 - B(\omega)|^2 |E(\omega)|^2 \, d\omega \qquad (6.5.20)$$

over the predictor coefficients $\{b_n\}$ yields the optimum predictor in the form

$$B(\omega) = 1 - \frac{G(\omega)}{g_0} \qquad (6.5.21)$$

where $G(\omega)$ is the solution to the spectral factorization

$$G(\omega)G^*(\omega) = \frac{1}{|E(\omega)|^2} \qquad (6.5.22)$$

and

$$G(\omega) = \sum_{n=0}^{\infty} g_n e^{-j\omega nT} \qquad (6.5.23)$$

The output of the infinite length linear predictor is a white noise sequence with power spectral density $1/g_0^2$ and the corresponding minimum MSE is given by (6.5.8). Therefore, the MSE performance of the infinite length predictive DFE is identical to the conventional DFE.

Although these two DFE structures result in equivalent performance if their lengths are infinite, the predictive DFE is suboptimum if the lengths of the two filters are finite. The reason for the optimality of the conventional DFE is relatively simple. The optimization of its tap coefficients in the feedforward and feedback filters is done jointly. Hence, it yields the minimum MSE. On the other hand, the optimizations of the feedforward filter and the feedback predictor in the predictive DFE are done separately. Hence, its MSE is at least as large as that of the conventional DFE. In spite of this suboptimality of the predictive DFE, it is suitable as an equalizer for trellis-coded signals, where the conventional DFE is not as suitable, as described in the following section.

6.5.2 Adaptive Equalization of Trellis-Coded Signals with a DFE

Bandwidth efficient trellis-coded modulation that was described in Sec. 5.4 is frequently used in digital communications over telephone channels to reduce the required SNR per bit for achieving a specified error rate. Channel distortion of the trellis-coded signal forces us to use adaptive equalization in order to reduce the intersymbol interference. The output of the equalizer is then fed to the Viterbi decoder which performs soft-decision decoding of the trellis-coded signal.

The question that arises in such a receiver is how do we adapt the equalizer in a data transmission mode? One possibility is to have the equalizer make its own decisions at its output solely for the purpose of generating an error signal for adjusting its tap coefficients as shown in the block diagram in Fig. 6.5.4. The problem with this approach is that such decisions are generally unreliable since the pre-decoding coded symbol SNR is relatively low. A high error rate would cause a significant degradation in the operation of the equalizer, which would ultimately affect the reliability of the decisions at the output of the decoder. The

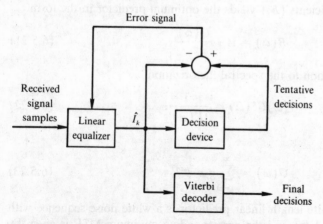

FIGURE 6.5.4
Adjustment of equalizer based on tentative decisions.

FIGURE 6.5.5
Adjustment of equalizer based on decisions from the Viterbi decoder.

more desirable alternative is to use the post-decoding decisions from the Viterbi decoder, which are much more reliable, to continuously adapt the equalizer. This approach is certainly preferable and viable when a linear equalizer is used prior to the Viterbi decoder. The decoding delay inherent in the Viterbi decoder can be overcome by introducing an identical delay D in the tap weight adjustment of the equalizer coefficients as shown in Fig. 6.5.5. The major price that must be paid for the added delay is that the step-size parameter in the LMS algorithm must be reduced, as described by Long et al. (1987, 1989), in order to achieve stability in the algorithm.

In channels with one or more in-band spectral nulls, the linear equalizer is no longer adequate for compensating the channel intersymbol interference. Instead, we would like to use a DFE. But the DFE requires reliable decisions in its feedback filter in order to cancel out the intersymbol interference from previously detected symbols. Tentative decisions prior to decoding would be highly unreliable and, hence, inappropriate. Unfortunately, the conventional DFE cannot be cascaded with the Viterbi algorithm in which post-decoding decisions from the decoder are fed back to the DFE.

The only alternative is to use the predictive DFE. In order to accommodate for the decoding delay as it affects the linear predictor, we introduce a periodic interleaver/deinterleaver pair that has the same delay as the Viterbi decoder and, thus, makes it possible to generate the appropriate error signal to the predictor as illustrated in the block diagram of Fig. 6.5.6. The novel way in which a predictive DFE can be combined with Viterbi decoding to equalize trellis-coded signals is described and analyzed by Eyuboglu (1988). This same idea has been carried over to the equalization of fading multipath channels by Zhou et al. (1987, 1989), but the structure of the DFE was modified to use recursive least-squares lattice-type filters which provide faster tracking for the time variations encountered in the channel.

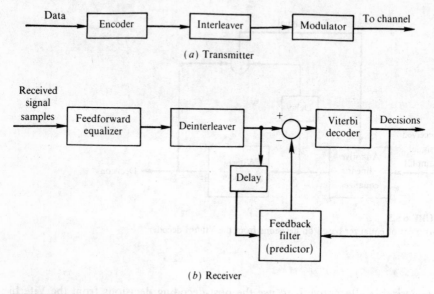

FIGURE 6.5.6
Use of predictive DFE with interleaving and trellis-coded modulation.

6.6 A PROBABILISTIC SYMBOL-BY-SYMBOL EQUALIZATION ALGORITHM

In the linear and decision-feedback equalization techniques discussed in Secs. 6.4 and 6.5, the estimate of the information symbol in any particular signaling interval is some linear combination of the received signal samples from the output of the matched filter. In the decision-feedback equalizer, there is the additional linear combination of previously detected symbols. The tap weights in these linear combinations are determined by minimizing either the MSE or the peak distortion. In any case, the adjustment of the tap weights does not require knowledge of the statistics of the signal and the additive noise. Consequently we may call these techniques "nonparametric."

In contrast, we now consider a symbol-by-symbol equalization algorithm, developed by Abend and Fritchman (1970), that is parametric in the sense that the probability density of the noise must be known. The algorithm is based on the computation of *a posteriori* probabilities and is optimum in the sense of minimizing the probability of a symbol error.

To illustrate the algorithm, suppose that it is desired to detect the information symbol I_k and let $v_1, v_2, \ldots, v_{k+D}$ be the observed received sequence, where the delay D is chosen to exceed the channel dispersion, i.e., $D \geq L$. On the basis of the received sequence one can compute the posterior probabilities

$$P(I_k = m | v_{k+D}, \ldots, v_1)$$

for the M possible symbol values and choose that symbol having the largest probability. Since

$$P(I_k = m|v_{k+D}, \ldots, v_1) = \frac{p(v_{k+D}, \ldots, v_1|I_k = m)P(I_k = m)}{p(v_{k+D}, v_{k+D-1}, \ldots, v_1)} \quad (6.6.1)$$

and since the denominator is common for all M probabilities, the maximum *a posteriori* probability (MAP) criterion is equivalent to choosing the value of I_k that maximizes the numerator $p(v_{k+D}, \ldots, v_1|I_k)P(I_k)$ in (6.6.1). Thus the criterion for deciding on the symbol I_k is

$$\bullet \quad \tilde{I}_k = \max_{I_k} p(v_{k+D}, \ldots, v_1|I_k)P(I_k) \quad (6.6.2)$$

When the symbols are equally probable, the probability $P(I_k)$ may be dropped from the computation.

The algorithm for computing the probabilities given in (6.6.2), recursively begins with the first symbol I_1. We have

$$\tilde{I}_1 = \max_{I_1} p(v_{1+D}, \ldots, v_1|I_1)P(I_1)$$

$$= \max_{I_1} \sum_{I_{1+D}} \cdots \sum_{I_2} p(v_{1+D}, \ldots, v_1|I_{1+D}, \ldots, I_1)P(I_{1+D}, \ldots, I_1)$$

$$= \max_{I_1} \sum_{I_{1+D}} \cdots \sum_{I_2} p_1(I_{1+D}, \ldots, I_2, I_1) \quad (6.6.3)$$

where $\tilde{I}_1$ denotes the decision on I_1 and, for mathematical convenience, we have defined

$$p_1(I_{1+D}, \ldots, I_2, I_1) \equiv p(v_{1+D}, \ldots, v_1|I_{1+D}, \ldots, I_1)P(I_{1+D}, \ldots, I_1) \quad (6.6.4)$$

The joint probability $P(I_{1+D}, \ldots, I_1)$ may be omitted if the symbols are equally probable and statistically independent. As a consequence of the statistical independence of the additive noise in the received sequence $v_1, v_2, \ldots, v_{1+D}$, the joint conditional probability density in (6.6.4) factors into a product (with $I_k = 0$ for $k \leq 0$). Thus

$$p(v_{1+D}, \ldots, v_1|I_{1+D}, \ldots, I_1)$$

$$= p(v_{1+D}|I_{1+D}, \ldots, I_{1+D-L})$$

$$\times p(v_D|I_D, \ldots, I_{D-L}) \cdots p(v_2|I_2, I_1)p(v_1|I_1) \quad (6.6.5)$$

For example, when the received samples $\{v_k\}$, the channel response $\{f_k\}$, and the symbols $\{I_k\}$ are all real-valued, the conditional probability density for gaussian additive noise is

$$p(v_{k+D}|I_{k+D}, \ldots, I_{k+D-L})$$

$$= \frac{1}{(2\pi N_0)^{1/2}} \exp\left[-\frac{1}{2N_0}\left(v_{k+D} - \sum_{j=0}^{L} f_j I_{k+D-j}\right)^2\right] \quad (6.6.6)$$

If $\{v_k\}$ is complex-valued then (6.6.6) must be modified to the joint conditional probability of the real and imaginary components of $\{v_k\}$.

For the detection of the symbol I_2, we have

$$\tilde{I}_2 = \max_{I_2} p(v_{2+D}, \ldots, v_1 | I_2) P(I_2)$$

$$= \max_{I_2} \sum_{I_{2+D}} \cdots \sum_{I_3} p(v_{2+D}, \ldots, v_1 | I_{2+D}, \ldots, I_2) P(I_{D+2}, \ldots, I_2) \quad (6.6.7)$$

The joint conditional probability in the multiple summation can be expressed as

$$p(v_{2+D}, \ldots, v_1 | I_{2+D}, \ldots, I_2)$$

$$= p(v_{2+D} | I_{2+D}, \ldots, I_{2+D-L}) p(v_{1+D}, \ldots, v_1 | I_{1+D}, \ldots, I_2) \quad (6.6.8)$$

Furthermore, the joint probability $p(v_{1+D}, \ldots, v_1 | I_{1+D}, \ldots, I_2) P(_{1+D}, \ldots, I_2)$ can be obtained as follows from the probabilities computed previously in the detection of I_1. That is,

$$p(v_{1+D}, \ldots, v_1 | I_{1+D}, \ldots, I_2) P(I_{1+D}, \ldots, I_2)$$

$$= \sum_{I_1} p(v_{1+D}, \ldots, v_1 | I_{1+D}, \ldots, I_1) P(I_{1+D}, \ldots, I_1)$$

$$= \sum_{I_1} p_1(I_{1+D}, \ldots, I_2, I_1) \quad (6.6.9)$$

Thus, by combining the expressions in (6.6.9) and (6.6.8) and then substituting into (6.6.7), we obtain

$$\tilde{I}_2 = \max_{I_2} \sum_{I_{2+D}} \cdots \sum_{I_3} p_2(I_{2+D}, \ldots, I_3, I_2) \quad (6.6.10)$$

where, by definition,

$$p_2(I_{2+D}, \ldots, I_3, I_2)$$

$$= p(v_{2+D} | I_{2+D}, \ldots, I_{2+D-L}) P(I_{2+D}) \sum_{I_1} p_1(I_{1+D}, \ldots, I_2, I_1) \quad (6.6.11)$$

In general, the recursive algorithm for detecting the symbol I_k is as follows. Upon reception of $v_{k+D}, \ldots, v_1$ we compute

$$\tilde{I}_k = \max_{I_k} p(v_{k+D}, \ldots, v_1 | I_k) P(I_k)$$

$$= \max_{I_k} \sum_{I_{k+D}} \cdots \sum_{I_{k+1}} P_k(I_{k+D}, \ldots, I_{k+1}, I_k) \quad (6.6.12)$$

where, by definition,

$$p_k(I_{k+D}, \ldots, I_{k+1}, I_k) = p(v_{k+D} | I_{k+D}, \ldots, I_{k+D-L}) P(I_{k+D})$$

$$\times \sum_{I_{k-1}} P_{k-1}(I_{k-1+D}, \ldots, I_{k-1}) \quad (6.6.13)$$

Thus the recursive nature of the algorithm is established by the relations in (6.6.12) and (6.6.13).

We observe, first of all, that the algorithm in (6.6.12) through its dependence on $\{f_k\}$ requires knowledge of the discrete-time channel response. For the case in which the $\{f_k\}$ are unknown or time-varying, an adaptive algorithm can be devised using the channel estimator described in Sec. 6.7. The channel estimator supplies the algorithm in (6.6.12) with estimates $\{\hat{f}_k\}$ that are used in place of the unknown parameters $\{f_k\}$. Thus an adaptive algorithm results.

Second, we observe that the averaging performed over the symbols $I_{k+D}, \ldots, I_{k+1}, I_{k-1}$ in the algorithm in (6.6.12) involves a large number of computations per received signal. In particular, the computations involve summations of exponential factors that depend on the received sequence $\{v_k\}$ and the information sequence $\{I_k\}$ as indicated in (6.6.6). Consequently the large computational burden is the major shortcoming of this algorithm.

The error rate performance of the algorithm in (6.6.12) has not been evaluated for general channel characteristics, primarily because the analysis is very difficult. A number of Monte Carlo simulation results have been published which indicate that the performance of this algorithm is superior to that of the decision-feedback equalizer and comparable to the performance of the Viterbi algorithm discussed in Sec. 6.7.

Symbol-by-symbol demodulation of CPM signals. We recall that CPM with a small modulation index is a bandwidth-efficient modulation technique that is especially suitable for bandwidth-constrained channels. In Chap. 4 we considered the demodulation of PSK and FSK signals by means of matched filtering or cross correlation based on independent symbol-by-symbol detection. This type of demodulation is also applicable to CPM and CPFSK signals. However, this demodulation ignores the memory that is inherent in a CPM and CPFSK signal due to the phase continuity. We shall now consider the benefit, expressed in terms of an improvement in performance obtained by employing a demodulator that exploits the phase continuity in the signal. White gaussian noise is assumed to be the only channel disturbance.

There are basically two types of optimum demodulation methods that can be used to exploit the memory in CPM. One is the maximum-likelihood sequence estimation (MLSE) technique that is described in the following section which may be realized efficiently by use of the Viterbi algorithm. A second type of optimum demodulation is the symbol-by-symbol detection algorithm described above which exploits the memory of CPM by performing matched filtering or cross correlation over several signaling intervals. Because of its computational complexity, however, this recursive algorithm has not been directly applied to the demodulation of CPM. Instead, two similar, albeit suboptimal, symbol-by-symbol detection methods have been described in the papers by deBuda (1972), Osborne and Luntz (1974), and Schonhoff (1976). One of these is functionally equivalent to the algorithm described above and the second is a suboptimum approximation of the first. We shall describe these two methods in the context of

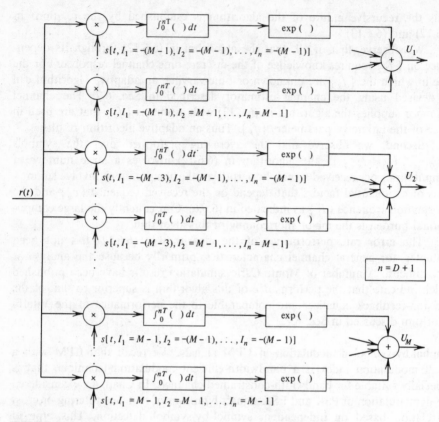

FIGURE 6.6.1
Block diagram of demodulator for coherent detection of CPFSK.

demodulation of CPFSK signals for which these detection algorithms have been applied directly.

To describe these methods we assume that the signal is observed over the present signaling interval and D signaling intervals into the future in deciding on the information symbol transmitted in the present signaling interval. A block diagram of the demodulator, implemented as a bank of cross correlators, is shown in Fig. 6.6.1. We recall that the transmitted CPFSK signal during the nth signaling interval is

$$s(t) = \text{Re}\left[v(t)e^{j2\pi f_c t}\right]$$

where

$$v(t) = \exp\left\{\frac{\pi h[t - (n-1)T]I_n}{T} + \pi h \sum_{k=0}^{n-1} I_k + \phi\right\}$$

$h = 2f_d T$ is the modulation index, f_d is the peak frequency deviation, and ϕ is the initial phase angle of the carrier.

In detecting the symbol I_1, the cross correlations shown in Fig. 6.6.1 are performed with the reference signals $s(t, I_1, I_2, \ldots, I_{1+D})$ for all M^{D+1} possible values of the symbols $I_1, I_2, \ldots, I_{1+D}$ transmitted over the $D + 1$ signaling intervals. But these correlations in effect generate the variables $v_1, v_2, \ldots, v_{1+D}$, which in turn are the arguments of the exponentials that occur in the pdf

$$p(v_1, v_2, \ldots, v_{1+D} | I_1, I_2, \ldots, I_{1+D})$$

Finally, the summations over the M_D possible values of the symbols $I_2, I_3, \ldots, I_{1+D}$ represent the averaging of

$$p(v_1, v_2, \ldots, v_{1+D} | I_1, I_2, \ldots, I_{1+D}) P(I_1, I_2, \ldots, I_{1+D})$$

over the M^D possible values of these symbols. The M outputs of the demodulator constitute the decision variables from which the largest is selected to form the demodulated symbol. Consequently the metrics generated by the demodulator shown in Fig. 6.6.1 are equivalent to the decision variables given by (6.6.3) on which the decision on I_1 is based.

Signals received in subsequent signaling intervals are demodulated in the same manner. That is, the demodulator cross-correlates the signal received over $D + 1$ signaling intervals with the M^{D+1} possible transmitted signals and forms the decision variables as illustrated in Fig. 6.6.1. Thus the decision made on the mth signaling interval is based on the cross correlations performed over the signaling intervals $m, m + 1, \ldots, m + D$. For coherent detection, the initial phase in the correlation interval of duration $(D + 1)T$ is assumed to be known. On the other hand, the algorithm described by (6.6.12) and (6.6.13) involves an additional averaging operation over the previously detected symbols. In this respect, the demodulator shown in Fig. 6.6.1 differs from the recursive algorithm described above. However, the difference is insignificant.

One suboptimum demodulation method that performs almost as well as the optimum method embodied in Fig. 6.6.1 bases its decision on the largest output from the bank of M^{D+1} cross correlators. Thus the exponential functions and the summations are eliminated. But this method is equivalent to selecting the symbol I_m for which the probability density function $p(v_m, v_{m+1}, \ldots, v_{m+D} | I_m, I_{m+1}, \ldots, I_{m+D})$ is a maximum.

The performance of the demodulator shown in Fig. 6.6.1 has been upperbounded and evaluated numerically. Figure 6.6.2 illustrates the performance of binary CPFSK with $n = D + 1$ as a parameter. The modulation index $h = 0.715$ used in generating these results minimizes the probability of error as shown by Schonhoff (1976). We note that an improvement of about 2.5 dB is obtained relative to orthogonal FSK ($n = 1$) by a demodulator that cross-correlates over two symbols. An additional gain of approximately 1.5 dB is obtained by extending the correlation time to three symbols. Further extension of the correlation time results in a relatively small additional gain.

Similar results are obtained with larger alphabet sizes. For example, Figs. 6.6.3 and 6.6.4 illustrate the performance improvements for quanternary and

FIGURE 6.6.2
Performance of binary CPFSK with coherent detection.

octal CPFSK, respectively. The modulation indices given in these graphs are the ones that minimize the probability of a symbol error.

Instead of performing coherent detection, which requires knowledge of the carrier phase ϕ, we may assume that ϕ is uniformly distributed over the interval 0 to 2π, and average over it in arriving at the decision variables. Thus coherent integration (cross correlation) is performed over the $n = D + 1$ signaling intervals, but the outputs of the correlators are envelope-detected. This is called *noncoherent detection* of CPFSK. In this detection scheme, performance is optimized by selecting n to be odd and making the decision on the middle symbol in

FIGURE 6.6.3
Performance of quaternary CPFSK with coherent detection.

the sequence of n symbols. The numerical results on the probability of error for noncoherent detection of CPFSK are similar to the results illustrated above for coherent detection. That is, a gain of 2 to 3 dB in performance is achieved by increasing the correlation interval from $n = 1$ to $n = 3$ and to $n = 5$.

As indicated above, CPM and CPFSK may also be demodulated by employing the maximum-likelihood sequence estimation criterion, implemented by means of the Viterbi algorithm, which is described in Sec. 6.7. Since the Viterbi algorithm is optimum in the sense of minimizing the probability of error for a sequence of information symbols, its performance should be approximately the same as that obtained with the detection methods described above.

FIGURE 6.6.4
Performance of octal CPFSK with coherent detection.

6.7 MAXIMUM-LIKELIHOOD SEQUENCE ESTIMATION FOR DETECTION OF ISI-CORRUPTED SIGNALS AND CPM SIGNALS

In Sec. 6.3 we derived the optimum receiver, based on the maximum-likelihood criterion, for estimating the symbol sequence $\{I_k\}$ from the received signal that is corrupted by intersymbol interference. The decision variables or metrics that are computed for the maximum-likelihood sequence estimate (MLSE) of the sequence $\{I_k\}$ are given by (6.3.7). It is easily demonstrated that the metrics in (6.3.7) can be computed recursively, according to the relation [see Ungerboeck

(1974) or MacKenchnie (1973)]

$$J_n(\mathbf{I}_n) = J_{n-1}(\mathbf{I}_{n-1}) + \text{Re}\left[I_n^*\left(2y_n - x_0 I_n - 2\sum_{m \le n-1} I_m x_{n-m}\right)\right] \quad (6.7.1)$$

Since $x_n = 0$ for $|n| > L$, (6.7.1) can be expressed in the form

$$J_n(\mathbf{I}_n) = J_{n-1}(\mathbf{I}_{n-1}) + \text{Re}\left[I_n^*\left(2y_n - x_0 I_n - 2\sum_{m=1}^{L} x_m I_{n-m}\right)\right] \quad (6.7.2)$$

If we transmit a sequence $\{I_k\}$ of N symbols, where each symbol is chosen from an M-symbol alphabet, there are M^N possible sequences and, hence, M^N metrics. The sequence that has the largest metric is selected as the detected sequence. However, not all M^N metrics need be computed. The reason is that the finite-duration discrete-time channel model shown in Fig. 6.3.2 is equivalent to a convolutional encoder operating on the information sequence. Hence the Viterbi algorithm described in Sec. 5.3.2 for decoding convolutional codes is the efficient method for optimum sequence estimation.

The Viterbi algorithm for MLSE of the information sequence $\{I_k\}$ is most easily described in terms of the received sequence $\{v_k\}$ at the output of the whitening filter. Thus, suppose that we have received the sequence $\mathbf{v}_N = (v_1, v_2, \ldots, v_N)$, where v_k is given by (6.3.15). We wish to estimate the sequence $\mathbf{I}_N = (I_1, I_2, \ldots, I_N)$. The decision, based on the maximum-likelihood criterion, is made in favor of the sequence $\mathbf{I}_N$ that maximizes the joint probability density function

$$p(\mathbf{v}_N|\mathbf{I}_N) \equiv p(v_N, v_{N-1}, \ldots, v_1|I_N, I_{N-1}, \ldots, I_1) \quad (6.7.3)$$

Since the additive noise sequence $\{\eta_k\}$ is white, the joint probability density $p(\mathbf{v}_N|\mathbf{I}_N)$ can be expressed as a product of marginal densities as indicated previously, i.e.,

$$p(\mathbf{v}_N|\mathbf{I}_N) = \prod_{k=1}^{N} p(v_k|I_k, I_{k-1}, \ldots, I_{k-L}) \quad (6.7.4)$$

where, by definition, $I_k = 0$ for $k \le 0$. When there is no intersymbol interference, the signal v_k depends only on the symbol I_k and, consequently, the maximum of $p(\mathbf{v}_N|\mathbf{I}_N)$ over the symbols $\mathbf{I}_N$ is equivalent to maximizing each of the marginal densities $p(v_k|I_k)$. In other words, sequence detection based on the maximum-likelihood criterion reduces to simple symbol-by-symbol detection. More important, however, is the fact that only M probabilities (metrics) are computed per signaling interval and, hence, MN probabilities need be computed for the detection of N symbols instead of M^N.

In the presence of intersymbol interference that spans L symbols (L interfering components), the maximization of $p(\mathbf{v}_N|\mathbf{I}_N)$ can be performed recursively and efficiently by computing NM^{L+1} probabilities or metrics as we now describe. We begin with the samples $v_1, v_2, \ldots, v_{L+1}$, from which we compute the

M^{L+1} joint probabilities

$$p(v_{L+1},\ldots,v_1|I_{L+1},\ldots,I_1) = \prod_{k=1}^{L+1} p(v_k|I_k,\ldots,I_{k-L}) \qquad (6.7.5)$$

corresponding to the M^{L+1} possible sequences for the symbols $I_{L+1},\ldots,I_1$. In practice the multiplication of probabilities indicated in (6.7.5) is avoided by taking the natural logarithm of both sides of the equation. It is observed that the subsequent signal samples beginning with v_{L+2} do not depend on I_1. This fact implies that we can discard some of the least-probable sequences with respect to I_1 according to the following procedure. The M^{L+1} possible sequences of $I_{L+1},\ldots,I_2,I_1$ are subdivided into M^L groups corresponding to the possible choices of the symbols $I_{L+1},\ldots,I_2$. In each of the M^L groups there are M sequences that differ only in the symbol I_1. From the M sequences in each group we select that sequence having the largest probability (with respect to I_1) and assign to the sequence the metric

$$\mu_1(\mathbf{I}_{L+1}) \equiv \mu_1(I_{L+1},\ldots,I_1) = \max_{I_1} \sum_{k=1}^{L+1} \ln p(v_k|I_k,\ldots,I_{k-L}) \qquad (6.7.6)$$

where $\ln(x)$ denotes the natural logarithm of x. The $M-1$ remaining sequences from each of the M^L groups are discarded. Thus we are left with M^L surviving sequences having the metrics $\mu_1(\mathbf{I}_{L+1})$ that correspond to the M^L possible choices of the symbols $I_{L+1},\ldots,I_2$. At this point, if all the M^L surviving sequences have the same value of I_1, that value constitutes the decision on I_1. Otherwise the decision on I_1 is deferred to a later time.

Upon reception of v_{L+2}, the probabilities for $v_{L+2},\ldots,v_1$, or, equivalently, the logarithms of the probabilities are computed as follows:

$$\ln p(v_{L+2},v_{L+1},\ldots,v_1|I_{L+2},\ldots,I_1)$$
$$= \ln p(v_{L+2}|I_{L+2},\ldots,I_2) + \sum_{k=1}^{L+1} \ln p(v_k|I_k,\ldots,I_{k-L}) \qquad (6.7.7)$$

Now (6.7.7) gives the probabilities of M^{L+2} sequences of the symbols $I_{L+2},\ldots,I_1$. It should be noted that the summation of probabilities on the right-hand side of (6.7.7) was computed previously in the reception of $v_{L+1},\ldots,v_1$. Since the subsequent signal samples beginning with v_{L+3} do not depend on I_1 and I_2, one may choose from the M^{L+2} sequences that terminate in the symbols $I_3,\ldots,I_{L+2}$ the more probable sequences with respect to I_1 and I_2 as follows. The M^{L+2} sequences are subdivided into M^L groups corresponding to the possible choices of the symbols $I_{L+2},\ldots,I_3$. Each of the M^L groups contains M^2 sequences that differ only in the symbols I_1 and I_2. From these M^2 sequences in each group we select that sequence having the largest probability. However, a little thought will convince the reader that the maximum of (6.7.7) over I_1 and I_2 must occur from the continuation of the surviving sequences obtained from (6.7.6) and that, in general, any continuation of a discarded sequence will always have a lower probability than the same continuation of the corresponding surviving sequence.

Consequently we need to consider only continuations of surviving sequences so that the M^L groups contain only M sequences instead of M^2. From each of the M sequences per group we select the one having the largest probability (with respect to I_2), thus reducing the number of surviving sequences down to M^L. Again, it is convenient to define the M^L metrics corresponding to the probabilities of the M^L surviving sequences at the end of the second stage as $\mu_2(\mathbf{I}_{L+2})$. It follows from the above discussion that

$$\mu_2(\mathbf{I}_{L+2}) = \max_{I_2} \left[\ln p(v_{L+2}|I_{L+2}, \ldots, I_2) + \mu_1(\mathbf{I}_{L+1}) \right] \qquad (6.7.8)$$

At this stage a decision is made on the symbol I_1 or the pair of symbols (I_1, I_2) if all the M^L surviving sequences terminating in the symbols $I_3, \ldots, I_{L+2}$ have the same value of I_1 or the same values of (I_1, I_2), respectively. Otherwise the decision is deferred to a later stage.

The procedure described continues with the reception of subsequent signal samples. In general, upon reception of v_{L+k} the metrics

$$\mu_k(\mathbf{I}_{L+k}) = \max_{I_k} \left[\ln p(v_{L+k}|I_{L+k}, \ldots, I_k) + \mu_{k-1}(\mathbf{I}_{L+k-1}) \right] \qquad (6.7.9)$$

which are computed give the probabilities of the M^L surviving sequences. Thus, as each signal sample is received, the Viterbi algorithm involves first the computation of the M^{L+1} probabilities

$$\ln p(v_{L+k}|I_{L+k}, \ldots, I_k) + \mu_{k-1}(\mathbf{I}_{L+k-1}) \qquad (6.7.10)$$

corresponding to the M^{L+1} sequences which form the continuations of the M^L surviving sequences from the previous stage of the process. Then the M^{L+1} sequences are subdivided into M^L groups, with each group containing M sequences that terminate in the same set of symbols $I_{L+k}, \ldots, I_{k+1}$ and differ in the symbol I_k. From each group of M sequences we select the one having the largest probability as indicated by (6.7.9) while the remaining $M - 1$ sequences are discarded. Thus we are left again with M^L sequences having the metrics $\mu_k(\mathbf{I}_{L+k})$.

In each stage of the Viterbi algorithm, M^{L+1} probabilities are computed. Hence a total of NM^{L+1} probabilities are computed in the detection of N symbols.

As indicated above, the delay in detecting each information symbol is variable. In practice the variable delay is avoided by truncating the surviving sequences to q most recent symbols where $q \gg L$, thus achieving a fixed delay. In case the M^L surviving sequences at time k disagree on the symbol I_{k-q}, the symbol in the most probable sequence may be chosen. The loss in performance resulting from this suboptimum decision procedure is negligible if $q \geq 5L$.

The MLSE technique described above may also be viewed as a problem in estimating the state of a discrete-time finite-state machine. The finite-state machine in this case is the channel with coefficients $\{f_k\}$ and its state at any

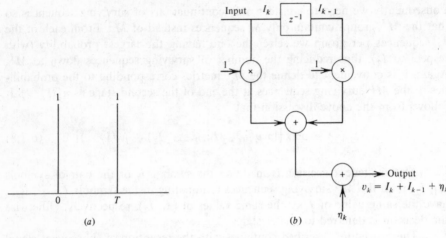

FIGURE 6.7.1
Equivalent discrete-time model for intersymbol interference resulting from a duobinary pulse.

instant in time is given by the L most recent inputs, i.e., the state at time k

$$S_k = (I_{k-1}, I_{k-2}, \ldots, I_{k-L})$$

where $I_k = 0$ for $k \leq 0$. Thus the channel has M^L states where M is the size of the symbol alphabet. Equivalently an M^L-state trellis may be used to describe the channel. In any case,the Viterbi algorithm described above tracks the state of the channel or the paths through the trellis and gives at stage k the M^L most probable sequences terminating in the symbols $I_{k-L}, I_{k-L+1}, \ldots, I_{k-1}$.

Example 6.7.1. The Viterbi algorithm is an optimum method for recovering the information sequence from a partial response signal. For illustrative purposes, suppose that a duobinary signal pulse is employed to transmit four-level ($M = 4$) PAM. Thus each symbol is a number selected from the set $\{-3, -1, 1, 3\}$. The controlled intersymbol interference in this partial response signal is represented by the equivalent discrete-time channel model shown in Fig. 6.7.1. Suppose we have received v_1 and v_2, where

$$v_1 = I_1 + \eta_1$$
$$v_2 = I_2 + I_1 + \eta_2 \qquad (6.7.11)$$

and $\{\eta_i\}$ is a sequence of statistically independent zero mean gaussian noise. We may now compute the 16 metrics

$$\mu_1(I_2, I_1) = -\sum_{k=1}^{2} \left(v_k - \sum_{j=0}^{1} I_{k-j} \right)^2 \qquad I_1, I_2 = \pm 1, \pm 3 \quad (6.7.12)$$

where $I_k = 0$ for $k \leq 0$.

We note that any subsequently received signals $\{v_i\}$ do not involve I_1. Hence, at this stage, we may discard 12 of the 16 possible pairs $\{I_1, I_2\}$. This step is illustrated by the tree diagram shown in Fig. 6.7.2. In other words, after computing

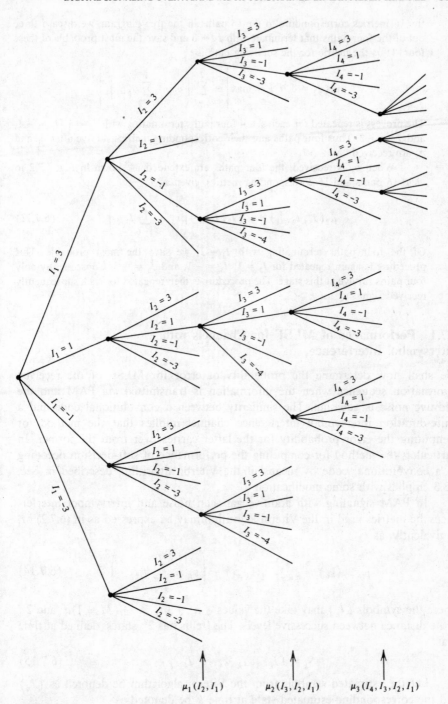

FIGURE 6.7.2
Tree diagram for Viterbi decoding of the duobinary pulse.

the 16 metrics corresponding to the 16 paths in the tree diagram, we discard three out of the four paths that terminate with $I_2 = 3$ and save the most probable of these four. Thus the metric for the surviving path is

$$\mu_1(I_2 = 3, I_1) = \max_{I_1}\left[-\sum_{k=1}^{2}\left(v_k - \sum_{j=0}^{1} I_{k-j} \right)^2 \right]$$

The process is repeated for each set of four paths terminating with $I_2 = 1$, $I_2 = -1$, and $I_2 = -3$. Thus four paths and their corresponding metrics survive after v_1 and v_2 are received.

When v_3 is received, the four paths are extended as shown in Fig. 6.7.2 to yield 16 paths and 16 corresponding metrics, given as

$$\mu_2(I_3, I_2, I_1) = \mu_1(I_2, I_1) - \left(v_3 - \sum_{j=0}^{1} I_{3-j} \right)^2 \tag{6.7.13}$$

Of the four paths terminating with $I_3 = 3$, we save the most probable. This procedure is again repeated for $I_3 = 1$, $I_3 = -1$, and $I_3 = -3$. Consequently only four paths survive at this stage. The procedure is then repeated for each subsequently received signal v_k for $k > 3$.

6.7.1 Performance of MLSE for Channels with Intersymbol Interference

We shall now determine the probability of error for MLSE of the received information sequence when the information is transmitted via PAM and the additive noise is gaussian. The similarity between a convolutional code and a finite-duration intersymbol interference channel implies that the method for computing the error probability for the latter carries over from the former. In particular, the method for computing the performance of soft-decision decoding of a convolutional code by means of the Viterbi algorithm, described in Sec. 5.3.3, applies with some modification.

In PAM signaling with additive gaussian noise and intersymbol interference, the metrics used in the Viterbi algorithm may be expressed as in (6.7.2) or, equivalently, as

$$\mu_{k-L}(\mathbf{I}_k) = \mu_{k-L-1}(\mathbf{I}_{k-1}) - \left(v_k - \sum_{j=0}^{L} f_j I_{k-j} \right)^2 \tag{6.7.14}$$

where the symbols $\{I_n\}$ may take the values $\pm d$, $\pm 3d, \ldots, \pm(M-1)d$, and $2d$ is the distance between successive levels. The trellis has 2^L states, defined at time k as

$$S_k = (I_{k-1}, I_{k-2}, \ldots, I_{k-L}) \tag{6.7.15}$$

Let the estimated symbols from the Viterbi algorithm be denoted as $\{\tilde{I}_n\}$ and the corresponding estimated state at time k be denoted as

$$\tilde{S}_k = \left(\tilde{I}_{k-1}, \tilde{I}_{k-2}, \ldots, \tilde{I}_{k-L} \right) \tag{6.7.16}$$

Now suppose that the estimated path through the trellis diverges from the correct path at time k and remerges with the correct path at time $k + l$. Thus $\tilde{S}_k = S_k$ and $\tilde{S}_{k+1} = S_{k+1}$, but $\tilde{S}_m \neq S_m$ for $k < m < k + l$. As in a convolutional code, we call this an *error event*. Since the channel spans $L + 1$ symbols, it follows that $l \geq L + 1$.

For such an error event, we have $\tilde{I}_k \neq I_k$ and $\tilde{I}_{k+l-L-1} \neq I_{k+l-L-1}$, but $\tilde{I}_m = I_m$ for $k - L \leq m \leq k - 1$ and $k + l - L \leq m \leq k + l - 1$. It is convenient to define an error vector $\boldsymbol{\varepsilon}$ corresponding to this error event as

$$\boldsymbol{\varepsilon} = (\varepsilon_k, \varepsilon_{k+1}, \ldots, \varepsilon_{k+l-L-1}) \tag{6.7.17}$$

where the components of $\boldsymbol{\varepsilon}$ are defined as

$$\varepsilon_j = \frac{1}{2d}(I_j - \tilde{I}_j) \qquad j = k, k+1, \ldots, k+l-L-1 \tag{6.7.18}$$

The normalization factor of $2d$ in (6.7.18) results in elements ε_j that take on the values $\pm 1, \pm 2, \pm 3, \ldots, \pm(M - 1)$. Moreover, the error vector is characterized by the properties that $\varepsilon_k \neq 0$, $\varepsilon_{k+l-L-1} \neq 0$, and there is no sequence of L consecutive elements that are zero. Associated with the error vector in (6.7.17) is the polynomial of degree $l - L - 1$,

$$\varepsilon(z) = \varepsilon_k + \varepsilon_{k+1}z^{-1} + \varepsilon_{k+2}z^{-2} + \cdots + \varepsilon_{k+l-L-1}z^{-(l-L-1)} \tag{6.7.19}$$

We wish to determine the probability of occurrence of the error event that begins at time k and is characterized by the error vector $\boldsymbol{\varepsilon}$ given in (6.7.17) or, equivalently, by the polynomial given in (6.7.19). To accomplish this we follow the procedure developed by Forney (1972). Specifically, for the error event $\boldsymbol{\varepsilon}$ to occur, the following three subevents $\mathscr{E}_1$, $\mathscr{E}_2$, and $\mathscr{E}_3$ must occur:

$\mathscr{E}_1$: At time k, $\tilde{S}_k = S_k$.

$\mathscr{E}_2$: The information symbols $I_k, I_{k+1}, \ldots, I_{k+l-L-1}$ when added to the scaled error sequence $2d(\varepsilon_k, \varepsilon_{k+1}, \ldots, \varepsilon_{k+l-L-1})$ must result in an allowable sequence, i.e., the sequence $\tilde{I}_k, \tilde{I}_{k+1}, \ldots, \tilde{I}_{k+l-L-1}$ must have values selected from $\pm d, \pm 3d, \pm \cdots \pm(M - 1)d$.

$\mathscr{E}_3$: For $k \leq m \leq k + l$, the sum of the branch metrics of the estimated path exceed the sum of the branch metrics of the correct path.

The probability of occurrence of $\mathscr{E}_3$ is

$$P(\mathscr{E}_3) = P\left[\sum_{i=k}^{k+l-1} \left(v_i - \sum_{j=0}^{L} f_j \tilde{I}_{i-j} \right)^2 < \sum_{i=k}^{k+l-1} \left(v_i - \sum_{j=0}^{L} f_j I_{i-j} \right)^2 \right] \tag{6.7.20}$$

But

$$v_i = \sum_{j=0}^{L} f_j I_{i-j} + \eta_i \tag{6.7.21}$$

where the $\{\eta_i\}$ are a real-valued white gaussian noise sequence. Substitution of

(6.7.21) into (6.7.20) yields

$$P(\mathscr{E}_3) = P\left[\sum_{i=k}^{k+l-1}\left(\eta_i + 2d\sum_{j=0}^{L}f_j\varepsilon_{i-j}\right)^2 < \sum_{i=k}^{k+l-1}\eta_i^2\right]$$

$$= P\left[4d\sum_{i=k}^{k+l-1}\eta_i\left(\sum_{j=0}^{L}f_j\varepsilon_{i-j}\right) < -4d^2\sum_{i=k}^{k+l-1}\left(\sum_{j=0}^{L}f_j\varepsilon_{i-j}\right)^2\right] \quad (6.7.22)$$

where $\varepsilon_j = 0$ for $j < k$ and $j > k + l - L - 1$. If we define

$$\alpha_i = \sum_{j=0}^{L}f_j\varepsilon_{i-j} \quad (6.7.23)$$

then (6.7.22) may be expressed as

$$P(\mathscr{E}_3) = P\left(\sum_{i=k}^{k+l-1}\alpha_i\eta_i < -d\sum_{i=k}^{k+l-1}\alpha_i^2\right) \quad (6.7.24)$$

where the factor of $4d$ common to both terms has been dropped. Now (6.7.24) is just the probability that a linear combination of statistically independent gaussian random variables is less than some negative number. Thus

$$P(\mathscr{E}_3) = \tfrac{1}{2}\,\text{erfc}\left(\sqrt{\frac{d^2}{N_0}\sum_{i=k}^{k+l-1}\alpha_i^2}\right) \quad (6.7.25)$$

For convenience we define

$$\delta^2(\varepsilon) = \sum_{i=k}^{k+l-1}\alpha_i^2 = \sum_{i=k}^{k+l-1}\left(\sum_{j=0}^{L}f_j\varepsilon_{i-j}\right)^2 \quad (6.7.26)$$

where $\varepsilon_j = 0$ for $j < k$ and $j > k + l - L - 1$. Note that the $\{\alpha_i\}$ resulting from the convolution of $\{f_j\}$ with $\{\varepsilon_j\}$ are the coefficients of the polynomial

$$\alpha(z) = F(z)\varepsilon(z)$$

$$= \alpha_k + \alpha_{k+1}z^{-1} + \cdots + \alpha_{k+l-1}z^{-(l-1)} \quad (6.7.27)$$

Furthermore, $\delta^2(\varepsilon)$ is simply equal to the coefficient of z^0 in the polynomial

$$\alpha(z)\alpha(z^{-1}) = F(z)F(z^{-1})\varepsilon(z)\varepsilon(z^{-1})$$

$$= X(z)\varepsilon(z)\varepsilon(z^{-1}) \quad (6.7.28)$$

We call $\delta^2(\varepsilon)$ the *Euclidean weight* of the error event ε.

An alternative method for representing the result of convolving $\{f_j\}$ with $\{\varepsilon_j\}$ is the matrix form

$$\alpha = \mathbf{e}\mathbf{f}$$

where α is an l-dimensional vector, $\mathbf{f}$ is an $(L + 1)$-dimensional vector, and $\mathbf{e}$ is an $l \times (L + 1)$ matrix, defined as

$$
\boldsymbol{\alpha} = \begin{bmatrix} \alpha_k \\ \alpha_{k+1} \\ \vdots \\ \alpha_{k+l-1} \end{bmatrix} \qquad \mathbf{f} = \begin{bmatrix} f_0 \\ f_1 \\ \vdots \\ f_L \end{bmatrix}
$$

$$
\mathbf{e} = \begin{bmatrix} \varepsilon_k & 0 & 0 & \cdots & 0 & \cdots & 0 \\ \varepsilon_{k+1} & \varepsilon_k & 0 & \cdots & 0 & \cdots & 0 \\ \varepsilon_{k+2} & \varepsilon_{k+1} & \varepsilon_k & & 0 & \cdots & 0 \\ \cdots & \cdots & \cdots & \cdots & \cdots & & \varepsilon_{k+l-L-1} \\ \varepsilon_{k+l-1} & \cdots & \cdots & \cdots & \cdots & & \end{bmatrix} \tag{6.7.29}
$$

Then

$$
\begin{aligned}
\delta^2(\boldsymbol{\varepsilon}) &= \boldsymbol{\alpha}'\boldsymbol{\alpha} \\
&= \mathbf{f}'\mathbf{e}'\mathbf{e}\mathbf{f} \\
&= \mathbf{f}'\mathbf{A}\mathbf{f}
\end{aligned} \tag{6.7.30}
$$

where $\mathbf{A}$ is an $(L + 1) \times (L + 1)$ matrix of the form

$$
\mathbf{A} = \mathbf{e}'\mathbf{e} = \begin{bmatrix} \beta_0 & \beta_1 & \beta_2 & \cdots & \beta_L \\ \beta_1 & \beta_0 & \beta_1 & \cdots & \beta_{L-1} \\ \beta_2 & \beta_1 & \beta_0 & \beta_1 & \beta_{L-2} \\ \cdots & \cdots & \cdots & \cdots & \cdots \\ \beta_L & \cdots & \cdots & \cdots & \beta_0 \end{bmatrix} \tag{6.7.31}
$$

and

$$
\beta_m = \sum_{i=k}^{k+l-1-m} \varepsilon_i \varepsilon_{i+m} \tag{6.7.32}
$$

We may use either (6.7.26) and (6.7.27) or (6.7.30), (6.7.31), and (6.7.32) in evaluating the error rate performance. We consider these computations later. For now we conclude that the probability of the subevent $\mathscr{E}_3$, given by (6.7.25), may be expressed as

$$
\begin{aligned}
P(\mathscr{E}_3) &= \tfrac{1}{2} \operatorname{erfc}\left(\sqrt{\frac{d^2}{N_0} \delta^2(\boldsymbol{\varepsilon})} \right) \\
&= \tfrac{1}{2} \operatorname{erfc}\left(\sqrt{\frac{3}{M^2 - 1} \gamma_{\mathrm{av}} \delta^2(\boldsymbol{\varepsilon})} \right)
\end{aligned} \tag{6.7.33}
$$

where we have used the relation

$$
d^2 = \frac{3}{M^2 - 1} TP_{\mathrm{av}} \tag{6.7.34}
$$

to eliminate d^2 and $\gamma_{\mathrm{av}} = TP_{\mathrm{av}}/N_0$. We note that in the absence of intersymbol interference $\delta^2(\boldsymbol{\varepsilon}) = 1$ and $P(\mathscr{E}_3)$ is proportional to the symbol error probability of M-ary PAM.

The probability of the subevent $\mathcal{E}_2$ depends only on the statistical properties of the input sequence. We assume that the information symbols are equally probable and that the symbols in the transmitted sequence are statistically independent. Then, for an error of the form $|\varepsilon_i| = j$, $j = 1, 2, \ldots, (M-1)$, there are $M - j$ possible values of I_i such that

$$I_i = \tilde{I}_i + 2d\varepsilon_i$$

Hence

$$P(\mathcal{E}_2) = \prod_{i=0}^{l-L-1} \frac{M - |i|}{M} \tag{6.7.35}$$

The probability of the subevent $\mathcal{E}_1$ is much more difficult to compute exactly because of its dependence on the subevent $\mathcal{E}_3$. That is, we must compute $P(\mathcal{E}_1|\mathcal{E}_3)$. However, $P(\mathcal{E}_1|\mathcal{E}_3) = 1 - P_M$, where P_M is the symbol error probability. Hence $P(\mathcal{E}_1|\mathcal{E}_3)$ is well approximated (and upper-bounded) by unity for reasonably low symbol error probabilities. Therefore the probability of the error event ε is well approximated and upper-bounded as

$$P(\varepsilon) \leq \tfrac{1}{2} \operatorname{erfc}\left[\sqrt{\frac{3}{M^2-1}\gamma_{av}\delta^2(\varepsilon)}\right] \prod_{i=0}^{l-L-1} \frac{M - |i|}{M} \tag{6.7.36}$$

Let E be the set of all error events ε starting at time k and let $w(\varepsilon)$ be the corresponding number of nonzero components (Hamming weight or number of symbol errors) in each error event ε. Then the probability of a symbol error is upper-bounded (union bound) as

$$P_M \leq \sum_{\varepsilon \in E} w(\varepsilon) P(\varepsilon)$$

$$\leq \tfrac{1}{2} \sum_{\varepsilon \in E} w(\varepsilon) \operatorname{erfc}\left[\sqrt{\frac{3}{M^2-1}\gamma_{av}\delta^2(\varepsilon)}\right] \prod_{i=0}^{l-L-1} \frac{M - |i|}{M} \tag{6.7.37}$$

Now let D be the set of all $\delta(\varepsilon)$. For each $\delta \in D$, let E_δ be the subset of error events for which $\delta(\varepsilon) = \delta$. Then (6.7.37) may be expressed as

$$P_M \leq \frac{1}{2} \sum_{\delta \in D} \operatorname{erfc} \sqrt{\frac{3}{M^2-1}\gamma_{av}\delta^2} \left[\sum_{\varepsilon \in E_\delta} w(\varepsilon) \prod_{i=0}^{l-L-1} \frac{M - |i|}{M}\right]$$

$$\leq \frac{1}{2} \sum_{\delta \in D} K_\delta \operatorname{erfc}\left(\sqrt{\frac{3}{M^2-1}\gamma_{av}\delta^2}\right) \tag{6.7.38}$$

where

$$K_\delta = \sum_{\varepsilon \in E_\delta} w(\varepsilon) \prod_{i=0}^{l-L-1} \frac{M - |i|}{M} \tag{6.7.39}$$

The expression for the error probability in (6.7.38) is similar to the form of the error probability for a convolutional code with soft-decision decoding given by (5.3.26). The weighting factors $\{K_\delta\}$ may be determined by means of the error state diagram, which is akin to the state diagram of a convolutional encoder. This approach has been illustrated by Forney (1972) and Viterbi and Omura (1979).

In general, however, the use of the error state diagram for computing P_M is tedious. Instead we may simplify the computation of P_M by focusing on the dominant term in the summation of (6.7.38). Due to the exponential dependence of each term in the sum, the expression P_M is dominated by the term corresponding to the minimum value of δ, denoted as δ_{min}. Hence the symbol error probability may be approximated as

$$P_M \approx \tfrac{1}{2} K_{\delta_{min}} \, \text{erfc} \left(\sqrt{ \frac{3}{M^2 - 1} \gamma_{av} \delta_{min}^2 } \right) \qquad (6.7.40)$$

where

$$K_{\delta_{min}} = \sum_{\varepsilon \in E_{\delta_{min}}} w(\varepsilon) \prod_{i=0}^{l-L-1} \frac{M - |i|}{M} \qquad (6.7.41)$$

In general, $\delta_{min}^2 \leq 1$. Hence $10 \log \delta_{min}^2$ represents the loss in SNR due to intersymbol interference.

The minimum value of δ may be determined either from (6.7.27) or from evaluation of the quadratic form in (6.7.30) for different error sequences. In the following two examples we use (6.7.27).

Example 6.7.2. Consider a two-path channel ($L = 1$) with arbitrary coefficients f_0 and f_1 satisfying the constraint $f_0^2 + f_1^2 = 1$. The channel characteristic is

$$F(z) = f_0 + f_1 z^{-1} \qquad (6.7.42)$$

For an error event of length n,

$$\varepsilon(z) = \varepsilon_0 + \varepsilon_1 z^{-1} + \cdots + \varepsilon_{n-1} z^{-(n-1)} \qquad n \geq 1 \qquad (6.7.43)$$

The product $\alpha(z) = F(z)\varepsilon(z)$ may be expressed as

$$\alpha(z) = \alpha_0 + \alpha_1 z^{-1} + \cdots + \alpha_n z^{-n} \qquad (6.7.44)$$

where $\alpha_0 = \varepsilon_0 f_0$ and $\alpha_n = f_1 \varepsilon_{n-1}$. Since $\varepsilon_0 \neq 0$, $\varepsilon_{n-1} \neq 0$, and

$$\delta^2(\varepsilon) = \sum_{k=0}^{n} \alpha_k^2 \qquad (6.7.45)$$

it follows that

$$\delta_{min}^2 \geq f_0^2 + f_1^2 = 1$$

Indeed, $\delta_{min}^2 = 1$ when a single error occurs, i.e. $\varepsilon(z) = \varepsilon_0$. Thus we conclude that there is no loss in SNR in maximum-likelihood sequence estimation of the information symbols when the channel dispersion has length 2.

Example 6.7.3. The controlled intersymbol interference in a partial response signal may be viewed as having been generated by a time-dispersive channel. Thus the intersymbol interference from a duobinary pulse may be represented by the (normalized) channel characteristic

$$F(z) = \frac{1}{\sqrt{2}} + \frac{1}{\sqrt{2}} z^{-1} \qquad (6.7.46)$$

Similarly the representation for a modified duobinary pulse is

$$F(z) = \frac{1}{\sqrt{2}} - \frac{1}{\sqrt{2}} z^{-2} \qquad (6.7.47)$$

The minimum distance $\delta_{\min}^2 = 1$ for any error event of the form

$$\varepsilon(z) = \pm(1 - z^{-1} - z^{-2} \cdots - z^{-(n-1)}) \qquad n \geq 1 \qquad (6.7.48)$$

for the channel given by (6.7.46) since

$$\alpha(z) = \pm \frac{1}{\sqrt{2}} \mp \frac{1}{\sqrt{2}} z^{-n}$$

Similarly, when

$$\varepsilon(z) = \pm(1 + z^{-2} + z^{-4} + \cdots + z^{-2(n-1)}) \qquad n \geq 1 \qquad (6.7.49)$$

$\delta_{\min}^2 = 1$ for the channel given by (6.7.47) since

$$\alpha(z) = \pm \frac{1}{\sqrt{2}} \mp \frac{1}{\sqrt{2}} z^{-2n}$$

Hence MLSE of these two partial response signals results in no loss in SNR. In contrast, the suboptimum symbol-by-symbol detection described previously resulted in a 2.1-dB loss.

The constant $K_{\delta_{\min}}$ is easily evaluated for these two signals. With precoding, the number of output symbol errors (Hamming weight) associated with the error events in (6.7.48) and (6.7.49) is two. Hence

$$K_{\delta_{\min}} = 2 \sum_{n=1}^{\infty} \left(\frac{M-1}{M} \right)^n = 2(M-1) \qquad (6.7.50)$$

On the other hand, without precoding, these error events result in n symbol errors and, hence,

$$K_{\delta_{\min}} = 2 \sum_{n=1}^{\infty} n \left(\frac{M-1}{M} \right)^n = 2M(M-1) \qquad (6.7.51)$$

As a final exercise, we consider the evaluation of $\delta_{\min}^2$ from the quadratic form in (6.7.30). The matrix $\mathbf{A}$ of the quadratic form is positive-definite; hence all its eigenvalues are positive. If $\{\mu_k(\varepsilon)\}$ are the eigenvalues and $\{\mathbf{v}_k(\varepsilon)\}$ are the corresponding orthonormal eigenvectors of $\mathbf{A}$ for an error event ε, then the quadratic form in (6.7.30) can be expressed as

$$\delta^2(\varepsilon) = \sum_{k=1}^{L+1} \mu_k(\varepsilon) [\mathbf{f}' \mathbf{v}_k(\varepsilon)]^2 \qquad (6.7.52)$$

In other words, $\delta^2(\varepsilon)$ is expressed as a linear combination of the squared projections of the channel vector $\mathbf{f}$ onto the eigenvectors of $\mathbf{A}$. Each squared projection in the sum is weighted by the corresponding eigenvalue $\mu_k(\varepsilon)$,

$k = 1, 2, \ldots, L + 1$. Then

$$\delta_{min}^2 = \min_{\varepsilon} \delta^2(\varepsilon) \tag{6.7.53}$$

It is interesting to note that the worst channel characteristic of a given length $L + 1$ can be obtained by finding the eigenvector corresponding to the minimum eigenvalue. Thus, if $\mu_{min}(\varepsilon)$ is the minimum eigenvalue for a given error event ε and $\mathbf{v}_{min}(\varepsilon)$ is the corresponding eigenvector, then

$$\mu_{min} = \min_{\varepsilon} \mu_{min}(\varepsilon)$$

$$\mathbf{f} = \min_{\varepsilon} \mathbf{v}_{min}(\varepsilon)$$

and

$$\delta_{min}^2 = \mu_{min}$$

Example 6.7.4. Let us determine the worst time-dispersive channel of length 3 ($L = 2$) by finding the minimum eigenvalue of $\mathbf{A}$ for different error events. Thus

$$F(z) = f_0 + f_1 z^{-1} + f_2 z^{-2}$$

where f_0, f_1, and f_2 are the components of the eigenvector of $\mathbf{A}$ corresponding to the minimum eigenvalue. An error event of the form

$$\varepsilon(z) = 1 - z^{-1}$$

results in a matrix

$$\mathbf{A} = \begin{bmatrix} 2 & -1 & 0 \\ -1 & 2 & -1 \\ 0 & -1 & 2 \end{bmatrix}$$

which has the eigenvalues $\mu_1 = 2$, $\mu_2 = 2 + \sqrt{2}$, $\mu_3 = 2 - \sqrt{2}$. The eigenvector corresponding to μ_3 is

$$\mathbf{v}_3' = \left[\frac{1}{2}, \frac{1}{\sqrt{2}}, \frac{1}{2} \right] \tag{6.7.54}$$

We may also consider the dual error event

$$\varepsilon(z) = 1 + z^{-1}$$

which results in the matrix

$$\mathbf{A} = \begin{bmatrix} 2 & 1 & 0 \\ 1 & 2 & 1 \\ 0 & 1 & 2 \end{bmatrix}$$

This matrix has the identical eigenvalues as the one for $\varepsilon(z) = 1 - z^{-1}$. The corresponding eigenvector for $\mu_3 = 2 - \sqrt{2}$ is

$$\mathbf{v}_3' = \left[-\frac{1}{2}, \frac{1}{\sqrt{2}}, -\frac{1}{2} \right] \tag{6.7.55}$$

Any other error events lead to larger values for μ_{min}. Hence $\mu_{min} = 2 - \sqrt{2}$ and the worst-case channel is either

$$\left[\frac{1}{2}, \frac{1}{\sqrt{2}}, \frac{1}{2} \right] \quad \text{or} \quad \left[-\frac{1}{2}, \frac{1}{\sqrt{2}}, -\frac{1}{2} \right].$$

TABLE 6.7.1
Maximum performance loss and corresponding channel characteristics

Channel length $(L + 1)$	Performance loss $-10 \log \delta^2_{\min}$, dB	Minimum distance channel
3	2.3	0.50, 0.71, 0.50
4	4.2	0.38, 0.60, 0.60, 0.38
5	5.7	0.29, 0.50, 0.58, 0.50, 0.29
6	7.0	0.23, 0.42, 0.52, 0.52, 0.42, 0.23

The loss in SNR from the channel is

$$-10 \log \delta^2_{\text{min}} = -10 \log \mu_{\min} = 2.3 \text{ dB}$$

Repetition of the above computation for channels with $L = 3, 4, 5$ yield the results given in Table 6.7.1.

We conclude this subsection on the performance of MLSE by comparing its performance against that of a decision-feedback equalizer. For the two-path channel with $f_0 = f_1 = 1/\sqrt{2}$, we have shown that MLSE suffers no SNR loss while the decision-feedback equalizer suffers a 3-dB loss. On channels with more distortion the SNR advantage of MLSE over decision-feedback equalization is even greater. Figure 6.7.3 illustrates a comparison of the error rate performance of these two equalization techniques, obtained via Monte Carlo simulation, for binary PAM and the channel characteristics shown in Fig. 6.4.7b and c. The error rate curves for the two methods have different slopes; hence the difference in SNR increases as the error probability decreases. As a benchmark, the error rate for the AWGN channel with no intersymbol interference is also shown in Fig. 6.7.3.

6.7.2 An Adaptive Channel Estimator

The MLSE criterion implemented via the Viterbi algorithm as embodied in the metric computation given by (6.7.9) and the probabilistic symbol-by-symbol detection algorithm described in Sec. 6.6 require knowledge of the equivalent discrete-time channel coefficients $\{f_k\}$. To accommodate a channel that is unknown or slowly time-varying, one may include a channel estimator connected in parallel with the detection algorithm, as shown in Fig. 6.7.4. The channel estimator, which is shown in Fig. 6.7.5, is identical in structure to the linear transversal equalizer discussed previously in Sec. 6.4. In fact, the channel estimator is a replica of the equivalent discrete-time channel filter that models the intersymbol interference. The estimated tap coefficients, denoted as $\{\hat{f}_k\}$, are adjusted recursively to minimize the MSE between the actual received sequence and the output of the estimator. For example, the steepest-descent algorithm in a

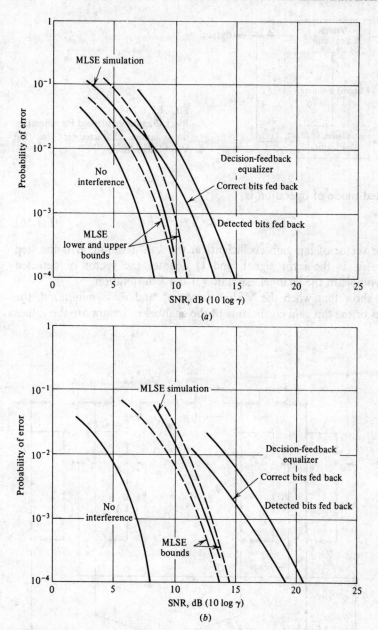

FIGURE 6.7.3
Comparison of performance between MLSE and decision-feedback equalization for channel characteristics shown in Fig. 6.4.7b and c.

FIGURE 6.7.4
Block diagram of method for estimating the channel characteristics for the Viterbi algorithm.

decision-directed mode of operation is

$$\hat{\mathbf{f}}_{k+1} = \hat{\mathbf{f}}_k + \Delta e_k \tilde{\mathbf{I}}_k^* \tag{6.7.56}$$

where $\hat{\mathbf{f}}_k$ is the vector of tap gain coefficients at the kth iteration, Δ is the step size, $e_k = v_k - \hat{v}_k$ is the error signal, and $\tilde{\mathbf{I}}_k$ denotes the vector of detected information symbols in the channel estimator at the kth iteration.

We now show that when the MSE between v_k and $\hat{v}_k$ is minimized, the resulting values of the tap gain coefficients of the channel estimator are the values

FIGURE 6.7.5
Adaptive transversal filter for estimating the channel dispersion.

of the equivalent discrete-time channel model. For mathematical tractability, we assume that the detected information sequence $\{\tilde{I}_k\}$ is correct, i.e., $\{\tilde{I}_k\}$ is identical to the transmitted sequence $\{I_k\}$. This is a reasonable assumption when the system is operating at a low probability of error. Thus the MSE between the received signal v_k and the estimate $\hat{v}_k$ is

$$J(\hat{\mathbf{f}}) = E\left(\left| v_k - \sum_{j=0}^{N-1} \hat{f}_j I_{k-j} \right|^2\right) \tag{6.7.57}$$

The tap coefficients $\{\hat{f}_k\}$ that minimize $J(\hat{\mathbf{f}})$ in (6.7.57) satisfy the set of N linear equations

$$\sum_{j=0}^{N-1} \hat{f}_j \phi_{kj} = d_k \qquad k = 0, 1, \ldots, N-1 \tag{6.7.58}$$

where

$$\phi_{kj} = E(I_k I_j^*) \qquad d_k = \sum_{j=0}^{N-1} f_j \phi_{kj} \tag{6.7.59}$$

From (6.7.58) and (6.7.59) we conclude that as long as the information sequence $\{I_k\}$ is uncorrelated, the optimum coefficients are exactly equal to the respective values of the equivalent discrete-time channel. It is also apparent that when the number of taps N in the channel estimator is greater than or equal to $L+1$, the optimum tap gain coefficients $\{\hat{f}_k\}$ are equal to the respective values of the $\{f_k\}$, even when the information sequence is correlated. Subject to the above conditions, the minimum MSE is simply equal to the noise variance N_0.

In the above discussion the estimated information sequence at the output of the Viterbi algorithm or the probabilistic symbol-by-symbol algorithm was used in making adjustments of the channel estimator. For startup operation one may send a short training sequence to perform the initial adjustment of the tap coefficients, as is usually done in the case of the linear transversal equalizer. In an adaptive mode of operation, the receiver simply uses its own decisions to form an error signal.

6.7.3 MLSE of CPM Signals

In Sec. 6.6 we observed the improvement in performance obtained from demodulation of CPFSK in a manner that exploits the phase continuity inherent in the modulated signal. This was accomplished by cross correlating the received signal over n signaling intervals with the possible transmitted waveforms and forming M decision variables corresponding to the M-ary CPFSK modulation. This detection method is basically a maximum likelihood symbol-by-symbol detection criterion as previously indicated in Sec. 6.6 and applies to general CPM signals.

As an alternative to this symbol-by-symbol detection we may use MLSE utilizing Euclidean metrics in the implementation of the Viterbi algorithm. Since the Viterbi algorithm is the optimum maximum likelihood sequence detector, we expect similar performance improvements to be achieved in the demodulation of CPM. The Viterbi algorithm based on a Euclidean distance metric, as was described in this section, is directly applicable to the detection of CPM signals. Let us first establish the general state trellis structure of CPM and then develop the metric computations. The reader should refer back to Sec. 3.3.2 for the general representation of CPM signals and to Sec. 3.4.2 for their spectral characteristics.

Recall that the carrier phase for a CPM signal may be expressed as

$$\phi(t; \mathbf{I}) = 2\pi h \sum_{k=-\infty}^{n} I_k q(t - hT)$$

$$= \pi h \sum_{k=-\infty}^{n-L} I_k + 2\pi h \sum_{k=n-L+1}^{n} I_k q(t - kT) \qquad nT \leq t \leq (n+1)T$$

$$= \theta_n + \theta(t; \mathbf{I}) \qquad nT \leq t \leq (n+1)T \qquad (6.7.60)$$

where we have assumed that $q(t) = 0$ for $t < 0$, $q(t) = 1/2$ for $t \geq LT$, and

$$q(t) = \int_0^t u(\tau) \, dt \qquad (6.7.61)$$

The signal pulse $u(t) = 0$ for $t < 0$ and $t \geq LT$. For $L = 1$ we have a full response CPM and for $L > 1$, where L is a positive integer, we have a partial response CPM signal.

Now, when h is rational, i.e., $h = m/p$ where m and p are relatively prime positive integers, the CPM scheme can be represented by a trellis. In this case, there are p phase states

$$\Theta_s = \left\{ 0, \frac{\pi m}{p}, \frac{2\pi m}{p}, \dots, \frac{(p-1)\pi m}{p} \right\} \qquad (6.7.62)$$

when m is even and $2p$ phase states

$$\Theta_s = \left\{ 0, \frac{\pi m}{p}, \dots, \frac{(2p-1)\pi m}{p} \right\} \qquad (6.7.63)$$

when m is odd. If $L = 1$, these are the only states in the trellis. On the other hand, if $L > 1$, we have an additional number of states due to the partial response character of the signal pulse $u(t)$. These additional states can be identified by expressing $\theta(t, \mathbf{I})$ given by (6.7.60) as

$$\theta(t; \mathbf{I}) = 2\pi h \sum_{k=n-L+1}^{n-1} I_k q(t - kT) + 2\pi h I_n q(t - nT) \qquad (6.7.64)$$

The first term on the right-hand side of (6.7.64) depends on the information symbols $(I_{n-1}, I_{n-2}, \dots, I_{n-L+1})$, which is called the *correlative state vector*, and represents the phase term corresponding to signal pulses that have not reached their final value. The second term in (6.7.64) represents the phase contribution due to the most recent symbol I_n. Hence, the state of the CPM signal (or the modulator) at time $t = nT$ may be expressed as the combined *phase state* and *correlative state*, denoted as

$$S_n = \{\theta_n, I_{n-1}, I_{n-2}, \ldots, I_{n-L+1}\} \qquad (6.7.65)$$

for a partial response signal pulse of length LT, where $L > 1$. In this case, the number of states is

$$N_s = \begin{cases} pM^{L-1} & m \text{ even} \\ 2pM^{L-1} & m \text{ odd} \end{cases} \qquad (6.7.66)$$

when $h = m/p$.

Now, suppose the state of the modulator at $t = nT$ is S_n. The effect of the new symbol in the time interval $nT \le t \le (n+1)T$ is to change the state from S_n to S_{n+1}. Hence, at $t = (n+1)T$ the state becomes

$$S_{n+1} = (\theta_{n+1}, I_n, I_{n+1}, \ldots, I_{n-L+2})$$

where

$$\theta_{n+1} = \theta_n + \pi h I_{n-L+1}$$

Example 6.7.5. Consider a CPM scheme with a modulation index $h = 3/4$ and a partial response pulse with $L = 2$. Let us determine the states S_n of the CPM scheme and sketch the phase tree and state trellis.

First, we note that there are $2p = 8$ phase states, namely,

$$\Theta_s = \left\{ 0, \pm\frac{\pi}{4}, \pm\frac{\pi}{2}, \pm\frac{3\pi}{4}, \pi \right\}$$

For each of these phase states, there are two states that result from the memory of the CPM scheme. Hence, the total number of states is $N_s = 16$, namely,

$$(0, +1), (0, -1), (\pi, 1), (\pi, -1), \left(\frac{\pi}{4}, 1\right), \left(\frac{\pi}{4}, -1\right), \left(\frac{\pi}{2}, 1\right), \left(\frac{\pi}{2}, -1\right),$$

$$\left(\frac{3\pi}{4}, 1\right), \left(\frac{3\pi}{4}, -1\right), \left(-\frac{\pi}{4}, 1\right), \left(-\frac{\pi}{4}, -1\right), \left(-\frac{\pi}{2}, 1\right), \left(-\frac{\pi}{2}, -1\right),$$

$$\left(-\frac{3\pi}{4}, 1\right), \left(-\frac{3\pi}{4}, -1\right)$$

If the system is in phase state $\theta_n = -\pi/4$ and $I_{n-1} = -1$, then

$$\theta_{n+1} = \theta_n + \pi h I_{n-1}$$

$$= -\pi/4 - \pi(3/4) = -\pi$$

The state trellis is illustrated in Fig. 6.7.6. A path through the state trellis corresponding to the sequence $(1, -1, -1, -1, 1, 1)$ is illustrated in Fig. 6.7.7.

In order to sketch the phase tree, we must know the signal pulse shape $u(t)$. Figure 6.7.8 illustrates the phase tree when $u(t)$ is a rectangular pulse of duration $2T$.

Having established the state trellis representation of CPM, let us now consider the metric computations performed in the Viterbi algorithm.

(θ_n, I_{n-1})

FIGURE 6.7.6
State trellis for partial response ($L = 2$) CPM with $h = 3/4$.

Metric computations. By referring back to the mathematical development for the derivation of the maximum likelihood demodulator given in Sec. 6.3, it is easy to show that the logarithm of the probability of the observed signal $r(t)$ conditioned on a particular sequence of transmitted symbols $\mathbf{I}$ is proportional to the cross correlation

$$J_n(\mathbf{I}) = \int_{-\infty}^{(n+1)T} r(t) \cos\left[\omega_c t + \phi(t; \mathbf{I})\right] dt$$

$$= J_{n-1}(\mathbf{I}) + \int_{nT}^{(n+1)T} r(t) \cos\left[\omega_c t + \theta(t; \mathbf{I}) + \theta_n\right] dt \quad (6.7.67)$$

The term $J_{n-1}(\mathbf{I})$ represents the metrics for the surviving sequences up to time

FIGURE 6.7.7
A signal signal path through the trellis.

nT, and the term

$$v_n(\mathbf{I}; \theta_n) = \int_{nT}^{(n+1)T} r(t) \cos\left[\omega_c t + \theta(t, \mathbf{I}) + \theta_n\right] dt \qquad (6.7.68)$$

represents the additional increments to the metrics contributed by the signal in the time interval $nT \leq t \leq (n+1)T$. Note that there are M^L possible sequences $\mathbf{I} = (I_n, I_{n-1}, \ldots, I_{n-L+1})$ of symbols and p (or $2p$) possible phase states $\{\theta_n\}$. Therefore, there are pM^L (or $2pM^L$) different values of $v_n(\mathbf{I}, \theta_n)$, computed in each signal interval, and each value is used to increment the metrics correspond-

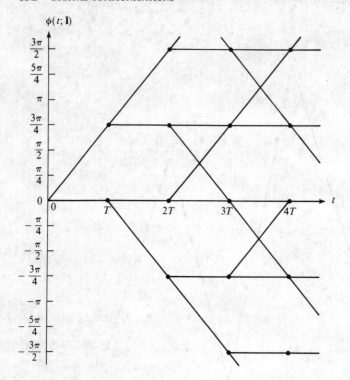

FIGURE 6.7.8
Phase tree for $L = 2$ partial response CPM with $h = 3/4$.

ing to the pM^{L-1} surviving sequences from the previous signaling interval. A general block diagram that illustrates the computations of $v_n(\mathbf{I}; \theta_n)$ for the Viterbi decoder is shown in Fig. 6.7.9.

Note that the number of surviving sequences at each state of the Viterbi decoding process is pM^{L-1} (or $2pM^{L-1}$). For each surviving sequence we have M new increments of $v_n(\mathbf{I}; \Theta_n)$ that are added to the existing metric to yield pM^L (or $2pM^L$) sequences with pM^L (or $2pM^L$) metrics. However, this number is then reduced back to pM^{L-1} (or $2pM^{L-1}$) survivors with corresponding metrics by selecting the most probable sequence of the M sequences merging at each node of the trellis and discarding the other $M - 1$ sequences.

Performance of CPM signals. In evaluating the performance of CPM signals achieved with MLSE, we must determine the minimum Euclidean distance of paths through the trellis that separate at the node at $t = 0$ and remerge at a later time at the same mode. The distance between two paths through the trellis is related to the corresponding signals as we now demonstrate.

Suppose that we have two signals $s_i(t)$ and $s_j(t)$ corresponding to two phase trajectories $\phi(t; \mathbf{I}_i)$ and $\phi(t; \mathbf{I}_j)$. The sequences $\mathbf{I}_i$ and $\mathbf{I}_j$ must be different

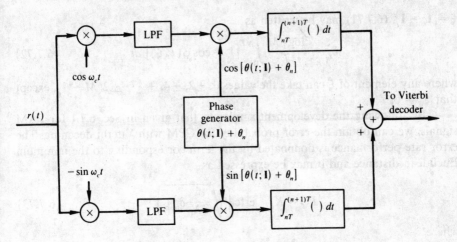

FIGURE 6.7.9
Computation of metric increments $v_n(\mathbf{I}; \theta_n)$.

in their first symbol. Then, the Euclidean distance between the two signals over an interval of length NT, where $1/T$ is the symbol rate, is defined as

$$
\begin{aligned}
d_{ij}^2 &= \int_0^{NT} \left[s_i(t) - s_j(t) \right]^2 dt \\
&= \int_0^{NT} s_i^2(t)\, dt + \int_0^{NT} s_j^2(t)\, dt - 2\int_0^{NT} s_i(t) s_j(t)\, dt \\
&= 2N\mathscr{E} - 2\left(\frac{2\mathscr{E}}{T}\right) \int_0^{NT} \cos\left[\omega_c t + \phi(t; \mathbf{I}_i)\right] \cos\left[\omega_c t + \phi(t; \mathbf{I}_j)\right] dt \\
&= 2N\mathscr{E} - \left(\frac{2\mathscr{E}}{T}\right) \int_0^{NT} \cos\left[\phi(t; \mathbf{I}_i) - \phi(t; \mathbf{I}_j)\right] dt \\
&= \frac{2\mathscr{E}}{T} \int_0^{NT} \left\{ 1 - \cos\left[\phi(t; \mathbf{I}_i) - \phi(t; \mathbf{I}_j)\right] \right\} dt \qquad (6.7.69)
\end{aligned}
$$

Hence the Euclidean distance is related to the phase difference between the paths in the state trellis according to (6.7.69).

It is desirable to express the distance d_{ij}^2 in terms of the bit energy. Since $\mathscr{E} = \mathscr{E}_b \log_2 M$, (6.7.69) may be expressed as

$$
d_{ij}^2 = 2\mathscr{E}_b \delta_{ij}^2 \qquad (6.7.70)
$$

where δ_{ij}^2 is defined as

$$
\delta_{ij}^2 = \frac{\log_2 M}{T} \int_0^{NT} \left\{ 1 - \cos\left[\phi(t; \mathbf{I}_i) - \phi(t; \mathbf{I}_j)\right] \right\} dt \qquad (6.7.71)
$$

Furthermore, we observe that $\phi(t; \mathbf{I}_i) - \phi(t; \mathbf{I}_j) = \phi(t; \mathbf{I}_i - \mathbf{I}_j)$, so that, with

$\xi = \mathbf{I}_i - \mathbf{I}_j$, (6.7.71) may be written as

$$\delta_{ij}^2 = \frac{\log_2 M}{T} \int_0^{NT} [1 - \cos \phi(t; \xi)] \, dt \qquad (6.7.72)$$

where any element of ξ can take the values $0, \pm 2, \pm 4, \pm \cdots \pm 2(M - 1)$, except that $\xi_0 \neq 0$.

By following the development similar to that given in Sec. 6.7.1 for PAM signals, we can obtain the error probability for CPM with Viterbi decoding. The error rate performance is dominated by the term corresponding to the minimum Euclidean distance and it may be expressed as

$$P_M = K_{\delta_{\min}} \, \text{erfc}\left(\sqrt{\frac{\mathcal{E}_b}{2 N_0} \delta_{\min}^2} \right) \qquad (6.7.73)$$

where

$$\delta_{\min}^2 = \lim_{N \to \infty} \min_{i, j} \delta_{ij}^2$$

$$= \lim_{N \to \infty} \min_{i, j} \left\{ \frac{\log_2 M}{T} \int_0^{NT} [1 - \cos \phi(t; \mathbf{I}_i - \mathbf{I}_j)] \, dt \right\} \qquad (6.7.74)$$

We note that for conventional binary PSK with no memory, $N = 1$ and $\delta_{\min}^2 = \delta_{12}^2 = 2$. Hence, (6.7.73) agrees with our previous result.

Since $\delta_{\min}^2$ characterizes the performance of CPM with MLSE, we can investigate the effect on $\delta_{\min}^2$ resulting from varying the alphabet size M, the modulation index h, and the length of the transmitted pulse in partial response CPM.

First, we consider full response $(L = 1)$ CPM. If we take $M = 2$ as a beginning, we note that the sequences

$$\mathbf{I}_i = +1, -1, I_2, I_3$$
$$\mathbf{I}_j = -1, +1, I_2, I_3 \qquad (6.7.75)$$

which differ for $k = 0, 1$ and agree for $k \geq 2$, result in two phase trajectories that merge after the second symbol. This corresponds to the difference sequences

$$\xi = \{2, -2, 0, 0, \dots\} \qquad (6.7.76)$$

The Euclidean distance for this sequence is easily calculated from (6.7.72) and provides an upper bound on $\delta_{\min}^2$. This upper bound for $M = 2$ is

$$d_B^2(h) = 2\left(1 - \frac{\sin 2\pi h}{2\pi h}\right) \qquad M = 2 \qquad (6.7.77)$$

For example, where $h = 1/2$, which corresponds to MSK, we have $d_B^2(1/2) = 2$, so that $\delta_{\min}^2(1/2) \leq 2$.

For $M > 2$ and full response CPM it is also easily seen that phase trajectories merge at $t = 2T$. Hence, an upper bound on $\delta_{\min}^2$ can be obtained by considering the phase difference sequence $\xi = \{\alpha, -\alpha, 0, 0, \dots\}$ where

FIGURE 6.7.10
The upper bound d_B^2 as a function of the modulation index h for full response CPM with rectangular pulses. [From Aulin and Sundberg (1984), © 1984, John Wiley Ltd. Reprinted with permission of the publisher.]

$\alpha = \pm 2, \pm 4, \ldots, \pm 2(M - 1)$. This sequence yields the upper bound

$$d_B^2(h) = \min_{1 \le k \le M-1} \left\{ (2 \log_2 M)\left(1 - \frac{\sin 2k\pi h}{2k\pi h}\right) \right\} \qquad (6.7.78)$$

The graphs of $d_B^2(h)$ versus h for $M = 2, 4, 8, 16$ are shown in Fig. 6.7.10. It is apparent from these graphs that large gains in performance can be achieved by increasing the alphabet size M. It must be remembered, however, that $\delta_{min}^2(h) \le d_B^2(h)$. That is, the upper bound may not be achievable for all values of h.

The minimum Euclidean distance $\delta_{min}^2(h)$ has been determined, by evaluating (6.7.72), for a variety of CPM signals by Aulin and Sundberg (1981). For example, Fig. 6.7.11 illustrates the dependence of the Euclidean distance for

$d^2(h)$

$N = 4$

MSK $\quad$ $N = 3$

$d_B^2(h)$

$N = 2$

$N = 1$

4 $\quad$ 3

2

FIGURE 6.7.11
Squared minimum Euclidean distance as a function of the modulation index for binary CPFSK. The upper bound is d_B^2. [From paper by Aulin and Sundberg (1981). © 1981 IEEE. Reprinted with permission.]

binary CPFSK as a function of the modulation index h, with the number N of bit observation (decision) intervals ($N = 1, 2, 3, 4$) as a parameter. Also shown is the upper bound $d_B^2(h)$ given by (6.7.77). In particular, we note that when $h = 1/2$, then $\delta_{\min}^2(1/2) = 2$, which is the same squared distance as PSK (binary or quaternary) with $N = 1$. On the other hand, the required observation interval for MSK is $N = 2$ intervals for which we have $\delta_{\min}^2(1/2) = 2$. Hence, the performance of MSK with MLSE is comparable to (binary or quaternary) PSK as we have previously observed.

We also note from Fig. 6.7.11 that the optimum modulation index for binary CPFSK is $h = 0.715$ when the observation interval is $N = 3$. This yields $\delta_{\min}^2(0.715) = 2.43$, or a gain of 0.85 dB relative to MSK.

Figure 6.7.12 illustrates the Euclidean distance as a function of h for $M = 4$ CPFSK, with the length of the observation interval N as a parameter. Also shown (as a dashed line where it is not reached) is the upper bound d_B^2 evaluated from (6.7.78). We note that $\delta_{\min}^2$ achieves the upper bound for several values of h for some N. In particular, note that the maximum value of d_B^2 which occurs at $h \approx 0.9$ is approximately reached for $N = 8$ observed symbol intervals.

FIGURE 6.7.12
Squared minimum Euclidean distance as a function of the modulation index for quaternary CPFSK. The upper bound is d_B^2. [From Aulin and Sundberg (1981). © 1981 IEEE. Reprinted with permission.]

The true maximum is achieved at $h = 0.914$ with $N = 9$. For this case, $\delta_{min}^2(0.914) = 4.2$, which represents a 3.2-dB gain over MSK. Also note that the Euclidean distance contains minima at $h = 1/3, 1/2, 2/3, 1$, etc. These values of h are called *weak modulation indices* and should be avoided. Similar results are available for larger values of M, and may be found in the paper by Aulin and Sundberg (1981) and the text by Anderson et al. (1986).

Large performance gains can also be achieved with MLSE of CPM by using partial response signals. For example, the distance bound $d_B^2(h)$ for partial

FIGURE 6.7.13
Upper bound d_B^2 on the minimum distance for partial response (raised cosine pulse) binary CPM. [From paper by Sundberg (1986). © 1986 IEEE. Reprinted with permission.]

response, raised cosine pulses given by

$$u(t) = \begin{cases} \dfrac{1}{2LT}\left(1 - \cos \dfrac{2\pi t}{LT}\right) & 0 \le t \le LT \\ 0 & \text{otherwise} \end{cases} \tag{6.7.79}$$

is shown in Fig. 6.7.13 for $M = 2$. Here, we note that as L increases, d_B^2 also achieves higher values. Clearly, the performance of CPM improves as the correlative memory L increases, but h must also be increased in order to achieve the larger values of d_B^2. Since a larger modulation index implies a larger bandwidth (for fixed L), while a larger memory length L (for fixed h) implies a smaller bandwidth, it is better to compare the Euclidean distance as a function of the normalized bandwidth $2WT_b$, where W is the 99 percent power bandwidth and T_b is the bit interval. Figure 6.7.14 illustrates this type of comparison with MSK used as a point of reference (0 dB). We note from this figure that there are

FIGURE 6.7.14

Power bandwidth tradeoff for partial response CPM signals with raised cosine pulses. W is the 99 percent in-band power bandwidth. [From paper by Sundberg (1986). © 1986 IEEE. Reprinted with permission.]

several decibels to be gained by using partial response signals and higher signaling alphabets. The major price to be paid for this performance gain is the added exponentially increasing complexity in the implementation of the Viterbi decoder.

The performance results shown in Fig. 6.7.14 illustrate that 3- to 4-dB gain relative to MSK can be easily obtained with relatively no increase in bandwidth by use of raised cosine partial response CPM and $M = 4$. Although these results are for raised cosine signal pulses, similar gains can be achieved with other partial response pulse shapes. We emphasize that this gain in SNR is achieved by introducing memory into the signal modulation and exploiting the memory in the demodulation of the signal. No redundancy through coding has been introduced. In effect the code has been built into the modulation and the trellis-type (Viterbi) decoding exploits the phase constraints in the CPM signal.

Additional gains in performance can be achieved by introducing additional redundancy through coding and increasing the alphabet size as a means of maintaining a fixed bandwidth. In particular trellis-coded CPM using relatively simple convolutional codes has been thoroughly investigated and many results are available in the technical literature. The Viterbi decoder for the convolutionally encoded CPM signal now exploits the memory inherent in the code and in the CPM signal. Performance gains of the order of 4 to 6 dB, relative to uncoded MSK with the same bandwidth, have been demonstrated by combining convolutional coding with CPM. Extensive numerical results for coded CPM are given by Lindell (1985).

Multi-h CPM. By varying the modulation index from one signaling interval to another it is possible to increase the minimum Euclidean distance δ_{min}^2 between pairs of phase trajectories and, thus, improve the performance gain over constant-h CPM. Usually, multi-h CPM employs a fixed number H of modulation indices that are varied cyclically in successive signaling intervals. Thus, the phase of the signal varies piecewise linearly.

Significant gains in SNR are achievable by using only a small number of different values of h. For example, with full response ($L = 1$) CPM and $H = 2$, it is possible to obtain a gain of 3 dB relative to binary or quaternary PSK. By increasing H to $H = 4$, a gain of 4.5 dB relative to PSK can be obtained. The performance gain can also be increased with an increase in the signal alphabet. Table 6.7.2 lists the performance gains achieved with $M = 2, 4, 8$ for several

TABLE 6.7.2
Maximum values of the upper bound d_B^2 for multi-h linear phase CPM†

M	H	Max d_B^2	dB gain compared to MSK	h_1	h_2	h_3	h_4	$\bar{h}$
2	1	2.43	0.85	0.715				0.715
2	2	4.0	3.0	0.5	0.5			0.5
2	3	4.88	3.87	0.620	0.686	0.714		0.673
2	4	5.69	4.54	0.73	0.55	0.73	0.55	0.64
4	1	4.23	3.25	0.914				0.914
4	2	6.54	5.15	0.772	0.772			0.772
4	3	7.65	5.83	0.795	0.795	0.795		0.795
8	1	6.14	4.87	0.964				0.964
8	2	7.50	5.74	0.883	0.883			0.883
8	3	8.40	6.23	0.879	0.879	0.879		0.879

† From Aulin and Sundberg (1982b). Reprinted with permission from the IEEE.

FIGURE 6.7.15
Upper bounds on minimum squared Euclidean distance for various M and H values. [From Aulin and Sundberg (1982b). © 1982 IEEE. Reprinted with permission.]

values of H. The upper bounds on the minimum Euclidean distance are also shown in Fig. 6.7.15 for several values of M and H. Note that the major gain in performance is obtained when H is increased from $H = 1$ to $H = 2$. For $H > 2$, the additional gain is relatively small for small values of $\{h_i\}$. On the other hand, significant performance gains are achieved by increasing the alphabet size M.

The results shown above hold for full response CPM. One can also extend the use of multi-h CPM to partial response in an attempt to further improve performance. It is anticipated that such schemes will yield some additional performance gains, but numerical results on partial response, multi-h CPM are limited. The interested reader is referred to the paper by Aulin and Sundberg (1982b).

Multiamplitude CPM. Multiamplitude CPM (MACPM) is basically a combined amplitude and phase digital modulation scheme that allows us to increase the signaling alphabet relative to CPM in another dimension and, thus, to achieve higher data rates on a band-limited channel. Simultaneously, the combination of multiple amplitude in conjunction with CPM results in a bandwidth-efficient modulation technique.

We have already observed the spectral characteristics of MACPM in Sec. 3.4. The performance characteristics of MACPM have been investigated by Mulligan (1988) for both uncoded and trellis-coded CPM. Of particular interest is the result that trellis-coded CPM with two amplitude levels achieves a gain of 3 to 4 dB relative to MSK without a significant increase in the signal bandwidth.

6.8 RECURSIVE LEAST-SQUARES ALGORITHMS FOR ADAPTIVE EQUALIZATION

The derivation of the algorithms for adjusting the coefficients of the linear equalizer and the decision-feedback equalizer to minimize the MSE was based on a statistical approach. The MSE was defined as the expected value of the squared difference between the desired information symbol and the estimated information symbol. The minimization of this performance index resulted in a set of linear equations for the coefficients which involved the statistical autocorrelation function of the received signal components and the cross-correlation function between the desired symbol and the received signal components. The solution of the linear equations was obtained by a steepest-descent (gradient) algorithm in which the true gradient vector was approximated by an estimate obtained directly from the data.

The major advantage of the steepest-descent algorithm lies in its computational simplicity. However, the price paid for the simplicity is slow convergence, especially when the channel characteristics result in an autocorrelation matrix Γ whose eigenvalues have a large spread, i.e., $\lambda_{max}/\lambda_{min} \gg 1$. Viewed in another way, the gradient algorithm has only a single adjustable parameter for controlling the convergence rate, namely, the parameter Δ. Consequently the slow convergence is due to this fundamental limitation.

In order to obtain faster convergence, it is necessary to devise more complex algorithms which involve additional parameters. In particular, if the matrix Γ is $N \times N$ and has eigenvalues $\lambda_1, \lambda_2, \ldots, \lambda_N$, we may use an algorithm that contains N parameters, one for each of the eigenvalues. The optimum selection of these parameters to achieve rapid convergence is a topic of this section.

In deriving faster converging algorithms we shall adopt a least-squares approach rather than the statistical approach used previously. That is, we shall deal directly with the received data in minimizing the quadratic performance index, whereas previously we minimized the expected value of the squared error. Put simply, this means that the performance index is expressed in terms of a time average instead of a statistical average.

It is convenient to express the recursive least-squares algorithms in matrix form. Hence we shall define a number of vectors and matrices that are needed in this development. In so doing, we shall change the notation slightly. Specifically, the estimate of the information symbol at time t, where t is an integer, from a linear equalizer is now expressed as

$$\hat{I}(t) = \sum_{j=-K}^{K} c_j(t-1)v_{t-j}$$

By changing the index j on $c_j(t-1)$ to run from $j=0$ to $j=N-1$ and simultaneously defining

$$y(t) = v_{t+K}$$

the estimate $\hat{I}(t)$ becomes

$$\hat{I}(t) = \sum_{j=0}^{N-1} c_j(t-1)y(t-j)$$

$$= \mathbf{C}'_N(t-1)\mathbf{Y}_N(t) \qquad (6.8.1)$$

where $\mathbf{C}_N(t-1)$ and $\mathbf{Y}_N(t)$ are, respectively, the column vectors of the equalizer coefficients $c_j(t-1)$, $j=0,1,\ldots,N-1$, and the input signals $y(t-j)$, $j=0,1,2,\ldots,N-1$.

Similarly, in the decision-feedback equalizer we have tap coefficients $c_j(t)$, $j=0,1,\ldots,N-1$, where the first K_1+1 are the coefficients of the feedforward filter and the remaining $K_2 = N - K_1 - 1$ are the coefficients of the feedback filter. The data in the estimate $\hat{I}(t)$ is $v_{t+K_1},\ldots,v_{t+1},v_t,\tilde{I}_{t-1},\ldots,\tilde{I}_{t-K_2}$, where $\tilde{I}_{t-j}$, $1 \le j \le K_2$ denote the decisions on previously detected symbols. In this development, we neglect the effect of decision errors in the algorithms. Hence we assume that $\tilde{I}_{t-j} = I_{t-j}$, $1 \le j \le K_2$. For notational convenience we also define

$$y(t-j) = \begin{cases} v_{t+K_1-j} & 0 \le j \le K_1 \\ I_{t+K_1-j} & K_1 < j \le N-1 \end{cases} \qquad (6.8.2)$$

Thus

$$\mathbf{Y}_N(t) = [y(t)\,y(t-1)\cdots y(t-N+1)]$$

$$= (v_{t+K_1} \cdots v_{t+1}\,v_t\,I_{t-1} \cdots I_{t-K_2}) \qquad (6.8.3)$$

6.8.1 Recursive Least-Squares (Kalman) Algorithm

The recursive least-squares (RLS) estimation of $\hat{I}(t)$ may be formulated as follows. Suppose we have observed the vectors $\mathbf{Y}_N(n)$, $n=0,1,\ldots,t$, and we wish to determine the coefficient vector $\mathbf{C}_N(t)$ of the equalizer (linear or decision-feedback) that minimizes the time-average weighted squared error

$$\mathscr{E}_N = \sum_{n=0}^{t} w^{t-n} |e_N(n, t)|^2 \tag{6.8.4}$$

where the error is defined as

$$e_N(n, t) = I(n) - \mathbf{C}_N'(t)\mathbf{Y}_N(n) \tag{6.8.5}$$

and w represents a weighting factor $0 < w < 1$. Thus we introduce exponential weighting into past data which is appropriate when the channel characteristics are time-variant. The minimization of $\mathscr{E}_N$ with respect to the coefficient vector $\mathbf{C}_N(t)$ yields the set of linear equations

$$\mathbf{R}_N(t)\mathbf{C}_N(t) = \mathbf{D}_N(t) \tag{6.8.6}$$

where $\mathbf{R}_N(t)$ is the signal correlation matrix defined as

$$\mathbf{R}_N(t) = \sum_{n=0}^{t} w^{t-n} \mathbf{Y}_N^*(n)\mathbf{Y}_N'(n) \tag{6.8.7}$$

and $\mathbf{D}_N(t)$ is the cross-correlation vector

$$\mathbf{D}_N(t) = \sum_{n=0}^{t} w^{t-n} I(n)\mathbf{Y}_N^*(n) \tag{6.8.8}$$

The solution of (6.8.6) is

$$\mathbf{C}_N(t) = \mathbf{R}_N^{-1}(t)\mathbf{D}_N(t) \tag{6.8.9}$$

The matrix $\mathbf{R}_N(t)$ is akin to the statistical autocorrelation matrix Γ_N while the vector $\mathbf{D}_N(t)$ is akin to the cross-correlation vector ξ_N, defined previously. We emphasize, however, that $\mathbf{R}_N(t)$ is not a Toeplitz matrix. We also should mention that for small values of t, $\mathbf{R}_N(t)$ may be ill conditioned; hence it is customary to add initially the matrix $\delta\mathbf{I}_N$ to $\mathbf{R}_N(t)$, where δ is a small positive constant and $\mathbf{I}_N$ is the identity matrix. With exponential weighting into the past, the effect of adding $\delta\mathbf{I}_N$ dissipates with time.

Now suppose we have the solution (6.8.9) for time $t - 1$, i.e., $\mathbf{C}_N(t - 1)$, and we wish to compute $\mathbf{C}_N(t)$. It is inefficient and, hence, impractical to solve the set of N linear equations for each new signal component that is received. To avoid this, we proceed as follows. First, $\mathbf{R}_N(t)$ may be computed recursively as

$$\mathbf{R}_N(t) = w\mathbf{R}_N(t - 1) + \mathbf{Y}_N^*(t)\mathbf{Y}_N'(t) \tag{6.8.10}$$

We call (6.8.10) the *time-update equation* for $\mathbf{R}_N(t)$.

Since the inverse of $\mathbf{R}_N(t)$ is needed in (6.8.9), we use the matrix inverse identity

$$\mathbf{R}_N^{-1}(t) = \frac{1}{w}\left[\mathbf{R}_N^{-1}(t - 1) - \frac{\mathbf{R}_N^{-1}(t - 1)\mathbf{Y}_N^*(t)\mathbf{Y}_N'(t)\mathbf{R}_N^{-1}(t - 1)}{w + \mathbf{Y}_N'(t)\mathbf{R}_N^{-1}(t - 1)\mathbf{Y}_N^*(t)}\right] \tag{6.8.11}$$

Thus $\mathbf{R}_N^{-1}(t)$ may be computed recursively according to (6.8.11).

For convenience, we define $\mathbf{P}_N(t) = \mathbf{R}_N^{-1}(t)$. It is also convenient to define an N-dimensional vector, called the *Kalman gain vector*, as

$$\mathbf{K}_N(t) = \frac{1}{w + \mu_N(t)} \mathbf{P}_N(t-1)\mathbf{Y}_N^*(t) \qquad (6.8.12)$$

where $\mu_N(t)$ is a scalar defined as

$$\mu_N(t) = \mathbf{Y}_N'(t)\mathbf{P}_N(t-1)\mathbf{Y}_N^*(t) \qquad (6.8.13)$$

With these definitions, (6.8.11) becomes

$$\mathbf{P}_N(t) = \frac{1}{w}\left[\mathbf{P}_N(t-1) - \mathbf{K}_N(t)\mathbf{Y}_N'(t)\mathbf{P}_N(t-1)\right] \qquad (6.8.14)$$

Suppose we postmultiply both sides of (6.8.14) by $\mathbf{Y}_N^*(t)$. Then

$$\mathbf{P}_N(t)\mathbf{Y}_N^*(t) = \frac{1}{w}\left[\mathbf{P}_N(t-1)\mathbf{Y}_N^*(t) - \mathbf{K}_N(t)\mathbf{Y}_N'(t)\mathbf{P}_N(t-1)\mathbf{Y}_N^*(t)\right]$$

$$= \frac{1}{w}\left\{[w + \mu_N(t)]\mathbf{K}_N(t) - \mathbf{K}_N(t)\mu_N(t)\right\}$$

$$= \mathbf{K}_N(t) \qquad (6.8.15)$$

Therefore the Kalman gain vector may also be defined as $\mathbf{P}_N(t)\mathbf{Y}_N(t)$.

Now we use the matrix inversion identity to derive an equation for obtaining $\mathbf{C}_N(t)$ from $\mathbf{C}_N(t-1)$. Since

$$\mathbf{C}_N(t) = \mathbf{P}_N(t)\mathbf{D}_N(t)$$

and

$$\mathbf{D}_N(t) = w\mathbf{D}_N(t-1) + I(t)\mathbf{Y}_N^*(t) \qquad (6.8.16)$$

we have

$$\mathbf{C}_N(t) = \frac{1}{w}\left[\mathbf{P}_N(t-1) - \mathbf{K}_N(t)\mathbf{Y}_N'(t)\mathbf{P}_N(t-1)\right]\left[w\mathbf{D}_N(t-1) + I(t)\mathbf{Y}_N^*(t)\right]$$

$$= \mathbf{P}_N(t-1)\mathbf{D}_N(t-1) + \frac{1}{w}I(t)\mathbf{P}_N(t-1)\mathbf{Y}_N^*(t)$$

$$\quad - \mathbf{K}_N(t)\mathbf{Y}_N'(t)\mathbf{P}_N(t-1)\mathbf{D}_N(t-1)$$

$$\quad - \frac{1}{w}I(t)\mathbf{K}_N(t)\mathbf{Y}_N'(t)\mathbf{P}_N(t-1)\mathbf{Y}_N^*(t)$$

$$= \mathbf{C}_N(t-1) + \mathbf{K}_N(t)\left[I(t) - \mathbf{Y}_N'(t)\mathbf{C}_N(t-1)\right] \qquad (6.8.17)$$

Note that $\mathbf{Y}_N'(t)\mathbf{C}_N(t-1)$ is the output of the equalizer at time t, i.e.,

$$\hat{I}(t) = \mathbf{Y}_N'(t)\mathbf{C}_N(t-1) \qquad (6.8.18)$$

and

$$e_N(t, t-1) = I(t) - \hat{I}(t) \equiv e_N(t) \qquad (6.8.19)$$

is the error between desired symbol and the estimate. Hence $\mathbf{C}_N(t)$ is updated recursively according to the relation

$$\mathbf{C}_N(t) = \mathbf{C}_N(t-1) + \mathbf{K}_N(t)e_N(t) \qquad (6.8.20)$$

The residual MSE resulting from this optimization is

$$\mathcal{E}_{N\min} = \sum_{n=0}^{t} w^{t-n}|I(n)|^2 - \mathbf{C}_N'(t)\mathbf{D}_N^*(t) \qquad (6.8.21)$$

To summarize, suppose we have $\mathbf{C}_N(t-1)$ and $\mathbf{P}_N(t-1)$. When a new signal component is received, we have $\mathbf{Y}_N(t)$. Then the recursive computation for the time update of $\mathbf{C}_N(t)$ and $\mathbf{P}_N(t)$ proceeds as follows:

Compute output: $\qquad \hat{I}(t) = \mathbf{Y}_N'(t)\mathbf{C}_N(t-1)$

Compute error: $\qquad e_N(t) = I(t) - \hat{I}(t)$

Compute Kalman gain vector:

$$\mathbf{K}_N(t) = \frac{\mathbf{P}_N(t-1)\mathbf{Y}_N^*(t)}{w + \mathbf{Y}_N'(t)\mathbf{P}_N(t-1)\mathbf{Y}_N^*(t)}$$

Update inverse of the correlation matrix:

$$\mathbf{P}_N(t) = \frac{1}{w}\left[\mathbf{P}_N(t-1) - \mathbf{K}_N(t)\mathbf{Y}_N'(t)\mathbf{P}_N(t-1)\right]$$

Update coefficients:

$$\begin{aligned}
\mathbf{C}_N(t) &= \mathbf{C}_N(t-1) + \mathbf{K}_N(t)e_N(t) \\
&= \mathbf{C}_N(t-1) + \mathbf{P}_N(t)\mathbf{Y}^*(t)e_N(t) \qquad (6.8.22)
\end{aligned}$$

The algorithm described by (6.8.22) is called the *RLS direct form* or *Kalman algorithm*. It is appropriate when the equalizer has a transversal (direct-form) structure.

We note that the equalizer coefficients change with time by an amount equal to the error $e_N(t)$ multiplied by the Kalman gain vector $\mathbf{K}_N(t)$. Since $\mathbf{K}_N(t)$ is N-dimensional, each tap coefficient in effect is controlled by one of the elements of $\mathbf{K}_N(t)$. Consequently rapid convergence is obtained. In contrast, the steepest-descent algorithm, expressed in our present notation, is

$$\mathbf{C}_N(t) = \mathbf{C}_N(t-1) + \Delta\mathbf{Y}_N^*(t)e_N(t) \qquad (6.8.23)$$

and the only variable parameter is the step size Δ.

Figure 6.8.1 illustrates the initial convergence rate of these two algorithms for a channel with fixed parameters $f_0 = 0.26$, $f_1 = 0.93$, $f_2 = 0.26$, and a linear equalizer which has 11 taps. The eigenvalue ratio for this channel is $\lambda_{\max}/\lambda_{\min} = 11$. All the equalizer coefficients were initialized to zero. The steepest-descent algorithm was implemented with $\Delta = 0.020$. The superiority of the Kalman algorithm is clearly evident. This is especially important in tracking a time-variant channel. For example, the time variations in the characteristics of an (iono-

FIGURE 6.8.1
Comparison of convergence rate for the Kalman and gradient algorithms.

spheric) high-frequency (HF) radio channel are too rapid to be equalized by the gradient algorithm, but the Kalman algorithm adapts sufficiently rapidly to track such variations.

In spite of its superior tracking performance, the Kalman algorithm described above has two disadvantages. One is its complexity. The second is its sensitivity to roundoff noise that accumulates due to the recursive computations. The latter may cause instabilities in the algorithm.

The number of computations or operations (multiplications, divisions, and subtractions) in computing the variables in (6.8.22) is proportional to N^2. Most of these operations are involved in the updating of $\mathbf{P}_N(t)$. This part of the computation is also susceptible to roundoff noise. To remedy that problem, algorithms have been developed which avoid the computation of $\mathbf{P}_N(t)$ according to (6.8.22). The basis of these algorithms lies in the decomposition of $\mathbf{P}_N(t)$ in the form

$$\mathbf{P}_N(t) = \mathbf{S}_N(t)\Lambda_N(t)\mathbf{S}'_N(t) \tag{6.8.24}$$

where $\mathbf{S}_N(t)$ is a lower triangular matrix and $\Lambda_N(t)$ is a diagonal matrix. Such a decomposition is called a *Cholesky factorization* or a *square-root factorization* [for reference, see Bierman (1977)]. The Cholesky factorization is described in Appendix 6A. In a square-root algorithm, $\mathbf{P}_N(t)$ is not updated as in (6.8.22) nor is it computed. Instead, the time updating is performed on $\mathbf{S}_N(t)$ and $\Lambda_N(t)$.

Square-foot algorithms are frequently used in control systems applications in which Kalman filtering is involved. In digital communications, the square-root Kalman algorithm has been implemented in a decision-feedback-equalized PSK modem designed to transmit at high speed over HF radio channels with a nominal 3-kHz bandwidth. This algorithm is described in the paper by Hsu (1982). It has a computational complexity of $1.5N^2 + 6.5N$ (complex-valued multiplications and divisions per output symbol). It is also numerically stable and exhibits good numerical properties as we will demonstrate below. For a detailed discussion of square-root algorithms in sequential estimation, the reader is referred to the book by Bierman (1977).

6.8.2 Linear Prediction and the Lattice Filter

The least-squares lattice algorithm to be described in Sec. 6.8.3 is intimately related to the topic of linear prediction of a signal, which was described in Chap. 2 in the context of speech encoding. In this section we shall establish the connection between linear prediction and a lattice filter.

The linear prediction problem may be stated as follows. Given a set of data $y(t-1), y(t-2), \ldots, y(t-p)$, predict the value of the next data point $y(t)$. The predictor of order p is

$$\hat{y}(t) = \sum_{k=1}^{p} a_{pk} y(t-k) \tag{6.8.25}$$

Minimization of the MSE, defined as

$$\mathscr{E}_p = E[y(t) - \hat{y}(t)]^2$$

$$= E\left[y(t) - \sum_{k=1}^{p} a_{pk} y(t-k)\right]^2 \tag{6.8.26}$$

with respect to the predictor coefficients $\{a_{pk}\}$ yields the set of linear equations

$$\sum_{k=1}^{p} a_{pk} \phi(k-l) = \phi(l) \qquad l = 1, 2, \ldots, p \tag{6.8.27}$$

where

$$\phi(l) = E[y(t) y(t+l)]$$

These are called the *normal equations* or the *Yule–Walker equations*.

The matrix Φ with elements $\phi(k-l)$ is a Toeplitz matrix and, hence, the Levinson–Durbin algorithm described in Appendix 2A provides an efficient means for solving the linear equations recursively, starting with a first-order predictor and proceeding recursively to the solution of the coefficients for the predictor of order p. The recursive relations for the Levinson–Durbin algorithm are

$$a_{11} = \frac{\phi(1)}{\phi(0)} \qquad \mathscr{E}_0 = \phi(0)$$

$$a_{mm} = \frac{\phi(m) - A'_m \Phi^r_{m-1}}{\mathscr{E}_{m-1}} \tag{6.8.28}$$

$$a_{mk} = a_{m-1\,k} - a_{mm} a_{m-1\,m-k}$$

$$\mathscr{E}_m = \mathscr{E}_{m-1}(1 - a^2_{mm})$$

for $m = 1, 2, \ldots, p$, where the vectors $\mathbf{A}_{m-1}$ and ϕ^r_{m-1} are defined as

$$\mathbf{A}'_{m-1} = [a_{m-11}a_{m-12} \cdots a_{m-1m-1}]$$
$$\phi^r_{m-1} = [\phi(m-1)\phi(m-2) \cdots \phi(1)]$$

The linear prediction filter of order m may be realized as a transversal filter with transfer function

$$A_m(z) = 1 - \sum_{k=1}^{m} a_{mk}z^{-k} \qquad (6.8.29)$$

Its input is the data $\{y(t)\}$ and its output is the error $e(t) = y(t) - \hat{y}(t)$. The prediction filter can also be realized in the form of a lattice, as we now demonstrate.

Our starting point is the use of the Levinson–Durbin algorithm for the predictor coefficients a_{mk} in (6.8.29). This substitution yields

$$A_m(z) = 1 - \sum_{k=1}^{m-1} \left(a_{m-1k} - a_{mm}a_{m-1m-k} \right) z^{-k} - a_{mm}z^{-m}$$

$$= 1 - \sum_{k=1}^{m-1} a_{m-1k}z^{-k} - a_{mm}z^{-m}\left(1 - \sum_{k=1}^{m-1} a_{m-1k}z^{k} \right)$$

$$= A_{m-1}(z) - a_{mm}z^{-m}A_{m-1}(z^{-1}) \qquad (6.8.30)$$

Thus we have the transfer function of the mth-order predictor in terms of the transfer function of the $(m-1)$st-order predictor.

Now suppose we define a filter with transfer function $G_m(z)$ as

$$G_m(z) = z^{-m}A_m(z^{-1}) \qquad (6.8.31)$$

Then (6.8.30) may be expressed as

$$A_m(z) = A_{m-1}(z) - a_{mm}z^{-1}G_{m-1}(z) \qquad (6.8.32)$$

We note that $G_{m-1}(z)$ represents a transversal filter with tap coefficients $(-a_{m-1m-1}, -a_{m-1m-2}, \ldots, -a_{m-11}, 1)$, while the coefficients of $A_{m-1}(z)$ are exactly the same except that they are given in reverse order.

More insight into the relationship between $A_m(z)$ and $G_m(z)$ can be obtained by computing the output of these two filters to an input sequence $y(t)$. Using z-transform relations, we have

$$A_m(z)Y(z) = A_{m-1}(z)Y(z) - a_{mm}z^{-1}G_{m-1}(z)Y(z) \qquad (6.8.33)$$

We define the outputs of the filters as

$$F_m(z) = A_m(z)Y(z)$$
$$B_m(z) = G_m(z)Y(z) \qquad (6.8.34)$$

Then (6.8.33) becomes

$$F_m(z) = F_{m-1}(z) - a_{mm}z^{-1}B_{m-1}(z) \qquad (6.8.35)$$

In the time domain, the relation in (6.8.35) becomes

$$f_m(t) = f_{m-1}(t) - a_{mm}b_{m-1}(t-1) \qquad m \geq 1 \qquad (6.8.36)$$

where

$$f_m(t) = y(t) - \sum_{k=1}^{m} a_{mk}y(t-k) \qquad (6.8.37)$$

$$b_m(t) = y(t-m) - \sum_{k=1}^{m} a_{mk}y(t-m+k) \qquad (6.8.38)$$

To elaborate, $f_m(t)$ in (6.8.37) represents the error of an mth-order forward predictor, while $b_m(t)$ represents the error of an mth-order backward predictor.

The relation in (6.8.36) is one of two that specifies a lattice filter. The second relation is obtained from $G_m(z)$ as follows:

$$\begin{aligned} G_m(z) &= z^{-m}A_m(z^{-1}) \\ &= z^{-m}\left[A_{m-1}(z^{-1}) - a_{mm}z^mA_{m-1}(z)\right] \\ &= z^{-1}G_{m-1}(z) - a_{mm}A_{m-1}(z) \end{aligned} \qquad (6.8.39)$$

Now, if we multiply both sides of (6.8.39) by $Y(z)$ and express the result in terms of $F_m(z)$ and $B_m(z)$ using the definitions in (6.8.34), we obtain

$$B_m(z) = z^{-1}B_{m-1}(z) - a_{mm}F_{m-1}(z) \qquad (6.8.40)$$

By transforming (6.8.40) into the time domain, we obtain the second relation that corresponds to the lattice filter, namely,

$$b_m(t) = b_{m-1}(t-1) - a_{mm}f_{m-1}(t) \qquad m \geq 1 \qquad (6.8.41)$$

The initial condition is

$$f_0(t) = b_0(t) = y(t) \qquad (6.8.42)$$

The lattice filter described by the recursive relations in (6.8.36) and (6.8.41) is illustrated in Fig. 6.8.2. Each stage is characterized by its own multiplication factor $\{a_{ii}\}$, $i = 1, 2, \ldots, m$, which is defined in the Levinson–Durbin algorithm. The forward and backward errors $f_m(t)$ and $b_m(t)$ are usually called the *residuals*. The mean-square value of these residuals is

$$\mathscr{E}_m = E[f_m^2(t)] = E[b_m^2(t)] \qquad (6.8.43)$$

$\mathscr{E}_m$ is given recursively, as indicated in the Levinson–Durbin algorithm, by

$$\begin{aligned} \mathscr{E}_m &= \mathscr{E}_{m-1}(1 - a_{mm}^2) \\ &= \mathscr{E}_0 \prod_{i=1}^{m}(1 - a_{ii}^2) \end{aligned} \qquad (6.8.44)$$

where $\mathscr{E}_0 = \phi(0)$.

FIGURE 6.8.2
A lattice filter.

The residuals $\{f_m(t)\}$ and $\{b_m(t)\}$ satisfy a number of interesting properties as described by Makhoul (1978). Most important of these are the orthogonality properties

$$E[b_m(t)b_n(t)] = \mathcal{E}_m \delta_{mn}$$

$$E[f_m(t+m)f_n(t+n)] = \mathcal{E}_m \delta_{mn}$$

(6.8.45)

Furthermore, the cross correlation between $f_m(t)$ and $b_n(t)$ is

$$E[f_m(t)b_n(t)] = \begin{cases} a_{nn}\mathcal{E}_m & m \geq n \\ 0 & m < n \end{cases} \qquad m, n \geq 0$$

(6.8.46)

As a consequence of the orthogonality properties of the residuals, the different sections of the lattice exhibit a form of independence which allows us to add or delete one or more of the last stages without affecting the parameters of the remaining stages. Since the residual mean-square error $\mathcal{E}_m$ decreases monotonically with the number of sections, $\mathcal{E}_m$ can be used as performance index in determining where the lattice should be terminated.

6.8.3 Recursive Least-Squares Lattice Algorithms

In Sec. 6.8.2 we showed that a linear prediction filter can be implemented either as a linear transversal filter or as a lattice filter. The lattice filter is order-recursive and, as a consequence, the number of sections it contains can be easily increased or decreased without affecting the parameters of the remaining sections. In contrast, the coefficients of a transversal filter obtained on the basis of the RLS criterion are interdependent. This means that an increase or a decrease in the size of the filter results in a change in all coefficients. Consequently the Kalman algorithm described in Sec. 6.8.1 is recursive in time but is not recursive in order.

We shall now describe a recursive least-squares lattice algorithm which is recursive in order and in time. The development in this section applies to a linear least-squares equalizer. The extention to a decision-feedback equalizer requires a two-dimensional formulation of the following derivation. This is a straightforward extension which is omitted for the sake of brevity, but the algorithm is discussed below.

Suppose we observe the signal $y(t - n)$, $n = 1, 2, \ldots, t$, and let us consider the problem of linear prediction based on least squares. Toward this end, let $f_m(n, t)$ denote the forward prediction error for an mth-order predictor, defined as

$$f_m(n, t) = y(n) + \mathbf{A}'_m(t)\mathbf{Y}_m(n - 1) \tag{6.8.47}$$

where the vector $\mathbf{A}'_m(t)$ consists of the forward predictor coefficients, i.e.,

$$\mathbf{A}'_m(t) = \begin{bmatrix} a_{m1}(t) & a_{m2}(t) \cdots a_{mm}(t) \end{bmatrix} \tag{6.8.48}$$

and the data vector $\mathbf{Y}'_m(n - 1)$ is

$$\mathbf{Y}'_m(n - 1) = \begin{bmatrix} y(n - 1)\, y(n - 2) \cdots y(n - m) \end{bmatrix} \tag{6.8.49}$$

The predictor coefficients $\mathbf{A}_m(t)$ are selected to minimize the average weighted squared error

$$\mathcal{E}_m^f(t) = \sum_{n=0}^{t} w^{t-n} |f_m(n, t)|^2 \tag{6.8.50}$$

The minimization of $\mathcal{E}_m^f(t)$ with respect to $\mathbf{A}_m(t)$ leads to the following set of linear equations:

$$\mathbf{R}_m(t - 1)\mathbf{A}_m(t) = -\mathbf{Q}_m(t) \tag{6.8.51}$$

where $\mathbf{R}_m(t)$ is defined by (6.8.7) and

$$\mathbf{Q}_m(t) = \sum_{n=0}^{t} w^{t-n} y(n)\mathbf{Y}_m^*(n - 1) \tag{6.8.52}$$

Thus

$$\begin{aligned} \mathbf{A}_m(t) &= -\mathbf{R}_m^{-1}(t - 1)\mathbf{Q}_m(t) \\ &= -\mathbf{P}_m(t - 1)\mathbf{Q}_m(t) \end{aligned} \tag{6.8.53}$$

The minimum value of $\mathcal{E}_m^f(t)$, obtained with the linear predictor specified by (6.8.53) and denoted as $r_m^f(t)$, is

$$\begin{aligned} r_m^f(t) &= \sum_{n=0}^{t} w^{t-n} y^*(n)\big[y(n) - \mathbf{A}'_m(t)\mathbf{Y}_m(n - 1) \big] \\ &= q(t) + \mathbf{A}'_m(t)\mathbf{Q}_m^*(t) \end{aligned} \tag{6.8.54}$$

where $q(t)$ is defined as

$$q(t) = \sum_{n=0}^{t} w^{t-n}|y(n)|^2 \tag{6.8.55}$$

The linear equations in (6.8.51) and the equation for $r_m^f(t)$ in (6.8.54) can be combined in a single matrix equation of the form

$$\begin{bmatrix} q(t) & \mathbf{Q}_m^{*\prime}(t) \\ \mathbf{Q}_m(t) & \mathbf{R}_m(t-1) \end{bmatrix} \begin{bmatrix} 1 \\ \mathbf{A}_m(t) \end{bmatrix} = \begin{bmatrix} r_m^f(t) \\ \mathbf{0}_m \end{bmatrix} \tag{6.8.56}$$

where $\mathbf{0}_m$ is the m-dimensional null vector. It is interesting to note that the $(m+1) \times (m+1)$ partitioned matrix in (6.8.56) is just $\mathbf{R}_{m+1}(t)$ since

$$\begin{aligned}
\mathbf{R}_{m+1}(t) &= \sum_{n=0}^{t} w^{t-n} \mathbf{Y}_{m+1}^*(n) \mathbf{Y}_{m+1}'(n) \\
&= \sum_{n=0}^{t} w^{t-n} \begin{bmatrix} y^*(n) \\ \mathbf{Y}_m^*(n-1) \end{bmatrix} [y(n) \quad \mathbf{Y}_m'(n-1)] \\
&= \begin{bmatrix} q(t) & \mathbf{Q}_m^{*\prime}(t) \\ \mathbf{Q}_m(t) & \mathbf{R}_m(t-1) \end{bmatrix}
\end{aligned} \tag{6.8.57}$$

In a completely parallel development to (6.8.47) through (6.8.57), we minimize the backward average weighted squared error for an mth-order backward predictor defined as

$$\mathscr{E}_m^b(t) = \sum_{n=0}^{t} w^{t-n}|b_m(n,t)|^2 \tag{6.8.58}$$

where the backward error is defined as

$$b_m(n,t) = y(n-m) + \mathbf{B}_m'(t)\mathbf{Y}_m(n) \tag{6.8.59}$$

and $\mathbf{B}_m(t)$ is the vector coefficients for the backward predictor. The minimization of $\mathscr{E}_m^b(t)$ leads to the equation

$$\mathbf{R}_m(t)\mathbf{B}_m(t) = -\mathbf{U}_m(t) \tag{6.8.60}$$

and, hence, to the solution

$$\begin{aligned}
\mathbf{B}_m(t) &= -\mathbf{R}_m^{-1}(t)\mathbf{U}_m(t) \\
&= -\mathbf{P}_m(t)\mathbf{U}_m(t)
\end{aligned} \tag{6.8.61}$$

where

$$\mathbf{U}_m(t) = \sum_{n=0}^{t} w^{t-n} y(n-m) \mathbf{Y}_m^*(n) \tag{6.8.62}$$

The minimum value of $\mathscr{E}_m^b(t)$, denoted as $r_m^b(t)$, is

$$r_m^b(t) = \sum_{n=0}^{t} w^{t-n} y^*(n-m)[y(n-m) - \mathbf{B}_m'(t)\mathbf{Y}_m(n)]$$

$$= u(t) + \mathbf{B}_m'(t)\mathbf{U}_m^*(t) \qquad (6.8.63)$$

where the scalar quantity $u(t)$ is defined as

$$u(t) = \sum_{n=0}^{t} w^{t-n}|y(n-m)|^2 \qquad (6.8.64)$$

If we combine (6.8.60) and (6.8.63) into a single matrix equation, we obtain

$$\begin{bmatrix} \mathbf{R}_m(t) & \mathbf{U}_m(t) \\ \mathbf{U}_m^{*\prime}(t) & u(t) \end{bmatrix} \begin{bmatrix} \mathbf{B}_m(t) \\ 1 \end{bmatrix} = \begin{bmatrix} \mathbf{0}_m \\ r_m^b(t) \end{bmatrix} \qquad (6.8.65)$$

Again we note that the $(m+1) \times (m+1)$ partioned matrix in (6.8.65) is $\mathbf{R}_{m+1}(t)$ since it can be expressed as

$$\mathbf{R}_{m+1}(t) = \sum_{n=0}^{t} w^{t-n} \begin{bmatrix} \mathbf{Y}_m^*(n) \\ y^*(n-m) \end{bmatrix} [\mathbf{Y}_m'(n) \quad y(n-m)]$$

$$= \begin{bmatrix} \mathbf{R}_m(t) & \mathbf{U}_m(t) \\ \mathbf{U}_m^{*\prime}(t) & u(t) \end{bmatrix} \qquad (6.8.66)$$

Thus we have obtained the equations for the forward and backward least-squares predictors of order m.

Next we derive the order-update equations for these predictors, which will lead us to the lattice filter structure. In deriving them for $\mathbf{A}_m(t)$ and $\mathbf{B}_m(t)$, we will make use of the two matrix inversion identities for a matrix of the form

$$\mathbf{M} = \begin{bmatrix} \mathbf{M}_{11} & \mathbf{M}_{12} \\ \mathbf{M}_{21} & \mathbf{M}_{22} \end{bmatrix} \qquad (6.8.67)$$

where $\mathbf{M}$, $\mathbf{M}_{11}$, and $\mathbf{M}_{22}$ are square matrices. The inverse of $\mathbf{M}$ is expressible in two different forms, namely,

$$\mathbf{M}^{-1} = \begin{bmatrix} \mathbf{M}_{11}^{-1} + \mathbf{M}_{11}^{-1}\mathbf{M}_{12}\tilde{\mathbf{M}}_{22}^{-1}\mathbf{M}_{21}\mathbf{M}_{11}^{-1} & -\mathbf{M}_{11}^{-1}\mathbf{M}_{12}\tilde{\mathbf{M}}_{22}^{-1} \\ -\tilde{\mathbf{M}}_{22}^{-1}\mathbf{M}_{21}\mathbf{M}_{11}^{-1} & \tilde{\mathbf{M}}_{22}^{-1} \end{bmatrix} \qquad (6.8.68)$$

and

$$\mathbf{M}^{-1} = \begin{bmatrix} \tilde{\mathbf{M}}_{11}^{-1} & -\tilde{\mathbf{M}}_{11}^{-1}\mathbf{M}_{12}\mathbf{M}_{22}^{-1} \\ -\mathbf{M}_{22}^{-1}\mathbf{M}_{21}\tilde{\mathbf{M}}_{11}^{-1} & \mathbf{M}_{22}^{-1}\mathbf{M}_{21}\tilde{\mathbf{M}}_{11}^{-1}\mathbf{M}_{12}\mathbf{M}_{22}^{-1} + \mathbf{M}_{22}^{-1} \end{bmatrix} \qquad (6.8.69)$$

where $\tilde{\mathbf{M}}_{11}$ and $\tilde{\mathbf{M}}_{22}$ are defined as

$$\tilde{\mathbf{M}}_{11} = \mathbf{M}_{11} - \mathbf{M}_{12}\mathbf{M}_{22}^{-1}\mathbf{M}_{21}$$

$$\tilde{\mathbf{M}}_{22} = \mathbf{M}_{22} - \mathbf{M}_{21}\mathbf{M}_{11}^{-1}\mathbf{M}_{12} \qquad (6.8.70)$$

Order-update recursions. Now suppose we use the formula in (6.8.68) to obtain the inverse of $\mathbf{R}_{m+1}(t)$ using the form in (6.8.66). In that case

$$\tilde{\mathbf{M}}_{22} = u(t) - \mathbf{U}_m^{*\prime}(t)\mathbf{R}_m^{-1}(t)\mathbf{U}_m(t)$$

$$= u(t) + \mathbf{B}_m'(t)\mathbf{U}_m^*(t) = r_m^b(t)$$

and

$$\mathbf{M}_{11}^{-1}\mathbf{M}_{12} = \left(\mathbf{M}_{21}\mathbf{M}_{11}^{-1}\right)' = \mathbf{R}_m^{-1}(t)\mathbf{U}_m(t) = -\mathbf{B}_m(t)$$

Hence

$$\mathbf{R}_{m+1}^{-1}(t) \equiv \mathbf{P}_{m+1}(t) = \begin{bmatrix} \mathbf{P}_m(t) + \dfrac{\mathbf{B}_m(t)\mathbf{B}_m^{*\prime}(t)}{r_m^b(t)} & \dfrac{\mathbf{B}_m(t)}{r_m^b(t)} \\[3mm] \dfrac{\mathbf{B}_m'(t)}{r_m^b(t)} & \dfrac{1}{r_m^b(t)} \end{bmatrix}$$

$$= \begin{bmatrix} \mathbf{P}_m(t) & 0 \\ 0 & 0 \end{bmatrix} + \frac{1}{r_m^b(t)}\begin{bmatrix} \mathbf{B}_m(t) \\ 1 \end{bmatrix}\begin{bmatrix} \mathbf{B}_m^{*\prime}(t) & 1 \end{bmatrix} \qquad (6.8.71)$$

By substituting $t - 1$ for t in (6.8.71) and postmultiplying the result by $\mathbf{Q}_{m+1}(t)$, we obtain the order update for $\mathbf{A}_m(t)$. That is,

$$\mathbf{A}_{m+1}(t) = -\mathbf{P}_{m+1}(t-1)\mathbf{Q}_{m+1}(t)$$

$$= \begin{bmatrix} \mathbf{P}_m(t-1) & 0 \\ 0 & 0 \end{bmatrix}\begin{bmatrix} \mathbf{Q}_m(t) \\ \cdots \end{bmatrix} - \frac{1}{r_m^b(t-1)}\begin{bmatrix} \mathbf{B}_m(t-1) \\ 1 \end{bmatrix}$$

$$\times \begin{bmatrix} \mathbf{B}_m^{*\prime}(t-1) & 1 \end{bmatrix}\mathbf{Q}_{m+1}(t)$$

$$= \begin{bmatrix} \mathbf{A}_m(t) \\ 0 \end{bmatrix} - \frac{k_{m+1}(t)}{r_m^b(t-1)}\begin{bmatrix} \mathbf{B}_m(t-1) \\ 1 \end{bmatrix} \qquad (6.8.72)$$

where the scalar quantity $k_{m+1}(t)$ is defined as

$$k_{m+1}(t) = \begin{bmatrix} \mathbf{B}_m^{*\prime}(t-1) & 1 \end{bmatrix}\mathbf{Q}_{m+1}(t) \qquad (6.8.73)$$

To obtain the corresponding order update for $\mathbf{B}_m(t)$, we use the inversion formula in (6.8.69) for the inverse of $\mathbf{R}_{m+1}(t)$, along with the form in (6.8.57). In this case, we have

$$\tilde{\mathbf{M}}_{11} = q(t) - \mathbf{Q}_m^{*\prime}(t)\mathbf{R}_m^{-1}(t-1)\mathbf{Q}_m(t)$$

$$= q(t) + \mathbf{A}_m'(t)\mathbf{Q}_m^*(t) = r_m^f(t)$$

and

$$\mathbf{M}_{22}^{-1}\mathbf{M}_{21} = \left(\mathbf{M}_{12}\mathbf{M}_{22}^{-1}\right) = \mathbf{R}_m^{-1}(t-1)\mathbf{Q}_m(t) = -\mathbf{A}_m(t)$$

Hence

$$\mathbf{P}_{m+1}(t) = \begin{bmatrix} \dfrac{1}{r_m^f(t)} & \dfrac{\mathbf{A}_m'(t)}{r_m^f(t)} \\[2ex] \dfrac{\mathbf{A}_m(t)}{r_m^f(t)} & \mathbf{P}_m(t-1) + \dfrac{1}{r_m^f(t)}\mathbf{A}_m(t)\mathbf{A}_m^{*'}(t) \end{bmatrix}$$

$$= \begin{bmatrix} 0 & 0 \\ 0 & \mathbf{P}_m(t-1) \end{bmatrix} + \frac{1}{r_m^f(t)}\begin{bmatrix} 1 \\ \mathbf{A}_m(t) \end{bmatrix}\begin{bmatrix} 1 & \mathbf{A}_m^{*'}(t) \end{bmatrix} \quad (6.8.74)$$

Now, if we postmultiply (6.8.74) by $-\mathbf{U}_{m+1}(t)$, we obtain

$$\mathbf{B}_{m+1}(t) = \begin{bmatrix} 0 & 0 \\ 0 & \mathbf{P}_m(t-1) \end{bmatrix}\begin{bmatrix} \cdots \\ \mathbf{U}_m(t-1) \end{bmatrix} - \frac{1}{r_m^f(t)}\begin{bmatrix} 1 \\ \mathbf{A}_m(t) \end{bmatrix}\begin{bmatrix} 1 & -\mathbf{A}_m'(t) \end{bmatrix}\mathbf{U}_{m+1}(t)$$

$$= \begin{bmatrix} 0 \\ \mathbf{B}_m(t-1) \end{bmatrix} - \frac{k_{m+1}^*(t)}{r_m^f(t)}\begin{bmatrix} 1 \\ \mathbf{A}_m(t) \end{bmatrix} \quad (6.8.75)$$

where

$$\begin{bmatrix} 1 & \mathbf{A}_m^{*'}(t) \end{bmatrix}\mathbf{U}_{m+1}(t) = \begin{bmatrix} \mathbf{B}_m'(t-1) & 1 \end{bmatrix}\mathbf{Q}_{m+1}^*(t) = k_{m+1}^*(t) \quad (6.8.76)$$

The proof that the two scalar quantities in (6.8.76) are identical is left as an exercise for the reader. Thus, (6.8.72) and (6.8.75) specify the order-update equations for $A_m(t)$ and $B_m(t)$, respectively.

The order-update equations for $r_m^f(t)$ and $r_m^b(t)$ may now be obtained. From the definition of $r_m^f(t)$ given in (6.8.54) we have

$$r_{m+1}^f(t) = q(t) + \mathbf{A}_{m+1}'(t)\mathbf{Q}_{m+1}^*(t) \quad (6.8.77)$$

Substituting from (6.8.72) for $\mathbf{A}_{m+1}(t)$ into (6.8.77), we obtain

$$r_{m+1}^f(t) = q(t) + \left\{ \begin{bmatrix} \mathbf{A}_m'(t) & 0 \end{bmatrix}\begin{bmatrix} \mathbf{Q}_m(t) \\ \cdots \end{bmatrix} - \frac{k_{m+1}(t)}{r_m^b(t-1)}\begin{bmatrix} \mathbf{B}_m'(t-1) & 1 \end{bmatrix}\mathbf{Q}_{m+1}^*(t) \right\}$$

$$= r_m^f(t) - \frac{|k_{m+1}(t)|^2}{r_m^b(t-1)} \quad (6.8.78)$$

Similarly, by using (6.8.63) and (6.8.75), we obtain the order update for $r_{m+1}^b(t)$ in the form

$$r_{m+1}^b(t) = r_m^b(t-1) - \frac{|k_{m+1}(t)|^2}{r_m^f(t)} \quad (6.8.79)$$

The lattice filter is specified by two coupled equations involving the forward and backward errors $f_m(t, t-1)$ and $b_m(t, t-1)$. From the definition of the forward error in (6.8.47) we have

$$f_{m+1}(t, t-1) = y(t) + \mathbf{A}_{m+1}'(t-1)\mathbf{Y}_{m+1}(t-1) \quad (6.8.80)$$

Substituting for $\mathbf{A}'_{m+1}(t-1)$ from (6.8.72) into (6.8.80) yields

$$f_{m+1}(t, t-1) = y(t) + \left[\mathbf{A}'_m(t-1) \quad 0\right]\left[\begin{matrix}\mathbf{Y}_m(t-1)\\ \cdots\end{matrix}\right] - \frac{k_{m+1}(t-1)}{r_m^b(t-2)}$$

$$\times \left[\mathbf{B}'_m(t-2) \quad 1\right]\mathbf{Y}_{m+1}(t-1)$$

$$= f_m(t, t-1) - \frac{k_{m+1}(t-1)}{r_m^b(t-2)}\left[y(t-m-1) + \mathbf{B}'_m(t-2)\mathbf{Y}_m(t-1)\right]$$

$$= f_m(t, t-1) - \frac{k_{m+1}(t-1)}{r_m^b(t-2)}b_m(t-1, t-2) \qquad (6.8.81)$$

To simplify the notation, we define

$$f_m(t) \equiv f_m(t, t-1)$$
$$b_m(t) \equiv b_m(t, t-1)$$

Then (6.8.81) may be expressed as

$$f_{m+1}(t) = f_m(t) - \frac{k_{m+1}(t-1)}{r_m^b(t-2)}b_m(t-1) \qquad (6.8.82)$$

Similarly, beginning with the definition of the backward error given in (6.8.59), we have

$$b_{m+1}(t, t-1) = y(t-m-1) + \mathbf{B}'_{m+1}(t-1)\mathbf{Y}_{m+1}(t)$$

Substituting for $\mathbf{B}_{m+1}(t)$ from (6.8.75) and simplifying the result, we obtain

$$b_{m+1}(t, t-1) = b_m(t-1, t-2) - \frac{k_{m+1}^*(t-1)}{r_m^f(t-1)}f_m(t, t-1) \qquad (6.8.83)$$

or, equivalently,

$$b_{m+1}(t) = b_m(t-1) - \frac{k_{m+1}^*(t-1)}{r_m^f(t-1)}f_m(t) \qquad (6.8.84)$$

The two recursive equations in (6.8.82) and (6.8.84) specify the lattice filter, illustrated in Fig. 6.8.3. The initial conditions on the order updates are

$$f_0(t) = b_0(t) = y(t)$$

$$r_0^f(t) = r_0^b(t) = \sum_{n=0}^{t} w^{t-n}|y(n)|^2$$

$$= wr_0^f(t-1) + |y(t)|^2 \qquad (6.8.85)$$

We note that (6.8.85) is also a time-update equation for $r_0^f(t)$ and $r_0^b(t)$. From these two quantities and from $r_0^b(t-1)$ and $k_{m+1}(t)$ we may evaluate $r_m^f(t)$ and $r_m^b(t)$ using the order-update equations in (6.8.78) and (6.8.79). However, we have yet to specify an update equation for $k_{m+1}(t)$. Below we show that $k_{m+1}(t)$ satisfies a simple time-update equation.

(a)

(b)

FIGURE 6.8.3
Least-squares lattice filter.

Time-update recursions. In order to determine the time-update equation for $k_{m+1}(t)$, we begin with the form

$$k_{m+1}(t) = U_{m+1}^{*\prime}(t) \begin{bmatrix} 1 \\ A_m(t) \end{bmatrix}$$

First we determine the time-update relations for $U_{m+1}(t)$ and $A_m(t)$. From (6.8.62), the time update for $U_{m+1}(t)$ is

$$U_{m+1}(t) = wU_{m+1}(t-1) + y(t-1-m)Y_{m+1}^*(t) \qquad (6.8.86)$$

The time-update for $A_m(t)$ is determined as follows. From (6.8.52), (6.8.53), and (6.8.14) we have

$$
\begin{aligned}
A_m(t) &= -P_m(t-1)Q_m(t) \\
&= -\frac{1}{w}\big[P_m(t-2) - K_m(t-1)Y_m^\prime(t-1)P_m(t-2)\big] \\
&\quad \times \big[wQ_m(t-1) + y(t)Y_m^*(t-1)\big] \\
&= A_m(t-1) - K_m(t-1)\big[y(t) + Y_m^\prime(t-1)A_m(t-1)\big] \quad (6.8.87)
\end{aligned}
$$

where $K_m(t-1)$ is the Kalman gain vector at $t-1$. But from (6.8.47) we have

$$y(t) + Y_m^\prime(t-1)A_m(t-1) = f_m(t, t-1) \equiv f_m(t)$$

Therefore the time update for $A_m(t)$ is

$$A_m(t) = A_m(t-1) - K_m(t-1)f_m(t) \qquad (6.8.88)$$

In the parallel development to the above, using (6.8.61), (6.8.62), and (6.8.14), we obtain the time update for the coefficients of the backward predictor in the form

$$\mathbf{B}_m(t) = \mathbf{B}_m(t-1) - \mathbf{K}_m(t)b_m(t) \tag{6.8.89}$$

Now, from (6.8.86) and (6.8.88), the time-update equation for $k_{m+1}(t)$ is

$$k_{m+1}(t) = \left[w\mathbf{U}_{m+1}^{*\prime}(t-1) + y^*(t-1-m)\mathbf{Y}_{m+1}'(t) \right]$$

$$\times \left\{ \begin{bmatrix} 1 \\ \mathbf{A}_m(t-1) \end{bmatrix} - \begin{bmatrix} 0 \\ \mathbf{K}_m(t-1)f_m(t) \end{bmatrix} \right\}$$

$$= wk_{m+1}(t-1) - w\mathbf{U}_{m+1}^{*\prime}(t-1) \begin{bmatrix} 0 \\ \mathbf{K}_m(t-1)f_m(t) \end{bmatrix}$$

$$+ y^*(t-1-m)\mathbf{Y}_{m+1}'(t) \begin{bmatrix} 1 \\ \mathbf{A}_m(t-1) \end{bmatrix}$$

$$- y^*(t-1-m)\mathbf{Y}_{m+1}'(t) \begin{bmatrix} 0 \\ \mathbf{K}_m(t-1)f_m(t) \end{bmatrix} \tag{6.8.90}$$

But

$$\mathbf{Y}_{m+1}'(t) \begin{bmatrix} 1 \\ \mathbf{A}_m(t-1) \end{bmatrix} = \begin{bmatrix} y(t) & \mathbf{Y}_m'(t-1) \end{bmatrix} \begin{bmatrix} 1 \\ \mathbf{A}_m(t-1) \end{bmatrix} = f_m(t) \tag{6.8.91}$$

Also

$$\mathbf{U}_{m+1}^{*\prime}(t-1) \begin{bmatrix} 0 \\ \mathbf{K}_m(t-1) \end{bmatrix} = \mathbf{U}_m^{*\prime}(t-2)\mathbf{K}_m(t-1)$$

$$= \frac{\mathbf{U}_m^{*\prime}(t-2)\mathbf{P}_m(t-2)\mathbf{Y}_m^*(t-1)}{w + \mu_m(t-1)}$$

$$= -\frac{\mathbf{B}_m^{*\prime}(t-2)\mathbf{Y}_m^*(t-1)}{w + \mu_m(t-1)}$$

$$= \frac{y^*(t-1-m) - b_m^*(t-1)}{w + \mu_m(t-1)} \tag{6.8.92}$$

Finally,

$$\mathbf{Y}_{m+1}'(t) \begin{bmatrix} 0 \\ \mathbf{K}_m(t-1) \end{bmatrix} = \frac{\mathbf{Y}_m'(t-1)\mathbf{P}_m(t-2)\mathbf{Y}_m^*(t-1)}{w + \mu_m(t-1)} = \frac{\mu_m(t-1)}{w + \mu_m(t-1)} \tag{6.8.93}$$

Substituting the results of (6.8.91), (6.8.92), and (6.8.93) into (6.8.90) and simplifying the resulting expression, we obtain the desired time-update equation in the form

$$k_{m+1}(t) = wk_{m+1}(t-1) + \frac{w}{w + \mu_m(t-1)} f_m(t)b_m^*(t-1) \tag{6.8.94}$$

It is convenient to define a new variable $\alpha_m(t)$ as

$$\alpha_m(t) = \frac{w}{w + \mu_m(t)} \qquad (6.8.95)$$

Then (6.8.94) becomes

$$k_{m+1}(t) = wk_{m+1}(t-1) + \alpha_m(t-1)f_m(t)b_m^*(t-1) \qquad (6.8.96)$$

Order-update for $\alpha_m(n)$. Although $\alpha_m(t)$ can be computed directly for each value of m and for each t, it is more efficient to use an order-update equation, which is determined as follows. First of all, from the definition of $\mathbf{K}_m(t)$ given in (6.8.12), it is easy to see that

$$\alpha_m(t) = 1 - \mathbf{Y}_m'(t)\mathbf{K}_m(t) \qquad (6.8.97)$$

To obtain an order-update equation for $\alpha_m(t)$, we need an order-update equation for the Kalman gain vector $\mathbf{K}_m(t)$. But $\mathbf{K}_{m+1}(t)$ may be expressed as

$$\mathbf{K}_{m+1}(t) = \mathbf{P}_{m+1}(t)\mathbf{Y}_{m+1}^*(t)$$

$$= \left\{ \begin{bmatrix} \mathbf{P}_m(t) & 0 \\ 0 & 0 \end{bmatrix} + \frac{1}{r_m^b(t)} \begin{bmatrix} \mathbf{B}_m(t) \\ 1 \end{bmatrix} \begin{bmatrix} \mathbf{B}_m'(t) & 1 \end{bmatrix} \right\} \begin{bmatrix} \mathbf{Y}_m^*(t) \\ y^*(t-m) \end{bmatrix}$$

$$= \begin{bmatrix} \mathbf{K}_m(t) \\ 0 \end{bmatrix} + \frac{b_m(t,t)}{r_m^b(t)} \begin{bmatrix} \mathbf{B}_m(t) \\ 1 \end{bmatrix} \qquad (6.8.98)$$

where we have used (6.8.59) and (6.8.71). The term $b_m(t,t)$ in (6.8.98) can be expressed as

$$b_m(t,t) = y(t-m) + \mathbf{B}_m'(t)\mathbf{Y}_m(t)$$

$$= y(t-m) + [\mathbf{B}_m'(t-1) - \mathbf{K}_m'(t)b_m(t)]\mathbf{Y}_m(t)$$

$$= y(t-m) + \mathbf{B}_m'(t-1)\mathbf{Y}_m(t) - \mathbf{K}_m'(t)\mathbf{Y}_m(t)b_m(t)$$

$$= \alpha_m(t)b_m(t) \qquad (6.8.99)$$

Hence the order-update equation for $\mathbf{K}_m(t)$ in (6.8.98) may be expressed as

$$\mathbf{K}_{m+1}(t) = \begin{bmatrix} \mathbf{K}_m(t) \\ 0 \end{bmatrix} + \frac{\alpha_m(t)b_m^*(t)}{r_m^b(t)} \begin{bmatrix} \mathbf{B}_m(t) \\ 1 \end{bmatrix} \qquad (6.8.100)$$

Using (6.8.100) and the relation in (6.8.97), we obtain the order-update equation for $\alpha_m(t)$ as follows:

$$\alpha_{m+1}(t) = 1 - \mathbf{Y}_{m+1}'(t)\mathbf{K}_{m+1}(t)$$

$$= 1 - [\mathbf{Y}_m'(t)\ y(t-m)] \left\{ \begin{bmatrix} \mathbf{K}_m(t) \\ 0 \end{bmatrix} + \frac{\alpha_m(t)b_m^*(t)}{r_m^b(t)} \begin{bmatrix} \mathbf{B}_m(t) \\ 1 \end{bmatrix} \right\}$$

$$= \alpha_m(t) - \frac{\alpha_m(t)b_m^*(t)}{r_m^b(t)} [y(t-m) - \mathbf{Y}_m'(t)\mathbf{B}_m(t)]$$

$$= \alpha_m(t) - \frac{\alpha_m(t)b_m^*(t)}{r_m^b(t)} b_m(t,t)$$

$$= \alpha_m(t) - \frac{\alpha_m^2(t)|b_m(t)|^2}{r_m^b(t)} \qquad (6.8.101)$$

Thus we have obtained the order-update and time-update relations for the basic least-squares lattice filter in Fig. 6.8.2. To summarize briefly, the errors $f_m(t)$ and $b_m(t)$ are computed via (6.8.82) and (6.8.84). The minimum forward and backward squared errors $r_m^f(t)$ and $r_m^b(t)$ are updated according to (6.8.78) and (6.8.79). The parameters $\{k_{m+1}(t)\}$ satisfy the time-update equation given by (6.8.96) and the parameters $\{\alpha_m(t)\}$ are updated according to the order-update equation given by (6.8.101). Initially we have

$$r_m^f(-1) = r_m^b(-1) = r_m^b(-2) = \varepsilon > 0$$

$$f_m(-1) = b_m(-1) = k_m(-1) = 0 \qquad (6.8.102)$$

$$\alpha_m(-1) = 1 \qquad \alpha_{-1}(t) = \alpha_{-1}(t-1) = 1$$

The latter (joint process) estimator. The last step in this derivation is to obtain the output of the least-squares lattice equalizer. Suppose the equalizer has $m + 1$ taps which are determined to minimize the average weighted squared error

$$\mathscr{E}_{m+1} = \sum_{n=0}^{t} w^{t-n} |e_{m+1}(n, t)|^2$$

where

$$e_{m+1}(n, t) = I(n) - \mathbf{C}'_{m+1}(t)\mathbf{Y}_{m+1}(n) \qquad (6.8.103)$$

From the results of Sec. 6.8.1, we have established that the equalizer coefficients that minimize $\mathscr{E}_{m+1}$ are given by the equation

$$\mathbf{C}_{m+1}(t) = \mathbf{P}_{m+1}(t)\mathbf{D}_{m+1}(t) \qquad (6.8.104)$$

We have also established that $\mathbf{C}_m(t)$ satisfies the time-update equation given in (6.8.20).

Next we obtain an order update for $\mathbf{C}_m(t)$. Using (6.8.71) in (6.8.104), we have

$$\mathbf{C}_{m+1}(t) = \begin{bmatrix} \mathbf{P}_m(t) & 0 \\ 0 & 0 \end{bmatrix} \begin{bmatrix} \mathbf{D}_m(t) \\ - \end{bmatrix} + \frac{1}{r_m^b(t)} \begin{bmatrix} \mathbf{B}_m(t) \\ 1 \end{bmatrix} [\mathbf{B}_m(t) \quad 1]\mathbf{D}_{m+1}(t)$$

$$(6.8.105)$$

We define a scalar, $d_m(t)$, as

$$d_m(t) = [\mathbf{B}'_m(t) \quad 1]\mathbf{D}_{m+1}(t) \qquad (6.8.106)$$

Then (6.8.105) may be expressed as

$$\mathbf{C}_{m+1}(t) = \begin{bmatrix} \mathbf{C}_m(t) \\ 0 \end{bmatrix} + \frac{d_m(t)}{r_m^b(t)} \begin{bmatrix} \mathbf{B}_m(t) \\ 1 \end{bmatrix} \qquad (6.8.107)$$

The scalar $d_m(t)$ satisfies a time-update equation which is obtained from the time-update equations for $\mathbf{B}_m(t)$ and $\mathbf{D}_m(t)$, given by (6.8.89) and (6.8.16), respectively. Thus

$$d_m(t) = \left[\mathbf{B}_m^{*\prime}(t-1) - \mathbf{K}_m^{*\prime}(t)b_m^*(t) \quad 1\right]\left[w\mathbf{D}_{m+1}(t-1) + I(t)\mathbf{Y}_{m+1}^*(t)\right]$$
$$= w\left[\mathbf{B}_m^{*\prime}(t-1) \quad 1\right]\mathbf{D}_{m+1}(t-1) + \left[\mathbf{B}_m'(t-1) \quad 1\right]I(t)\mathbf{Y}_{m+1}^*(t)$$
$$- wb_m^*(t)\left[\mathbf{K}_m^{*\prime}(t) \quad 0\right]\mathbf{D}_{m+1}(t-1) - I(t)b_m^*(t)\left[\mathbf{K}_m^{*\prime}(t) \quad 0\right]\mathbf{Y}_{m+1}^*(t)$$
$$(6.8.108)$$

But

$$\left[\mathbf{B}_m'(t-1) \quad 1\right]\mathbf{Y}_{m+1}^*(t) = y^*(t-m) - \mathbf{B}_m^{*\prime}(t-1)\mathbf{Y}_m^*(t) = b_m^*(t) \quad (6.8.109)$$

Also

$$\left[\mathbf{K}_m^{*\prime}(t) \quad 0\right]\mathbf{D}_{m+1}(t-1) = \frac{1}{w + \mu_m(t)}\left[\mathbf{Y}_m'(t)\mathbf{P}_m(t-1) \quad 0\right]\begin{bmatrix}\mathbf{D}_m(t-1)\\ \cdots\end{bmatrix}$$
$$= \frac{1}{w + \mu_m(t)}\mathbf{Y}_m'(t)\mathbf{C}_m(t-1) \quad (6.8.110)$$

The last term in (6.8.108) may be expressed as

$$\left[\mathbf{K}_m^{*\prime}(t) \quad 0\right]\begin{bmatrix}\mathbf{Y}_m^*(t)\\ \cdots\end{bmatrix} = \frac{1}{w + \mu_m(t)}\mathbf{Y}_m'(t)\mathbf{P}_m(t-1)\mathbf{Y}_m^*(t) = \frac{\mu_m(t)}{w + \mu_m(t)}$$
$$(6.8.111)$$

Substituting the expressions in (6.8.109), (6.8.110), and (6.8.111) into (6.8.108) and collecting the terms, we obtain

$$d_m(t) = wd_m(t-1) + \alpha_m(t)b_m^*(t)e_m(t) \quad (6.8.112)$$

Order-update equations for $\alpha_m(t)$ and $b_m(t)$ have already been derived. With $e_0(t) = I(t)$, the order-update equation for $e_m(t)$ is obtained as follows:

$$e_m(t) \equiv e_m(t, t-1) = I(t) - \mathbf{C}_m'(t-1)\mathbf{Y}_m(t)$$
$$= I(t) - \left[\mathbf{C}_{m-1}'(t-1) \quad 0\right]\begin{bmatrix}\mathbf{Y}_{m-1}(t)\\ \cdots\end{bmatrix}$$
$$- \frac{d_{m-1}(t-1)}{r_{m-1}^b(t-1)}\left[\mathbf{B}_{m-1}'(t-1) \quad 1\right]\mathbf{Y}_m(t)$$
$$= I(t) - \mathbf{C}_{m-1}'(t-1)\mathbf{Y}_{m-1}(t) - \frac{d_{m-1}(t-1)}{r_{m-1}^b(t-1)}$$
$$\times\left[y(t-m+1) + \mathbf{B}_{m-1}'(t-1)\mathbf{Y}_{m-1}(t)\right]$$
$$= e_{m-1}(t) - \frac{d_{m-1}(t-1)}{r_{m-1}^b(t-1)}b_{m-1}(t) \quad (6.8.113)$$

Finally, the output of the lattice equalizer at time t is

$$\hat{I}(t) = \mathbf{C}_{m+1}'(t-1)\mathbf{Y}_{m+1}(t) \quad (6.8.114)$$

FIGURE 6.8.4
Least-squares lattice equalizer.

But $C'_m(t - 1)$ is not computed explicitly. By repeated use of the order-update equation given by (6.8.107) in (6.8.114), we obtain the desired expression for $I(t)$ in the form

$$\hat{I}(t) = \sum_{k=0}^{m} \frac{d_k(t-1)}{r_k^b(t-1)} b_k(t) \qquad (6.8.115)$$

The recursive least-squares lattice equalizer is shown in Fig. 6.8.4. The corresponding algorithm is summarized in Table 6.8.1. It has a computational complexity of $13N$ complex multiplications and additions. This linear growth in complexity makes this lattice equalizer very attractive in practical applications.

From the structure of the equalizer shown in Fig. 6.8.4 we note that it consists of two parts, a lattice filter and a *ladder filter*. The ladder part is sometimes called a *joint process estimator*. It is that part of the equalizer that forms a weighted linear combination of the backward errors $\{b_m(t)\}$ to yield the equalizer output.

We also emphasize that the RLS lattice-ladder structure of the linear equalizer, as shown in Fig. 6.8.4, is mathematically equivalent to the RLS transversal (direct form) equalizer structure described in Sec. 6.8.1. Hence, the performance of these two structures is identical provided that the computations are performed with infinite precision. However, the numerical properties of these structures and their corresponding algorithms are different when fixed-point arithmetic is used as we show below.

The RLS lattice-ladder algorithm given in Table 6.8.1 is by no means unique. Modification can be made to some of the equations without affecting the optimality of the algorithm. However, some modifications result in algorithms that are most robust numerically when fixed-point arithmetic is used in the implementation of the algorithm. Of particular interest is the modification in the manner in which the reflection coefficients in the lattice and the ladder gains are computed. In the conventional (indirect) method that is given in Table 6.8.1 the reflection coefficients and the ladder gains are computed according to the relations

$$\mathscr{K}_{m+1}^{f}(t) = -\frac{k_{m+1}(t)}{r_m^b(t-1)}$$

$$\mathscr{K}_{m+1}^{b}(t) = -\frac{k_{m+1}^*(t)}{r_m^f(t)} \qquad (6.8.116)$$

$$\xi_m(t) = -\frac{d_m(t)}{r_m^b(t)}$$

where the numerator and denominator terms in the reflection coefficients $\{\mathscr{K}_{m+1}^{f}(t)\}$ and $\{\mathscr{K}_{m+1}^{b}(t)\}$ and in the ladder gains $\{\xi_m(t)\}$ are updated recursively.

Alternatively, $\mathscr{K}_{m+1}^{f}(t)$, $\mathscr{K}_{m+1}^{b}(t)$, and $\xi_m(t)$ may be computed directly as shown by Ling et al. (1986) according to the equations

$$\mathscr{K}_{m+1}^{f}(t) = \mathscr{K}_{m+1}^{f}(t-1) - \frac{\alpha_m(t-1)f_{m+1}(t)b_m^*(t-1)}{r_m^b(t-1)}$$

$$\mathscr{K}_{m+1}^{b}(t) = \mathscr{K}_{m+1}^{b}(t-1) - \frac{\alpha_m(t-1)f_m^*(t)b_{m+1}(t)}{r_m^f(t)} \qquad (6.8.117)$$

$$\xi_m(t) = \xi_m(t-1) - \frac{\alpha_m(t)b_m^*(t)e_{m+1}(t)}{r_m^b(t)}$$

An important characteristic of these direct-update equations is that the forward and backward residuals (errors) are fed back to time-update the reflection coefficients in the lattice stages and $e_{m+1}(t)$ fed back to update the ladder gain. The RLS algorithm, which is given in Table 6.8.2, is called the *error-feedback form*.

RLS lattice decision-feedback equalizer. A RLS lattice-ladder DFE algorithm can be derived by following the procedure given above for the linear equalizer, but modifying the data vector so that two new inputs (the signal and the past decision) enter the data vector in each iteration. The general procedure is described in the paper by Ling and Proakis (1984) and the RLS lattice-ladder DFE algorithm is given in the paper by Ling and Proakis (1985).

The structure of this DFE is illustrated in Fig. 6.8.5. This equalizer consists of a multichannel lattice predictor and a ladder (joint process) estimator. The multichannel lattice predictor has $M = N_1 - N_2$ single-channel lattice states followed by $N_2 - 1$ two-channel lattice stages. This is equivalent to the RLS transversal (direct form) DFE with N_1 feedforward taps and N_2 feedback taps. Note the two inputs into the DFE as indicated above. It should also be noted that this structure (and algorithm) does not allow us to change the number of

TABLE 6.8.1
RLS lattice-ladder algorithm

Lattice predictor: Begin with $t = 0$ and compute the order updates
for $m = 0, 1, \ldots, N - 2$

$$k_{m+1}(t) = wk_{m+1}(t - 1) + \alpha_m(t - 1)f_m(t)b_m^*(t - 1)$$

$$\mathscr{K}_{m+1}^f(t) = -\frac{k_{m+1}(t)}{r_m^b(t - 1)}$$

$$\mathscr{K}_{m+1}^b(t) = -\frac{k_{m+1}^*(t)}{r_m^f(t)}$$

$$f_{m+1}(t + 1) = f_m(t + 1) + \mathscr{K}_{m+1}^f(t)b_m(t)$$

$$b_{m+1}(t + 1) = b_m(t) + \mathscr{K}_{m+1}^b(t)f_m(t + 1)$$

$$r_{m+1}^f(t) = r_m^f(t) - \frac{|k_{m+1}(t)|^2}{r_m^b(t + 1)}$$

$$r_{m+1}^b(t) = r_m^b(t - 1) - \frac{|k_{m+1}(t)|^2}{r_m^f(t)}$$

$$\alpha_{m+1}(t) = \alpha_m(t) - \frac{\alpha_m^2(t)|b_m(t)|^2}{r_m^b(t)}$$

Ladder filter: Begin with $t = 0$ and compute the order updates
for $m = 0, 1, \ldots, N - 1$

$$d_m(t) = wd_m(t - 1) + \alpha_m(t)b_m^*(t)e_m(t)$$

$$\xi_m(t) = -\frac{d_m(t)}{r_m^b(t)}$$

$$e_{m+1}(t + 1) = e_m(t + 1) + \xi_m(t)b_m(t + 1)$$

Initialization

$$\alpha_0(t) = 1 \qquad e_0(t + 1) = I(t + 1) \qquad f_0(t + 1) = b_0(t + 1) = y(t + 1)$$

$$r_0^f(t + 1) = r_0^b(t + 1) = wr_0^f(t) + |y(t + 1)|^2$$

$$\alpha_m(-1) = 1 \qquad k_m(-1) = 0$$

$$r_m^b(-1) = r_m^f(0) = \varepsilon > 0 \qquad d_m(-1) = 0$$

---

feedforward and feedback stages individually after it is started. A two-channel stage can be added or dropped during operation, thus keeping the difference between the number of feedforward and feedback stages unchanged. The average output error can be monitored to determine if two-channel stages should be added or dropped.

The version of the lattice-ladder DFE algorithm given in the paper by Ling and Proakis (1985) is the error-feedback form. Its computational complexity is

TABLE 6.8.2
Direct-update (error feedback) form of the RLS lattice-ladder algorithm

Lattice predictor: Begin with $t = 0$ and compute the order updates
for $m = 0, 1, \ldots, N - 2$

$$\mathcal{K}^f_{m+1}(t) = \mathcal{K}^f_{m+1}(t-1) - \frac{\alpha_m(t-1)f_{m+1}(t)b^*_m(t-1)}{r^b_m(t-1)}$$

$$\mathcal{K}^b_{m+1}(t) = \mathcal{K}^b_{m+1}(t-1) - \frac{\alpha_m(t-1)f^*_m(t)b_{m+1}(t)}{r^f_m(t-1)}$$

$$f_{m+1}(t+1) = f_m(t+1) + \mathcal{K}^f_{m+1}(t)b_m(t)$$

$$b_{m+1}(t+1) = b_m(t) + \mathcal{K}^b_{m+1}(t)f_m(t+1)$$

$$r^f_{m+1}(t) = wr^f_{m+1}(t-1) + \alpha_{m+}(t-1)|f_{m+1}(t)|^2$$

$$\alpha_{m+1}(t) = \alpha_m(t) - \frac{\alpha^2_m(t)|b_m(t)|^2}{r^b_m(t)}$$

$$r^b_{m+1}(t) = wr^b_{m+1}(t-1) + \alpha_{m+1}(t)|b_m(t)|^2$$

Ladder filter: Begin with $t = 0$ and compute the order updates
$m = 0, 1, \ldots, N - 1$

$$\xi_m(t) = \xi_m(t-1) - \frac{\alpha_m(t)b^*_m(t)e_{m+1}(t)}{r^b_m(t)}$$

$$e_{m+1}(t+1) = e_m(t+1) + \xi_m(t)b_m(t+1)$$

Initialization

$$\alpha_0(t) = 1 \qquad e_0(t+1) = I(t+1) \qquad f_0(t+1) = b_0(t+1) = y(t+1)$$

$$r^f_0(t+1) = r^b_0(t+1) = w^f_0 r(t) + |y(t+1)|^2$$

$$\alpha_m(-1) = 1 \qquad \mathcal{K}^f_m(-1) = \mathcal{K}^b_m(-1) = 0$$

$$r^b_m(-1) = r^f_m(0) = \varepsilon > 0$$

$18N_1 + 39N_2 - 39$ complex multiplications and divisions. Hence, its complexity is also linear in N. Its numerical properties are described below.

The RLS lattice linear equalizer derived above may also be incorporated into the predictive form of the DFE as the feedforward filter. Recall that the predictive DFE is particularly attractive in applications of decision-feedback equalization where trellis-coded modulation is used for transmitting the data. In rapidly fading channels such as HF ionospheric propagation, there is a need for the RLS lattice to provide rapid convergence. The noise predictor in this DFE is best realized by using another rapidly converging algorithm, called the modified recursive Gram–Schmidt algorithm, described in the paper by Ling et al. (1986). The interested reader may refer to the papers by Zhou and Proakis (1988) and Ling and Qureshi (1989) on the use of the predictive DFE for adaptive equalization of fading channels.

FIGURE 6.8.5
Structure of RLS lattice DFE

6.8.4 Gradient Lattice Equalizer

The RLS lattice-ladder algorithms described in the preceding section are significantly more complicated than the LMS algorithm. However, they do result in superior performance as we shall observe below.

In an attempt to simplify the computational complexity of the RLS algorithms but yet retain most of their optimal properties, we shall consider a lattice-ladder equalizer in which the number of parameters is significantly reduced. The equalizer structure is shown in Fig. 6.8.6. Each stage of the lattice is characterized by the output-input relations

$$f_m(t) = f_{m-1}(t) + k_m(t)b_{m-1}(t-1)$$
$$b_m(t) = b_{m-1}(t-1) + k_m^*(t)f_{m-1}(t) \qquad (6.8.118)$$

where $k_m(t)$ is the reflection coefficient of the mth stage of the lattice and the forward and backward errors $f_m(t)$ and $b_m(t)$ are defined by (6.8.37) and (6.8.38), respectively. In the present notation, $a_{mm} = k_m(t)$.

This form of the lattice filter is identical to that obtained from the Levinson–Durbin algorithm, except that now $k_m(t)$ is allowed to vary with time, so that the lattice filter adapts to the time variations in the channel. In compari-

FIGURE 6.8.6
Gradient lattice equalizer.

son with the least-squares lattice filter described in Sec. 6.8.3, this lattice structure is more restrictive in that the forward predictor and the backward predictor have identical reflection coefficients (except in reverse order).

The lattice filter parameters $\{k_m\}$ may be optimized according to a MSE criterion or by employing the method of least squares. Suppose we adopt the MSE criterion and select the parameters to minimize the sum of the mean-square forward and backward errors. In other words, the MSE is given as

$$
\begin{aligned}
\mathscr{E}_m &= E\left[|f_m(t)|^2 + |b_m(t)|^2\right] \\
&= E\{|f_{m-1}(t) - k_m b_{m-1}(t-1)|^2 \\
&\quad + |b_{m-1}(t-1) - k_m f_{m-1}(t)|^2\}
\end{aligned}
\tag{6.8.119}
$$

where we have dropped the time dependence on the parameters $\{k_m\}$ for this optimization, since they are now assumed to be time-invariant. Differentiation of $\mathscr{E}_m$ with respect to k_m yields the solution

$$
k_m = \frac{-2E\left[f_{m-1}(t)b_{m-1}^*(t-1)\right]}{E|f_{m-1}(t)|^2 + E|b_{m-1}(t-1)|^2}
\tag{6.8.120}
$$

Alternatively, suppose we adopt a least-squares criterion. The performance index to be minimized is

$$
\begin{aligned}
\mathscr{E}_m^{\mathrm{LS}} &= \sum_{n=0}^{t} w^{t-n}\left[|f_m(n)|^2 + |b_m(n)|^2\right] \\
&= \sum_{n=0}^{t} w^{t-n}\{|f_{m-1}(n) + k_m(t)b_{m-1}(n-1)|^2 \\
&\quad + |b_{m-1}(n-1) + k_m^*(t)f_{m-1}(n)|^2\}
\end{aligned}
\tag{6.8.121}
$$

Minimization of $\mathscr{E}_m^{\mathrm{LS}}$ with respect to $k_m(t)$ yields the solution

$$
k_m(t) = \frac{-2\sum_{n=0}^{t} w^{t-n}f_{m-1}(n)b_{m-1}^*(n-1)}{\sum_{n=0}^{t} w^{t-n}\left[|f_{m-1}(n)|^2 + |b_{m-1}(n-1)|^2\right]}
\tag{6.8.122}
$$

Although (6.8.120) and (6.8.122) are equivalent, the expression in (6.8.122) is the appropriate one for adaptive filtering applications. In particular, the numerator and denominator in (6.8.122) may be updated recursively in time as follows:

$$
\begin{aligned}
u_m(t) &= wu_m(t-1) + 2f_{m-1}(t)b_{m-1}^*(t-1) \\
v_m(t) &= wv_m(t-1) + |f_{m-1}(t)|^2 + |b_{m-1}(t-1)|^2
\end{aligned}
\tag{6.8.123}
$$

Then

$$
k_m(t) = -\frac{u_m(t)}{v_m(t)}
$$

Equivalently, $k_m(t)$ may be updated recursively in time according to the relation

$$k_m(t) = k_m(t-1) + \frac{f_{m-1}(t-1)b_m^*(t-1) + b_{m-1}^*(t-2)f_m(t-1)}{v_m(t-1)}$$

(6.8.124)

The output of the lattice equalizer, which is the estimate of the desired information symbol, is a linear combination of the backward errors $\{b_m(t)\}$, i.e.,

$$\hat{I}(t) = \sum_{k=0}^{N-1} g_k(t)b_k(t)$$

(6.8.125)

where $\{g_k(t)\}$ are the weighting coefficients. The optimum values of the weighting coefficients are obtained by minimizing the MSE between the desired symbol $I(t)$ and the estimate. Let $e_{m+1}(t)$ be the error between the desired symbol and the estimate at the output of an m-stage lattice. Then with $e_0(t) = I(t)$, we have

$$\begin{aligned} e_{m+1}(t) &= I(t) - \sum_{k=0}^{m} g_k(t)b_k(t) \\ &= I(t) - \mathbf{G}'_{m+1}(t)\boldsymbol{\beta}_{m+1}(t) \\ &= e_m(t) - g_m(t)b_m(t) \end{aligned}$$

(6.8.126)

where $\mathbf{G}_{m+1}(t)$ is the $(m+1)$-dimensional vector of weighting coefficients and $\boldsymbol{\beta}_{m+1}(t)$ is the $(m+1)$-dimensional vector of backward errors.

The vector of weighting coefficients is selected to minimize the mean-square value of $e_N(t)$. Assuming that the statistics of the noise and signal components are stationary, we may drop the time dependence of $\mathbf{G}_N$ and apply the orthogonality principle. Thus $\mathbf{G}_N$ is selected to satisfy the orthogonality condition

$$E[e_N(t)\boldsymbol{\beta}_N^*(t)] = \mathbf{0}_N$$

(6.8.127)

Substituting for $e_N(t)$, we obtain

$$\begin{aligned} E\{[I(t) - \mathbf{G}'_N\boldsymbol{\beta}_n(t)]\boldsymbol{\beta}_N^*(t)\} &= \mathbf{0} \\ \mathbf{G}_N &= \{E[\boldsymbol{\beta}_n^*(t)\boldsymbol{\beta}_N'(t)]\}^{-1}E[I(t)\boldsymbol{\beta}_N^*(t)] \end{aligned}$$

(6.8.128)

But the backward residuals are orthogonal, i.e.,

$$E[b_k(t)b_j^*(t)] = \mathcal{E}_k^{bs}\delta_{kj}$$

and, hence, the matrix $E[\boldsymbol{\beta}_N(t)\boldsymbol{\beta}_N'(t)]$ is diagonal. Consequently the optimum coefficients are given as

$$g_k = \frac{1}{\mathcal{E}_k^b}E[I(t)b_k(t)] \qquad k = 0,1,\dots,N-1$$

(6.8.129)

Due to the orthogonality of the residuals, the $\{g_k\}$ are decoupled from one another.

The coefficient vector $\mathbf{G}_N$ with components given by (6.8.129) is easily related to the coefficient vector $\mathbf{C}_N$ in a transversal equalizer. First we note that the expression for the backward error, given by (6.8.38), may be expressed in the matrix form

$$
\begin{bmatrix}
1 & 0 & 0 & 0 & \cdots & 0 \\
-a_{11} & 1 & 0 & 0 & \cdots & 0 \\
-a_{22} & -a_{21} & 1 & 0 & \cdots & 0 \\
\multicolumn{6}{c}{\cdots\cdots\cdots\cdots\cdots\cdots\cdots\cdots\cdots\cdots\cdots\cdots} \\
-a_{mm} & -a_{mm-1} & \cdots & \cdots & -a_{m1} & 1
\end{bmatrix}
\begin{bmatrix}
y(t) \\
y(t-1) \\
\vdots \\
y(t-m)
\end{bmatrix}
=
\begin{bmatrix}
b_0(t) \\
b_1(t) \\
\vdots \\
b_m(t)
\end{bmatrix}
$$

or, equivalently,

$$
\mathbf{H}_{m+1}\mathbf{Y}_{m+1}(t) = \boldsymbol{\beta}_{m+1}(t) \tag{6.8.130}
$$

Using (6.8.130) to substitute for $\boldsymbol{\beta}_N(t)$ in (6.8.128), we obtain

$$
\begin{aligned}
\mathbf{G}_N &= \left\{ E\left[\mathbf{H}_N\mathbf{Y}_N(t)\mathbf{Y}_N'(t)\mathbf{H}_N' \right] \right\}^{-1} E\left[I(t)\mathbf{H}_N\mathbf{Y}(t) \right] \\
&= \left(\mathbf{H}_N\boldsymbol{\Gamma}_N\mathbf{H}_N' \right)^{-1}\mathbf{H}_N\boldsymbol{\xi}_N = (\mathbf{H}^{-1})\boldsymbol{\Gamma}_N^{-1}\boldsymbol{\xi}_N
\end{aligned}
$$

where $\boldsymbol{\Gamma}_N = E[\mathbf{Y}_N(t)\mathbf{Y}_N'(t)]$ and $\boldsymbol{\xi}_N = E[I(t)\mathbf{Y}_N(t)]$. But $\mathbf{C}_N = \boldsymbol{\Gamma}_N^{-1}\boldsymbol{\xi}_N$. Hence

$$
\mathbf{G}_N = \left(\mathbf{H}_N^{-1} \right)'\mathbf{C}_N \tag{6.2.131}
$$

Furthermore, the output of the lattice equalizer is

$$
\begin{aligned}
\hat{I}(t) &= \mathbf{G}_N'\boldsymbol{\beta}_N(t) \\
&= \mathbf{C}_N'\mathbf{H}_N^{-1}\mathbf{H}_N\mathbf{Y}_N(t) \\
&= \mathbf{C}_N'\mathbf{Y}_N(t)
\end{aligned}
$$

This establishes the relationship between the tap coefficients in the transversal equalizer and the weighting coefficients in the lattice equalizer.

There remains the problem of adjusting the coefficients $\{g_k\}$ adaptively. Since the desired $\{g_k\}$ minimize the MSE between $I(t)$ and $\hat{I}(t)$, the error will be orthogonal to the $\{b_k(t)\}$ in $\hat{I}(t)$. This suggests a gradient algorithm of the form

$$
g_k(t+1) = g_k(t) + \frac{e_k(t)b_k^*(t)}{\hat{\mathscr{E}}_k^b(t)} \tag{6.8.132}
$$

where $\hat{\mathscr{E}}_k^b(t)$ is an estimate of $\mathscr{E}_k^b$, which may be computed recursively as

$$
\hat{\mathscr{E}}_k^b(t) = w\hat{\mathscr{E}}_k^b(t-1) + |b_k(t)|^2 \tag{6.8.133}
$$

However, the computation in (6.8.133) can be avoided. Since the forward and backward errors have identical mean-square values, the variable $v_m(t)$ in (6.8.123), which represents the combined residual noise power in $f_m(t)$ and $b_m(t)$, is an estimate of $2\mathscr{E}_m^b$. Hence (6.8.132) is simply replaced by

$$
g_k(t+1) = g_k(t) + \frac{2e_k(t)b_k^*(t)}{v_k(t)} \tag{6.8.134}
$$

and, thus, the computation in (6.8.133) is avoided. Note that (6.8.134) is an

TABLE 6.8.3
Gradient lattice-ladder aglorithm

$$v_m(t) = wv_m(t-1) + |f_{m-1}(t)|^2 + |b_{m-1}(t-1)|^2$$

$$k_m(t) = k_m(t-1) + \frac{f_{m-1}(t-1)b_m^*(t-1) + b_{m-1}^*(t-2)f_m(t-1)}{v_m(t-1)}$$

$$f_m(t) = f_{m-1}(t) - k_m(t)b_{m-1}(t-1)$$

$$b_m(t) = b_{m-1}(t) - k_m^*(t)f_{m-1}(t)$$

$$e_{m+1}(t) = e_m(t) - b_m(t)g_m(t)$$

$$\hat{I}(t) = \beta_N(t)\mathbf{G}_N(t)$$

$$g_m(t+1) = g_m(t) + \frac{2e_m(t)b_m^*(t)}{v_m(t)}$$

Initialization

$$f_0(t) = b_0(t) = y(t) \qquad f_m(-1) = b_m(-1) = b_m(-2) = 0$$

$$e_0(t) = I(t) \qquad e_m(0) = 0 \qquad m > 1$$

$$v_m(-1) = \varepsilon > 0$$

$$g_m(0) = 0$$

$$k_m(-1) = 0$$

LMS-type (gradient) algorithm with a step size of $2/v_k(t)$. For this reason, the equalizer is called a *gradient lattice-ladder equalizer*. The algorithm is listed in Table 6.8.3. It has a computational complexity of $11N$ complex multiplications and divisions. However, the convergence rate of the gradient lattice is a little slower than that of the least-squares lattice equalizer.

Simulation results in Figs. 6.8.7 and 6.8.8, developed by Satorius and Pack (1981), illustrate the convergence rates of the least-squares lattice equalizer, the gradient lattice equalizer, and the transversal equalizer with the gradient algorithm for adjusting the coefficients, for the two time-dispersive channels shown in Fig. 6.8.9. The first channel has an eigenvalue spread $\lambda_{\max}/\lambda_{\min} = 11$, while the second has a spread of 21. The signal-to-noise ratio is 30 dB. The linear transversal equalizer had 11 taps and a step-size parameter $\Delta = 0.02$ was used to adjust its coefficients. This choice of Δ results in approximately the same minimum MSE as that of the lattice equalizers. Both lattice equalizers had 10 sections, which is equivalent to an 11-tap equalizer. Since the channels are time-invariant, the weighting factor w is selected as unity.

The results of these simulations show that the recursive least-squares equalizer converges in about 40 to 50 iterations, while the gradient lattice equalizer converges in about 120 iterations, independent of the eigenvalue spread. On the other hand, the linear transversal equalizer with coefficients adjusted adaptively by means of the LMS gradient algorithm given in (6.8.23) requires

DIGITAL SIGNALING OVER A BANDWIDTH-CONSTRAINED LINEAR FILTER CHANNEL **673**

FIGURE 6.8.7
Comparison of the convergence rate of lattice equalizers and the gradient LMS algorithm for an eigenvalue ratio of 11.

FIGURE 6.8.8
Comparison of the convergence rate of lattice equalizers and the gradient LMS algorithm for an eigenvalue ratio of 21.

(a)

(b)

FIGURE 6.8.9
Equivalent discrete-time channel characteristics used to generate the convergence results in Figs. 6.8.6 and 6.8.7.

approximately 600 to 700 iterations to converge when $\lambda_{max}/\lambda_{min} = 11$ and approximately 1000 to 1100 iterations to converge when $\lambda_{max}/\lambda_{min} = 21$. Therefore, as previously indicated, the convergence rate of the gradient algorithm is not only slow, but it also depends on the channel characteristics. In some applications, such as digital communications over telephone channels which exhibit very slow time variations, this convergence rate is adequate. However, on a number of radio channels the time variations are sufficiently fast that only recursive least-squares algorithms and the gradient lattice algorithms are sufficiently fast to track the channel characteristics.

A gradient lattice algorithm can also be devised for a DFE structure. This has been done and its performance has been described in the paper by Ling and Proakis (1985).

6.8.5 Fast (Kalman) RLS Algorithm

In Sec. 6.8.1 we described transversal (direct form) linear and decision-feedback equalizers based on the method of recursive least squares and derived a RLS (Kalman) algorithm for updating the tap coefficients of the equalizer. An important variation of this RLS algorithm is the square-root RLS algorithm which has a computational complexity of $1.5N^2 + 6.5N$. However, this algorithm is still computationally intensive. The reason for such a large computational burden is due to the matrix multiplication in calculating the Kalman gain vector $\mathbf{K}_N(t)$ and in updating the inverse covariance matrix $\mathbf{P}_N(t)$. On the other hand, the recursive least-squares lattice and gradient lattice algorithms described in Secs. 6.8.3 and 6.8.4 are more efficient. They avoid matrix multiplications and, hence, the number of computations per symbol is proportional to N.

It is possible to modify the RLS Kalman algorithm so as to avoid the matrix multiplications involved in computing $\mathbf{K}_N(t)$ and $\mathbf{P}_N(t)$. The modification is based on the use of the forward and backward prediction relations, previously derived for the least-squares lattice algorithm, to arrive at a time update for the Kalman gain vector. The resulting algorithm has a complexity that is proportional to N, and, hence, it is called a *fast RLS Kalman algorithm*. The proportionality factor is smaller than that of the recursive least-squares lattice equalizer and comparable to that of the gradient lattice equalizer.

Let us develop a fast RLS Kalman algorithm for the linear equalizer by using the relations derived in Secs. 6.8.1 and 6.8.3. The extension to the decision-feedback equalizer requires a two-channel formulation of the linear prediction relations given in Sec. 6.8.3. For the sake of brevity, this derivation is omitted. The fast RLS Kalman DFE algorithm is given in the paper by Proakis and Ling (1984).

We restate, in the proper sequence in which the computations are performed, the following relations given previously in Secs. 6.8.1 and 6.8.3:

$$f_{N-1}(t) = y(t) + \mathbf{A}'_{N-1}(t-1)\mathbf{Y}_{N-1}(t-1) \tag{6.8.135}$$

$$b_{N-1}(t) = y(t-N+1) + \mathbf{B}'_{N-1}(t-1)\mathbf{Y}_{N-1}(t) \tag{6.8.136}$$

$$\mathbf{A}_{N-1}(t) = \mathbf{A}_{N-1}(t-1) - \mathbf{K}_{N-1}(t-1)f_{N-1}(t) \tag{6.8.137}$$

$$f_{N-1}(t,t) = \alpha_{N-1}(t-1)f_{N-1}(t) \tag{6.8.138}$$

Next we need a time-update equation for $r^f_{N-1}(t)$. Beginning with the definition of $r^f_{N-1}(t)$ given by (6.8.54) and substituting (6.8.137) for $\mathbf{A}_{N-1}(t)$ and (6.8.52) for $\mathbf{Q}_{N-1}(t)$, we obtain

$$
\begin{aligned}
r^f_{N-1}(t) &= q(t) + \mathbf{A}'_{N-1}(t)\mathbf{Q}^*_{N-1}(t) \\
&= wq(t-1) + |y(t)|^2 + \left[\mathbf{A}'_{N-1}(t-1) - \mathbf{K}'_{N-1}(t-1)f_{n-1}(t)\right] \\
&\quad \times \left[w\mathbf{Q}^*_{N-1}(t-1) + y^*(t)\mathbf{Y}_{N-1}(t-1)\right] \\
&= wr^f_{N-1}(t-1) + \alpha_{N-1}(t-1)|f_{N-1}(t)|^2 \\
&= wr^f_{N-1}(t-1) + f_{N-1}(t,t)f_{n-1}(t) \tag{6.8.139}
\end{aligned}
$$

The time-update equation for the Kalman gain vector is obtained by substituting for $\mathbf{P}_N(t)$ in (6.8.15) using the update equation in (6.8.74). Thus

$$
\begin{aligned}
\mathbf{K}_N(t) &= \mathbf{P}_N(t)\mathbf{Y}^*_N(t) \\
&= \begin{bmatrix} 0 & 0 \\ 0 & \mathbf{P}_{N-1}(t-1) \end{bmatrix}\begin{bmatrix} y^*(t) \\ \mathbf{Y}^*_{N-1}(t-1) \end{bmatrix} \\
&\quad + \frac{1}{r^f_{N-1}(t)}\begin{bmatrix} 1 \\ \mathbf{A}_{N-1}(t) \end{bmatrix}\begin{bmatrix} 1 & \mathbf{A}^{*\prime}_{N-1}(t) \end{bmatrix}\begin{bmatrix} y^*(t) \\ \mathbf{Y}^*_{N-1}(t-1) \end{bmatrix} \\
&= \begin{bmatrix} 0 \\ \mathbf{K}_{N-1}(t-1) \end{bmatrix} + \frac{f^*_{N-1}(t,t)}{r^f_{N-1}(t)}\begin{bmatrix} 1 \\ \mathbf{A}_{N-1}(t) \end{bmatrix} \tag{6.8.140}
\end{aligned}
$$

It is convenient to partition $\mathbf{K}_N(t)$ as

$$\mathbf{K}_N(t) = \begin{bmatrix} \mathbf{M}_{N-1}(t) \\ k_{NN}(t) \end{bmatrix} \tag{6.8.141}$$

where $\mathbf{M}_{N-1}(t)$ is an $(N-1)$-dimensional vector and $k_{NN}(t)$ is a scalar. To perform the time-update computation in (6.8.140), we need the expression for $\mathbf{K}_{N-1}(t-1)$. This can be obtained from (6.8.98), which relates $\mathbf{K}_N(t)$ to the backward error. That is,

$$\mathbf{K}_N(t) = \begin{bmatrix} \mathbf{K}_{N-1}(t) \\ 0 \end{bmatrix} + \frac{b_{N-1}^*(t,t)}{r_{N-1}^b(t)} \begin{bmatrix} \mathbf{B}_{N-1}(t) \\ 1 \end{bmatrix} = \begin{bmatrix} \mathbf{M}_{N-1}(t) \\ k_{NN}(t) \end{bmatrix} \tag{6.8.142}$$

Clearly

$$\frac{b_{N-1}^*(t,t)}{r_{N-1}^b(t)} = k_{NN}(t) \tag{6.8.143}$$

where $k_{NN}(t)$ has already been determined from (6.8.140). Thus

$$\mathbf{K}_{N-1}(t) + k_{NN}(t)\mathbf{B}_{N-1}(t) = \mathbf{M}_{N-1}(t)$$

$$\mathbf{K}_{N-1}(t) + k_{NN}(t)\left[\mathbf{B}_{N-1}(t-1) + \mathbf{K}_{N-1}(t)b_{N-1}(t)\right] = \mathbf{M}_{N-1}(t)$$

$$\mathbf{K}_{N-1}(t) = \frac{\mathbf{M}_{N-1}(t) - k_{NN}(t)\mathbf{B}_{N-1}(t-1)}{1 - k_{NN}(t)b_{N-1}(t)} \tag{6.8.144}$$

This is the time-update equation for $\mathbf{K}_{N-1}(t)$.

Having $\mathbf{K}_{N-1}(t)$, we may now update $\mathbf{B}_{N-1}(t)$. That is,

$$\mathbf{B}_{N-1}(t) = \mathbf{B}_{N-1}(t-1) - \mathbf{K}_{N-1}(t)b_{N-1}(t) \tag{6.8.145}$$

We must also determine a time-update equation for $\alpha_{N-1}(t)$. From (6.8.101) we have

$$\alpha_N(t) = \alpha_{N-1}(t) - \frac{\alpha_{N-1}^2(t)|b_{N-1}(t)|^2}{r_{N-1}^b(t)}$$

$$= \alpha_{N-1}(t)\left[1 - k_{NN}(t)b_{N-1}(t)\right] \tag{6.8.146}$$

A second relation is obtained by using (6.8.140) to eliminate $\mathbf{K}_{N-1}(t)$ in the expression for $\alpha_N(t)$. Thus

$$\alpha_N(t) = 1 - \mathbf{Y}_N'(t)\mathbf{K}_N(t)$$

$$= 1 - \left[y(t)\mathbf{Y}_{N-1}'(t-1)\right]$$

$$\times \left\{ \begin{bmatrix} 0 \\ \mathbf{K}_{N-1}(t-1) \end{bmatrix} + \frac{f_{N-1}^*(t,t)}{r_{N-1}^f(t)} \begin{bmatrix} 1 \\ \mathbf{A}_{N-1}(t) \end{bmatrix} \right\}$$

$$= \alpha_{N-1}(t-1) - \frac{\alpha_{N-1}(t-1)f_{N-1}^*(t,t)f_{N-1}(t)}{r_{N-1}^f(t)} \tag{6.8.147}$$

Equating the two expressions for $\alpha_N(t)$ in (6.8.146) and (6.8.147) and solving for

$\alpha_{N-1}(t)$, we obtain

$$\alpha_{N-1}(t) = \alpha_{N-1}(t-1) \frac{1 - \dfrac{f_{N-1}^*(t,t)f_{N-1}(t)}{r_{N-1}^f(t)}}{1 - k_{NN}(t)b_{N-1}(t)} \qquad (6.8.148)$$

Thus we have the desired time-update equation for $\alpha_{N-1}(t)$.

The output of the equalizer and the corresponding error are

$$\hat{I}(t) = \mathbf{C}_N'(t-1)\mathbf{Y}_N(t) \qquad (6.8.149)$$

$$e_N(t) = I(t) - \hat{I}(t) \qquad (6.8.150)$$

The final step is to update the equalizer coefficient vector. This equation is

$$\mathbf{C}_N(t) = \mathbf{C}_N(t-1) + \mathbf{K}_N(t)e_N(t) \qquad (6.8.151)$$

This algorithm is summarized in Table 6.8.4. The computational complexity of this algorithm is $9N + 3$ complex multiplications and divisions per output signal sample. In this respect it is slightly better than the gradient lattice algorithm.

Other versions of the fast RLS (Kalman) algorithm that have a computational complexity of $7N$ have been given in the literature, but these algorithms are considerably more sensitive to roundoff noise than the algorithm given in Table 6.8.4. The interested reader is referred to the papers by Carayannis et al. (1983, 1986) and Cioffi and Kailath (1984).

6.8.6 Properties of the RLS Algorithms

The RLS algorithms that we have described in Secs. 6.8.1, 6.8.3, 6.8.4, and 6.8.5 have a number of desirable properties. In this section we consider the properties of these algorithms and compare them with the corresponding properties of the LMS algorithm.

Convergence rate. All the RLS algorithms (lattice and direct form) basically have the same convergence rate. The gradient lattice algorithm is only an approximation to an RLS algorithm and, hence, its convergence rate is slightly slower (by about a factor of 2) as we have already demonstrated. All the RLS algorithms and the gradient lattice algorithm provide significantly faster convergence than the LMS algorithm. In fact, it is this basic limitation of the LMS algorithm that has led us to consider faster converging algorithms.

Computational requirements. The RLS lattice and the fast RLS (Kalman) algorithms have a computational complexity proportional to N. In contrast, the RLS square-root algorithms have a complexity proportional to N^2. Table 6.8.5 lists the computational complexity of these algorithms for the DFE in terms in complex multiplications and divisions. The corresponding graphs of the number of computations as a function of the number of equalizer coefficients are shown in

TABLE 6.8.4
Fast RLS (Kalman) algorithm

$$f_{N-1}(t) = y(t) + \mathbf{A}'_{N-1}(t-1)\mathbf{Y}_{N-1}(t-1)$$

$$b_{n-1}(n) = y(t - N + 1) + \mathbf{B}'_{N-1}(t-1)\mathbf{Y}_N(t)$$

$$\mathbf{A}_{N-1}(t) = \mathbf{A}_{N-1}(t-1) - \mathbf{K}_{N-1}(t-1)f_{N-1}(t)$$

$$f_{N-1}(t, t) = \alpha_{N-1}(t-1)f_{N-1}(t)$$

$$r^f_{N-1}(t) = wr^f_{N-1}(t-1) + \alpha_{N-1}(t-1)|f_{N-1}(t)|^2$$

$$\begin{bmatrix} \mathbf{M}_{N-1}(t) \\ k_{NN}(t) \end{bmatrix} \equiv \mathbf{K}_N(t) = \begin{bmatrix} 0 \\ \mathbf{K}_{N-1}(t-1) \end{bmatrix} + k_{NN}(t)\begin{bmatrix} 1 \\ \mathbf{A}_{N-1}(t) \end{bmatrix}$$

$$\mathbf{K}_{N-1}(t) = \frac{\mathbf{M}_{N-1}(t) - k_{NN}(t)\mathbf{B}_{N-1}(t-1)}{1 - k_{NN}(t)b_{N-1}(t)}$$

$$\mathbf{B}_{N-1}(t) = \mathbf{B}_{N-1}(t-1) - \mathbf{K}_{N-1}(t)b_{N-1}(t)$$

$$\alpha_{N-1}(t) = \alpha_{N-1}(t-1)\left[\frac{1 - \dfrac{f_{N-1}(t)f^*_{N-1}(t,t)}{r^f_{N-1}(t)}}{1 - k_{NN}(t)b_{N-1}(t)} \right]$$

$$\hat{I}(t) = \mathbf{C}'_N(t-1)\mathbf{Y}_N(t)$$

$$e_N(t) = I(t) - \hat{I}(t)$$

$$\mathbf{C}_N(t) = \mathbf{C}_N(t-1) + \mathbf{K}(t)e_N(t)$$

Initialization

$$\mathbf{A}_{N-1}(-1) = \mathbf{B}_{N-1}(-1) = \mathbf{0}$$

$$\mathbf{K}_{N-1}(-1) = \mathbf{0} \qquad \mathbf{C}_N(-1) = \mathbf{0}$$

$$r^f_{N-1}(-1) = \varepsilon > 0$$

TABLE 6.8.5
Computational complexity of adaptive DFE algorithms

Algorithm	Total number of complex operations	Number of divisions
LMS DFE	$2N + 1$	0
Fast RLS (Kalman) DFE	$20N + 5$	3
Conventional RLS (Kalman) DFE	$2.5N^2 + 4.5N$	2
Square-root RLS (Kalman) DFE	$1.5N^2 + 6.5N$	N
Gradient lattice DFE	$13N_1 + 33N_2 - 36$	$2N_1$
RLS lattice DFE	$18N_1 + 39N_2 - 39$	$2N_1$

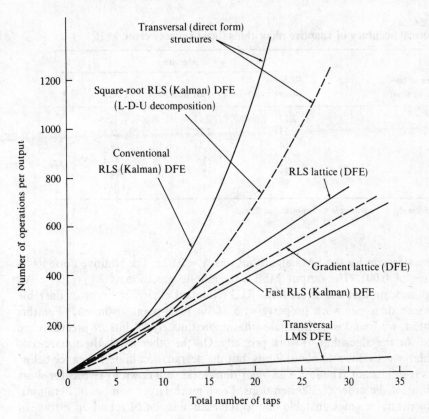

FIGURE 6.8.10
Computational complexity of adaptive DFE's.

Fig. 6.8.10. For the lattice equalizer, we have assumed that $N_1 = N_2 = N/2$. Note that for equalizers with more than 10 taps coefficients, the lattice structures and the fast RLS (Kalman) algorithm are much more efficient than the square-root or conventional RLS (Kalman) algorithms. On the other hand, for small equalizers there is little difference in the computational complexity of the rapidly converging algorithms.

Numerical properties. The numerical properties of the RLS algorithms have been investigated by several researchers, including Ling and Proakis (1984a,b), Gardner (1984), Ljung and Ljung (1985), and Cioffi (1987).

For illustrative purposes, Table 6.8.6 includes simulation results on the steady state (time-averaged) square error for the RLS square-root (Kalman) algorithm, the fast RLS (Kalman) algorithm, and the LMS algorithm for different word lengths. The simulation was performed with a linear adaptive equalizer having $M = 11$ coefficients. The channel had an eigenvalue ratio of $\lambda_{max}/\lambda_{min} = 11$. The exponential weighting factor used in the RLS algorithms was $w = 0.975$

TABLE 6.8.6
Numerical accuracy of adaptive algorithms (least-squares error $\times\ 10^{-3}$)

Number of bits (including sign)	Algorithm		
	RLS square root	Fast RLS	LMS
16	2.17	2.17	2.30
13	2.33	2.21	2.30
11	6.14	3.34	19.0
10	17.6	†	77.2
9	75.3	†	311.0
8	†	†	1170.0

† Algorithm does not converge to optimum coefficients.

and the step size for the LMS algorithm was $\Delta = 0.025$. The additive noise has a variance of 0.001. The output MSE with infinite precision is 2.1×10^{-3}. We should indicate that the direct-form RLS (Kalman) algorithm becomes unstable and hence does not work properly with 16-bit fixed-point arithmetic. For this algorithm, we found experimentally that approximately 24 bits of precision are needed for the algorithm to work properly. On the other hand, the square-root algorithm works down to about 9 bits, but the degradation in performance below 11 bits is significant. The fast RLS algorithm works well down to 11 bits for short durations of the order of 500 iterations. For a much larger number of iterations, the algorithm becomes unstable due to the accumulation of round-off errors. In such a case several methods have been proposed for preventing overflow in the filter coefficients. The interested reader may refer to the papers of Eleftheriou and Falconer (1987), Stock and Kailath (1988), Cioffi and Kailath (1984), and Hsu (1982). We also observe from the results of Table 6.8.6 that the LMS algorithm is quite robust to round-off noise. It deteriorates as expected with a decrease in the precision of the filter coefficients, but we did not observe any catastrophic failure (instability) with 8 or 9 bits of precision. However, the degradation in performance below 12 bits is significant.

The RLS and gradient lattice-ladder algorithms are particularly attractive because of their numerical robustness. First of all, these lattice-ladder algorithms are numerically stable as demonstrated by Ling et al. (1985) and Ling and Proakis (1984). By numerical stability we mean that the output estimation error from the computations is bounded when a bounded error signal is introduced at the input (bounded-input, bounded-output stability). Second, the numerical accuracy of the optimum solution (in steady state operation) is also relatively good. For purposes of comparison, we illustrate in Table 6.8.7 the steady state average squared error obtained by means of computer simulation of the RLS and gradient lattice algorithms and the transversal (direct form) algorithms. The striking result in Table 6.8.7 is the superior performance obtained with the RLS lattice-ladder algorithm in which the reflection coefficients and the ladder gains

TABLE 6.8.7
Numerical accuracy, in terms of output MSE for channel with $\lambda_{max} / \lambda_{min} = 11$ and $w = 0.975$, MSE $\times 10^{-3}$

Number bits (including sign)	Algorithm				
	RLS square root	Fast RLS	Conventional RLS lattice	Error feedback RLS lattice	LMS
16	2.17	2.17	2.16	2.16	2.30
13	2.33	2.21	3.09	2.22	2.30
11	6.14	3.34	25.2	3.09	19.0
9	75.3	†	365	31.6	311

† Algorithm did not converge.

are updated directly according to (6.8.117). This is the error-feedback form of the RLS lattice algorithm. It is clear that the direct updating of these coefficients is significantly more robust to roundoff errors then all the other adaptive algorithms, including the LMS algorithm. It is also apparent that the two-step process used in the conventional RLS lattice algorithm to estimate the reflection coefficients is not as accurate. Furthermore the estimation errors that are generated in the coefficient at each stage propagate from stage to stage, causing additional errors.

In the gradient lattice algorithm, the reflection coefficients and the ladder gains are also updated directly. Consequently, the numerical accuracy of the gradient lattice algorithm is comparable to that obtained with the direct update form of the RLS lattice.

Analytical and simulation results on numerical stability and numerical accuracy in fixed-point implementation of these algorithms can be found in the papers by Ling and Proakis (1984a, b), Ling et al. (1986a, b), Ljung and Ljung (1985), and Gardner (1984).

Error rate performance. The rapid convergence properties of the RLS algorithms make it possible to track time variations in time-variant dispersive channels that cause intersymbol interference. An example of such a channel is when the ionosphere is used for long distance transmission at frequencies below 30 MHz. Let us consider modelling such a channel by the filter characteristic

$$H(z) = a_0(t) + a_1(t)z^{-1} + a_2(t)z^{-2}$$

where z^{-1} represents a unit of delay equal to one symbol. Thus, we have a discrete-time channel with three time-variant tap coefficients $\{a_i(t)\}$. These coefficients are generated by passing white gaussian noise through lowpass filters of a specified bandwidth. Let us assume that we have a nominal 3 kHz channel, a signaling rate of 2400 symbols per second, and noise filtered by a two-pole Butterworth filter having a 3-dB bandwidth of 0.5 Hz. The graphs of the tap

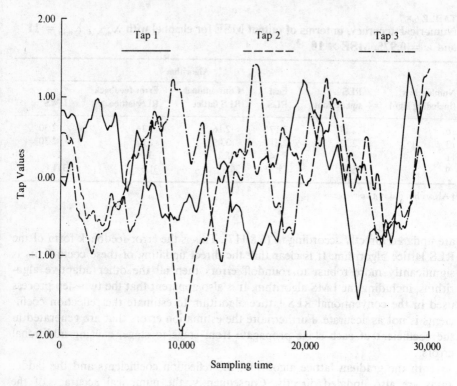

FIGURE 6.8.11
Tap values of time-variant channel.

values obtained by passing the white gaussian noise through the Butterworth filters are shown in Fig. 6.8.11.

The performance obtained with a DFE is illustrated in Fig. 6.8.12 for the LMS and RLS lattice algorithm. Also shown for comparison is the performance of the optimum DFE which is based on exact knowledge of the channel coefficients. Note that the LMS DFE performs poorly because it cannot track the time variations in the channel.

The results in Fig. 6.8.12 were based on the correct symbols being fed back in the DFE. In Fig. 6.8.13 we illustrate the performance with decision-directed operation. To help the equalizer during times when the channel is in a deep fade (a_0, a_1, a_2 are all small), a small percentage of the transmitted bits were training bits known to the equalizer. The percentage of bits used for training is also noted in Fig. 6.8.13. These results demonstrate the capability of the RLS algorithms to track time variations inherent in a number of radio channels.

Implementation considerations. As we have observed, the lattice filter structure is highly modular and allows for the computations to be pipelined. Because of the high degree of modularity, the RLS and gradient lattice algorithms are particu-

FIGURE 6.8.12
Error rate performance of DFE's on time-variant channel (correct symbols fed back).

larly suitable for implementation in VLSI (very large scale integration). As a result of this advantage in implementation and the desirable properties of stability, excellent numerical accuracy, and fast convergence, we anticipate that in the near future, more and more adaptive equalizers will be implemented as lattice-ladder structures.

6.9 ECHO CANCELLATION IN DATA TRANSMISSION OVER TELEPHONE CHANNELS

In the transmission of data over telephone channels, modems (modulator/demodulator) are used to provide an interface between the digital data sequence and the analog channel. Shown in Fig. 6.9.1 is a block diagram of a communications system in which two terminals, labeled A and B, transmit data by using

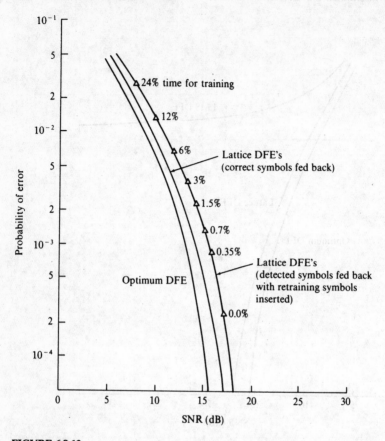

FIGURE 6.8.13
Error rate performance of lattice DFE's on time-variant channel (detected symbols fed back with training symbols inserted).

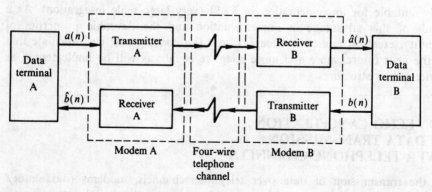

FIGURE 6.9.1
Full-duplex data transmission over telephone channels.

modems A and B to interface to a telephone channel. As shown, a digital sequence $a(n)$ is transmitted from terminal A to terminal B while another digital sequence $b(n)$ is transmitted from terminal B to terminal A. This simultaneous transmission in both directions is called *full-duplex transmission*.

When a subscriber leases a private line from a telephone company for the purpose of transmitting data between terminals A and B, the telephone line provided is a four-wire line, which is equivalent to having two dedicated telephone (two-wire) channels, one (pair of wires) for transmitting data in one direction and one (pair of wires) for receiving data from the other direction. In such a case the two transmission paths are isolated and, consequently, there is no "cross talk" or mutual interference between the two signal paths. Channel distortion is compensated by use of an adaptive equalizer, as described above, at the receiver of each modem.

The major problem with the system shown in Fig. 6.9.1 is the cost of leasing a four-wire telephone channel. If the volume of traffic is high and the telephone channel is used either continuously or frequently, as in banking transactions systems or airline reservation systems, the system pictured in Fig. 6.9.1 may be cost effective. Otherwise, it is not.

An alternative solution for low volume, infrequent transmission of data is to use the dial-up switched telephone network. In this case, the local communication link between the subscriber and the local central telephone office is a two-wire line, called the *local loop*. At the central office, the subscriber two-wire line is connected to the main four-wire telephone channels that interconnect different central offices, called *trunk lines*, by a device called a *hybrid*. By using transformer coupling, the hybrid is tuned to provide isolation between the transmit and receive channels in full duplex operation. However, due to impedance mismatch between the hybrid and the telephone channel, the level of isolation is often insufficient and, consequently, some of the signal on the transmit side leaks back and corrupts the signal on the receiver side, causing an "echo" that is often heard in voice communications over telephone channels.

To mitigate the echoes in voice transmission, the telephone companies employ a device called an *echo suppressor*. In data transmission, the solution is to use an *echo canceller* within each modem. The echo cancellers are implemented as adaptive filters with automatically adjustable coefficients, just as in the case of transversal equalizers.

With the use of hybrids to couple a two-wire channel to a four-wire channel and echo cancellers at each modem to estimate and subtract out the echoes, the data communications system for the dial-up switched network takes the form shown in Fig. 6.9.2. A hybrid is needed at each modem to isolate the transmitter from the receiver and to couple to the two-wire local loop. Hybrid A is physically located at the central office of subscriber A while hybrid B is located at the central office to which subscriber B is connected. The two central offices are connected by a four-wire line, one pair used for transmission from A to B and the other pair used for transmission in the reverse direction, from B to A. An

FIGURE 6.9.2
Block diagram model of a digital communication system that uses echo cancellers in the modems.

echo at terminal A due to the hybrid A is called a *near-end echo*, while an echo at terminal A due to the hybrid B is termed a *far-end echo*. Both types of echoes are usually present in data transmission and must be removed by the echo canceller.

Suppose we neglect the channel distortion for purposes of this discussion, and let us deal with the echoes only. The signal received at modem A may be expressed as

$$s_{RA}(t) = A_1 s_B(t) + A_2 s_A(t - d_1) + A_3 s_A(t - d_2)$$

where $s_B(t)$ is the desired signal to be demodulated at modem A, $s_A(t - d_1)$ is the near-end echo due to hybrid A, $s_A(t - d_2)$ is the far-end echo due to hybrid B, A_i, $i = 1, 2, 3$, are the corresponding amplitudes of the three signal components, and (d_1, d_2) are the delays associated with the echo components. A further disturbance that corrupts the received signal is additive noise, so that the received signal at modem A is

$$r_A(t) = s_{RA}(t) + w(t)$$

where $w(t)$ represents the additive process.

The adaptive echo canceller attempts to estimate adaptively the two echo components. If its coefficients are $h(n)$, $n = 0, 1, \ldots, N - 1$, its output is

$$\hat{s}_A(n) = \sum_{k=0}^{N-1} h(k) a(n - k)$$

which is an estimate of the echo signal components. This estimate is subtracted from the sampled received signal and the resulting error signal can be minimized in the least-squares sense to optimally adjust the coefficients of the echo canceller.

There are several possible configurations for placement of the echo canceller in the modem and for forming the corresponding error signal. Figure 6.9.3 illustrates one configuration in which the canceller output is subtracted from the sampled output of the demodulator. Figure 6.9.4 illustrates a second configuration, in which the echo canceller is generating samples at the Nyquist rate instead of the symbol rate. In this case the error signal used to adjust the coefficients is simply the difference between the sampled received signal and the canceller output. Finally, Fig. 6.9.5 illustrates the canceller operating in combination with an adaptive equalizer.

The adaptive LMS and RLS algorithms that we have described in this chapter can be applied to the adaptation of the echo canceller. In telephone channel communications the LMS algorithm is more than adequate for tracking purposes and there is no need for the more complex RLS algorithms.

The major problem with adapting the echo canceller is that the near- and far-end echo signals may be several orders of magnitude (10 to 40 dB) larger than the received signal. Consequently, the amount of echo suppression that must be provided by the canceller is very large. Otherwise the received signal will be masked by the echo. A large suppression requires that the coefficients of the echo

FIGURE 6.9.3
Symbol-rate echo canceller.

canceller must be precisely set and the dynamic range of the coefficients (number of bits of accuracy) must be significantly higher than one would need in an adaptive equalizer. Furthermore, the initial adjustment period for the echo canceller is usually longer than that for an equalizer.

In the adjustment of the coefficients of the echo canceller, the desired signal is basically the near- and the far-end echo signals. As indicated above, the canceller attempts to synthesize a signal that matches the echo according to the optimization criterion, e.g., minimum mean square error. However, once the echo has been sufficiently suppressed so that the incoming received data signal to the

FIGURE 6.9.4
Nyquist rate echo canceller.

FIGURE 6.9.5
Modem with adaptive equalizer and echo canceller.

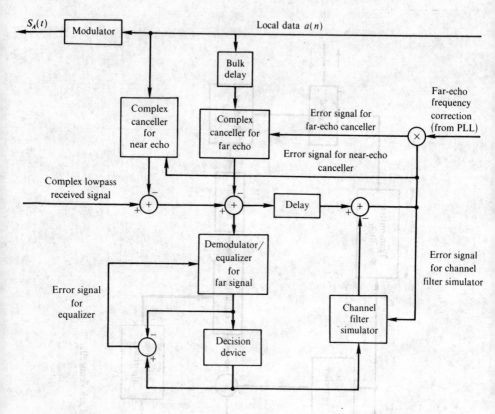

FIGURE 6.9.6
Adaptive cancellation of near- and far-end echoes with suppression of far-end received signal.

demodulator is larger than the echo, the error signal to the echo canceller is dominated by the incoming data signal which acts as an additive noise, thus, making it difficult for the echo canceller to adjust its coefficients very precisely.

To remedy this problem, we can eliminate the received data signal from the error signal to the echo canceller by regenerating the received signal after demodulation and subtracting it from the error signal. The general configuration for this scheme is shown in Fig. 6.9.6. Note that a bulk delay is used in the suppression of the far-end echo. This is desirable in order to keep the echo canceller from becoming too large and having many idle (zero weight) taps. Computer simulations of this technique given in the paper by Quatieri and O'Leary (1986) demonstrate that the suppression of the far- and near-end echos is significantly better after the received data signal is subtracted from the error signal to the echo canceller.

The far-end echo signal usually contains a frequency offset of a few hertz, which must be estimated along with the echo signal. The estimate of the frequency offset may be used to frequency-shift the echo signal prior to cancellation.

To date, modems with echo cancellers have made it possible to transmit data on two-wire telephone circuits in both directions simultaneously at speeds up to 9600 bits per second.

The design of echo cancellers has received considerable attention in the technical literature. We cite for reference the papers by Koll and Weinstein (1973), Weinstein (1979), Mueller (1979), Falconer and Mueller (1979), Sondhi and Berkley (1980), Falconer (1982), Salz (1983), and Messerschmidt (1984).

6.10 BIBLIOGRAPHICAL NOTES AND REFERENCES

The pioneering work on signal design for bandwidth-constrained channels was done by Nyquist (1928). The use of binary partial response signals was originally proposed by Lender (1963) and was later generalized by Kretzmer (1966). Other early work on problems dealing with intersymbol interference (ISI) and transmitter and receiver optimization with constraints on ISI was done by Gerst and Diamond (1961), Tufts (1965), Smith (1965), and Berger and Tufts (1967). "Faster than Nyquist" transmission has been studied by Mazo (1975) and Foschini (1984).

Adaptive equalization for digital communications was developed by Lucky (1965, 1966). His algorithm was based on the peak distortion criterion and led to the zero-forcing algorithm. Lucky's work was a major breakthrough which led to the rapid development of high speed modems within five years of publication of his work. Concurrently, the LMS algorithm was devised by Widrow (1966) and its use for adaptive equalization for complex-valued (in-phase and quadrature components) signals was described and analyzed in a tutorial paper by Proakis and Miller (1969).

Decision-feedback equalization was proposed and analyzed by Austin (1967). The adaptive DFE was developed and applied to the equalization of fading dispersive channels by Monsen (1971). An analysis of the performance of the DFE is given in the papers by George et al. (1971), Price (1972), Salz (1973), and Duttweiler et al. (1974).

The Viterbi algorithm was devised by Viterbi (1967) for the purpose of decoding convolutional codes. Its use as the optimal maximum likelihood sequence estimator for symbols corrupted with ISI was proposed and analyzed by Forney (1972) and Omura (1971). Its use for carrier-modulated signals was considered by Ungerboeck (1974). It was subsequently applied to the demodulation of CPM by Aulin and Sunberg (1981a, b) and others.

A tutorial treatment of adaptive equalization algorithms that were developed during the period 1965 through 1975 is given by Proakis (1975). A more recent tutorial treatment of adaptive equalization is given in the paper by Qureshi (1985). The major breakthrough in adaptive equalization techniques, beginning with the work of Lucky in 1965 coupled with the development of trellis-coded modulation which was proposed by Ungerboeck and Csajka (1976), has led to the development of commercially available high speed modems with a capability of speeds between 9600 and 19,200 bits per second on telephone channels.

The use of a faster converging algorithm for adaptive equalization was proposed by Godard (1974). Our derivation of the RLS (Kalman) algorithm, described in Sec. 6.8.1 follows the approach outlined by Picinbono (1978). RLS lattice algorithms for general signal estimation applications were developed by Morf et al. (1977a, b, c, d, 1979). The applications of these algorithms have been investigated by several researchers, including Makhoul (1978), Satorius and Pack (1981), Satorius and Alexander (1979), and Ling and Proakis (1982, 1984a, b, c, 1985). The fast RLS Kalman algorithm for adaptive equalization was first described by Falconer and Ljung (1978). Our derivation of this algorithm was based on fixing the order of the RLS lattice algorithm and obtaining a time-update equation for the Kalman gain vector. The above references are just a few of the important papers that have been published on RLS algorithms for adaptive equalization and other applications.

APPENDIX 6A
THE CHOLESKY FACTORIZATION

Consider the solution of the set of linear equations

$$\mathbf{R}_N \mathbf{C}_N = \mathbf{U}_N \tag{6A.1}$$

where $\mathbf{R}_N$ is an $N \times N$ positive-definite symmetric matrix, $\mathbf{C}_N$ is an N-dimensional vector of coefficients to be determined, and $\mathbf{U}_N$ is an arbitrary N-dimensional vector. The equations in (6A.1) can be solved efficiently by expressing $\mathbf{R}_N$ in the factored form

$$\mathbf{R}_N = \mathbf{S}_N \mathbf{D}_N \mathbf{S}'_N \tag{6A.2}$$

where $\mathbf{S}_N$ is a lower triangular matrix with elements $\{s_{ik}\}$ and $\mathbf{D}_N$ is a diagonal matrix with diagonal elements $\{d_k\}$. The diagonal elements of $\mathbf{S}_N$ are set to unity, i.e., $s_{ii} = 1$. Then we have

$$r_{ij} = \sum_{k=1}^{j} s_{ik} d_k s_{jk} \qquad 1 \le j \le i-1 \qquad i \ge 2$$
$$r_{11} = d_1 \tag{6A.3}$$

where $\{r_{ij}\}$ are the elements of $\mathbf{R}_N$. Consequently the elements $\{s_{ik}\}$ and $\{d_k\}$ are determined from (6A.3) according to the equations

$$d_1 = r_{11}$$
$$s_{ij} d_j = r_{ij} - \sum_{k=1}^{j-1} s_{ik} d_k s_{jk} \qquad i \le j \le i-1 \qquad 2 \le i \le N$$
$$d_i = r_{ii} - \sum_{k=1}^{j-1} s_{ik}^2 d_k \qquad 2 \le i \le N \tag{6A.4}$$

Thus (6A.4) define $\mathbf{S}_N$ and $\mathbf{D}_N$ in terms of the elements of $\mathbf{R}_N$.

The solution to (6A.1) is performed in two steps. With (6A.2) substituted into (6A.1), we have

$$\mathbf{S}_N \mathbf{D}_N \mathbf{S}'_N \mathbf{C}_N = \mathbf{U}_N$$

Let

$$\mathbf{Y}_N = \mathbf{D}_N \mathbf{S}'_N \mathbf{C}_N \tag{6A.5}$$

Then

$$\mathbf{S}_N \mathbf{Y}_N = \mathbf{U}_N \tag{6A.6}$$

First we solve (6A.6) for $\mathbf{Y}_N$. Because of the triangular form of $\mathbf{S}_N$, we have

$$y_1 = u_1$$

$$y_i = u_i - \sum_{j=1}^{i-1} s_{ij} y_j \qquad 2 \le i \le N \tag{6A.7}$$

Having obtained $\mathbf{Y}_N$, the second step is to compute $\mathbf{C}_N$. That is,

$$\mathbf{D}_N \mathbf{S}'_N \mathbf{C}_N = \mathbf{Y}_N$$

$$\mathbf{S}'_N \mathbf{C}_N = \mathbf{D}_N^{-1} \mathbf{Y}_N$$

Beginning with

$$c_N = \frac{y_N}{d_N} \tag{6A.8}$$

the remaining coefficients of $\mathbf{C}_N$ are obtained recursively as follows:

$$c_i = \frac{y_i}{d_i} - \sum_{j=i+1}^{N} s_{ji} c_j \qquad 1 \le i \le N - 1 \tag{6A.9}$$

The number of multiplications and divisions required to perform the factorization of $\mathbf{R}_N$ is proportional to N^3. The number of multiplications and divisions required to compute $\mathbf{C}_N$, once $\mathbf{S}_N$ is determined, is proportional to N^2. In contrast, when $\mathbf{R}_N$ is Toeplitz, the Levinson–Durbin algorithm should be used to determine the solution of (6A.1) since the number of multiplications and divisions is proportional to N^2. On the other hand, in a recursive least-squares formulation, $\mathbf{S}_N$ and $\mathbf{D}_N$ are not computed as in (6A.3) but they are updated recursively. The update is accomplished with N^2 operations (multiplications and divisions). Then the solution for the vector $\mathbf{C}_N$ follows the steps (6A.5) through (6A.9). Consequently the computational burden of the recursive least-squares formulation is proportional to N^2.

PROBLEMS

6.1. A channel is said to be distortionless if the response $y(t)$ to an input $x(t)$ is $Kx(t - t_0)$, where K and t_0 are constants. Show that if the frequency response of the channel is $A(f)e^{j\theta(f)}$, where $A(f)$ and $\theta(f)$ are real, the necessary and sufficient conditions for distortionless transmission are $A(f) = K$ and $\theta(f) = 2\pi f t_0 \pm n\pi$, $n = 0, 1, 2, \ldots$.

6.2. The raised-cosine spectral characteristic is given by (6.2.23).

(a) Show that the corresponding impulse response is

$$x(t) = \frac{\sin \pi t/T}{\pi t/T} \frac{\cos \beta \pi t/T}{1 - 4\beta^2 t^2/T^2}$$

(b) Determine the Hilbert transform of $x(t)$ when $\beta = 1$.

(c) Does $\hat{x}(t)$ possess the desirable properties of $x(t)$ that make it appropriate for data transmission? Explain.

(d) Determine the envelope of the SSB suppressed-carrier signal generated from $x(t)$.

6.3. (a) Show that (Poisson sum formula)

$$x(t) = \sum_{k=-\infty}^{\infty} g(t)h(t - kT) \Rightarrow X(f) = \frac{1}{T} \sum_{n=-\infty}^{\infty} H\left(\frac{n}{T}\right)G\left(f - \frac{n}{T}\right)$$

Hint: Make a Fourier-series expansion of the periodic factor

$$\sum_{k=-\infty}^{\infty} h(t - kT)$$

(b) Using the result in (a), verify the following versions of the Poisson sum:

(i) $\displaystyle\sum_{k=-\infty}^{\infty} h(kT) = \frac{1}{T} \sum_{n=-\infty}^{\infty} H\left(\frac{n}{T}\right)$

(ii) $\displaystyle\sum_{k=-\infty}^{\infty} h(t - kT) = \frac{1}{T} \sum_{n=-\infty}^{\infty} H\left(\frac{n}{T}\right) \exp\frac{j2\pi nt}{T}$

(iii) $\displaystyle\sum_{k=-\infty}^{\infty} h(kT) \exp(-j2\pi kTf) = \frac{1}{T} \sum_{n=-\infty}^{\infty} H\left(f - \frac{n}{T}\right)$

(c) Derive the condition for no intersymbol interference (Nyquist criterion) by using the Poisson sum formula.

6.4. Suppose a digital communications system employs gaussian-shaped pulses of the form

$$x(t) = \exp(-\pi a^2 t^2)$$

To reduce the level of intersymbol interference to a relatively small amount, we impose the condition that $x(T) = 0.01$, where T is the symbol interval. The bandwidth W of the pulse $x(t)$ is defined as that value of W for which $X(W)/X(0) = 0.01$, where $X(f)$ is the Fourier transform of $x(t)$. Determine the value of W and compare this value to that of raised-cosine spectrum with 100 percent rolloff.

6.5. A band-limited signal having bandwidth W can be represented as

$$x(t) = \sum_{n=-\infty}^{\infty} x_n \frac{\sin 2\pi W(t - n/2W)}{2\pi W(t - n/2W)}$$

(a) Determine the spectrum $X(f)$ and plot $|X(f)|$ for the following cases:

(i) $x_0 = 2$ (ii) $x_{-1} = -1$
    $x_1 = 1$ $x_0 = 2$
    $x_2 = -1$ $x_1 = -1$
    $x_n = 0$ $n \neq 0,1,2$ $x_n = 0$ $n \neq -1,0,1$

(b) Plot $x(t)$ for these two cases.

(c) If these signals are used for binary signal transmission, determine the number of received levels possible at the sampling instants $t = nT = n/2W$, and the probabilities of occurrence of the received levels. Assume that the binary digits at the transmitter are equally probable.

6.6. Consider the problem of equalizing the discrete-time equivalent channel shown in Fig. P6.6. The information sequence $\{I_n\}$ is binary (± 1) and uncorrelated. The additive noise $\{v_n\}$ is white and real-valued, with variance N_0. The received sequence $\{y_n\}$ is processed by a linear three-tap equalizer which is optimized on the basis of the MSE criterion.

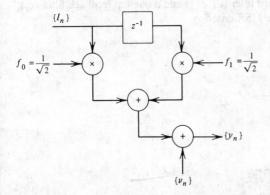

FIGURE P6.6

(a) Determine the optimum coefficients of the equalizer as a function of N_0.

(b) Determine the three eigenvalues $\lambda_1, \lambda_2, \lambda_3$ of the covariance matrix Γ and the corresponding (normalized to unit length) eigenvectors v_1, v_2, v_3.

(c) Determine the minimum MSE for the three-tap equalizer as a function of N_0.

(d) Determine the output SNR for the three-tap equalizer as a function of N_0. How does this compare with the output SNR for the infinite-tap equalizer? For example, evaluate the output SNR for these two equalizers when $N_0 = 0.1$.

(e) Determine the maximum value of Δ that can be used to ensure that the equalizer coefficients converge during operation in the adaptive mode.

(f) What is the variance of the self-noise generated by the three-tap equalizer when operating in an adaptive mode, as a function of Δ? Suppose it is desired to limit the variance of the self-noise to 10 percent of the minimum MSE for the three-tap equalizer when $N_0 = 0.1$. What value of Δ would you select?

(g) If the optimum coefficients of the equalizer are computed recursively by the method of steepest descent, the recursive equation can be expressed in the form

$$\mathbf{C}_{(n+1)} = (\mathbf{I} - \Delta\Gamma)\mathbf{C}_{(n)} + \Delta\xi$$

where $\mathbf{I}$ is the identity matrix. The above represents a set of three coupled first-order difference equations. They can be decoupled by a linear transformation that diagonalizes the matrix Γ. That is, $\Gamma = \mathbf{U}\Lambda\mathbf{U}^T$ where Λ is the diagonal matrix having the eigenvalues of Γ as its diagonal elements and $\mathbf{U}$ is the (normalized) modal matrix that can be obtained from your answer in (b). Let $\mathbf{C}' = \mathbf{U}^T\mathbf{C}$ and determine the steady-state solution for $\mathbf{C}'$. From this evaluate $\mathbf{C} = (\mathbf{U}^T)^{-1}\mathbf{C}' = \mathbf{U}\mathbf{C}'$ and, thus, show that your answer agrees with the result obtained in (a).

6.7. Use the orthogonality principle to derive the equations for the coefficients in a decision-feedback equalizer based on the MSE criterion and given by (6.5.3) and (6.5.5).

6.8. Suppose the discrete-time model for the intersymbol interference is characterized by the tap coefficients $f_0, f_1, \ldots, f_L$. From the equations for the tap coefficients of a decision-feedback equalizer (DFE), show that only L taps are needed in the feedback filter of the DFE. That is, if $\{c_k\}$ are the coefficients of the feedback filter, then $c_k = 0$ for $k \geq L + 1$.

6.9. Consider the channel model shown in Fig. P6.9. $\{v_n\}$ is a real-valued white-noise sequence with zero mean and variance N_0. Suppose the channel is to be equalized by DFE having a two-tap feedforward filter (c_0, c_{-1}) and a one-tap feedback filter (c_1). The $\{c_i\}$ are optimized using the MSE criterion.

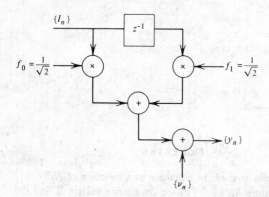

FIGURE P6.9

(a) Determine the optimum coefficients and their approximate values for $N_0 \ll 1$.
(b) Determine the exact value of the minimum MSE and a first-order approximation appropriate for the case $N_0 \ll 1$.
(c) Determine the exact value of the output SNR for the three-tap equalizer as a function of N_0 and a first-order approximation appropriate for the case $N_0 \ll 1$.
(d) Compare the results in (b) and (c) with the performance of the infinite-tap DFE.
(e) Evaluate and compare the exact values of the output SNR for the three-tap and infinite-tap DFE in the special cases where $N_0 = 0.1$ and 0.01. Comment on how well the three-tap equalizer performs relative to the infinite-tap equalizer.

6.10. A pulse and its (raised-cosine) spectral characteristic are shown in Fig. P6.10a. This pulse is used for transmitting digital information over a band-limited channel at a rate $1/T$ symbols/s.
(a) What is the rolloff factor β?
(b) What is the pulse rate?
(c) The channel distorts the signal pulses. Suppose the sampled values of the filtered received pulse $x(t)$ are as shown in Fig. P6.10c. It is obvious that there are five interfering signal components. Give the sequence of +1s and −1s that will cause the largest (destructive or constructive) interference and the corresponding value of the interference (the peak distortion).
(d) What is the probability of occurrence of the worst sequence obtained in (c), assuming that all binary digits are equally probable and independent?

(c)

FIGURE P6.10

6.11. A time-dispersive channel having an impulse response $h(t)$ is used to transmit four-phase PSK at a rate $R = 1/T$ symbols/s. The equivalent discrete-time channel is shown in Fig. P6.11. The sequence $\{\eta_k\}$ is a white-noise sequence having zero mean and variance $\sigma^2 = N_0$.

(a) What is the sampled autocorrelation function sequence $\{x_k\}$ defined by

$$x_k = \int_{-\infty}^{\infty} h^*(t) h(t + kT) \, dt$$

for this channel?

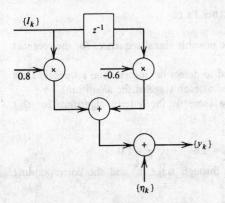

FIGURE P6.11

(b) The minimum MSE performance of a linear equalizer and a decision-feedback equalizer having an infinite number of taps depends on the *folded spectrum of the channel*

$$\frac{1}{T} \sum_{n=-\infty}^{\infty} \left| H\left(\omega + \frac{2\pi n}{T}\right) \right|^2$$

where $H(\omega)$ is the Fourier transform of $h(t)$. Determine the *folded spectrum of the channel* given above.

(c) Use your answer in (b) to express the minimum MSE of a linear equalizer in terms of the folded spectrum of the channel. (You may leave your answer in integral form.)

(d) Repeat (c) for an infinite-tap decision-feedback equalizer.

6.12. Consider a four-level PAM system with possible transmitted levels $3, 1, -1, -3$. The channel through which the data are transmitted introduces intersymbol interference over two successive symbols. The equivalent discrete-time channel model is shown in Fig. P6.12. The $\{\eta_k\}$ is a sequence of real-valued independent zero mean gaussian noise variables with variance $\sigma^2 = N_0$. The received sequence is

$$y_1 = 0.8I_1 + n_1$$
$$y_2 = 0.8I_2 - 0.6I_1 + n_2$$
$$y_3 = 0.8I_3 - 0.6I_2 + n_3$$
$$\vdots$$
$$y_k = 0.8I_k - 0.6I_{k-1} + n_k$$

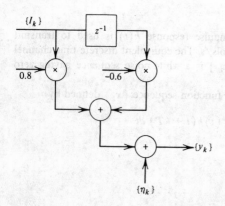

FIGURE P6.12

(a) Sketch the tree structure showing the possible signal sequences for the received signals y_1, y_2, y_3.

(b) Suppose the Viterbi algorithm is used to detect the information sequence. How many probabilities must be computed at each stage of the algorithm?

(c) How many surviving sequences are there in the Viterbi algorithm for this channel?

(d) Suppose that the received signals are

$$y_1 = 0.5 \qquad y_2 = 2.0 \qquad y_3 = -1.0$$

Determine the surviving sequences through stage y_3 and the corresponding metrics.

(e) Give a tight upper bound for the probability of error for four-level PAM transmitted over this channel.

6.13. A transversal equalizer with K taps has an impulse response

$$e(t) = \sum_{k=0}^{K-1} c_k \delta(t - kT)$$

where T is the delay between adjacent taps, and a transfer function

$$E(z) = \sum_{k=0}^{K-1} c_k z^{-k}$$

The *discrete Fourier transform* (DFT) of the equalizer coefficients $\{c_k\}$ is defined as

$$|E_n \equiv E(z)|_{z=e^{j2\pi n/k}} = \sum_{k=0}^{K-1} c_k e^{-j2\pi n/K} \qquad n = 0, 1, \ldots, K-1$$

The *inverse DFT* is defined as

$$b_k = \frac{1}{K} \sum_{n=0}^{K-1} E_n e^{j2\pi nk/K} \qquad k = 0, 1, \ldots, K-1$$

(a) Show that $b_k = c_k$ by substituting for E_n in the above expression.

(b) From the relations given above, derive an equivalent filter structure having the z transform

$$E(z) = \underbrace{\frac{(1 - z^{-K})}{K}}_{E_1(z)} \sum_{n=0}^{K-1} \underbrace{\frac{E_n}{1 - e^{j2\pi n/K}z^{-1}}}_{E_2(z)}$$

(c) If $E(z)$ is considered as two separate filters $E_1(z)$ and $E_2(z)$ in cascade, sketch a block diagram for each of the filters, using z^{-1} to denote a unit of delay.

(d) In the transversal equalizer the adjustable parameters are the coefficients $\{c_k\}$. What are the adjustable parameters of the equivalent filter in (b) and how are they related to $\{c_k\}$? Can you suggest an algorithm for the automatic adjustment of these coefficients in the equivalent filter?

6.14. An equivalent discrete-time channel with white gaussian noise is shown in Fig. P6.14.

(a) Suppose we use a linear equalizer to equalize the channel. Determine the tap coefficients c_{-1}, c_0, c_1 of a three-tap equalizer. To simplify the computation, let the AWGN be zero.

(b) The tap coefficients of the linear equalizer in (a) are determined recursively via the algorithm

$$\mathbf{C}_{k+1} = \mathbf{C}_k - \Delta\mathbf{g}_k \qquad \mathbf{C}_k = \begin{bmatrix} c_{1k} \\ c_{0k} \\ c_{1k} \end{bmatrix}$$

where $\mathbf{g}_k = \mathbf{\Gamma}\mathbf{C}_k - \mathbf{b}$ is the gradient vector and Δ is the step size. Determine the range of values of Δ to ensure convergence of the recursive algorithm. To simplify the computation, let the AWGN be zero.

(c) Determine the tap weights of a DFE with two feedforward taps and one feedback tap. To simplify the computation, let the AWGN be zero.

FIGURE P6.14

6.15. Determine the phase states, the correlative states, and the combined phase correlative states S_n for CPM with $L = 2$ and $h = 2/3$.

6.16. In Example 6.7.5 suppose that $u(t)$ is the rectangular pulse, defined as

$$u(t) = \begin{cases} \dfrac{1}{4T} & 0 \leq t \leq 2T \\ 0 & \text{otherwise} \end{cases}$$

Sketch the phase tree and identify the phase trajectory corresponding to the sequence $(-1, -1, -1, -1, 1, 1, -1, 1)$ when the modulation $h = 3/4$.

6.17. Recall that MSK can be represented as a four-phase offset PSK modulation having the lowpass equivalent form

$$v(t) = \sum_k \left[I_k u(t - 2kT_b) + j J_k u(t - 2kT_b - T_b) \right]$$

where

$$u(t) = \begin{cases} \sin \dfrac{\pi t}{2T_b} & 0 \leq t \leq 2T_b \\ 0 & \text{otherwise} \end{cases}$$

and $\{I_k\}$ and $\{J_k\}$ are sequences of information symbols (± 1).
(a) Sketch the block diagram of an MSK demodulator for offset QPSK.
(b) Evaluate the performance of the four-phase demodulator if no account is taken of the memory in the modulation.
(c) Compare the performance obtained in (b) with that for Viterbi decoding of the MSK signal.
(d) The MSK signal is also equivalent to binary FSK. Determine the performance of noncoherent detection of the MSK symbol. Compare your result with (b) and (c).

6.18. When a periodic pseudo-random sequence of length N is used to adjust the coefficients of an N-tap linear equalizer, the computations can be performed efficiently in the frequency domain by use of the discrete Fourier transform (DFT). Suppose that $\{y_n\}$ is a sequence of N received samples (taken at the symbol rate) at the equalizer input. Then the computation of the equalizer coefficients is performed as follows:

(*a*) Compute the DFT of one period of the equalizer input sequence $\{y_n\}$, i.e.,

$$Y_k = \sum_{n=0}^{N-1} y_n e^{-j2\pi nk/N}$$

(*b*) Compute the desired equalizer spectrum

$$C_k = \frac{X_k Y_k^*}{|Y_k|^2} \qquad k = 0, 1, \ldots, N-1$$

(*c*) Compute the inverse DFT of $\{C_k\}$ to obtain the equalizer coefficients $\{c_n\}$. Show that this procedure in the absence of noise yields an equalizer whose frequency response is equal to the frequency response of the inverse folded channel spectrum at the N uniformly spaced frequencies $f_k = k/NT$, $k = 0, 1, \ldots, N-1$.

CHAPTER
7

DIGITAL SIGNALING OVER FADING MULTIPATH CHANNELS

The previous chapters have described the design and performance of digital communications systems operating over either the classical AWGN channel or a linear filter channel with AWGN. We observed that the distortion inherent in linear filter channels required special signal design techniques and rather sophisticated adaptive equalization algorithms in order to achieve good performance.

In this chapter we consider the problems of signal design, receiver structure, and receiver performance for more complex channels, namely, channels having randomly time-variant impulse responses. This characterization serves as a model for signal transmission over many radio channels such as shortwave ionospheric radio communication in the 3- to 30-MHz frequency band (HF), tropospheric scatter (beyond-the-horizon) radio communications in the 300- to 3000-MHz frequency band (UHF) and 3000- to 30,000-MHz frequency band (SHF), and ionospheric forward scatter in the 30- to 300-MHz frequency band (VHF). The time-variant impulse responses of these channels are a consequence of the constantly changing physical characteristics of the media. For example, the ions in the ionospheric layers that reflect the signals transmitted in the HF frequency band are always in motion. To the user of the channel the motion of the ions appears to be random. Consequently, if the same signal is transmitted at HF in two widely separated time intervals, the two received signals will not only be

702

different, but the difference will be random rather than deterministic. The random variations that occur are treated in statistical terms.

We shall begin our treatment of digital signaling over fading multipath channels by first developing a statistical characterization of the channel. Then we shall evaluate the performance of several basic digital signaling techniques for communication over such channels. The performance results will demonstrate the severe penalty in SNR that must be paid as a consequence of the fading characteristics of the received signal. We shall then show that the penalty in SNR can be dramatically reduced by means of efficient modulation/coding and demodulation/decoding techniques.

7.1 CHARACTERIZATION OF FADING MULTIPATH CHANNELS

If we transmit an extremely short pulse, ideally an impulse, over a time-varying multipath channel, the received signal might appear as a train of pulses, as shown in Fig. 7.1.1. Hence one characteristic of a multipath medium is the time spread introduced in the signal which is transmitted through the channel.

A second characteristic is due to the time variations in the structure of the medium. As a result of such time variations the nature of the multipath varies with time. That is, if we repeat the pulse-sounding experiment over and over, we would observe changes in the received pulse train, which will include changes in the sizes of the individual pulses, changes in the relative delays among the pulses, and, quite often, changes in the number of pulses observed in the received pulse train as shown in Fig. 7.1.1. Moreover, the time variations appear to be unpredictable to the user of the channel. Therefore it is reasonable to characterize the time-variant multipath channel statistically. Toward this end let us examine the effects of the channel on a transmitted signal that is represented in general as

$$s(t) = \text{Re}\left[u(t)e^{j2\pi f_c t}\right] \qquad (7.1.1)$$

We assume that there are multiple propagation paths. Associated with each path is a propagation delay and an attenuation factor. Both the propagation delays and the attenuation factors are time-variant as a result of changes in the structure of the medium. Thus the received bandpass signal may be expressed in the form

$$x(t) = \sum_n \alpha_n(t)s\left[t - \tau_n(t)\right] \qquad (7.1.2)$$

where $\alpha_n(t)$ is the attenuation factor for the signal received on the nth path and $\tau_n(t)$ is the propagation delay for the nth path. Substitution for $s(t)$ from (7.1.1) into (7.1.2) yields the result

$$x(t) = \text{Re}\left(\left\{\sum_n \alpha_n(t)e^{-j2\pi f_c \tau_n(t)}u\left[t - \tau_n(t)\right]\right\}e^{j2\pi f_c t}\right) \qquad (7.1.3)$$

It is apparent from observation of (7.1.3) that the equivalent lowpass received signal is

$$r(t) = \sum_n \alpha_n(t)e^{-j2\pi f_c \tau_n(t)}u\left[t - \tau_n(t)\right] \qquad (7.1.4)$$

Transmitted signal　　　Received signal

FIGURE 7.1.1
Example of the response of a time-variant multipath channel to a very narrow pulse.

Since $r(t)$ is the response of an equivalent lowpass channel to the equivalent lowpass signal $u(t)$, it follows that the equivalent lowpass channel is described by the time-variant impulse response

$$c(\tau; t) = \sum_n \alpha_n(t) e^{-j2\pi f_c \tau_n(t)} \delta[\tau - \tau_n(t)] \qquad (7.1.5)$$

For some channels, such as the tropospheric scatter channel, it is more appropriate to view the received signal as consisting of a continuum of multipath components. In such a case, the received signal $x(t)$ is expressed in the integral form

$$x(t) = \int_{-\infty}^{\infty} \alpha(\tau; t) s(t - \tau) \, d\tau \qquad (7.1.6)$$

where $\alpha(\tau; t)$ denotes the attenuation of the signal components at delay τ and at time instant t. Now substitution for $s(t)$ from (7.1.1) into (7.1.6) yields

$$x(t) = \text{Re}\left\{\left[\int_{-\infty}^{\infty} \alpha(\tau; t) e^{-j2\pi f_c \tau} u(t - \tau) \, d\tau\right] e^{j2\pi f_c t}\right\} \quad (7.1.7)$$

Since the integral in (7.1.7) represents the convolution of $u(t)$ with an equivalent lowpass time-variant impulse response $c(\tau; t)$, it follows that

$$c(\tau; t) = \alpha(\tau; t) e^{-j2\pi f_c \tau} \quad (7.1.8)$$

where $c(\tau; t)$ represents the response of the channel at time t due to an impulse applied at time $t - \tau$. Thus (7.1.8) is the appropriate definition of the equivalent lowpass impulse response when the channel results in continuous multipath and (7.1.5) is appropriate for a channel that contains discrete multipath components.

Now let us consider the transmission of an unmodulated carrier at frequency f_c. Then $u(t) = 1$ for all t and, hence, the received signal for the case of discrete multipath, given by (7.1.4), reduces to

$$r(t) = \sum_n \alpha_n(t) e^{-j2\pi f_c \tau_n(t)}$$

$$= \sum_n \alpha_n(t) e^{-j\theta_n(t)} \quad (7.1.9)$$

where $\theta_n(t) = 2\pi f_c \tau_n(t)$. Thus the received signal consists of the sum of a number of time-variant vectors (phasors) having amplitudes $\alpha_n(t)$ and phases $\theta_n(t)$. We note that large dynamic changes in the medium are required for $\alpha_n(t)$ to change sufficiently to cause a significant change in the received signal. On the other hand, $\theta_n(t)$ will change by 2π rad whenever τ_n changes by $1/f_c$. But $1/f_c$ is a small number and, hence, θ_n can change by 2π rad with relatively small motions of the medium. We also expect the delays $\tau_n(t)$ associated with the different signal paths to change at different rates and in an unpredictable (random) manner. This implies that the received signal $r(t)$ in (7.1.9) can be modeled as a random process. When there are a large number of paths, the central limit theorem can be applied. That is, $r(t)$ can be modeled as a complex-valued gaussian random process. This means that the time-variant impulse response $c(\tau; t)$ is a complex-valued gaussian random process in the t variable.

The multipath propagation model for the channel embodied in the received signal $r(t)$, given in (7.1.9), results in signal fading. The fading phenomenon is primarily a result of the time variations in the phases $\{\theta_n(t)\}$. That is, the randomly time-variant phases $\{\theta_n(t)\}$ associated with the vectors $\{\alpha_n e^{-j\theta_n}\}$ at times result in the vectors adding destructively. When that occurs, the resultant received signal $r(t)$ is very small or practically zero. At other times the vectors $\{\alpha_n e^{-j\theta_n}\}$ add constructively, so that the received signal is large. Thus the amplitude variations in the received signal, termed *signal fading*, are due to the time-variant multipath characteristics of the channel.

When the impulse response $c(\tau; t)$ is modeled as a zero mean complex-valued gaussian process, the envelope $|c(\tau; t)|$ at any instant t is Rayleigh-distributed. In this case the channel is said to be a *Rayleigh fading channel*. In the event that there are fixed scatterers or signal reflectors in the medium, in addition to randomly moving scatterers, $c(\tau; t)$ can no longer be modeled as having a zero mean. In this case, the envelope $|c(\tau; t)|$ has a Rice distribution and the channel is said to be a *Ricean fading channel*. In our treatment of fading channels, we consider only the Rayleigh-distributed envelope statistics. This model appears to be a realistic one. Rayleigh-distributed envelope fading has been observed often on HF and troposcatter channels and, as a consequence, this channel model is widely accepted.

We shall now develop a number of useful correlation functions and power spectral density functions that define the characteristics of a fading multipath channel. Our starting point is the equivalent lowpass impulse response $c(\tau; t)$, which is characterized as a complex-valued zero mean gaussian random process in the t variable. We assume that $c(\tau; t)$ is wide-sense-stationary. Then we define the autocorrelation function of $c(\tau; t)$ as

$$\phi_c(\tau_1, \tau_2; \Delta t) = \tfrac{1}{2} E\left[c^*(\tau_1; t) c(\tau_2; t + \Delta t) \right] \tag{7.1.10}$$

In most radio transmission media, the attenuation and phase shift of the channel associated with path delay τ_1 is uncorrelated with the attenuation and phase shift associated with path delay τ_2. This is usually called *uncorrelated scattering*. We make the assumption that the scattering at two different delays is uncorrelated and incorporate it into (7.1.10) to obtain

$$\tfrac{1}{2} E\left[c^*(\tau_1; t) c(\tau_2; t + \Delta t) \right] = \phi_c(\tau_1; \Delta t) \delta(\tau_1 - \tau_2) \tag{7.1.11}$$

If we let $\Delta t = 0$, the resulting autocorrelation function $\phi_c(\tau; 0) \equiv \phi_c(\tau)$ is simply the average power output of the channel as a function of the time delay τ. For this reason, $\phi_c(\tau)$ is called the *multipath intensity profile* or *the delay power spectrum* of the channel. In general, $\phi_c(\tau; \Delta t)$ gives the average power output as a function of the time delay τ and the difference Δt in observation time.

In practice, the function $\phi_c(\tau; \Delta t)$ is measured by transmitting very narrow pulses or, equivalently, a wideband signal and cross-correlating the received signal with a delayed version of itself. Typically, the measured function $\phi_c(\tau)$ may appear as shown in Fig. 7.1.2. The range of values of τ over which $\phi_c(\tau)$ is

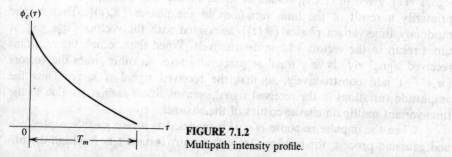

FIGURE 7.1.2
Multipath intensity profile.

essentially nonzero is called the *multipath spread of the channel* and is denoted by T_m.

A completely analogous characterization of the time-variant multipath channel begins in the frequency domain. By taking the Fourier transform of $c(\tau; t)$, we obtain the time-variant transfer function $C(f; t)$, where f is the frequency variable. Thus

$$C(f; t) = \int_{-\infty}^{\infty} c(\tau; t) e^{-j2\pi f\tau} \, d\tau \qquad (7.1.12)$$

Since $c(\tau; t)$ is a complex-valued zero mean gaussian random process in the t variable, it follows that $C(f; t)$ also has the same statistics. Thus, under the assumption that the channel is wide-sense-stationary, we define the autocorrelation function

$$\phi_C(f_1, f_2; \Delta t) = \tfrac{1}{2} E\left[C^*(f_1; t) C(f_2; t + \Delta t) \right] \qquad (7.1.13)$$

Since $C(f; t)$ is the Fourier transform of $c(\tau; t)$, it is not surprising to find that $\phi_C(f_1, f_2; \Delta t)$ is related to $\phi_C(\tau; \Delta t)$ by the Fourier transform. The relationship is easily established by substituting (7.1.12) into (7.1.13). Thus

$$\phi_C(f_1, f_2; \Delta t) = \frac{1}{2} \int_{-\infty}^{\infty} \int_{-\infty}^{\infty} E\left[c^*(\tau_1; t) c(\tau_2; t + \Delta t) \right] e^{j2\pi(f_1\tau_1 - f_2\tau_2)} \, d\tau_1 \, d\tau_2$$

$$= \int_{-\infty}^{\infty} \int_{-\infty}^{\infty} \phi_c(\tau_1; \Delta t) \delta(\tau_1 - \tau_2) e^{j2\pi(f_1\tau_1 - f_2\tau_2)} \, d\tau_1 \, d\tau_2$$

$$= \int_{-\infty}^{\infty} \phi_c(\tau_1; \Delta t) e^{j2\pi(f_1 - f_2)\tau_1} \, d\tau_1$$

$$= \int_{-\infty}^{\infty} \phi_c(\tau_1; \Delta t) e^{-j2\pi\Delta f\tau_1} \, d\tau \equiv \phi_C(\Delta f; \Delta t) \qquad (7.1.14)$$

where $\Delta f = f_2 - f_1$. From (7.1.14) we observe that $\phi_C(\Delta f; \Delta t)$ is the Fourier transform of the multipath intensity profile. Furthermore, the assumption of uncorrelated scattering implies that the autocorrelation function of $C(f; t)$ in frequency is a function of only the frequency difference $\Delta f = f_2 - f_1$. Therefore it is appropriate to call $\phi_C(\Delta f; \Delta t)$ the *spaced-frequency spaced-time correlation function of the channel*. It can be measured in practice by transmitting a pair of sinusoids separated by Δf and cross-correlating the two separately received signals with a relative delay Δt.

Suppose we set $\Delta t = 0$ in (7.1.14). Then, with $\phi_C(\Delta f; 0) \equiv \phi_C(\Delta f)$ and $\phi_c(\tau; 0) \equiv \phi_c(\tau)$, the transform relationship is simply

$$\phi_C(\Delta f) = \int_{-\infty}^{\infty} \phi_c(\tau) e^{-j2\pi\Delta f\tau} \, d\tau \qquad (7.1.15)$$

The relationship is graphically depicted in Fig. 7.1.3. Since $\phi_C(\Delta f)$ is an autocorrelation function in the frequency variable, it provides us with a measure of the frequency coherence of the channel. As a result of the Fourier transform relationship between $\phi_C(\Delta f)$ and $\phi_c(\tau)$, the reciprocal of the multipath spread is

FIGURE 7.1.3
Relationship between $\phi_C(\Delta f)$ and $\phi_c(\tau)$.

a measure of the *coherence bandwidth of the channel*. That is,

$$(\Delta f)_c \approx \frac{1}{T_m} \tag{7.1.16}$$

where $(\Delta f)_c$ denotes the coherence bandwidth. Thus two sinusoids with frequency separation greater than $(\Delta f)_c$ are affected differently by the channel. When an information-bearing signal is transmitted through the channel, if $(\Delta f)_c$ is small in comparison to the bandwidth of the transmitted signal, the channel is said to be *frequency-selective*. In this case the signal is severely distorted by the channel. On the other hand, if $(\Delta f)_c$ is large in comparison to the bandwidth of the transmitted signal the channel is said to be *frequency-nonselective*.

We now focus our attention on the time variations of the channel as measured by the parameter Δt in $\phi_C(\Delta f; \Delta t)$. The time variations in the channel are evidenced as a Doppler broadening and, perhaps, in addition as a Doppler shift of a spectral line. In order to relate the Doppler effects to the time variations of the channel, we define the Fourier transform of $\phi_C(\Delta f; \Delta t)$ with respect to the variable Δt to be the function of $S_C(\Delta f; \lambda)$. That is,

$$S_C(\Delta f; \lambda) = \int_{-\infty}^{\infty} \phi_C(\Delta f; \Delta t) e^{-j2\pi\lambda\,\Delta t}\, d\Delta t \tag{7.1.17}$$

With Δf set to zero and $S_C(0; \lambda) \equiv S_C(\lambda)$, the relation in (7.1.17) becomes

$$S_C(\lambda) = \int_{-\infty}^{\infty} \phi_C(\Delta t) e^{-j2\pi\lambda\,\Delta t}\, d\Delta t \tag{7.1.18}$$

The function $S_C(\lambda)$ is a power spectrum that gives the signal intensity as a function of the Doppler frequency λ. Hence we call $S_C(\lambda)$ the *Doppler power spectrum of the channel*.

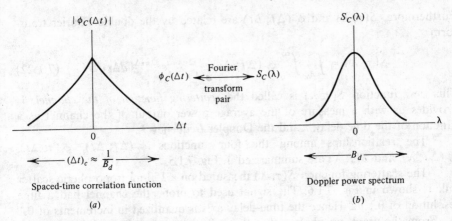

FIGURE 7.1.4

Relationship between $\phi_C(\Delta t)$ and $S_C(\lambda)$.

From (7.1.18) we observe that if the channel is time-invariant, $\phi_C(\Delta t) = 1$ and $S_C(\lambda)$ becomes equal to the delta function $\delta(\lambda)$. Therefore, when there are no time variations in the channel, there is no spectral broadening observed in the transmission of a pure frequency tone.

The range of values of λ over which $S_C(\lambda)$ is essentially nonzero is called the *Doppler spread* B_d of the channel. Since $S_C(\lambda)$ is related to $\phi_C(\Delta t)$ by the Fourier transform, the reciprocal of B_d is a measure of the coherence time of the channel. That is,

$$(\Delta t)_c \approx \frac{1}{B_d} \tag{7.1.19}$$

where $(\Delta t)_c$ denotes the *coherence time*. Clearly a slowly changing channel has a large coherence time or, equivalently, a small Doppler spread. Figure 7.1.4 illustrates the relationship between $\phi_C(\Delta t)$ and $S_C(\lambda)$.

We have now established a Fourier transform relationship between $\phi_C(\Delta f; \Delta t)$ and $\phi_c(\tau; \Delta t)$ involving the variables $(\tau, \Delta f)$, and a Fourier transform relationship between $\phi_C(\Delta f; \Delta t)$ and $S_C(\Delta f; \lambda)$ involving the variables $(\Delta t, \lambda)$. There are two additional Fourier transform relationships that we can define which serve to relate $\phi_c(\tau; \Delta t)$ to $S_C(\Delta f; \lambda)$ and, thus, close the loop. The desired relationship is obtained by defining a new function, denoted as $S(\tau; \lambda)$, to be the Fourier transform of $\phi_c(\tau; \Delta t)$ in the Δt variable. That is,

$$S(\tau; \lambda) = \int_{-\infty}^{\infty} \phi_c(\tau; \Delta t) e^{-j2\pi\lambda \Delta t} \, d\Delta t \tag{7.1.20}$$

It follows that $S(\tau; \lambda)$ and $S_C(\Delta f; \lambda)$ are a Fourier transform pair. That is,

$$S(\tau; \lambda) = \int_{-\infty}^{\infty} S_C(\Delta f; \lambda) e^{j2\pi\tau\Delta f} \, d\Delta f \tag{7.1.21}$$

Furthermore, $S(\tau; \lambda)$ and $\phi_C(\Delta f; \Delta t)$ are related by the double Fourier transform

$$S(\tau; \lambda) = \int_{-\infty}^{\infty} \int_{-\infty}^{\infty} \phi_C(\Delta f; \Delta t) e^{-j2\pi\lambda \Delta t} e^{j2\pi\tau\Delta f} \, d\Delta t \, d\Delta f \qquad (7.1.22)$$

This new function $S(\tau; \lambda)$ is called the *scattering function of the channel*. It provides us with a measure of the average power output of the channel as a function of the time delay τ and the Doppler frequency λ.

The relationships among the four functions $\phi_C(\Delta f; \Delta t)$, $\phi_c(\tau; \Delta t)$, $S_C(\Delta f; \lambda)$, and $S(\tau; \lambda)$ are summarized in Fig. 7.1.5.

The scattering function $S(\tau; \lambda)$ measured on a 150-mi tropospheric scatter link is shown in Fig. 7.1.6. The signal used to probe the channel had a time resolution of 0.1 μs. Hence the time-delay axis is quantized in increments of 0.1 μs. From the graph we observe that the multipath spread $T_m = 0.7$ μs. On the other hand, the Doppler spread, which may be defined as the 3-dB bandwidth of the power spectrum for each signal path, appears to vary with each signal path. For example, in one path it is less than 1 Hz while in some other paths the Doppler spread is several hertz. For our purposes, we shall take the largest of these 3-dB bandwidths of the various paths and call that the *Doppler spread*.

There are many radio channels in which fading is encountered that are basically line-of-sight (LOS) communication links with multipath components arising from secondary reflections, or signal paths, from surrounding terrain. In such channels the number of multipath components is small and, hence, the channel may be modeled in a somewhat simpler form. We cite two channel models as examples.

As the first example, let us consider an airplane to ground communication link in which there is the direct path and a single multipath component at a delay t_0 relative to the direct path. The impulse response of such a channel may be modeled as

$$c(\tau; t) = a\delta(\tau) + \beta(t)\delta(\tau - \tau_0(t)) \qquad (7.1.23)$$

where α is the attenuation factor of the direct path and $\beta(t)$ represents the time-variant multipath signal component resulting from terrain reflections. Often, $\beta(t)$ can be characterized as a zero mean gaussian random process. The transfer function for this channel model may be expressed as

$$C(f; t) = \alpha + \beta(t)e^{-j2\pi f\tau_0(t)} \qquad (7.1.24)$$

This channel fits the Ricean fading model defined previously. The direct path with attenuation α represents the specular component and $\beta(t)$ represents the Rayleigh fading component.

A similar model has been found to hold for microwave LOS radio channels used for long distance voice and video transmission by telephone companies throughout the world. For such channels Rummler (1979) has developed a three path model based on channel measurements performed on typical LOS links in the 6-GHz frequency band. The differential delay on the two multipath compo-

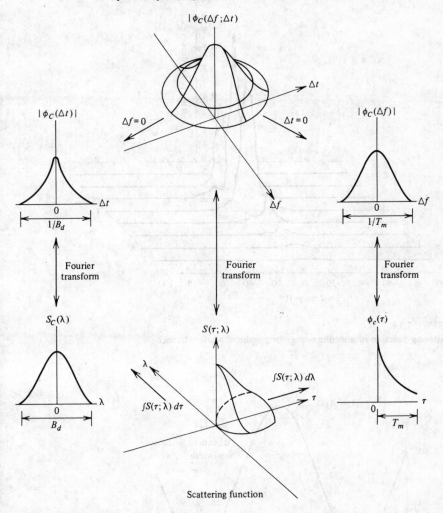

Spaced–frequency, Spaced–time correlation function

FIGURE 7.1.5
Relationships among the channel correlation functions and power spectra (from P. E. Green, Jr. [1962], with permission).

nents is relatively small and, hence, the model developed by Rummler is one that has a channel transfer function

$$C(f) = \alpha\left[1 - \beta e^{-j2\pi(f-f_0)\tau_0}\right] \qquad (7.1.25)$$

where α is the overall attenuation parameter, β is called a shape parameter which is due to the multipath components, f_0 is the frequency of the fade minimum, and τ_0 is the relative time delay between the direct and the multipath compo-

FIGURE 7.1.6
Scattering function of a medium-range tropospheric scatter channel.

FIGURE 7.1.7
Magnitude-square frequency response of LOS channel model.

nents. This simplified model was used to fit data derived from channel measurements.

Rummler found that the parameters α and β may be characterized as random variables which for practical purposes are nearly statistically independent. From the channel measurements, he found that the distribution of β has the form $(1 - \beta)^{2.3}$. The distribution of α is well modeled by the log normal distribution, i.e., $-\log \alpha$ is gaussian. For $\beta > 0.5$ the mean of $-20 \log \alpha$ was found to be 25 dB and the standard deviation was 5 dB. For smaller values of β, the mean decreases to 15 dB. The delay parameter determined from the measurements was $\tau_0 = 6.3$ ns. The magnitude square response of $C(f)$, which is

$$|C(f)|^2 = \alpha^2 \left[1 + \beta^2 - 2\beta \cos 2\pi (f - f_0)\tau_0 \right] \qquad (7.1.26)$$

is plotted in Fig. 7.1.7 as a function of the frequency $(f - f_0)$ for $\tau_0 = 6.3$ ns. Note that the effect of the multipath component is to create a deep attenuation at $f = f_0$ and at multiples of $1/\tau \approx 159$ MHz. By comparison, the typical channel bandwidth is 30 MHz. This model was used by Lundgren and Rummler (1979) to determine the error rate performance of a digital radio system.

7.2 THE EFFECT OF SIGNAL CHARACTERISTICS ON THE CHOICE OF A CHANNEL MODEL

Having discussed the statistical characterization of time-variant multipath channels generally in terms of the correlation functions described in Sec. 7.1, we now consider the effect of signal characteristics on the selection of a channel model that is appropriate for the specified signal. Thus, let $u(t)$ be the equivalent lowpass signal transmitted over the channel and let $U(f)$ denote its frequency content. Then the equivalent lowpass received signal, exclusive of additive noise, may be expressed either in terms of the time domain variables $c(\tau; t)$ and $u(t)$ as

$$r(t) = \int_{-\infty}^{\infty} c(\tau; t) u(t - \tau) \, d\tau \qquad (7.2.1)$$

or in terms of the frequency functions $C(f; t)$ and $U(f)$ as

$$r(t) = \int_{-\infty}^{\infty} C(f; t) U(f) e^{j2\pi ft} \, df \qquad (7.2.2)$$

Suppose we are transmitting digital information over the channel by modulating (either in amplitude, or in phase, or both) the basic pulse $u(t)$ at a rate $1/T$, where T is the signaling interval. It is apparent from (7.2.2) that the time-variant channel characterized by the transfer function $C(f; t)$ distorts the signal $U(f)$. If $U(f)$ has a bandwidth W greater than the coherence bandwidth $(\Delta f)_c$ of the channel, $U(f)$ is subjected to different gains and phase shifts across the band. In such a case the channel is said to be frequency-selective. Additional distortion is caused by the time variations in $C(f; t)$. This type of distortion is evidenced as a variation in the received signal strength and has been termed *fading*. It should be emphasized that the frequency selectivity and fading are viewed as two different types of distortion. The former depends on the multipath spread or, equivalently, on the coherence bandwidth of the channel relative to the transmitted signal bandwidth W. The latter depends on the time variations of the channel, which are grossly characterized by the coherence time $(\Delta t)_c$ or, equivalently, by the Doppler spread B_d.

The effect of the channel on the transmitted signal $u(t)$ is a function of our choice of signal bandwidth and signal duration. For example, if we select the signaling interval T to satisfy the condition $T \gg T_m$, the channel introduces a negligible amount of intersymbol interference. If the bandwidth of the signal pulse $u(t)$ is $W \approx 1/T$, the condition $T \gg T_m$ implies that

$$W \ll \frac{1}{T_m} \approx (\Delta f)_c \qquad (7.2.3)$$

That is, the signal bandwidth W is much smaller than the coherence bandwidth of the channel. Hence the channel is frequency-nonselective. In other words, all of the frequency components in $U(f)$ undergo the same attenuation and phase shift in transmission through the channel. But this implies that, within the bandwidth occupied by $U(f)$, the time-variant transfer function $C(f; t)$ of the channel is a complex-valued constant in the frequency variable. Since $U(f)$ has its frequency content concentrated in the vicinity of $f = 0$, then $C(f; t) = C(0; t)$. Consequently (7.2.2) reduces to

$$r(t) = C(0; t) \int_{-\infty}^{\infty} U(f) e^{j2\pi f t} \, df$$

$$= C(0; t) u(t) \qquad (7.2.4)$$

Thus, when the signal bandwidth W is much smaller than the coherence bandwidth $(\Delta f)_c$ of the channel, the received signal is simply the transmitted signal multiplied by a complex-valued gaussian random process $C(0; t)$, which represents the time-variant characteristics of the channel. In this case, we say that the multipath components in the received are not resolvable because $W \ll (\Delta f)_c$.

The transfer function $C(0; t)$ for a frequency-nonselective channel may be expressed in the form

$$C(0; t) = \alpha(t) e^{-j\phi(t)} \qquad (7.2.5)$$

where $\alpha(t)$ represents the envelope and $\phi(t)$ represents the phase of the equivalent lowpass channel. When $C(0; t)$ is modeled as a zero mean complex-valued gaussian random process, the envelope $\alpha(t)$ is Rayleigh-distributed for any fixed value of t and $\phi(t)$ is uniformly distributed over the interval $(-\pi, \pi)$. The rapidity of the fading on the frequency-nonselective channel is determined either from the correlation function $\phi_C(\Delta t)$ or from the Doppler power spectrum $S_C(\lambda)$. Alternatively, either of the channel parameters $(\Delta t)_c$ or B_d can be used to characterize the rapidity of the fading.

For example, suppose it is possible to select the signal bandwidth W to satisfy the condition $W \ll (\Delta f)_c$ and the signaling interval T to satisfy the

TABLE 7.2.1
Multipath spread, Doppler spread, and spread factor
for several time-variant multipath channels

Type of channel	Multipath duration	Doppler spread	Spread factor
Shortwave ionospheric propagation (HF)	10^{-3}–10^{-2}	10^{-1}–1	10^{-4}–10^{-2}
Ionospheric propagation under disturbed auroral conditions (HF)	10^{-3}–10^{-2}	10–100	10^{-2}–1
Ionospheric forward scatter (VHF)	10^{-4}	10	10^{-3}
Tropospheric scatter (SHF)	10^{-6}	10	10^{-5}
Orbital scatter (X band)	10^{-4}	10^{3}	10^{-1}
Moon at max. libration ($f_0 = 0.4$ kmc)	10^{-2}	10	10^{-1}

condition $T \ll (\Delta t)_c$. Since T is smaller than the coherence time of the channel, the channel attenuation and phase shift are essentially fixed for the duration of at least one signaling interval. When this condition holds, we call the channel a *slowly fading channel*. Furthermore, when $W \approx 1/T$, the conditions that the channel is frequency-nonselective and slowly fading imply that the product of T_m and B_d must satisfy the condition $T_m B_d < 1$.

The product $T_m B_d$ is called the *spread factor* of the channel. If $T_m B_d < 1$, the channel is said to be *underspread*; otherwise, it is *overspread*. The multipath spread, the Doppler spread, and the spread factor are listed in Table 7.2.1 for several channels. We observe from this table that several radio channels, including the moon when used as a passive reflector, are underspread. Consequently it is possible to select the signal $u(t)$ such that these channels are frequency-non-selective and slowly fading. The slow-fading condition implies that the channel characteristics vary sufficiently slowly that they can be measured.

In Sec. 7.3 we shall determine the error rate performance for binary signaling over a frequency-nonselective slowly fading channel. This channel model is, by far, the simplest to analyze. More important, it yields insight into the performance characteristics for digital signaling on a fading channel and serves to suggest the type of signal waveforms that are effective in overcoming the fading caused by the channel.

Since the multipath components in the received signal are not resolvable when the signal bandwidth W is less than the coherence bandwidth $(\Delta f)_c$ of the channel, the received signal appears to arrive at the receiver via a single fading path. On the other hand, we may choose $W \gg (\Delta f)_c$, so that the channel becomes frequency-selective. We shall show later that under this condition the multipath components in the received signal are resolvable with a resolution in time delay of $1/W$. Thus we shall illustrate that the frequency-selective channel can be modeled as a tapped delay line (transversal) filter with time-variant tap coefficients. We shall then derive the performance of binary signaling over such a frequency-selective channel model.   RAKE!

7.3   BINARY SIGNALING OVER A FREQUENCY-NONSELECTIVE, SLOWLY FADING CHANNEL

In this section we derive the error rate performance of binary PSK and binary FSK when these signals are transmitted over a frequency-nonselective, slowly fading channel. As described in Sec. 7.2, the frequency-nonselective channel results in multiplicative distortion of the transmitted signal $u(t)$. Furthermore, the condition that the channel fades slowly implies that the multiplicative process may be regarded as a constant during at least one signaling interval. Consequently, if the transmitted signal is $u(t)$, the received equivalent lowpass signal in one signaling interval is

$$r(t) = \alpha e^{-j\phi} u(t) + z(t) \qquad 0 \le t \le T \tag{7.3.1}$$

where $z(t)$ represents the complex-valued white gaussian noise process corrupting the signal.

Let us assume that the channel fading is sufficiently slow to that the phase shift ϕ can be estimated from the received signal without error. In that case we can achieve ideal coherent detection of the received signal. Thus the received signal can be processed by passing it through a matched filter in the case of binary PSK or through a pair of matched filters in the case of binary FSK. One method that we can use to determine the performance of the binary communications systems is to evaluate the decision variables and from these determine the probability of error. However, we have already done this for a fixed (time-invariant) channel. That is, for a fixed attenuation α, we have previously derived the probability of error for binary PSK and binary FSK. From (4.2.40), the expression for the error rate of binary PSK as a function of the received SNR γ_b is

$$P_2(\gamma_b) = \tfrac{1}{2} \operatorname{erfc}\left(\sqrt{\gamma_b}\right) \tag{7.3.2}$$

where $\gamma_b = \alpha^2 \mathscr{E}_b / N_0$. The expression for the error rate of binary FSK, detected coherently, is given by (4.2.42) as

$$P_2(\gamma_b) = \tfrac{1}{2} \operatorname{erfc}\left(\sqrt{\tfrac{\gamma_b}{2}}\right) \tag{7.3.3}$$

We view (7.3.2) and (7.3.3) as conditional error probabilities, where the condition is that α is fixed. To obtain the error probabilities when α is random, we must average $P_2(\gamma_b)$, given in (7.3.2) and (7.3.3), over the probability density function of γ_b. That is, we must evaluate the integral

$$P_2 = \int_0^\infty P_2(\gamma_b) p(\gamma_b) \, d\gamma_b \tag{7.3.4}$$

where $p(\gamma_b)$ is the probability density function of γ_b when α is random.

Since α is Rayleigh-distributed, α^2 has a chi-square probability distribution with 2 degrees of freedom. Consequently γ_b also is chi-square-distributed. It is easily shown that

$$p(\gamma_b) = \frac{1}{\bar{\gamma}_b} e^{-\gamma_b/\bar{\gamma}_b} \qquad \gamma_b \geq 0 \tag{7.3.5}$$

where $\bar{\gamma}_b$ is the average signal-to-noise ratio, defined mathematically as

$$\bar{\gamma}_b = \frac{\mathcal{E}_b}{N_0} E(\alpha^2) \tag{7.3.6}$$

The term $E(\alpha^2)$ is simply the average value of α^2.

Now we can substitute (7.3.5) into (7.3.4) and carry out the integration for $P_2(\gamma_b)$ as given by (7.3.2) and (7.3.3). The result of this integration for binary PSK is

$$P_2 = \frac{1}{2}\left[1 - \sqrt{\frac{\bar{\gamma}_b}{1 + \bar{\gamma}_b}}\right] \tag{7.3.7}$$

If we repeat the integration with $P_2(\gamma_b)$ given by (7.3.3), we obtain the probability of error for binary FSK, detected coherently, in the form

$$P_2 = \frac{1}{2}\left[1 - \sqrt{\frac{\bar{\gamma}_b}{2 + \bar{\gamma}_b}}\right] \tag{7.3.8}$$

In arriving at the error rate results in (7.3.7) and (7.3.8), we assumed that the estimate of the channel phase shift, obtained in the presence of slow fading, is noiseless. Such an ideal condition may not hold in practice. In such a case, the expressions in (7.3.7) and (7.3.8) should be viewed as representing the best achieveable performance in the presence of Rayleigh fading. In Appendix 7A we consider the problem of estimating the phase in the presence of noise and we evaluate the error rate performance of binary and multiphase PSK.

On channels for which the fading is sufficiently rapid to preclude the estimation of a stable phase reference by averaging the received signal phase over many signaling intervals, DPSK is an alternative signaling method. Since DPSK requires phase stability over only two consecutive signaling intervals, this modulation technique is quite robust in the presence of signal fading. In deriving the performance of binary DPSK for a fading channel, we begin again with the error probability for a nonfading channel, which is

$$P_2(\gamma_b) = \tfrac{1}{2}e^{-\gamma_b} \tag{7.3.9}$$

This expression is substituted into the integral in (7.3.4) along with $p(\gamma_b)$ obtained from (7.3.5). Evaluation of the resulting integral yields the probability of error for binary DPSK, in the form

$$P_2 = \frac{1}{2(1 + \bar{\gamma}_b)} \tag{7.3.10}$$

If we choose not to estimate the channel phase shift at all, but instead we employ a noncoherent (envelope or square-law) detector with binary, orthogonal

FIGURE 7.3.1
Performance of binary signaling on a Rayleigh fading channel.

FSK signals, the error probability for a nonfading channel is

$$P_2(\gamma_b) = \tfrac{1}{2}e^{-\gamma_b/2} \tag{7.3.11}$$

When we average $P_2(\gamma_b)$ over the Rayleigh fading channel attenuation, the resulting error probability is

$$P_2 = \frac{1}{2 + \bar{\gamma}_b} \tag{7.3.12}$$

The error probabilities in (7.3.7), (7.3.8), (7.3.10), and (7.3.12) are illustrated in Fig. 7.3.1. In comparing the performance of the four binary signaling systems, we focus our attention on the probabilities of error for large SNR, i.e., $\bar{\gamma}_b \gg 1$.

Under this condition the error rates in (7.3.7), (7.3.8), (7.3.10), and (7.3.12) simplify to

$$
P_2 \approx
\begin{cases}
\dfrac{1}{4\gamma_b} & \text{coherent PSK} \\[2mm]
\dfrac{1}{2\gamma_b} & \text{coherent, orthogonal FSK} \\[2mm]
\dfrac{1}{2\gamma_b} & \text{DPSK} \\[2mm]
\dfrac{1}{\gamma_b} & \text{noncoherent, orthogonal FSK}
\end{cases}
\tag{7.3.13}
$$

From (7.3.13) we observe that coherent PSK is 3 dB better than DPSK and 6 dB better than noncoherent FSK. More striking, however, is the observation that the error rates decrease only inversely with SNR. In contrast, the decrease in error rate on a nonfading channel is exponential with SNR. This means that, on a fading channel, the transmitter must transmit a large amount of power in order to obtain a low probability of error. In many cases, a large amount of power is not possible, technically and/or economically. An alternative solution to the problem of obtaining acceptable performance on a fading channel is the use of redundancy, which can be obtained by means of diversity techniques, as discussed in Sec. 7.4.

7.4 DIVERSITY TECHNIQUES FOR FADING MULTIPATH CHANNELS

Diversity techniques are based on the notion that errors occur in reception when the channel attenuation is large, i.e., when the channel is in a deep fade. If we can supply to the receiver several replicas of the same information signal transmitted over independently fading channels, the probability that all the signal components will fade simultaneously is reduced considerably. That is, if p is the probability that any one signal will fade below some critical value, then p^L is the probability that all L independently fading replicas of the same signal will fade below the critical value. There are several ways that we can provide the receiver with L independently fading replicas of the same information-bearing signal.

One method is to employ *frequency diversity*. That is, the same information-bearing signal is transmitted on L carriers where the separation between successive carriers equals or exceeds the coherence bandwidth $(\Delta f)_c$ of the channel.

A second method for achieving L independently fading versions of the same information-bearing signal is to transmit the signal in L different time slots, where the separation between successive time slots equals or exceeds the coherence time $(\Delta t)_c$ of the channel. This method is called *time diversity*.

We note that the fading channel fits the model of a bursty error channel. Furthermore, we may view the transmission of the same information either at different frequencies or in different time slots (or both) as a simple form of repetition coding. The separation of the diversity transmissions in time by $(\Delta t)_c$ or in frequency by $(\Delta f)_c$ is basically a form of block interleaving the bits in the repetition code in an attempt to break up the error bursts and, thus, to obtain independent errors. Later in the chapter, we shall demonstrate that, in general, repetition coding is wasteful of bandwidth when compared to nontrivial coding.

Another commonly used method for achieving diversity employs multiple antennas. For example, we may employ a single transmitting antenna and multiple receiving antennas. The receiving antennas must be spaced sufficiently far apart so that the multipath components in the signal have significantly different propagation delays at the antennas. Usually a separation of at least 10 wavelengths is required between two antennas in order to obtain signals that fade independently.

A more sophisticated method for obtaining diversity is based on the use of a signal having a bandwidth much greater than the coherence bandwidth $(\Delta f)_c$ of the channel. Such a signal with bandwidth W will resolve the multipath components and, thus, provide the receiver with several independently fading signal paths. The time resolution is $1/W$. Consequently, with a multipath spread of T_m seconds, there are $T_m W$ resolvable signal components. Since $T_m \approx 1/(\Delta f)_c$, the number of resolvable signal components may also be expressed as $W/(\Delta f)_c$. Thus the use of a wideband signal may be viewed as just another method for obtaining frequency diversity of order $L \approx W/(\Delta f)_c$. The optimum receiver for processing the wideband signal will be derived in Sec. 7.5. It is called a *RAKE correlator* or a *RAKE matched filter* and was invented by Price and Green (1958).

There are other diversity techniques that have received some consideration in practice, such as angle-of-arrival diversity and polarization diversity. However, these have not been as widely used as the ones described above.

We shall now determine the error rate performance for a binary digital communications system with diversity. We begin by describing the mathematical model for the communications system with diversity. First of all, we assume that there are L diversity channels, carrying the same information-bearing signal. Each channel is assumed to be frequency-nonselective and slowly fading with Rayleigh-distributed envelope statistics. The fading processes among the L diversity channels are assumed to be mutually statistically independent. The signal in each channel is corrupted by an additive zero mean white gaussian noise process. The noise processes in the L channels are assumed to be mutually statistically independent, with identical autocorrelation functions. Thus the equivalent low-pass received signals for the L channels can be expressed in the form

$$r_k(t) = \alpha_k e^{-j\phi_k} u_{km}(t) + z_k(t) \qquad \begin{matrix} k = 1,2,\ldots,L \\ m = 1,2 \end{matrix} \qquad (7.4.1)$$

where $\{\alpha_k e^{-j\phi_k}\}$ represent the attenuation factors and phase shifts for the L

channels, $u_{km}(t)$ denotes the mth signal transmitted on the kth channel, and $z_k(t)$ denotes the additive white gaussian noise on the kth channel. All signals in the set $\{u_{km}(t)\}$ have the same energy.

The optimum demodulator for the signal received from the kth channel consists of two matched filters, one having the impulse response

$$b_{k1}(t) = u_{k1}^*(T - t) \tag{7.4.2}$$

and the other having the impulse response

$$b_{k2}(t) = u_{k2}^*(T - t) \tag{7.4.3}$$

Of course, if binary PSK is the modulation method used to transmit the information, then $u_{k1}(t) = -u_{k2}(t)$. Consequently only a single matched filter is required for binary PSK. Following the matched filters is a combiner which forms the two decision variables. The combiner that achieves the best performance is one in which each matched filter output is multiplied by the corresponding complex-valued (conjugate) channel gain $\alpha_k e^{j\phi_k}$. The effect of this multiplication is to compensate for the phase shift in the channel and to weight the signal by a factor that is proportional to the signal strength. Thus a strong signal carries a larger weight than a weak signal. After the complex-valued weighting operation is performed, two sums are formed. One consists of the real parts of the weighted outputs from the matched filters corresponding to a transmitted 0. The second consists of the real part of the outputs from the matched filters corresponding to a transmitted 1. This optimum combiner is called a *maximal ratio combiner* by Brennan (1959). Of course the realization of this optimum combiner is based on the assumption that the channel attenuations $\{\alpha_k\}$ and the phase shifts $\{\phi_k\}$ are known perfectly. That is, the estimates of the parameters $\{\alpha_k\}$ and $\{\phi_k\}$ contain no noise. (The effect of noisy estimates on the error rate performance of multiphase PSK is considered in Appendix 7A.)

A block diagram illustrating the model for the binary digital communications system described above is shown in Fig. 7.4.1.

Let us first consider the performance of binary PSK with Lth-order diversity. The output of the maximal ratio combiner can be expressed as a single decision variable in the form

$$U = \text{Re}\left(2\mathscr{E}\sum_{k=1}^{L}\alpha_k^2 + \sum_{k=1}^{L}\alpha_k N_k\right)$$

$$= 2\mathscr{E}\sum_{k=1}^{L}\alpha_k^2 + \sum_{k=1}^{L}\alpha_k N_{kr} \tag{7.4.4}$$

where N_{kr} denotes the real part of the complex-valued gaussian noise variable

$$N_k = e^{j\phi_k}\int_0^T z_k(t)u_k^*(t)\,dt \tag{7.4.5}$$

We follow the approach used in Sec. 7.3 in deriving the probability of error. That is, the probability of error conditioned on a fixed set of attenuation factors $\{\alpha_k\}$

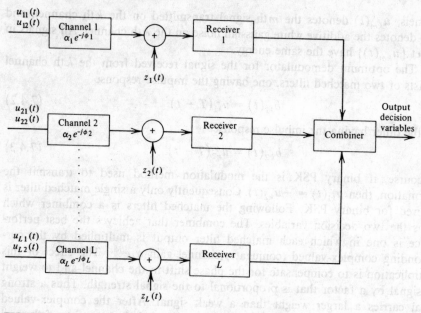

FIGURE 7.4.1
Model of binary digital communications system with diversity.

is obtained first. Then the conditional probability of error is averaged over the probability density function of the $\{\alpha_k\}$.

For a fixed set of $\{\alpha_k\}$ the decision variable U is gaussian with mean

$$E(U) = 2\mathscr{E} \sum_{k=1}^{L} \alpha_k^2 \tag{7.4.6}$$

and variance

$$\sigma_U^2 = 2\mathscr{E} N_0 \sum_{k=1}^{L} \alpha_k^2 \tag{7.4.7}$$

For these values of the mean and variance, the probability that U is less than zero is simply

$$P_2(\gamma_b) = \tfrac{1}{2} \operatorname{erfc}\left(\sqrt{\gamma_b}\right) \tag{7.4.8}$$

where the SNR per bit γ_b is given as

$$\gamma_b = \frac{\mathscr{E}}{N_0} \sum_{k=1}^{L} \alpha_k^2$$

$$= \sum_{k=1}^{L} \gamma_k \tag{7.4.9}$$

where $\gamma_k = \mathscr{E}\alpha_k^2/N_0$ is the instantaneous SNR on the kth channel. Now we must determine the probability density function $p(\gamma_b)$. This function is most easily determined via the characteristic function of γ_b. First of all, we note that for $L = 1$, $\gamma_b \equiv \gamma_1$ has a chi-square probability density function given previously in (7.3.5). The characteristic function of γ_1 is easily shown to be

$$\psi_{\gamma_1}(jv) = E(e^{jv\gamma_1})$$

$$= \frac{1}{1 - jv\bar{\gamma}_c} \tag{7.4.10}$$

where $\bar{\gamma}_c$ is the average SNR per channel, which is assumed to be identical for all channels. That is,

$$\bar{\gamma}_c = \frac{\mathscr{E}}{N_0}E(\alpha_k^2) \tag{7.4.11}$$

independent of k. This assumption applies for the results throughout this section. Since the fading on the L channels is mutually statistically independent, the $\{\gamma_k\}$ are statistically independent and, hence, the characteristic function for the sum γ_b is simply the result in (7.4.10) raised to the Lth power, i.e.,

$$\psi_{\gamma_b}(jv) = \frac{1}{(1 - jv\bar{\gamma}_c)^L} \tag{7.4.12}$$

But this is the characteristic function of a chi-square-distributed random variable with $2L$ degrees of freedom. It follows from (1.1.107) that the probability density function $p(\gamma_b)$ is

$$p(\gamma_b) = \frac{1}{(L - 1)!\bar{\gamma}_c^L}\gamma_b^{L-1}e^{-\gamma_b/\bar{\gamma}_c} \tag{7.4.13}$$

The final step in this derivation is to average the conditional error probability given in (7.4.8) over the fading channel statistics. Thus we evaluate the integral

$$P_2 = \int_0^\infty P_2(\gamma_b)p(\gamma_b)\,d\gamma_b \tag{7.4.14}$$

There is a closed-form solution for (7.4.14), which can be expressed as

$$P_2 = \left(\frac{1-\mu}{2}\right)^L \sum_{k=0}^{L-1}\binom{L-1+k}{k}\left(\frac{1+\mu}{2}\right)^k \tag{7.4.15}$$

where, by definition,

$$\mu = \sqrt{\frac{\bar{\gamma}_c}{1 + \bar{\gamma}_c}} \tag{7.4.16}$$

When the average SNR per channel $\bar{\gamma}_c$ satisfies the condition $\bar{\gamma}_c \gg 1$, the term

$(1 + \mu)/2 \approx 1$ and the term $(1 - \mu)/2 \approx 1/4\bar{\gamma}_c$. Furthermore,

$$\sum_{k=0}^{L-1} \binom{L-1+k}{k} = \binom{2L-1}{L} \qquad (7.4.17)$$

Therefore, when $\bar{\gamma}_c$ is sufficiently large (greater than 10 dB), the probability of error in (7.4.15) can be approximated as

$$P_2 \approx \left(\frac{1}{4\bar{\gamma}_c}\right)^L \binom{2L-1}{L} \qquad (7.4.18)$$

We observe from (7.4.18) that the probability of error varies as $1/\bar{\gamma}_c$ raised to the Lth power. Thus, with diversity, the error rate decreases inversely with the Lth power of the SNR.

Having obtained the performance of binary PSK with diversity, we now turn our attention to binary, orthogonal FSK that is detected coherently. In this case the two decision variables at the output of the maximal ratio combiner may be expressed as

$$U_1 = \text{Re}\left(2\mathscr{E}\sum_{k=1}^{L} \alpha_k^2 + \sum_{k=1}^{L} \alpha_k N_{k1}\right)$$

$$U_2 = \text{Re}\left(\sum_{k=1}^{L} \alpha_k N_{k2}\right) \qquad (7.4.19)$$

where we have assumed that signal $u_{k1}(t)$ was transmitted and where $\{N_{k1}\}$ and $\{N_{k2}\}$ are the two sets of noise components at the output of the matched filters. The probability of error is simply the probability that U_2 exceeds U_1. This computation is similar to the one performed for PSK, except that we now have twice the noise power. Consequently, when the $\{\alpha_k\}$ are fixed, the conditional probability of error is

$$P_2(\gamma_b) = \tfrac{1}{2} \text{erfc}\left(\sqrt{\frac{\gamma_b}{2}}\right) \qquad (7.4.20)$$

We use (7.4.13) to average $P_2(\gamma_b)$ over the fading. It is not surprising to find that the result given in (7.4.15) still applies, with $\bar{\gamma}_c$ replaced by $\bar{\gamma}_c/2$. That is, (7.4.15) is the probability of error for binary, orthogonal FSK with coherent detection, where the parameter μ is defined as

$$\mu = \sqrt{\frac{\bar{\gamma}_c}{2 + \bar{\gamma}_c}} \qquad (7.4.21)$$

Furthermore, for large values of $\bar{\gamma}_c$ the performance P_2 can be approximated as

$$P_2 \approx \left(\frac{1}{2\bar{\gamma}_c}\right)^L \binom{2L-1}{L} \qquad (7.4.22)$$

In comparing (7.4.22) with (7.4.18), we observe that the 3-dB difference in performance between PSK and orthogonal FSK with coherent detection, which exists in a nonfading, nondispersive channel, is the same also in a fading channel.

In the above discussion of binary PSK and FSK, detected coherently, we assumed that noiseless estimates of the complex-valued channel parameters $\{\alpha_k e^{-j\phi_k}\}$ were used at the receiver. Since the channel is time-variant, the parameters $\{\alpha_k e^{-j\phi_k}\}$ cannot be estimated perfectly. In fact, on some channels the time variations may be sufficiently fast to preclude the implementation of coherent detection. In such a case, we should consider using either DPSK or FSK with noncoherent detection.

Let us consider DPSK first. In order for DPSK to be a viable digital signaling method, the channel variations must be sufficiently slow so that the channel phase shifts $\{\phi_k\}$ do not change appreciably over two consecutive signaling intervals. In our analysis, we assume that the channel parameters $\{\alpha_k e^{-j\phi_k}\}$ remain constant over two successive signaling intervals. Thus the combiner for binary DPSK will yield as an output the decision variable

$$U = \text{Re}\left[\sum_{k=1}^{L}\left(2\mathscr{E}\alpha_k e^{-j\phi_k} + N_{k2}\right)\left(2\mathscr{E}\alpha_k e^{j\phi_k} + N_{k1}^*\right)\right] \qquad (7.4.23)$$

where $\{N_{k1}\}$ and $\{N_{k2}\}$ denote the received noise components at the output of the matched filters in the two consecutive signaling intervals. The probability of error is simply the probability that U is less than zero. Since U is a special case of the general quadratic form in complex-valued gaussian random variables treated in Appendix 4B, the probability of error can be obtained directly from the results given in that appendix. Alternatively, we may use the error probability given in (4.4.13), which applies to binary DPSK transmitted over L time-invariant channels, and average it over the Rayleigh fading channel statistics. Thus we have the conditional error probability

$$P_2(\gamma_b) = \frac{1}{2^{2L-1}}e^{-\gamma_b}\sum_{k=0}^{L-1}b_k\gamma_b^k \qquad (7.4.24)$$

where γ_b is given by (7.4.9) and

$$b_k = \frac{1}{k!}\sum_{n=0}^{L-1-k}\binom{2L-1}{n} \qquad (7.4.25)$$

The average of $P_2(\gamma_b)$ over the fading channel statistics given by $p(\gamma_b)$ in (7.4.13) is easily shown to be

$$P_2 = \frac{1}{2^{2L-1}(L-1)!(1+\bar{\gamma}_c)^L}\sum_{k=0}^{L-1}b_k(L-1+k)!\left(\frac{\bar{\gamma}_c}{1+\bar{\gamma}_c}\right)^k \qquad (7.4.26)$$

We indicate that the result in (7.4.26) can be manipulated into the form given in (7.4.15), which applies also to coherent PSK and FSK. For binary DPSK, the parameter μ in (7.4.15) is defined as (see Appendix 7A)

$$\mu = \frac{\bar{\gamma}_c}{1+\bar{\gamma}_c} \qquad (7.4.27)$$

For $\bar{\gamma}_c \gg 1$, the error probability in (7.4.26) can be approximated by the expression

$$P_2 \approx \left(\frac{1}{2\bar{\gamma}_c}\right)^L \binom{2L-1}{L} \tag{7.4.28}$$

Orthogonal FSK with noncoherent detection is the final signaling technique that we consider in this section. It is appropriate for both slow and fast fading. However, the analysis of the performance presented below is based on the assumption that the fading is sufficiently slow so that the channel parameters $\{\alpha_k e^{j\phi_k}\}$ remain constant for the duration of the signaling interval. The combiner for the multichannel signals is a square-law combiner. Its output consists of the two decision variables

$$
\begin{aligned}
U_1 &= \sum_{k=1}^{L} |2\mathscr{E}\alpha_k e^{-j\phi_k} + N_{k1}|^2 \\
U_2 &= \sum_{k=1}^{L} |N_{k2}|^2
\end{aligned}
\tag{7.4.29}
$$

where U_1 is assumed to contain the signal. Consequently the probability of error is the probability that U_2 exceeds U_1.

As in DPSK, we have a choice of two approaches in deriving the performance of FSK with square-law combining. In Sec. 4.4 we indicated that the expression for the error probability for square-law combined FSK is the same as that for DPSK with γ_b replaced by $\gamma_b/2$. That is, the FSK system requires 3 dB of additional SNR to achieve the same performance on a time-invariant channel. Consequently the conditional error probability for DPSK given in (7.4.24) applies to square-law-combined FSK when γ_b is replaced by $\gamma_b/2$. Furthermore, the result obtained by averaging (7.4.24) over the fading, which is given by (7.4.26), must also apply to FSK with $\bar{\gamma}_c$ replaced by $\bar{\gamma}_c/2$. But we also stated previously that (7.4.26) and (7.4.15) are equivalent. Therefore the error probability given in (7.4.15) also applies to square-law-combined FSK with the parameter μ defined as

$$\mu = \frac{\bar{\gamma}_c}{2 + \bar{\gamma}_c} \tag{7.4.30}$$

An alternative derivation used by Pierce (1958) to obtain the probability that the decision variable U_2 exceeds U_1 is just as easy as the method described above. It begins with the probability density functions $p(U_1)$ and $p(U_2)$. Since the complex-valued random variables $\{\alpha_k e^{-j\phi_k}\}$, $\{N_{k1}\}$, and $\{N_{k2}\}$ are zero mean gaussian-distributed, the decision variables U_1 and U_2 are distributed according to a chi-square probability distribution with $2L$ degrees of freedom. That is,

$$p(U_1) = \frac{1}{(2\sigma_1^2)^L (L-1)!} U_1^{L-1} e^{-U_1/2\sigma_1^2} \tag{7.4.31}$$

where

$$\sigma_1^2 = \tfrac{1}{2}E\left(|2\mathscr{E}\alpha_k e^{-j\phi_k} + N_{k1}|^2\right)$$
$$= 2\mathscr{E}N_0(1 + \bar{\gamma}_c) \tag{7.4.32}$$

Similarly,

$$p(U_2) = \frac{2}{(2\sigma_2^2)^L (L-1)!} U_2^{L-1} e^{-U_2/2\sigma_2^2} \tag{7.4.33}$$

where

$$\sigma_2^2 = 2\mathscr{E}N_0 \tag{7.4.34}$$

FIGURE 7.4.2
Performance of binary signals with diversity.

The probability of error is just the probability that U_2 exceeds U_1. It is left as an exercise for the reader to show that this probability is given by (7.4.15), where μ is defined by (7.4.30).

When $\bar{\gamma}_c \gg 1$, the performance of square-law-detected FSK can be simplified as we have done for the other binary multichannel systems. In this case, the error rate is well approximated by the expression

$$P_2 \approx \left(\frac{1}{\bar{\gamma}_c} \right)^L \binom{2L-1}{L} \tag{7.4.35}$$

The error rate performance of PSK, DPSK, and square-law-detected orthogonal FSK is illustrated in Fig. 7.4.2 for $L = 1, 2, 4$. The performance is plotted as a function of the average SNR per bit $\bar{\gamma}_b$, which is related to the average SNR per channel $\bar{\gamma}_c$ by the formula

$$\bar{\gamma}_b = L\bar{\gamma}_c \tag{7.4.36}$$

The results in Fig. 7.4.2 clearly illustrate the advantage of diversity as a means for overcoming the severe penalty in SNR caused by fading.

7.5 DIGITAL SIGNALING OVER A FREQUENCY-SELECTIVE, SLOWLY FADING CHANNEL

When the spread factor of the channel satisfies the condition $T_m B_d \ll 1$, it is possible, we showed, to select signals having a bandwidth $W \ll (\Delta f)_c$ and a signaling duration $T \ll (\Delta t)_c$. Thus the channel is frequency-nonselective and slowly fading. In such a channel, diversity techniques can be employed to overcome the severe consequences of fading.

When a bandwidth $W \gg (\Delta f)_c$ is available to the user, the channel can be subdivided into a number of frequency division multiplexed (FDM) subchannels having a mutual separation in center frequencies of at least $(\Delta f)_c$. Then the same signal can be transmitted on the FDM subchannels and, thus, frequency diversity is obtained.

As we shall now demonstrate, a more direct method for achieving basically the same result is to employ a wideband signal covering the bandwidth W. The channel is still assumed to be slowly fading by virtue of the assumption that $T \ll (\Delta t)_c$. Now suppose that W is the bandwidth occupied by the real bandpass signal. Then the band occupancy of the equivalent lowpass signal $u(t)$ is $|f| \leq W/2$. Since $u(t)$ is band-limited to $|f| \leq W/2$, application of the sampling theorem results in the signal representation

$$u(t) = \sum_{n=-\infty}^{\infty} u\left(\frac{n}{W} \right) \frac{\sin \pi W(t - n/W)}{\pi W(t - n/W)} \tag{7.5.1}$$

The Fourier transform of $u(t)$ is

$$U(f) = \begin{cases} \dfrac{1}{W} \displaystyle\sum_{n=-\infty}^{\infty} u\left(\dfrac{n}{W}\right) e^{-j2\pi fn/W} & |f| \le W/2 \\ 0 & |f| > W/2 \end{cases} \quad (7.5.2)$$

The noiseless received signal from a frequency-selective channel was previously expressed in the form

$$r(t) = \int_{-\infty}^{\infty} C(f; t) U(f) e^{j2\pi ft} \, df \quad (7.5.3)$$

where $C(f; t)$ is the time-variant transfer function. Substitution for $U(f)$ from (7.5.2) into (7.5.3) yields

$$r(t) = \frac{1}{W} \sum_{n=-\infty}^{\infty} u\left(\frac{n}{W}\right) \int_{-\infty}^{\infty} C(f; t) e^{j2\pi f(t-n/W)} \, df$$

$$= \frac{1}{W} \sum_{n=-\infty}^{\infty} u\left(\frac{n}{W}\right) c\left(t - \frac{n}{W}; t\right) \quad (7.5.4)$$

where $c(\tau; t)$ is the time-variant impulse response. We observe that (7.5.4) has the form of a convolution sum. Hence it can also be expressed in the alternative form

$$r(t) = \frac{1}{W} \sum_{n=-\infty}^{\infty} u\left(t - \frac{n}{W}\right) c\left(\frac{n}{W}; t\right) \quad (7.5.5)$$

It is convenient to define a set of time-variable channel coefficients as

$$c_n(t) = \frac{1}{W} c\left(\frac{n}{W}; t\right) \quad (7.5.6)$$

Then (7.5.5) expressed in terms of these channel coefficients becomes

$$r(t) = \sum_{n=-\infty}^{\infty} c_n(t) u\left(t - \frac{n}{W}\right) \quad (7.5.7)$$

The form for the received signal in (7.5.7) implies that the time-variant frequency-selective channel can be modeled or represented as a tapped delay line with tap spacing $1/W$ and tap weight coefficients $\{c_n(t)\}$. In fact, we deduce from (7.5.7) that the lowpass impulse response for the channel is

$$c(\tau; t) = \sum_{n=-\infty}^{\infty} c_n(t) \delta\left(\tau - \frac{n}{W}\right) \quad (7.5.8)$$

and the corresponding time-variant transfer function is

$$C(f; t) = \sum_{n=-\infty}^{\infty} c_n(t) e^{-j2\pi fn/W} \quad (7.5.9)$$

Thus, with an equivalent lowpass signal having a bandwidth $W/2$, where $W \gg (\Delta f)_c$, we achieve a resolution of $1/W$ in the multipath delay profile. Since the

FIGURE 7.5.1
Tapped delay line model of frequency-selective channel.

total multipath spread is T_m, for all practical purposes the tapped delay line model for the channel can be truncated at $L = [T_m W] + 1$ taps. Then the noiseless received signal can be expressed in the form

$$r(t) = \sum_{n=1}^{L} c_n(t) u\left(t - \frac{n}{W}\right) \qquad (7.5.10)$$

The truncated tapped delay line model is shown in Fig. 7.5.1. Consistent with the statistical characterization of the channel presented in Sec. 7.1, the time-variant tap weights $\{c_n(t)\}$ are zero mean complex-valued stationary gaussian random processes. The magnitudes $|c_n(t)| \equiv \alpha_n(t)$ are Rayleigh-distributed and the phases $\phi_n(t)$ are uniformly distributed. Since the $\{c_n(t)\}$ represent the tap weights corresponding to the L different delays $\tau = n/W$, $n = 1, 2, \ldots, L$, the uncorrelated scattering assumption made in Sec. 7.1 implies that the $\{c_n(t)\}$ are mutually uncorrelated. But the $\{c_n(t)\}$ are gaussian random processes; hence they are statistically independent.

We now consider the problem of digital signaling over a frequency-selective channel that is modeled by a tapped delay line with statistically independent time-variant tap weights $\{c_n(t)\}$. It is apparent at the outset, however, that the tapped delay line model with statistically independent tap weights provides us with L replicas of the same transmitted signal at the receiver. Hence a receiver that processes the received signal in an optimum manner will achieve the performance of an equivalent Lth-order diversity communications system.

Let us consider binary signaling over the channel. We have two equal-energy signals $u_1(t)$ and $u_2(t)$ which are either antipodal or orthogonal. Their time duration T is selected to satisfy the condition $T \gg T_m$. Thus we may neglect any intersymbol interference due to multipath. Since the bandwidth of the signal exceeds the coherent bandwidth of the channel, the received signal is expressed as

$$r(t) = \sum_{k=1}^{L} c_k(t) u_i\left(t - \frac{k}{W}\right) + z(t)$$

$$= v_i(t) + z(t) \qquad 0 \le t \le T \quad i = 1, 2 \qquad (7.5.11)$$

where $z(t)$ is a complex-valued zero mean white gaussian noise process. Assume for the moment that the channel tap weights are known. Then the optimum receiver consists of two filters matched to $v_1(t)$ and $v_2(t)$, followed by samplers and a decision circuit that selects the signal corresponding to the largest output. An equivalent optimum receiver employs cross correlation instead of matched

INPUT SIGNAL FROM CHANNEL

FIGURE 7.5.2
Optimum demodulator for wideband binary signals (delayed reference configuration).

FIGURE 7.5.3
Optimum demodulator for wideband binary signals (delayed received signal configuration).

filtering. In either case, the decision variables for coherent detection of the binary signals can be expressed as

$$U_m = \operatorname{Re} \int_0^T r(t) v_m^*(t)\, dt$$

$$= \operatorname{Re} \sum_{k=1}^L \int_0^T r(t) c_k^*(t) u_m^*\!\left(t - \frac{k}{W}\right) dt \qquad m = 1, 2 \qquad (7.5.12)$$

Figure 7.5.2 illustrates the operations involved in the computation of the decision variables. In this realization of the optimum receiver, the two reference signals are delayed and correlated with the received signal $r(t)$.

An alternative realization of the optimum receiver employs a single delay line through which is passed the received signal $r(t)$. The signal at each tap is correlated with $c_k(t) u_m^*(t)$, where $k = 1, 2, \ldots, L$ and $m = 1, 2$. This receiver structure is shown in Fig. 7.5.3. In effect, the tapped delay line receiver attempts to collect the signal energy from all the received signal paths that fall within the span of the delay line and carry the same information. Its action is somewhat analogous to an ordinary garden rake and, consequently, the name "RAKE receiver" has been coined for this receiver structure by Price and Green (1958).

We shall now evaluate the performance of the RAKE receiver under the condition that the fading is sufficiently slow to allow us to estimate $c_k(t)$ perfectly (without noise). Furthermore, within any one signaling interval, $c_k(t)$ is treated as a constant and denoted as c_k. Thus the decision variables in (7.5.12) may be expressed in the form

$$U_m = \text{Re}\left[\sum_{k=1}^{L} c_k^* \int_0^T r(t) u_m^*\left(t - \frac{k}{W}\right) dt\right] \qquad m = 1, 2 \qquad (7.5.13)$$

Suppose the transmitted signal is $u_1(t)$; then the received signal is

$$r(t) = \sum_{n=1}^{L} c_n u_1^*\left(t - \frac{n}{W}\right) + z(t) \qquad 0 \le t \le T \qquad (7.5.14)$$

Substitution of (7.5.14) into (7.5.13) yields

$$U_m = \text{Re}\left[\sum_{k=1}^{L} c_k^* \sum_{n=1}^{L} c_n \int_0^T u_1\left(t - \frac{n}{W}\right) u_m^*\left(t - \frac{k}{W}\right) dt\right]$$

$$+ \text{Re}\left[\sum_{k=1}^{L} c_k^* \int_0^T z(t) u_m^*\left(t - \frac{k}{W}\right) dt\right] \qquad m = 1, 2 \qquad (7.5.15)$$

Usually the wideband signals $u_1(t)$ and $u_2(t)$ are generated from pseudo-random sequences, which results in signals that have the property

$$\int_0^T u_i\left(t - \frac{n}{W}\right) u_i^*\left(t - \frac{k}{W}\right) dt \approx 0 \qquad \text{for } k \ne n \quad i = 1, 2 \qquad (7.5.16)$$

If we assume that our binary signals are designed to satisfy this property, then (7.5.15) simplifies to†

$$U_m = \text{Re}\left[\sum_{k=1}^{L} |c_k|^2 \int_0^T u_1\left(t - \frac{k}{W}\right) u_m^*\left(t - \frac{k}{W}\right) dt\right]$$

$$+ \text{Re}\left[\sum_{k=1}^{L} c_k^* \int_0^T z(t) u_m^*\left(t - \frac{k}{W}\right) dt\right] \qquad m = 1, 2 \qquad (7.5.17)$$

When the binary signals are antipodal, a single decision variable suffices. In this case (7.5.17) reduces to

$$U_1 = \text{Re}\left(2\mathscr{E}\sum_{k=1}^{L} \alpha_k^2 + \sum_{k=1}^{L} \alpha_k N_k\right) \qquad (7.5.18)$$

---

† Although the orthogonality property specified by (7.5.16) can be satisfied by proper selection of the pseudo-random sequences, the cross correlation of $u_1(t - n/W)$ with $u_i^*(t - k/W)$ gives rise to a signal-dependent self-noise which ultimately limits the performance. For simplicity, we do not consider the self-noise term in the following calculations. Consequently the performance results presented below should be considered as lower bounds (ideal RAKE). An approximation to the performance of the RAKE can be obtained by treating the self-noise as an additional gaussian noise component with noise power equal to its variance.

where $\alpha_k = |c_k|$ and

$$N_k = e^{j\phi_k} \int_0^T z(t) u^* \left(t - \frac{k}{W} \right) dt \tag{7.5.19}$$

But (7.5.18) is identical to the decision variable given in (7.4.4), which corresponds to the output of a maximal ratio combiner in a system with Lth-order diversity. Consequently the RAKE receiver with perfect (noiseless) estimates of the channel tap weights is equivalent to a maximal ratio combiner in a system with Lth-order diversity. Thus, when all the tap weights have the same mean-square value, i.e., $E(\alpha_k^2)$ is the same for all k, the error rate performance of the RAKE receiver is given by (7.4.15) and (7.4.16). On the other hand, when the mean-square values $E(\alpha_k^2)$ are not identical for all k, the derivation of the error rate performance must be repeated since (7.4.15) no longer applies.

We shall derive the probability of error for binary antipodal and orthogonal signals under the condition that the mean-square values are distinct. We begin with the conditional error probability

$$P_2(\gamma_b) = \tfrac{1}{2} \operatorname{erfc} \left(\sqrt{\frac{\gamma_b}{2} (1 - \rho_r)} \right) \tag{7.5.20}$$

where $\rho_r = -1$ for antipodal signals, $\rho_r = 0$ for orthogonal signals, and

$$\gamma_b = \frac{\mathscr{E}}{N_0} \sum_{k=1}^L \alpha_k^2$$

$$= \sum_{k=1}^L \gamma_k \tag{7.5.21}$$

Each of the $\{\gamma_k\}$ is distributed according to a chi-squared distribution with 2 degrees of freedom. That is,

$$p(\gamma_k) = \frac{1}{\bar{\gamma}_k} e^{-\gamma_k / \bar{\gamma}_k} \tag{7.5.22}$$

where $\bar{\gamma}_k$ is the average SNR for the kth path, defined as

$$\bar{\gamma}_k = \frac{\mathscr{E}}{N_0} E(\alpha_k^2) \tag{7.5.23}$$

Furthermore, from (7.4.10) we know that the characteristic function of γ_k is

$$\psi_{\gamma_k}(jv) = \frac{1}{1 - jv\bar{\gamma}_k} \tag{7.5.24}$$

Since γ_b is the sum of L statistically independent components $\{\gamma_k\}$, the characteristic function of γ_b is

$$\psi_{\gamma_b}(jv) = \prod_{k=1}^L \frac{1}{(1 - jv\bar{\gamma}_k)} \tag{7.5.25}$$

The inverse Fourier transform of the characteristic function in (7.5.25) yields the probability density function of γ_b in the form

$$p(\gamma_b) = \sum_{k=1}^L \frac{\pi_k}{\bar{\gamma}_k} e^{-\gamma_b / \bar{\gamma}_k} \qquad \gamma_b \geq 0 \tag{7.5.26}$$

where π_k is defined as

$$\pi_k = \prod_{\substack{i=1 \\ i \neq k}}^{L} \frac{\bar{\gamma}_k}{\bar{\gamma}_k - \bar{\gamma}_i} \tag{7.5.27}$$

When the conditional error probability in (7.5.20) is averaged over the probability density function given in (7.5.26), the result is

$$P_2 = \frac{1}{2} \sum_{k=1}^{L} \pi_k \left[1 - \sqrt{\frac{\bar{\gamma}_k (1 - \rho_r)}{2 + \bar{\gamma}_k (1 - \rho_r)}} \right] \tag{7.5.28}$$

This error probability can be approximated as ($\bar{\gamma}_k \gg 1$)

$$P_2 \approx \binom{2L-1}{L} \prod_{k=1}^{L} \frac{1}{[2\bar{\gamma}_k (1 - \rho_r)]} \tag{7.5.29}$$

By comparing (7.5.29) for $\rho_r = -1$ with (7.4.18), we observe that the same type of asymptotic behavior is obtained for the case of unequal SNR per path and the case of equal SNR per path.

In the derivation for the error rate performance of the RAKE receiver, we assumed that the estimates of the channel tap weights are perfect. In practice, relatively good estimates can be obtained if the channel fading is sufficiently slow, e.g., $(\Delta t)_c / T \geq 100$, where T is the signaling interval. Figure 7.5.4 illustrates a method for estimating the tap weights when the binary signaling waveforms are orthogonal. The estimate is the output of the lowpass filter. At any one instant in time, the incoming signal is either $u_1(t)$ or $u_2(t)$. Hence the input to the lowpass filter used to estimate $c_k(t)$ contains signal-plus-noise from one of the correlators and noise only from the other correlator. This method for channel estimation is not appropriate for antipodal signals because the addition of the two correlator outputs results in signal cancellation. Instead, a single correlator can be employed for antipodal signals. Its output is fed to the input of the lowpass filter after the information-bearing signal is removed. To accomplish this, we must introduce a delay of one signaling interval into the channel estimation procedure, as illustrated in Fig. 7.5.5. That is, first the receiver must decide whether the information in the received signal is $+1$ or -1 and, then, it uses the decision to remove the information from the correlator output prior to feeding it to the lowpass filter.

If we choose not to estimate the tap weights of the frequency-selective channel, we may use either DPSK signaling or noncoherently detected orthogonal signaling. The RAKE receiver structure for DPSK is illustrated in Fig. 7.5.6. It is apparent that when the transmitted signal waveform $u(t)$ satisfies the orthogonality property given in (7.5.16), the decision variable is identical to that given in (7.4.23) for an Lth-order diversity system. Consequently the error rate performance of the RAKE receiver for a binary DPSK is identical to that given in (7.4.15) with $\mu = \bar{\gamma}_c / (1 + \bar{\gamma}_c)$, when all the signal paths have the same SNR $\bar{\gamma}_c$. On the other hand, when the SNR's $\{\bar{\gamma}_k\}$ are distinct, the error probability can be obtained by averaging (7.4.24), which is the probability of error conditioned on a time-invariant channel, over the probability density function of γ_b given by (7.5.26). The result of this integration is

$$P_2 = \frac{1}{2^{2L-1}} \sum_{m=0}^{L-1} m! b_m \sum_{k=1}^{L} \frac{\pi_k}{\bar{\gamma}_k} \left(\frac{\bar{\gamma}_k}{1 + \bar{\gamma}_k} \right)^{m+1} \tag{7.5.30}$$

where π_k was previously defined in (7.5.27) and b_m was defined in (7.4.25).

Finally, we consider binary orthogonal signaling over the frequency-selective channel with square-law detection at the receiver. This type of signal is appropriate when either the fading is rapid enough to preclude a good estimate of the channel tap weights or the cost of implementing the channel estimators is high. The RAKE receiver with square-law combining of the signal from each tap is illustrated in Fig. 7.5.7. In computing its performance, we again assume that the orthogonality property given in (7.5.16) holds. Then the decision variables at the output of the RAKE are

$$\begin{aligned} U_1 &= \sum_{k=1}^{L} |2\mathcal{E}c_k + N_{k1}|^2 \\ U_2 &= \sum_{k=1}^{L} |N_{k2}|^2 \end{aligned} \tag{7.5.31}$$

FIGURE 7.5.4
Channel tap weight estimation with binary orthogonal signals.

FIGURE 7.5.5
Channel tap weight estimation with binary antipodal signals.

where we have assumed that $u_1(t)$ was the transmitted signal. Again we observe that the decision variables are identical to the ones given in (7.4.29), which apply to orthogonal signals with Lth-order diversity. Therefore the performance of the RAKE receiver for square-law-detected orthogonal signals is given by (7.4.15) with $\mu = \bar{\gamma}_c/(2 + \bar{\gamma}_c)$ when all the signal paths have the same SNR. If the SNR's are distinct, we can average the conditional error probability given by (7.4.24), with γ_b replaced by $\gamma_b/2$, over the probability density function $p(\gamma_b)$ given in (7.5.26). The result of this averaging is given by (7.5.30) with $\bar{\gamma}_k$ replaced by $\bar{\gamma}_k/2$.

In the above analysis, the RAKE demodulator shown in Fig. 7.5.7 for square-law combining of orthogonal signals is assumed to contain a signal component at each delay. If that is not the case, its performance will be degraded, since some of the tap correlators will contribute only noise. As a demonstration of the performance degradation, we show in Fig. 7.5.8 the error rate versus the average SNR per bit when the RAKE demodulator contains either 4 or 10 or 20 or 50 taps, but only $L = 4$ of these taps contain signal components. The others contain noise. From these results we observe that a degradation of about 4 dB occurs when the square-law combining encompasses all 50 taps but only four contain a signal. Most of this degradation can be recovered by use of

FIGURE 7.5.6
RAKE demodulator for DPSK signals.

adaptive algorithms which exclude low-level noise-only contributions from the tap correlators in the combining process, as shown by Chyi et al. (1988).

An additional degradation in the square-law combined output is due to the signal dependent self-noise generated by the correlators, as mentioned earlier. Figure 7.5.9 illustrates the effect of this noise when we have only four multipath signal components (with equal average strength) and 46 noise-only contributions in the RAKE demodulator. When only the outputs from the active signal taps are square-law combined, the error rate saturates at about 10^{-5} as a result of the self-noise. On the other hand, if all 50 taps are used, the error rate saturates at about 2.5×10^{-2}. This is a very significant degradation and results in unacceptably poor performance. It also emphasizes the need to exclude noise-only correlator outputs from the square-law combiner.

This concludes our discussion of signaling over a frequency-selective channel. The configurations of the RAKE receiver presented in this section can be easily generalized to multilevel signaling. In fact, if M-ary PSK or DPSK is chosen, the RAKE structures presented in this section remain unchanged. Only the PSK and DPSK detectors that follow the RAKE correlator are different.

FIGURE 7.5.7
RAKE demodulator for square-law combining of orthogonal signals.

7.6 MULTIPHASE SIGNALING OVER A FREQUENCY-NONSELECTIVE, SLOWLY FADING CHANNEL

Multiphase signaling over a Rayleigh fading channel is the topic presented in some detail in Appendix 7A. Our main purpose in this section is to cite the general result for the probability of a symbol error in M-ary PSK and DPSK systems and the probability of a bit error in four-phase PSK and DPSK.

The general result for the probability of a symbol error in M-ary PSK and DPSK is

$$P_M = \frac{(-1)^{L-1}(1-\mu^2)^L}{\pi(L-1)!} \left(\frac{\partial^{L-1}}{\partial b^{L-1}} \left\{ \frac{1}{b-\mu^2} \left[\frac{\pi}{M}(M-1) \right. \right. \right.$$
$$\left. \left. \left. - \frac{\mu \sin(\pi/M)}{\sqrt{b-\mu^2 \cos^2(\pi/M)}} \cot^{-1} \frac{-\mu \cos(\pi/M)}{\sqrt{b-\mu^2 \cos^2(\pi/M)}} \right] \right\} \right)_{b=1} \quad (7.6.1)$$

FIGURE 7.5.8
Performance of equal-weight square-law combining of signals and noise from a RAKE receiver with L taps of which only l_1 taps contain a signal.

where

$$\mu = \sqrt{\frac{\bar{\gamma}_c}{1 + \bar{\gamma}_c}} \qquad (7.6.2)$$

for coherent PSK and

$$\mu = \frac{\bar{\gamma}_c}{1 + \bar{\gamma}_c} \qquad (7.6.3)$$

for DPSK. Again, $\bar{\gamma}_c$ is the average received SNR per channel. The SNR per bit is $\bar{\gamma}_b = L\bar{\gamma}_c/k$.

The bit error rate for four-phase PSK and DPSK is derived on the basis that the pair of information bits is mapped into the four phases according to a

FIGURE 7.5.9
Performance of the equal-weight square-law combining with and without correlator self-noise.

Gray code. The expression for the bit error rate derived in Appendix 7A is

$$P_b = \frac{1}{2}\left[1 - \frac{\mu}{\sqrt{2 - \mu^2}} \sum_{k=0}^{L-1}\binom{2k}{k}\left(\frac{1 - \mu^2}{4 - 2\mu^2}\right)^k\right] \tag{7.6.4}$$

where μ is again given by either (7.6.2) or (7.6.3) for PSK and DPSK, respectively.

Figure 7.6.1 illustrates the probability of a symbol error of DPSK and coherent PSK for $M = 2$, 4, and 8 with $L = 1$. We note that the difference in performance between DPSK and coherent PSK is approximately 3 dB for all three values of M. In fact, when $\bar{\gamma}_b \gg 1$ and $L = 1$, (7.6.1) is well approximated as

$$P_M \approx \frac{M - 1}{(M \log_2 M)\left[\sin^2(\pi/M)\right]\bar{\gamma}_b} \tag{7.6.5}$$

FIGURE 7.6.1
Probability of symbol error for PSK and DPSK.

for DPSK and as

$$P_M \approx \frac{M-1}{(M \log_2 M)\left[\sin^2(\pi/M)\right]2\bar{\gamma}_b} \tag{7.6.6}$$

for PSK. Hence, at high SNR, coherent PSK is 3 dB better than DPSK on a Rayleigh fading channel. This difference also holds as L is increased.

FIGURE 7.6.2
Probability of a bit error for DPSK with diversity.

Bit error probabilities are depicted in Fig. 7.6.2 for two-phase, four-phase, and eight-phase DPSK signaling with $L = 1$, 2, and 4. The expression for the bit error probability of eight-phase DPSK with Gray encoding has not been given here, but it is available in the paper by Proakis (1968). In this case, we observe that the performance for two-phase and four-phase DPSK is (approximately) the same, while that for eight-phase DPSK is about 3 dB poorer. Although we have not shown the bit error probability for coherent PSK, it can be demonstrated that two-phase and four-phase coherent PSK also yield approximately the same performance.

7.7 *M*-ARY ORTHOGONAL SIGNALING OVER A FREQUENCY-NONSELECTIVE, SLOWLY FADING CHANNEL

In this section we determine the performance of *M*-ary orthogonal signals transmitted over a Rayleigh fading channel and we assess the advantages of higher-order signal alphabets relative to a binary alphabet. The orthogonal signals may be viewed as *M*-ary FSK with a minimum frequency separation of an integer multiple of $1/T$, where T is the signaling interval. The same information-bearing signal is transmitted on L diversity channels. Each diversity channel is assumed to be frequency-nonselective and slowly fading and the fading processes on the L channels are assumed to be mutually statistically independent. An additive white Gaussian noise process corrupts the signal on each diversity channel. We assume that the additive noise processes are mutually statistically independent.

Although it is relatively easy to formulate the structure and analyze the performance of a maximal ratio combiner for the diversity channels in the *M*-ary communication system, it is more likely that a practical system would employ noncoherent detection. Consequently we confine our attention to square-law combining of the diversity signals. The output of the combiner containing the signal is

$$U_1 = \sum_{k=1}^{L} |2\mathscr{E}\alpha_k e^{-j\phi_k} + N_{k1}|^2 \tag{7.7.1}$$

while the outputs of the remaining $M - 1$ combiners are

$$U_m = \sum_{k=1}^{L} |N_{km}|^2 \qquad m = 2, 3, 4, \ldots, M \tag{7.7.2}$$

The probability of error is simply 1 minus the probability that U_1 exceeds all the $\{U_m\}$ for $m = 2, 3, \ldots, M$. Since the signals are orthogonal and the additive noise processes are mutually statistically independent, the random variables $U_1, U_2, \ldots, U_M$ are also mutually statistically independent. The probability density function of U_1 was given previously in (7.4.31). On the other hand, $U_2, \ldots, U_M$ are identically distributed and described by the marginal probability density function in (7.4.33). With U_1 fixed, the joint probability $P(U_2 < U_1, U_3 < U_1, \ldots, U_m < U_1)$ is equal to $P(U_2 < U_1)$ raised to the $M - 1$ power. Now

$$P(U_2 < U_1) = \int_0^{U_1} p(U_2)\, dU_2$$

$$= 1 - e^{-U_1/2\sigma_2^2} \sum_{k=0}^{L-1} \frac{(U_1/2\sigma_2^2)^k}{k!} \tag{7.7.3}$$

where $\sigma_2^2 = 2\mathscr{E}N_0$. The $M - 1$ power of this probability is then averaged over the probability density function of U_1 to yield the probability of a correct decision. If we subtract this result from unity, we obtain the probability of error in the form given by Hahn (1962)

$$P_M = 1 - \int_0^\infty \frac{1}{(2\sigma_1^2)^L (L-1)!} U_1^{L-1} e^{-U_1/2\sigma_2^2}$$

$$\times \left[1 - e^{-U_1/2\sigma_2^2} \sum_{k=0}^{L-1} \frac{(U_1/2\sigma_2^2)^k}{k!} \right]^{M-1} dU_1$$

$$= 1 - \int_0^\infty \frac{1}{(1+\bar{\gamma}_c)^L (L-1)!} U_1^{L-1} e^{-U_1/(1+\bar{\gamma}_c)}$$

$$\times \left(1 - e^{-U_1} \sum_{k=0}^{L-1} \frac{U_1^k}{k!} \right)^{M-1} dU_1 \qquad (7.7.4)$$

where $\bar{\gamma}_c$ is the average SNR per diversity channel. The average SNR per bit is $\bar{\gamma}_b = L\bar{\gamma}_c/\log_2 M = L\bar{\gamma}_c/k$.

The integral in (7.7.4) can be expressed in closed form as a double summation. This can be seen if we write

$$\left(\sum_{k=0}^{L-1} \frac{U_1^k}{k!} \right)^m = \sum_{k=0}^{m(L-1)} \beta_{km} U_1^k \qquad (7.7.5)$$

where β_{km} is the set of coefficients in the above expansion. Then it follows that (7.7.4) reduces to

$$P_M = \frac{1}{(L-1)!} \sum_{m=1}^{M-1} \frac{(-1)^{m+1}\binom{M-1}{m}}{(1+m+m\bar{\gamma}_c)^L}$$

$$\times \sum_{k=0}^{m(L-1)} \beta_{km}(L-1+k)! \left(\frac{1+\bar{\gamma}_c}{1+m+m\bar{\gamma}_c} \right)^k \qquad (7.7.6)$$

When there is no diversity ($L = 1$), the error probability in (7.7.6) reduces to the simple form

$$P_M = \sum_{m=1}^{M-1} \frac{(-1)^{m+1}\binom{M-1}{m}}{1+m+m\bar{\gamma}_c} \qquad (7.7.7)$$

The symbol error rate P_M may be converted to an equivalent bit error rate by multiplying P_M with $(2^{k-1}/2^k - 1)$.

Although the expression for P_M given in (7.7.6) is in closed form, it is computationally cumbersome to evaluate for large values of M and L. An alternative is to evaluate P_M by numerical integration using the expression in (7.7.4). The results illustrated in the following graphs were generated from (7.7.4).

First of all, let us observe the error rate performance of M-ary orthogonal signaling with square-law combining as a function of the order of diversity. Figures 7.7.1 through 7.7.4 illustrate the characteristics of P_M for $M + 2, 4, 8$, and 16 as a function of L when the total SNR, defined as $\bar{\gamma}_t = L\bar{\gamma}_c$, remains fixed. These results indicate that there is an optimum order of diversity for each

FIGURE 7.7.1
Performance of square-law-detected binary orthogonal signals as a function of diversity.

$\bar{\gamma}_t$. That is, for any $\bar{\gamma}_t$, there is a value of L for which P_M is a minimum. A careful observation of these graphs reveals that the minimum in P_M is obtained when $\bar{\gamma}_c = \bar{\gamma}_t/L \approx 3$. This result appears to be independent of the alphabet size M.

Second, let us observe the error rate P_M as a function of the average SNR per bit, defined as $\bar{\gamma}_b = L\bar{\gamma}_c/k$. (If we interpret M-ary orthogonal FSK as a form of coding† and the order of diversity as the number of times a symbol is repeated in a repetition code, then $\bar{\gamma}_b = \bar{\gamma}_c/R_c$ where $R_c = k/L$ is the code rate.) The

---

† In Sec. 7.8 we show that M-ary orthogonal FSK with diversity may be viewed as a block orthogonal code.

FIGURE 7.7.2
Performance of square-law-detected $M = 4$ orthogonal signals as a function of diversity.

graphs of P_M versus $\bar{\gamma}_b$ for $M = 2, 4, 8, 16, 32$ and $L = 1, 2, 4$ are shown in Fig. 7.7.5. These results illustrate the gain in performance as M increases and L increases. First we note that a significant gain in performance is obtained by increasing L. Second, we note that the gain in performance obtained with an increase in M is relatively small when L is small. However, as L increases, the gain achieved by increasing M also increases. Since an increase in either parameter results in an expansion of bandwidth, i.e.,

$$B_e = \frac{LM}{\log_2 M}$$

the results illustrated in Fig. 7.7.5 indicate that an increase in L is more efficient

FIGURE 7.7.3
Performance of square-law-detected $M = 8$ orthogonal signals as a function of diversity.

than a corresponding increase in M. As we shall see in Sec. 7.8, coding is a bandwidth-effective means for obtaining diversity in the signal transmitted over the fading channel.

Chernoff bound. Before concluding this section, we develop a Chernoff upper bound on the error probability of binary orthogonal signaling with Lth-order diversity, which will be useful in our discussion of coding for fading channels, the topic of Sec. 7.8. Our starting point is the expression for the two decision variables U_1 and U_2 given by (7.4.29), where U_1 consists of the square-law-com-

FIGURE 7.7.4
Performance of square-law-detected $M = 16$ orthogonal signals as a function of diversity.

bined signal-plus-noise terms and U_2 consists of square-law-combined noise terms. The binary probability of error, denoted here as $P_2(L)$, is

$$P_2(L) = P(U_2 - U_1 > 0)$$
$$= P(X > 0) = \int_0^\infty p(x)\, dx \qquad (7.7.8)$$

where the random variable X is defined as

$$X = U_2 - U_1 = \sum_{k=1}^{L} \left(|N_{k2}|^2 - |2\mathscr{E}\alpha_k + N_{k1}|^2 \right) \qquad (7.7.9)$$

FIGURE 7.7.5
Performance of orthogonal signaling with M and L as parameters.

The phase terms $\{\phi_k\}$ in U_1 have been dropped since they do not affect the performance of the square-law detector.

Let $S(X)$ denote the unit step function. Then the error probability in (7.7.8) can be expressed in the form

$$P_2(L) = E[S(X)] \tag{7.7.10}$$

Following the development in Sec. 1.1.5, the Chernoff bound is obtained by overbounding the unit step function by an exponential function. That is,

$$S(X) \le e^{\zeta X} \qquad \zeta \ge 0 \tag{7.7.11}$$

where the parameter ζ is optimized to yield a tight bound. Thus we have

$$P_2(L) = E[S(X)] \le E(e^{\zeta X}) \tag{7.7.12}$$

Upon substituting for the random variable X from (7.7.9) and noting that the random variables in the summation are mutually statistically independent, we obtain the result

$$P_2(L) \le \prod_{k=1}^{L} E\left(e^{\zeta |N_{k2}|^2}\right) E\left(e^{-\zeta |2\mathcal{E}\alpha_k + N_{k1}|^2}\right) \tag{7.7.13}$$

But

$$E\left(e^{\zeta |N_{k2}|^2}\right) = \frac{1}{1 - 2\zeta\sigma_2^2} \qquad \zeta < \frac{1}{2\sigma_2^2} \tag{7.7.14}$$

and

$$E\left(e^{-\zeta |2\mathcal{E}\alpha_k + N_{k1}|^2}\right) = \frac{1}{1 + 2\zeta\sigma_1^2} \qquad \zeta > \frac{-1}{2\sigma_1^2} \tag{7.7.15}$$

where $\sigma_2^2 = 2\mathcal{E}N_0$, $\sigma_1^2 = 2\mathcal{E}N_0(1 + \bar{\gamma}_c)$, and $\bar{\gamma}_c$ is the average SNR per diversity channel. Note that σ_1^2 and σ_2^2 are independent of k, i.e., the additive noise terms on the L diversity channels as well as the fading statistics are identically distributed. Consequently (7.7.13) reduces to

$$P_2(L) \le \left[\frac{1}{(1 - 2\zeta\sigma_2^2)(1 + 2\zeta\sigma_1^2)}\right]^L \qquad 0 \le \zeta \le \frac{1}{2\sigma_2^2} \tag{7.7.16}$$

By differentiating the right-hand side of (7.7.16) with respect to ζ, we find that the upper bound is minimized when

$$\zeta = \frac{\sigma_1^2 - \sigma_2^2}{4\sigma_1^2\sigma_2^2} \tag{7.7.17}$$

Substitution of (7.7.17) for ζ into (7.7.16) yields the Chernoff upper bound in the form

$$P_2(L) \le \left[\frac{4(1 + \bar{\gamma}_c)}{(2 + \bar{\gamma}_c)^2}\right]^L \tag{7.7.18}$$

It is interesting to note that (7.7.18) may also be expressed as

$$P_2(L) \le [4p(1 - p)]^L \tag{7.7.19}$$

where $p = 1/(2 + \bar{\gamma}_c)$ is the probability of error for binary orthogonal signaling on a fading channel without diversity.

A comparison of the Chernoff bound in (7.7.18) with the exact error probability for binary orthogonal signaling and square-law combining of the L

FIGURE 7.7.6
Comparison of Chernoff bound with exact error probability.

diversity signals, which is given by the expression

$$P_2(L) = \left(\frac{1}{1+\bar{\gamma}_c}\right)^L \sum_{k=0}^{L-1} \binom{L-1+k}{k} \left(\frac{1+\bar{\gamma}_c}{2+\bar{\gamma}_c}\right)^k$$

$$= p^L \sum_{k=0}^{L-1} \binom{L-1+k}{k} (1-p)^k \qquad (7.7.20)$$

reveals the tightness of the bound. Figure 7.7.6 illustrates this comparison. We observe that the Chernoff upper bound is approximately 6 dB from the exact error probability for $L = 1$, but as L increases it becomes tighter. For example, the difference between the bound and the exact error probability is about 2.5 dB when $L = 4$.

Finally we mention that the error probability for M-ary orthogonal signaling with diversity can be upper bounded by means of the union bound

$$P_M \le (M - 1)P_2(L) \tag{7.7.21}$$

where we may use either the exact expression given in (7.7.20) or the Chernoff bound in (7.7.18) for $P_2(L)$.

7.8 CODED WAVEFORMS FOR FADING CHANNELS

Up to this point we have demonstrated that diversity techniques are very effective in overcoming the detrimental effects of fading caused by the time-variant dispersive characteristics of the channel. Time and/or frequency diversity techniques may be viewed as a form of repetition (block) coding of the information sequence. From this point of view, the combining techniques described previously represent soft-decision decoding of the repetition code. Since a repetition code is a trivial form of coding, we shall now consider the additional benefits derived from more efficient types of codes. In particular, we demonstrate that coding provides an efficient means for obtaining diversity on a fading channel. The amount of diversity provided by a code is directly related to its minimum distance.

As explained in Sec. 7.4, time diversity is obtained by transmitting the signal components carrying the same information in multiple time intervals mutually separated by an amount equal to or exceeding the coherence time $(\Delta t)_c$ of the channel. Similarly, frequency diversity is obtained by transmitting the signal components carrying the same information in multiple frequency slots mutually separated by an amount of least equal to the coherence bandwidth $(\Delta f)_c$ of the channel. Thus the signal components carrying the same information undergo statistically independent fading.

To extend these notions to a coded information sequence, we simply require that the signal waveform corresponding to a particular code bit or code symbol fade independently from the signal waveform corresponding to any other code bit or code symbol. This requirement may result in inefficient utilization of the available time-frequency space, with the existence of large unused portions in this two-dimensional signaling space. To reduce the inefficiency, a number of code words may be interleaved either in time or in frequency or both, in such a manner that the waveforms corresponding to the bits or symbols of a given code word fade independently. Thus we assume that the time-frequency signaling space is partitioned into nonoverlapping time-frequency cells. A signal waveform corresponding to a code bit or code symbol is transmitted within such a cell.

In addition to the assumption of statistically independent fading of the signal components of a given code word, we also assume that the additive noise components corrupting the received signals are white gaussian processes which

FIGURE 7.8.1
Model of communications system with FSK modulation/demodulation and encoding/decoding.

are statistically independent and identically distributed among the cells in the time-frequency space. Also, we assume that there is sufficient separation between adjacent cells so that intercell interference is negligible.

An important issue is the modulation technique that is used to transmit the coded information sequence. If the channel fades slowly enough to allow the establishment of a phase reference, then PSK or DPSK may be employed. If this is not possible, then FSK modulation with noncoherent detection at the receiver is appropriate. In our treatment, we assume that it is not possible to establish a phase reference or phase references for the signals in the different cells occupied by the transmitted signal. Consequently we choose FSK modulation with noncoherent detection.

A model of the digital communications system for which the error rate performance will be evaluated is shown in Fig. 7.8.1. The encoder may be binary, nonbinary, or a concatenation of a nonbinary encoder with a binary encoder. Furthermore, the code generated by the encoder may be a block code, a convolutional code, or, in the case of concatenation, a mixture of a block code and a convolutional code.

In order to explain the modulation, demodulation, and decoding for FSK-type (orthogonal) signals, consider a linear binary block code in which k information bits are encoded into a block of n bits. For simplicity and without loss of generality, let us assume that all n bits of a code word are transmitted simultaneously over the channel on multiple frequency cells. A code word $\mathbf{C}_i$ having bits $\{c_{ij}\}$ is mapped into FSK signal waveforms in the following way. If $c_{ij} = 0$, the tone f_{0j} is transmitted, and if $c_{ij} = 1$, the tone f_{1j} is transmitted. This means that $2n$ tones or cells are available to transmit the n bits of the code word, but only n tones are transmitted in any signaling interval. Since each code word conveys k bits of information, the bandwidth expansion factor for FSK is $B_e = 2n/k$.

The demodulator for the received signal separates the signal into $2n$ spectral components corresponding to the available tone frequencies at the transmitter. Thus the demodulator can be realized as a bank of $2n$ filters, where each filter is matched to one of the possible transmitted tones. The outputs of the $2n$ filters are detected noncoherently. Since the Rayleigh fading and the additive white gaussian noises in the $2n$ frequency cells are mutually statistically independent and identically distributed random processes, the optimum maximum-likelihood soft-decision decoding criterion requires that these filter responses be square-law-detected and appropriately combined for each code word to form the $M = 2^k$ decision variables. The code word corresponding to the maximum of the decision variables is selected. If hard-decision decoding is employed, the optimum maximum-likelihood decoder selects the code word having the smallest Hamming distance relative to the received code word.

Although the discussion above assumed the use of a block code, a convolutional encoder can be easily accommodated in the block diagram shown in Fig. 7.8.1. For example, if a binary convolutional code is employed, each bit in the output sequence may be transmitted by binary FSK. The maximum-likelihood soft-decision decoding criterion for the convolutional code can be efficiently implemented by means of the Viterbi algorithm, in which the metrics for the surviving sequences at any point in the trellis consist of the square-law-combined outputs for the corresponding paths through the trellis. On the other hand, if hard-decision decoding is employed, the Viterbi algorithm is implemented with Hamming distance as the metric.

7.8.1 Probability of Error for Soft-Decision Decoding of Linear Binary Block Codes

Consider the decoding of a linear binary (n, k) code transmitted over a Rayleigh fading channel, as described above. The optimum soft-decision decoder, based on the maximum-likelihood criterion, forms the $M = 2^k$ decision variables

$$
\begin{aligned}
U_i &= \sum_{j=1}^{n} \left[\left(1 - c_{ij}\right)|y_{0j}|^2 + c_{ij}|y_{1j}|^2 \right] \\
&= \sum_{j=1}^{n} \left[|y_{0j}|^2 + c_{ij}\left(|y_{1j}|^2 - |y_{0j}|^2\right) \right] \qquad i = 1, 2, \dots, 2^k \quad (7.8.1)
\end{aligned}
$$

where $|y_{r_j}|^2$, $j = 1, 2, \dots, n$, and $r = 0, 1$ represent the squared envelopes at the outputs of the $2n$ filters that are tuned to the $2n$ possible transmitted tones. A decision is made in favor of the code word corresponding to the largest decision variable of the set $\{U_i\}$.

Our objective in this section is the determination of the error rate performance of the soft-decision decoder. Toward this end, let us assume that the all-zero code word $\mathbf{C}_1$ is transmitted. The average received signal-to-noise ratio per tone (cell) is denoted as $\bar{\gamma}_c$. The total received SNR for the n tones in $n\bar{\gamma}_c$ and, hence, the average SNR per bit is

$$
\begin{aligned}
\bar{\gamma}_b &= \frac{n}{k}\bar{\gamma}_c \\
&= \frac{\bar{\gamma}_c}{R_c} \qquad (7.8.2)
\end{aligned}
$$

where R_c is the code rate.

The decision variable U_1 corresponding to the code word $\mathbf{C}_1$ is given by (7.8.1) with $c_{ij} = 0$ for all j. The probability that a decision is made in favor of the mth code word is just

$$P_2(m) = P(U_m > U_1) = P(U_1 - U_m < 0)$$

$$= P\left[\sum_{j=1}^{n} (c_{1j} - c_{mj})(|y_{1j}|^2 - |y_{0j}|^2) < 0 \right]$$

$$= P\left[\sum_{j=1}^{w_m} (|y_{0j}|^2 - |y_{1j}|^2) < 0 \right] \tag{7.8.3}$$

where w_m is the weight of the mth code word. But the probability in (7.8.3) is just the probability of error for square-law combining of binary orthogonal FSK with w_m-order diversity. That is,

$$P_2(m) = p^{w_m} \sum_{k=0}^{w_m - 1} \binom{w_m - 1 + k}{k}(1 - p)^k \tag{7.8.4}$$

$$\leq p^{w_m} \sum_{k=0}^{w_m - 1} \binom{w_m - 1 + k}{k} = \binom{2w_m - 1}{w_m} p^{w_m} \tag{7.8.5}$$

where

$$p = \frac{1}{2 + \bar{\gamma}_c} = \frac{1}{2 + R_c\bar{\gamma}_b} \tag{7.8.6}$$

As an alternative, we may use the Chernoff upper bound derived in Sec. 7.7, which in the present notation is

$$P_2(m) \leq [4p(1 - p)]^{w_m} \tag{7.8.7}$$

The sum of the binary error events over the $M - 1$ nonzero-weight code words gives an upper bound on the probability of error. Thus

$$P_M \leq \sum_{m=2}^{M} P_2(m) \tag{7.8.8}$$

Since the minimum distance of the linear code is equal to the minimum weight, it follows that

$$\frac{1}{(1 + R_c\bar{\gamma}_b)^{w_m}} \leq \frac{1}{(2 + R_c\bar{\gamma}_b)^{d_{\min}}}$$

The use of this relation in conjunction with (7.8.5) and (7.8.8) yields a simple, albeit looser, upper bound which may be expressed in the form

$$P_M < \frac{\sum_{m=2}^{M} \binom{2w_m - 1}{w_m}}{(2 + R_c\bar{\gamma}_b)^{d_{\min}}} \tag{7.8.9}$$

This simple bound indicates that the code provides an effective order of diversity equal to $d_{\min}$. An even simpler bound is the union bound

$$P_M < (M-1)\left[4p(1-p)\right]^{d_{\min}} \qquad (7.8.10)$$

which is obtained from the Chernoff bound given in (7.8.7).

As an example which serves to illustrate the benefits of coding for a Rayleigh fading channel, we have plotted in Fig. 7.8.2 the performance obtained with the extended Golay $(24, 12)$ code and the performance of binary FSK and quarternary FSK each with dual diversity. Since the extended Golay code requires a total of 48 cells and $k = 12$, the bandwidth expansion factor $B_e = 4$.

FIGURE 7.8.2
Example of performance obtained with conventional diversity versus coding for $B_e = 4$.

This is also the bandwidth expansion factor for binary and quaternary FSK with $L = 2$. Thus the three types of waveforms are compared on the basis of the same bandwidth expansion factor. We note that at $P_b = 10^{-4}$ the Golay code outperforms quaternary FSK by more than 6 dB and at $P_b = 10^{-5}$ the difference is approximately 10 dB.

The reason for the superior performance of the Golay code is due to its large minimum distance ($d_{\min} = 8$), which translates into an equivalent eighth-order ($L = 8$) diversity. In contrast, the binary and quaternary FSK signals have only second-order diversity. Hence the code makes more efficient use of the available channel bandwidth. The price that we must pay for the superior performance of the code is the increase in decoding complexity.

7.8.2   Probability of Error for Hard-Decision Decoding of Linear Binary Block Codes

Bounds on the performance obtained with hard-decision decoding of a linear binary (n, k) code have already been given in Sec. 5.2.5. These bounds are applicable to a general binary-input binary-output memoryless (binary symmetric) channel and, hence, they apply without modification to a Rayleigh fading AWGN channel with statistically independent fading of the symbols in the code word. The probability of a bit error needed to evaluate these bounds when binary FSK with noncoherent detection is used as the modulation and demodulation technique is given by (7.8.6).

A particularly interesting result is obtained when we use the Chernoff upper bound on the error probability for hard-decision decoding given by (5.2.89). That is,

$$P_2(m) \le \left[4p(1 - p)\right]^{w_m/2} \tag{7.8.11}$$

and P_M is upper-bounded by (7.8.8). In comparison, the Chernoff upper bound for $P_2(m)$ when soft-decision decoding is employed is given by (7.8.7). We observe that the effect of hard-decision decoding is a reduction in the distance between any two code words by a factor of 2. When the minimum distance of a code is relatively small, the reduction of the distances by a factor of 2 is much more noticeable in a fading channel than in a nonfading channel.

For illustrative purposes we have plotted in Fig. 7.8.3 the performance of the Golay (23, 12) code when hard-decision and soft-decision decoding are used. The difference in performance at $P_b = 10^{-5}$ is approximately 6 dB. This is a significant difference in performance compared to the 2-dB difference between soft-decision and hard-decision decoding in a nonfading AWGN channel. We also note that the difference in performance increases as P_b decreases. In short, these results indicate the benefits of soft-decision decoding over hard-decision decoding on a Rayleigh fading channel.

FIGURE 7.8.3
Comparison of performance between hard-decision and soft-decision decoding.

7.8.3 Upper Bounds on the Performance of Convolutional Codes for a Rayleigh Fading Channel

In this subsection we derive the performance of binary convolutional codes when used on a Rayleigh fading AWGN channel. The encoder accepts k binary digits at a time and puts out n binary digits at a time. Thus the code rate is $R_c = k/n$. The binary digits at the output of the encoder are transmitted over the Rayleigh fading channel by means of binary FSK which is square-law-detected at the receiver. The decoder for either soft-decision or hard-decision decoding performs maximum-likelihood sequence estimation, which is efficiently implemented by means of the Viterbi algorithm.

First we consider soft-decision decoding. In this case the metrics computed in the Viterbi algorithm are simply sums of square-law-detected outputs from the demodulator. Suppose the all-zero sequence is transmitted. Following the procedure outlined in Sec. 5.3.3, it is easily shown that the probability of error in a pairwise comparison of the metrics corresponding to the all-zero sequence with the metric corresponding to another sequence that merges for the first time at the all-zero state is

$$P_2(d) = p^d \sum_{k=0}^{d-1} \binom{d-1+k}{k} (1-p)^k \qquad (7.8.12)$$

where d is the number of bit positions in which the two sequences differ and p is given by (7.8.6). That is, $P_2(d)$ is just the probability of error for binary FSK with square-law detection and dth-order diversity. Alternatively, we may use the Chernoff bound in (7.8.7) for $P_2(d)$. In any case, the bit error probability is upperbounded, as shown in Sec. 5.3.3, by the expression

$$P_b < \frac{1}{k} \sum_{d=d_{\text{free}}}^{\infty} \beta_d P_2(d) \qquad (7.8.13)$$

where the weighting coefficients $\{\beta_d\}$ in the summation are obtained from the expansion of the first derivative of the transfer function $T(D, N)$, given by (5.3.25).

When hard-decision decoding is performed at the receiver, the bounds on the error rate performance for binary convolutional codes derived in Sec. 5.3.4 apply. That is, P_b is again upper-bounded by the expression in (7.8.13), where $P_2(d)$ is defined by (5.3.28) for d odd and by (5.3.29) for d even, or upper-bounded (Chernoff bound) by (5.3.31), and p is defined by (7.8.6).

As in the case of block coding, when the respective Chernoff bounds are used for $P_2(d)$ with hard-decision and soft-decision decoding, it is interesting to note that the effect of hard-decision decoding is to reduce the distances (diversity) by a factor of 2 relative to soft-decision decoding.

The following numerical results illustrate the error rate performance of binary, rate $1/n$, maximal free distance convolutional codes for $n = 2$, 3, and 4 with soft-decision Viterbi decoding. First of all, Fig. 7.8.4 shows the performance of the rate $1/2$ convolutional codes for constraint lengths 3, 4, and 5. The bandwidth expansion factor for binary FSK modulation is $B_e = 2n$. Since an increase in the constraint length results in an increase in the complexity of the decoder to go along with the corresponding increase in the minimum free distance, the system designer can weigh these two factors in the selection of the code.

Another way to increase the distance without increasing the constraint length of the code is to repeat each output bit m times. This is equivalent to reducing the code rate by a factor of m or expanding the bandwidth by the same factor. The result is a convolutional code that has a minimum free distance of

FIGURE 7.8.4
Performance of rate $1/2$ binary convolutional codes with soft-decision decoding.

md_{free} where d_{free} is the minimum free distance of the original code without repetitions. Such a code is almost as good, from the viewpoint of minimum distance, as a maximal free distance, rate $1/mn$ code. The error rate performance with repetitions is upper-bounded by

$$P_b < \frac{1}{k} \sum_{d=d_{\text{free}}}^{\infty} \beta_d P_2(md) \qquad (7.8.14)$$

where $P_2(md)$ is given by (7.8.12). Figures 7.8.5 and 7.8.6 illustrate the performance of the rate $1/2$ codes with repetitions ($m = 1, 2, 3, 4$) for constraint lengths 3 and 5, respectively.

FIGURE 7.8.5
Performance of rate $1/2m$, constraint length 3, binary convolutional codes with soft-decision decoding.

Figure 7.8.7 shows the performance of the rate $1/3$, maximal free distance code for constraint lengths 3 and 4 and $m = 2$ and 4. Finally, Fig. 7.8.8 shows the performance of the rate $1/4$ code for constraint lengths 3 and 4 and $m = 1$ and 2.

7.8.4 Use of Constant-Weight Codes and Concatenated Codes for a Fading Channel

Our treatment of coding for a Rayleigh fading channel to this point was based on the use of binary FSK as the modulation technique for transmitting each of the binary digits in a code word. For this modulation technique, all the 2^k code words

FIGURE 7.8.6
Performance of rate $1/2m$, constraint length 5, binary convolutional codes with soft-decision decoding.

in the (n, k) code have identical transmitted energy. Furthermore, under the condition that the fading on the n transmitted tones is mutually statistically independent and identically distributed, the average received signal energy for the $M = 2^k$ possible code words is also identical. Consequently, in a soft-decision decoder, the decision is made in favor of the code word having the largest decision variable.

The condition that the received code words have identical average SNR has an important ramification in the implementation of the receiver. If the received code words do not have identical average SNR, the receiver must provide bias compensation for each received code word so as to render it equal energy. In

FIGURE 7.8.7
Performance of rate $1/3m$, binary convolutional codes with soft-decision decoding.

general, the determination of the appropriate bias terms is difficult to implement because it requires the estimation of the average received signal power; hence the equal-energy condition on the received code words considerably simplifies the receiver processing.

There is an alternative modulation method for generating equal-energy waveforms from code words when the code is constant weight, i.e., when every code word has the same number of 1s. Note that such a code is nonlinear. Nevertheless, suppose we assign a single tone or cell to each bit position of the 2^k code words. Thus an (n, k) binary block code has n tones assigned. Waveforms are constructed by transmitting the tone corresponding to a particular bit in a

FIGURE 7.8.8
Performance of rate $1/4m$, binary convolutional codes with soft-decison decoding.

code word if that bit is a 1; otherwise that tone is not transmitted for the duration of the interval. This modulation technique for transmitting the coded bits is called *on-off keying* (OOK). Since the code is constant weight, say w, every coded waveform consists of w transmitted tones which depend on the positions of the 1s in each of the code words.

As in FSK, all tones in the OOK signal that are transmitted over the channel are assumed to fade independently across the frequency band and in time from one code word to another. The received signal envelope for each tone is described statistically by the Rayleigh distribution. Statistically independent additive white gaussian noise is assumed to be present in each frequency cell.

The receiver employs maximum-likelihood (soft-decision) decoding to map the received waveform into one of the M possible transmitted code words. For this purpose, n matched filters are employed, each matched to one of the n frequency tones. For the assumed statistical independence of the signal fading for the n frequency cells and additive white gaussian noise, the envelopes of the matched filter outputs are squared and combined to form the M decision variables

$$U_i = \sum_{j=1}^{n} c_{ij}|y_j|^2 \qquad i = 1, 2, \ldots, 2^k \qquad (7.8.15)$$

where $|y_j|^2$ corresponds to the squared envelope of the filter corresponding to the jth frequency, where $j = 1, 2, \ldots, n$.

It may appear that the constant-weight condition severely restricts our choice of codes. This is not the case, however. To illustrate this point, we briefly describe some methods for constructing constant-weight codes. This discussion is by no means exhaustive.

Method 1: nonlinear transformation of a linear code. In general, if in each word of an arbitrary binary code we substitute one binary sequence for every occurrence of a 0 and another sequence for each 1, a constant-weight binary block code will be obtained if the two substitution sequences are of equal weights and lengths. If the length of the sequence is ν and the original code is an (n, k) code, then the resulting constant-weight code will be an $(\nu n, k)$ code. The weight will be n times the weight of the substitution sequence, and the minimum distance will be the minimum distance of the original code times the distances between the two substitution sequences. Thus the use of complementary sequences when ν is even results in a code with minimum distance $\nu d_{\min}$ and weight $\nu n/2$.

The simplest form of this method is the case $\nu = 2$, in which every 0 is replaced by the pair 01 and every 1 is replaced by the complementary sequence 10 (or vice versa). As an example, we take as the initial code the (24, 12) extended Golay code. The parameters of the original and the resultant constant-weight code are given in Table 7.8.1.

We note that this substitution process can be viewed as a separate encoding. This secondary encoding clearly does not alter the information content of a

TABLE 7.8.1
Example of constant-weight code formed by method 1

Code parameters	Original Golay	Constant weight
n	24	48
k	12	12
M	4096	4096
$d_{\min}$	8	16
w	variable	24

TABLE 7.8.2
Examples of constant-weight codes formed by expurgation

Parameters	Original	Constant weight No. 1	Constant weight No. 2
n	24	24	24
k	12	9	11
M	4096	759	2576
d_{min}	8	≥ 8	≥ 8
w	variable	8	12

code word—it merely changes the form in which it is transmitted. Since the new
code word is composed of pairs of bits, one "on" and one "off," the use of OOK
transmission of this code word produces a waveform that is identical to that
obtained by binary FSK modulation for the underlying linear code.

Method 2: expurgation. In this method we start with an arbitrary binary block
code and select from it a subset consisting of all words of a certain weight.
Several different constant-weight codes can be obtained from one initial code by
varying the choice of the weight w. Since the code words of the resulting
expurgated code can be viewed as a subset of all possible permutations of any
one code word in the set, the term "binary expurgated permutation modulation"
(BEXPERM) has been used by Gaarder (1971) to describe such a code. In fact,
the constant-weight binary block codes constructed by the other methods may
also be viewed as BEXPERM codes. This method of generating constant-weight
codes is in a sense opposite to the first method in that the word length n is held
constant and the code size M is changed. The minimum distance for the
constant-weight subset will clearly be no less than that of the original code. As an
example, we consider the Golay $(24, 12)$ code and form the two different
constant-weight codes shown in Table 7.8.2.

Method 3: Hadamard matrices. This method might appear to form a constant-
weight binary block code directly, but it actually is a special case of the method
of expurgation. In this method, a Hadamard matrix is formed as described in
Sec. 5.2.2, and a constant-weight code is created by selection of rows (code
words) from this matrix. Recall that a Hadamard matrix is an $n \times n$ matrix
(n an even integer) of 1s and 0s with the property that any row differs from any
other row in exactly $n/2$ positions. One row of the matrix is normally chosen as
being all 0s.

In each of the other rows, half of the elements are 0s and the other half are
1s. A Hadamard code of size $2(n - 1)$ code words is obtained by selecting these
$n - 1$ rows and their complements. By selecting $M = 2^k \leq 2(n - 1)$ of these
code words, we obtain a Hadamard code, which we denote as $H(n, k)$, where
each code word conveys k information bits. The resulting code has constant
weight equal to $n/2$ and minimum distance $d_{min} = n/2$.

Since n frequency cells are used to transmit k information bits, the bandwidth expansion factor for the Hadamard $H(n, k)$ code is defined as

$$B_e = \frac{n}{k} \text{ cells/information bit}$$

which is simply the reciprocal of the code rate. Also, the average signal-to-noise (SNR) per bit, denoted as $\bar{\gamma}_b$, is related to the average SNR per cell $\bar{\gamma}_c$, by the expression

$$\bar{\gamma}_c = \frac{k}{n/2}\bar{\gamma}_b$$

$$= 2\left(\frac{k}{n}\right)\bar{\gamma}_b = 2R_c\bar{\gamma}_b = \frac{2\bar{\gamma}_b}{B_e} \tag{7.8.16}$$

Let us compare the performance of the constant-weight Hadamard codes under a fixed bandwidth constraint with a conventional M-ary orthogonal set of waveforms where each waveform has diversity L. The M orthogonal waveforms with diversity are equivalent to a block orthogonal code having a block length $n = LM$ and $k = \log_2 M$. For example, if $M = 4$ and $L = 2$, the code words of the block orthogonal code are

$$C_1 = [1 \quad 1 \quad 0 \quad 0 \quad 0 \quad 0 \quad 0 \quad 0]$$
$$C_2 = [0 \quad 0 \quad 1 \quad 1 \quad 0 \quad 0 \quad 0 \quad 0]$$
$$C_3 = [0 \quad 0 \quad 0 \quad 0 \quad 1 \quad 1 \quad 0 \quad 0]$$
$$C_4 = [0 \quad 0 \quad 0 \quad 0 \quad 0 \quad 0 \quad 1 \quad 1]$$

To transmit these code words using OOK modulation requires $n = 8$ cells, and since each code word conveys $k = 2$ bits of information, the bandwidth expansion factor $B_e = 4$. In general, we denote the block orthogonal code as $O(n, k)$. The bandwidth expansion factor is

$$B_e = \frac{n}{k} = \frac{LM}{k} \tag{7.8.17}$$

Also, the SNR per bit is related to the SNR per cell by the expression

$$\bar{\gamma}_c = \frac{k}{L}\bar{\gamma}_b$$

$$= M\left(\frac{k}{n}\right)\bar{\gamma}_b = M\frac{\bar{\gamma}_b}{B_e} \tag{7.8.18}$$

Now we turn our attention to the performance characteristics of these codes. First, the exact probability of a code word (symbol) error for M-ary orthogonal signaling over a Rayleigh fading channel with diversity was given in closed form in Sec. 7.7. As previously indicated, this expression is rather cumbersome to evaluate, especially if either L or M or both are large. Instead, we shall use a union bound which is very convenient. That is, for a set of M orthogonal waveforms, the probability of a symbol error can be upper-bounded as

$$P_M \le (M - 1)P_2(L)$$
$$= (2^k - 1)P_2(L) < 2^k P_2(L) \tag{7.8.19}$$

where $P_2(L)$, the probability of error for two orthogonal waveforms, each with diversity L, is given by (7.8.12) with $p = 1/(2 + \bar{\gamma}_c)$. The probability of bit error is obtained by multiplying P_M by $2^{k-1}/(2^k - 1)$, as explained previously.

A simple upper (union) bound on the probability of a code word error for the Hadamard $H(n, k)$ code is obtained by noting the probability of error in deciding between the transmitted code word and any other code word is bounded from above by $P_2(d_{\min}/2)$, where $d_{\min}$ is the minimum distance of the code. Therefore an upper bound on P_M is

$$P_M \leq (M - 1) P_2\left(\frac{d_{\min}}{2}\right) < 2^k P_2\left(\frac{d_{\min}}{2}\right) \qquad (7.8.20)$$

Thus the "effective order of diversity" of the code for OOK modulation is

FIGURE 7.8.9
Performance of Hadamard codes.

FIGURE 7.8.10
Performance of block orthogonal codes.

$d_{\min}/2$. The bit error probability may be approximated as $P_M/2$, or slightly overbounded by multiplying P_M by the factor $(2^{k-1})(2^k - 1)$, which is the factor used above for orthogonal codes. The latter was selected for the error probability computations given below.

Figures 7.8.9 and 7.8.10 illustrate the error rate performance of a selected number of Hadamard codes and block orthogonal codes, respectively, for several bandwidth expansion factors. The advantage resulting from an increase in the seize (M) of the alphabet (or k, since $k = \log_2 M$) and an increase in the bandwidth expansion factor is apparent from observation of these curves. Note, for example, that the $H(20,5)$ code when repeated twice results in a code that is

FIGURE 7.8.11
Comparison of performance between Hadamard codes and block orthogonal codes.

denoted as $_2H(20, 5)$ and has a bandwidth expansion factor $B_e = 8$. Figure 7.8.11 shows the performance of the two types of codes compared on the basis of equal bandwidth expansion factors. It is observed that the error rate curves for the Hadamard codes are steeper than the corresponding curves for the block orthogonal codes. This characteristic behavior is due simply to the fact that, for the same bandwidth expansion factor, the Hadamard codes provide more diversity than block orthogonal codes. Alternatively, one may say that Hadamard codes provide better bandwidth efficiency than block orthogonal codes. It should be mentioned, however, that at low SNR a lower-density code outperforms a higher-diversity code as a consequence of the fact that, on a Rayleigh fading channel, there is an optimum distribution of the total received SNR among the diversity signals. Therefore the curves for the block orthogonal codes will cross over the curves for the Hadamard codes at the low-SNR (high-error-rate) region.

Method 4: concatenation. In this method we begin with two codes, one binary and the other nonbinary. The binary code is the inner code and is an (n, k) constant-weight (nonlinear) block code. The nonbinary code, which may be linear, is the outer code. To distinguish it from the inner code, we use uppercase letters, e.g., an (N, K) code, where N and K are measured in terms of symbols from a q-ary alphabet. The size q of the alphabet over which the outer code is defined cannot be greater than the number of words in the inner code. The outer code, when defined in terms of the binary inner code words rather than q-ary symbols, is the new code.

An important special case is obtained when q is equal to 2^k and the inner code size is chosen to be 2^k. Then the number of words is $M = 2^{kK}$ and the concatenated structure is an (nN, kK) code. The bandwidth expansion factor of this concatenated code is the product of the bandwidth expansions for the inner and outer codes.

Now we shall demonstrate the performance advantages obtained on a Rayleigh fading channel by means of code concatenation. Specifically, we construct a concatenated code in which the outer code is a dual-k (nonbinary) convolutional code and the inner code is either a Hadamard code or a block orthogonal code. That is, we view the dual-k code with M-ary ($M = 2^k$) orthogonal signals for modulation as a concatenated code. In all cases to be considered, soft-decision demodulation and Viterbi decoding are assumed.

The error rate performance of the dual-k convolutional codes is obtained from the derivation of the transfer function given by (5.3.39). For a rate $1/2$, dual-k code with no repetitions, the bit error probability, appropriate for the case in which each k-bit output symbol from the dual-k encoder is mapped into one of $M = 2^k$ orthogonal code words, is upper-bounded as

$$P_b < \frac{2^{k-1}}{2^k - 1} \sum_{m=4}^{\infty} \beta_m P_2(m) \qquad (7.8.21)$$

where $P_2(m)$ is given by (7.8.12).

For example, a rate $1/2$, dual-2 code may employ a 4-ary orthogonal code $O(4, 2)$ as the inner code. The bandwidth expansion factor of the resulting concatenated code is, of course, the product of the bandwidth expansion factors of the inner and outer codes. Thus, in this example, the rate of the outer code is $1/2$ and the inner code is $1/2$. Hence $B_e = (4/2)(2) = 4$.

Note that if every symbol of the dual-k code is repeated r times, this is equivalent to using an orthogonal code with diversity $L = r$. If we select $r = 2$ in the example given above, the resulting orthogonal code is denoted as $O(8, 2)$ and the bandwidth expansion factor for the rate $1/2$, dual-2 code becomes $B_e = 8$.

Consequently the term $P_2(m)$ in (7.8.21) must be replaced by $P_2(mL)$ when the orthogonal code has diversity L. Since a Hadamard code has an "effective diversity" $d_{min}/2$, it follows that when a Hadamard code is used as the inner code with a dual-k outer code, the upper bound on the bit error probability of the resulting concatenated code given by (7.8.21) still applies if $P_2(m)$ is replaced by $P_2(md_{min}/2)$. With these modifications, the upper bound on the bit error probability given by (7.8.21) has been evaluated for rate 1/2, dual-k convolutional codes with either Hadamard codes or block orthogonal codes as inner codes. Thus the resulting concatenated code has a bandwidth expansion factor equal to twice the bandwidth expansion factor of the inner code.

FIGURE 7.8.12

Comparison of performance between block orthogonal codes and dual-k with block orthogonal inner codes.

FIGURE 7.8.13
Comparison of performance between Hadamard codes and dual-k codes with Hadamard inner codes.

First we consider the performance gains due to code concatenation. Figure 7.8.12 illustrates the performance of dual-k codes with block orthogonal inner codes compared to the performance of block orthogonal codes for bandwidth expansion factors $B_e = 4, 8, 16,$ and 32. The performance gains due to concatenation are very impressive. For example, at an error rate of 10^{-6} and $B_e = 8$, the dual-k code outperforms the orthogonal block code by 7.5 dB. In short, this gain may be attributed to the increased diversity (increase in minimum distance) obtained via code concatenation. Similarly, Fig. 7.8.13 illustrates the performance of two dual-k codes with Hadamard inner codes compared to the performance of the Hadamard codes alone for $B_e = 8$ and 12. It is observed that the performance

FIGURE 7.8.14
Performance of dual-k codes with either Hadamard or block orthogonal inner code for $B_e = 8$.

gains due to code concatenation are still significant, but certainly not as impressive as those illustrated in Fig. 7.8.12. The reason is that the Hadamard codes alone yield a large diversity, so that the increased diversity arising from concatenation does not result in as large a gain in performance for the range of error rates covered in Fig. 7.8.13.

Next we compare the performance for the two types of inner codes used with dual-k outer codes. Figure 7.8.14 shows the comparison for $B_e = 8$. Note that the $_2H(4, 2)$ inner code has a $d_{\min} = 4$ and, hence, it has an effective order of diversity equal to 2. But this dual diversity is achieved by transmitting four frequencies per code word. On the other hand, the orthogonal code $O(8, 2)$ also gives dual diversity, but this is achieved by transmitting only two frequencies per

FIGURE 7.8.15
Performance of dual-k codes with either Hadamard or block orthogonal inner code for $B_e = 16$.

code word. Consequently the $O(8, 2)$ code is 3 dB better than the $_2H(4, 2)$. This difference in performance is maintained when the two codes are used as inner codes in conjunction with dual-2 code. On the other hand, for $B_e = 8$, one can use the $H(20, 5)$ as the inner code of a dual-5 code and its performance is significantly better than that of the dual-2 code at low error rates. This improvement in performance is achieved at the expense of an increase in decoding complexity. Similarly, in Fig. 7.8.15 we compare the performance of the dual-k codes with two types of inner codes for $B_e = 16$. We note that the $_3H(8, 3)$ inner code has a $d_{\min} = 12$ and, hence, it yields an effective diversity of 6. This diversity is achieved by transmitting 12 frequencies per code word. The orthogonal inner code $O(24, 3)$ gives only third-order diversity, which is achieved by transmitting three frequencies per code word. Consequently the $O(24, 3)$ inner code is more efficient at low SNR, that is, for the range of error rates covered in Fig. 7.8.15. At large SNR the dual-3 code with the Hadamard $_3H(8, 3)$ inner code outperforms its counterpart with the $O(24, 3)$ inner code due to the large diversity provided by the Hadamard code. For the same bandwidth expansion factor $B_e = 16$, one may use a dual-6 code with a $H(48, 6)$ code to achieve an improvement over the dual-3 code with the $_3H(8, 3)$ inner code. Again, this improvement in performance, which in this case is not as impressive as that shown in Fig. 7.8.14, must be weighed against the increased decoding complexity inherent in the dual-6 code.

The numerical results given above illustrate the performance advantages in using codes with good distance properties and soft-decision decoding on a Rayleigh fading channel as an alternative to conventional M-ary orthogonal signaling with diversity. In addition, the results illustrate the benefits of code concatenation on such a channel, using a dual-k convolutional code as the outer code and either a Hadamard code or a block orthogonal code as the inner code. Although dual-k codes were used for the outer code, similar results are obtained when a Reed–Solomon code is used for the outer code. There is an even greater choice in the selection of the inner code.

The important parameter in the selection of both the outer and the inner codes is the minimum distance of the resultant concatenated code required to achieve a specified level of performance. Since many codes will meet the performance requirements, the ultimate choice is made on the basis of decoding complexity and bandwidth requirements.

7.8.5 System Design Based on the Cutoff Rate

In the treatment of coded waveforms given above we have demonstrated the effectiveness of various codes for fading channels. In particular, we have observed the benefits of soft-decision decoding and code concatenation as a means for increasing the minimum distance and, hence, the amount of diversity in the coded waveforms. In this section, we consider randomly selected code words and derive an upper (union) bound on the error probability which depends on the cutoff rate parameter for the Rayleigh fading channel.

Let us consider the model for the communication system illustrated in Fig. 7.8.1. The modulator has an q-ary orthogonal FSK alphabet. Code words of block length n are mapped into waveforms by selecting n tones from the alphabet of q tones. The demodulation is performed by passing the signal through a bank of q matched filters followed by square-law detectors. The decoding is assumed to be soft-decision. Thus, the square-law detected outputs from the demodulator are appropriately combined (added) with equal weighting to form M decision variables corresponding to the M possible transmitted code words.

To evaluate the union bound on the probability of error in a Rayleigh fading channel with AWGN, we first evaluate the binary error probability involving the decision variable U_1, which corresponds to the transmitted code word, and any of the other $M - 1$ decision variables corresponding to the other code words. Let U_2 be the other decision variable and suppose that U_1 and U_2 have l tones in common. Hence, the contributions to U_1 and U_2 from these l tones are identical and, therefore, cancel out when we form the difference $U_1 - U_2$. Since the two decision variables differ in $n - l$ tones, the probability of error is simply that for a binary orthogonal FSK system with $n - l$ order diversity. The exact form for this probability of error is given by (7.8.4) where $p = 1/(2 + \bar{\gamma}_c)$, and $\bar{\gamma}_c$ is the average SNR per tone. For simplicity, we choose to use the Chernoff bound for this binary event error probability, given by (7.8.7), i.e.,

$$P_2(U_1, U_2|l) = [4p(1 - p)]^{n-l} \qquad (7.8.22)$$

Now, let us average over the ensemble of binary communication systems. There are q^n possible code words from which we randomly select two code words. Thus, each code word is selected with equal probability. Then, the probability that two randomly selected code words have l tones in common is

$$P(l) = \binom{n}{l}\left(\frac{1}{q}\right)^l\left(1 - \frac{1}{q}\right)^{n-l} \qquad (7.8.23)$$

When we average (7.8.22) over the probability distribution of l given by (7.8.23) we obtain

$$\begin{aligned}
\overline{P_2(U_1, U_2)} &= \sum_{l=0}^{n} P_2(U_1, U_2|l)P(l) \\
&\leq \sum_{l=0}^{n} \binom{n}{l}\left(\frac{1}{q}\right)^l\left[4\left(1 - \frac{1}{q}\right)p(1 - p)\right]^{n-l} \\
&\leq \left\{\frac{1}{q}[1 + 4(q - 1)p(1 - p)]\right\}^n \qquad (7.8.24)
\end{aligned}$$

Finally, the union bound for communication systems which use $M = 2^k$ randomly selected code words is simply

$$\overline{P}_M \leq (M - 1)\overline{P_2(U_1, U_2)} < M\overline{P_2(U_1, U_2)} \qquad (7.8.25)$$

By combining (7.8.24) with (7.8.25) we obtain the upper bound on the symbol error probability as

$$\overline{P}_M < 2^{-n(R_0 - R_c)} \qquad (7.8.26)$$

where $R_c = k/n$ is the code rate and R_0 is the cutoff rate defined as

$$R_0 = \log_2 \frac{q}{1 + 4(q-1)p(1-p)} \qquad (7.8.27)$$

where

$$p = \frac{1}{2 + \bar{\gamma}_c} \qquad (7.8.28)$$

FIGURE 7.8.16
Cutoff rate as a function of $\bar{\gamma}_c$ for Rayleigh fading channel.

Graphs of R_0 as a function of $\bar{\gamma}_c$ are shown in Fig. 7.8.16 for $q = 2, 4,$ and 8.

A more interesting form of (7.8.26) is obtained if we express $\overline{P}_M$ in terms of the SNR per bit. In particular (7.8.26) may be expressed as

$$\overline{P}_M < 2^{-k[\bar{\gamma}_b g(q, \bar{\gamma}_c) - 1]} \tag{7.8.29}$$

where, by definition,

$$g(q, \bar{\gamma}_c) = \frac{R_0}{\bar{\gamma}_c}$$

$$= \frac{1}{\bar{\gamma}_c} \log_2 \frac{q}{1 + 4(q - 1)p(1 - p)} \tag{7.8.30}$$

Graphs of $g(q, \bar{\gamma}_c)$ as a function of $\bar{\gamma}_c$ are plotted in Fig. 7.8.17 with q as a parameter. First we note there is an optimum $\bar{\gamma}_c$ for each value of q, which minimizes the probability of error. For large q, this value is approximately $\bar{\gamma}_c = 3$ (5 dB), which is consistent with our previous observation for ordinary square-law diversity combining. Furthermore, as $q \to \infty$, the function $g(q, \bar{\gamma}_c)$ approaches a limit which is

$$\lim_{q \to \infty} g(q, \bar{\gamma}_c) = g_\infty(\bar{\gamma}_c) = \frac{1}{\bar{\gamma}_c} \log_2 \frac{(2 + \bar{\gamma}_c)^2}{4(1 + \bar{\gamma}_c)} \tag{7.8.31}$$

The value of $g_\infty(\bar{\gamma}_c)$ evaluate at $\bar{\gamma}_c = 3$ is

$$g_\infty(3) = \max_{\bar{\gamma}_c} g_\infty(\bar{\gamma}_c)$$

$$= 0.215 \tag{7.8.32}$$

Therefore, the error probability in (7.8.29) for this optimum division of total SNR is

$$\overline{P}_M < 2^{-0.215k(\bar{\gamma}_b - 4.65)} \tag{7.8.33}$$

This result indicates that the probability of error can be made arbitrarily small with optimum SNR per code chip, if the average SNR per bit $\bar{\gamma}_b > 4.65$ (6.7 dB). Even a relatively modest value of $q = 20$ comes close to this minimum value. As seen from Fig. 7.8.17, $g(20, 3) = 0.2$, so that $P_M \to 0$ provided $\bar{\gamma}_b > 5$ (7 dB). On the other hand, if $q = 2$, the maximum value of $g(2, \bar{\gamma}_c) \approx 0.096$ and the corresponding minimum SNR per bit is 10.2 dB.

In the case of binary FSK waveforms ($q = 2$), we may easily compare the cutoff rate for the unquantized (soft-decision) demodulator output with the cutoff rate for binary quantization, for which

$$R_Q = 1 - \log \left[1 + \sqrt{4p(1 - p)} \right] \qquad Q = 2$$

as was previously given in (5.2.104). Figure 7.8.18 illustrates the graphs for R_0

FIGURE 7.8.17
Graph of function $g(q, \bar{\gamma}_c)$.

and R_Q. We note that the difference between R_0 and R_Q is approximately 3 dB for rates below 0.3 and the difference increases rapidly at high rates. This loss may be reduced significantly by increasing the number of quantization levels to $Q = 8$ (three bits).

Similar comparisons in the relative performance between unquantized soft-decision decoding and quantized decision decoding can also be made for $q > 2$.

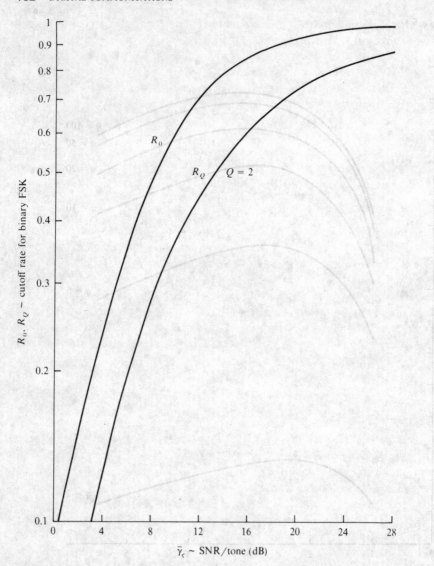

FIGURE 7.8.18
Cutoff rate for (unquantized) soft-decision and hard-decision decoding of coded binary FSK.

7.9 BIBLIOGRAPHICAL NOTES AND REFERENCES

In this chapter we considered a number of topics concerned with digital communications over a fading multipath channel. We began with a statistical characterization of the channel and then we described the ramifications of the channel characteristics on the design of digital signals and on their performance. We observed that the reliability of the communications system is enhanced by the use of diversity transmission and reception. Finally, we demonstrated that channel encoding and soft-decision decoding provide a bandwidth-efficient means for obtaining diversity over such channels.

The pioneering work on the characterization of fading multipath channels and on signal and receiver design for reliable digital communications over such channels was done by Price (1954, 1956). This work was followed by additional significant contributions from Price and Green (1958, 1960), Kailath (1960, 1961), and Green (1962). Diversity transmission and diversity combining techniques under a variety of channel conditions have been considered in the papers by Pierce (1958), Brennan (1959), Turin (1961, 1962), Pierce and Stein (1960), Barrow (1963), Bello and Nelin (1962a, b, 1963), Price (1962a, b), and Lindsey (1964).

Our treatment of coding for fading channels has relied on contributions from a number of researchers. In particular, the use of dual-k codes with M-ary orthogonal FSK was proposed in publications by Viterbi and Jacobs (1975) and Odenwalder (1976). The importance of coding for digital communications over a fading channel was also emphasized in a paper by Chase (1976). The benefits derived from concatenated coding with soft-decision decoding for a fading channel were demonstrated by Pieper et al. (1978). In that paper, a Reed–Solomon code was used for the outer code and a Hadamard code was selected as the inner code. The performance of dual-k codes with either block orthogonal codes or Hadamard codes as inner codes were investigated by Proakis and Rahman (1979). The error rate performance of maximal free distance binary convolutional codes was evaluated by Rahman (1981). Finally, the derivation of the cutoff rate for Rayleigh fading channels is due to Wozencraft and Jacobs (1965).

Our treatment of digital communications over fading channels was necessarily restricted to the Rayleigh fading channel model. For the most part, this is due to the wide acceptance of this model for describing the fading effects on many radio channels and to its mathematical tractability. Although other statistical models, such as a Ricean fading model, may be more appropriate for characterizing fading on some real channels, the general approach in the design of reliable communications presented in this chapter carries over.

APPENDIX 7A
PROBABILITIES OF ERROR FOR ADAPTIVE
RECEPTION OF M-PHASE SIGNALS

In this appendix we derive probabilities of error for two- and four-phase signaling over an L-diversity-branch time-invariant additive gaussian noise channel and for M-phase signaling over an L-diversity-branch Rayleigh fading additive gaussian noise channel. Both channels corrupt the signaling waveforms transmitted through them by introducing additive white gaussian noise and an unknown or random multiplicative gain and phase shift in the transmitted signal. The receiver processing consists of cross-correlating the signal plus noise received over each diversity branch by a noisy reference signal, which is derived either from the previously received information-bearing signals or from the transmission and reception of a pilot signal, and adding the outputs from all L-diversity branches to form the decision variable.

7A.1 MATHEMATICAL MODEL FOR AN M-PHASE SIGNALING COMMUNICATIONS SYSTEM

In the general case of M-phase signaling, the signaling waveforms at the transmitter are†

$$s_n(t) = \text{Re}\left[u_n(t)e^{j2\pi f_c t}\right]$$

where

$$u_n(t) = u(t)\exp\left[j\frac{2\pi}{M}(n-1)\right] \qquad n = 1, 2, \ldots, M \quad 0 \le t \le T \quad (7A.1)$$

and T is the time duration of the signaling interval.

Consider the case in which one of these M waveforms is transmitted, for the duration of the signaling interval, over L channels. Assume that each of the channels corrupts the signaling waveform transmitted through it by introducing a multiplicative gain and phase shift, represented by the complex-valued number g_k, and an additive noise $z_k(t)$. Thus, when the transmitted waveform is $u_n(t)$, the waveform received over the kth channel is

$$r_k(t) = g_k u_n(t) + z_k(t) \qquad 0 \le t \le T \quad k = 1, 2, \ldots, L \quad (7A.2)$$

The noises $\{z_k(t)\}$ are assumed to be sample functions of a stationary white gaussian random process with zero mean and autocorrelation function $\phi_n(\tau) = N_0\delta(\tau)$, where N_0 is the value of the spectral density. These sample functions are assumed to be mutually statistically independent.

At the demodulator, $r_k(t)$ is passed through a filter whose impulse response is matched to the waveform $u(t)$. The output of this filter, sampled at time $t = T$, is denoted as

$$X_k = 2\mathscr{E}g_k\exp\left[j\frac{2\pi}{M}(n-1)\right] + N_k \qquad (7A.3)$$

where $\mathscr{E}$ is the transmitted signal energy per channel and N_k is the noise sample from the kth filter. In order for the demodulator to decide which of the M phases was transmitted in the signaling interval $0 \le t \le T$, it attempts to undo the phase shift introduced by each channel. In practice, this is accomplished by multiplying the matched filter output X_k by the complex conjugate of an

---

† The complex representation of real signals is used throughout. Complex conjugation is denoted by an asterisk.

estimate $\hat{g}_k$ of the channel gain and phase shift. The result is a weighted and phase-shifted sampled output from the kth-channel filter, which is then added to the weighted and phase-shifted sampled outputs from the other $L-1$ channel filters.

The estimate $\hat{g}_k$ of the gain and phase shift of the kth channel is assumed to be derived either from the transmission of a pilot signal or by undoing the modulation on the information-bearing signals received in previous signaling intervals. As an example of the former, suppose that a pilot signal, denoted by $u_{pk}(t)$, $0 \le t \le T$, is transmitted over the kth channel for the purpose of measuring the channel gain and phase shift. The received waveform is

$$g_k u_{pk}(t) + z_{pk}(t) \qquad 0 \le t \le T$$

where $z_{pk}(t)$ is a sample function of a stationary white gaussian random process with zero mean and autocorrelation function $\phi_p(\tau) = N_0 \delta(\tau)$. This signal plus noise is passed through a filter matched to $u_{pk}(t)$. The filter output is sampled at time $t = T$ to yield the random variable $X_{pk} = 2\mathscr{E}_p g_k + N_{pk}$, where $\mathscr{E}_p$ is the energy in the pilot signal, which is assumed to be identical for all channels, and N_{pk} is the additive noise sample. An estimate of g_k is obtained by properly normalizing X_{pk}, i.e., $\hat{g}_k = g_k + N_{pk}/2\mathscr{E}_p$.

On the other hand, an estimate of g_k can be obtained from the information-bearing signal as follows. If one knew the information component contained in the matched filter output, then an estimate of g_k is obtained by properly normalizing this output. For example, the information component in the filter output given by (7A.3) is $2\mathscr{E} g_k \exp[j(2\pi/M)(n-1)]$ and, hence, the estimate is

$$\hat{g}_k = \frac{X_k}{2\mathscr{E}} \exp\left[-j\frac{2\pi}{M}(n-1)\right] = g_k + \frac{N_k'}{2\mathscr{E}}$$

where $N_k' = N_k \exp[-j(2\pi/M)(n-1)]$ and the pdf of N_k' is identical to the pdf of N_k. An estimate that is obtained from the information-bearing signal in this manner is called a *clairvoyant estimate*. Although a physically realizable receiver does not possess such clairvoyance, it can approximate this estimate by employing a time delay of one signaling interval and by feeding back the estimate of the transmitted phase in the previous signaling interval.

Whether the estimate of g_k is obtained from a pilot signal or from the information-bearing signal, the estimate can be improved by extending the time interval over which it is formed to include several prior signaling intervals in a way that has been described by Price (1962a, b). As a result of extending the measurement interval, the signal-to-noise ratio in the estimate of g_k is increased. In the general case where the estimation interval is the infinite past, the normalized *pilot signal estimate* is

$$\hat{g}_k = g_k + \frac{\sum_{i=1}^{\infty} c_i N_{pki}}{2\mathscr{E}_p \sum_{i=1}^{\infty} c_i} \tag{7A.4}$$

where c_i is the weighting coefficient on the subestimate of g_k derived from the ith prior signal interval and N_{pki} is the sample of additive gaussian noise at the output of the filter matched to $u_{pk}(t)$ in the ith prior signaling interval. Similarly, the clairvoyant estimate that is obtained from the information-bearing signal by undoing the modulation over the infinite past is

$$\hat{g}_k = g_k + \frac{\sum\limits_{i=1}^{\infty} c_i N_{ki}}{2\mathscr{E}\left(\sum\limits_{i=1}^{\infty} c_i\right)} \tag{7A.5}$$

As indicated, the demodulator forms the product between $\hat{g}_k^*$ and X_k and adds this to the products of the other $L-1$ channels. The random variable that results is

$$z = \sum_{k=1}^{L} X_k \hat{g}_k^* = \sum_{k=1}^{L} X_k Y_k^*$$

$$= z_r + jz_i \tag{7A.6}$$

where, by definition, $Y_k = \hat{g}_k$, $z_r = \mathrm{Re}\{z\}$, and $z_i = \mathrm{Im}\{z\}$. The phase of z is the decision variable. This is simply

$$\theta = \tan^{-1}\frac{z_i}{z_r} = \tan^{-1}\frac{\mathrm{Im}\left(\sum\limits_{k=1}^{L} X_k Y_k^*\right)}{\mathrm{Re}\left(\sum\limits_{k=1}^{L} X_k Y_k^*\right)} \tag{7A.7}$$

7A.2 CHARACTERISTIC FUNCTION AND PROBABILITY DENSITY FUNCTION OF THE PHASE θ

The following derivation is based on the assumption that the transmitted signal phase is zero, i.e., $n = 1$. If desired, the pdf of θ conditional on any other transmitted signal phase can be obtained by translating $p(\theta)$ by the angle $2\pi(n-1)/M$. We also assume that the complex-valued numbers $\{g_k\}$, which characterize the L channels, are mutually statistically independent and identically distributed zero mean gaussian random variables. This characterization is appropriate for slowly Rayleigh fading channels. As a consequence, the random variables (X_k, Y_k) are correlated, complex-valued, zero mean, gaussian, and statistically independent but identically distributed with any other other pair (X_i, Y_i).

The method that has been used in evaluating the probability density $p(\theta)$ in the general case of diversity reception is as follows. First, the characteristic function of the joint probability distribution function of z_r and z_i, where z_r and z_i are two components that make up the decision variable θ, is obtained. Second, the double Fourier transform of the characteristic function is performed and yields the density $p(z_r, z_i)$. Then the transformation

$$r = \sqrt{z_r^2 + z_i^2}$$

$$\theta = \tan^{-1}\left(\frac{z_i}{z_r}\right)$$

(7A.8)

yields the joint pdf of the envelope r and the phase θ. Finally, integration of this joint pdf over the random variable r yields the pdf of θ.

The joint characteristic function of the random variables z_r and z_i can be expressed in the form

$$\psi(jv_1, jv_2) = \left[\cfrac{\cfrac{4}{m_{xx}m_{yy}\left(1 - |\mu|^2\right)}}{\left(v_1 - j\cfrac{2|\mu|\cos\varepsilon}{\sqrt{m_{xx}m_{yy}}\left(1 - |\mu|^2\right)}\right)^2}\right.$$

$$\left. + \left(v_2 - j\frac{2|\mu|\sin\varepsilon}{\sqrt{m_{xx}m_{yy}}\left(1 - |\mu|^2\right)}\right)^2 + \frac{4}{m_{xx}m_{yy}\left(1 - |\mu|^2\right)^2}\right]^L$$

(7A.9)

where, by definition,

$$
\begin{aligned}
m_{xx} &= E\left(|X_k|^2\right) &&\text{identical for all } k \\
m_{yy} &= E\left(|Y_k|^2\right) &&\text{identical for all } k \\
m_{xy} &= E\left(X_k Y_k^*\right) &&\text{identical for all } k \\
\mu &= \frac{m_{xy}}{\sqrt{m_{xx}m_{yy}}} = |\mu|e^{-j\varepsilon}
\end{aligned}
$$

(7A.10)

The result of Fourier-transforming the function $\psi(jv_1, jv_2)$ with respect to the variables v_1 and v_2 is

$$p(z_r, z_i) = \frac{\left(1 - |\mu|^2\right)^L}{(L-1)!\pi 2^L}\left(\sqrt{z_r^2 + z_i^2}\right)^{L-1}$$

$$\times \exp\left[|\mu|(z_r\cos\varepsilon + z_i\sin\varepsilon)\right]$$

$$\times K_{L-1}\left(\sqrt{z_r^2 + z_i^2}\right)$$

(7A.11)

where $K_n(x)$ is the modified Hankel function of order n. At this point, the transformation of random variables, as indicated in (7A.8), yields the joint pdf of the envelope r and the phase θ in the form

$$p(r, \theta) = \frac{(1 - |\mu|^2)^L}{(L - 1)! \pi 2^L} r^L \exp\left[|\mu| r \cos(\theta - \varepsilon)\right] K_{L-1}(r) \quad (7A.12)$$

Now integration over the variable r yields the marginal pdf of the phase θ. We have evaluated the integral to obtain $p(\theta)$ in the form

$$p(\theta) = \frac{(-1)^{L-1}(1 - |\mu|^2)^L}{2\pi(L - 1)!}$$

$$\times \left\{ \frac{\partial^{L-1}}{\partial b^{L-1}} \left[\frac{1}{b - |\mu|^2 \cos^2(\theta - \varepsilon)} + \frac{|\mu| \cos(\theta - \varepsilon)}{\left(b - |\mu|^2 \cos^2(\theta - \varepsilon)\right)^{3/2}} \right. \right.$$

$$\left. \left. \times \cos^{-1}\left(-\frac{|\mu| \cos(\theta - \varepsilon)}{b^{1/2}}\right)\right] \right\}\Bigg|_{b=1} \quad (7A.13)$$

In this equation, the notation

$$\frac{\partial^L}{\partial b^L} f(b, \mu)\Bigg|_{b=1}$$

denotes the Lth partial derivative of the function $f(b, \mu)$ evaluated at $b = 1$.

7A.3 ERROR PROBABILITIES FOR SLOWLY RAYLEIGH FADING CHANNELS

In this section, the probability of a character error and the probability of a binary digit error are derived for M-phase signaling. The probabilities are evaluated via the probability density function and the probability distribution function of θ.

The probability distribution function of the phase. In order to evaluate the probability of error, we need to evaluate the definite integral

$$P(\theta_1 \leq \theta \leq \theta_2) = \int_{\theta_1}^{\theta_2} p(\theta) \, d\theta$$

where θ_1 and θ_2 are limits of integration and $p(\theta)$ is given by (7A.13). All subsequent calculations are made for a real cross-correlation coefficient μ. A real-valued μ implies that the signals have symmetric spectra. This is the usual situation encountered. Since a complex-valued μ causes a shift of ε in the pdf of θ, i.e., ε is simply a bias term, the results that are given for real μ can be altered in a trivial way to cover the more general case of complex-valued μ.

In the integration of $p(\theta)$, only the range $0 \leq \theta \leq \pi$ is considered, because $p(\theta)$ is an even function. Furthermore, the continuity of the integrand and its derivatives and the fact that the limits θ_1 and θ_2 are independent of b allow for the interchange of integration and differentiation. When this is done, the resulting integral can be evaluated quite readily and it can be expressed as follows:

$$\int_{\theta_1}^{\theta_2} p(\theta)\, d\theta = \frac{(-1)^{L-1}(1-\mu^2)^L}{2\pi(L-1)!} \frac{\partial^{L-1}}{\partial b^{L-1}}$$

$$\times \left(\frac{1}{b-\mu^2} \left\{ \frac{\mu\sqrt{1-\left[(b/\mu^2)-1\right]x^2}}{b^{1/2}} \cot^{-1} x \right.\right.$$

$$\left.\left. -\cot^{-1} \frac{xb^{1/2}/\mu}{\sqrt{1-\left[(b/\mu^2)-1\right]x^2}} \right\} \right) \Bigg|_{x_1}^{x_2} \Bigg|_{b=1} \qquad (7A.14)$$

where, by definition,

$$x_i = \frac{-\mu \cos\theta_i}{\sqrt{b-\mu^2 \cos\theta_i}} \qquad i = 1,2 \qquad (7A.15)$$

Probability of a symbol error. The probability of a symbol error for any M-phase signaling system is

$$P_M = 2\int_{\pi/M}^{\pi} p(\theta)\, d\theta$$

When (7A.14) is evaluated at these two limits, the result is

$$P_M = \frac{(-1)^{L-1}(1-\mu^2)^L}{\pi(L-1)!}$$

$$\times \left(\frac{\partial^{L-1}}{\partial b^{L-1}} \left\{ \frac{1}{b-\mu^2} \left[\frac{\pi}{M}(M-1) - \frac{\mu \sin(\pi/M)}{\sqrt{b-\mu^2 \cos^2(\pi/M)}} \right.\right.\right.$$

$$\left.\left.\left. \times \cot^{-1} \frac{-\mu \cos(\pi/M)}{\sqrt{b-\mu^2 \cos^2(\pi/M)}} \right] \right\} \right)_{b=1} \qquad (7A.16)$$

Probabilities of a binary digit error. First let us consider two-phase signaling. In this case, the probability of a binary digit error is obtained by integrating the pdf $p(\theta)$ over the ranges $\pi/2 < \theta < 3\pi/2$. Since $p(\theta)$ is an even function and the signals are *a priori* equally likely, this probability can be written as

$$P_2 = 2\int_{\pi/2}^{\pi} p(\theta)\, d\theta$$

It is easily verified that $\theta_1 = \pi/2$ implies $x_1 = 0$ and $\theta_2 = \pi$ implies $x_2 = \mu/\sqrt{b - \mu^2}$. Thus

$$P_2 = \frac{(-1)^{L-1}(1 - \mu^2)^L}{2(L-1)!} \frac{\partial^{L-1}}{\partial b^{L-1}} \left[\frac{1}{b - \mu^2} - \frac{\mu}{b^{1/2}(b - \mu^2)} \right]\Bigg|_{b=1} \qquad (7A.17)$$

After performing the differentiation indicated in (7A.17) and evaluating the resulting function at $b = 1$, the probability of a binary digit error is obtained in the form

$$P_2 = \frac{1}{2} \left[1 - \mu \sum_{k=0}^{L-1} \binom{2k}{k} \left(\frac{1 - \mu^2}{4} \right)^k \right] \qquad (7A.18)$$

Next we consider the case of four-phase signaling in which a Gray code is used to map pairs of bits into phases. Assuming again that the transmitted signal is $u_1(t)$, it is clear that a single error is committed when the received phase is $\pi/4 < \theta < 3\pi/4$, and a double error is committed when the received phase is $3\pi/4 < \theta < \pi$. That is, the probability of a binary digit error is

$$P_{4b} = \int_{\pi/4}^{3\pi/4} p(\theta) \, d\theta + 2 \int_{3\pi/4}^{\pi} p(\theta) \, d\theta \qquad (7A.19)$$

It is easily established from (7A.14) and (7A.19) that

$$P_{4b} = \frac{(-1)^{L-1}(1 - \mu^2)^L}{2(L-1)!} \left\{ \frac{\partial^{L-1}}{\partial b^{L-1}} \left[\frac{1}{b - \mu^2} - \frac{\mu}{(b - \mu^2)(2b - \mu^2)^{1/2}} \right] \right\}_{b=1}$$

Hence the probability of a binary digit error for four-phase signaling is

$$P_{4b} = \frac{1}{2} \left[1 - \frac{\mu}{\sqrt{2 - \mu^2}} \sum_{k=0}^{L-1} \binom{2k}{k} \left(\frac{1 + \mu^2}{4 - 2\mu^2} \right)^k \right] \qquad (7A.20)$$

Notice that if one defines the quantity $\rho = \mu/\sqrt{2 - \mu^2}$, the expression for P_{4b} in terms of ρ is

$$P_{4b} = \frac{1}{2} \left[1 - \rho \sum_{k=0}^{L-1} \binom{2k}{k} \left(\frac{1 - \rho^2}{4} \right)^k \right] \qquad (7A.21)$$

In other words, P_{4b} has the same form as P_2 given in (7A.18). Furthermore, note that ρ, just like μ, can be interpreted as a cross-correlation coefficient since the range of ρ is $0 \le \rho \le 1$ for $0 \le \mu \le 1$. This simple fact will be used later in Sec. 7A.4.

The above procedure for obtaining the bit error probability for an M-phase signal with a Gray code can be used to generate results for $M = 8, 16$, etc., as shown by Proakis (1968).

Evaluation of the cross-correlation coefficient. The expressions for the probabilities of error given above depend on a single parameter, namely, the cross-correlation coefficient μ. The clairvoyant estimate is given by (7A.5) and the matched filter output, when signal waveform $u_1(t)$ is transmitted, is $X_k = 2\mathscr{E}g_k + N_k$. Hence the cross-correlation coefficient is

$$\mu = \frac{\sqrt{\nu}}{\sqrt{\left(\dfrac{1}{\bar{\gamma}_c} + 1\right)\left(\dfrac{1}{\bar{\gamma}_c} + \nu\right)}} \tag{7A.22}$$

where, by definition

$$\nu = \frac{\left|\displaystyle\sum_{i=1}^{\infty} c_i\right|^2}{\displaystyle\sum_{i=1}^{\infty} |c_i|^2} \tag{7A.23}$$

$$\bar{\gamma}_c = \frac{\mathscr{E}}{N_0} E\left(|g_k|^2\right) \qquad k = 1, 2, \ldots, L$$

The parameter ν represents the effective number of signaling intervals over which the estimate is formed, and $\bar{\gamma}_c$ is the average SNR per channel.

In the case of differential phase signaling, the weighting coefficients are $c_1 = 1$, $c_i = 0$ for $i \neq 1$. Hence $\nu = 1$ and $\mu = \bar{\gamma}_c/(1 + \bar{\gamma}_c)$.

When $\nu = \infty$, the estimate is perfect and

$$\lim_{\nu \to \infty} \mu = \sqrt{\frac{\bar{\gamma}_c}{\bar{\gamma}_c + 1}}$$

Finally, in the case of a pilot signal estimate, given by (7A.4), the cross-correlation coefficient is

$$\mu = \frac{1}{\sqrt{\left(1 + \dfrac{r+1}{r\bar{\gamma}_t}\right)\left(1 + \dfrac{r+1}{\nu\bar{\gamma}_t}\right)}} \tag{7A.24}$$

where, by definition,

$$\bar{\gamma}_t = \frac{\mathscr{E}_t}{N_0} E\left(|g_k|^2\right)$$

$$\mathscr{E}_t = \mathscr{E} + \mathscr{E}_p$$

$$r = \frac{\mathscr{E}}{\mathscr{E}_p}$$

The values of μ given above are summarized in Table 7A.1.

TABLE 7A.1
Rayleigh fading channel

Type of estimate	Cross-correlation coefficient μ
Clairvoyant estimate	$\dfrac{\sqrt{\nu}}{\sqrt{\left(1 + \dfrac{1}{\bar{\gamma}_c}\right)\left(\nu + \dfrac{1}{\bar{\gamma}_c}\right)}}$
Pilot signal estimate	$\dfrac{\sqrt{r\nu}}{(r+1)\sqrt{\left(\dfrac{1}{\bar{\gamma}_t} + \dfrac{r}{r+1}\right)\left(\dfrac{1}{\bar{\gamma}_t} + \dfrac{\nu}{r+1}\right)}}$
Differential phase signaling	$\dfrac{\bar{\gamma}_c}{\bar{\gamma}_c + 1}$
Perfect estimate	$\sqrt{\dfrac{\bar{\gamma}_c}{\bar{\gamma}_c + 1}}$

7A.4 ERROR PROBABILITIES FOR TIME-INVARIANT AND RICEAN FADING CHANNELS

In Sec. 7A.2, the complex-valued channel gains $\{g_k\}$ were characterized as zero mean gaussian random variables, which is appropriate for Rayleigh fading channels. In this section, the channel gains $\{g_k\}$ are assumed to be nonzero mean gaussian random variables. Estimates of the channel gains are formed by the demodulator and are used as described in Sec. 7A.1. Moreover, the decision variable θ is defined again by (7A.7). However, in this case the gaussian random variables X_k, Y_k, which denote the matched filter output and the estimate, respectively, for the kth channel, have nonzero means that are denoted by $\bar{X}_k$ and $\bar{Y}_k$. Furthermore, the second moments are

$$m_{xx} = E\left(|X_k - \bar{X}_k|^2\right) \qquad \text{identical for all channels}$$

$$m_{yy} = E\left(|Y_k - \bar{Y}_k|^2\right) \qquad \text{identical for all channels}$$

$$m_{xy} = E\left[\left(X_k - \bar{X}_k\right)\left(Y_k^* - \bar{Y}_k^*\right)\right] \quad \text{identical for all channels}$$

and the normalized covariance is defined as

$$\mu = \frac{m_{xy}}{\sqrt{m_{xx}m_{yy}}}$$

Error probabilities are given below only for two- and four-phase signaling with this channel model. We are interested in the special case in which the fluctuating component of each of the channel gains $\{g_k\}$ is zero, so that the channels are time-invariant. If, in addition to this time invariance, the noises between the estimate and the matched filter output are uncorrelated, then $\mu = 0$.

In the general case, the probability of error for two-phase signaling over L statistically independent channels characterized in the manner described above can be obtained from the results in Appendix 4B. In its most general form, the expression for the binary error rate is

$$
P_2 = Q(a, b) - I_0(a) \exp\left(-\frac{a^2 + b^2}{2}\right) + \frac{I_0(ab) \exp\left[-(a^2 + b^2)/2\right]}{[2/(1 - \mu)]^{2L-1}}
$$

$$
\times \sum_{k=0}^{L-1} \binom{2L-1}{k}\left(\frac{1+\mu}{1-\mu}\right)^k + \frac{\exp\left[-(a^2 + b^2)/2\right]}{[2/(1-\mu)]^{2L-1}}
$$

$$
\times \sum_{n=1}^{L-1} I_n(ab)\left\{ \sum_{k=0}^{L-1-n} \binom{2L-1}{k}\left[\left(\frac{b}{a}\right)^n\left(\frac{1+\mu}{1-\mu}\right)^k - \left(\frac{a}{b}\right)^n\left(\frac{1+\mu}{1-\mu}\right)^{2L-1-k}\right]\right\}
$$

$$
L \geq 2
$$

$$
P_2 = Q(a, b) - \frac{1+\mu}{2}I_0(ab)\exp\left(-\frac{a^2 + b^2}{2}\right) \qquad L = 1 \tag{7A.25}
$$

where, by definition,

$$
a = \left(\frac{1}{2} \sum_{k=1}^{L} \left| \frac{\overline{X}_k}{\sqrt{m_{xx}}} - \frac{\overline{Y}_k}{\sqrt{m_{yy}}} \right|^2 \right)^{1/2}
$$

$$
b = \left(\frac{1}{2} \sum_{k=1}^{L} \left| \frac{\overline{X}_k}{\sqrt{m_{xx}}} + \frac{\overline{Y}_k}{\sqrt{m_{yy}}} \right|^2 \right)^{1/2} \tag{7A.26}
$$

$$
Q(a, b) = \int_b^\infty x \exp\left(-\frac{a^2 + x^2}{2}\right) I_0(ab)\, dx
$$

$I_n(x)$ is the modified Bessel function of the first kind and of order n.

Let us evaluate the constants a and b when the channel is time-invariant, $\mu = 0$, and the channel gain and phase estimates are those given in Sec. 7A.1. Recall that when signal $u_1(t)$ is transmitted, the matched filter output is $X_k = 2\mathscr{E}g_k + N_k$. The clairvoyant estimate is given by (7A.5). Hence, for this estimate, the moments are $\overline{X}_k = 2\mathscr{E}g_k$, $\overline{Y}_k = g_k$, $m_{xx} = 4\mathscr{E}N_0$, and $m_{yy} = N_0/\mathscr{E}\nu$, where $\mathscr{E}$ is the signal energy, N_0 is the value of the noise spectral density, and ν is defined in (7A.23). Substitution of these moments into (7A.26) results in the following expressions for a and b:

$$
a = \sqrt{\frac{\gamma_b}{2}}\,|\sqrt{\nu} - 1|
$$

$$
b = \sqrt{\frac{\gamma_b}{2}}\,|\sqrt{\nu} + 1| \tag{7A.27}
$$

$$
\gamma_b = \frac{\mathscr{E}}{N_0} \sum_{k=1}^{L} |g_k|^2
$$

This is a result originally derived by Price (1962).

The probability of error for differential phase signaling can be obtained by setting $\nu = 1$ in (7A.27).

Next, consider a pilot signal estimate. In this case, the estimate is given by (7A.4) and the matched filter output is again $X_k = 2\mathscr{E}g_k + N_k$. When the moments are calculated and these are substituted into (7A.26), the following expressions for a and b are obtained:

$$
\begin{aligned}
a &= \sqrt{\frac{\gamma_t}{2}} \left| \sqrt{\frac{\nu}{r+1}} - \sqrt{\frac{r}{r+1}} \right| \\
b &= \sqrt{\frac{\gamma_t}{2}} \left(\sqrt{\frac{\nu}{r+1}} + \sqrt{\frac{r}{r+1}} \right)
\end{aligned}
\tag{7A.28}
$$

where

$$
\gamma_t = \frac{\mathscr{E}_t}{N_0} \sum_{k=1}^{L} |g_k|^2
$$

$$
\mathscr{E}_t = \mathscr{E} + \mathscr{E}_p
$$

$$
r = \frac{\mathscr{E}}{\mathscr{E}_p}
$$

TABLE 7A.2
Time-invariant channel

Type of estimate	a	b
Two-phase signaling		
Clairvoyant estimate	$\sqrt{\dfrac{\gamma_b}{2}} \, \lvert\sqrt{\nu} - 1\rvert$	$\sqrt{\dfrac{\gamma_b}{2}} \, (\sqrt{\nu} + 1)$
Differential phase signaling	0	$\sqrt{2\gamma_b}$
Pilot signal estimate	$\sqrt{\dfrac{\gamma_t}{2}} \left\lvert \sqrt{\dfrac{\nu}{r+1}} - \sqrt{\dfrac{r}{r+1}} \right\rvert$	$\sqrt{\dfrac{\gamma_t}{2}} \left(\sqrt{\dfrac{\nu}{r+1}} + \sqrt{\dfrac{r}{r+1}} \right)$
Four-phase signaling		
Clairvoyant estimate	$\sqrt{\dfrac{\gamma_b}{2}} \left\lvert \sqrt{\nu + 1 + \sqrt{\nu^2 + 1}} - \sqrt{\nu + 1 - \sqrt{\nu^2 + 1}} \right\rvert$	$\sqrt{\dfrac{\gamma_b}{2}} \left(\sqrt{\nu + 1 + \sqrt{\nu^2 + 1}} + \sqrt{\nu + 1 - \sqrt{\nu^2 + 1}} \right)$
Differential phase signaling	$\sqrt{\dfrac{\gamma_b}{2}} \left(\sqrt{2 + \sqrt{2}} - \sqrt{2 - \sqrt{2}} \right)$	$\sqrt{\dfrac{\gamma_b}{2}} \left(\sqrt{2 + \sqrt{2}} + \sqrt{2 - \sqrt{2}} \right)$
Pilot signal estimate	$\sqrt{\dfrac{\gamma_t}{4(r+1)}} \left\lvert \sqrt{\nu + r + \sqrt{\nu^2 + r^2}} - \sqrt{\nu + r - \sqrt{\nu^2 + r^2}} \right\rvert$	$\sqrt{\dfrac{\gamma_t}{4(r+1)}} \left(\sqrt{\nu + r + \sqrt{\nu^2 + r^2}} + \sqrt{\nu + r - \sqrt{\nu^2 + r^2}} \right)$

Finally, we consider the probability of a binary digit error for four-phase signaling over a time-invariant channel for which the condition $\mu = 0$ obtains. One approach that can be used to derive this error probability is to determine the pdf of θ and then to integrate the pdf over the appropriate range of values of θ. Unfortunately, this approach proves to be intractable mathematically. Instead, a simpler, albeit roundabout, method may be used that involves the Laplace transform. In short, the integral in (7.4.14) of the text that relates the error probability $P_2(\gamma_b)$ in an AWGN channel to the error probability P_2 in a Rayleigh fading channel is a Laplace transform. Since the bit error probabilities P_2 and P_{4b} for a Rayleigh fading channel, given by (7A.18) and (7A.21), respectively, have the same form but differ only in the correlation coefficient, it follows that the bit error probabilities for the time-invariant channel also have the same form. That is, (7A.25) with $\mu = 0$ is also the expression for the bit error probability of a four-phase signaling system with the parameters a and b modified to reflect the difference in the correlation coefficient. The detailed derivation may be found in the paper by Proakis (1968). The expressions for a and b are given in Table 7A.2.

PROBLEMS

7.1. The scattering function $S(\tau; \lambda)$ for a fading multipath channel is nonzero for the range of values $0 \leq \tau \leq 1$ ms and $-0.1 \leq \lambda \leq 0.1$ Hz. Assume that the scattering function is approximately uniform in the two variables.

  (*a*) Give numerical values for the following parameters:
  (*i*) Multipath spread of the channel
  (*ii*) Doppler spread of the channel
  (*iii*) Coherence time of the channel
  (*iv*) Coherence bandwidth of the channel
  (*v*) Spread factor of the channel.

  (*b*) *Explain* the meaning of the following terms, taking into consideration the answers given in (*a*):
  (*i*) The channel is frequency-nonselective.
  (*ii*) The channel is slowly fading.
  (*iii*) The channel is frequency-selective.

  (*c*) Suppose we have a frequency allocation (bandwidth) of 10 kHz and we wish to transmit at a rate of 100 bits/s over this channel. Design a binary communications system with frequency diversity. In particular, specify (1) the type of modulation, (2) the number of subchannels, (3) the frequency separation between adjacent carriers, and (4) the signaling interval used in your design. Justify your choice of parameters.

7.2. Consider a binary communications system for transmitting a binary sequence over a fading channel. The modulation is orthogonal FSK with third-order frequency diversity ($L = 3$). The demodulator consists of matched filters followed by square-law detectors. Assume that the FSK carriers fade independently and identically according to a Rayleigh envelope distribution. The additive noises on the diversity signals are zero mean gaussian with autocorrelation function $\frac{1}{2}E[z_k^*(t)z_k(t + \tau)] = N_0\delta(\tau)$. The noise processes are mutually statistically independent.

(a) The transmitted signal may be viewed as binary FSK with square-law detection, generated by a repetition code of the form

$$1 \to \mathbf{C}_1 = [1 \quad 1 \quad 1] \qquad 0 \to \mathbf{C}_0 = [0 \quad 0 \quad 0]$$

Determine the error rate performance P_{2h} for a hard-decision decoder following the square-law-detected signals.

(b) Evaluate P_{2h} for $\bar{\gamma}_c = 100$ and 1000.

(c) Evaluate the error rate P_{2s} for $\bar{\gamma}_c = 100$ and 1000 is the decoder employs soft-decision decoding.

(d) Consider the generalization of the result in (a). If a repetition code of block length L (L odd) is used, determine the error probability P_{2h} of the hard-decision decoder and compare that with P_{2s}, the error rate of the soft-decision decoder. Assume $\bar{\gamma} \gg 1$.

7.3. Suppose the binary signal $u(t)$ is transmitted over a fading channel and the received signal is

$$r(t) = \pm a u(t) + z(t) \qquad 0 \le t \le T$$

where $z(t)$ is zero mean white gaussian noise with autocorrelation function

$$\phi_{zz}(\tau) = N_0 \delta(\tau)$$

The energy in the transmitted signal is $\mathscr{E} = \frac{1}{2} \int_0^T |u(t)|^2 \, dt$. The channel gain a is specified by the probability density function

$$p(a) = 0.1\delta(a) + 0.9\delta(a - 2)$$

(a) Determine the average probability of error P_e for the demodulator which employs a filter matched to $u(t)$.

(b) What value does P_e approach as $\mathscr{E}/N_0$ approaches infinity.

(c) Suppose the same signal is transmitted on two statistically *independently fading* channels with gains a_1 and a_2 where

$$p(a_k) = 0.1\delta(a_k) + 0.9\delta(a_k - 2) \qquad k = 1, 2$$

The noises on the two channels are statistically independent and identically distributed. The demodulator employs a matched filter for each channel and simply adds the two filter outputs to form the decision variable. Determine the average P_e.

(d) For the case in (c) what value does P_e approach as $\mathscr{E}/N_0$ approaches infinity.

7.4. A multipath fading channel has a multipath spread of $T_m = 1$ s and a Doppler spread $B_d = 0.01$ Hz. The total channel bandwidth at bandpass available for signal transmission is $W = 5$ Hz. To reduce the effects of intersymbol interference, the signal designer selects a pulse duration $T = 10$ s.

(a) Determine the coherence bandwidth and the coherence time.

(b) Is the channel frequency selective? Explain.

(c) Is the channel fading slowly or rapidly? Explain.

(d) Suppose the channel is used to transmit binary data via (antipodal) coherently detected PSK in a frequency diversity mode. Explain how you would use the available channel bandwidth to obtain frequency diversity and determine how much diversity is available.

(e) For the case in (d), what is the *approximate* SNR required per diversity to achieve an error probability of 10^{-6}?

(f) Suppose a wideband signal is used for transmission and a RAKE-type receiver is used for demodulation. How many taps would you use in the RAKE receiver?

(g) Explain whether or not the RAKE receiver can be implemented as a coherent receiver with predetection combining.

(h) If binary orthogonal signals are used for the wideband signal with square-law postdetection combining in the RAKE receiver, what is the *approximate* SNR required to achieve an error probability of 10^{-6}? (Assume all taps have the same SNR.)

7.5. In the binary communications system shown in Fig. P7.5, $z_1(t)$ and $z_2(t)$ are statistically independent white gaussian noise processes with zero mean and identical autocorrelation functions $\phi_{zz}(\tau) = N_0\delta(\tau)$. The sampled values U_1 and U_2 represent the *real parts* of the matched filter outputs. For example, if $u(t)$ is transmitted, then we have

$$U_1 = 2\mathscr{E} + N_1$$
$$U_2 = N_1 + N_2$$

where $\mathscr{E}$ is the transmitted signal energy and

$$N_k = \text{Re}\left[\int_0^T u^*(t)z_k(t)\,dt\right] \qquad k = 1,2$$

It is apparent that U_1 and U_2 are correlated gaussian variables while N_1 and N_2 are independent gaussian variables. Thus

$$p(n_1) = \frac{1}{\sqrt{2\pi}\,\sigma}\exp\left(-\frac{1}{2\sigma^2}n_1^2\right)$$
$$p(n_2) = \frac{1}{\sqrt{2\pi}\,\sigma}\exp\left(-\frac{1}{2\sigma^2}n_2^2\right)$$

where σ^2, the variance of N_k, is $2\mathscr{E}N_0$.

(*a*) Show that the joint probability density function for U_1 and U_2 is

$$p(U_1, U_2) = \frac{1}{2\pi\sigma^2}\exp\left\{-\frac{1}{2\sigma^2}\left[(U_1 - 2\mathscr{E})^2 - U_2(U_1 - 2\mathscr{E}) + \frac{U_2^2}{2}\right]\right\}$$

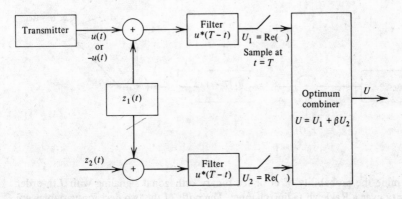

FIGURE P7.5

if $u(t)$ is transmitted and

$$p(U_1, U_2) = \frac{1}{2\pi\sigma^2} \exp\left\{-\frac{1}{2\sigma^2}\left[(U_1 + 2\mathscr{E})^2 - U_2(U_1 + 2\mathscr{E}) + \frac{U_2^2}{2}\right]\right\}$$

if $-u(t)$ is transmitted.

(b) Based on the likelihood ratio, show that the optimum combining of U_1 and U_2 results in the decision variable

$$U = U_1 + \beta U_2$$

where β is a constant. What is the optimum value of β?

(c) Suppose that $u(t)$ is transmitted. What is the probability density function of U?

(d) What is the probability of error assuming that $u(t)$ was transmitted? Express your answer as a function for the SNR $\mathscr{E}/N_0$.

(e) What is the loss in performance if only $U = U_1$ is the decision variable?

7.6. Consider the model for a binary communications system with diversity as shown in Fig. P7.6. The channels have fixed attenuations and phase shifts. The $\{z_k(t)\}$ are complex-valued white gaussian noise processes with zero means and autocorrelation functions.

$$\phi_{zz}(t) = \frac{1}{2}E\left[z_k^*(t)z_k(t + \tau)\right] = N_{0k}\delta(\tau)$$

(Note that the spectral densities $\{N_{0k}\}$ are all different.) Also, the noise processes $\{z_k(t)\}$ are mutually statistically independent. The $\{\beta_k\}$ are complex-valued weighting factors to be determined. The decision variable from the combiner is

$$U = \text{Re}\left(\sum_{k=1}^{L} \beta_k U_k\right) \underset{-1}{\overset{1}{\gtrless}} 0$$

(a) Determine the pdf $p(U)$ when $a + 1$ is transmitted.

(b) Determine the probability of error P_2 as a function of the weights $\{\beta_k\}$.

(c) Determine the values of $\{\beta_k\}$ that minimize P_2.

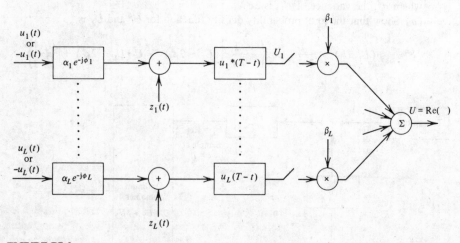

FIGURE P7.6

7.7. Determine the probability of error for binary orthogonal signaling with Lth-order diversity over a Rayleigh fading channel. The pdf's of the two decision variables are given by (7.4.31) and (7.4.33).

7.8. The rate $1/3$, $L = 3$, binary convolutional code with transfer function given by (5.3.5) is used for transmitting data over a Rayleigh fading channel via binary PSK.

(a) Determine and plot the probability of error for hard-decision decoding. Assume that the transmitted waveforms corresponding to the coded bits fade independently.

(b) Determine and plot the probability of error for soft-decision decoding. Assume that the waveforms corresponding to the coded bits fade independently.

7.9. A binary sequence is transmitted via binary antipodal signaling over a Rayleigh fading channel with Lth-order diversity. When $u(t)$ is transmitted, the received equivalent lowpass signals are

$$r_k(t) = \alpha_k e^{-j\phi_k} u(t) + z_k(t) \qquad k = 1, 2, \ldots, L$$

The fading among the L subchannels is statistically independent. The additive noise terms $\{z_k(t)\}$ are zero mean, statistically independent, and identically distributed white gaussian noise processes with autocorrelation function $\phi_{zz}(\tau) = N_0(\tau)$. Each of the L signals is passed through a filter matched to $u(t)$ and the output is phase-corrected to yield

$$U_k = \text{Re}\left\{ e^{j\phi_k} \int_0^T r_k(t) u^*(t) \, dt \right\} \qquad k = 1, 2, \ldots, L$$

The $\{U_k\}$ are combined by a linear combiner to form the decision variable

$$U = \sum_{k=1}^{L} U_k$$

(a) Determine the pdf of U conditional on fixed values for the $\{a_k\}$.

(b) Determine the expression for the probability of error when the $\{\alpha_k\}$ are statistically independent and identically distributed Rayleigh random variables.

7.10. The Chernoff bound for the probability of error for binary FSK with diversity L in Rayleigh fading was shown to be

$$P_2(L) < [4p(1-p)]^L = \left[4\frac{1 + \bar{\gamma}_c}{(2 + \bar{\gamma}_c)^2} \right]^L$$

$$< 2^{-\bar{\gamma}_b g(\bar{\gamma}_c)}$$

where

$$g(\bar{\gamma}_c) = \frac{1}{\bar{\gamma}_c} \log_2 \frac{(2 + \bar{\gamma}_c)^2}{4(1 + \bar{\gamma}_c)}$$

(a) Plot $g(\bar{\gamma}_c)$ and determine approximately its maximum value and the value of $\bar{\gamma}_c$ where the maximum occurs.

(b) For a given $\bar{\gamma}_b$, determine the optimum order of diversity.

(c) Compare $P_2(L)$, under the condition that $g(\bar{\gamma}_c)$ is maximized (optimal diversity), with the error probability for binary FSK in AWGN with no fading, which is

$$P_2 = \tfrac{1}{2} e^{-\gamma_b/2}$$

and determine the penalty in SNR due to fading and noncoherent (square-law) combining.

CHAPTER
8

SPREAD SPECTRUM SIGNALS FOR DIGITAL COMMUNICATIONS

Spread spectrum signals used for the transmission of digital information are distinguished by the characteristic that their bandwidth W is much greater than the information rate R in bits per second. That is, the bandwidth expansion factor $B_e = W/R$ for a spread spectrum signal is much greater than unity. The large redundancy inherent in spread spectrum signals is required to overcome the severe levels of interference that are encountered in the transmission of digital information over some radio and satellite channels. Since coded waveforms are also characterized by a bandwidth expansion factor that is greater than unity and since coding is an efficient method for introducing redundancy, it follows that coding is an important element in the design of spread spectrum signals.

A second important element employed in the design of spread spectrum signals is pseudo-randomness, which makes the signals appear similar to random noise and difficult to demodulate by receivers other than the intended ones. This element is intimately related with the application or purpose of such signals.

To be specific, spread spectrum signals are used for the purpose of (1) combatting or suppressing the detrimental effects of interference due to jamming, interference arising from other users of the channel, and self-interference due to multipath propagation, (2) hiding a signal by transmitting it at a low power and, thus, making it difficult for an unintended listener to detect in the presence of background noise, and (3) achieving message privacy in the presence of other listeners. In applications other than communications, spread spectrum signals are used to obtain accurate range (time delay) and range rate (velocity) measurements in radar and navigation. For the sake of brevity, we shall limit our discussion to digital communications applications.

In combatting intentional interference (jamming), it is important to the communicators that the jammer who is trying to disrupt the communication does not have prior knowledge of the signal characteristics except for the overall channel bandwidth and the type of modulation, (PSK, FSK, etc.) being used. If the digital information is just encoded as described in Chap. 5, a sophisticated jammer can easily mimic the signals emitted by the transmitter and, thus, confuse the receiver. To circumvent this possibility, the transmitter introduces an element of unpredictability or randomness (pseudo-randomness) in each of the transmitted coded signal waveforms which is known to the intended receiver but not to the jammer. As a consequence, the jammer must synthesize and transmit an interfering signal without knowledge of the pseudo-random pattern.

Interference from other users arises in multiple-access communications systems in which a number of users share a common channel bandwidth. At any given time, a subset of these users may transmit information simultaneously over the common channel to corresponding receivers. Assuming that all the users employ the same code for the encoding and decoding of their respective information sequences, the transmitted signals in this common spectrum may be distinguished from one another by superimposing a different pseudo-random pattern, also called a *code*, in each transmitted signal. Thus a particular receiver can recover the transmitted information intended for it by knowing the pseudo-random pattern, i.e., the key, used by the corresponding transmitter. This type of communication technique, which allows multiple users to simultaneously use a common channel for transmission of information, is called *code division multiple access* (CDMA). CDMA will be considered in Secs. 8.2 and 8.3.

Resolvable multipath components resulting from time-dispersive propagation through a channel may be viewed as a form of self-interference. This type of interference may also be suppressed by the introduction of a pseudo-random pattern in the transmitted signal, as will be described below.

A message may be hidden in the background noise by spreading its bandwidth with coding and transmitting the resultant signal at a low average power. Because of its low power level, the transmitted signal is said to be "covert." It has a low probability of being intercepted (detected) by a casual listener and, hence, is also called a *low-probability of intercept* (LPI) signal.

Finally, message privacy may be obtained by superimposing a pseudo-random pattern on a transmitted message. The message can be demodulated by the intended receivers that know the pseudo-random pattern or key used at the transmitter but not by any other receivers that do not have knowledge of the key.

In the following sections we shall describe a number of different types of spread spectrum signals, their characteristics, and their application. The emphasis will be on the use of spread spectrum signals for combatting, jamming (antijam or AJ signals), for CDMA, and for LPI. Before discussing the signal design problem, however, we shall briefly describe the types of channel characteristics assumed for the applications cited above.

8.1 MODEL OF SPREAD SPECTRUM DIGITAL COMMUNICATIONS SYSTEM

The block diagram shown in Fig. 8.1.1 illustrates the basic elements of a spread spectrum digital communications with a binary information sequence at its input at the transmitting end and at its output at the receiving end. The channel encoder and decoder and the modulator and demodulator are basic elements of the system which were treated in Chaps. 4, 5, 6, and 7. In addition to these elements, we have two identical pseudo-random pattern generators, one which interfaces with the modulator at the transmitting end and the second which interfaces with the demodulator at the receiving end. The generators generate a pseudo-random or pseudo-noise (PN) binary-valued sequence which is impressed on the transmitted signal at the modulator and removed from the received signal at the demodulator.

Synchronization of the PN sequence generated at the receiver with the PN sequence contained in the incoming received signal is required in order to demodulate the received signal. Initially, prior to the transmission of information, synchronization may be achieved by transmitting a fixed pseudo-random bit pattern which the receiver will recognize in the presence of interference with a high probability. After time synchronization of the generators is established, the transmission of information may commence.

Interference is introduced in the transmission of the information-bearing signal through the channel. The characteristics of the interference depend to a large extent on its origin. It may be categorized as being either broadband or narrowband relative to the bandwidth of the information-bearing signal, and either continuous in time or pulsed (discontinuous) in time. For example, a jamming signal may consist of one or more sinusoids in the bandwidth used to transmit the information. The frequencies of the sinusoids may remain fixed or they may change with time according to some rule. As a second example, the interference generated in CDMA by other users of the channel may be either broadband or narrowband depending on the type of spread spectrum signal that

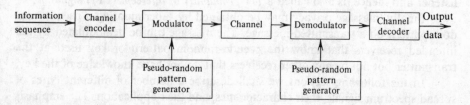

FIGURE 8.1.1
Model of spread spectrum digital communications system.

is employed to achieve multiple access. If it is broadband, it may be characterized as an equivalent additive white gaussian noise. We shall consider these types of interference and some others in the following sections.

Our treatment of spread spectrum signals will focus on the performance of the digital communications system in the presence of narrowband and broadband interference. Two types of modulation are considered, PSK and FSK. PSK is appropriate in applications where phase coherence between the transmitted signal and the received signal can be maintained over a time interval that is relatively long compared to the reciprocal of the transmitted signal bandwidth. On the other hand, FSK modulation is appropriate in applications where such phase coherence cannot be maintained due to time-variant effects on the communications link. This may be the case in a communications link between two high-speed aircraft or between a high-speed aircraft and a ground terminal.

The PN sequence generated at the modulator is used in conjunction with the PSK modulation to shift the phase of the PSK signal pseudo-randomly as described in Sec. 8.2. The resulting modulated signal is called a *direct sequence* (DS) or a *pseudo-noise* (PN) spread spectrum signal. When used in conjunction with binary or *M*-ary ($M > 2$) FSK, the pseudo-random sequence selects the frequency of the transmitted signal pseudo-randomly. The resulting signal is called a *frequency-hopped* (FH) spread spectrum signal. Although a number of other types of spread spectrum signals will be briefly described, the emphasis of our treatment will be on PN and FH spread spectrum signals.

8.2 DIRECT SEQUENCE SPREAD SPECTRUM SIGNALS

In the model shown in Fig. 8.1.1, we assume that the information rate at the input to the encoder is R bits/s and the available channel bandwidth is W Hz. The modulation is assumed to be binary PSK. In order to utilize the entire available channel bandwidth, the phase of the carrier is shifted pseudo-randomly according to the pattern from the PN generator at a rate W times/s. The reciprocal of W, denoted as T_c, defines the duration of a rectangular pulse which is called a *chip* and its time duration T_c is called the *chip interval*. The pulse is the basic element in a DS spread spectrum signal.

If we define $T_b = 1/R$ to be the duration of a rectangular pulse corresponding to the transmission time of an information bit, the bandwidth expansion factor W/R may be expressed as

$$B_c = \frac{W}{R} = \frac{T_b}{T_c} \qquad (8.2.1)$$

In practical systems the ratio T_b/T_c is an integer,

$$L_c = \frac{T_b}{T_c} \qquad (8.2.2)$$

which is the number of chips per information bit. That is, L_c is the number of

(a) PN and data signals

(b) DS-QPSK modulator

FIGURE 8.2.1
(a) The PN and data signals and (b) the QPSK modulator for a DS spread spectrum system.

phase shifts that occur in the transmitted signal during the bit duration $T_b = 1/R$. Figure 8.2.1a illustrates the relationships between the PN signal and the data signal.

Suppose that the encoder takes k information bits at a time and generates a binary linear (n, k) block code. The time duration available for transmitting the n code elements is kT_b seconds. The number of chips that occur in this time interval is kL_c. Hence we may select the block length of the code as $n = kL_c$. If the encoder generates a binary convolutional code of rate k/n, the number chips in the time interval kT_b is also $n = kL_c$. Therefore the following discussion applies to both block codes and convolutional codes.

One method for impressing the PN sequence on the transmitted signal is to alter directly the coded bits by modulo-2 addition with the PN sequence.† Thus each coded bit is altered by its addition with a bit from the PN sequence. If b_i represents the ith bit of the PN sequence and c_i is the corresponding bit from the encoder, the modulo-2 sum is

$$a_i = b_i \oplus c_i \qquad (8.2.3)$$

Hence $a_i = 1$ if either $b_i = 1$ and $c_i = 0$ or $b_i = 0$ and $c_i = 1$; also $a_i = 0$ is either $b_i = 1$ and $c_i = 1$ or $b_i = 0$ and $c_i = 0$. We may say that $a_i = 0$ when $b_i = c_i$ and $a_i = 1$ when $b_i \neq c_i$. The sequence $\{a_i\}$ is mapped into a binary PSK signal of the form $s(t) = \pm \mathrm{Re}[u(t)e^{j2\pi f_c t}]$ according to the convention

$$\begin{aligned} a_i = 0: \quad & u_i(t) = u(t - iT_c) \\ a_i = 1: \quad & u_i(t) = -u(t - iT_c) \end{aligned} \qquad (8.2.4)$$

where $u(t)$ represents a pulse of duration T_c seconds and arbitrary shape.

The modulo-2 addition of the coded sequence $\{c_i\}$ and the sequence $\{b_i\}$ from the PN generator may also be represented as a multiplication of two waveforms. To demonstrate this point, suppose that the elements of the coded sequence are mapped into a binary PSK signal according to the relation

$$c_i(t) = (2c_i - 1)u(t - T_c) \qquad (8.2.5)$$

Similarly, we define a waveform $p_i(t)$ as

$$p_i(t) = (2b_i - 1)p(t - iT_c) \qquad (8.2.6)$$

where $p(t)$ is a rectangular pulse of duration T_c. Then the equivalent lowpass transmitted signal corresponding to the ith coded bit is

$$\begin{aligned} u_i(t) &= p_i(t)c_i(t) \\ &= (2b_i - 1)(2c_i - 1)u(t - iT_c) \end{aligned} \qquad (8.2.7)$$

This signal is identical to the one given by (8.2.4) which is obtained from the sequence $\{a_i\}$. Consequently modulo-2 addition of the coded bits with the PN sequence followed by a mapping which yields a binary PSK signal is equivalent

†When four-phase PSK is desired, one PN sequence is added to the information sequence carried on the in-phase signal component and a second PN sequence is added to the information sequence carried on the quadrature component. In many PN-spread spectrum systems, the same binary information sequence is added to the two PN sequences to form the two quadrature components. Thus, a four-phase PSK signal is generated with a binary information stream.

to multiplying a binary PSK signal generated from the coded bits with a sequence of unit amplitude rectangular pulses, each of duration T_c, and with a polarity which is determined from the PN sequence according to (8.2.6). Although it is easier to implement modulo-2 addition followed by PSK modulation instead of waveform multiplication, it is convenient, for purposes of demodulation, to consider the transmitted signal in the multiplicative form given by (8.2.7). A functional block diagram of a four-phase PSK DS spread spectrum modulator is shown in Fig. 8.2.1*b*.

The received signal for the *i*th code element is†

$$r_i(t) = p_i(t)c_i(t) + z(t) \qquad iT_c \leq t \leq (i+1)T_c$$
$$= (2b_i - 1)(2c_i - 1)u(t - iT_c) + z(t) \qquad (8.2.8)$$

where $z(t)$ represents the interference or jamming signal corrupting the information-bearing signal. The interference is assumed to be a stationary random process with zero mean.

If $z(t)$ is a sample function from a complex-valued gaussian process, the optimum demodulator may be implemented either as a filter matched to the waveform $u(t)$ or as a correlator, as illustrated by the block diagrams in Fig. 8.2.2. In the matched filter realization, the sampled output from the matched filter is multiplied by $(2b_i - 1)$, which is obtained from the PN generator at the demodulator when the PN generator is properly synchronized. Since $(2b_i - 1)^2 = 1$ when $b_i = 0$ and $b_i = 1$, the effect of the PN sequence on the received coded bits is thus removed.

In Fig. 8.2.2 we also observe that the cross correlation can be accomplished in either one of two ways. The first, illustrated in Fig. 8.2.2*b*, involves premultiplying $r(t)$ with the waveform $p_i(t)$ generated from the output of the PN generator and then cross-correlating with $u^*(t)$ and sampling the output in each chip interval. The second method, illustrated in Fig. 8.2.2*c*, involves cross correlation with $u^*(t)$ first, sampling the output of the correlator and, then, multiplying this output with $(2b_i - 1)$, which is obtained from the PN generator.

If $z(t)$ is not a gaussian random process, the demodulation methods illustrated in Fig. 8.2.2 are no longer optimum. Nevertheless, we may still use any of these three demodulator structures to demodulate the received signal. When the statistical characteristics of the interference $z(t)$ are unknown *a priori*, this is certainly one possible approach. An alternative method, which is described later, utilizes an adaptive filter prior to the matched filter or correlator to suppress narrowband interference. The rationale for this second method is also described later.

---

†For simplicity, we assume that the channel attenuation $\alpha = 1$ and the phase shift of the channel is zero. Since coherent PSK detection is assumed, any arbitrary channel phase shift is compensated for in the demodulation.

FIGURE 8.2.2
Possible demodulator structures for PN spread spectrum signals.

In Sec. 8.2.1 we derive the error rate performance of the DS spread spectrum system in the presence of wideband and narrowband interference. The derivations are based on the assumption that the demodulator is any of the three equivalent structures shown in Fig. 8.2.2.

8.2.1 Error Rate Performance of the Decoder

Let the unquantized output of the demodulator be denoted as y_j, $1 \leq j \leq n$. First we consider a linear binary (n, k) block code and, without loss of generality, we assume that the all-zero code word is transmitted.

A decoder that employs soft-decision decoding computes the decision variables

$$U_i = \sum_{j=1}^{n} (2c_{ij} - 1) y_j \qquad i = 1, 2, \ldots, 2^k \qquad (8.2.9)$$

where c_{ij} denotes the bit in the jth position of the ith code word. The decision variable corresponding to the all-zero code word is

$$U_1 = 2n\mathscr{E}_c + \sum_{j=1}^{n} (2c_{1j} - 1)(2b_j - 1)\nu_j$$

$$= 2n\mathscr{E}_c - \sum_{j=1}^{n} (2b_j - 1)\nu_j \qquad (8.2.10)$$

where ν_j, $1 \le j \le n$, is the additive noise term corrupting the jth coded bit and $\mathscr{E}_c$ is the chip energy. It is defined as

$$\nu_j = \text{Re}\left\{ \int_0^{T_c} u^*(t)z[t + (j-1)T_c]\, dt \right\} \qquad j = 1, 2, \ldots, n \quad (8.2.11)$$

Similarly, the decision variable corresponding to code word $\mathbf{C}_m$ having weight w_m is

$$U_m = 2\mathscr{E}_c n \left(1 - \frac{2w_m}{n}\right) + \sum_{j=1}^{n} (2c_{mj} - 1)(2b_j - 1)\nu_j \qquad (8.2.12)$$

Following the procedure used previously in Sec. 5.2.4, we shall determine the probability that U_m exceeds U_1. The difference between U_1 and U_m is

$$D = U_1 - U_m$$

$$= 4\mathscr{E}_c w_m - 2 \sum_{j=1}^{n} c_{mj}(2b_j - 1)\nu_j \qquad (8.2.13)$$

Since the code word $\mathbf{C}_m$ has weight w_m, there are w_m nonzero components in the summation of noise terms contained in (8.2.13). We shall assume that the minimum distance of the code is sufficiently large so that we can invoke the central limit theorem to the summation of noise components. This assumption is valid for PN spread spectrum signals which have a bandwidth expansion of 20 or more.† Thus the summation of noise components is modeled as a gaussian random variable. Since $E(2b_j - 1) = 0$ and $E(\nu_j) = 0$, the mean of the second term in (8.2.13) is also zero.

The variance is

$$\sigma_m^2 = 4 \sum_{j=1}^{n} \sum_{i=1}^{n} c_{mi}c_{mj}E\left[(2b_j - 1)(2b_i - 1)\right] E(\nu_i\nu_j) \qquad (8.2.14)$$

The sequence of binary digits from the PN generator are assumed to be uncorrelated. Hence

$$E\left[(2b_j - 1)(2b_i - 1)\right] = \delta_{ij} \qquad (8.2.15)$$

---

†Typically the bandwidth expansion factor in a spread spectrum signal is of the order of 100 and higher.

and

$$\sigma_m^2 = 4w_m E(v^2) \tag{8.2.16}$$

where $E(v^2)$ is the second moment of any one element from the set $\{v_j\}$. This moment is easily evaluated to yield

$$E(v^2) = \int_0^{T_c} \int_0^{T_c} u^*(t) u(\tau) \phi_{zz}(t-\tau) \, dt \, d\tau$$

$$= \int_{-\infty}^{\infty} |U(f)|^2 \Phi_{zz}(f) \, df \tag{8.2.17}$$

where $\phi_{zz}(\tau) = \frac{1}{2} E[z^*(t) z(t+\tau)]$ is the autocorrelation function and $\Phi_{zz}(f)$ is the power spectral density of the interference $z(t)$.

We observe that when the interference is spectrally flat within the bandwidth† occupied by the transmitted signal, i.e.,

$$\Phi_{zz}(f) = J_0 \qquad |f| \le \frac{W}{2} \tag{8.2.18}$$

the second moment in (8.2.17) is $E(v^2) = 2\mathscr{E}_c J_0$ and, hence, the variance of the interference term in (8.2.16) becomes

$$\sigma_m^2 = 8\mathscr{E}_c J_0 w_m \tag{8.2.19}$$

In this case the probability that $D < 0$ is

$$P_2(m) = \frac{1}{2} \operatorname{erfc}\left(\sqrt{\frac{\mathscr{E}_c}{J_0}} w_m\right) \tag{8.2.20}$$

But the energy per coded bit $\mathscr{E}_c$ may be expressed in terms of the energy per information bit $\mathscr{E}_b$ as

$$\mathscr{E}_c = \frac{k}{n} \mathscr{E}_b = R_c \mathscr{E}_b \tag{8.2.21}$$

With this substitution, (8.2.20) becomes

$$P_2(m) = \frac{1}{2} \operatorname{erfc}\left(\sqrt{\frac{\mathscr{E}_b}{J_0} R_c w_m}\right)$$

$$= \frac{1}{2} \operatorname{erfc}\left(\sqrt{\gamma_b R_c w_m}\right) \tag{8.2.22}$$

where $\gamma_b = \mathscr{E}_b / J_0$ is the SNR per information bit. Finally, the code word error probability may be upper-bounded by the union bound as

$$P_m \le \frac{1}{2} \sum_{m=2}^{M} \operatorname{erfc}\left(\sqrt{\gamma_b R_c w_m}\right) \tag{8.2.23}$$

---

†If the bandwidth of the bandpass channel is W, the bandwidth of the equivalent low-pass channel is $W/2$.

where $M = 2^k$. Note that this expression is identical to the probability of a code word error for soft-decision decoding of a linear binary block code in an AWGN channel.

Although we have considered a binary block code in the derivation given above, the procedure is similar for an (n, k) convolutional code. The result of such a derivation is the following upper bound on the equivalent bit error probability:

$$P_b \le \frac{1}{2k} \sum_{d=d_{\text{free}}}^{\infty} \beta_d \, \text{erfc}\left(\sqrt{\gamma_b R_c d}\right) \tag{8.2.24}$$

The set of coefficients $\{\beta_d\}$ are obtained from an expansion of the derivative of the transfer function $T(D, N)$, as previously described in Sec. 5.3.3.

Next, we consider a narrowband interference centered at the carrier (at dc for the equivalent lowpass signal). We may fix the total (average) jamming power to $J_{\text{av}} = J_0 W$, where J_0 is the value of the power spectral density of an equivalent wideband interference (jamming signal). The narrowband interference is characterized by the power spectral density

$$\Phi_{zz}(f) = \begin{cases} \dfrac{J_{\text{av}}}{W_1} = \dfrac{J_0 W}{W_1} & |f| \le \dfrac{W_1}{2} \\[2ex] 0 & |f| > \dfrac{W_1}{2} \end{cases} \tag{8.2.25}$$

where $W \gg W_1$.

Substitution of (8.2.25) for $\Phi_{zz}(f)$ into (8.2.17) yields

$$E(\nu^2) = \frac{J_{\text{av}}}{W_1} \int_{-W_1/2}^{W_1/2} |U(f)|^2 \, df \tag{8.2.26}$$

The value of $E(\nu^2)$ depends on the spectral characteristics of the pulse $u(t)$. In the following examples we consider two special cases.

Example 8.2.1. Suppose that $u(t)$ is a rectangular pulse as shown in Fig. 8.2.3a and $|U(f)|^2$ is the corresponding energy density spectrum shown in Fig. 8.2.3b. For the narrowband interference given by (8.2.26), the variance of the total interference is

$$\sigma_m^2 = 4 w_m E(\nu^2)$$

$$= \frac{8 \mathscr{E}_c w_m T_c J_{\text{av}}}{W_1} \int_{-W_1/2}^{W_1/2} \left(\frac{\sin \pi f T_c}{\pi f T_c}\right)^2 df$$

$$= \frac{8 \mathscr{E}_c w_m J_{\text{av}}}{W_1} \int_{-\beta/2}^{\beta/2} \left(\frac{\sin \pi x}{\pi x}\right)^2 dx \tag{8.2.27}$$

where $\beta = W_1 T_c$. Figure 8.2.4 illustrates the value of this integral for $0 \le \beta \le 1$. In any case, an upper bound on the value of the integral is $\beta = W_1 T_c$. Hence $\sigma_m^2 \le 8 \mathscr{E}_c w_m T_c J_{\text{av}}$.

In the limit as W_1 becomes zero, the interference becomes an impulse at the carrier. In this case the interference is a pure frequency tone and it is usually called

(a)

(b)

FIGURE 8.2.3
Rectangular pulse and its energy density spectrum.

FIGURE 8.2.4
Plot of the value of the integral in (8.2.27).

a *CW jamming signal*. The power spectral density is

$$\Phi_{zz}(f) = J_{av}\delta(f) \qquad (8.2.28)$$

and the corresponding variance for the decision variable $D = U_1 - U_m$ is

$$\sigma_m^2 = 4w_m J_{av}|U(0)|^2$$
$$= 8w_m \mathscr{E}_c T_c J_{av} \qquad (8.2.29)$$

The probability of a code word error for CW jamming is upper-bounded as

$$P_M \le \frac{1}{2}\sum_{m=2}^{M}\text{erfc}\left(\sqrt{\frac{\mathscr{E}_c}{J_{av}T_c}w_m}\right) \qquad (8.2.30)$$

But $\mathscr{E}_c = R_c\mathscr{E}_b$. Furthermore, $T_c \approx 1/W$ and $J_{av}/W = J_0$. Therefore (8.2.30) may be expressed as

$$P_M \le \frac{1}{2}\sum_{m=2}^{M}\text{erfc}\left(\sqrt{\frac{\mathscr{E}_b}{J_0}R_c w_m}\right) \qquad (8.2.31)$$

which is the result obtained previously for a broad band interference. This result indicates that a CW jammer has the same effect on performance as an equivalent broadband jammer. This equivalence is discussed further below.

Example 8.2.2. Let us determine the performance of the DS spread spectrum system in the presence of a CW jammer of average power J_{av} when the transmitted signal pulse $u(t)$ is one-half cycle of a sinusoid as illustrated in Fig. 8.2.5, i.e.,

$$u(t) = \sqrt{\frac{4\mathscr{E}_c}{T_c}}\sin\frac{\pi t}{T_c} \qquad 0 \le t \le T_c \qquad (8.3.32)$$

The variance of the interference of this pulse is

$$\sigma_m^2 = 4w_m J_{av}|U(0)|^2$$
$$= \frac{64}{\pi^2}\mathscr{E}_c T_c J_{av} w_m \qquad (8.2.33)$$

Hence the upper bound on the code word error probability is

$$P_M \le \frac{1}{2}\sum_{m=2}^{M}\text{erfc}\left(\sqrt{\frac{\pi^2}{8}\frac{\mathscr{E}_b}{J_{av}T_c}R_c w_m}\right) \qquad (8.2.34)$$

FIGURE 8.2.5
A sinusoidal signal pulse.

We observe that the performance obtained with this pulse is 0.9 dB better than that obtained with a rectangular pulse. Recall that this pulse shape when used in offset QPSK results in an MSK signal. MSK modulation is frequently used in DS spread spectrum systems.

The processing gain and the jamming margin. An interesting interpretation of the performance characteristics for the DS spread spectrum signal is obtained by expressing the signal energy per bit $\mathscr{E}_b$ in terms of the average power. That is, $\mathscr{E}_b = S_{av}T_b$, where S_{av} is the average received signal power and T_b is the bit interval. Let us consider the performance obtained in the presence of CW jamming for the rectangular pulse treated in Example 8.2.1. When we substitute for $\mathscr{E}_b$ and J_0 in (8.2.31), we obtain

$$P_M \leq \frac{1}{2} \sum_{m=2}^{M} \text{erfc}\left(\sqrt{\frac{S_{av}}{J_{av}}\frac{T_b}{T_c}R_c w_m}\right) = \frac{1}{2}\sum_{m=2}^{M}\text{erfc}\left(\sqrt{\frac{S_{av}}{J_{av}}L_c R_c w_m}\right) \quad (8.2.35)$$

where L_c is the number of chips per information bit and S_{av}/J_{av} is the signal-to-jamming power ratio.

An identical result is obtained with broadband jamming for which the performance is given by (8.2.23). For the signal energy per bit we have

$$\mathscr{E}_b = S_{av}T_b = \frac{S_{av}}{R} \quad (8.2.36)$$

where R is the information rate in bits per second. The value of the power spectral density for the jamming signal may be expressed as

$$J_0 = \frac{J_{av}}{W} \quad (8.2.37)$$

Using the relations in (8.2.36) and (8.2.37), the ratio $\mathscr{E}_b/J_0$ may be expressed as

$$\frac{\mathscr{E}_b}{J_0} = \frac{S_{av}/R}{J_{av}/W} = \frac{W/R}{J_{av}/S_{av}} \quad (8.2.38)$$

The ratio J_{av}/S_{av} is the jamming-to-signal power ratio, which is usually less than unity. The ratio $W/R = T_b/T_c = B_e = L_c$ is just the bandwidth expansion factor, or, equivalently, the number of chips per information bit. This ratio is usually called the *processing gain* of the DS spread spectrum system. It represents the advantage gained over the jammer that is obtained by expanding the bandwidth of the transmitted signal. If we interpret $\mathscr{E}_b/J_0$ as the SNR required to achieve a specified error rate performance and W/R as the available bandwidth expansion factor, the ratio J_{av}/S_{av} is called the *jamming margin* of the DS spread spectrum system. In other words, the jamming margin is the largest value that the ratio J_{av}/S_{av} can take and still satisfy the specified error probability.

The performance of a soft-decision decoder for a linear (n, k) binary code, expressed in terms of the processing gain and the jamming margin, is

$$P_m \leq \frac{1}{2}\sum_{m=2}^{M}\text{erfc}\left(\sqrt{\frac{W/R}{J_{av}/S_{av}}R_c w_m}\right) \leq \frac{M-1}{2}\text{erfc}\left(\sqrt{\frac{W/R}{J_{av}/S_{av}}R_c d_{min}}\right) \quad (8.2.39)$$

In addition to the processing gain W/R and J_{av}/S_{av}, we observe that the performance depends on a third factor, namely, $R_c w_m$. This factor is the *coding gain*. A lower bound for this factor is $R_c\, d_{min}$. Thus the jamming margin achieved by the DS spread spectrum signal depends in the processing gain and the coding gain.

Uncoded DS spread spectrum signals. The performance results given above for DS spread spectrum signals generated by means of an (n, k) code may be specialized to a trivial type of code, namely, a binary repetition code. For this case, $k = 1$ and the weight of the nonzero code word is $w = n$. Thus $R_c w = 1$ and, hence, the performance of the binary signaling system reduces to

$$P_2 = \tfrac{1}{2}\,\text{erfc}\left(\sqrt{\frac{\mathscr{E}_b}{J_0}}\right)$$

$$= \tfrac{1}{2}\,\text{erfc}\left(\sqrt{\frac{W/R}{J_{av}/S_{av}}}\right) \tag{8.2.40}$$

We note that the trivial (repetition) code gives no coding gain. It does result in a processing gain of W/R.

Example 8.2.3. Suppose that we wish to achieve an error rate performance of 10^{-6} or less with an uncoded DS spread spectrum system. The available bandwidth expansion factor is $W/R = 1000$. Let us determine the jamming margin.

The $\mathscr{E}_b/J_0$ required to achieve a bit error probability of 10^{-6} with uncoded binary PSK is 10.5 dB. The processing gain is $10\log_{10} 1000 = 30$ dB. Hence the maximum jamming-to-signal power that can be tolerated, i.e., the jamming margin, is

$$10\log_{10}\frac{J_{av}}{S_{av}} = 30 - 10.5 = 19.5\text{ dB}$$

Since this is the jamming margin achieved with an uncoded DS spread spectrum system, it may be increased by coding the information sequence.

There is another way to view the modulation and demodulation processes for the uncoded (repetition code) DS spread spectrum system. At the modulator, the signal waveform generated by the repetition code with rectangular pulses, for example, is identical to a unit amplitude rectangular pulse $s(t)$ of duration T_b or its negative, depending on whether the information bit is 1 or 0, respectively. This may be seen from (8.2.7) where the coded chips $\{c_i\}$ within a single information bit are either all 1s or all 0s. The PN sequence multiplies either $s(t)$ or $-s(t)$. Thus, when the information bit is a 1, the L_c PN chips generated by the PN generator are transmitted with the same polarity. On the other hand, when the information bit is a 0, the L_c PN chips when multiplied by $-s(t)$ are reversed in polarity.

The demodulator for the repetition code, implemented as a correlator, is illustrated in Fig. 8.2.6. We observe the integration interval in the integrator is

FIGURE 8.2.6
Correlation-type demodulator for a repetition code.

the bit interval T_b. Thus the decoder for the repetition code is eliminated and its function is subsumed in the demodulator.

Now let us assess qualitatively the effect of this demodulation process on the interference $z(t)$. The multiplication of $z(t)$ by the output of the PN generator, which is expressed as

$$w(t) = \sum_i (2b_i - 1) p(t - iT_c)$$

yields

$$v(t) = w(t) z(t)$$

The waveforms $w(t)$ and $z(t)$ are statistically independent random processes each with zero mean and autocorrelation functions $\phi_{ww}(\tau)$ and $\phi_{zz}(\tau)$, respectively. The product $v(t)$ is also a random process having an autocorrelation function equal to the product of $\phi_{ww}(\tau)$ with $\phi_{zz}(\tau)$. Hence the power spectral density of the process $v(t)$ is equal to the convolution of the power spectral density of $w(t)$ with the power spectral density of $z(t)$.

The effect of convolving the two spectra is to spread the power in bandwidth. Since the bandwidth of $w(t)$ occupies the available channel bandwidth W, the result of convolution of the two spectra is to spread the power spectral density of $z(t)$ over the frequency band of width W. If $z(t)$ is a narrowband process, i.e., its power spectral density has a width much less than W, the power spectral density of the process $v(t)$ will occupy a bandwidth equal to at least W.

The integrator used in the cross correlation shown in Fig. 8.2.6 has a bandwidth approximately equal to $1/T_b$. Since $1/T_b \ll W$, only a fraction of the total interference power appears at the output of the correlator. This fraction is approximately equal to the ratio of bandwidths $1/T_b$ to W. That is,

$$\frac{1/T_b}{W} = \frac{1}{WT_b} = \frac{T_c}{T_b} = \frac{1}{L_c}$$

In other words, the multiplication of the interference with the signal from the PN generator spreads the interference to the signal bandwidth W and the narrowband integration following the multiplication sees only the fraction $1/L_c$ of the total interference. Thus the performance of the uncoded DS spread spectrum system is enhanced by the processing gain L_c.

Linear code concatenated with a binary repetition code. As illustrated above, a binary repetition code provides a margin against an interference or jamming signal but yields no coding gain. To obtain an improvement in performance, we may use a linear (n_1, k) block or convolutional code, where $n_1 \le n = kL_c$. One possibility is to select $n_1 < n$ and to repeat each code bit n_2 times such that $n = n_1 n_2$. Thus we can construct a linear (n_1, k) code by concatenating the (n_1, k) code with a binary $(n_2, 1)$ repetition code. This may be viewed as a trivial form of code concatenation where the outer code is the (n_1, k) code and the inner code is the repetition code.

Since the repetition code yields no coding gain, the coding gain achieved by the combined code must reduce to that achieved by the (n_1, k) outer code. It is easily demonstrated that this is indeed the case. The coding gain of the overall combined code is

$$R_c w_m = \frac{k}{n} w_m \qquad m = 2, 3, \ldots, 2^k$$

But the weights $\{w_m\}$ for the combined code may be expressed as

$$w_m = n_2 w_m^0$$

where $\{w_m^0\}$ are the weights of the outer code. Therefore the coding gain of the combined code is

$$R_c w_m = \frac{k}{n_1 n_2} n_2 w_m^0 = \frac{k}{n_1} w_m^0 = R_c^0 w_m^0 \qquad (8.2.41)$$

which is just the coding gain obtained from the outer code.

A coding gain is also achieved if the (n_1, k) outer code is decoded using hard decisions. The probability of a bit error obtained with the $(n_2, 1)$ repetition code (based on soft-decision decoding) is

$$p = \tfrac{1}{2} \operatorname{erfc}\left(\sqrt{\frac{n_2 \mathcal{E}_c}{J_0}}\right) = \tfrac{1}{2} \operatorname{erfc}\left(\sqrt{\frac{\mathcal{E}_b}{J_0} R_c^0}\right)$$

$$= \tfrac{1}{2} \operatorname{erfc}\left(\sqrt{\frac{W/R}{J_{av}/S_{av}} R_c^0}\right) \qquad (8.2.42)$$

Then the code word error probability for a linear (n_1, k) block code is upper-bounded as

$$P_M \le \sum_{m=t+1}^{n_1} \binom{n_1}{m} p^m (1-p)^{n_1 - m} \qquad (8.2.43)$$

where $t = (d_{\min} - 1)/2$ or as

$$P_M \le \sum_{m=2}^{M} [4p(1-p)]^{w_m^0/2} \qquad (8.2.44)$$

where the latter is a Chernoff bound. For an (n_1, k) binary convolutional code, the upper bound on the bit error probability is

$$P_b \le \sum_{d=d_{\text{free}}}^{\infty} \beta_d P_2(d) \qquad (8.2.45)$$

where $P_2(d)$ is defined by (5.3.28) for d odd and by (5.3.29) for d even.

Concatenated coding for DS spread spectrum systems. It is apparent from the discussion given above that an improvement in performance can be obtained by replacing the repetition code by a more powerful code which will yield a coding gain in addition to the processing gain. Basically, the objective in a DS spread spectrum system is to construct a long, low rate code having a large minimum distance. This may be best accomplished by using code concatenation. When binary PSK is used in conjunction with DS spread spectrum, the elements of a concatenated code word must be expressed in binary form.

Best performance is obtained when soft-decision decoding is used on both the inner and outer code. However, an alternative, which usually results in a reduced complexity for the decoder, is to employ soft-decision decoding on the inner code and hard-decision decoding on the outer code. The expressions for the error rate performance of these decoding schemes depends, in part, on the type of codes (block or convolutional) selected for the inner and outer codes. For example, the concatenation of two block codes may be viewed as an overall long binary (n, k) block code having a performance given by (8.2.39). The performance of other code combinations may also be readily derived. For the sake of brevity, this is left as an exercise for the reader.

8.2.2 Some Applications For DS Spread Spectrum Signals

In this subsection we shall briefly consider the use of coded DS spread spectrum signals for four specific applications. One application is concerned with providing immunity against a jamming signal. In the second, a communication signal is hidden in the background noise by transmitting the signal at a very low power level. The third application is concerned with accommodating a number of simultaneous signal transmissions on the same channel, i.e., CDMA. Finally, the fourth deals with the use of DS spread spectrum for resolving the multipath in a time-dispersive radio channel.

Antijamming application. In Sec. 8.2.1 we derived the error rate performance for a DS spread spectrum signal in the presence of either a narrowband or a wideband jamming signal. As examples to illustrate the performance of a digital communications system in the presence of a jamming signal, we shall select three codes. One is the Golay (24, 12), which is characterized by the weight distribution given in Table 5.2.1 and has a minimum distance $d_{\text{min}} = 8$. The second code is an expurgated Golay (24, 11) obtained by selecting 2048 code words of constant weight 12. Of course this expurgated code is nonlinear. These two codes will be used in conjunction with a repetition code. The third code to be considered is a maximum-length shift-register code.

The error rate performance of the Golay $(24, 12)$ with soft-decision decoding is

$$P_M \leq \frac{1}{2} \left[759 \operatorname{erfc} \left(\sqrt{4 \frac{W/R}{J_{av}/S_{av}}} \right) + 2576 \operatorname{erfc} \left(\sqrt{6 \frac{W/R}{J_{av}/S_{av}}} \right) \right.$$

$$\left. + 759 \operatorname{erfc} \left(\sqrt{8 \frac{W/R}{J_{av}/S_{av}}} \right) + \operatorname{erfc} \left(\sqrt{12 \frac{W/R}{J_{av}/S_{av}}} \right) \right] \quad (8.2.46)$$

where W/R is the processing gain and J_{av}/S_{av} is the jamming margin. Since $n = n_1 n_2 = 12 W/R$ and $n_1 = 24$, each coded bit is, in effect, repeated $n_2 = W/2R$ times. For example, if $W/R = 100$ (a processing gain of 20 dB), the block length of the repetition code is $n_2 = 50$.

If hard-decision decoding is used, the probability of error for a coded bit is

$$p = \frac{1}{2} \operatorname{erfc} \left(\sqrt{\frac{W/2R}{J_{av}/S_{av}}} \right) \quad (8.2.47)$$

and the corresponding probability of a code word error is upper-bounded as

$$P_M \leq \sum_{m=4}^{24} \binom{24}{m} p^m (1-p)^{24-m} \quad (8.2.48)$$

As an alternative, we may use the Chernoff bound for hard-decision decoding, which is

$$P_M \leq 759[4p(1-p)]^4 + 2576[4p(1-p)]^6$$

$$+ 759[4p(1-p)]^8 + [4p(1-p)]^{12} \quad (8.2.49)$$

Figure 8.2.7 illustrates the performance of the Golay $(24, 12)$ as a function of the jamming margin J_{av}/S_{av}, with the processing gain as a parameter. The Chernoff bound was used to compute the error probability for hard-decision decoding. The error probability for soft-decision decoding is dominated by the term

$$\frac{759}{2} \operatorname{erfc} \left(\sqrt{4 \frac{W/R}{J_{av}/S_{av}}} \right)$$

and that for hard-decision decoding is dominated by the term $759[4p(1-p)]^4$. Hence the coding gain for soft-decision decoding† is at most $10 \log 4 = 6$ dB. We

---

†The coding gain is less than 6 dB due to the multiplicative factor of 759 which increases the error probability relative to the performance of the binary uncoded system.

FIGURE 8.2.7
Performance of Golay codes used in DS spread spectrum signal.

note that the two curves corresponding to $W/R = 1000$ (30 dB) are identical in shape to the ones for $W/R = 100$ (20 dB), except that the latter are shifted by 10 dB to the right relative to the former. This shift is simply the difference in processing gain between these two DS spread spectrum signals.

The error rate performance of the expurgated Golay $(24, 11)$ is upper-bounded as

$$P_M \leq \frac{2047}{2} \, \text{erfc} \left(\sqrt{\frac{W/R}{J_{av}/S_{av}} \frac{11}{2}} \right) \qquad (8.2.50)$$

for soft-decision decoding and as†

$$P_M \leq 2047[4p(1-p)]^6 \qquad (8.2.51)$$

for hard-decision decoding, where p is given as

$$p = \frac{1}{2} \operatorname{erfc}\left(\sqrt{\frac{W/R}{J_{av}/S_{av}} \frac{11}{24}}\right) \qquad (8.2.52)$$

The performance characteristics of this code are also plotted in Fig. 8.2.7 for $W/R = 100$. We observe that this expurgated Golay $(24, 11)$ code performs about 1 dB better than the Golay $(24, 12)$ code.

Instead of using a block code concatenated with a low-rate $(1/n_2)$ repetition code, let us consider using a single low-rate code. A particularly suitable set of low-rate codes is the set of maximum-length shift-register codes described in Sec. 5.2.3. We recall that for this set of codes

$$\begin{aligned}
(n, k) &= (2^m - 1, m) \\
d_{\min} &= 2^{m-1}
\end{aligned} \qquad (8.2.53)$$

All code words except the all-zero code word have an identical weight of 2^{m-1}. Hence the error rate for soft-decision decoding is upper-bounded as‡

$$\begin{aligned}
P_M &\leq \frac{M-1}{2} \operatorname{erfc}\left(\sqrt{\frac{W/R}{J_{av}/S_{av}} R_c d_{\min}}\right) \\
&\leq 2^{m-1} \operatorname{erfc}\left(\sqrt{\frac{W/R}{J_{av}/S_{av}} \frac{m2^{m-1}}{2^m - 1}}\right) \leq 2^{m-1} \exp\left(-\frac{W/R}{J_{av}/S_{av}} \frac{m2^{m-1}}{2^m - 1}\right) \quad (8.2.54)
\end{aligned}$$

For moderate values of m, $R_c d_{\min} \approx m/2$ and, hence, $(8.2.54)$ may be expressed as

$$P_M \leq 2^{m-1} \operatorname{erfc}\left(\sqrt{\frac{W/R}{J_{av}/S_{av}} \frac{m}{2}}\right) \leq 2^{m-1} \exp\left(-\frac{m}{2} \frac{W/R}{J_{av} S_{av}}\right) \quad (8.2.55)$$

Hence the coding gain is at most $10 \log m/2$.

For example, if we select $m = 10$, then $n = 2^{10} - 1 = 1023$. Since $n = kW/R = mW/R$, it follows that $W/R \approx 102$. Thus, we have a processing gain of about 20 dB and a coding gain of 7 dB. This performance is comparable to that obtained with the expurgated Golay $(24, 11)$ code. Higher coding gains can be achieved with larger values of m.

---

†We remind the reader that the union bound is not very tight for large signal sets.

‡The $M = 2^m$ waveforms generated by a maximum-length shift-register code form a simplex set (see Prob. 5.18). The exact expression for the error probability, given in Sec. 4.2.3, may be used for large values of M where the union bound becomes very loose.

If hard-decision decoding is used for the maximum-length shift-register codes, the error rate is upper-bounded by the Chernoff bound as

$$P_M \le (M-1)[4p(1-p)]^{d_{min}/2} = (2^m - 1)[4p(1-p)]^{2^{m-2}} \quad (8.2.56)$$

where p is given as

$$p = \tfrac{1}{2} \operatorname{erfc} \left(\sqrt{\frac{W/R}{J_{av}/S_{av}} R_c} \right) = \tfrac{1}{2} \operatorname{erfc} \left(\sqrt{\frac{W/R}{J_{av}/S_{av}} \frac{m}{2^m - 1}} \right) \quad (8.2.57)$$

For $m = 10$, the code word error rate P_M is comparable to that obtained with the expurgated Golay $(24, 11)$ code for hard-decision decoding.

The results given above illustrate the performance that can be obtained with a single level of coding. Greater coding gains can be achieved with concatenated codes.

Low-detectability signal transmission. In this application the signal is purposely transmitted at a very low power level relative to the background channel noise and thermal noise that is generated in the front end of the receiver. If the DS spread spectrum signal occupies a bandwidth W and the spectral density of the additive noise is N_0 watts per hertz, the average noise power in the bandwidth W is $N_{av} = WN_0$.

The average received signal power at the intended receiver is S_{av}. If we wish to hide the presence of the signal from receivers that are in the vicinity of the intended receiver, the signal is transmitted at a low power level such that $S_{av}/N_{av} \ll 1$. The intended receiver can recover the information-bearing signal with the aid of the processing gain and the coding gain. However, any other receiver which has no prior knowledge of the PN sequence is unable to take advantage of the processing gain and the coding gain. Hence the presence of the information-bearing signal is difficult to detect. We say that the signal has a *low probability* of being *intercepted* (LPI) and it is called an *LPI signal*.

The probability of error results given in Sec. 8.2.1 apply as well to the demodulation and decoding of LPI signals at the intended receiver.

Code division multiple access. The enhancement in performance obtained from a DS spread spectrum signal through the processing gain and coding gain can be used to enable many DS spread spectrum signals to occupy the same channel bandwidth provided that each signal has its own distinct PN sequence. Thus it is possible to have several users transmit messages simultaneously over the same channel bandwidth. This type of digital communication in which each user (transmitter-receiver pair) has a distinct PN code for transmitting over a common channel bandwidth is called either *code division multiple access* (CDMA) or *spread spectrum multiple access* (SSMA).

In the demodulation of each PN signal, the signals from the other simultaneous users of the channel appear as an additive interference. The level of interference varies depending on the number of users at any given time. A major advantage of CDMA is that a large number of users can be accommodated if each transmits messages for a short period of time. In such a multiple access system it is relatively easy either to add new users or to decrease the number of users without disrupting the system.

Let us determine the number of simultaneous signals that can be supported in a CDMA system. For simplicity, we assume that all signals have identical average powers. Thus, if there are N_u simultaneous users, the desired signal-to-noise interference power ratio at a given receiver is

$$\frac{S_{av}}{J_{av}} = \frac{S_{av}}{(N_u - 1)S_{av}} = \frac{1}{N_u - 1} \qquad (8.2.58)$$

Hence the performance for soft-decision decoding at the given receiver is upper-bounded as

$$P_M \le \frac{1}{2} \sum_{m=2}^{M} \text{erfc}\left(\sqrt{\frac{W/R}{N_u - 1} R_c w_m}\right) \le \frac{M-1}{2} \text{erfc}\left(\sqrt{\frac{W/R}{N_u - 1} R_c d_{min}}\right)$$

$$(8.2.59)$$

As an example, suppose that the desired level of performance (error probability of 10^{-6}) is achieved when

$$\frac{W/R}{N_u - 1} R_c d_{min} = 20$$

Then the maximum number of users that can be supported in the CDMA system is

$$N_u = \frac{W/R}{20} R_c d_{min} + 1 \qquad (8.2.60)$$

If $W/R = 100$ and $R_c d_{min} = 4$, as obtained with the Golay (24, 12) code, the maximum number is $N_u = 21$. If $W/R = 1000$ and $R_c d_{min} = 4$, this number becomes $N_u = 201$.

In determining the maximum number of simultaneous users of the channel, we implicitly assumed that the PN code sequences are mutually orthogonal and the interference from other users adds on a power basis only. However, orthogonality among a number of PN code sequences is not easily achieved, especially if the number of PN code sequences required is large. In fact, the selection of a good set of PN sequences for a CDMA system is an important problem that has received considerable attention in the technical literature. We shall briefly discuss this problem in Sec. 8.2.3.

Communication over channels with multipath. As a final application of DS spread spectrum signals, we consider the problem of transmitting information over time-invariant multipath channels. We treated this problem previously in Chap. 7, and we observed that by transmitting a signal having a bandwidth W that exceeds the coherence bandwidth of the channel, we can resolve the channel multipath with a time resolution of $1/W$.

The RAKE receiver described in Sec. 7.5 represents the realization of the matched filter for the received signal components. The wideband signals used to transmit the information are usually DS spread spectrum signals. If orthogonal signaling is employed, two PN sequences are generated, one corresponding to a 0 and the other corresponding to a 1. On the other hand, if PSK or DPSK is employed as the modulation technique, only a single PN sequence is required. The cross correlations performed at each tap of the RAKE receiver, as illustrated in Figs. 7.5.3 through 7.5.7, are simply cross correlations of the received DS spread spectrum signal with the locally generated PN reference waveforms. Thus PN spread spectrum signals are used to resolve the multipath introduced by the channel. By demodulating and combining the signals in the resolvable components as performed with a RAKE correlator, the error rate of the system is reduced.

8.2.3 Effect of Pulsed Interference on DS Spread Spectrum Systems

Thus far we have considered the effect of continuous interference or jamming on a DS spread spectrum signal. We have observed that the processing gain and coding gain provide a means for overcoming the detrimental effects of this type of interference. However, there is a jamming threat that has a dramatic effect on the performance of a DS spread spectrum system. That jamming signal consists of pulses of spectrally flat noise that covers the entire signal bandwidth W. This is usually called *pulsed interference* or *partial-time jamming*.

Suppose the jammer has an average power J_{av} in the signal bandwidth W. Hence $J_0 = J_{av}/W$. Instead of transmitting continuously, the jammer transmits pulses at a power J_{av}/α for α percent of the time, i.e., the probability that the jammer is transmitting at a given instant in α. For simplicity, we assume that an interference pulse spans an integral number of signaling intervals and, thus, it affects an integral number of bits. When the jammer is not transmitting, the transmitted bits are assumed to be received error-free, and when the jammer is transmitting, the probability of error for an uncoded DS spread spectrum system is $\frac{1}{2}\,\text{erfc}\,(\sqrt{\alpha\mathscr{E}_b/J_0}\,)$. Hence the average probability of a bit error is

$$P_2(\alpha) = \frac{\alpha}{2}\,\text{erfc}\left(\sqrt{\alpha\mathscr{E}_b/J_0}\right) = \frac{\alpha}{2}\,\text{erfc}\left(\sqrt{\frac{\alpha W/R}{J_{av}/S_{av}}}\right) \quad (8.2.61)$$

The jammer selects the duty cycle α to maximize the error probability. Upon differentiating (8.2.61) with respect to α, we find the worst-case pulse jamming occurs when

$$\alpha^* = \begin{cases} \dfrac{0.71}{\mathscr{E}_b/J_0} & \dfrac{\mathscr{E}_b}{J_0} \geq 0.71 \\[3mm] 1 & \dfrac{\mathscr{E}_b}{J_0} < 0.71 \end{cases} \quad (8.2.62)$$

and the corresponding error probability is

$$
P_2 = \begin{cases} \dfrac{0.083}{\mathscr{E}_b/J_0} = \dfrac{0.083 J_{av}/S_{av}}{W/R} & \dfrac{\mathscr{E}_b}{J_0} > 0.71 \\[3mm] \dfrac{1}{2}\,\text{erfc}\,\sqrt{\dfrac{W/R}{J_{av}/S_{av}}} & \dfrac{\mathscr{E}_b}{J_0} < 0.71 \end{cases} \tag{8.2.63}
$$

The error rate performance given by (8.2.61) for $\alpha = 1.0$, 0.1, and 0.01 along with the worst-case performance based on α^* is plotted in Fig. 8.2.8. By comparing the error rate for continuous gaussian noise jamming with worst case pulse jamming we observe a large difference in performance, which is approximately 40 dB at an error rate of 10^{-6}.

We should point out that the analysis given above applies when the jammer pulse duration is equal to or larger than the bit duration. In addition, we should indicate that practical considerations may prohibit the jammer from achieving small values of α at a high peak power. Nevertheless, the error probability given by (8.2.63) serves as an upper bound on the performance of the uncoded binary PSK in worst-case pulse jamming. Clearly, the performance of the DS spread spectrum system in the presence of such jamming is extremely poor.

If we simply add coding to the DS spread spectrum system, the improvement over the uncoded system is the coding gain. Thus, $\mathscr{E}_b/J_0$ is reduced by the coding gain, which in most cases is limited to less than 10 dB. The reason for the poor performance is that the jamming signal pulse duration may be selected to affect many consecutive coded bits when the jamming signal is turned on. Consequently, the code word error probability is high due to the burst characteristics of the jammer.

In order to improve the performance we should interleave the coded bits prior to transmission over the channel. The effect of the interleaving, as previously discussed in Sec. 5.2.10, is to make the coded bits that are hit by the jammer statistically independent.

The block diagram of the digital communication system that includes interleaving/deinterleaving is shown in Fig. 8.2.9. Also shown in this figure is the possibility that the receiver knows the jammer state, i.e., it knows when the jammer is on or off. Knowledge of the jammer state (called *side information*) is sometimes available from channel measurements of noise power levels in adjacent frequency bands. In our treatment, we consider two extreme cases, namely, no knowledge of the jammer state or complete knowledge of the jammer state. In any case, the random variable ζ representing the jammer state is characterized by the probabilities

$$
P(\zeta = 1) = \alpha \qquad P(\zeta = 0) = 1 - \alpha
$$

When the jammer is on, the channel is modeled as an AWGN with power spectral $N_0 = J_0/\alpha = J_{av}/\alpha W$, and when the jammer is off there is no noise in the

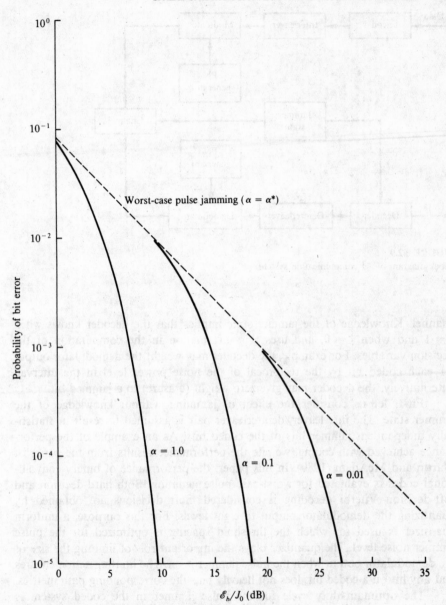

FIGURE 8.2.8
Performance of DS binary PSK with pulse jamming.

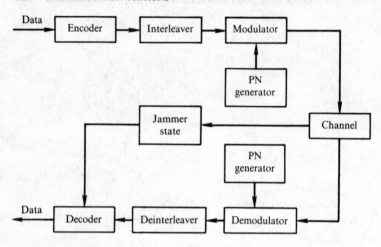

FIGURE 8.2.9
Block diagram of AJ communication system.

channel. Knowledge of the jammer state implies that the decoder knows when $\zeta = 1$ and when $\zeta = 0$, and uses this information in the computation of the decision variables. For example, the decoder may weight the demodulator output for each coded bit by the reciprocal of the noise power level in the interval. Alternatively, the decoder may give zero weight (erasure) to a jammed bit.

First, let us consider the effect of jamming without knowledge of the jammer state. The interleaver/deinterleaver pair is assumed to result in statistically independent jammer hits of the coded bits. As an example of the performance achieved with coding, we cite the performance results from the paper of Martin and McAdam (1980). In this paper, the performance of binary convolutional codes is evaluated for worst-case pulse jamming. Both hard-decision and soft-decision Viterbi decoding is considered. Soft decisions are obtained by quantizing the demodulator output to eight levels. For this purpose, a uniform quantizer is used for which the threshold spacing is optimized for the pulse jammer noise level. The quantizer plays the important role of limiting the size of the demodulator output when the pulse jammer is on. The limiting action ensures that any hit on a coded bit does not heavily bias the corresponding path metrics.

The optimum duty cycle for the pulse jammer in the coded system is generally inversely proportional to the SNR, but its value is different from that given by (8.2.62) for the uncoded system. Figure 8.2.10 graphically illustrates the optimal jammer duty cycle for both hard- and soft-decision decoding of the rate 1/2 convolutional codes. The corresponding error rate results for this worst-case pulse jammer are illustrated in Figs. 8.2.11 and 8.2.12 for rate 1/2 codes with constraint lengths $3 \leq L \leq 9$. For example, we note that at $P_2 = 10^{-6}$, the $L = 7$ convolutional code with soft-decision decoding requires a $\mathscr{E}_b/J_0$ of 7.6 dB,

Pulse jamming
Binary phase shift keying

Hard decisions

Soft decisions (union bound)

FIGURE 8.2.10
Optimal duty cycle for pulse jammer. [From paper by Martin and McAdam (1980). © 1980 IEEE. Reprinted with permission.]

whereas hard-decision decoding requires an $\mathscr{E}_b/J_0$ of 11.7 dB. This 4.1-dB difference in SNR is relatively large and is reminiscent of the similarly large differences between soft- and hard-decision decoding that we observed on a fading channel. With continuous gaussian noise, the corresponding SNR's for an error rate of 10^{-6} are 5 dB for soft-decision decoding and 7 dB for hard-decision decoding. Hence, the worst-case pulse jammer has degraded the performance by 2.6 dB for soft-decision decoding and by 4.7 dB for hard-decision decoding. These levels of degradation increase as the constraint length of the convolutional code is decreased. The important point, however, is that the loss in SNR due to jamming has been reduced from 40 dB for the uncoded system to less than 5 dB for the coded system based on an $L = 7$, rate $1/2$ convolutional code.

A simpler method for evaluating the performance of a coded AJ communication system is to use the cutoff rate parameter R_0 as proposed by Omura and

FIGURE 8.2.11
Performance of rate 1/2 convolutional codes with hard-decision Viterbi decoding binary PSK with optimal pulse jamming. [From paper by Martin and McAdam (1980). © 1980 IEEE. Reprinted with permission.]

Levitt (1982). For example, with binary-coded modulation the cutoff rate may be expressed as

$$R_0 = 1 - \log(1 + D) \tag{8.2.64}$$

where the factor D depends on the channel noise characteristics and the decoder processing. Recall that for binary PSK in an AWGN channel and soft-decision decoding

$$D = e^{-\mathscr{E}_c/N_0} \tag{8.2.65}$$

where $\mathscr{E}_c$ is the energy per coded bit, and for hard-decision decoding

$$D = \sqrt{4p(1 - p)} \tag{8.2.66}$$

where p is the probability of a coded bit error. Here, we have $N_0 \equiv J_0$.

For a coded binary PSK, with pulse jamming, Omura and Levitt (1982) have shown that

$$D = \alpha e^{-\alpha \mathcal{E}_c / N_0} \qquad \begin{array}{l} \text{soft-decision decoding with} \\ \text{knowledge of jammer state} \end{array} \qquad (8.2.67)$$

$$D = \min_{\lambda \geq 0} \left\{ \left[\alpha \exp \left(\lambda^2 \mathcal{E}_c N_0 / \alpha \right) + 1 - \alpha \right] \exp \left(-2\lambda \mathcal{E}_c \right) \right\}$$

$$\begin{array}{l} \text{soft-decision decoding with} \\ \text{no knowledge of jammer state} \end{array} \qquad (8.2.68)$$

$$D = \alpha \sqrt{4p(1 - p)} \qquad \begin{array}{l} \text{hard-decision decoding with} \\ \text{knowledge of the jammer state} \end{array} \qquad (8.2.69)$$

$$D = \sqrt{4\alpha p (1 - \alpha p)} \qquad \begin{array}{l} \text{hard-decision decoding with no} \\ \text{knowledge of the jammer state} \end{array} \qquad (8.2.70)$$

where the probability of error for hard-decision decoding of binary PSK is

$$p = \frac{1}{2} \operatorname{erfc} \left(\sqrt{\frac{\alpha \mathcal{E}_c}{N_0}} \right)$$

The graphs for R_0 as a function of $\mathcal{E}_c / N_0$ are illustrated in Fig. 8.2.13, for the cases given above. Note that these graphs represent the cutoff rate for the worst-case value of $\alpha = \alpha^*$ which maximizes D (minimizes R_0) for each value of $\mathcal{E}_c / N_0$. Furthermore, we note that with soft-decision decoding and no knowledge of the jammer state, $R_0 = 0$. This situation results from the fact that the demodulator output is not quantized.

The graphs in Fig. 8.2.13 may be used to evaluate the performance of coded systems. To demonstrate the procedure, suppose that we wish to determine the SNR required to achieve an error probability of 10^{-6} with coded binary PSK in worst-case pulse jamming. To be specific, we assume that we have a rate $1/2$, $L = 7$ convolutional code. We begin with the performance of the rate $1/2$, $L = 7$ convolutional code with soft-decision decoding in an AWGN channel. At $P_2 = 10^{-6}$, the SNR required is found from Fig. 5.3.21 to be

$$\mathcal{E}_b / N_0 = 5 \text{ dB}$$

Since the code is rate $1/2$, we have

$$\mathcal{E}_c / N_0 = 2 \text{ dB}$$

Now, we go to the graphs in Fig. 8.2.13 and find that for the AWGN channel (reference system) with $\mathcal{E}_c / N_0 = 2$ dB, the corresponding value of the cutoff rate is

$$R_0 = 0.74 \text{ bits/symbol}$$

If we have another channel with different noise characteristics (a worst-case pulse noise channel) but with the same value of the cutoff rate R_0, then the upper bound on the bit error probability is the same, i.e., 10^{-6} in this case. Conse-

830   DIGITAL COMMUNICATIONS

FIGURE 8.2.12
Performance of rate 1/2 convolutional codes with soft-decision Viterbi decoding binary PSK with optimal pulse jamming. [From paper by Martin and McAdam (1980). © 1980 IEEE. Reprinted with permission.]

quently, we can use this rate to determine the SNR required for the worst-case pulse jammer channel. From the graphs in Fig. 8.2.13 we find that

$$
\frac{\mathscr{E}_c}{J_0} = \begin{cases} 10\text{ dB} & \text{hard-decision decoding with} \\ & \text{no knowledge of jammer state} \\ 5\text{ dB} & \text{hard-decision decoding with} \\ & \text{knowledge of jammer state} \\ 3\text{ dB} & \text{soft-decision decoding with} \\ & \text{knowledge of jammer state} \end{cases}
$$

Therefore, the corresponding values of $\mathscr{E}_b/J_0$ for the rate 1/2, $L = 7$ convolutional are 13, 8, and 6 dB, respectively.

Key

(0) Soft-decision decoding in AWGN ($\alpha = 1$)

(1) Soft-decision with jammer state information

(2) Hard-decision with jammer state information

(3) Soft decision with no jammer state information

(4) Hard decision with no jammer state information

FIGURE 8.2.13

The cutoff rate for coded DS binary PSK modulation. [From paper by Omura and Levitt (1982). © 1982 IEEE. Reprinted with permission.]

This general approach may be used to generate error rate graphs for coded binary signals in a worst-case pulse jamming channel by using corresponding error rate graphs for the AWGN channel. The approach we describe above is easily generalized to M-ary coded signals as indicated by Omura and Levitt (1982).

By comparing the cutoff rate for coded DS binary PSK modulation shown in Fig. 8.2.13, we note that for rates below 0.7 there is no penalty in SNR with soft-decision decoding and jammer state information compared with the performance on the AWGN channel ($\alpha = 1$). On the other hand, at $R_0 = 0.7$ there is a 6-dB difference in performance between the SNR in an AWGN channel and that required for hard-decision decoding with no jammer state information. At rates below 0.4, there is no penalty in SNR with hard-decision decoding if the jammer state is unknown. However, there is the expected 2-dB loss in hard-decision decoding compared with soft-decision decoding in the AWGN channel.

8.2.4 Generation of PN Sequences

The generation of PN sequences for spread spectrum applications is a topic that has received considerable attention in the technical literature. We shall briefly discuss the construction of some PN sequences and present a number of

FIGURE 8.2.14
General m-stage shift register with linear feedback.

important properties of the autocorrelation and cross-correlation functions of such sequences. For a comprehensive treatment of this subject, the interested reader may refer to the book by Golomb (1967).

By far the most widely known binary PN sequences are the maximum-length shift-register sequences introduced in Sec. 5.2.3 in the context of coding and suggested again in Sec. 8.2.2 for use as low-rate codes. A maximum-length shift-register sequence, or m-sequence for short, has length $n = 2^m - 1$ bits and is generated by an m-stage shift register with linear feedback as illustrated in Fig. 8.2.14. The sequence is periodic with period n. Each period of the sequence contains 2^{m-1} ones and $2^{m-1} - 1$ zeros.

In DS spread spectrum applications the binary sequence with elements $\{0, 1\}$ is mapped into a corresponding sequence of positive and negative pulses according to the relation

$$p_i(t) = (2b_i - 1)p(t - iT)$$

where $p_i(t)$ is the pulse corresponding to the element b_i in the sequence with elements $\{0, 1\}$. Equivalently, we may say that the binary sequence with elements $\{0, 1\}$ is mapped into a corresponding binary sequence with elements $\{-1, 1\}$. We shall call the equivalent sequence with elements $\{-1, 1\}$ a *bipolar sequence* since it results in pulses of positive and negative amplitudes.

An important characteristic of a periodic PN sequence is its periodic autocorrelation function, which is usually defined in terms of the bipolar sequence as

$$\phi(j) = \sum_{i=1}^{n} (2b_i - 1)(2b_{i+j} - 1) \qquad 0 \leq j \leq n - 1 \qquad (8.2.71)$$

where n is the period. Clearly $\phi(j + rn) = \phi(j)$ for any integer value r.

Ideally, a pseudo-random sequence should have an autocorrelation function with the property that $\phi(0) = n$ and $\phi(j) = 0$ for $1 \leq j \leq n - 1$. In the case of m sequences, the periodic autocorrelation function is

$$\phi(j) = \begin{cases} n & j = 0 \\ -1 & 1 \leq j \leq n - 1 \end{cases} \qquad (8.2.72)$$

For large values of n, i.e., for long m sequences, the size of the off-peak values of $\phi(j)$ relative to the peak value $\phi(j)/\phi(0) = -1/n$ is small and, from a practical viewpoint, inconsequential. Therefore, m sequences are almost ideal when viewed in terms of their autocorrelation function.

In antijamming applications of PN spread spectrum signals, the period of the sequence must be large in order to prevent the jammer from learning the feedback connections of the PN generator. However, this requirement is impractical in most cases because the jammer can determine the feedback connections by observing only $2m$ chips from the PN sequence. This vulnerability of the PN sequence is due to the linearity property of the generator. To reduce the vulnerability to a jammer, the output sequences from several stages of the shift register or the outputs from several distinct m sequences are combined in a nonlinear way to produce a nonlinear sequence which is considerably more difficult for the jammer to learn. Further reduction in vulnerability is achieved by frequently changing the feedback connections and/or the number of stages in the shift register according to some prearranged plan formulated between the transmitter and the intended receiver.

In some applications the cross-correlation properties of PN sequences are as important as the autocorrelation properties. For example, in CDMA each user is assigned a particular PN sequence. Ideally the PN sequences among users should be mutually orthogonal so that the level of interference experienced by any one user from transmissions of other users adds on a power basis. However, the PN sequences used in practice exhibit some correlation.

To be specific, we consider the class of m sequences. It is known [for reference, see Sarwate and Pursley (1980)] that the periodic cross-correlation function between any pair of m sequences of the same period can have relatively large peaks. Table 8.2.1 lists the peak magnitude ϕ_{max} for the periodic cross correlation between pairs of m sequences for $3 \le m \le 12$. Also listed in Table

TABLE 8.2.1
Peak cross correlations of m sequences and Gold sequences

m	$n = 2^m - 1$	Number of m sequences	Peak cross correlation ϕ_{max}	$\phi_{max}/\phi(0)$	$t(m)$	$t(m)/\phi(0)$
3	7	2	5	0.71	5	0.71
4	15	2	9	0.60	9	0.60
5	31	6	11	0.35	9	0.29
6	63	6	23	0.36	17	0.27
7	127	18	41	0.32	17	0.13
8	255	16	95	0.37	33	0.13
9	511	48	113	0.22	33	0.06
10	1023	60	383	0.37	65	0.06
11	2047	176	287	0.14	65	0.03
12	4095	144	1407	0.34	129	0.03

834 DIGITAL COMMUNICATIONS

8.2.1 is the number of m sequences of length $n = 2^m - 1$ for $3 \leq m \leq 12$. As we can see from this table, the number of m sequences of length n increases rapidly with m. We also observe that for most sequences the peak magnitude $\phi_{\max}$ of the cross-correlation function is a large percentage of the peak value of the autocorrelation function.

Such high values for the cross correlations are unacceptable in CDMA. Although it is possible to select a small subset of m sequences that have relatively smaller cross-correlation peak values, the number of sequences in the set is usually too small for CDMA applications.

PN sequences with better periodic cross-correlation properties than m sequences have been given by Gold (1967, 1968) and Kasami (1966). They are derived from m sequences as described below.

Gold and Kasami proved that certain pairs of m sequences of length n exhibit a three-valued cross-correlation function with values $\{-1, -t(m), t(m) - 2\}$ where

$$t(m) = \begin{cases} 2^{(m+1)/2} + 1 & m \text{ odd} \\ 2^{(m+2)/2} + 1 & m \text{ even} \end{cases} \tag{8.2.73}$$

For example, if $m = 10$, $t(10) = 2^6 + 1 = 65$ and the three possible values of the periodic cross-correlation function are $\{-1, -65, 63\}$. Hence the maximum cross correlation for the pair of m sequences is 65 while the peak for the family of 60 possible sequences generated by a 10-stage shift register with different feedback connections is $\phi_{\max} = 383$, about a sixfold difference in peak values. Two m sequences of length n with a periodic cross-correlation function that takes on the possible values $\{-1, -t(m), t(m) - 2\}$ are called *preferred sequences*.

From a pair of preferred sequences, say $\mathbf{a} = [a_1 a_2 \cdots a_n]$ and $\mathbf{b} = [b_1 b_2 \cdots b_n]$, we construct a set of sequences of length n by taking the modulo-2 sum of $\mathbf{a}$ with the n cyclicly shifted versions of $\mathbf{b}$ or vice versa. Thus we obtain n new periodic sequences† with period $n = 2^m - 1$. We may also include the original sequences $\mathbf{a}$ and $\mathbf{b}$ and, thus, we have a total of $n + 2$ sequences. The $n + 2$ sequences constructed in this manner are called *Gold sequences*.

Example 8.2.4. Let us consider the generation of Gold sequences of length $n = 31 = 2^5 - 1$. As indicated above for $m = 5$, the cross-correlation peak is

$$t(5) = 2^3 + 1 = 9$$

Two preferred sequences that may be obtained from Peterson and Weldon (1972) are described by the polynomials

$$g_1(p) = p^5 + p^2 + 1$$
$$g_2(p) = p^5 + p^6 + p^2 + p + 1$$

---

†An equivalent method for generating the n new sequences is to employ a shift register of length $2m$ with feedback connections specified by the polynomial $h(p) = g_1(p)g_2(p)$, where $g_1(p)$ and $g_2(p)$ are the polynomials which specify the feedback connections of the m-stage shift registers that generate the m sequences $\mathbf{a}$ and $\mathbf{b}$.

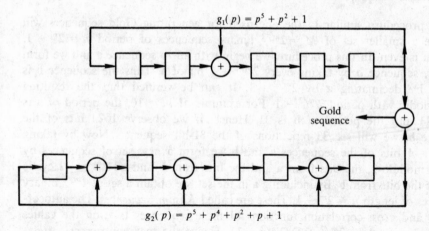

$$g_1(p) = p^5 + p^2 + 1$$

Gold sequence

$$g_2(p) = p^5 + p^4 + p^2 + p + 1$$

FIGURE 8.2.15
Generation of Gold sequences of length 31.

The shift registers for generating the two m sequences and the corresponding Gold sequences are shown in Fig. 8.2.15. In this case, there are 33 different sequences, corresponding to the 33 relative phases of the two m sequences. Of these, 31 sequences are non-maximal-length sequences.

With the exception of the sequences **a** and **b**, the set of Gold sequences are not maximum-length shift-register sequences of length n. Hence their autocorrelation functions are not two-valued. Gold (1968) has shown that the cross-correlation function for any pair of sequences from the set of $n + 2$ Gold sequences is three-valued with possible values $\{-1, -t(m), t(m) - 2\}$, where $t(m)$ is given by (8.2.73). Similarly, the off-peak autocorrelation function for a Gold sequence takes on values from the set $\{-1, -t(m), t(m) - 2\}$. Hence the off-peak values of the autocorrelation function are upper-bounded by $t(m)$.

The values of the off-peak, autocorrelation function and the peak cross-correlation function, i.e., $t(m)$, for Gold sequences is listed in Table 8.2.1. Also listed are the values normalized by $\phi(0)$.

It is interesting to compare the peak cross-correlation value of Gold sequences with a known lower bound on the cross correlation between any pair of binary sequences of period n in a set of M sequences. A lower bound developed by Welch (1974) for $\phi_{\max}$ is

$$\phi_{\max} \geq n\sqrt{\frac{M-1}{Mn-1}} \tag{8.2.74}$$

which, for large values of n and M, is well approximated as $\sqrt{n}$. For Gold sequences, $n = 2^m - 1$ and, hence, the lower bound is $\phi_{\max} \approx 2^{m/2}$. This bound is lower by $\sqrt{2}$ for m odd and by 2 for m even relative to $\phi_{\max} = t(m)$ for Gold sequences.

A procedure similar to the one used for generating Gold sequences will generate a smaller set of $M = 2^{m/2}$ binary sequences of period $n = 2^m - 1$, where m is even. In this procedure, we begin with an m sequence **a** and we form a binary sequence **b** by taking every $2^{m/2} + 1$ bit of **a**. Thus the sequence **b** is formed by decimating **a** by $2^{m/2} + 1$. It can be verified that the resulting **b** is periodic with period $2^{m/2} - 1$. For example, if $m = 10$, the period of **a** is $n = 1023$ and the period of **b** is 31. Hence, if we observe 1023 bits of the sequence **b**, we will see 33 repetitions of the 31-bit sequence. Now by taking $n = 2^m - 1$ bits of the sequences **a** and **b** we form a new set of sequences by adding, modulo-2, the bits from **a** and the bits from **b** and all $2^{m/2} - 2$ cyclic shifts of the bits from **b**. By including **a** in the set, we obtain a set of $2^{m/2}$ binary sequences of length $n = 2^m - 1$. These are called *Kasami sequences*. The autocorrelation and cross-correlation functions of these sequences take on the values from the set $\{-1, -(2^{m/2} + 1), 2^{m/2} - 1\}$. Hence the maximum cross-correlation value for any pair of sequences from the set is

$$\phi_{\max} = 2^{m/2} + 1 \qquad (8.2.75)$$

This value of $\phi_{\max}$ satisfies the Welch lower bound for a set of $2^{m/2}$ sequences of length $n = 2^m - 1$. Hence the Kasami sequences are optimal.

Besides the well-known Gold sequences and Kasami sequences, there are other binary sequences that are appropriate for CDMA applications. The interested reader may refer to the work of Scholtz (1979), Olsen (1977), and Sarwate and Pursley (1980).

Finally, we wish to indicate that, although we discussed the periodic cross-correlation function between pairs of periodic sequences, many practical CDMA systems may use an information bit duration that encompasses only a fraction of a periodic sequence. In such a case, it is the partial period cross correlation between two sequences that is important. A number of papers in the literature deal with this problem, among them the papers by Lindholm (1968), Wainberg and Wolf (1970), Fredricsson (1975), Bekir et al. (1978), and Pursley (1979).

8.2.5 Excision of Narrowband Interference in DS Spread Spectrum Systems

We have shown that DS spread spectrum signals reduce the effects of interference due to other users of the channel and intentional jamming. When the interference in narrowband, the cross correlation of the received signal with the replica of the PN code sequence reduces the level of the interference by spreading it across the frequency band occupied by the PN signal. Thus the interference is rendered equivalent to a lower-level noise with a relatively flat spectrum. Simultaneously the cross correlation operation collapses the desired signal to the bandwidth occupied by the information signal prior to spreading.

The interference immunity of a DS spread spectrum communications system corrupted by narrowband interference can be further improved by filter-

FIGURE 8.2.16
Demodulator for PN spread spectrum signal corrupted by narrowband interference.

ing the signal prior to cross correlation, where the objective is to reduce the level of the interference at the expense of introducing some distortion on the desired signal. This filtering can be accomplished by exploiting the wideband spectral characteristics of the desired DS signal and the narrowband characteristic of the interference as described below.

To be specific, we consider the demodulator illustrated in Fig. 8.2.16. The received signal is passed through a filter matched to the chip pulse $u(t)$. The output of this filter is synchronously sampled every T_c seconds to yield

$$r_j = 2\mathscr{E}_c(2b_j - 1)(2c_{ij} - 1) + \nu_j \qquad j = 1, 2, \ldots \qquad (8.2.76)$$

Following the sampler is a linear discrete-time transversal filter with impulse response $h_k, 0 \le k \le K - 1$. The values $\{h_k\}$ are determined as described below. After the sequence $\{r_j\}$ is filtered, it is cross-correlated with the PN sequence and decoded.

The additive noise term ν_j will be assumed to consist of two terms, one corresponding to a broadband noise (usually thermal noise) and the other to narrowband interference. Consequently we may express r_j^* as

$$r_j = s_j + i_j + n_j \qquad (8.2.77)$$

where s_j denotes the signal component, i_j the narrowband interference, and n_j the broadband noise.

The interference component i_j can be estimated from the received signal and suppressed by passing it through the linear transversal filter. Computationally efficient algorithms based on linear prediction may be used to estimate the interference. Basically, in this method the narrowband interference is modeled as having been generated by passing white noise through an all-pole filter. Linear prediction is used to estimate the coefficients of the all-pole model. The estimated coefficients specify an appropriate noise-whitening all-zero (transversal) filter through which the received signal is passed for the purpose of suppressing the narrowband interference. Linear prediction algorithms are described below.

Interference estimation and suppression based on linear prediction. An estimate of the interference $\{i_j\}$ is formed from $\{r_j\}$. Let us assume for the moment that the statistics of the sequence $\{i_j\}$ are known and that they are stationary. Then, due to the narrowband characteristics of $\{i_j\}$, we can predict i_j from $r_{j-1}, r_{j-2}, \ldots, r_{j-m}$. That is,

$$\hat{i}_j = \sum_{l=1}^{m} a_{ml} r_{j-l} \tag{8.2.78}$$

where $\{a_{ml}\}$ are the coefficients of an mth-order linear predictor. It should be emphasized that (8.2.78) predicts the interference but not the signal s_j, because the PN chips are uncorrelated and, hence, s_j is uncorrelated with $r_{j-l}, l = 1, 2, \ldots, m$, where m is less than the length of the PN sequence.

The coefficients in (8.2.78) are determined by minimizing the mean-square error between r_j and $\hat{i}_j$, which is defined as

$$\mathscr{E}(m) = E\left[\left(r_j - \hat{i}_j\right)^2\right]$$

$$= E\left[\left(r_j - \sum_{l=1}^{m} a_{ml} r_{j-1}\right)^2\right] \tag{8.2.79}$$

Minimization of $\mathscr{E}(m)$ with respect to the predictor coefficients leads to the familiar set of linear equations, which were first encountered in Chap. 2, namely,

$$\sum_{l=1}^{m} a_{ml} \phi(k - l) = \phi(k) \qquad k = 1, 2, \ldots, m \tag{8.2.80}$$

where

$$\phi(k) = E(r_j r_{j+k}) \tag{8.2.81}$$

is the autocorrelation function of the received signal $\{r_j\}$.

The solution of (8.2.80) for the coefficients of the prediction filter requires knowledge of the autocorrelation function $\phi(k)$. In practice, the autocorrelation function of i_j and, hence, r_j is unknown and it may also be slowly varying in time. Consequently we must consider methods for obtaining the predictor coefficients directly from the sequence $\{r_j\}$. This may be accomplished in a number of ways. Three different methods are described below. In all cases, the predictor coefficients are obtained by using a block of N samples of $\{r_j\}$.

The first method is simply based on the direct estimation of $\phi(k)$ from the block of N samples. The estimate of $\phi(k)$ is

$$\hat{\phi}(k) = \sum_{l=0}^{N-k} r(l) r(l + k) \qquad k = 0, 1, \ldots, m \tag{8.2.82}$$

The estimate $\hat{\phi}(k)$ may then be substituted in (8.2.80) in place of $\phi(k)$ and the Levinson–Durbin algorithm described in Appendix 2A can be used to solve the equations efficiently.

The second method that may be used for obtaining the prediction coefficients is the Burg algorithm [for reference, see Burg (1967)]. Basically the Burg algorithm may be viewed as an order-recursive least-squares algorithm in which the Levinson recursion is used in each iteration. The performance index used in the Burg algorithm is

$$\mathscr{E}_B(m) = \sum_{j=m+1}^{N} \left[f_m^2(j) + b_m^2(j) \right] \tag{8.2.83}$$

where $f_m(j)$ and $b_m(j)$ are the forward and backward errors in an mth-order predictor, which are defined as

$$f_m(j) = r_j - \sum_{k=1}^{m} a_{mk} r_{j-k}$$
$$b_m(j) = r_{j-m} - \sum_{k=1}^{m} a_{mk} r_{j-m+k} \tag{8.2.84}$$

The predictor coefficients a_{mk} for $1 \le k \le m-1$ ar forced to satisfy the Levinson–Durbin recursion given in Appendix 2A. As a consequence of this constraint, the forward and backward errors satisfy the recursive relations

$$f_m(j) = f_{m-1}(j) - a_{mm} b_{m-1}(j-1)$$
$$b_m(j) = b_{m-1}(j-1) - a_{mm} f_{m-1}(j) \tag{8.2.85}$$
$$f_0(j) = b_0(j) = r_j$$

The relations in (8.2.85) are substituted into (8.2.83) and $\mathscr{E}_B(m)$ is minimized with respect to a_{mm}. The result of this minimization is

$$a_{mm} = \frac{2 \sum_{j=m+1}^{N} f_{m-1}(j) b_{m-1}(j-1)}{\sum_{j=m+1}^{N} \left[f_{m-1}^2(j) + b_{m-1}^2(j-1) \right]} \qquad m \ge 1 \tag{8.2.86}$$

The relations given in (8.2.85) and (8.2.86) along with the Levinson–Durbin recursion for $\{a_{mk}\}$, $1 \le k \le m-1$, constitute the Burg algorithm for obtaining the prediction coefficients directly from the data.

The Burg algorithm is basically a least-squares lattice algorithm with the added constraint that the predictor coefficients satisfy the Levinson recursion. As a result of this constraint, an increase in the order of the predictor requires only a single parameter optimization at each stage. In contrast to this method, we may use an unconstrained least-squares algorithm to determine the predictor coefficients. Specifically, this approach is based on minimizing (8.2.83) over the entire set of predictor coefficients. This minimization yields the set of equations

$$\sum_{k=1}^{m} a_{mk} \phi(l,k) = \phi(l,0) \qquad l = 1, 2, \dots, m \tag{8.2.87}$$

where

$$\phi(l,k) = \sum_{j=m+1}^{N} \left(r_{j-k} r_{j-l} + r_{j-m+k} r_{j-m+l} \right) \tag{8.2.88}$$

A computationally efficient algorithm for solving the equations in (8.2.76) has been given by Marple (1980).

The three methods described above for determining the prediction coefficients are intimately related to the least-squares algorithms described in Sec. 6.8 in the context of adaptive equalization. Specifically, the algorithms given above are formulated in terms of block processing of N received signal samples from the chip matched filter and are recursive in order only. It is also possible to make them recursive in time. In particular, the time-recursive version of the unconstrained least-squares equations given above is just the algorithm described in detail in Sec. 6.8.3. For the sake of brevity, time-recursive versions of these algorithms will not be considered in this subsection. Instead, we consider narrowband interference that remains fixed over the time duration spanned by a transmitted code word.

FIGURE 8.2.17
Equivalent structures for the interference suppression filter.

FIGURE 8.2.18
Frequency response characteristics for the interference suppression filter.

Once the prediction coefficients are determined by any one of the methods described above, the estimate $\hat{i}_j$ of the interference, given by (8.2.78), is subtracted from r_j and the difference signal is processed further in order to extract the digital information. Consequently the filter for suppressing the interference has the transfer function

$$A_m(z) = 1 - \sum_{k=1}^{m} a_{mk} z^{-k} \tag{8.2.89}$$

where z^{-1} denotes a unit of (one-chip) delay. Two equivalent structures for this filter are shown in Fig. 8.2.17.

Example 8.2.5. Suppose the narrowband interference is confined to 20 percent of the spectral band occupied by the PN spread spectrum signal. The average power of the interference is 20 dB above the average power of the signal. The average power of the broadband noise is 20 dB below the average power of the signal. Figure 8.2.18 illustrates the frequency characteristics of the filter with transfer function given by (8.2.89) when the coefficients of a fourth-order predictor are obtained by means of the three algorithms described above. We observe only minor differences in the spectral characteristics of the filters. These results were generated by Monte Carlo simulation on a digital computer. Good estimates of the coefficients were obtained with as few as 50 samples (PN chips) and predictors that varied in length from

FIGURE 8.2.19
Frequency response characteristics of 16- and 29-tap filters for four bands of interference.

fourth to fifteenth order. In general, the Burg algorithm and the unconstrained least-squares algorithm yield good estimates of the predictor coefficients provided the number of signal samples is at least twice the length of the prediction filter.

Figure 8.2.19 illustrates an example in which narrowband interference occupying 20 percent of the signal band is split equally into four frequency bands. The spectral characteristics of a 16- and a 29-tap filter are shown in this example. It is apparent that the 29-tap filter has better spectral characteristics. In general, the number of taps in the filter should be about four times the number of interference bands for adequate suppression.

It is apparent from the results given in Example 8.2.5 that the interference suppression filter acts as a notch filter. In effect, it attempts to flatten or whiten the total noise plus interference, so that the power spectral density of these components at its output is approximately flat. While suppressing the interference, the filter also distorts the desired signal by spreading it in time.

Since the noise plus interference at the output of the suppression filter is spectrally flat, the matched filtering or cross correlation following the suppression filter should be performed with the distorted signal (see Prob. 4.15). This may be accomplished by having a filter matched to the interference suppression filter, i.e., a discrete-time filter impulse response $\{-a_{mm}, -a_{mm-1}, \ldots, -a_{m1}, 1\}$ followed by the PN correlator. In fact, we can combine the interference suppression filter and its matched filter into a single filter having an impulse response

$$h_0 = -a_{mm}$$

$$h_k = -a_{mm-k} + \sum_{l=0}^{k-1} a_{mm-l} a_{mk-l} \qquad 1 \leq k \leq m-1$$

$$h_m = 1 + \sum_{l=1}^{m} a_{ml}^2 \qquad\qquad (8.2.90)$$

$$h_{m+k} = h_{m-k} \qquad\qquad 0 \leq k \leq m$$

The combined filter is a linear phase (symmetric) transversal filter with $K = 2m + 1$ taps. The impulse response may be normalized by dividing every term by h_m. Thus the center tap is normalized to unity.

Performance improvement from interference suppression. In order to demonstrate the effectiveness of the interference suppression filter, we shall compare the performance of the decoder with and without the suppression filter. The output SNR is a convenient performance index for this purpose. Since the output of the decoder is characterized as gaussian, there is a one-to-one correspondence between the SNR and the probability of error.

We assume that the all-zero code word $\mathbf{C}_1$ is transmitted. We shall compute the SNR at the output of the decoder for the decision variable U_1 with and without the suppression filter.

Without the suppression filter, the decoder output has a mean $2\mathcal{E}_c n$ and a variance $n[2\mathcal{E}_c N_0 + \phi_{ii}(0)]$ where $\phi_{ii}(k)$ is the autocorrelation function of the sequence $\{i_j\}$, defined as

$$\phi_{ii}(k) = E(i_m i_{m+k})$$

and

$$i_j = \operatorname{Re}\left\{ \int_0^{T_c} u^*(t) i[t - (j-1)T_c] \, dt \right\}$$

It follows that

$$\phi_{ii}(k) = \int_{-\infty}^{\infty} |U(f)|^2 \Phi_{ii}(f) e^{j2\pi f k T_c} \, df \qquad (8.2.91)$$

where $\Phi_{ii}(f)$ is the power spectral density of the narrowband interference.

The output SNR is defined as the ratio of the square of the mean to twice the variance. Hence the SNR without the suppression filter is

$$\text{SNR}_{no} = \frac{\mathcal{E}_c n}{N_0 + \phi_{ii}(0)/2\mathcal{E}_c} \qquad (8.2.92)$$

For example, if the narrowband interference is a CW tone, $\phi_{ii}(0) = 2\mathcal{E}_c T_c J_{av}$ as shown in Example 8.2.1. Then the output SNR becomes

$$\text{SNR}_{no} = \frac{\mathscr{E}_c n}{N_0 + J_{av} T_c} = \frac{\mathscr{E}_c n}{N_0 + J_0} \tag{8.2.93}$$

which is consistent with our previous results.

    With an interference suppression filter having a symmetric impulse response as defined in (8.2.90) and normalized such that the center tap is unity, the mean value of U_1 is also $2\mathscr{E}_c n$. However, the variance of U_1 now consists of three terms. One corresponds to the additive wideband noise, the second to the residual narrowband interference, and the third to a self-noise caused by the time dispersion introduced by the suppression filter. The expression for the variance can be shown to be

$$\text{var}(U_1) = 2n\mathscr{E}_c N_0 \sum_{k=0}^{K} h_k^2 + n \sum_{k=0}^{K} \sum_{l=0}^{K} h(l)h(k)\phi_{ii}(k - l)$$

$$= 4n\mathscr{E}_c^2 \sum_{k=0}^{K/2-1} \left(2 - \frac{k}{n}\right) h_k^2 \tag{8.2.94}$$

Hence the output SNR with the filter is

$$\text{SNR}_0 = \frac{\mathscr{E}_c n}{N_0 \displaystyle\sum_{k=0}^{K} h_k^2 + \frac{1}{2\mathscr{E}_c} \displaystyle\sum_{k=0}^{K} \sum_{l=0}^{K} h(k)h(l)\phi_{ii}(k - l) + 2\mathscr{E}_c \displaystyle\sum_{k=0}^{K/2-1} \left(2 - \frac{k}{n}\right) h_k^2} \tag{8.2.95}$$

    The ratio of the SNR in (8.2.95) to the SNR in (8.2.92) represents the improvement in performance due to the use of the interference suppression filter. This ratio, denoted by η, is

$$\eta = \frac{N_0 + \phi_{ii}(0)/2\mathscr{E}_c}{N_0 \displaystyle\sum_{k=0}^{K} h_k^2 + \frac{1}{2\mathscr{E}_c} \displaystyle\sum_{k=0}^{K} \sum_{l=0}^{K} h(k)h(l)\phi_{ii}(k - l) + 2\mathscr{E}_c \displaystyle\sum_{k=0}^{K/2-1} \left(2 - \frac{k}{n}\right) h_k^2} \tag{8.2.96}$$

This ratio is called the improvement factor resulting from interference suppression. It may be plotted against the normalized SNR per chip without filtering, defined as

$$\frac{\text{SNR}_{no}}{n} = \frac{\mathscr{E}_c}{N_0 + \phi_{ii}(0)/2\mathscr{E}_c} \tag{8.2.97}$$

The resulting graph of η versus SNR_{no}/n is universal in the sense that it applies to any PN spread spectrum system with arbitrary processing gain for a given $\mathscr{E}_c$, N_0, and $\phi_{ii}(0)$.

    As an example, the improvement factor in (decibels) is plotted against SNR_{no}/n in Fig. 8.2.20 for a single-band equal-amplitude randomly phased

FIGURE 8.2.20
Improvement factor for interference suppression filter in cascade with its matched filter.

sinusoids covering 20 percent of the frequency band occupied by the DS spread spectrum signal. The interference suppression filter consists of a nine-tap suppression filter which corresponds to a fourth-order predictor. These numerical results indicate that the notch filter is very effective in suppressing the interference prior to PN correlation and decoding. As a consequence, the jamming margin of the system is increased or, equivalently, the processing gain (bandwidth expansion) of the DS spread spectrum system can be reduced.

8.3 FREQUENCY-HOPPED SPREAD SPECTRUM SIGNALS

In a frequency-hopped (FH) spread spectrum communications system the available channel bandwidth is subdivided into a large number of contiguous frequency slots. In any signaling interval, the transmitted signal occupies one or more of the available frequency slots. The selection of the frequency slot(s) in each signaling interval is made pseudo-randomly according to the output from a PN generator. Figure 8.3.1 illustrates a particular frequency-hopped pattern in the time-frequency plane.

A block diagram of the transmitter and receiver for a frequency-hopped spread spectrum system is shown in Fig. 8.3.2. The modulation is usually either binary of M-ary FSK. For example, if binary FSK is employed, the modulator selects one of two frequencies corresponding to the transmission of either a 1 or a 0. The resulting FSK signal is translated in frequency by an amount that is determined by the output sequence from the PN generator which, in turn, is used

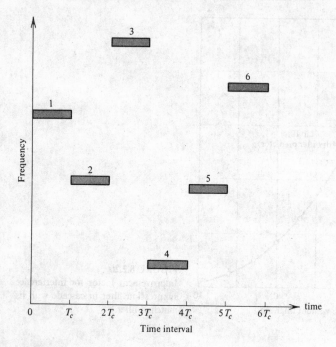

FIGURE 8.3.1
An example of a frequency-hopped pattern.

to select a frequency that is synthesized by the frequency synthesizer. This frequency is mixed with the output of the modulator and the resultant frequency-translated signal is transmitted over the channel. For example, m bits from the PN generator may be used to specify $2^m - 1$ possible frequency translations.

At the receiver, we have an identical PN generator, synchronized with the received signal, which is used to control the output of the frequency synthesizer. Thus the pseudo-random frequency translation introduced at the transmitter is removed at the receiver by mixing the synthesizer output with the received signal. The resultant signal is demodulated by means of an FSK demodulator. A signal for maintaining synchronism of the PN generator with the frequency-translated received signal is usually extracted from the received signal.

Although PSK modulation gives better performance than FSK in an AWGN channel, it is difficult to maintain phase coherence in the synthesis of the frequencies used in the hopping pattern and, also, in the propagation of the signal over the channel as the signal is hopped from one frequency to another over a wide bandwidth. Consequently FSK modulation with noncoherent detection is usually employed with FH spread spectrum signals.

In the frequency-hopping system depicted in Fig. 8.3.2, the carrier frequency is pseudo-randomly hopped in every signaling interval. The M informa-

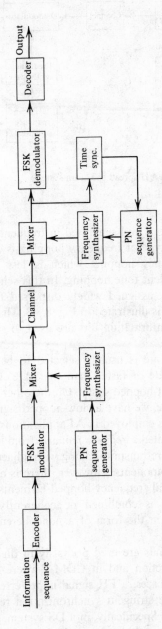

FIGURE 8.3.2
Block diagram of a FH spread spectrum system.

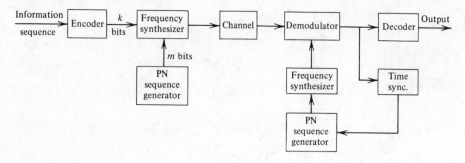

FIGURE 8.3.3
Block diagram of an independent tone FH spread spectrum system.

tion-bearing tones are contiguous and separated in frequency by $1/T_c$, where T_c is the signaling interval. This type of frequency hopping is called *block hopping*.

Another type of frequency hopping which is less vulnerable to some jamming strategies is independent tone hopping. In this scheme, the M possible tones from the modulator are assigned widely dispersed frequency slots. One method for accomplishing this is illustrated in Fig. 8.3.3. That is, the m bits from the PN generator and the k information bits are used to specify the frequency slots for the transmitted signal.

The frequency-hopping rate is usually selected to be either equal to the (coded or uncoded) symbol rate or faster than that rate. If there are multiple hops per symbol, we have a fast-hopped signal. On the other hand, if the hopping is performed at the symbol rate, we have a slow-hopped signal.

Fast frequency hopping is employed in AJ applications when it is necessary to prevent a type of jammer, called a *follower jammer*, from having sufficient time to intercept the frequency and retransmit it along with adjacent frequencies so as to create interfering signal components. However, there is a penalty incurred in subdividing a signal into several frequency-hopped elements because the energy from these separate elements is combined noncoherently. Consequently the demodulator incurs a penalty in the form of a noncoherent combining loss as described in Sec. 4.4.

FH spread spectrum signals are used primarily in digital communications systems that require AJ projection and in CDMA where many users share a common bandwidth. In most cases, a FH signal is preferred over a DS spread spectrum signal because of the stringent synchronization requirements inherent in DS spread spectrum signals. Specifically, in a DS system, timing and synchronization must be established to within a fraction of the chip interval $T_c \approx 1/W$. On the other hand, in an FH system, the chip interval is the time spent in transmitting a signal in a particular frequency slot of bandwidth $B \ll W$. But this interval is approximately $1/B$, which is much larger than $1/W$. Hence the timing requirements in a FH system are not a stringent as in a PN system.

In Secs. 8.3.2 and 8.3.3 we shall focus on the AJ and CDMA applications of FH spread spectrum signals. First we shall determine the error rate performance of an uncoded and a coded FH signal in the presence of a broadband AWGN interference. Then we shall consider a more serious type of interference that arises in AJ and CDMA applications, called *partial-band interference*. The benefits obtained from coding for this type of interference are determined. We conclude the discussion in Sec. 8.3.3 with an example of a FH CDMA system that was designed for use by mobile users with a satellite serving as the channel.

8.3.1 Performance of FH Spread Spectrum Signals in AWGN

Let us consider the performance of a FH spread spectrum signal in the presence of broadband interference characterized statistically as AWGN with power spectral density J_0. For binary orthogonal FSK with noncoherent detection and slow frequency hopping (1 hop/bit), the probability of error, derived in Sec. 4.3.1, is

$$P_2 = \tfrac{1}{2}e^{-\gamma_b/2} \qquad (8.3.1)$$

where $\gamma_b = \mathscr{E}_b/J_0$. On the other hand, if the bit interval is subdivided into L subintervals and FH binary FSK is transmitted in each subinterval, we have a fast FH signal. With square-law combining of the output signals from the corresponding matched filters for the L subintervals, the error rate performance of the FH signal, obtained from the results in Sec. 4.4, is

$$P_2(L) = \frac{1}{2^{2L-1}} e^{-\gamma_b/2} \sum_{i=0}^{L-1} K_i \left(\frac{\gamma_b}{2}\right)^i \qquad (8.3.2)$$

where the SNR per bit is $\gamma_b = \mathscr{E}_b/J_0 = L\gamma_c$, γ_c is the SNR per chip in the L-chip symbol, and

$$K_i = \frac{1}{i!} \sum_{r=0}^{L-1-i} \binom{2L-1}{r} \qquad (8.3.3)$$

We recall that, for a given SNR per bit γ_b, the error rate obtained from (8.3.2) is larger than that obtained from (8.3.1). The difference in SNR for a given error rate and a given L is called the *noncoherent combining loss*, which was previously described and illustrated in Sec. 4.4.

Coding improves the performance of the FH spread spectrum system by an amount, which we call the *coding gain*, that depends on the code parameters. Suppose we use a linear binary (n, k) block code and binary FSK modulation with one hop per coded bit for transmitting the bits. With soft-decision decoding of the square-law-demodulated FSK signal, the probability of a code word error is upper-bounded as

$$P_M \leq \sum_{m=2}^{M} P_2(m) \tag{8.3.4}$$

where $P_2(m)$ is the error probability in deciding between the mth code word and the all-zero code word when the latter has been transmitted. The expression for $P_2(m)$ is derived in Sec. 5.2.4 and has the same form as (8.3.2) and (8.3.3), with L being replaced by w_m and γ_b by $\gamma_b R_c w_m$, where w_m is the weight of the mth code word and R_c is the code rate. The product $R_c w_m$, which is no less than $R_c d_{\min}$, represents the coding gain. Thus we have the performance of a block coded FH system with slow frequency hopping in broadband interference.

The probability of error for fast frequency hopping with n_2 hops/coded bit is obtained by reinterpreting the binary event probability $P_2(m)$ in (8.3.4). The n_2 hops/coded bit may be interpreted as a repetition code, which when combined with a nontrivial (n_1, k) binary linear code having weight distribution $\{w_m\}$ yields an $(n_1 n_2, k)$ binary linear code with weight distribution $\{n_2 w_m\}$. Hence $P_2(m)$ has the form given in (8.3.2), with L replaced by $n_2 w_m$ and γ_b by $\gamma_b R_c n_2 w_m$, where $R_c = k/n_1 n_2$. Note that $\gamma_b R_c n_2 w_m = \gamma_b w_m k/n_1$, which is just the coding gain obtained from the nontrivial (n_1, k) code. Consequently the use of the repetition code will result in an increase in the noncoherent combining loss.

With hard-decision decoding and slow frequency hopping, the probability of a coded bit error at the output of the demodulator for noncoherent detection is

$$p = \tfrac{1}{2} \exp\left(-\frac{\gamma_b R_c}{2}\right) \tag{8.3.5}$$

The code word error probability is easily upper-bounded, by use of the Chernoff bound, as

$$P_M \leq \sum_{m=2}^{M} \left[4p(1-p)\right]^{w_m/2} \tag{8.3.6}$$

However, if fast frequency hopping is employed with n_2 hops/coded bit, and the square-law-detected outputs from the corresponding matched filters for the n_2 hops are added as in soft-decision decoding to form the two decision variables for the coded bits, the bit error probability p is also given by (8.3.2), with L replaced by n_2 and $\gamma_b R_c n_2$, where R_c is the rate of the nontrivial (n_1, k) code. Consequently the performance of the fast FH system in broadband interference is degraded relative to the slow FH system by an amount equal to the noncoherent combining loss of the signals received from the n_2 hops.

We have observed that for both hard-decision and soft-decision decoding, the use of the repetition code in a fast frequency-hopping system yields no coding gain. The only coding gain obtained comes from the (n_1, k) block code. Hence the repetition code is inefficient in a fast FH system with noncoherent combining. A more efficient coding method is one in which either a single low-rate binary code or a concatenated code is employed. Additional improvements in performance may be obtained by using nonbinary codes in conjunction with M-ary

FSK. Bounds on the error probability for this case may be obtained from the results given in Secs. 4.3 and 4.4.

Although we have evaluated the performance of linear block codes only in the above discussion, it is relatively easy to derive corresponding performance results for binary convolutional codes. We leave as an exercise for the reader the derivation of the bit error probability for soft-decision Viterbi decoding and hard-decision Viterbi decoding of FH signals corrupted by broadband interference.

Finally, we observe that $\mathscr{E}_b$, the energy per bit, can be expressed as $\mathscr{E}_b = S_{av}/R$, where R is the information rate in bits per second and $J_0 = J_{av}/W$. Therefore γ_b may be expressed as

$$\gamma_b = \frac{\mathscr{E}_b}{J_0} = \frac{W/R}{J_{av}/S_{av}} \tag{8.3.7}$$

In this expression, we recognize W/R as the processing gain and J_{av}/S_{av} as the jamming margin for the FH spread spectrum signal.

8.3.2 Performance of FH Spread Spectrum Signals in Partial-Band Interference

The partial-band interference considered in this subsection is modeled as a zero mean gaussian random process with a flat power spectral density over a fraction α of the total bandwidth W and zero elsewhere. In the region or regions where the power spectral density is nonzero, its value is $\Phi_{zz}(f) = J_0/\alpha$, $0 < \alpha \le 1$. This model of the interference may be applied to a jamming signal or to interference from other users in a FH CDMA system.

Suppose that the partial-band interference comes from a jammer who may select α to optimize the effect on the communications system. In an uncoded pseudo-randomly hopped (slow-hopping) FH system with binary FSK modulation and noncoherent detection, the received signal will be jammed with probability α and it will not be jammed with probability $1 - \alpha$. When it is jammed, the probability of error is $\frac{1}{2} \exp(-\mathscr{E}_b \alpha/2J_0)$ and when it is not jammed the demodulation is error-free. Consequently the average probability of error is

$$P_2(\alpha) = \frac{\alpha}{2} \exp\left(-\frac{\alpha \mathscr{E}_b}{2J_0}\right) \tag{8.3.8}$$

where $\mathscr{E}_b/J_0$ may also be expressed as $(W/R)/(J_{av}/S_{av})$.

Figure 8.3.4 illustrates the error rate as a function of $\mathscr{E}_b/J_0$ for several values of α. The jammer's optimum strategy is to select the value of α that maximizes the error probability. By differentiating $P_2(\alpha)$ and solving for the extremum with the restriction that $0 \le \alpha \le 1$ we find that

$$\alpha^* = \begin{cases} \dfrac{1}{\mathscr{E}_b/2J_0} = 2\dfrac{J_{av}/S_{av}}{W/R} & \dfrac{\mathscr{E}_b}{J_0} \ge 2 \\[2ex] 1 & \dfrac{\mathscr{E}_b}{J_0} < 2 \end{cases} \tag{8.3.9}$$

FIGURE 8.3.4
Performance of binary FSK with partial-band interference.

The corresponding error probability for the worst-case partial-band jammer is

$$P_2 = \frac{e^{-1}}{\mathscr{E}_b/J_0} = \frac{1}{e\left(\dfrac{W/R}{J_{av}/S_{av}}\right)} \tag{8.3.10}$$

Whereas the error probability decreases exponentially for full-band jamming, we now find that the error probability decreases only inversely with $\mathscr{E}_b/J_0$ for the worst-case partial-band jamming. This result is similar to the error rate performance of binary FSK in a Rayleigh fading channel and to the uncoded DS spread spectrum system corrupted by worst-case pulse jamming.

In our discussion of signal design for efficient and reliable communication over a fading channel, we found that diversity obtained by means of coding provides a significant improvement in performance relative to uncoded signals. It should not be surprising that the same type of signal waveforms that were described for signaling over a fading channel are also effective on partial-band interference channels.

To illustrate the benefits of diversity in a FH spread spectrum signal with partial-band interference, we assume that the same information symbol is transmitted by binary FSK on L independent frequency hops. This may be accomplished by subdividing the signaling interval into L subintervals, as described previously for fast frequency hopping. After the hopping pattern is removed, the signal is demodulated by passing it through a pair of matched filters whose outputs are square-law-detected and sampled at the end of each subinterval. The square-law-detected signals corresponding to the L frequency hops are weighted and summed to form the decision variables U_1 and U_2.

When the decision variable U_1 contains the signal components, U_1 and U_2 may be expressed as

$$U_1 = \sum_{k=1}^{L} \beta_k \left| 2\mathcal{E}_c + N_{1k} \right|^2$$

$$U_2 = \sum_{k=1}^{L} \beta_k \left| N_{2k} \right|^2 \tag{8.3.11}$$

where $\{\beta_k\}$ represent the weighting coefficients, $\mathcal{E}_c$ is the signal energy per chip in the L-chip symbol, and $\{N_{jk}\}$ represent the additive gaussian noise terms at the output of the matched filters.

The coefficients are optimally selected to prevent the jammer from saturating the combiner should the transmitted frequencies be successfully hit in one or more hops. Ideally β_k is selected to be equal to the reciprocal of the variance of the corresponding noise terms $\{N_k\}$. Thus the noise variance for each chip is normalized to unity by this weighting and the corresponding signal is also scaled accordingly. This means that when the signal frequencies on a particular hop are jammed, the corresponding weight is very small. In the absence of jamming on a given hop, the weight is relatively large. In practice, for partial-bound noise jamming the weighting may be accomplished by use of an AGC having a gain that is set on the basis of noise power measurements obtained from frequency bands adjacent to the transmitted tones. This is equivalent to having side information (knowledge of jammer state) at the decoder.

Suppose that we have broadband gaussian noise with power spectral density N_0 and partial-band interference, over αW of the frequency band, which is also gaussian with power spectral density J_0/α. In the presence of partial-band interference, the second moments of the noise terms N_{1k} and N_{2k} are

$$\sigma_k^2 = \tfrac{1}{2} E\left(|N_{1k}|^2 \right) = \tfrac{1}{2} E\left(|N_{2k}|^2 \right) = 2\mathcal{E}_c \left(N_0 + \frac{J_0}{\alpha} \right) \tag{8.3.12}$$

In this case, we select $\beta_k = 1/\sigma_k^2 = 1/2\mathscr{E}_c(N_0 + J_0/\alpha)$. In the absence of partial-band interference, $\sigma_k^2 = 2\mathscr{E}_c N_0$ and, hence, $\beta_k = 1/2\mathscr{E}_c N_0$. Note that β_k is a random variable.

An error occurs in the demodulation if $U_2 > U_1$. Although it is possible to determine the exact error probability, we shall resort to the Chernoff bound, which yields a result that is much easier to evaluate and interpret. Specifically, the Chernoff (upper) bound in the error probability is

$$P_2 = P(U_2 - U_1 > 0) \leq E\{\exp[\nu(U_2 - U_1)]\}$$

$$= E\left\{\exp\left[-\nu \sum_{k=1}^{L} \beta_k\left(|2\mathscr{E} + N_{1k}|^2 - |N_{2k}|^2\right)\right]\right\} \quad (8.3.13)$$

where ν is a variable which is optimized to yield the tightest possible bound.

The averaging in (8.3.13) is performed with respect to the statistics of the noise components and the statistics of the weighting coefficients $\{\beta_k\}$ which are random as a consequence of the statistical nature of the interference. Keeping the $\{\beta_k\}$ fixed and averaging over the noise statistics first, we obtain

$$P_2(\boldsymbol{\beta}) = E\left[\exp\left(-\nu \sum_{k=1}^{L} \beta_k |2\mathscr{E}_c + N_{1k}|^2 + \nu \sum_{k=1}^{L} \beta_k |N_{2k}|^2\right)\right]$$

$$= \prod_{k=1}^{L} E\left[\exp\left(-\nu\beta_k |2\mathscr{E}_c + N_{1k}|^2\right)\right] E\left[\exp\left(\nu\beta_k |N_{2k}|^2\right)\right]$$

$$= \prod_{k=1}^{L} \frac{1}{1 - 4\nu^2} \exp\left(\frac{-4\mathscr{E}_c^2 \beta_k \nu}{1 + 2\nu}\right) \quad (8.3.14)$$

Since the FSK tones are jammed with probability α, it follows that $\beta_k = 1/2\mathscr{E}(N_0 + J_0/\alpha)$ with probability α and $1/2\mathscr{E}_c N_0$ with probability $1 - \alpha$. Hence the Chernoff bound is

$$P_2 \leq \prod_{k=1}^{L} \left\{\frac{\alpha}{1 - 4\nu^2} \exp\left[\frac{-2\mathscr{E}_c\nu}{(N_0 + J_0/\alpha)(1 + 2\nu)}\right] + \frac{1 - \alpha}{1 - 4\nu^2} \exp\left[\frac{-2\mathscr{E}_c\nu}{N_0(1 + 2\nu)}\right]\right\}$$

$$= \left\{\frac{\alpha}{1 - 4\nu^2} \exp\left[\frac{-2\mathscr{E}_c\nu}{(N_0 + J_0/\alpha)(1 + 2\nu)}\right] + \frac{1 - \alpha}{1 - 4\nu^2} \exp\left[\frac{-2\mathscr{E}_c\nu}{N_0(1 + 2\nu)}\right]\right\}^L \quad (8.3.15)$$

The next step is to optimize the bound in (8.3.15) with respect to the variable ν. In its present form, however, the bound is messy to manipulate. A significant simplification occurs if we assume that $J_0/\alpha \gg N_0$, which renders the second term in (8.3.15) negligible compared to the first. Alternatively, we let $N_0 = 0$ so that the bound on P_2 reduces to

$$P_2 \leq \left\{\frac{\alpha}{1 - 4\nu^2} \exp\left[\frac{-2\alpha\nu\mathscr{E}_c}{J_0(1 + 2\nu)}\right]\right\}^L \quad (8.3.16)$$

The minimum value of this bound with respect to ν and the maximum with respect to α (worst-case partial-band interference) is easily shown to occur when $\alpha = 3J_0/\mathscr{E}_c \leq 1$ and $\nu = \frac{1}{4}$. For these values of the parameters, (8.3.16) reduces to

$$P_2 \leq P_2(L) = \left(\frac{4}{e\gamma_c}\right)^L = \left(\frac{1.47}{\gamma_c}\right)^L, \qquad \gamma_c = \frac{\mathscr{E}_c}{J_0} = \frac{\mathscr{E}_b}{LJ_0} \geq 3 \quad (8.3.17)$$

where γ_c is the SNR per chip in the L-chip symbol. Equivalently

$$P_2 \leq \left[\frac{1.47(J_{av}/S_{av})}{W/R}\right]^L \qquad \frac{W/R}{L(J_{av}/S_{av})} \geq 3 \qquad (8.3.18)$$

The result in (8.3.17) was first derived by Viterbi and Jacobs (1975).

We observe that the probability of error for the worst-case partial-band interference decreases exponentially with an increase in the SNR per chip γ_c in a manner that is very similar to the performance characteristics of diversity techniques for Rayleigh fading channels. In fact, we may express the right-hand side of (8.3.17) in the form

$$P_2(L) = \exp\left[-\gamma_b h(\gamma_c)\right] \qquad (8.3.19)$$

where the function $h(\gamma_c)$ is defined as

$$h(\gamma_c) = -\frac{1}{\gamma_c}\left[\ln\left(\frac{4}{\gamma_c}\right) - 1\right] \qquad (8.3.20)$$

A plot of $h(\gamma_c)$ is given in Fig. 8.3.5. We observe that the function has a maximum value of $\frac{1}{4}$ at $\gamma_c = 4$. Consequently there is an optimum SNR per chip

FIGURE 8.3.5
Graph of the function $h(\gamma_c)$.

of $10 \log \gamma_c = 6$ dB. At the optimum SNR, the error rate is upper-bounded as

$$P_2 \le P_2(L_{\text{opt}}) = \exp\left(-\frac{\gamma_b}{4}\right) \qquad (8.3.21)$$

When we compare the error probability bound in (8.3.21) with the error probability for binary FSK in spectrally flat noise, which is given by (8.3.1), we see that the combined effect of worst-case partial-band interference and the noncoherent combining loss in the square-law combining of the L chips is 3 dB. We emphasize, however, that for a given $\mathscr{E}_b/J_0$, the loss is greater when the order of diversity is not optimally selected.

Coding provides a means for improving the performance of the frequency-hopped system corrupted by partial-band interference. In particular, if a block orthogonal code is used, with $M = 2^k$ code words and Lth-order diversity per code word, the probability of a code word error is upper-bounded as

$$P_M \le (2^k - 1)P_2(L) = (2^k - 1)\left(\frac{1.47}{\gamma_c}\right)^L = (2^k - 1)\left(\frac{1.47}{k\gamma_b/L}\right)^L \quad (8.3.22)$$

and the equivalent bit error probability is upper-bounded as

$$P_b \le 2^{k-1}\left(\frac{1.47}{k\gamma_b/L}\right)^L \qquad (8.3.23)$$

Figure 8.3.6 illustrates the probability of a bit error for $L = 1, 2, 4, 8$ and $k = 1, 3$. With an optimum choice of diversity, the upper bound can be expressed as

$$P_b \le 2^{k-1}\exp\left(-\frac{k}{4}\gamma_b\right) = \tfrac{1}{2}\exp\left[-k\left(\frac{\gamma_b}{4} - \ln 2\right)\right] \qquad (8.3.24)$$

Thus we have an improvement in performance by an amount equal to $10 \log k(1 - 2.77/\gamma_b)$. For example, when $\gamma_b = 10$ and $k = 3$ (octal modulation), the gain is 3.4 dB, while if $k = 5$ the gain is 5.6 dB.

Additional gains can be achieved by employing the bandwidth-efficient coding techniques described in Sec. 7.8 for communication over Rayleigh fading channels in conjunction with soft-decision decoding. Due to their relatively large minimum distance, concatenated codes are particularly effective. The examples described in Sec. 7.8.4, in which a dual-k convolutional code serves as the outer code and an orthogonal code or a Hadamard code is the inner code, are especially effective on the channel with partial-band interference.

Example 8.3.1. Suppose we use a Hadamard $H(n, k)$ constant weight code with on-off keying (OOK) modulation. The minimum distance of the code is $d_{\min} = n/2$ and, hence, the effective order of diversity obtained with OOK modulation is $d_{\min}/2 = n/4$. There are $n/2$ frequency-hopped tones transmitted per code word. Hence

$$\gamma_c = \frac{k}{n/2}\gamma_b = 2R_c\gamma_b \qquad (8.3.25)$$

FIGURE 8.3.6
Performance of binary and octal FSK with L-order diversity for a channel with worst-case partial-band interference.

when this code is used alone. The bit error rate performance for soft-decision decoding of these codes for the partial-band interference channel is upper-bounded as

$$P_b \leq 2^{k-1} P_2\left(\frac{d_{\min}}{2}\right) = 2^{k-1}\left(\frac{1.47}{2R_c\gamma_b}\right)^{n/4} \qquad (8.3.26)$$

Now if a Hadamard (n, k) code is used as the inner code and a rate $1/2$ dual-k convolutional code is the outer code, the bit error performance in the

FIGURE 8.3.7
Performance of dual-k codes concatenated with Hadamard codes for a channel with worst-case partial-band interference.

presence of worst-case partial-band interference is

$$P_b \leq \frac{2^{k-1}}{2^k - 1} \sum_{m=4}^{\infty} \beta_m P_2\left(\frac{md_{\min}}{2}\right) = \frac{2^{k-1}}{2^k - 1} \sum_{m=4}^{\infty} \beta_m P_2\left(\frac{mn}{4}\right) \quad (8.3.27)$$

where $P_2(L)$ is given by (8.3.17) with

$$\gamma_c = \frac{k}{n}\gamma_b = R_c\gamma_b \quad (8.3.28)$$

Figure 8.3.7 illustrates the performance of the dual-k for $k = 5, 4, 3$ concatenated with the Hadamard $H(20,5)$, $H(16,4)$, and $H(12,3)$ codes, respectively.

In the above discussion we focused on soft-decision decoding. As in a fading channel, the performance achieved with hard-decision decoding is significantly (several decibels) poorer than that obtained with soft-decision decoding. In a concatenated coding scheme, however, a mixture involving soft-decision decoding of the inner code and hard-decision decoding of the outer code represents a reasonable compromise between decoding complexity and performance.

Finally, we wish to indicate that another serious threat in a FH spread spectrum system is partial-band multitone jamming. This type is similar in effect to partial-band spectrally flat noise jamming. Diversity obtained through coding is an effective means for improving the performance of the FH system. An additional improvement is achieved by properly weighting the demodulator outputs so as to suppress the effect of the jammer.

8.3.3 A CDMA System Based on FH Spread Spectrum Signals

In Sec. 8.2.2 we considered a CDMA system based on use of DS spread spectrum signals. As previously indicated, it is also possible to have a CDMA system based on FH spread spectrum signals. Each transmitter-receiver pair in such a system is assigned its own pseudo-random frequency-hopping pattern. Aside from this distinguishing feature, the transmitters and receivers of all the users may be identical in that they may have identical encoders, decoders, modulators, and demodulators.

CDMA systems based on FH spread spectrum signals are particularly attractive for mobile (land, air, sea) users because timing requirements are not as stringent as in a PN spread spectrum signal. In addition, frequency synthesis techniques and associated hardware have been developed that make it possible to frequency-hop over bandwidths that are significantly larger than those currently possible with DS spread spectrum systems. Consequently larger processing gains are possible with FH. The capacity of CDMA with FH is also relatively high. Viterbi (1978) has shown that with dual-k codes and M-ary FSK modulation it is possible to accommodate up to $\frac{3}{8} W/R$ simultaneous users who transmit at an information rate R bits/s over a channel with bandwidth W.

A CDMA system based on FH coded spread spectrum signals was built and successfully tested during the late 1960s. It was developed to provide multiple-access tactical satellite communications for small mobile (land, sea, air) terminals each of which transmitted relatively short messages over the channel intermittently. The system was called *Tactical Transmission System* (TATS) and it is described in a paper by Drouilhet and Bernstein (1969).

An octal Reed–Solomon (7, 2) code is used in the TATS system. Thus two 3-bit information symbols from the input to the encoder are used to generate a seven-symbol code word. Each 3-bit coded symbol is transmitted by means of octal FSK modulation. The eight possible frequencies are spaced at $1/T_c$ Hz apart, where T_c is the time (chip) duration of a single frequency transmission. In addition to the seven symbols in a code word, an eighth symbol is included. That symbol and its corresponding frequency is fixed and is transmitted at the beginning of each code word for the purpose of providing timing and frequency synchronization† at the receiver. Consequently each code word is transmitted in $8T_c$ seconds.

TATS was designed to transmit at information rates of 75 bits/s and 2400 bits/s. Hence $T_c = 10$ ms and 312.5 μs, respectively. Each frequency tone corresponding to a code symbol is frequency-hopped. Hence the hopping rate is 100 hops/s at the 75-bits/s rate and at 3200 hops/s at the 2400-bits/s rate.

There are $M = 2^6 = 64$ code words in the Reed–Solomon $(7, 2)$ code and the minimum distance of the code is $d_{min} = 6$. This means that the code provides an effective order of diversity equal to 6.

At the receiver, the received signal is first dehopped and then demodulated by passing it through a parallel bank of eight matched filters, where each filter is tuned to one of the eight possible frequencies. Each filter output is envelope-detected, quantized to 4 bits (one of 16 levels), and fed to the decoder. The decoder takes the 56 filter outputs corresponding to the reception of each seven-symbol code word and forms 64 decision variables corresponding to the 64 possible code words in the $(7, 2)$ code by linearly combining the appropriate envelope detected outputs. A decision is made in favor of the code word having the largest decision variable.

By limiting the matched filter outputs to 16 levels, interference (cross talk) from other users of the channel causes a relatively small loss in performance (0.75 dB with strong interference on one chip and 1.5 dB with strong interference on two chips out of the seven). The AGC used in TATS has a time constant greater than the chip interval T_c, so that no attempt is made to perform optimum weighting of the demodulator outputs as described in Sec. 8.3.2.

The derivation of the error probability for the TATS signal in AWGN and worst-case partial-band interference is left as an exercise for the reader.

8.4 OTHER TYPES OF SPREAD SPECTRUM SIGNALS

DS and FH are the most common forms of spread spectrum signals used in practice. However, other methods may be used to introduce pseudo-randomness in a spread spectrum signal. One method, which is analogous to FH, is time hopping (TH). In TH, a time interval, which is selected to be much larger than the reciprocal of the information rate, is subdivided into a large number of time slots. The coded information symbols are transmitted in a pseudo-randomly selected time slot as a block of one or more code words. PSK modulation may be used to transmit the coded bits.

---

†Since mobile users are involved, there is a Doppler frequency offset associated with each transmission. This frequency offset must be tracked and compensated for in the demodulation of the signal. The sync symbol is used for this purpose.

For example, suppose that a time interval A is subdivided into 1000 time slots of width ... 1000 each. With ... information bit rate of ... bits/s, the number of bits to be transmitted is T seconds (A, T) ... time slots, and that number of (T_s, K) ... where T_s is the ... so we ... a time interval of $T \times 100$ seconds, we must ... modulation method, the ... approximately $R = 1000$ ...

A block diagram of a complete ... ... TH spread spectrum system is shown in Fig. 8.4.1. ... ... of the transmitted signal, and an ... rate below ... ... a TH system, as shown in Fig. 8.4.1. A buffer in ... ... to provide a uniform data stream to the user ...

... as usual, band interference degrades an optical TH spread spectrum system, narrow-time (pulsed) interference has a ... ... a TH spread spectrum with coding and interleaving are effective ... for combating this type of interference ... we have already demonstrated for DS and TH systems. Perhaps the major disadvantage of a TH system is the stringent timing requirements compared ... for DS but also to DS.

A combined ... spread spectrum signal can be obtained by combining DS, FH, and TH. For example, we may have a hybrid DS/FH, which means that a PN sequence is used in conjunction with frequency hopping. The signal transmitted on a single hop consists of a DS spread spectrum signal, which is demodulated coherently. However, the received signals on different hops are combined noncoherently (envelope or square-law detection). Since coherent detection is performed within a hop, the ... ... attained relative to a pure FH system. Therefore, the price paid for the gain in performance is an increase in complexity, greater cost, and more stringent ...

A hybrid DS/TH spread spectrum signal ... ... DS/FH. This does not seem to be as practical as DS/FH, primarily because ... ... to system complexity and more stringent timing requirements.

8.5 SYNCHRONIZATION OF SPREAD SPECTRUM SYSTEMS

Time synchronization ... the receiver to the received spread spectrum signal may be separated into two steps. There is an initial acquisition phase and a tracking phase after the signal timing has been initially acquired.

Acquisition in a direct sequence spread spectrum system ... the PN code within a fraction of the chip interval $T_c = 1/W$. The problem of initial synchronization may be viewed ... which we attempt to synchronize in time the local PN code to the transmitter in each possibly overlapping

FIGURE 8.4.1
Block diagram of a TH spread spectrum system.

For example, suppose that a time interval T is subdivided into 1000 time slots of width $T/1000$ each. With an information bit rate of R bits/s, the number of bits to be transmitted in T seconds is RT. Coding increases this number to RT/R_c bits, where R_c is the coding rate. Consequently, in a time interval of $T/1000$ seconds, we must transmit RT/R_c bits. If binary PSK is used as the modulation method, the bit rate is $1000R/R_c$ and the bandwidth required is approximately $W = 1000R/R_c$.

A block diagram of a transmitter and a receiver for a TH spread spectrum system is shown in Fig. 8.4.1. Due to the burst characteristics of the transmitted signal, buffer storage must be provided at the transmitter in a TH system, as shown in Fig. 8.4.1. A buffer may also be used at the receiver to provide a uniform data stream to the user.

Just as partial-band interference degrades an uncoded FH spread spectrum system, partial-time (pulsed) interference has a similar effect on a TH spread spectrum system. Coding and interleaving are effective means for combatting this type of interference, as we have already demonstrated for FH and TH systems. Perhaps the major disadvantage of a TH system is the stringent timing requirements compared not only to FH but, also, to DS.

Other types of spread spectrum signals can be obtained by combining DS, FH, and TH. For example, we may have a hybrid DS/FH, which means that a PN sequence is used in combination with frequency hopping. The signal transmitted on a single hop consists of a DS spread spectrum signal which is demodulated coherently. However, the received signals from different hops are combined noncoherently (envelope or square-law combining). Since coherent detection is performed within a hop, there is an advantage obtained relative to a pure FH system. However, the price paid for the gain in performance is an increase in complexity, greater cost, and more stringent timing requirements.

Another possible hybrid spread spectrum signal is DS/TH. This does not seem to be as practical as DS/FH, primarily because of an increase in system complexity and more stringent timing requirements.

8.5 SYNCHRONIZATION OF SPREAD SPECTRUM SYSTEMS

Time synchronization of the receiver to the received spread spectrum signal may be separated into two phases. There is an initial acquisition phase and a tracking phase after the signal timing has been initially acquired.

Acquisition. In a direct sequence spread spectrum system the PN code must be time synchronized to within a small fraction of the chip interval $T_c \approx 1/W$. The problem of initial synchronization may be viewed as one in which we attempt to synchronize in time the receiver clock to the transmitter clock. Usually, extremely accurate and stable time clocks are used in spread spectrum systems. Consequently, accurate time clocks result in a reduction of the time uncertainty between the receiver and the transmitter. However, there is always an initial timing uncertainty that is due to range uncertainty between the transmitter and the receiver. This is especially a problem when communication is taking place between two mobile users. In any case, the usual procedure for establishing initial synchronization is for the transmitter to send a known pseudo-random data sequence to the receiver. The receiver is continuously in a search mode looking for this sequence in order to establish initial synchronization.

Let us suppose that the initial timing uncertainty is T_u seconds and the chip duration is T_c. If initial synchronization is to take place in the presence of additive noise and other interference, it is necessary to dwell for $T_d = NT_c$ seconds in order to test synchronism at each time instant. If we search over the time uncertainty interval in (coarse) time steps if $T_c/2$, then the time required to establish initial synchronization is

$$T_{\text{init sync}} = \frac{T_u}{T_c/2} NT_c = 2NT_u \text{ s} \qquad (8.5.1)$$

Clearly, the synchronization sequence transmitted to the receiver must be at least as long as $2NT_c$ in order for the receiver to have sufficient time to perform the necessary search in a serial fashion.

In principle, matched filtering or cross correlation are optimum methods for establishing initial synchronization. A filter matched to the known data waveform generated from the known pseudo-random sequence continuously looks for exceedence of a predetermined threshold. When this occurs, initial synchronization is established and the demodulator enters the "data receive" mode.

Alternatively, we may use a *sliding correlator* as shown in Fig. 8.5.1. The correlator cycles through the time uncertainty, usually in discrete time intervals of $T_c/2$ seconds, and correlates the received signal with the known synchronization sequence. The cross correlation is performed over the time interval

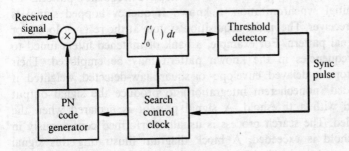

FIGURE 8.5.1
A sliding correlator for DS signal acquisition.

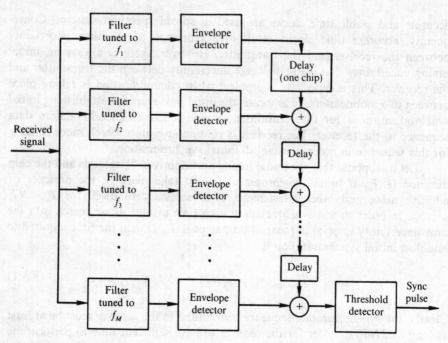

FIGURE 8.5.2
System for acquisition of a FH signal.

NT_c (N chips) and the correlator output is compared with a threshold to determine if the known signal sequence is present. If the threshold is not exceeded, the known reference sequence is advanced in time by $T_c/2$ seconds and the correlation process is repeated. These operations are performed until a signal is detected or until the search has been performed over the time uncertainty interval T_u. In the latter case, the search process is then repeated.

A similar process may also be used for FH signals. In this case the problem is to synchronize the PN code that controls the hopped frequency pattern. To accomplish this initial synchronization, a known frequency hopped signal is transmitted to the receiver. The initial acquisition system at the receiver looks for this known FH signal pattern. For example, a bank of matched filters tuned to the transmitted frequencies in the known pattern may be employed. Their outputs must be properly delayed, envelope- or square-law-detected, weighted, if necessary, and added (noncoherent integration) to produce the signal output which is compared with a threshold. A signal present is declared when the threshold is exceeded. The search process is usually performed continuously in time until a threshold is exceeded. A block diagram illustrating this signal acquisition scheme is given in Fig. 8.5.2. As an alternative, a single matched-filter-envelope detector pair may be used preceded by a frequency hopping

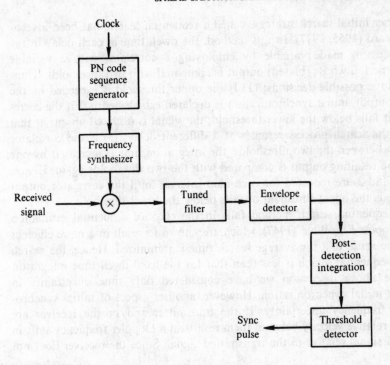

FIGURE 8.5.3
Alternative system for acquisition of a FH signal.

pattern generator and followed by a post-detection integrator and a threshold detector. This configuration, shown in Fig. 8.5.3, is based on a serial search and is akin to the sliding correlator for DS spread spectrum signals.

The sliding correlator for the DS signals or its counterpart shown in Fig. 8.5.3 for FH signals basically perform a serial search that is generally time consuming. As an alternative, one may introduce some degree of parallelism by having two or more such correlators operating in parallel and searching over nonoverlapping time slots. In such a case, the search time is reduced at the expense of a more complex and costly implementation. Figure 8.5.2 represents such a parallel realization for the FH signals.

During the search mode there may be false alarms that occur at the designed false alarm rate of the system. To handle the occasional false alarms, it is necessary to have an additional method or circuit that checks to confirm that the received signal at the output of the correlator remains above the threshold. With such a detection strategy, a large noise pulse that causes a false alarm will cause only a temporary exceedence of the threshold. On the other hand, when a signal is present the correlator or matched filter output will stay above the threshold for the duration of the transmitted signal. Thus, if confirmation fails, the search is resumed.

Another initial search strategy, called a sequential search, has been investigated by Ward (1965, 1977). In this method, the dwell time at each delay in the search process is made variable by employing a correlator with a variable integration period whose (biased) output is compared with two thresholds. Thus, there are three possible decisions: (1) If the upper threshold is exceeded by the correlator output, initial synchronization is declared established; (2) if the correlator output falls below the lower threshold, the signal is declared absent at that delay and the search process resumes at a different delay; (3) if the correlator output falls between the two thresholds, the integration time is increased by one chip and the resulting output is compared with the two thresholds again. Hence, steps 1, 2, and 3 are repeated for each chip interval until the correlator output either exceeds the upper threshold or falls below the lower threshold.

The sequential search method falls in the class of sequential estimation methods proposed by Wald (1947), which are known to result in a more efficient search in the sense that the average search time is minimized. Hence, the search time for a sequential search is less than that for the fixed dwell time integrator.

In the above discussion we have considered only time uncertainty in establishing initial synchronization. However, another aspect of initial synchronization is frequency uncertainty. If the transmitter and/or the receiver are mobile, the relative velocity between them results in a Doppler frequency shift in the received signal relative to the transmitted signal. Since the receiver does not

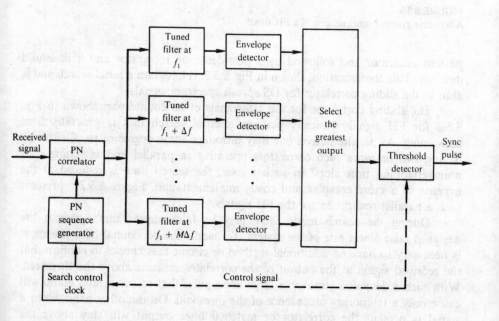

FIGURE 8.5.4
Initial search for Doppler frequency offset in a DS system.

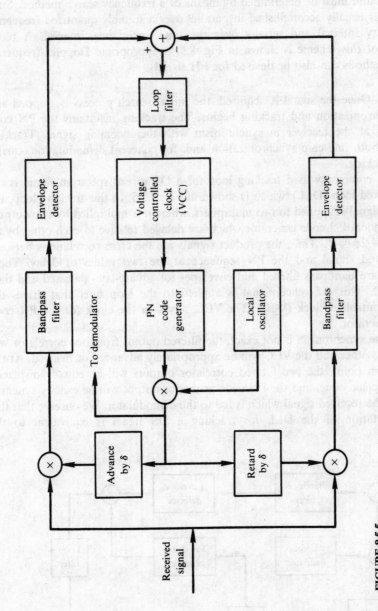

FIGURE 8.5.5
Delay-locked loop (DLL) for PN code tracking.

Received signal

Envelope detector

Voltage controlled clock (VCC)

Envelope detector

Bandpass filter

PN code generator

Local oscillator

Bandpass filter

Loop filter

To demodulator

Advance by δ

Retard by δ

×

×

×

+

−

867

usually know the relative velocity, *a priori*, the Doppler frequency shift is unknown and must be determined by means of a frequency search method. Such a search is usually accomplished in parallel over a suitably quantized frequency uncertainty interval and serially over the time uncertainty interval. A block diagram of this scheme is shown in Fig. 8.5.4. Appropriate Doppler frequency search methods can also be devised for FH signals.

Tracking. Once the signal is acquired, the initial search process is stopped and fine synchronization and tracking begins. The tracking maintains the PN code generator at the receiver in synchronism with the incoming signal. Tracking includes both fine chip synchronization and, for coherent demodulation, carrier phase tracking.

The commonly used tracking loop for a DS spread spectrum signal is the delay-locked loop (DLL), which is shown in Fig. 8.5.5. In this tracking loop, the received signal is applied to two multipliers, where it is multiplied by two outputs from the local PN code generator which are delayed relative to each other by an amount of $2\delta \leq T_c$. Thus, the product signals are the cross correlations between the received signal and the PN sequence at the two values of delay. These products are bandpass filtered and envelope- (or square-law) detected and then subtracted. This difference signal is applied to the loop filter that drives the voltage controlled clock (VCC). The VCC serves as the clock for the PN code signal generator.

If the synchronism is not exact, the filtered output from one correlator will exceed the other and the VCC will be appropriately advanced or delayed. At the equilibrium point, the two filtered correlator outputs will be equally displaced from the peak value, and the PN code generator output will be exactly synchronized to the received signal which is fed to the demodulator. We observe that this implementation of the DLL for tracking a DS signal is equivalent to the

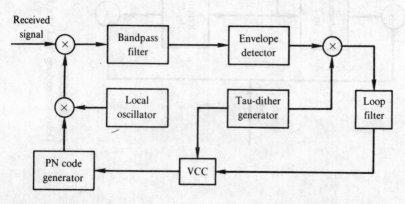

FIGURE 8.5.6
Tau-dither loop (TDL).

early-late gate bit tracking synchronizer previously discussed in Sec. 4.5.2 and shown in Fig. 4.5.11.

An alternate method for time tracking a DS signal is to use a tau-dither loop (TDL), illustrated by the block diagram in Fig. 8.5.6. The TDL employs only a single "arm" instead of the two "arms" shown in Fig. 8.5.5. By providing a suitable gating waveform it is possible to make this single "arm" implementation appear to be equivalent to the two "arm" realization. In this case, the cross correlation is regularly sampled at two values of delay, by stepping the code clock forward or backward in time by an amount δ. The envelope of the cross correlation that is sampled at $\pm\delta$ has an amplitude modulation whose phase relative to the tau-dither modulator determines the sign of the tracking error.

A major advantage of the TDL is the less costly implementation resulting from elimination of one of the two arms that are employed in the conventional DLL. A second and less apparent advantage is that the TDL does not suffer from performance degradation that is inherent in the DLL when the amplitude gain in the two arms is not properly balanced.

The DLL (and its equivalent, the TDL) generate an error signal by sampling the signal correlation function at $\pm\delta$ off the peak as shown in Fig.

(a) Autocorrelation function

(b) Tracking error signal

FIGURE 8.5.7
Autocorrelation function and tracking error signal for DLL.

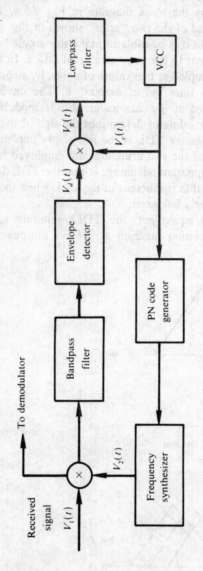

FIGURE 8.5.8
Tracking method for FH signals [from paper by Pickholtz et al. (1982). © 1982 IEEE. Reprinted with permission.]

(a) Tracking loop for FH signals

Received signal $V_1(t)$

To demodulator

$V_2(t)$

Frequency synthesizer

PN code generator

Bandpass filter

Envelope detector $V_d(t)$

$V_c(t)$

VCC

Lowpass filter

$V_g(t)$

FIGURE 8.5.8 (Continued).

(b) Waveforms for tracking an FH signal

Received frequency: f_0 f_1 f_2

Local FH: f_0 f_1 f_2

τ

$V_d(t)$: 1, 0

$V_c(t)$: +1, −1

$V_g(t)$

Timing offset $= \tau$

871

8.5.7a. This generates an error signal as shown in Fig. 8.5.7b. The analysis of the performance of the DLL is similar to that for the phase-locked loop (PLL) carried out in Sec. 4.5. If it were not for the envelope detectors in the two arms of the DLL, the loop resembles a Costas loop. In general, the variance of the time estimation error in the DLL is inversely proportional to the loop SNR, which depends on the input SNR to the loop and the loop bandwidth. Its performance is somewhat degraded as in the squaring PLL by the nonlinearities inherent in the envelope detectors, but this degradation is relatively small.

A typical tracking technique for FH spread spectrum signals is illustrated in Fig. 8.5.8a. This method is also based on the premise that, although initial acquisition has been achieved, there is a small timing error between the received signal and the receiver clock. The bandpass filter is tuned to a single intermediate frequency and its bandwidth is of the order of $1/T_c$, where T_c is the chip interval. Its output is envelope detected and then multiplied by the clock signal to produce a three-level signal, as shown in Fig. 8.5.8b, which drives the loop filter. Note that when the chip transitions from the locally generated sinusoidal waveform do not occur at the same time as the transitions in the incoming signal, the output of the loop filter will be either negative or positive, depending on whether the VCC is lagging or advanced relative to the timing of the input signal. This error signal from the loop filter will provide the control signal for adjusting the VCC timing signal so as to drive the frequency synthesized pulsed sinusoid to proper synchronism with the received signal.

8.6 BIBLIOGRAPHICAL NOTES AND REFERENCES

The introductory treatment on spread spectrum signals and their performance that we have given in this chapter is necessarily brief. Detailed and more specialized treatments of signal acquisition techniques, code tracking methods, and hybrid spread spectrum systems as well as other general topics on spread spectrum signals and systems can be found in the vast body of technical literature that now exists on the subject.

Historically, the primary application of spread spectrum communications has been in the development of secure (AJ) digital communication systems for military use. In fact, prior to 1970 most of the work on the design and development of spread spectrum communications was classified. In the last two decades, however, this trend has been reversed. The open literature now contains numerous publications on all aspects of spread spectrum signal analysis and design. Moreover, we have recently seen publications dealing with the application of spread spectrum signaling techniques to commercial communications such as interoffice radio communications [for reference, see paper by Pahlavan (1985)] and mobile-user radio communications [for reference, see paper by Yue (1983)].

A historical perspective on the development of spread spectrum communication systems covering the period 1920 to 1960 is given in a paper by Scholtz (1982). Tutorial treatments focusing on the basic concepts are found in the papers by Scholtz (1977) and Pickholtz et al. (1982). These papers also contain a large number of references to previous work. In addition, there are two papers by Viterbi (1979, 1985) which provide a basic review of the performance characteristic of DS and FH signaling techniques.

Comprehensive treatments of various aspects of analysis and design of spread spectrum signals and systems, including synchronization techniques are now available in the texts by Simon et al. (1985), Ziemer and Peterson (1985), and Holmes (1982). In addition to these texts, there are two special issues of the *IEEE Transactions on Communications* devoted to spread spectrum communications (August 1977 and May 1982). These issues contain a collection of papers devoted to a variety of topics, including multiple access techniques, synchronization techniques, and performance analyses with various types of interference. A number of important papers that have been published in IEEE journals have also been reprinted in book form by the IEEE Press [see Dixon (1976), ed. and Cook et al. (1983)]. Finally, we recommend the book by Golomb (1967) as a basic reference on shift register sequences for the reader who wishes to delve deeper into this topic.

PROBLEMS

8.1. Following the procedure outlined in Example 8.2.2, determine the error rate performance of a DS spread spectrum system in the presence of CW jamming when the signal pulse is

$$u(t) = \sqrt{\frac{16\mathscr{E}_c}{3T_c}} \cos^2 \frac{\pi}{T_c}\left(t - \frac{T_c}{2}\right) \qquad 0 \le t \le T_c$$

8.2. The sketch in Fig. P8.2 illustrates the power density spectra of a PN spread spectrum signal and a narrowband interference in an uncoded (trivial repetition

FIGURE P8.2

code) digital communications systems. Referring to Fig. 8.2.6 of the text, which shows the demodulator for this signal, sketch the (approximate) spectral characteristics of the signal and the interference after the multiplication of $r(t)$ with the output of the PN generator. Determine the fraction of the total interference which appears at the output of the correlator when the number of PN chips per bit is L_c.

8.3. Consider the concatenation of a Reed–Solomon $(31, 3)$ ($q = 32$-ary alphabet) as the outer code with a Hadamard $(16, 5)$ binary code as the inner code in a DS spread spectrum system. Assume that soft-decision decoding is performed on both codes. Determine an upper (union) bound on the probability of a bit error based on the minimum distance of the concatenated code.

8.4. The Hadamard $(n, k) = (2^m, m + 1)$ are low-rate codes with $d_{min} = 2^{m-1}$. Determine the performance of this class of codes for DS spread spectrum signals with binary PSK modulation and either soft-decision or hard-decision decoding.

8.5. Determine the error probability for a FH spread spectrum signal in which a binary convolutional code is used in combination with binary FSK. The interference on the channel is AWGN. The FSK demodulator outputs are square-law-detected and passed to the decoder, which performs optimum soft-decision Viterbi decoding as described in Sec. 5.3. Assume that the hopping rate is 1 hop/coded bit.

8.6. Repeat Prob. 8.5 for hard-decision Viterbi decoding.

8.7. Repeat Prob. 8.5 when fast frequency hopping is performed at a hopping rate of L hops/coded bit.

8.8. Repeat Prob. 8.5 when fast frequency hopping is performed with L hops/coded bit and the decoder is a hard-decision Viterbi decoder. The L chips/coded bit are square-law-detected and combined prior to the hard decision.

8.9. The TATS signal described in Sec. 8.3.3 is demodulated by a parallel bank of eight matched filters (octal FSK), and each filter output is square-law-detected. The eight outputs obtained in each of seven signal intervals (56 total outputs) are used to form the 64 possible decision variables corresponding to the Reed–Solomon $(7, 2)$ code. Determine an upper (union) bound of the code word error probability for AWGN and soft-decision decoding.

8.10. Repeat Prob. 8.9 for the worst-case partial-band interference channel.

8.11. Derive the results in (8.3.8) and (8.3.9) from (8.3.10).

8.12. Show that (8.3.14) follows from (8.3.13).

8.13. Derive (8.3.17) from (8.3.16).

8.14. The generator polynomials for constructing Gold code sequences of length $n = 7$ are

$$g_1(p) = p^3 + p + 1$$

$$g_2(p) = p^3 + p^2 + 1$$

Generate all the Gold codes of length 7 and determine the cross correlations of one sequence with each of the others.

8.15. In Sec. 8.2.3 we demonstrated techniques for evaluating the error of a coded system with interleaving in pulse interference by using the cutoff rate parameter R_0. Use the error probability curves given in Fig. P8.15 for rate 1/2 and 1/3 convolutional codes with soft-decision Viterbi decoding to determine the corresponding error rates for a coded system in pulse interference. Perform this computation for $L = 3$, 5, and 7.

Union bound
Continuous gaussian noise
Binary phase shift keying
Rate 1/3 convolutional code
  with Viterbi decoding
Soft decisions

(a)

FIGURE P8.15

8.16. In coded and interleaved DS binary PSK modulation with pulse jamming and soft-decision decoding, the cutoff rate is

$$R_0 = 1 + \log_2\left[1 + \alpha e^{-\alpha \mathscr{E}_c/N_0}\right]$$

where α is the fraction of the time the system is being jammed, $\mathscr{E}_c = \mathscr{E}_b R$, R is the bit rate, and $N_0 \equiv J_0$.
(a) Show that the SNR per bit $\mathscr{E}_b/N_0$ can be expressed as

$$\frac{\mathscr{E}_b}{N_0} = \frac{1}{\alpha R}\ln\frac{\alpha}{2^{1-R_0}-1}$$

$\mathscr{E}_b/N_0$ (dB)

(b)

FIGURE P8.15 (*Continued*).

(b) Determine the value of α that maximizes the required $\mathscr{E}_b/N_0$ (worst-case pulse jamming) and the resulting maximum value of $\mathscr{E}_b/N_0$.

(c) Plot the graph of $10 \log \mathscr{E}_b/r N_0$ versus R_0, where $r = R_0/R$, for worst-case pulse jamming and for AWGN ($\alpha = 1$). What conclusions do you reach regarding the effect of worst-case pulse jamming?

8.17. In a coded and interleaved frequency-hopped q-ary FSK modulation with partial band jamming and coherent demodulation with soft-decision decoding, the cutoff rate is

$$R_0 = \log_2 \frac{q}{1 + (q-1)\alpha e^{-\alpha \mathscr{E}_c/2N_0}}$$

where α is the fraction of the band being jammed, $\mathscr{E}_c$ is the chip (or tone) energy, and $N_0 = J_0$.

(*a*) Show that the SNR per bit can be expressed as

$$\frac{\mathscr{E}_b}{N_0} = \frac{2}{\alpha R} \ln \frac{(q-1)\alpha}{q 2^{-R_0} - 1}$$

(*b*) Determine the value of α that maximizes the required $\mathscr{E}_b/N_0$ (worst-case partial band jamming) and the resulting maximum value of $\mathscr{E}_b/N_0$.

(*c*) Define $r = R_0/R$ in the result for $\mathscr{E}_b/N_0$ from (*b*), and plot $10 \log \mathscr{E}_b/rN_0$ versus the normalized cutoff rate $R_0/\log_2 q$ for $q = 2, 4, 8, 16, 32$. Compare these graphs with the results of Prob. 8.16(*c*). What conclusions do you reach regarding the effect of worst-case partial band jamming? What is the effect of increasing the alphabet size q? What is the penalty in SNR between the results in Prob. 8.16(*c*) and q-ary FSK as $q \to \infty$?

REFERENCES
AND BIBILIOGRAPHY

Abend, K. and Fritchman, B. D. (1970). "Statistical Detection for Communication Channels with Intersymbol Interference," *Proc. IEEE*, pp. 779–785, May.

Abramson, N. (1963). *Information Theory and Coding*, McGraw-Hill, New York.

Anderberg, M. R. (1973). *Cluster Analysis for Applications*, Academic, New York.

Anderson J. B., Aulin, T., and Sundberg, C. W. (1986). *Digital Phase Modulation*, Plenum, New York.

Anderson R. R. and Salz, J. (1965). "Spectra of Digital FM," *Bell Sys. Tech. J.*, vol. 44 pp. 1165–1189, July–August.

Ash, R. B. (1965). *Information Theory*, Interscience, New York.

Aulin, T. (1980). "Viterbi Detection of Continuous Phase Modulated Signals," *Nat. Telecommun. Conf.*, *Conf. Record*, pp. 14.2.1–14.2.7, Houston, Tex., November.

Aulin, T., Rydbeck N., and Sundberg, C. W. (1981). "Continuous Phase Modulation—Part II: Partial Response Signaling," *IEEE Trans. Commun.*, vol. COM-29, pp. 210–225, March.

Aulin T. and Sundberg, C. W. (1981). "Continuous Phase Modulation—Part I: Full Response Signaling," *IEEE Trans. Commun.*, vol. COM-29, pp. 196–209, March.

Aulin T. and Sundberg, C. W. (1982a). "On the Minimum Euclidean Distance for a Class of Signal Space Codes," *IEEE Trans. Inform Theory*, vol. IT-28, pp. 43–55, January.

Aulin, T. and Sundberg, C. W. (1982b). "Minimum Euclidean Distance and Power Spectrum for a Class of Smoothed Phase Modulation Codes with Constant Envelope," *IEEE Trans. Commun.*, vol. COM-30, pp. 1721–1729, July.

Aulin, T. and Sundberg, C. W. (1984). "CPM—An Efficient Constant Aplitude Modulation Scheme," *Int. J. Satellite Commun.*, vol. 2, pp. 161–186.

Austin, M. E. (1967). "Decision-Feedback Equalization for Digital Communication Over Dispersive Channels," MIT Lincoln Laboratory, Lexington, Mass., Tech. Report No. 437, August.

Barrow, B. (1963). "Diversity Combining of Fading Signals with Unequal Mean Strengths," *IEEE Trans. Commun. Syst.*, vol. CS-11. pp. 73–78, March.

Bekir, N. E., Scholtz, R. A., and Welch, L. R. (1978). "Partial-Period Correlation Properties of PN Sequences," *1978 National Telecommunications Conf. Record*, Birmingham, Alabama, pp. 35.1.1–35.1.4, November.

Belfiore, C. A. and Park, J. H., Jr. (1979). "Decision-Feedback Equalization," *Proc. IEEE*, vol. 67, pp. 1143–1156, August.

Bello, P. A. and Nelin, B. D. (1962a). "Predetection Diversity Combining with Selectivity Fading Channels," *IRE Trans. Commun. Syst.*, vol. CS-10, pp. 32–42, March.

Bello, P. A. and Nelin, B. D. (1962b). "The Influence of Fading Spectrum on the Binary Error Probabilities of Incoherent and Differentially Coherent Matched Filter Receivers," *IRE Trans. Commun. Syst.*, vol. CS-10, pp. 160–168, June.

Bello, P. A. and Nelin, B. D. (1963). "The Effect of Frequency Selective Fading on the Binary Error Probabilities of Incoherent and Differentially Coherent Matched Filter Receivers," *IEEE Trans. Commun. Syst.*, vol. CS-11, pp. 170–186, June.

Bennett W. R. and Davey, J. R. (1965). *Data Transmission*, McGraw-Hill, New York.

Bennett, W. R. and Rice, S. O. (1963). "Spectral Density and Autocorrelation Functions Associated with Binary Frequency-Shift Keying," *Bell Syst. Tech. J.*, vol. 42, pp. 2355–2385, September.

Benveniste, A. and Goursat, M. (1984). "Blind Equalizers," *IEEE Trans. Commun.*, vol. COM-32, pp. 871–883, August.

Berger, T. (1971). *Rate Distortion Theory*, Prentice-Hall, Englewood Cliffs, N.J.

Berger, T. and Tufts, D. W. (1967). "Optimum Pulse Amplitude Modulation, Part I: Transmitter-Receiver Design and Bounds from Information Theory," *IEEE Trans. Inform. Theory*, vol. IT-13, pp. 196–208.

Berlekamp, E. R. (1968). *Algebraic Coding Theory*, McGraw-Hill, New York.

Berlekamp, E. R. (1972). *Key Papers in the Development of Coding Theory*, IEEE Press, New York.

Berlekamp, E. R. (1973). "Goppa Codes," *IEEE Trans. Inform. Theory*, vol. IT-19, pp. 590–592.

Bierman, G. J. (1977). *Factorization Methods for Discrete Sequential Estimation*, Academic, New York.

Blahut, R. E. (1983). *Theory and Practice of Error Control Codes*, Addison-Wesley, Reading, Mass.

Blahut, R. E. (1987). *Principles and Practice of Information Theory*, Addison-Wesley, Reading, Mass.

Blanco, M. (1989). "Probability of Symbol Error for M-ary FSK with Diversity," *IEEE Trans. Commun.*, vol. COM-37.

Bose, R. C. and Ray-Chaudhuri, D. K. (1960a). "On a Class of Error Correcting Binary Group Codes," *Inform. Control*, vol. 3, pp. 68–79, March.

Bose R. C. and Ray-Chaudhuri, D. K. (1960b). "Further Results in Error Correcting Binary Group Codes," *Inform. Control*, vol. 3, pp. 279–290, September.

Brennan, D. G. (1959). "Linear Diversity Combining Techniques," *Proc. IRE*, vol. 47, pp. 1075–1102, June.

Bucher, E. A. (1980). "Coding Options for Efficient Communications on Non-Stationary Channels," *Rec. IEEE Int. Conf. Comm.*, pp. 4.1.1–4.1.7.

Burg, J. P. (1967). "Maximum Entropy Spectral Analysis," *Proc. 37th Meeting of the Society of Exploration Geophysicists, 1967*; Also reprinted in *Modern Spectrum Analysis*, D. G. Childers (ed.), pp. 34–41, IEEE Press, New York, 1978.

Burton, H. O. (1969). "A Class of Asymptotically Optimal Burst Correcting Block Codes," *Proc. ICCC*, Boulder, Col., June.

Buzo, A., Gray, A. H., Jr., Gray, R. M., and Markel, J. D. (1980). "Speech Coding Based Upon Vector Quantization," *IEEE Trans Acoust., Speech, Signal Processing*, vol. ASSP-28 pp. 562–574, October.

Cahn, C. R. (1960). "Combined Digital Phase and Amplitude Modulation Communication Systems," *IRE Trans. Commun. Syst.*, vol. CS-8, pp. 150–155, September.

Calderbank, A.R. and Sloane, N. J. A. (1987). "New Trellis Codes Based on Lattices and Cosets," *IEEE Trans. Inform. Theory*, vol. IT-33, pp. 177–195, March.

Campopiano, C. N. and Glazer, B. G. (1962). "A Coherent Digital Amplitude and Phase Modulation Scheme," *IRE Trans. Commun. Syst.*, vol. CS-10, pp. 90–95, June.

Caraiscos, C. and Liu, B. (1984). "A Roundoff Error Analysis of the LMS Adaptive Algorithm," *IEEE Trans. Acoust., Speech, Signal Processing*, vol. ASSP-32, pp. 34–41, January.

Carayannis, G., Manolakis, D. G., and Kalouptsidis, N. (1983). "A Fast Sequential Algorithm for Least-Squares Filtering and Prediction," *IEEE Trans. Acoust., Speech, Signal Processing*, vol. ASSP-31, pp. 1394–1402, December.

Carayannis, G. Manolakis, D. G., and Kalouptsidis, N. (1986). "A Unified View of Parametric Processing Algorithms for Prewindowed Signals," *Signal Processing*, vol. 10, pp. 335–368, June.

Chang, D. Y., Gersho, A., Ramamurthi, B., and Shohan, Y. (1984). "Fast Search Algorithms for Vector Quantization and Pattern Matching," *Proc. IEEE Int. Conf. Acoust. Speech, Signal Processing*, San Diego, Calif., March, paper 9.11.

Chase, D. (1972). "A Class of Algorithms for Decoding Block Codes With Channel Measurement Information," *IEEE Trans. Inform. Theory*, vol. IT-18, pp. 170–182, January.

Chase, D. (1976). "Digital Signal Design Concepts for a Time-Varying Ricean Channel," *IEEE Trans. Commun.*, vol. COM-24, pp. 164–172, February.

Chien, R. T. (1964). "Cyclic Decoding Procedures for BCH Codes," *IEEE Trans. Inform. Theory*, vol. IT-10, pp 357–363, October.

Chyi, G. T., Proakis, J. G., and Keller, C.M. (1988). "Diversity Selection/Combining Schemes with Excess Noise-Only Diversity Receptions Over a Rayleigh-Fading Multipath Channel," *Proc. Conf. Inform. Sciences Syst.*, Princeton University, Princeton, N.J. March.

Cioffi, J. M. (1987). "Limited Precision Effects in Adaptive Filtering," *IEEE Trans. Circuits Systems*, vol. CAS-34, pp. 821–833, July.

Cioffi, J. M. and Kailath, T. (1984). "Fast Recursive-Least Squares Transversal Filters for Adaptive Filtering," *IEEE Trans. Acoust., Speech, Signal Processing*, vol. ASSP-32, pp. 304–337, April.

Cook, C. E., Ellersick, F. W., Milstein, L. B., and Schilling, D. L. (1983). *Spread Spectrum Communications*, IEEE Press, New York.

Costas, J. P. (1956). "Synchronous Communications," *Proc. IRE*, vol. 44, pp. 1713–1718, December.

Cramér, H. (1946). *Mathematical Methods of Statistics*, Princeton University Press, Princeton, N.J.

Daut, D. G., Modestino, J. W., and Wismer, L. D. (1982). "New Short Constraint Length Convolutional Code Construction for Selected Rational Rates," *IEEE Trans. Inform. Theory*, vol. IT-28, pp. 793–799, September.

Davenport, W. B., Jr. (1970). *Probability and Random Processes*, McGraw-Hill, New York.

Davenport, W. B., Jr. and Root, W. L. (1958). *Random Signals and Noise*, McGraw-Hill, New York.

deBuda, R. (1972). "Coherent Demodulation of Frequency Shift Keying with Low Deviation Ratio," *IEEE Trans. Commun.*, vol. COM-20, pp. 429–435, June.

Dixon, R. C. (1976). *Spread Spectrum Techniques*, IEEE Press, New York.

Drouilhet, P. R., Jr. and Bernstein, S. L. (1969). "TATS—A Bandspread Modulation-Demodulation System for Multiple Access Tactical Satellite Communication," *1969 IEEE Electronics and Aerospace Systems (EASCON) Conv. Record*, Washington, D.C., pp. 126–132, October 27–29.

Duffy, F. P. and Tratcher, T. W. (1971). "Analog Transmission Performance on the Switched Telecommunications Network," *Bell System Tech. J.*, vol. 50, pp. 1311–1347, April.

Durbin, J. (1959). "Efficient Estimation of Parameters in Moving-Average Models," *Biometrika*, vol. 46, parts 1 and 2. pp. 306–316.

Duttweiler, D. L., Mazo, J. E., and Messerschmitt, D. G. (1974). "Error Propagation in Decision-Feedback Equalizers," *IEEE Trans. Inform. Theory*, vol. IT-20, pp. 490–497, July.

Eleftheriou, E. and Falconer, D. D. (1987). "Adaptive Equalization Techniques for HF Channels," *IEEE J. Selected Areas Commun.*, vol. SAC-5 pp. 238–247, February.

Elias, P. (1954). "Error-Free Coding," *IRE Trans. Inform. Theory*, vol. IT-4, pp. 29–37, September.

Elias, P. (1955). "Coding for Noisy Channels," *IRE Convention Record*, vol. 3, part 4, pp. 37–46.

Eyuboglu, V. M. (1988). "Detection of Coded Modulation Signals on Linear, Severely Distorted Channels Using Decision-Feedback Noise Prediction with Interleaving." *IEEE Trans. Commun.*, vol. COM-36, pp. 401–409, April.

Falconer, D. D. (1976). "Jointly Adaptive Equalization and Carrier Recovery in Two-Dimensional Digital Communication Systems," *Bell Syst. Tech. J.*, vol. 55, pp. 317–334, March.

Falconer, D. D. (1982). "Adaptive Reference Echo Cancellation," *IEEE Trans. Commun.*, vol. COM-30, pp. 2083–2094, September.

Falconer, D. D. and Ljung, L. (1978). "Application of Fast Kalman Estimation to Adaptive Equalization," *IEEE Trans. Commun.*, vol. COM-26, pp. 1439–1446, October.

Falconer, D. D. and Mueller, K. H. (1979). "Adaptive Echo Cancellation/AGC Structures for Two-Wire, Full-Duplex Data Transmission," *Bell Syst. Tech. J.*, vol. 58, pp. 1593–1616, September.

Falconer, D. D. and Salz, J. (1977). "Optimal Reception of Digital Data Over the Gaussian Channel with Unknown Delay and Phase Jitter," *IEEE Trans. Inform. Theory*, vol. IT-23, pp. 117–126, January.

Fano, R. M. (1961) *Transmission of Information*, MIT Press, Cambridge, Mass.

Fano, R. M. (1963) "A Heuristic Discussion of Probabilistic Coding," *IEEE Trans. Inform. Theory*, vol. IT-9, pp. 64–74, April.

Feinstein, A. (1958). *Foundations of Information Theory*, McGraw-Hill, New York.

Fire, P. (1959). "A Class of Multiple-Error-Correcting Binary Codes for Non-Independent Errors," Sylvania Report No. RSL-E-2, Sylvania Electronic Defense Laboratory, Mountain View, Calif., March.

Flanagan, J. L., et al. (1979). "Speech Coding," *IEEE Trans. Commun.* vol. COM-27, pp. 710–736, April.

Forney, G. D., Jr. (1965). "One Decoding BCH Codes," *IEEE Trans. Inform. Theory*, vol. IT-11, pp. 549–557, October.

Forney, G. D., Jr. (1966a). *Concatenated Codes*, MIT Press, Cambridge, Mass.

Forney, G. D., Jr. (1966b). "Generalized Minimum Distance Decoding," *IEEE Trans. Inform. Theory*, vol. IT-12, pp. 125–131, April.

Forney, G. D., Jr. (1970). "Coding and Its Application in Space Communications," *IEEE Spectrum*, vol. 7, pp. 47–58, June.

Forney, G. D., Jr. (1971). "Burst Correcting Codes for the Classic Bursty Channel," *IEEE Trans. Commun. Tech.*, vol. COM-19, pp. 772–781, October.

Forney, G. D., Jr. (1972). "Maximum-Likelihood Sequence Estimation of Digital Sequences in the Presence of Intersymbol Interference," *IEEE Trans. Inform. Theory*, vol. IT-18, pp. 363–378, May.

Forney, G. D., Jr. (1974). "Convolutional Codes III: Sequential Decoding," *Inform. Control*, vol. 25, pp. 267–297, July.

Forney, G. D. Jr. (1988). "Coset Codes I: Introduction and Geometrical Classification," *IEEE Trans. Inform. Theory*, vol. IT-34, pp. 671–680, September.

Forney, G. D., Jr., Gallager, R. G., Lang, G. R., Longstaff, F. M., and Qureshi, S. U. (1984). "Efficient Modulation for Band-Limited Channels," *IEEE J. Select Areas Commun.*, vol. SAC-2, pp. 632–647, September.

Foschini, G. J. (1984). "Contrasting Performance of Faster-Binary Signaling with QAM," *Bell Syst. Tech. J.*, vol. 63, pp. 1419–1445, October.

Foschini, G. J. (1985). "Equalizing Without Altering or Detecting Data, *Bell Syst. Tech. J.*, vol. 64, pp. 1885–1911, October.

Foschini, G. J., Gitlin, R. D., and Weinstein, S. B. (1974). "Optimization of Two-Dimensional Signal Constellations in the Presence of Gaussian Noise," *IEEE Trans. Commun*, vol. COM-22, pp. 28–38, January.

Franks, L. E. (1969). *Signal Theory*, Prentice-Hall, Englewood Cliff, N.J.

Franks, L. E. (1980). "Carrier and Bit Synchronization in Data Communication—A Tutorial Review," *IEEE Trans. Commun.*, vol. COM-28, pp. 1107–1121, August.

Franks, L. E. (1981). "Synchronization Subsystems: Analysis and Design," in *Digital Communications, Satellite/Earth Station Engineering*, (K. Feher), Prentice-Hall, Englewood Cliffs, N.J.

Fredricsson, S. (1975). "Pseudo-Randomness Properties of Binary Shift Register Sequences," *IEEE Trans. Inform. Theory*, vol. IT-21, pp. 115–120, January.

Gaarder, N. T. (1971). "Signal Design for Fast-Fading Gaussian Channels," *IEEE Trans. Inform. Theory*, vol. IT-17, pp. 247–256, May.

Gallager, R. G. (1968). *Information Theory and Reliable Communication*, Wiley, New York.

Gardner, F. M. (1968). *Phaselock Techniques*, Wiley, New York.

Gardner, W. A. (1984). "Learning Characteristics of Stochastic-Gradient Descent Algorithms: A General Study, Analysis, and Critique," *Signal Processing*, vol. 6, pp. 113–133, April.

George, D. A., Bowen, R. R., and Storey, J. R. (1971). "An Adaptive Decision-Feedback Equalizer," *IEEE Trans. Commun. Tech.*, vol. COM-19, pp. 281–293, June.

Gersho, A. (1982). "On the Structure of Vector Quantizers," *IEEE Trans. Inform. Theory*, vol. IT-28, pp. 157–166, March.

Gersho, A. and Lawerence, V. B. (1984). "Multidimensional Signal Constellations for Voiceband Data Transmission," *IEEE J. Select. Areas Commun.*, vol. SAC-2, pp. 687–702, September.

Gerst, I. and Diamond, J. (1961). "The Elimination of Intersymbol Interference by Input Pulse Shaping," *Proc. IRE*, vol. 53, July.

Gilbert, E. N. (1952). "A Comparison of Signalling Alphabets," *Bell Syst. Tech. J.*, vol. 31, pp. 504–522, May.

Gitlin, R. D. Meadors, H. C., and Weinstein, S. B. (1982). "The Tap Leakage Algorithm: An Algorithm for the Stable Operation of a Digitally Implemented Fractionally Spaced, Adaptive Equalizer," *Bell Syst. Tech. J.*, vol. 61, pp. 1817–1839, October.

Gitlin, R. D. and Weinstein, S. B. (1979). "On the Required Tap-Weight Precision for Digitally Implemented Mean-Squared Equalizers," *Bell Syst. Tech. J.*, vol. 58, pp. 301–321, February.

Gitlin, R. D. and Weinstein, S. B. (1981). "Fractionally-Spaced Equalization: An Improved Digital Transversal Equalizer," *Bell Syst. Tech. J.*, vol. 60, pp. 275–296, February.

Glave, F. E. (1972). "An Upper Bound on the Probability of Error due to Intersymbol Interference for Correlated Digital Signals," *IEEE Trans. Inform. Theory*, vol. IT-18, pp. 356–362, May.

Goblick, T. J., Jr. and Holsinger, J. L. (1967). "Analog Source Digitization: A Comparison of Theory and Practice," *IEEE Trans Inform. Theory*, vol. IT-13, pp. 323–326, April.

Godard, D. N. (1974). "Channel Equalization Using a Kalman Filter for Fast Data Transmission," *IBM J. Res. Develop.*, vol 18, pp. 267–273, May.

Godard, D. N. (1980). "Self-Recovering Equalization and Carrier Tracking in Two-Dimensional Data Communications Systems," *IEEE Trans. Commun.*, vol. COM-28, pp. 1867–1875, November.

Golay, M. J. E. (1949). "Notes on Digital Coding," *Proc. IRE*, vol. 37, p. 657, June.

Gold, R. (1967). "Optimal Binary Sequences for Spread Spectrum Multiplexing," *IEEE Trans. Inform. Theory*, vol. IT-13, pp. 619–621, October.

Gold, R. (1968). "Maximal Recursive Sequences with 3-Valued Recursive Cross Correlation Functions," *IEEE Trans. Inform. Theory*, vol. IT-14, pp. 154–156, January.

Golomb. S. W. (1967). *Shift Register Sequences*, Holden-Day, San Francisco, Calif.

Goppa, V. D. (1970). "New Class of Linear Correcting Codes," *Probl. Peredach. Inform.*, vol. 6, pp. 24–30.

Goppa, V. D. (1971). "Rational Presentation of Codes and (L, g)-Codes," *Probl. Peredach. Inform.*, vol. 7, pp. 41–49.

Greefkes, J. A. (1970). "A Digitally Companded Delta Modulation Modem for Speech Transmission," *Proc. IEEE Int. Conf. on Communications*, pp. 7.33–7.48, June.

Green, P. E., Jr. (1962). "Radar Astronomy Measurement Techniques," MIT Lincoln Laboratory, Lexington, Mass., Tech. Report No. 282, December.

Gronemeyer, S. A. and McBride, A. L. (1976). "MSK and Offset QPSK Modulation," *IEEE Trans. Commun.*, vol. COM-24, pp. 809–820, August.

Gupta, S. C. (1975). "Phase-Locked Loops," *Proc. IEEE*, vol. 63, pp. 291–306, February.

Hahn, P. M. (1962). "Theoretical Diversity Improvement in Multiple Frequency Shift Keying," *IRE Trans. Commun. Syst.* vol. CS-10, pp. 177–184, June.

Hamming, R. W. (1950). "Error Detecting and Error Correcting Codes," *Bell Syst. Tech. J.*, vol. 29, pp. 147–160, April.

Hamming, R. W. (1986). *Coding and Information Theory*, Prentice-Hall, Englewood Cliffs, N.J.

Hancock, J. C. and Lucky, R. W. (1960). "Performance of Combined Amplitude and Phase-Modulated Communication Systems," *IRE Trans. Commun. Syst.*, vol. CS-8, pp. 232–237, December.

Hecht, M. and Guida, A. (1969). "Delay Modulation," *Proc. IEEE*, vol. 57, pp. 1314–1316, July.

Heller, J. A. (1968). "Short Constraint Length Convolutional Codes," Jet Propulsion Laboratory, California Institute of Technology, Pasadena, Calif., *Space Program Summary* 37–54, vol. 3, pp. 171–174, December.

Heller, J. A. (1975). "Feedback Decoding of Convolutional Codes," in *Advances in Communication Systems*, vol. 4, A. J. Viterbi (ed.), Academic, New York.

Heller, J. A. and Jacobs, I. M. (1971). "Viterbi Decoding for Satellite and Space Communication," *IEEE Trans. Commun. Tech.*, vol. COM-19, pp. 835–848, October.

Helstrom, C. W. (1955). "The Resolution of Signals in White Gaussian Noise," *Proc. IRE*, vol. 43, pp. 1111–1118, September.

Helstrom, C. W. (1968, 1960). *Statistical Theory of Signal Detection*, Pergamon, London.

Helstrom, C. W. (1984). *Probability and Stochastic Processes for Engineers*, Macmillan, New York.

Hildebrand, F. B. (1960). *Methods of Applied Mathematics*, Prentice-Hall, Englewood Cliffs, N.J.

Ho, E. Y. and Yeh, Y. S. (1970). "A New Approach for Evaluating the Error Probability in the Presence of Intersymbol Interference and Additive Gaussian Noise," *Bell Syst. Tech. J.*, vol. 49, pp. 2249–2265, November.

Hocquenghem, A. (1959). "Codes Correcteurs d'Erreurs," *Chiffres*, vol. 2, pp. 147–156.

Holmes, J. K. (1982). *Coherent Spread Spectrum Systems*, Wiley-Interscience, New York.

Hsu, F. M. (1982). "Square-Root Kalman Filtering for High-Speed Data Received over Fading Dispersive HF Channels," *IEEE Trans. Inform. Theory*, vol. IT-28, pp. 753–763, September.

Hsu, F. M. and Giordano A. A. (1978). "Digital Whitening Techniques for Improving Spread Spectrum Communications Performance in the Presence of Narrowband Jamming and Interference," *IEEE Trans. Commun.*, vol. COM-26, pp. 209–216, February.

Huffman, D. A. (1952). "A Method for the Construction of Minimum Redundancy Codes" *Proc*, *IRE*, vol. 40, pp. 1098–1101, September.

Itakura, F. (1975). "Minimum Prediction Residual Principle Applied to Speech Recognition," *IEEE Trans. Acoust., Speech, Signal Processing*, vol. ASSP-23, pp. 67–72, February.

Itakura, F. and Saito, S. (1968). "Analysis Synthesis Telephony Based on the Maximum-Likelihood Method," *Proc. 6th Int. Congr. Acoust.*, Tokyo, Japan, pp. C17–C20.

Jacobs, I. M. (1974). "Practical Applications of Coding," *IEEE Trans. Inform. Theory*, vol. IT-20, pp. 305–310, May.

Jayant, N. S. (1970). "Adaptive Delta Modulation with a One-Bit Memory," *Bell Syst. Tech. J.*, pp. 321–342, March.

Jayant, N. S. (1974). "Digital Coding of Speech Waveforms: PCM, DPCM, and DM Quantizers," *Proc. IEEE*, vol. 62, pp. 611–632, May.

Jayant, N. J. (1976). *Waveform Quantization and Coding*, IEEE Press, New York.

Jayant, N. S. and Noll, P. (1984). *Digital Coding of Waveforms*, Prentice-Hall, Englewood Cliffs, N.J.

Jelinek, F. (1968). *Probabilistic Information Theory*, McGraw-Hill, New York.

Jelinek, F. (1969). "Fast Sequential Decoding Algorithm Using a Stack," *IMB J. Res. Dev.*, vol. 13, pp. 675–685, November.

Jones, S. K., Cavin, R. K., and Reed, W. M. (1982). "Analysis of Error-Gradient Adaptive Linear Equalizers for a Class of Stationary-Dependent Processes," *IEEE Trans. Inform. Theory*, vol. IT-28, pp. 318–329, March.

Jordan, K. L., Jr. (1966). "The Performance of Sequential Decoding in Conjunction with Efficient Modulation," *IEEE Trans. Commun. Syst.*, vol. COM-14, pp. 283–287, June.

Justesen, J. (1972). "A Class of Constructive Asymptotically Good Algebraic Codes," *IEEE Trans. Inform. Theory*, vol. IT-18, pp. 652–656, September.

Kailath, T. (1960). "Correlation Detection of Signals Perturbed by a Random Channel," *IRE Trans. Inform. Theory*, vol. IT-6, pp. 361–366, June.

Kailath, T. (1961). "Channel Characterization: Time-Variant Dispersive Channels," in *Lectures on Communication System Theory*, Chap. 6, E. Baghdady, (ed.), McGraw-Hill, New York.

Kasami, T. (1966). "Weight Distribution Formula for Some Class of Cyclic Codes," Coordinated Science Laboratory, University of Illinois, Urbana, Ill., Tech. Report No. R-285, April.

Kelly, E. J., Reed, I. S., and Root, W. L. (1960). "The Detection of Radar Echoes in Noise, Pt. I," *J. SIAM*, vol. 8, pp. 309–341, September.

Kobayashi, H. (1971). "Simultaneous Adaptive Estimation and Decision Algorithm for Carrier Modulated Data Transmission Systems," *IEEE Trans. Commun. Tech.*, vol. COM-19, pp. 268–280, June.

Koll, V. G. and Weinstein, S. B. (1973). "Simultaneous Two-Way Data Transmission Over a Two-Wire Circuit," *IEEE Trans. Commun.*, vol. COM-21, pp. 143–147, February.

Kretzmer, E. R. (1966). "Generalization of a Technique for Binary Data Communication," *IEEE Trans. Commun. Tech.*, vol. COM-14, pp. 67–68, February.

Larsen, K. J. (1973). "Short Convolutional Codes with Maximal Free Distance for Rates 1/2, 1/3, and 1/4," *IEEE Trans. Inform. Theory*, vol. IT-19, pp. 371–372, May.

Lender, A. (1963). "The Duobinary Technique for High Speed Data Transmission," *AIEE Trans. Commun. Electronics*, vol. 82, pp. 214–218.

Levinson, N. (1947). "The Wiener RMS (Root Mean Square) Error Criterion in Filter Design and Prediction," *J. Math. Phys.*, vol. 25, pp. 261–278.

Lin, S. and Costello, D. J., Jr. (1983). *Error Control Coding: Fundamentals and Applications*, Prentice-Hall, Englewood Cliffs, N.J.

Linde, Y., Buzo, A., and Gray, R. M. (1980). "An Algorithm for Vector Quantizer Design," *IEEE Trans. Commun.*, vol. COM-28, pp. 84–95, January.

Lindell, G. (1985). "On Coded Continuous Phase Modulation," Ph.d. Dissertation, Telecommunication Theory, University of Lund, Lund, Sweden, May.

Lindholm, J. H. (1968). "An Analysis of the Pseudo-Randomness Properties of Subsequences of Long *m*-Sequences," *IEEE Trans. Inform. Theory*, vol. IT-14, pp. 569–576, July.

Lindsey, W. C. (1964). "Error Probabilities for Ricean Fading Multichannel Reception of Binary and *N*-ary Signals," *IEEE Trans. Inform. Theory*, vol. IT-10, pp. 339–350, October.

Lindsey, W. C. (1972). *Synchronization Systems in Communications*, Prentice-Hall, Englewood Cliffs, N.J.

Lindsey, W. C. and Chie, C. M. (1981). "A Survey of Digital Phase-Locked Loops," *Proc. IEEE*, vol. 69, pp. 410–432.

Lindsey, W. C. and Simon, M. K. (1973). *Telecommunication Systems Engineering*, Prentice-Hall, Englewood Cliffs, N.J.

Ling, F. (1987). "On the Convergence Speed of Fractionally-Spaced Equalizer Using Intersymbol Interpolation," *Proc. Int. Conf. Commun. Tech.*, Nanjing, China, November.

Ling, F. (1988). "Convergence Characteristics of LMS and LS Adaptive Algorithms for Signals with Rank-Deficient Correlation Matrices," *Proc. Int. Conf. Acoust., Speech, Signal Processing*, New York, 25.D.4.7, April.

Ling, F., Manolakis, D. G., and Proakis, J. G. (1986a). "Finite Word-Length Effects in Recursive Least Squares Algorithms with Application to Adaptive Equalization," *Annales des Telecommunications*, vol. 41, pp. 1–9, May–June.

Ling F., Manolakis, D. G., and Proakis, J. G. (1986b). "Numerically Robust Least-Squares Lattice-Ladder Algorithms with Direct Updating of the Reflection Coefficients," *IEEE Trans. Acoust., Speech, Signal Processing*, vol. ASSP-34, pp. 837–845, August.

Ling F. and Proakis, J. G. (1982). "Generalized Least Squares Lattice and Its Applications to DFE," *Proc. 1982, IEEE Int. Conf. on Acoustics, Speech and Sig. Proc.*, Paris, France, May.

Ling F. and Proakis, J. G. (1984a). "Numerical Accuracy and Stability: Two Problems of Adaptive Estimation Algorithms Caused by Round-Off Error, " *Proc. Int. Conf. Acoust., Speech, Signal Processing*, pp. 30.3.1–30.3.4, San Diego, Calif. March.

Ling F. and Proakis, J. G. (1984b). "Nonstationary Learning Characteristics of Least Squares Adaptive Estimation Algorithms," *Proc. Int. Conf. Acoust., Speech, Signal Processing*, pp. 3.7.1–3.7.4, San Diego, California, March.

Ling, F. and Proakis, J. G. (1984c). "A Generalized Multichannel Least-Squares Lattice Algorithm with Sequential Processing Stages," *IEEE Trans. Acoust., Speech, Signal Processing*, vol ASSP-32, pp. 381–389, April.

Ling, F. and Proakis, J. G. (1985). "Adaptive Lattice Decision-Feedback Equalizers-Their Performance and Application to Time-Variant Multipath Channels" *IEEE Trans. Commun.*, vol. COM-33, pp. 348–356, April.

Ling F. and Proakis, J. G. (1986). "A Recursive Modified Gram–Schmidt Algorithm," *IEEE Trans. Acoust., Speech, Signal Processing*, vol. ASSP-34, pp. 829–836, August.

Ling, F. and Quershi, S. U. H. (1989). "Lattice Predictive Decision-Feedback Equalizer for Digital Communication Over Fading Multipath Channels," *IEEE Trans. Commun.*, vol. COM-37, March.

Ljung, S. and Ljung, L. (1985). "Error Propagation Properties of Recursive Least-Squares Adaptation Algorithms," *Automatica*, vol. 21, pp. 159–167.

Lloyd, S. P. (1982). "Least Squares Quantization in PCM," *IEEE Trans. Inform. Theory*, vol. IT-28, pp. 129–137, March.

Loeve, M. (1955). *Probability Theory*, Van Nostrand, Princeton, N.J.

Long, G., Ling, F., and Proakis, J. G. (1988a). "Fractionally-Spaced Equalizers Based on Singular-Value Decomposition," *Proc. Int. Conf., Acoust., Speech, Signal Processing*, New York, 25.D.4.10, April.

Long, G., Ling, F., and Proakis, J. G. (1988b). "Applications of Fractionally-Spaced Decision-Feedback Equalizers to HF Fading Channels," *Proc. MILCOM*, San Diego, Calif., October.

Long, G., Ling, F., and Proakis, J. G. (1987). "Adaptive Transversal Filters with Delayed Coefficient Adaptation," *Proc. Int. Conf. Acoust., Speech, Signal Processing*, Dallas, Tex., March.

Long, G., Ling, F., and Proakis, J. G. (1989). "The LMS Algorithm with Delayed Coefficient Adaptation," *IEEE Trans. Acoust., Speech, Signal Processing*, vol. ASSP-37, March.

Lucky, R. W. (1965). "Automatic Equalization for Digital Communications, *Bell Syst. Tech. J.*, vol. 44, pp. 547–588, April.

Lucky, R. W. (1966). "Techniques for Adaptive Equalization of Digital Communication," *Bell Syst. Tech. J.*, vol. 45, pp. 255–286.

Lucky, R. W. and Hancock, J. C. (1962). "On the Optimum Performance of N-ary Systems Having Two Degrees of Freedom," *IRE Trans. Commun. Syst.*, vol. CS-10, pp. 185–192, June.

Lucky, R. W., Salz, J., and Weldon, E. J., Jr. (1968). *Principles of Data Communication*, McGraw-Hill, New York.

Lugannani, R. (1969). "Intersymbol Interference and Probability of Error in Digital Systems," *IEEE Trans. Inform. Theory*, vol. IT-15, pp. 682–688, November.

Lundgren, C. W. and Rummler, W. D. (1979). "Digital Radio Outage Due to Selective Fading-Observation vs. Prediction from Laboratory Simulation," *Bell Syst. Tech. J.*, vol. 58, pp. 1074–1100, May–June.

MacKenchnie, L. R. (1973). "Maximum Likelihood Receivers for Channels Having Memory," Ph.D. dissertation, Department of Electrical Engineering, University of Notre Dame, Notre Dame, Ind., January.

MacWilliams, F.J. and Sloane, J.J. (1977). *The Theory of Error Correcting Codes*, North Holland, New York.

Magee, F. R. and Proakis, J. G. (1973). "Adaptive Maximum-Likelihood Sequence Estimation for Digital Signaling in the Presence of Intersymbol Interference," *IEEE Trans. Inform. Theory*, vol. IT-19, pp. 120–124, January.

Makhoul, J. (1978). "A Class of All-Zero Lattice Digital Filters: Properties and Applications," *IEEE Trans. Acoustics, Speech, and Signal Processing*, vol. ASSP-26, pp. 304–314, August.

Makhoul, J., Roucos, S., and Gish, H. (1985). "Vector Quantization in Speech Coding," *Proc. IEEE*, vol. 73, pp. 1551–1587, November.

Marple, S. L., Jr. (1980). "A New Autoregressive Spectrum Analysis Algorithm," *IEEE Trans. Acoustics, Speech, and Signal Processing*, vol. ASSP-28, pp. 441–454, August.

Marple, S. L., Jr. (1982). "Fast Algorithm for Linear Prediction and System Identification Filters with Linear Phase," *IEEE Trans. Acoust. Speech, Signal Processing*, vol. ASSP-30, pp. 942–953, December.

Martin, D. R. and McAdam, P. L. (1980). "Convolutional Code Performance with Optimal Jamming," *Conf. Rec. Int. Conf. Commun.*, pp. 4. 3. 1–4. 3. 7, May.

Massey, J. L. (1963). *Threshold Decoding*, MIT Press, Cambridge, Mass.

Massey, J. L. (1965). "Step-by-Step Decoding of the BCH Codes," *IEEE Trans. Inform. Theory*, vol. IT-11, pp. 580–585, October.

Massey, J. L. and Sain, M. (1968). "Inverses of Linear Sequential Circuits," *IEEE Trans. Comput.*, vol. C-17, pp. 330–337, April.

Matis, K. R. and Modestino, J. W. (1982). "Reduced-State Soft-Decision Trellis Decoding of Linear Block Codes," *IEEE Trans. Inform. Theory*, vol. IT-28, pp. 61–68, January.

Max, J. (1960). "Quantizing for Minimum Distortion," *IRE Trans. Inform. Theory*, vol. IT-6, pp. 7–12, March.

Mazo, J. E. (1975). "Faster-Than-Nyquist Signaling," *Bell Syst. Tech. J.*, vol. 54, pp. 1451–1462, October.

Mazo, J. E. (1979). "On the Independence Theory of Equalizer Convergence," *Bell Syst. Tech. J.*, vol. 58, pp. 963–993, May.

Mengali, U. (1977). "Joint Phase and Timing Acquisition in Data Transmission," *IEEE Trans. Commun.*, vol. COM-25, pp. 1174–1185, October.

Messerschmitt, D.G. (1984). "Echo Cancellation in Speech and Data Transmission," *IEEE J. Selected Areas Commun.*, vol. SAC-2, pp. 283–297, March.

Meyers, M. H. and Franks, L. E. (1980). "Joint Carrier Phase and Symbol Timing for PAM Systems," *IEEE Trans. Commun.*, vol. COM-28, pp. 1121–1129, August.

Miller, K. S. (1964). *Multidimensional Gaussian Distributions*, Wiley, New York.

Monsen, P. (1971). "Feedback Equalization for Fading Dispersive Channels,"*IEEE Trans. Inform. Theory*, vol. IT-17, pp. 56–64. January.

Morf, M. (1977). "Ladder Forms in Estimation and System Identification," *Proc. 11th Annual Asilomar Conf. on Circuits, Systems and Computers*, Monterey, Calif., Nov. 7–9.

Morf, M., Dickinson, B., Kailath, T., and Vieira, A. (1977). "Efficient Solution of Covariance Equations for Linear Prediction," *IEEE Trans. Acoustics, Speech, and Signal Processing*, vol. ASSP-25, pp. 429–433, October.

Morf, M. and Lee, D. (1979). "Recursive Least Squares Ladder Forms for Fast Parameter Tracking," *Proc. 1978 IEEE Conf. on Decision and Control*, San Diego, Calif., pp. 1362–1367, January 12.

Morf, M., Lee, D., Nickolls, J., and Vieira, A. (1977). "A Classification of Algorithms for ARMA Models and Ladder Realizations," *Proc. 1977 IEEE Int. Conf. on Acoustics, Speech, and Sig. Proc.*, Hartford, Conn., pp. 13–19, May.

Morf, M., Vieira, A., and Lee, D. (1977). "Ladder Forms for Identification and Speech Processing," *Proc. 1977 IEEE Conf. on Decision and Control*, New Orleans, La., pp. 1074–1078, December.

Mueller, K. H. (1979). "Combining Echo Cancellation and Decision-Feedback Equalization," *Bell Syst. Tech. J.*, vol. 58, pp. 491–500, February.

Muller, D. E. (1954). "Application of Boolean Algebra to Switching Circuit Design and to Error Detection," *IRE Trans. Electronic Comput.*, vol. EC-3, pp. 6–12, September.

Mulligan, M. G. (1988). "Multi-Amplitude Continuous Phase Modulation with Convolutional Coding," Ph.D. Dissertation, Department of Electrical and Computer Engineering, Northeastern University, June.

Natali, F. D. and Walbesser, W. J. (1969). "Phase-Locked Loop Detection of Binary PSK Signals Utilizing Decision Feedback," *IEEE Trans. Aerospace Electronic Syst.*, vol. AES-5, pp. 83–90, January.

Neyman, J. and Pearson, E. S. (1933). "On the Problem of the Most Efficient Tests of Statistical Hypotheses," *Phil. Trans. Roy. Soc. London, Series A*, vol. 231, pp. 289–337.

North, D. O. (1943). "An Analysis of the Factors Which Determine Signal/Noise Discrimination in Pulse-Carrier Systems," RCA Tech. Report No. 6 PTR-6C.

Nyquist, H. (1928). "Certain Topics in Telegraph Transmission Theory," *AIEE Trans.*, vol. 47, pp. 617–644.

Odenwalder, J. P. (1970). "Optimal Decoding of Convolutional Codes," Ph.D. Dissertation, Department of Systems Sciences, School of Engineering and Applied Sciences, University of California, Los Angeles.

Odenwalder, J. P. (1976). "Dual-*k* Convolutional Codes for Noncoherently Demodulated Channels," *Proc. Int. Telemetering Conf.*, vol. 12, pp. 165–174, September.

Olsen, J. D. (1977). "Nonlinear Binary Sequences with Asymptotically Optimum Periodic Cross Correlation," Ph.D. Dissertation, University of Southern California, December.

Omura, J. (1971). "Optimal Receiver Design for Convolutional Codes and Channels with Memory Via Control Theoretical Concepts," *Inform. Sci.*, vol. 3, pp. 243–266.

Omura, J. K. and Levitt, B. K. (1982). "Coded Error Probability Evaluation for Antijam Communication Systems," *IEEE Trans. Commun.*, vol. COM-30, pp. 896–903, May.

Oppenheim, A. V. and Schafer, R. W. (1975). *Digital Signal Processing*, Prentice-Hall, Englewood Cliffs, N.J.

Osborne, W. P. and Luntz, M. B. (1974). "Coherent and Noncoherent Detection of CPFSK," *IEEE Trans. Commun.*, vol. COM-22, pp. 1023–1036, August.

Paaske, E. (1974). "Short Binary Convolutional Codes with Maximal Free Distance for Rates 2/3 and 3/4," *IEEE Trans. Inform. Theory*, vol. IT-20, pp. 683–689, September.

Paez, M. D. and Glisson, T. H. (1972). "Minimum Mean Squared Error Quantization in Speech PCM and DPCM Systems," *IEEE Trans. Commun.*, vol. COM-20, pp. 225–230, April.

Pahlavan, K. (1985). "Wireless Communications for Office Information Networks," *IEEE Commun. Mag.*, vol. 23, pp. 18–27, June.

Papoulis, A. (1984, 1965). *Probability, Random Variables, and Stochastic Processes*, McGraw-Hill, New York.

Paul, D. B. (1983). "An 800 bps Adaptive Vector Quantization Vocoder Using a Perceptual Distance Measure," *Proc. IEEE Int. Conf. Acoust., Speech, Signal Processing*, Boston, Mass. pp. 73–76, April.

Pearson, K. (1965). *Tables of the Incomplete Γ-Function*, Cambridge University Press, London.

Peebles, P. Z. (1987, 1980). *Probability, Random Variables, and Random Signal Principles*, McGraw-Hill, New York.

Peterson, W. W. (1960). "Encoding and Error-Correction Procedures for Bose–Chaudhuri Codes," *IRE Trans. Inform. Theory*, vol. IT-6, pp. 459–470, September.

Peterson, W. W. and Weldon, E. J., Jr. (1972). *Error-Correcting Codes*, 2d ed. MIT Press, Cambridge, Mass.

Picci, G. and Prati, G. (1987). "Blind Equalization and Carrier Recovery Using a Stop-and-Go Decision Directed Algorithm," *IEEE Trans. Commun.*, vol. COM-35, pp. 877–887, September.

Picinbono, B. (1978). "Adaptive Signal Processing for Detection and Communication," in *Communication Systems and Random Process Theory*, J. K. Skwirzynski (ed.), Sijthoff & Nordhoff, Alphen aan den Rijn, The Netherlands.

Pickholtz, R. L., Schilling, D. L., and Milstein, L. B. (1982). "Theory of Spread Spectrum Communications—A Tutorial," *IEEE Trans. Commun.*, vol. COM-30, pp. 855–884, May.

Pieper, J. F., Proakis, J. G., Reed, R. R., and Wolf, J. K. (1978). "Design of Efficient Coding and Modulation for a Rayleigh Fading Channel," *IEEE Trans. Inform. Theory*, vol. IT-24, pp. 457–468, July.

Pierce, J. N. (1958). "Theoretical Diversity Improvement in Frequency-Shift Keying," *Proc. IRE*, vol. 46, pp. 903–910, May.

Pierce, J. N. and Stein, S. (1960). "Multiple Diversity with Non-Independent Fading," *Proc. IRE*, vol. 48, pp. 89–104, January.

Plotkin, M. (1960). "Binary Codes with Specified Minimum Distance," *IRE Trans. Inform. Theory*, vol. IT-6, pp. 445–450, September.

Price, R. (1954). "The Detection of Signals Perturbed by Scatter and Noise," *IRE Trans. Inform. Theory*, vol. PGIT-4, pp. 163–170, September.

Price, R. (1956). "Optimum Detection of Random Signals in Noise, with Application to Scatter-Multipath Communication," *IRE Trans. Inform. Theory*, vol. IT-2, pp. 125–135, December.

Price, R. (1962a). "Error Probabilities for Adaptive Multichannel Reception of Binary Signals," MIT Lincoln Laboratory, Lexington, Mass., Tech. Report No. 258, July.

Price, R. (1962b). "Error Probabilities for Adaptive Multichannel Reception of Binary Signals," *IRE Trans. Inform. Theory*, vol. IT-8, pp. 305–316, September.

Price, R. (1972). "Nonlinearly Feedback-Equalized PAM vs. Capacity," *Proc. 1972 IEEE Int. Conf. on Communications*, Philadelphia, Penn., pp. 22.12–22.17, June.

Price, R. and Green, P. E., Jr. (1958). "A Communication Technique for Multipath Channels," *Proc. IRE*, vol. 46, pp. 555–570, March.

Price, R. and Green, P. E., Jr. (1960). "Signal Processing in Radar Astronomy—Communication via Fluctuating Multipath Media," MIT Lincoln Labortory, Lexington, Mass., Tech. Report No. 234, October.

Proakis, J. G. (1968). "Probabilities of Error for Adaptive Reception of M-Phase Signals," *IEEE Trans. Commun. Tech.*, vol. COM-16, pp. 71–81, February.

Proakis, J. G. (1975). "Advances in Equalization for Intersymbol Interference," in *Advances in Communication Systems*, vol. 4, A. J. Viterbi (ed.), Academic, New York.

Proakis, J. G., Drouilhet, P. R., Jr., and Price, R. (1964). "Performance of Coherent Detection Systems Using Decision-Directed Channel Measurement," *IEEE Trans. Commun. Syst.*, vol. CS-12, pp. 54–63, March.

Proakis, J. G. and Ling, F. (1984). "Recursive Least Squares Algorithms for Adaptive Equalization of Time-Variant Multipath Channels," *Proc. Int. Conf. Commun*, Amsterdam, The Netherlands, May.

Proakis, J. G. and Manolakis, D. G. (1988). *Introduction to Digital Processing*, Macmillan, New York.

Proakis, J. G. and Miller J. H. (1969). "Adaptive Receiver for Digital Signaling through Channels with Intersymbol Interference," *IEEE Trans. Inform. Theory*, vol. IT-15, pp. 484–497, July.

Proakis, J. G. and Rahman, I. (1979). "Performance of Concatenated Dual-k Codes on a Rayleigh Fading Channel with a Bandwidth Constraint," *IEEE Trans. Commun.*, vol. COM-27, pp. 801–806, May.

Pursley, M. B. (1979). "On the Mean-Square Partial Correlation of Periodic Sequences," *Proc. 1979 Conf. Inform. Science and Systems*, Johns Hopkins University, Baltimore, Md., pp. 377–379, March.

Quatieri, T. F. and O'Leary, G. C. (1986). "Decision-Directed Echo Cancellation for Full-Duplex Data Transmission at 4800 bps," *Proc. Int. Conf. Acoust., Speech, Signal Processing*, Tokyo, Japan, April.

Qureshi, S. U. H. (1976). "Timing Recovery for Equalized Partial Response Systems, " *IEEE Trans. Communications*, vol. COM-24, pp. 1326–1331, December.

Qureshi, S. U. H. (1985). "Adaptive Equalization," *Proc. IEEE*, vol. 53, pp. 1349–1387, September.

Qureshi, S. U. H. and Forney, G. D., Jr. (1977). "Performance and Properties of a $T/2$ Equalizer," *Natl. Telecom. Conf. Record*, pp. 11.1.1–11.1.14, Los Angeles, Calif., December.

Rabiner, L. R. and Schafer, R. W. (1978). *Digital Processing of Speech Signals*, Prentice-Hall, Englewood Cliffs, N.J.

Rahman, I. (1981). "Bandwidth Constrained Signal Design for Digital Communication over Rayleigh Fading Channels and Partial Band Interference Channels," Ph.D. Dissertation, Department of Electrical Engineering, Northeastern University, Boston, Mass.

Ramsey, J. L. (1970). "Realization of Optimum Interleavers," *IEEE Trans. Inform. Theory*, vol. IT-16, pp. 338–345.

Reed, I. S. (1954). "A Class of Multiple-Error Correcting Codes and the Decoding Scheme," *IRE Trans. Inform.*, vol. IT-4, pp. 38–49, September.

Reed, I. S. and Solomon, G. (1960). "Polynomial Codes Over Certain Finite Fields," *SIAM J.*, vol. 8, pp. 300–304, June.

Roucos, S., Schwartz, R., and Makhoul, J. (1982). "Segment Quantization for Very-Low-Rate Speech Coding," *Proc. IEEE Int. Conf. Acoust., Speech, Signal Processing*, Paris, France, pp. 1565–569. May.

Rowe, H. E. and Prabhu, V. K. (1975). "Power Spectrum of a Digital Frequency Modulation Signal," *Bell Syst. Tech. J.*, vol. 54, pp. 1095–1125, July–August.

Rummler, W. D. (1979). "A New Selective Fading Model: Application to Propagation Data," *Bell Syst. Tech. J.*, vol. 58, pp. 1037–1071, May–June.

Saltzberg, B. R. (1968). "Intersymbol Interference Error Bounds with Application to Ideal Bandlimited Signaling," *IEEE Trans. Inform. Theory*, vol. IT-14, pp. 563–568, July.

Salz, J. (1973). "Optimum, Mean-Square Decision Feedback Equalization," *Bell Syst. Tech. J.*, vol. 52, pp. 1341–1373, October.

Salz, J. (1983). "On the Start-up Problem in Digital Echo Cancellers,," *Bell. Syst. Tech. J.*, vol. 62, pp. 1353–1364, July–August.

Salz, J., Sheehan, J. R., and Paris, D. J. (1971). "Data Transmission by Combined AM and PM," *Bell Syst. Tech. J.*, vol. 50, pp. 2399–2419, September.

Sarwate, D. V. and Pursley, M. B. (1980). "Crosscorrelation Properties of Pseudorandom and Related Sequences," *Proc. IEEE*, vol. 68, pp. 593–619, May.

Sato, Y. (1975). "A Method of Self-Recovering Equalization for Multilevel Amplitude-Modulation Systems," *IEEE Trans. Commun.*, vol. COM-23, pp. 679–682, June.

Satorius, E. H. and Alexander, S. T. (1979). "Channel Equalization Using Adaptive Lattice Algorithms," *IEEE Trans. Communications*, vol. COM-27, pp. 899–905, June.

Satorius, E. H. and Pack, J. D. (1981). "Application of Least Squares Lattice Algorithms to Adaptive Equalization," *IEEE Trans. Commun.*, vol. COM-29, pp. 136–142, February.

Savage, J. E. (1966). "Sequential Decoding—The Computation Problem," *Bell Syst. Tech. J.*, vol. 45, pp. 149–176. January.

Scholtz, R. A. (1977). "The Spread Spectrum Concept," *IEEE Trans. Commun.*, vol. COM-25, pp. 748–755, August.

Scholtz, R. A. (1979). "Optimal CDMA Codes," *1979 National Telecommunications Conf. Record*, Washington, D.C., pp. 54.2.1–54.2.4, November.

Scholtz, R. A. (1982). "The Origins of Spread Spectrum," *IEEE Trans. Commun.*, vol. COM-30, pp. 822–854, May.

Schonhoff, T. A. (1976). "Symbol Error Probabilities for M-ary CPFSK: Coherent and Noncoherent Detection," *IEEE Trans. Commun.*, vol. COM-24, pp. 644–652, June.

Shannon, C. E. (1948a). "A Mathematical Theory of Communication," *Bell Syst. Tech. J.*, vol. 27, pp. 379–423, July.

Shannon, C. E. (1948b). "A Mathematical Theory of Communication," *Bell Syst. Tech. J.*, vol. 27, pp. 623–656, October.

Shannon, C. E. (1959a). "Coding Theorems for a Discrete Source with a Fidelity Criterion," *IRE Nat. Conv. Rec.*, pt. 4, pp. 142–163, March.

Shannon, C. E. (1959b). "Probability of Error for Optimal Codes in a Gaussian Channel," *Bell Syst. Tech. J.*, vol. 38, pp. 611–656, May.

Shimbo, O. and Celebiler, M. (1971). "The Probability of Error due to Intersymbol Interference and Gaussian Noise in Digital Communication Systems," *IEEE Trans. Commun. Tech.*, vol. COM-19, pp. 113–119, April.

Simon, M. K., Omura, J. K., Scholtz, R. A., and Levitt, B. K. (1985). *Spread Spectrum Communications Vol. I, II, III*, Computer Science Press, Rockville, Md.

Simon, M. K. and Smith, J. G. (1973). "Hexagonal Multiple Phase-and-Amplitude-Shift Keyed Signal Sets," *IEEE Trans. Commun.*, COM-21, pp. 1108–1115, October.

Slepian, D. (1956). "A Class of Binary Signaling Alphabets," *Bell Syst. Tech. J.*, vol. 35, pp. 203–234, January.

Smith, J. W. (1965). "The Joint Optimization of Transmitted Signal and Receiving Filter for Data Transmission Systems," *Bell Syst. Tech. J.*, vol. 44, pp. 1921–1942, December.

Sondhi, M. M. and Berkley, D. A. (1980). "Silencing Echoes on the Telephone Network," *Proc. IEEE*, vol. 68, pp. 948–963, August.

Stenbit, J. P. (1964). "Table of Generators for BCH Codes," *IEEE Trans. Inform. Theory*, vol. IT-10, pp. 390–391, October.

Stiffler, J. J. (1971). *Theory of Synchronous Communications*, Prentice-Hall, Englewood Cliffs, N.J.

Stock, D. T. M. amd Kailath, T. (1988). "Numerically Stable Fast Recursive Least-Squares Transversal Filters," *Proc. Int. Conf. Acoust., Speech, Signal Processing*, pp. 1365–1368, New York, April.

Sundberg, C. E. (1986). "Continuous Phase Modulation," *IEEE Commun. Mag.*, vol. 24, pp. 25–38, April.

Thomas, C. M., Weidner, M. Y., and Durrani, S. H. (1974). "Digital Amplitude-Phase-Keying with M-ary Alphabets," *IEEE Trans. Commun.*, vol. COM-22, pp. 168–180, February.

Tufts, D. W. (1965). "Nyquist's Problem—The Joint Optimization of Transmitter and Receiver in Pulse Amplitude Modulation," *Proc. IEEE*, vol. 53, pp. 248–259, March.

Turin, G. L. (1961). "On Optimal Diversity Reception," *IRE Trans. Inform. Theory*, vol. IT-7, pp. 154–166, July.

Turin, G. L. (1962). "On Optimal Diversity Reception II," *IRE Trans. Commun. Syst.*, vol. CS-12, pp. 22–31, March.

Ulrych, T. J. and Bishop, T. N. (1975). "Maximum Entropy Spectral Analysis and Autoregressive Decomposition," *Rev. Geophys. Space Phys.*, vol. 13, pp. 183–200, February.

Ungerboeck, G. (1972). "Theory on the Speed of Convergence in Adaptive Equalizers for Digital Communication," *IBM J. Rec. Devel.*, vol. 16, pp. 546–555, November.

Ungerboeck, G. (1974). "Adaptive Maximum-Likelihood Receiver for Carrier-Modulated Data-Transmission Systems," *IEEE Trans. Commun.*, vol. COM-22, pp. 624–636, May.

Ungerboeck, G. (1976). "Fractional Tap-Spacing Equalizer and Consequences for Clock Recovery in Data Modems," *IEEE Trans. Commun.*, vol. COM-24, pp. 856–864, August.

Ungerboeck, G. (1982). "Channel Coding with Multilevel/Phase Signals," *IEEE Trans. Inform. Theory*, vol. IT-28, pp. 55–67, January.

Ungerboeck, G. (1987). "Trellis-Coded Modulation with Redundant Signal Sets, Parts I and II," *IEEE Commun. Mag.*, vol. 25, pp. 5–21, February.

Ungerboeck, G. and Csajka, I. (1976). "On Improving Data-Link Performance by Increasing the Channel Alphabet and Introducing Sequence Coding, *1976 Inter. Conf. Inform. Theory*, *Ronneby, Sweden*, June.

Van Trees, H. L. (1968). *Detection, Estimation, and Modulation Theory. Part I*, Wiley, New York.

Varsharmov. R. R. (1957). "Estimate of the Number of Signals in Error Correcting Codes," *Doklady Akad, Nauk, S.S.S.R.*, vol. 117, pp. 739–741.

Viterbi, A. J. (1966). *Principles of Coherent Communication*, McGraw-Hill, New York.

Viterbi, A. J. (1967). "Error Bounds for Convolutional Codes and an Asymptotically Optimum Decoding Algorithm," *IEEE Trans. Inform. Theory*, vol. IT-13, pp. 260–269, April.

Viterbi, A. J. (1971). "Convolutional Codes and Their Performance in Communication Systems," *IEEE Trans. Commun. Tech.* vol. COM-19, pp. 751–772, October.

Viterbi, A. J. (1978). "A Processing Satellite Transponder for Multiple Access by Low-Rate Mobile Users," *Proc. Fourth Int. Conf. on Digital Satellite Communications*, Montreal, Canada, pp. 166–174, October.

Viterbi, A. J. (1979). "Spread Spectrum Communication—Myths and Realities," *IEEE Commun. Mag.*, vol. 17, pp. 11–18, May.

Viterbi, A. J. (1985), "When Not to Spread Spectrum—A Sequel," *IEEE Commun. Mag.*, vol. 23, pp. 12–17, April.

Viterbi, A. J. and Jacobs, I. M. (1975). "Advances in Coding and Modulation for Noncoherent Channels Affected by Fading, Partial Band, and Multiple-Access Interference," in *Advances in Communication Systems*, vol. 4, A. J. Viterbi (ed.), Academic, New York.

Viterbi, A. J. and Omura, J. K. (1979). *Principles of Digital Communication and Coding*, McGraw-Hill, New York.

Wainberg, S. and Wolf, J. K. (1970). "Subsequences of Pseudo-Random Sequences," *IEEE Trans. Commun. Tech.*, vol. COM-18, pp. 606–612, October.

Wainberg, S. and Wolf, J. K. (1973). "Algebraic Decoding of Block Codes Over a q-ary Input, Q-ary Output Channel, $Q > q$," *Inform. and Control*, vol. 22, pp. 232–247, April.

Wald, A. (1974). *Sequential Analysis*, Wiley, New York.

Ward, R. B. (1965). "Acquisition of Pseudonoise Signals by Sequential Estimation," *IEEE Trans. Commun. Technol.*, vol. COM-13, pp. 474–483, December.

Ward, R. B. and Yiu, K. P. (1977). "Acquisition of Pseudonoise Signals by Recursion-Aided Sequential Estimation," *IEEE Trans. Commun.*, vol. COM-25, pp. 784–794, August.

Weber, W. J., III, Stanton, P. H., and Sumida, J. T. (1978). "A Bandwidth Compressive Modulation System Using Multi-Amplitude Minimum-Shift Keying (MAMSK)," *IEEE Trans. Commun.*, vol. COM-26, pp. 543–551, May.

Wei, L. F. (1987). "Trellis-Coded Modulation with Multi-Dimensional Constellations," *IEEE Trans. Inform. Theory*, vol. IT-33, pp. 483–501, July.

Weinstein, S. B. (1977). "A Passband Data-Driven Echo Canceller for Full-Duplex Transmission on Two-Wire Circuits," *IEEE Trans. Commun.*, vol. COM-25, pp. 654–666, July.

Welch, L. R. (1974). "Lower Bounds on the Maximum Cross Correlation of Signals," *IEEE Trans. Inform. Theory*, vol. IT-20, pp. 397–399, May.

Weldon, E. J., Jr. (1971). "Decoding Binary Block Codes on Q-ary Output Channels," *IEEE Trans. Inform. Theory*, vol. IT-17, pp. 713–718, November.

Widrow, B. (1966). "Adaptive Filters, I: Fundamentals," Stanford Electronics Laboratory, Stanford University, Stanford, Calif., Tech. Report No. 6764-6, December.

Widrow, B. and Hoff, M. E., Jr. (1960). "Adaptive Switching Circuits," *IRE WESCON Conc. Rec.*, pt. 4, pp. 96–104.

Wolf, J. K., (1978). "Efficient Maximum Likelihood Decoding of Linear Block Codes Using a Trellis," *IEEE Trans. Inform. Theory*, vol. IT-24, pp. 76–81, January.

Wozencraft, J. M. (1957). "Sequential Decoding for Reliable Communication," *IRE Natl. Conv. Rec.*, vol. 5, pt. 2, pp. 11–25.

Wozencraft, J. M. and Jacobs, I. M. (1965). *Principles of Communication Engineering*, Wiley, New York.

Wozencraft, J. M. and Kennedy, R. S. (1966). "Modulation and Demodulation for Probabilistic Decoding," *IEEE Trans. Inform. Theory*, vol. IT-12, pp. 291–297, July.

Wozencraft, J. M. and Reiffen, B. (1961). *Sequential Decoding*, MIT Press, Cambridge, Mass.

Yao, K. (1972). "On Minimum Average Probability of Error Expression for Binary Pulse-Communication System with Intersymbol Interference," *IEEE Trans. Inform. Theory*, vol. IT-18, pp. 528–531, July.

Yao, K. and Tobin, R. M. (1976). "Moment Space Upper and Lower Error Bounds for Digital Systems with Intersymbol Interference," *IEEE Trans. Inform. Theory*, vol. IT-22, pp. 65–74, January.

Yue, O. (1983). "Spread Spectrum Mobile Radio 1977–1982," *IEEE Trans. Vehicular Technology*, vol. VT-32, pp. 98–105, February.

Zhou, K. and Proakis, J. G. (1988). "Coded Reduced-Bandwidth QAM with Decision-Feedback Equalization," *Conf. Rec. IEEE Int. Conf. Commun.*, Philadelphia, Penn., pp. 12.6.1–12.6.5, June.

Zhou, K., Proakis, J. G., and Ling, F. (1987). "Decision-Feedback Equalization of Fading Dispersive Channels with Trellis-Coded Modulation," *Int. Conf. Commun. Tech.*, Nanjing, China, November.

Ziemer, R. E. and Peterson, R. L. (1985). *Digital Communications and Spread Spectrum Systems*, Macmillan, New York.

Zigangirov, K. S. (1966). "Some Sequential Decoding Procedures," *Probl. Peredach. Inform.*, vol. 2, pp. 13–25.

INDEX

DIGITAL COMMUNICATIONS